Game Theory Framework Applied to Wireless Communication Networks

Chungang Yang
Xidian University, China

Jiandong Li
Xidian University, China

A volume in the Advances in Wireless Technologies and Telecommunication (AWTT) Book Series

Information Science REFERENCE
An Imprint of IGI Global

Managing Director: Lindsay Johnston
Managing Editor: Keith Greenberg
Director of Intellectual Property & Contracts: Jan Travers
Acquisitions Editor: Kayla Wolfe
Production Editor: Christina Henning
Development Editor: Brandon Carbaugh
Typesetter: Mike Brehm
Cover Design: Jason Mull

Published in the United States of America by
Information Science Reference (an imprint of IGI Global)
701 E. Chocolate Avenue
Hershey PA, USA 17033
Tel: 717-533-8845
Fax: 717-533-8661
E-mail: cust@igi-global.com
Web site: http://www.igi-global.com

Library of Congress Cataloging-in-Publication Data

Game theory framework applied to wireless communication networks / Chungang Yang and Jiandong Li, editors.
pages cm
Includes bibliographical references and index.
ISBN 978-1-4666-8642-7 (hardcover) -- ISBN 978-1-4666-8643-4 (ebook) 1. Wireless communication systems--Computer simulation. 2. Cell phone systems--Computer simulation. 3. Computer networks--Computer simulation. 4. Telecommunication systems--Computer simulation. 5. Game theory. I. Yang, Chungang, 1982- II. Li, Jiandong, 1962-
TK5103.2.G3485 2016
004.601'13--dc23
2015015653

This book is published in the IGI Global book series Advances in Wireless Technologies and Telecommunication (AWTT) (ISSN: 2327-3305; eISSN: 2327-3313)

British Cataloguing in Publication Data
A Cataloguing in Publication record for this book is available from the British Library.

Advances in Wireless Technologies and Telecommunication (AWTT) Book Series

Xiaoge Xu
The University of Nottingham Ningbo China

ISSN: 2327-3305
EISSN: 2327-3313

MISSION

The wireless computing industry is constantly evolving, redesigning the ways in which individuals share information. Wireless technology and telecommunication remain one of the most important technologies in business organizations. The utilization of these technologies has enhanced business efficiency by enabling dynamic resources in all aspects of society.

The **Advances in Wireless Technologies and Telecommunication Book Series** aims to provide researchers and academic communities with quality research on the concepts and developments in the wireless technology fields. Developers, engineers, students, research strategists, and IT managers will find this series useful to gain insight into next generation wireless technologies and telecommunication.

COVERAGE

- Wireless Sensor Networks
- Digital Communication
- Mobile Web Services
- Global Telecommunications
- Network Management
- Radio Communication
- Wireless Broadband
- Wireless Technologies
- Cellular Networks
- Broadcasting

IGI Global is currently accepting manuscripts for publication within this series. To submit a proposal for a volume in this series, please contact our Acquisition Editors at Acquisitions@igi-global.com or visit: http://www.igi-global.com/publish/.

The Advances in Wireless Technologies and Telecommunication (AWTT) Book Series (ISSN 2327-3305) is published by IGI Global, 701 E. Chocolate Avenue, Hershey, PA 17033-1240, USA, www.igi-global.com. This series is composed of titles available for purchase individually; each title is edited to be contextually exclusive from any other title within the series. For pricing and ordering information please visit http://www.igi-global.com/book-series/advances-wireless-technologies-telecommunication-awtt/73684. Postmaster: Send all address changes to above address.

Titles in this Series

For a list of additional titles in this series, please visit: www.igi-global.com

Enabling Real-Time Mobile Cloud Computing through Emerging Technologies
Tolga Soyata (University of Rochester, USA)
Information Science Reference • copyright 2015 • 400pp • H/C (ISBN: 9781466686625) • US $195.00 (our price)

Emerging Perspectives on the Design, Use, and Evaluation of Mobile and Handheld Devices
Joanna Lumsden (School of Engineering and Applied Science, Aston University, UK)
Information Science Reference • copyright 2015 • 334pp • H/C (ISBN: 9781466685833) • US $205.00 (our price)

Technological Breakthroughs in Modern Wireless Sensor Applications
Hamid Sharif (University of Nebraska – Lincoln, USA) and Yousef S. Kavian (Shahid Chamran University of Ahvaz, Iran)
Information Science Reference • copyright 2015 • 417pp • H/C (ISBN: 9781466682511) • US $200.00 (our price)

Handbook of Research on Software-Defined and Cognitive Radio Technologies for Dynamic Spectrum Management
Naima Kaabouch (University of North Dakota, USA) and Wen-Chen Hu (University of North Dakota, USA)
Information Science Reference • copyright 2015 • 927pp • H/C (ISBN: 9781466665712) • US $505.00 (our price)

Interdisciplinary Mobile Media and Communications Social, Political, and Economic Implications
Xiaoge Xu (The University of Nottingham Ningbo China, China)
Information Science Reference • copyright 2014 • 409pp • H/C (ISBN: 9781466661660) • US $205.00 (our price)

Cognitive Radio Sensor Networks Applications, Architectures, and Challenges
Mubashir Husain Rehmani (Department of Electrical Engineering, COMSATS Institute of Information Technology, Pakistan) and Yasir Faheem (Department of Computer Science, COMSATS Institute of Information Technology, Pakistan)
Information Science Reference • copyright 2014 • 313pp • H/C (ISBN: 9781466662124) • US $235.00 (our price)

Game Theory Applications in Network Design
Sungwook Kim (Sogang University, South Korea)
Information Science Reference • copyright 2014 • 500pp • H/C (ISBN: 9781466660502) • US $225.00 (our price)

Convergence of Broadband, Broadcast, and Cellular Network Technologies
Ramona Trestian (Middlesex University, UK) and Gabriel-Miro Muntean (Dublin City University, Ireland)
Information Science Reference • copyright 2014 • 333pp • H/C (ISBN: 9781466659780) • US $235.00 (our price)

Table of Contents

Detailed Table of Contents

New decision-making paradigms addressing the requirements of flexibility, adaptability and intelligence are needed for future wireless networks. Moreover, mutual interactions should be captured when all the devices are autonomous and smart. Game theory is a powerful tool to study such interactions. However, since it is a branch of applied mathematic and mainly studied in economic, some featured challenges should be addressed when applied in wireless networks. This chapter bridges game theory and practical wireless applications, by focusing on the incomplete, dynamic and uncertain information constraints. Four kinds of distributed learning algorithms including stochastic learning automata, payoff-based log-linear learning, learning by trial and error, and no-regret learning are discussed. The learning procedures and basic theoretical results are presented, and their applications in wireless networks are reviewed. Contrastive analysis on environment dynamics, solution concepts, synchrony, convergence, and convergent results is discussed, and some future research directions are given.

Improving capacity and coverage is one of the main issues in next-generation wireless communication. Heterogeneous networks (HetNets), which is currently investigated in LTE-Advanced standard, is a promising solution to enhance capacity and eliminate coverage holes in a cost-efficient manner. A HetNet is composed of existing macrocells and various types of small cells. By deploying small cells into the existing network, operators enhance the users' quality of service which are suffering from severe signal degradation at cell edges or coverage holes. Nevertheless, there are numerous challenges in integrating small cells into the existing cellular network due to the characteristics: unplanned deployment, intercell interference, economic potential, etc. Recently, game theory has been shown to be a powerful tool for investigating the challenges in HetNets. Several game-theoretic approaches have been proposed to model the distributed deployment and self-organization feature of HetNets. In this chapter, the authors first give an overview of the challenges in HetNets. Subsequently, the authors illustrate how game theory can be applied to solve issues related to HetNets.

With the introduction of Small Cell into current cellular structure, the ever-growing demand for mobile traffic gets the opportunity to be fulfilled. But the overlapped dense deployment caused by the Small Cell Heterogeneous Network (HetNet) also arouses the interference problems. In order to solve those problems, this chapter focuses on Game Theory based uplink power control and downlink power allocation strategies for the interference mitigation of the heterogeneous Small Cell Network (SCN). For the uplink scenario, this chapter proposes the non-cooperative game model based power control algorithm, which can optimize the initial transmission power of both Macro Cell and Small Cell users through the Nash Equilibrium solution. For the downlink scenario with multiple service types in the SCN, the non-cooperative game model based scheme is proposed to optimize the transmission power allocation with constraints of different Quality of Service (QoS) requirements. The simulation results show the merits of the proposed strategies over current works.

To optimize radio resource allocation, the game theory is utilized as a powerful tool because its characteristic can be adaptive to the distribution characteristics of in heterogeneous small cell networks (HSCNs). This chapter summarizes the recent achievements for the game theory based radio resource allocation in HSCNs, where macro base stations (MBSs) and dense small cell base stations (SBSs) share the same frequency spectrum and interfere with each other. Two kinds of game models are introduced to optimize the radio resource allocation, namely the non-cooperative Stackelberg and the cooperative coalition. System models, optimization problem formulation, problem solution, and simulation results for these two kinds of game models are presented. Particularly, the Stackelberg models for HSCNs are presented with the Stackelberg equilibrium and the closed-form expressions. The coalition formations for traditional HCSNs, cloud small cell networks, and heterogeneous cloud small cell networks are introduced. Simulation results are shown to demonstrate the proposed game theory based radio resource optimization strategies converged and efficient.

Focusing on the coverage optimization of small cell networks (SCN), this chapter starts with a detailed analysis on various coverage problems, based on which the coverage optimization problem is formulated. Then centralized and distributed coverage optimization methods based on game theory are described. Firstly, considering the coverage optimization with a control center, a modified particle swarm optimization (MPSO) is presented for the self-optimization of SCN, which employs a heuristic power control scheme to search for the global optimum solution. Secondly, distributed optimization using game theory (DGT) without a control center is concerned. Considering both throughput and interference, a utility function is formulated. Then a power control scheme is proposed to find the Nash Equilibrium (NE). Simulation results show that MPSO and DGT significantly outperform conventional schemes. Moreover, compared with MPSO, DGT uses much less overhead. Finally, further research directions are discussed and conclusions are drawn.

In the Long Term Evolution-Advanced (LTE-A) system, Device-to-device (D2D) communication underlying cellular networks can bring some advantages such as high data rates, low delay, low power consumption by reusing frequency bands with cellular user equipments (UEs). While at the same time, inter-cell and intra-cell interference is inevitable. The D2D users and cellular UEs will compete or cooperate to coordinate interference and share resources which incurs game theory an effective mathematical tool. This chapter proposed a Stackelberg game based algorithm to jointly allocate power and resources when the uplink frequency is shared with LTE-A users. In the game, the evolved NodeB (eNB) and D2D UEs are grouped to form the seller-buyer pair and the eNB sets prices to reduce the interference that it suffers meanwhile maximizes its revenue. For given specified prices, the D2D users compete for the resources to communicate with each other and reach their individual utility maximization. Simulation results prove that satisfying performance can be achieved by using the proposed mechanisms.

Game theoretical approaches have been used to develop distributed resource allocation technologies for cognitive heterogeneous networks. In this chapter, we present a novel distributed resource allocation strategy for cognitive small cell networks based on orthogonal frequency-division multiple access. In particular, we consider a heterogeneous network consisting of macrocell networks overlaid with cognitive small cells that opportunistically access the available spectrum. We focus on a regret-matching game approach, aiming at maximizing the total throughput of the small cell network subject to cross-tier interference and quality of service (QoS) constraints. The regret-matching game approach exploits a regret procedure to learn the optimal resource allocation strategy from the regrets of the actions of cognitive users. Furthermore, the regret-matching game approach is extended to the joint resource allocation and user admission control problem. Numerical results are presented to demonstrate the effectiveness of the proposed regre-matching approaches.

Dynamic spectrum access is envisioned as a promising paradigm for addressing the spectrum under-utilization problem. According to the recent ruling of Federal Communications Commission (FCC) for white-space spectrum access, white-space devices are required to query a geo-location database to determine the spectrum availability. This chapter adopts a game theoretic approach for the self-organizing white-space spectrum access network design. This chapter first models the distributed channel selection problem among the devices as a distributed spectrum access game, and shows that the game is a potential game. This chapter then designs a self-organizing spectrum access algorithm which can achieve a Nash equilibrium of the game without any information exchange among the devices. Numerical results demonstrate that the proposed algorithm is efficient and can adapt to the dynamical network context changing.

This chapter addresses the complex scenario of multi cell OFDMA network resource allocation and interference control by auction theory. An auction framework that meets the properties of efficiency and incentive compatibility is proposed. We consider a multi-cell OFDMA network with Fractional Frequency Reuse (FFR) implementation. An auction is presented to allocate subcarriers. According to the proposed auction framework, users will avoid bidding for the subcarriers where they have a relatively low chance of winning. Optimal bidding strategy based on Bayesian Nash Equilibrium (BNE) is obtained in which users are maximizing their net profit. A focal distance which classifies the users into cell-center and cell-edge users is characterized. The proposed approach maximizes the total auctioneer revenue, while balancing a tradeoff between system performance in terms of total system throughput and quality of service provisioning for cell-edge users.

Section 3
Advanced Game Theory for Promising Wireless Techniques and Services

This chapter provides a comprehensive review of the domain of game theory based physical layer security in wireless communications. By exploiting the wireless channel characteristic and secure cooperation of nodes, physical layer security is to enable the exchange of confidential messages over a wireless medium in the presence of unauthorized eavesdroppers, without relying on higher-layer encryption. However, the selfness of nodes seriously affects the secure cooperation; game theory can model the influence of the selfness on physical layer security. This chapter firstly describes the physical layer security issues in the wireless networks and the role of game theory in the research on physical layer security. And then the typical applications of game theory in physical layer security are subsequently covered, including zero-sum game, Stackelberg game, auction theory, coalition game. Finally, the chapter concludes with observations on potential research directions in this area.

This chapter investigates the interplay between cooperative device-to-device (D2D) communications and green communications in LTE heterogeneous networks (HetNets). Two game theoretic concepts are studied and analyzed in order to perform dynamic HetNet base station (BS) on/off switching. The first approach is a coalition-based method whereas the second is based on the Nash bargaining solution. Afterwards, a method for coupling the BS on/off switching approach with D2D collaborative communications is presented and shown to lead to increased energy efficiency. The savings are additionally increased when a portion of the small cell BSs in a HetNet are powered by renewable energy sources. Different utility functions, modeling the game theoretic framework governing the energy consumption balance between the cellular network and the mobile terminals (MTs), are proposed and compared, and their impact on MT quality of service (QoS) is analyzed.

The emerging traffic demand has fueled the rapid densification of cellular networks. The increased number of Base Stations (BSs) leads to augmented energy consumption and expenditures for the Mobile Network Operators (MNOs), especially during low traffic, when many of the BSs remain underutilized. Hence, the MNOs are encouraged to provide "green" and cost effective solutions for their networks. In this chapter, an innovative algorithm for infrastructure sharing in two-operator environments is proposed, based on BSs switching off during low traffic periods. Motivated by the conflicting interests of the operators, the problem is formulated in a game theoretic framework that enables the MNOs to act individually to estimate the switching off probabilities that reduce their financial cost. The authors analytically and experimentally estimate the potential energy and cost savings that can be accomplished. The obtained results show a significant reduction in both energy consumption and expenditures, thus giving the operators the necessary incentives for infrastructure sharing.

This chapter provides an overview of game theoretic solutions in a wireless application of particular interest for multimedia and social interactions in modern telecommunication systems. In particular, coalitional games, bargaining solutions and fairness and stability issues will be investigated within wireless cooperative content-sharing. According to this paradigm, users download portions of data of common interest over long-range cellular links while exchanging the downloaded portions over short-range radio links. Expected benefits, which are natural incentives to cooperation, may be in terms of content price, energy consumption, and transfer delay reduction. Significant research activity has been conducted to design strategies that simultaneously exploit the multiple radio interfaces of modern wireless devices and maximize the gain. A valid solution for the cooperative content-sharing application should be based on fairness in the utility distribution among the involved nodes which can be eased by exploiting the game theoretic approaches presented in this chapter.

This chapter presents a mechanism for cooperative video transmission based on game theory for heterogeneous devices during broadcasting. Broadcasting is a multipoint delivery of transmission that sends data from a source to multiple destinations. The terminal is involved in cooperative transmission when the station broadcasts video data. To enhance performance, the heterogeneity and forwarding capabilities should be considered. This work studies power control and allocation in a collaborative transmission based on game theory, which provides an effective strategy when network resources are limited. First, a novel power-allocation model of the base station (BS) based on noncooperative game theory and bidding is presented in this study. Additionally, we also propose a utility function of Signal-to-Noise Ratios (SNRs) along with Signal-to-Interference Ratio (SIRs). Subsequently, based on such noncooperative game theory with a utility function, the model of the power distribution of terminals in cooperative transmission can be built. Experiments on the System-in-the-Loop (SITL) mode in OPNETs have proven the correctness of the designed model and superiority, verifying the effectiveness of the proposed power-control idea.

Foreword

Heterogeneous small cell networks can potentially handle the challenge of the cost-effectively supporting a 1000-fold increase in traffic demand for future 5G networks. However, the interference mitigation, resource management, coverage optimization, and self-organizing white-space spectrum access should be paid much attention from academia, industry, and standardization. *Game theory can characterize, analyze, and model the strategic interactions and the rational behaviors of macrocell eNBs and small cell eNBs, so that it facilitates to design the distributed management and control algorithms with capacity enhancement and energy consumption into consideration.*

*This book investigates game-theoretic approaches to challenged techniques in h*eterogeneous small cell networks *with the following sections: In session 1, game theoretic models are surveyed including complete information-supported non-cooperative potential games (Chapter 1) and cooperative coalition games (Chapter 2), and games with incomplete information (Chapter 3). Session 2 from Chapter 4 to Chapter11 is the main part, where* interference mitigation, power control, coverage optimization, etc. *are investigated via various game models. In Session 3, we employ advanced game theory to variety of application scenarios: the physical layer security (Chapter 12), infrastructure sharing (Chapter 13) and renewable energy (Chapter 14) for green communications, and then we concentrate on the challenges and application of game theory for promising wireless services including multimedia and social interactions (Chapter 15) and cooperative video transmission by broadcasting (Chapter 16).*

The topic is timely and interesting. The editors carefully chosen the promising topic, and excellently selected the up-to-date chapters addressing these technical aspects. The surveyed game models are novel. This book covers many types of games and their applications in wireless networks. The targeted audiences of the book are either students or researchers to get valuable insights into their own research. I am sure it will be an important reference for many years in the field of game theory for wireless engineers.

Zhu Han
University of Houston, USA

Zhu Han *received the B.S. degree in electronic engineering from Tsinghua University, in 1997, and the M.S. and Ph.D. degrees in electrical engineering from the University of Maryland, College Park, in 1999 and 2003, respectively. Currently, he is a Full Professor in Electrical and Computer Engineering Department at University of Houston, Texas. His research interests include wireless resource allocation and management, wireless communications and networking, game theory, wireless multimedia, and security. Dr. Han is an NSF CAREER award recipient 2010, and he is an Associate Editor of IEEE Transactions on Wireless Communications since 2010. Dr. Han is the winner of the 2011 IEEE Communications Society Fred W. Ellersick Prize, and he is the coauthor for the papers that won eight IEEE best conference paper awards. Dr. Han is an IEEE Fellow since 2014, and IEEE distinguished lecturer since 2015.*

Preface

Mobile wireless communication has experienced explosive growth over the past decade, which is fueled by the popularity of smart phones and various spectrum hungry services. The wireless industry has taken the challenge of cost-effectively supporting a 1000-fold increase in traffic demand over the next decade. Recently, to enhance the network capacity and improve the quality of user experience, various emerging wireless communication network architectures and technologies have been extensively researched, for instance, heterogeneous and small cell networks, device-to-device communication, self-organizing networks, green communication, spectrum aggregation, massive MIMO and COMP, and so on. Additionally, the techniques including interference mitigation, resource management, hybrid access decision, revenue maximization should be poured more research attention with novel modeling, analyzing, and designing methodologies.

Game theory can well characterize the emerging typical features of the mobile wireless networks, including the selfish behaviors, cross-tier cooperation, multiple dimensions of resource and fairness requirements. It can be used to model, analyze and design various issues, for instance, resource allocation, interference management, and energy saving, and so on. Therefore, game theory has been widely recognized as a useful mathematical tool. In the past decade, we have witnessed a huge explosion of interests that intersect wireless networks with game theory. Researchers are still working hard on employing game theory to much wider applications. Therefore, it is necessary to call for them to propose their novel ideas with advanced game theory to study wireless communication network and techniques. It will help to further develop the powerful mathematical tool for future challenges in wireless industry.

As we known, game theory itself is an evergreen evolving discipline. More advanced game models, effective equilibrium concepts, and differential games and mean field games, learning mechanisms in games and so on should be explored and exploited for better understanding and design wireless networks. This book looks to discuss and address the applications, mechanism designs, and challenges that the advanced game theoretic framework for the emerging wireless communication networks techniques. This book is therefore both timely and apposite.

This comprehensive and timely publication aims to be an essential reference source, built on the available literature in the field of game-theoretic framework for wireless communication networks, in particular, the advanced game model, equilibrium, and mechanism for the emerging network architectures and technologies. The book will provide the novel and potential way to solve the encountered problems for wireless network researcher and engineer. It also will help various kinds of networking engineers for understanding and designing the suitable wireless networks.

Researchers, advanced-level students, technology developers, and engineers will find this text useful in furthering their research on game theory for networks. This text will assist both economics and wireless engineers in furthering their own research efforts and interests in developing game theory for wireless communication and networks. Economics researchers who are interested in the engineering field; wireless engineers who are using game theory.

After careful and rigorous review, finally we accept 16 chapters according to the review results, and content, concentration, and objective of this book; and then we divide them into three sections according to their internal logic:

SECTION 1: ADVANCED GAME THEORY FOR WIRELESS COMMUNICATIONS

In Section 1, various advanced game theoretic models are surveyed in the wireless communications and networks including complete information-supported games of non-cooperative potential games (Chapter 1) and cooperative coalition games (Chapter 2), and games with incomplete information (Chapter 3). In detail, authors in chapter 1 first introduce the concept of potential games and give definitions of different types of potential games. Then, they provide some important results and properties of potential games. After that, the authors introduce three algorithms to achieve Nash equilibrium of a potential game. Next, chapter 1 will show how to apply potential game theory to design efficient algorithms for wireless network optimization. One application of potential game theory is to study the power control problem in wireless networks. Another example is the multimode precoding design in multi-input multi-output multiple access channels. Chapter 1 also shows how to apply potential games for joint resource allocation in a relay network where multiple users can conduct bidirectional communications through a relay node over multiple available channels. As it is known that with the emergence of cooperation as a new communication paradigm, and there needs for self-organizing, distributed networks. Therefore, it is necessary to seek suitable game theoretical tools that allow to analyze and study the behavior and interactions of the nodes in future communication networks. In this context, chapter 2 introduces the coalition formation game theory, and its potential applications in communication and wireless networks. Specifically, it presents the fundamental components, the key properties, the mathematical techniques, the solution concepts, and describes the methodologies for applying these games in several applications drawn from the state-of-the-art research in communications. Furthermore, chapter 3 bridges game theory and practical wireless applications, by focusing on the incomplete, dynamic and uncertain information constraints. Four kinds of distributed learning algorithms including stochastic learning automata, payoff-based log-linear learning, learning by trial and error, and no-regret learning are discussed. The learning procedures and basic theoretical results are presented, and their applications in wireless networks are reviewed. Contrastive analysis on environment dynamics, solution concepts, synchrony, convergence, and convergent results is discussed, and some future research directions are given.

We can see that these surveyed game models of chapter 1, 2, and 3 in Section 1 are relatively novel and with typical characteristics. Therefore, this book covers all types of games and their applications appearing in the game theory, which will provide a full understanding of game theory for wireless communications. Section 1 will help understand the following chapters since their further application of the survey games.

SECTION 2: GAME THEORY IN EMERGING HETEROGENEOUS SMALL CELL NETWORKS

Section 2 is the main part of this book, where we focus on the heterogeneous small cell networks since they can be recognized as the most powerful technology to provide exponential capacity growth for data traffic. Basics for interference mitigation, resource management, hybrid access decision, revenue maximization challenges of heterogeneous small cell networks, and applications of various game theoretic models in them including strategic game approach, potential game, Bayesian game, Stackelberg game, and mechanism design, and future research directions are provided in Chapter 4. In detail, improving capacity and coverage is one of the main issues in next-generation wireless communication. Chapter 4 concentrates on heterogeneous networks (HetNets), which is currently investigated in LTE-Advanced standard, and is a promising solution to enhance capacity and eliminate coverage holes in a cost-efficient manner. A HetNet is composed of existing macrocells and various types of small cells. By deploying small cells into the existing network, operators enhance the users' quality of service which are suffering from severe signal degradation at cell edges or coverage holes. Nevertheless, there are numerous challenges in integrating small cells into the existing cellular network due to the characteristics: unplanned deployment, intercell interference, economic potential, etc. Recently, game theory has been shown to be a powerful tool for investigating the challenges in HetNets. Several game-theoretic approaches have been proposed to model the distributed deployment and self-organization feature of HetNets. In chapter 4, the authors first give an overview of the challenges in HetNets. Subsequently, the authors illustrate how game theory can be applied to solve issues related to HetNets.

Similarly, in order to solve interference problems, chapter 5 focuses on Game Theory based uplink power control and downlink power allocation strategies for the interference mitigation of the heterogeneous Small Cell Network (SCN). For the uplink scenario, chapter 5 proposes the non-cooperative game model based power control algorithm, which can optimize the initial transmission power of both Macro Cell and Small Cell users through the Nash Equilibrium solution. For the downlink scenario with multiple service types in the SCN, the non-cooperative game model based scheme is proposed to optimize the transmission power allocation with constraints of different Quality of Service (QoS) requirements. The simulation results show the merits of the proposed strategies over current works.

To optimize radio resource allocation, the game theory is utilized as a powerful tool because its characteristic can be adaptive to the distribution characteristics of in heterogeneous small cell networks (HSCNs). Chapter 6 summarizes the recent achievements for the game theory based radio resource allocation in HSCNs, where macro base stations (MBSs) and dense small cell base stations (SBSs) share the same frequency spectrum and interfere with each other. Two kinds of game models are introduced to optimize the radio resource allocation, namely the non-cooperative Stackelberg and the cooperative coalition. System models, optimization problem formulation, problem solution, and simulation results for these two kinds of game models are presented. Particularly, the Stackelberg models for HSCNs are presented with the Stackelberg equilibrium and the closed-form expressions. The coalition formations for traditional HCSNs, cloud small cell networks, and heterogeneous cloud small cell networks are introduced. Simulation results are shown to demonstrate the proposed game theory based radio resource optimization strategies converged and efficient.

Chapter 7 focuses on the coverage optimization of small cell networks (SCN), and starts with a detailed analysis on various coverage problems, based on which the coverage optimization problem is formulated. Then centralized and distributed coverage optimization methods based on game theory are described. Firstly, considering the coverage optimization with a control center, a modified particle swarm optimization (MPSO) is presented for the self-optimization of SCN, which employs a heuristic power control scheme to search for the global optimum solution. Secondly, distributed optimization using game theory (DGT) without a control center is concerned. Considering both throughput and interference, a utility function is formulated. Then a power control scheme is proposed to find the Nash Equilibrium (NE). Simulation results show that MPSO and DGT significantly outperform conventional schemes. Moreover, compared with MPSO, DGT uses much less overhead. Finally, further research directions are discussed and conclusions are drawn.

In the Long Term Evolution-Advanced (LTE-A) system, Device-to-Device (D2D) communication underlying cellular networks can bring some advantages such as high data rates, low delays, low power consumption by reusing frequency band with the cellular user equipments (UEs). While at the same time, inter-cell and intra-cell interference is inevitable. The D2D users and cellular UEs will compete or cooperate to coordinate interference and share resources which incurs game theory an effective mathematical tool. Chapter 8 proposed a Stackelberg game based algorithm to jointly allocate power and resources when uplink frequency is shared with LTE-A users. In the game, the evolved NodeB (eNB) and D2D UE are grouped to form the seller-buyer pair and the eNB sets prices to reduce the interference that it suffers meanwhile maximizes its revenue. For given specified prices, the D2D users compete for the resources to communicate with each other and reach their individual utility maximization. Simulation results prove that satisfying performance can be achieved by using the proposed mechanisms.

Game theoretical approaches have been used to develop distributed resource allocation technologies for cognitive heterogeneous networks. In chapter 9, the authors present a novel distributed resource allocation strategy for cognitive small cell networks based on orthogonal frequency-division multiple access. In particular, teh authors consider a heterogeneous network consisting of macrocell networks overlaid with cognitive small cells that opportunistically access the available spectrum. A regret-matching game approach is proposed, aiming at maximizing the total throughput of the small cell network subject to cross-tier interference and quality of service (QoS) constraints. The regret-matching game approach exploits a regret procedure to learn the optimal resource allocation strategy from the regrets of the actions of cognitive users. Furthermore, the regret-matching game approach is extended to the joint resource allocation and user admission control problem. Numerical results are presented to demonstrate the effectiveness of the proposed regret-matching approaches.

Dynamic spectrum access is envisioned as a promising paradigm for addressing the spectrum under-utilization problem. According to the recent ruling of Federal Communications Commission (FCC) for white-space spectrum access, white-space devices are required to query a geo-location database to determine the spectrum availability. Chapter 10 adopts a game theoretic approach for the self-organizing white-space spectrum access network design. Chapter 10 first models the distributed channel selection problem among the devices as a distributed spectrum access game, and show that the game is a potential game. Chapter 10 then designs a self-organizing spectrum access algorithm which can achieve a Nash equilibrium of the game without any information exchange among the devices. Numerical results demonstrate that the proposed algorithm is efficient and can adapt to the dynamical network context changing.

Chapter 11 addresses the complex scenario of multi cell OFDMA network resource allocation and interference control by auction theory. An auction framework that meets the properties of efficiency and incentive compatibility is proposed. We consider a multi-cell OFDMA network with Fractional Frequency Reuse (FFR) implementation. An auction is presented to allocate subcarriers. According to the proposed auction framework, users will avoid bidding for the subcarriers where they have a relatively low chance of winning. Optimal bidding strategy based on Bayesian Nash Equilibrium (BNE) is obtained in which users are maximizing their net profit. A focal distance which classifies the users into cell-center and cell-edge users is characterized. The proposed approach maximizes the total auctioneer revenue, while balancing a tradeoff between system performance in terms of total system throughput and quality of service provisioning for cell-edge users.

In summary, in section 2 game theory based uplink power control and downlink power allocation strategies for the interference mitigation are investigated in Chapter 5. Two kinds of game models are introduced in Chapter 6 to optimize the radio resource allocation, namely the non-cooperative Stackelberg and the cooperative coalition. The centralized and distributed coverage optimization methods based on game theory are described in Chapter 7. Chapter 8 proposed a Stackelberg game based algorithm to jointly allocate power and resources when the uplink frequency is shared with LTE-A users. A heterogeneous network consisting of macrocell networks overlaid with cognitive small cells that opportunistically access the available spectrum is considered in Chapter 9. Further, the regret-matching procedure in game theory is exploited to solve a competition problem. Chapter 10 adopts a game theoretic approach for the self-organizing white-space spectrum access network design. Optimal bidding strategy based on Bayesian Nash Equilibrium (BNE) is obtained in Chapter 11 in which users are maximizing their net profit.

SECTION 3: ADVANCED GAME THEORY FOR PROMISING WIRELESS TECHNIQUES AND SERVICES

In Section 3, we summarize advanced game theory for the promising wireless techniques including physical layer security (Chapter 12), infrastructure sharing (Chapter 13) and renewable energy (Chapter 14) for green communications, and then concentrate on the challenges and application of game theory for promising wireless services including multimedia and social interactions (Chapter 15) and cooperative video transmission by broadcasting (Chapter 16).

Chapter 12 provides a comprehensive review of the domain of game theory based physical layer security in wireless communications. By exploiting the wireless channel characteristic and secure cooperation of nodes, physical layer security is to enable the exchange of confidential messages over a wireless medium in the presence of unauthorized eavesdroppers, without relying on higher-layer encryption. However, the selfishness of nodes seriously affects the secure cooperation; game theory can model the influence of the selfishness on physical layer security. Chapter 12 firstly describes the physical layer security issues in the wireless networks and the role of game theory in the research on physical layer security. And then the typical applications of game theory in physical layer security are subsequently covered, including zero-sum game, Stackelberg game, auction theory, coalition game. Finally, Chapter 12 concludes with observations on potential research directions in this area.

Chapter 13 investigates the interplay between cooperative device-to-device (D2D) communications and green communications in LTE heterogeneous networks (HetNets). Two game theoretic concepts are studied and analyzed in order to perform dynamic HetNet base station (BS) on/off switching. The first approach is a coalition-based method whereas the second is based on the Nash bargaining solution. Afterwards, a method for coupling the BS on/off switching approach with D2D collaborative communications is presented and shown to lead to increased energy efficiency. The savings are additionally increased when a portion of the small cell BSs in a HetNet are powered by renewable energy sources.

Different utility functions, modeling the game theoretic framework governing the energy consumption balance between the cellular network and the mobile terminals (MTs), are proposed and compared, and their impact on MT quality of service (QoS) is analyzed.

The emerging traffic demand has fueled the rapid densification of cellular networks. The increased number of Base Stations (BSs) leads to augmented energy consumption and expenditures for the Mobile Network Operators (MNOs), especially during low traffic, when many of the BSs remain underutilized. Hence, the MNOs are encouraged to provide "green" and cost effective solutions for their networks. In chapter 14, an innovative algorithm for infrastructure sharing in two-operator environments is proposed, based on BSs switching off during low traffic periods. Motivated by the conflicting interests of the operators, the problem is formulated in a game theoretic framework that enables the MNOs to act individually to estimate the switching off probabilities that reduce their financial cost. The authors inchapter 14 analytically and experimentally estimate the potential energy and cost savings that can be accomplished. The obtained results show a significant reduction in both energy consumption and expenditures, thus giving the operators the necessary incentives for infrastructure sharing.

Chapter 15 provides an overview of game theoretic solutions in a wireless application of particular interest for multimedia and social interactions in modern telecommunication systems. In particular, coalitional games, bargaining solutions and fairness and stability issues will be investigated within wireless cooperative content-sharing. According to this paradigm, users download portions of data of common interest over long-range cellular links while exchanging the downloaded portions over short-range radio links. Expected benefits, which are natural incentives to cooperation, may be in terms of content price, energy consumption, and transfer delay reduction. Significant research activity has been conducted to design strategies that simultaneously exploit the multiple radio interfaces of modern wireless devices and maximize the gain. A valid solution for the cooperative content-sharing application should be based on fairness in the utility distribution among the involved nodes which can be eased by exploiting the game theoretic approaches presented in chapter 15.

Chapter 16 presents a cooperative video transmission mechanism based on game theory for heterogeneous devices during broadcasting. Broadcasting is a multi-point delivery of transmission that sends data from a source to multiple destinations. The terminal is involved in cooperative transmission when the station broadcast video data. To enhance performance, the heterogeneity and forwarding capabilities should be considered. This work studies power control and allocation in a collaborative transmission based on game theory, which provides an effective strategy when network resources are limited. First, a novel power allocation model of the base station (BS) based on non-cooperative game theory and bidding is presented in this study. Additionally, the authors also propose the utility function of Signal-to-Noise Ratios (SNRs) along with Signal-to-Interference Ratio (SIRs).

In summary, in Section 3 the typical applications of game theory in physical layer security are subsequently covered in Chapter 12, including zero-sum game, Stackelberg game, auction theory, coalition game. Two game theoretic concepts are studied and analyzed in Chapter 13 in order to perform dynamic HetNet base station (BS) on/off switching. The first approach is a coalition-based method whereas the second is based on the Nash bargaining solution. Afterwards, a method for coupling the BS on/off switching approach with D2D collaborative communications is presented and shown to lead to the increased energy efficiency. The authors in Chapter 14 analytically and experimentally estimate the potential energy and cost savings that can be accomplished. Chapter 15 provided an overview of game theoretic solutions in a wireless application of particular interest for multimedia and social interactions in modern telecommunication systems. Chapter 16 presented a cooperative video transmission mechanism based on game theory for heterogeneous devices during broadcasting.

This comprehensive and timely publication aims to be an essential reference source, building on the available literature in the field of game-theoretic framework for wireless communication networks, in particular, the advanced game model, equilibrium, and mechanism for the emerging network architectures and technologies.

The book will provide the novel and potential way to solve the encountered problems for wireless network researcher and engineer. It also will help various kinds of networking engineers for understanding and designing the suitable wireless networks.

Section 1
Advanced Game Theory for Wireless Communication

Chapter 1
Potential Games and Its Applications to Wireless Networks

Wei Zhong
PLA University of Science and Technology, China

Jiaheng Wang
Southeast University, China

Meixia Tao
Shanghai Jiao Tong University, China

ABSTRACT

Potential games are a subclass of noncooperative games that bear some special properties and are often exploited in wireless network designs. In this chapter, we first introduce the concept of potential games and give definitions of different types of potential games. Then, we provide some important results and properties of potential games. After that, we introduce three algorithms to achieve Nash equilibrium of a potential game. Next, this chapter will show how to apply potential game theory to design efficient algorithms for wireless network optimization. One application of potential game theory is to study the power control problem in wireless networks. Another example is the multimode precoding design in multi-input multi-output multiple access channels. We also show how to apply potential games for joint resource allocation in a relay network where multiple users can conduct bidirectional communications through a relay node over multiple available channels.

INTRODUCTION

Noncooperative game theory constitutes a powerful framework for analyzing conflicts among interacting decision-makers, usually called players, each of whom wishes to maximize his own payoff (Monderer & Shapley, 1996a; Myerson, 1991; MacKenzie & DaSilva, 2007; Hossain, Niyato, & Han, 2009; Saraydar, Mandayam, & Goodman, 2002). Such a framework has been widely applied to designing wireless

DOI: 10.4018/978-1-4666-8642-7.ch001

networks, where users compete limited resources and meanwhile generate interference to each other. By formulating a network design as a noncooperative game, one can devise proper algorithms to achieve the solution of the game, i.e., the Nash equilibrium, at which no player wishes to unilaterally change his strategy. However, the overall network performance may not be improved at a Nash equilibrium due to selfish behaviors of players.

Potential games (Monderer & Shapley, 1996a) are a subclass of noncooperative games with nice properties that can lead to an enhancement of global performance at the equilibrium point. Particularly, in potential games, the incentive of all players to change their strategies can be expressed in a single global function, called potential function. That is, players in a potential game can serve the greater good by furthering their own interests. Consequently, each user's selfish optimization would contribute to optimizing the potential function, thus improving the global system performance, which is particularly important from a whole network perspective.

In light of global optimality improvement, potential games have received great attention and have been introduced to transmission designs in, for example, multiple access or interference channels. When using potential games in wireless networks, an important issue is to devise distributed algorithms, using local information and processing abilities, to reach a Nash equilibrium of the potential game. The property of potential game can guarantee the convergence of local search algorithms, of which two algorithms, namely, best response dynamic and better response dynamic, are most commonly used. Combined with appropriate learning mechanisms, these local search algorithms can achieve globally optimal or nearly globally optimal performance.

Potential games has been used to study the wireless networks in many previous works. For examples, the authors proposed a game theoretic framework for the power control problem in wireless networks based on potential game in the work of Scutari, Barbarossa, & Palomar, (2006). The work of Ai, Srinivasan, & Tham, (2008) applied potential game theory to study the coverage game in wireless sensor networks. The work of Menon, MacKenzie, Buehrer, & Reed, (2009) established a potential game theoretic framework for the interference avoidance in wireless communication systems. Xu, Wang, Wu, Anpalagan, & Yao, (2012) used potential game theory to study the opportunistic spectrum access in cognitive radio networks. Zhong, Chen, Jin, & Wong, (2014) formulated the relay selection and discrete power control problem for cognitive relay systems as a potential game and analyze the price of anarchy of the proposed game. Uykan & Jantti, (2014). analyzed the transmission order (TO) optimization problem in the presence of channel allocation (CA), i.e., joint CA and the TO optimization problem from a game theoretic perspective, and prove that the joint optimization problem can be formulated as an exact potential game. Among existing works, one important application of potential game theory is to design proper resource allocation strategies for wireless networks.

In this chapter, we will introduce the concept and definition of potential game, important properties of potential game, and algorithms to achieve Nash equilibrium of potential game. Then, we would use three examples to show applications of potential game theory in wireless networks. The first one is the application of potential games to a power control problem (Scutari, Barbarossa, & Palomar, 2006), the second one is the multimode precoding design in multi-input multi-output (MIMO) multiple access channels (Zhong, Xu, Tao, & Cai, 2010), and the third one is the joint resource management for multichannel dual-mode relay networks.

A BRIEF INTRODUCTION OF POTENTIAL GAMES

Potential games, first introduced by Monderer and Shapley in their pioneering work (Monderer & Shapley, 1996a), are a class of noncooperative games for which a pure strategy Nash equilibrium always exists. For a potential game, selfish behaviors of rational players could be characterized by a potential function and contribute to some global measure. Due to these favorable features over conventional noncooperative games, potential games have been widely applied to wireless communication system designs in recent years. In this section, we will briefly introduce the definition of a potential game and then show important properties of potential games.

The first to use potential functions for games in strategic form was Rosenthal, (1973). Rosenthal, (1973) defined the class of congestion games and proved, by explicitly constructing a potential function, that every game in this class possesses a pure-strategy equilibrium (Monderer & Shapley, 1996a). The class of congestion games is narrow compared to the potential games. Any game where a collection of homogeneous agents have to choose from a finite set of alternatives, and where the payoff of a player depends on the number of players choosing each alternative, is a congestion game (Monderer & Shapley, 1996a). It was shown that the class of congestion games coincides with the class of finite potential games.

1. Definition of Potential Games

Consider a noncooperative game where the set of players is denoted as $\mathcal{N} = \{1, 2, \cdots, n\}$, the action set is denoted as $\mathcal{A}_i$ for $i \in \mathcal{N}$, and the payoff function is denoted as $u_i : \mathcal{A}_i \to \mathbb{R}$ for $i \in \mathcal{N}$. We introduce the notations $\mathcal{A} = \prod_{i \in \mathcal{N}} \mathcal{A}_i$, $\mathcal{A}_{-i} = \prod_{j \neq i} \mathcal{A}_j$, and $\mathbf{a}_{-i} = (a_1, \cdots, a_{i-1}, a_{i+1}, \cdots, a_n) \in \mathcal{A}_{-i}$. Potential games were first formally introduced by Monderer and Shapley in [1]. In the following, we adopt the definitions in [1] and introduce several important concepts in potential games.

Definition 1 A game $\mathcal{G} = [\mathcal{N}, \mathcal{A}, \{u_i\}_{i \in \mathcal{N}}]$ is an exact potential game if there exists a function $\Phi : \mathcal{A} \to \mathbb{R}$ such that for all $i \in \mathcal{N}$, all $\mathbf{a} \in \mathcal{A}$, and all $a_i' \in \mathcal{A}_i$,

$$\Phi(a_i, \mathbf{a}_{-i}) - \Phi(a_i', \mathbf{a}_{-i}) = u_i(a_i, \mathbf{a}_{-i}) - u_i(a_i', \mathbf{a}_{-i}). \tag{1}$$

The function Φ is called an exact potential (function) for the game $\mathcal{G}$.

The definition of an ordinal potential game is given as follows.

Definition 2 A game $\mathcal{G} = [\mathcal{N}, \mathcal{A}, \{u_i\}_{i \in \mathcal{N}}]$ is an ordinal potential game if there exists a function $\Phi : \mathcal{A} \to \mathbb{R}$ such that for all $i \in \mathcal{N}$, all $\mathbf{s} \in \mathcal{S}$, and all $a_i' \in \mathcal{A}_i$,

$$\Phi(a_i, \mathbf{a}_{-i}) - \Phi(a_i', \mathbf{a}_{-i}) > 0 \Leftrightarrow u_i(a_i, \mathbf{a}_{-i}) - u_i(a_i', \mathbf{a}_{-i}) > 0. \tag{2}$$

The function Φ is called an ordinal potential function for the game $\mathcal{G}$.

Another definition of potential game is given as follows.

Definition 3 Let $\omega = (\omega_i)_{i \in \mathcal{N}}$ be a vector of positive numbers called weights. A function $\Phi : \mathcal{A} \to \mathbb{R}$ is an ω-potential for $\mathcal{G}$, if for every $i \in \mathcal{N}$ and for all $\mathbf{a} \in \mathcal{A}$, and all $a_i' \in \mathcal{A}_i$,

$$\Phi(a_i, \mathbf{a}_{-i}) - \Phi(a'_i, \mathbf{a}_{-i}) = \omega_i(u_i(a_i, \mathbf{a}_{-i}) - u_i(a'_i, \mathbf{a}_{-i})) . \quad (3)$$

$\mathcal{G}$ is called an ω-potential game if it admits an ω-potential function.

From the above definitions of three types of potential games, one may find that the potential function reflects the change in utility for any unilaterally deviating player. Any individual player's interest is aligned with the group's interest, i.e., the potential function, and any player choosing a better strategy given all other players' current strategies will necessarily lead to an improvement in the value of potential function. Note that the potential function of a potential game may not be unique. It is easily seen that every exact potential (or ω-potential) game is also an ordinal potential game with the same potential function Φ, but the inverse generally does not hold. In the sense of the particular conditions in Definitions 1-3, potential games seem to be rare. Nevertheless, potential games appear with surprising regularity when real-world situations are modeled.

2. Properties of Potential Games

In the following, we introduce some useful properties of potential games. The definition of a potential game provides a basic way to identify whether a noncooperative game is a potential game. Such a way hinges on finding the potential function of a possible potential game, which, however, is usually not straightforward. The following result provides an easy way to identify an exact potential game with continuously differentiable utility functions.

Theorem 1 (Monderer & Shapley, 1996a) A game $\mathcal{G} = [\mathcal{N}, \mathcal{A}, \{u_i\}_{i\in\mathcal{N}}]$ is an exact potential game with a potential function $\Phi(\cdot)$

if and only if

$$\frac{\partial^2 u_i}{\partial a_i \partial a_j} = \frac{\partial^2 u_j}{\partial a_i \partial a_j}, \forall i, j \in \mathcal{N}, a_i \in \mathcal{A}_i, a_j \in \mathcal{A}_j \quad (5)$$

provided that $\mathcal{A}_i$ is an interval of real numbers and u_i is continuously differentiable for all $i \in \mathcal{N}$.

Theorem 1 can be regarded as an equivalent definition of exact potential games with continuous strategies. The advantage of this equivalent definition is that one could identify an exact potential game without knowing its potential function in advance.

Given a noncooperative game, one basic question is whether it admits a Nash equilibrium as its solution. In general, the existence of a Nash equilibrium is not always guaranteed for any noncooperative game, but requires some condition, e.g., the quasi-concavity of each utility function [3]. A nice property of potential games is that the existence of a pure-strategy Nash equilibrium is always guaranteed as follows.

Theorem 2 (Monderer & Shapley, 1996a) Every finite ordinal potential game possesses a pure-strategy equilibrium.

Theorem 3 (MacKenzie & DaSilva, 2007) Let $\mathcal{G}$ be a continuous potential game with compact strategy sets. Then $\mathcal{G}$ possesses a pure-strategy equilibrium point.

As a potential game is closely related to (in fact defined on) a potential function, one may wonder what is the relation between the solution of a potential game, i.e., Nash equilibrium, and its potential function. The following result establishes such a connection.

Theorem 4 (MacKenzie & DaSilva, 2007) If Φ is an ordinal potential function of the game $\mathcal{G}$ and $\mathbf{a}^* \in \{\arg\max_{\mathbf{a}' \in \mathcal{A}} \Phi(\mathbf{a}')\}$ is a maximizer of the potential function, then $\mathbf{a}^*$ is a Nash Equilibrium of the game.

Since an exact potential (or ω-potential) game is also an ordinal potential game, Theorem 4 also applies to exact potential (or ω-potential) games. Theorem 2 indicates that every maximizer of the potential function is also a Nash equilibrium of the potential game, but the inverse may not be true. Nevertheless, if the potential game admits a unique Nash equilibrium, it must be the unique maximizer of the potential function.

The connection between a potential game and its potential function implies that even though a player is myopic and care about only his own interest, his unilateral action may still contribute to the overall wellbeing. To study how a potential is played, we introduce fictitious play process (Monderer & Shapley, 1996b). In fictitious play process, players observe the actions taken in previous stages, update their beliefs about their opponents' strategies, and choose myopic pure best responses against these beliefs. Note that every player (except for player i) takes the empirical distribution of player i's actions to be his belief about player i's mixed strategy in fictitious play process (Monderer & Shapley, 1996b). The fictitious play property of potential game is studied in the following theorem.

Theorem 5 (Monderer & Shapley, 1996a) Every finite weighted potential game has the Fictitious Play property.

We say that a game has the fictitious play property if every fictitious play process converges in beliefs to equilibrium (Monderer & Shapley, 1996b). Therefore, Theorem 5 shows that one can use the fictitious play process to find the equilibrium for every finite weighted potential game.

An improvement path is a path generated by myopic players. We say that a game has the finite improvement property (FIP) if every improvement path is finite (Monderer & Shapley, 1996a). The FIP of potential games is investigated as the following theorem.

Theorem 6 (Monderer & Shapley, 1996a) All finite ordinal potential games have the finite improvement path property.

Obviously, for finite ordinal potential game with the FIP, every maximal improvement path must terminate in an equilibrium point. That is, the myopic learning process converges to the equilibrium set (Monderer & Shapley, 1996a). However, we should note having the FIP is not equivalent to having an ordinal potential.

GAME THEORECTIC NASH EQUILIRIUM ALGORITHM

After introducing the definitions and important properties of potential games, now we would like to introduce several popular algorithms to achieve the solution, i.e., Nash equilibrium, of a potential game.

1. Best Response Dynamic

Definition 4 (MacKenzie & DaSilva, 2007) In the best response dynamic, the players will iteratively update their strategies. At each iteration, once player i has an opportunity to update its strategy, it will choose

$$a_i^{m+1} \in \arg\max_{a_i' \in \mathcal{A}_i} u_i(a_i', \mathbf{a}_{-i}^m). \tag{6}$$

In other words, whenever a player has an opportunity to revise its strategy, it will choose a strategy that maximizes its payoff, given the current strategies of the other players. The best response dynamic requires that players know the instantaneous payoffs available for each of their strategies $a_i \in \mathcal{A}_i$.

2. Random Better Response Dynamic

Definition 5 (MacKenzie & DaSilva, 2007) Random Better Reply Dynamics: At each time period, one player $i \in \mathcal{N}$ is randomly selected with all players having the same selection probability; player *i* randomly samples action $a_i^E \in \mathcal{A}_i - a_i$ with probability $(K_i - 1)$ divided equally among the elements of and switches his status quo action to a_i^E if and only if $u_i(a_i^E, a_{-i}) > u_i(a_i, a_{-i})$.

In the random better response dynamic, whenever player *i* has an opportunity to revise its strategy, it will choose a_i^{m+1} at random from the set $\left\{a_i' \in \mathcal{A}_i \mid u_i(a_i', a_{-i}^m) > u_i(a^m)\right\}$ unless this set is empty, in which case it will choose $a_i^{m+1} = a_i^m$. In other words, whenever a player has an opportunity to revise its strategy, it will choose a strategy with a higher payoff than its current payoff.

This strategy can be implemented by randomly sampling another strategy until one is found with a higher payoff than the current strategy. Hence, it does not require that the player know the utility of every strategy in its strategy space. (Friedman & Mezzetti, 2001).

Then, we will study the convergence of the above two algorithms in the following two theorems.

Theorem 7 (MacKenzie & DaSilva, 2007) Let $\mathcal{G}$ be a finite ordinal potential game. Then both the best response dynamic and the random better response dynamic will (almost surely) converge to a Nash Equilibrium in a finite number of steps.

Theorem 8 (MacKenzie & DaSilva, 2007) Let $\mathcal{G}$ be an ordinal potential game with a compact action space $\mathcal{A}$ and a continuous potential function Φ. Then the best response dynamic will (almost surely) either converge to a Nash Equilibrium or every limit point of the sequence will be a Nash Equilibrium.

The above two theorems show that both the best response dynamic algorithm and the random better response dynamic algorithm can be used to find the Nash equilibrium of the potential games.

3. Refined Algorithm

In potential games, it is known that the maximizer of the potential function is a Nash equilibrium. However, the potential games may have multiple Nash equilibriums. Thus, a Nash equilibrium may not be the maximizer of the potential function. Although the above best response dynamic and better response dynamic can obtain a Nash equilibrium of the potential games, however, we can not know which Nash equilibrium can be obtained by these algorithms. Then, if we want to obtain the maximizer of the potential function or get a solution with similar performance of the maximizer, these algorithms need to be refined. Here, we introduce a refined algorithm based on best response dynamic which is first given in [16]. The details of the algorithm are given as follows.

Step 1: $k = 0; K_{min} = \min(\{K_i\}_{i\in\mathcal{N}})$, where K_{min} is the minimum number of the strategies of players.

Step 2: $k \le K_{min}$

Implement best response dynamic using the k-th round strategies as the initial starting strategies of all players, save the strategy profile obtained by best response dynamic and its corresponding payoff;

$k = k + 1$;

End

Step 3: Return a strategy profile with the highest common payoff from the set of the strategy profiles obtained from step 2.

Although this algorithm cannot guarantee to find the maximizer of the potential function, however, it can find a Nash equilibrium with near optimal or optimal value of the potential function. In other words, this algorithm refines the pure strategy Nash equilibrium obtained by this algorithm.

THE APPLICATION OF POTENTIAL GAME TO POWER CONTROL IN WIRELESS SYSTEMS

Consider a single-cell CDMA system with total bandwidth W Hz and unspread bandwidth BHz, supporting n users. The signal-to-interference plus noise ratio (SINR) of the i-th user at the receiver is given by (Scutari, Barbarossa, & Palomar, 2006)

$$SIR_i(\mathbf{p}) = \frac{W}{B}\frac{|h_i|^2 p_i}{1+\sum_{j\neq i}|h|^2 p_j} \tag{7}$$

where $\mathbf{p} = [p_1, ..., p_n]^T$, p_i is the transmit power of the i-th user, $\{h_i\}_{i\in\mathcal{N}}$ is the set of channels from the users to Base Station (BS) and the thermal noise power is normalized to one, where $\mathcal{N} = \{1,\cdots,n\}$.

A general formulation for standard power control problems is given as below (Scutari, Barbarossa, & Palomar, 2006)

minimize p_i
subject to $f_{ji}(SIR_i(\mathbf{p})) \le \gamma_{ji}, j = 1,\cdots,n_i,\ p_i \in [0, P_i],\ i \in \mathcal{N}$. (8)

where P_i is the total power budget of each transmitter, and is a continuous function on $\mathbb{R}^+$. Note that based on the choice of $f_{ji}(\bullet)$'s, different types of QoS constraints from individual users, either in fixed/slow-fading or fast-fading Rayleigh channels, can be taken into account. However, QoS constraints give rise to a coupling among the individual powers, which makes problem (8) in general harder to study.

To overcome the above shortages, (8) can be equivalently formulated as the following normal form noncooperative game (Scutari, Barbarossa, & Palomar, 2006)

$$\mathcal{G} = \left[\mathcal{N}, \mathcal{A}, \{\log(p_i)\}_{i\in\mathcal{N}}\right] \tag{9}$$

with $\mathcal{A} = \{\mathbf{p} \in \mathbb{R}_+^n : f_{ji}(SIR_i(\mathbf{p})) \le \gamma_{ji}, p_i \le P_i, \forall i, \forall j \in \mathcal{N}\}$.

It is easy to verify that the game $\mathcal{G}$ defined in (9) is an exact potential game with potential function [6]

$$P(\mathbf{p}) = \sum_{i \in \mathcal{N}} log(p_i). \tag{10}$$

If the original problem (8) is feasible, then the set defined in (9) is compact. Moreover, multiple equilibriums may exist. If, in addition, the set is also convex, then the Nash equilibrium of the game (and thus the solution of (8)) is unique and it can be reached using the algorithms given the previous section (Scutari, Barbarossa, & Palomar, 2006). Furthermore, if the Nash equilibrium is unique, then it is also the maximizer of $P(\mathbf{p})$ based on the Theorems 2-4.

THE MULTIMODE PRECODING GAME IN MIMO MULTIPLE ACCESS CHANNELS

In this subsection, we introduce the multimode precoding strategy selection game with limited feedback for MIMO multiple access channels (Zhong, Xu, Tao, & Cai, 2010). We first established the game model of multimode precoding strategy selection in MIMO multiple access channels. Then we show that the proposed game is a potential game.

1. System Model

Consider n-user MIMO MAC equipped with M_r antennas at the receiver and M_{t_i} transmit antennas for user $i \in \mathcal{N}$, with $\mathcal{N} = \{1, \cdots, n\}$. Each user i has M_i radio frequency (RF) chains, with $1 \le M_i \le M_{t_i}$ and $\sum_{i=1}^{n} M_i \le M_r$. $L_i \in \mathcal{L}_i$ denote the number of transmission data substreams (i.e. mode) of user i, where $\mathcal{L}_i \subseteq \{0, 1, \cdots, M_i\}$ is the set of the supported mode values of user i.The MIMO MAC with multimode precoding strategy selection can be modeled as (Zhong, Xu, Tao, & Cai, 2010)

$$\mathbf{y} = \sum_{i=1}^{n} \mathbf{H}_i \mathbf{W}_i \mathbf{x}_i + \mathbf{z} = \sum_{i=1}^{n} \widetilde{\mathbf{H}}_i \mathbf{x}_i + \mathbf{z} \tag{11}$$

where $\mathbf{y}$ ($M_r \times 1$) is the channel output with multimode precoding strategy selection, $\mathbf{x}_i$ ($L_i \times 1$) is the channel input of user i, $\mathbf{z}$ ($M_r \times 1$) is a vector of additive white Gaussian noise (AWGN) with zero mean and covariance matrix equal to $\mathbf{I}_{M_r}$, $\mathbf{H}_i$ denote the $M_r \times M_{t_i}$ channel matrix of user i with each entry modeled as independent identically distributed (i.i.d.) zero-mean unit-variance circularly symmetric complex Gaussian. We assume that the number of users and the channel matrices $\{\mathbf{H}_i\}_{i \in \mathcal{N}}$ are fixed during each transmission block and change independently from one block to another according to ergodic random processes. Let $\widetilde{\mathbf{H}}_i = \mathbf{H}_i \mathbf{W}_i$ denote the $M_r \times L_i$ matrix associated with the selected matrix W_i of user i.

2. Game Theoretic Formulation

We define the payoff u_i for user i as (Zhong, Xu, Tao, & Cai, 2010)

$$u_i = \frac{1}{2}\log_2\left|\frac{\rho_i}{L_i}\widetilde{\mathbf{H}}_i\widetilde{\mathbf{H}}_i + \mathbf{N}_i\right| - \frac{1}{2}\log_2\left|\mathbf{N}_i\right| \tag{12}$$

where $\mathbf{N}_i = \sum_{j\in\mathcal{N}, j\neq i} \frac{\rho_j}{L_j}\widetilde{\mathbf{H}}_j\widetilde{\mathbf{H}}_j^{\dagger} + \mathbf{I}_{M_r}$.

Thus, the users along with their multimode precoding strategies and associated payoffs constitute a discrete noncooperative game denoted as $\mathcal{G}^E = [\mathcal{N}, \{\mathcal{W}_i\}_{i\in\mathcal{N}}, \{u_i\}_{i\in\mathcal{N}}]$, where the rational users are regarded as the players, the codebook $\mathcal{W}_i$ is the pure strategy (i.e. action) set of user i, and the individual mutual information of each user is its payoff. Formally, the game $\mathcal{G}^E$ is expressed as (Zhong, Xu, Tao, & Cai, 2010)

$$(\mathcal{G}^E): \max_{\mathbf{W}_i\in\mathcal{W}_i} u_i\left(\mathbf{W}_i, \mathbf{W}_{-i}\right), \quad for\, all\, i \in \mathcal{N} \tag{13}$$

where $\mathbf{W}_{-i}$ denotes the multimode precoding strategy selection strategies (matrices) of all the users excluding the ith user.

Assume that $\mathbf{W}_i^{'} \in \mathcal{W}_i$ is an alternate strategy of user i. We define $\widetilde{\mathbf{H}}_i^{'}$ as $\widetilde{\mathbf{H}}_i^{'} = \mathbf{H}_i\mathbf{W}_i^{'}$ and let $L_i^{'}$ denote the mode corresponding to the strategy $\mathbf{W}_i^{'}$. Then, we have (Zhong, Xu, Tao, & Cai, 2010)

$$u_i\left(\mathbf{W}_i, \mathbf{W}_{-i}\right) - u_i\left(\mathbf{W}_i^{'}, \mathbf{W}_{-i}\right) = \frac{1}{2}\log_2\left|\frac{\rho_i}{L_i}\widetilde{\mathbf{H}}_i\widetilde{\mathbf{H}}_i + \mathbf{N}_i\right| - \frac{1}{2}\log_2\left|\frac{\rho_i}{L_i}\widetilde{\mathbf{H}}_i^{'}\widetilde{\mathbf{H}}_i^{'} + \mathbf{N}_i\right|$$

$$= R_{sum}\left(\mathbf{W}_i, \mathbf{W}_{-i}\right) - R_{sum}\left(\mathbf{W}_i^{'}, \mathbf{W}_{-i}\right)$$

where $R_{sum} = \frac{1}{2}\log_2\left|\sum_{i=1}^{n}\frac{\rho_i}{L_i}\tilde{\mathbf{H}}_i\tilde{\mathbf{H}}_i^{\dagger} + \mathbf{I}_{M_r}\right|$. In the above equation, replace the sum rate R_{sum} by potential function Φ, we obtain equation (14). Thus the game $\mathcal{G}^E$ satisfies potential game definition (Monderer & Shapley, 1996a):

$$\forall \mathbf{W}_i, \mathbf{W}_i^{'}, \quad u_i\left(\mathbf{W}_i, \mathbf{W}_{-i}\right) - u_i\left(\mathbf{W}_i^{'}, \mathbf{W}_{-i}\right) = \Phi\left(\mathbf{W}_i, \mathbf{W}_{-i}\right) - \Phi\left(\mathbf{W}_i^{'}, \mathbf{W}_{-i}\right). \tag{14}$$

That is, $\mathcal{G}^E$ is a potential game with the sum rate R_{sum} as the potential function Φ.

Based on the above theoretic analysis, one can obtain the Nash equilibrium of the game $\mathcal{G}^E$ through the best response dynamic and the random better response dynamic.

THE APPLICATION OF POTENTIAL GAME TO JOINT RESOURCE ALLOCATION FOR MULTI-CHANNEL DUAL-MODE RELAY NETWORKS

1. System Models

Consider a wireless relay network depicted in Figure 2 where $N/2$ (N is an even number) user pairs attempt to exchange information (e.g., U1 attempts to exchange information with U2, U3 attempts to exchange information with U4, ..., and UN-1 attempts to exchange information with UN, where Ui denotes the user i in Figure 1), via a relay node through M available channels. U2n-1 and U2n denote the two users of the n-th user pair and $n \in \{1, 2, 3, ..., N/2\}$, respectively. Such a system is referred as a multi-channel relay network in this chapter. The relay node is assumed to know the intended targets of the communication pairs, which, in practice, can be realized by letting the users send indication signals to the relay node that stores these indication information into a table.

Suppose that M channels correspond to M frequency spectra with each channel occupying a bandwidth B, i.e., FDMA. Note that the multi-channel relay system can also work in the TDMA or CDMA manner with the channels being time slots or spread codes instead. In addition, we use h_{im} to denote the channel coefficient of the user i on channel m, with $i \in \mathcal{N} = \{1, 2, \cdots, N\}$ and $m \in \mathcal{M} = \{1, 2, \cdots, M\}$.

Some similar models can be found in previous works, e.g., the multi-way relay channel in Gunduz, Yener, Goldsmith, & Poor, (2013) and the multi-pair two-way relay channel in Yilmaz, Zakhour, Gesbert,

Figure 1. System model of MIMO multiple access channels

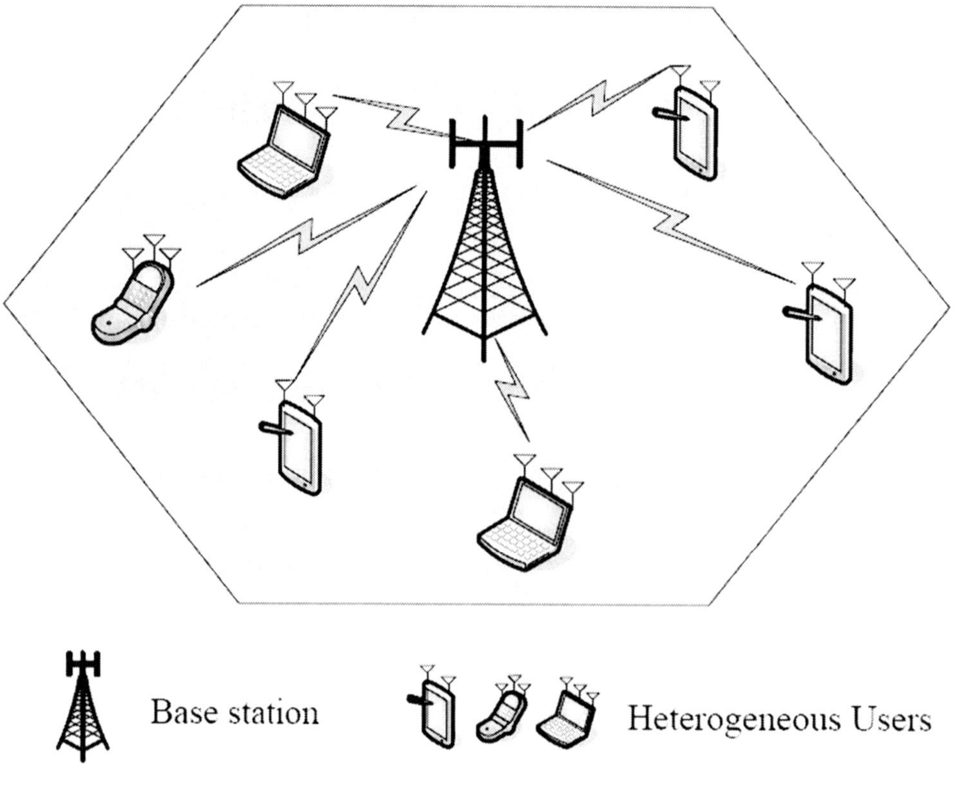

Figure 2. The system model

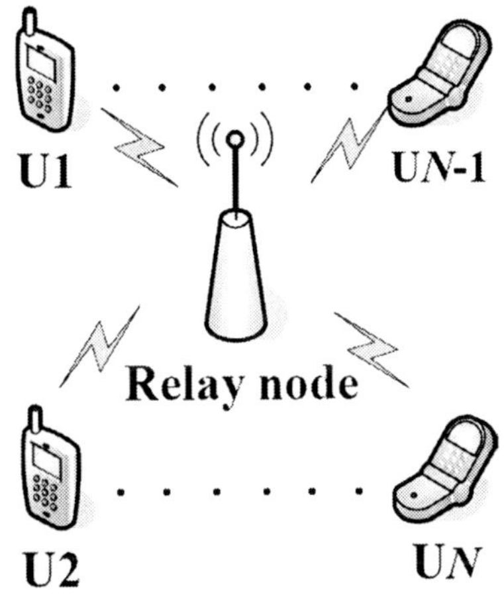

& Knopp, (2010). However, the users in Gunduz, Yener, Goldsmith, & Poor, (2013) and Yilmaz, Zakhour, Gesbert, & Knopp, (2010) are assumed to share the same channel and will interference with each other.

The relay node generally operates in the half-duplex mode, i.e., it cannot transmit and receive at the same time. The traffic pattern for relaying is divided into two phases (or time slots). The channels are assumed to be constant over both phases. We consider that the relay node does not attempt to decode the information of users and the amplify-and-forward (AF) relaying is adopted due to its simplicity.

Based on the emerging PNC (Zhang, Liew, & Lam, 2006; Katti, Gollakota, & Katabi, 2007) that allows two users exchange their information through one relay simultaneously, we propose a novel *dual-mode relay scheme* where the relay can intellectually choose between the traditional one-way or unidirectional relay mode and the two-way relay or bidirectional mode with the help of the PNC.

To be more exact, if Un and Un+1 select the same channel, then the relay node will exchange the signals from Un and Un+1 using the physical-layer network coding technique in the bidirectional mode.

Otherwise, if Un and Un+1 select different channels, the relay node will relay the signals individually on different channels in the unidirectional mode. From a practical point of view, the users from different communication pairs can not select the same channel (*this constraint is denoted as Co.*1), e.g., U1 and U3 can not choose the same channel to exchange their signals. Otherwise, collusion may happen. How to avoid the collusion will be discussed later in this chapter.

Moreover, we assume that each user can select only one channel to forward signals and the transmission of two phases of relaying will use the same selected channel, which is reasonable since spectral resources are scarce and there are often more users than available channels. We note that the model in this chapter can be extended to the cases where multiple channels can be selected by each user and different channels can be used by the two phases of relaying.

1.1 Bidirectional Mode

Suppose n is an odd number and $n \in \{1, 3, ..., N-1\}$. If Un and Un+1 select the same channel η. Then, at phase 1, the signal received at the relay node by using channel η is given by

$$y_{R_\eta} = \sqrt{P_n} h_{n\eta} x_n + \sqrt{P_{n+1}} h_{n+1\eta} x_{n+1} + z_{R_\eta} \quad (15)$$

where x_n and x_{n+1} denote the signals to be exchanged from U_n to U_{n+1}, and from U_n to U_{n+1}, respectively, P_n and P_{n+1} denote the transmit power of U_n and U_{n+1}, respectively, and z_{R_η} denotes the additive Gaussian noise that follows the distribution $\mathcal{CN}(0,\sigma^2)$. The transmission powers of Un and $Un+1$ are limited by the power budgets $\bar{P}_n$ and $\bar{P}_{n+1}$ respectively, so $0 \le P_n \le \bar{P}_n, 0 \le P_n \le \bar{P}_{n+1}$.

At the relay node, the received signal y_{R_η} of the channel η is amplified to $\mu_\eta y_{R_\eta}$, where μ_η is the amplification coefficient. Thus, we have

$$\mu_\eta^2 P_n |h_{n\eta}|^2 + \mu_\eta^2 P_{n+1} |h_{n+1\eta}|^2 + \mu_\eta^2 \sigma^2 \le P_{R_\eta}. \quad (16)$$

In phase 2, the relay node broadcasts the signal $\mu_\eta y_{R_\eta}$ to both U_n and U_{n+1}. The signals received at U_n and U_{n+1} can be expressed as

$$\tilde{y}_n = h_{n\eta} \mu_\eta y_{R_\eta} + z_{n\eta} = \underbrace{\mu_\eta h_{n\eta}^2 \sqrt{P_n} x_n}_{self\ interference} + \mu_\eta h_{n\eta} h_{n+1\eta} \sqrt{P_{n+1}} x_{n+1} + \mu h_{n\eta} z_{R_\eta} + z_{n\eta} \quad (17)$$

$$\tilde{y}_{n+1} = h_{n+1\eta} \mu_\eta y_{R_\eta} + z_{n+1\eta} = \underbrace{\mu_\eta h_{n+1\eta}^2 \sqrt{P_{n+1}} x_{n+1}}_{self\ interference} + \mu_\eta h_{n\eta} h_{n+1\eta} \sqrt{P_n} x_n + \mu_\eta h_{n+1\eta} z_{R_\eta} + z_{j\eta} \quad (18)$$

where $z_{n\eta}$ and $z_{n+1\eta}$ are additive Gaussian noise at U_n and U_{n+1}, respectively.

The principle of PNC is then applied at U_n and U_{n+1} for decoding. Assume that the channel state information (CSI) are perfectly known at both U_n and U_{n+1}, the self interference components can be completely canceled out, yielding

$$y_n = \mu_\eta h_{n\eta} h_{n+1\eta} \sqrt{P_{n+1}} x_{n+1} + \mu_\eta h_{n\eta} z_{R_\eta} + z_{n\eta} \quad (19)$$

$$y_{n+1} = \mu_\eta h_{n\eta} h_{n+1\eta} \sqrt{P_n} x_n + \mu_\eta h_{n+1\eta} z_{R_\eta} + z_{n+1\eta}. \quad (20)$$

1.2 Unidirectional Mode

If U_n and U_{n+1} selects different channels, e.g., U_n selects channel l and U_{n+1} selects channel d, the signals received at the selected channels at the relay node can be given by

$$y_{R_l} = \sqrt{P_n} h_{nl} x_n + z_{R_l} \quad (21)$$

$$y_{R_d} = \sqrt{P_{n+1}} h_{n+1d} x_{n+1} + z_{R_d} \quad (22)$$

where z_{R_l} and z_{R_d} denote the received Gaussian noise and follow the same distribution $\mathcal{CN}(0,\sigma^2)$.

Furthermore, the transmission power of U_n and U_{n+1} should satisfy $P_n \le \min(\bar{P}_n, P_l^{th} / g_{nl})$ and $P_{n+1} \le \bar{P}_n$.

In phase 2, the channel *l* and the channel *d* transmit the amplified received signals $\mu_l y_{R_l}$ and $\mu_d y_{R_d}$ to U_n and U_{n+1} respectively on their own channels without interfering with each other. Note that, the power of $\mu_l y_{R_l}$ and $\beta_d y_{R_d}$ need to satisfy the power budget P_{R_l} and P_{R_d} interfering to each other respectively, i.e.,

$$\mu_l^2 P_n |h_{nl}|^2 + \mu_l^2 \sigma^2 \le P_{R_l} \tag{23}$$

$$\mu_d^2 P_{n+1} |h_{n+1d}|^2 + \mu_d^2 \sigma^2 \le P_{R_d}. \tag{24}$$

The signals received at U*n* and U*n*+1 can be then expressed as

$$y_n = \mu_d h_{nd} \left(\sqrt{P_{n+1}} h_{n+1d} x_{n+1} + z_{R_d} \right) + z_{nd} \tag{25}$$

$$y_{n+1} = \mu_l h_{n+1l} \left(\sqrt{P_n} h_{nl} x_n + z_{R_l} \right) + z_{n+1l} \tag{26}$$

where μ_l and μ_d denote the amplification coefficients at channel *l* and channel *d*, respectively, z_{nd} and z_{n+1l} denote the received Gaussian noises and follow the same distribution $\mathcal{CN}(0, \sigma^2)$.

2. Optimal Resource Allocation Problem

This section studies global optimization problem to determine the optimal spectrum access strategy (i.e., the optimal channel, relay mode, and discrete power selection strategies of the users) such that the overall performance of the dual-mode multi-channel relay network is maximized under the stated constraints. It is assumed that all additive noises are independent and $0 < P_{R_m} \le P_R, \forall m \in \mathcal{M} = \{1, 2, \cdots, M\}$.

Assume that all additive noises are independent and $0 < P_{R_m} \le P_R, \forall m \in \mathcal{M} = \{1, 2, \cdots, M\}$. Considering that the power level in practical wireless communication systems is usually quantized into discrete values (Xing & Chandramouli, 2008), the transmission power of U*i*, $\forall i \in \mathcal{N} = \{1, 2, \cdots, N\}$ is chosen from the finite set $\mathcal{P}_i$, where $\mathcal{P}_i = \{P_{i1}, P_{i2}, \cdots, P_{i\Omega_i}\}$, Ω_i is the cardinality of $\mathcal{P}_i$, and $P_{i1} = 0 < P_{i2} < \cdots < P_{i\Omega_i} = \bar{P}_i$. We introduce the integer variable

$$\mathbf{a}_i = \begin{cases} (\mathrm{m}_i, P_i), P_i > 0 \\ \mathbf{0}, P_i = 0 \end{cases} \tag{27}$$

to denote a resource management strategy, where m_i denotes the index of the channel selected by user *i*, and $\mathcal{A}_i = \mathcal{M} \times \mathcal{P}_i$ denotes the set of all possible resource management strategies. For $\forall i \in \mathcal{N}$, the cardinality of $\mathcal{A}_i$ is $K_i = \Omega_i M$. Then, a resource management is essentially a joint channel, relay mode, and discrete power selection problem.

In order to guarantee the quality of service (QoS) of the users, the *minimum transmission rate constraint* should be considered (this constraint is denoted as Co.2). That is, each user shall satisfy $c_i \geq c_i^{min}, \forall i \in \mathcal{N}$, where c_i denotes the rate of user *i* and c_i^{min} denotes the minimum required transmission rate. Constrains Co.1 and Co.2 are summarized in Table 1.

Suppose that $n \in \{1,3,\cdots,N-1\}$, U*n* selects channel m_n and U_{n+1} selects channel m_{n+1}. If $m_n = m_{n+1} = m$, then the transmission rates of U_n and U_{n+1} are given by (28) and (29), following the equations (19) and (20), respectively.

$$c_n = B\log\left(1+\frac{\mu_m^2 P_n |h_{nm}|^2 |h_{n+1m}|^2}{\mu_m^2 |h_{n+1m}|^2 \sigma^2 + \sigma^2}\right) \tag{28}$$

$$c_{n+1} = B\log\left(1+\frac{\mu_m^2 P_{n+1} |h_{nm}|^2 |h_{n+1m}|^2}{\mu_m^2 |h_{nm}|^2 \sigma^2 + \sigma^2}\right) \tag{29}$$

where

$$\mu_m^2 = \frac{P_{R_m}}{P_n |h_{nm}|^2 + P_{n+1} |h_{n+1m}|^2 + \sigma^2}. \tag{30}$$

If $m_n \neq m_{n+1}$, then the relay node uses the unidirectional mode, and based on the equations (25) and (26), the transmission rates are given by

$$c_n = B\log\left(1+\frac{\mu_{m_n}^2 P_n |h_{nm_n}|^2 |h_{n+1m_n}|^2}{\mu_{m_n}^2 |h_{n+1m_n}|^2 \sigma^2 + \sigma^2}\right) \tag{31}$$

$$c_{n+1} = B\log\left(1+\frac{\mu_{m_{n+1}}^2 P_{n+1} |h_{nm_{n+1}}|^2 |h_{n+1m_{n+1}}|^2}{\mu_{m_{n+1}}^2 |h_{nm_{n+1}}|^2 \sigma^2 + \sigma^2}\right) \tag{32}$$

where

$$\mu_{m_n}^2 = \frac{P_{R_{m_1}}}{P_n |h_{nm_n}|^2 + \sigma^2},\ \mu_{m_{n+1}}^2 = \frac{P_{R_{m_{n+1}}}}{P_{n+1} |h_{n+1m_{n+1}}|^2 + \sigma^2}. \tag{33}$$

Furthermore, if all possible pure strategies of a user can not satisfy the stated constraints, then this user should be shut down and the power level of this user is set to be 0.

For multi-channel bidirectional relay network, the optimization objective is naturally related to the total information bits that the users can transfer. On the other hand, since users are usually energy limited, the power consumption is also a major concern. However, there is a fundamental tradeoff between the rate efficiency and power efficiency, since decreasing the transmit power usually degrades the transmission rate. Thus, to take into account both issues in the joint resource management, we consider the following optimization objective:

$$(Network\ utility) \quad U = c_{sum} - f(P_1, P_2, \cdots, P_N) \tag{34}$$

where $c_{sum} = \sum_{i \in \mathcal{N}} c_i$ denotes the sum rate of the system whose unit is *bit/s*, $f(P_1, P_2, ..., P_N)$ is a monotonic positive non-decreasing pricing function of $(P_1, P_2, ..., P_N)$ which reflects the *total energy consumption* of the system. Note that, this *energy-aware* optimization objective is also called *network utility* (Liang & Dandekar, 2007; Xiao & Shroff, 2001).

In this chapter, we would consider a linear pricing function $f(P_1, P_2, \cdots, P_N) = \sum_{i \in \mathcal{N}} \varepsilon_i P_i$ for convenience, where $\varepsilon_i > 0$ is the pricing factor of the power level of user i. The larger the ε_i is, the more aware of the energy consumption the user i is. Furthermore, the pricing factor ε_i is considered to have units *bits*/($W \bullet s$) so that it is consistent with the units of U in *bits/s*, $\forall i \in \mathcal{N}$.

The network utility c_{sum} will increase as the powers of the users become larger, which, on the other hand, will also increases the pricing function. Thus, the proposed network utility function will induce the system to achieve an appropriate resource management strategy profile that can balance the transmission rate and power consumption.

In summary, the optimal resource management problem in a bidirectional relay network with Co.1 and Co.2 can be expressed as:

$$\text{(OP.1)} \ \max_{\{\mathbf{a}_i \in \mathcal{S}_i\}_{i \in \mathcal{N}}} U \tag{35}$$

where

$$S_i = \{\mathbf{a}_i \in \mathcal{A}_i, \mathbf{a}_i \neq 0 \,|\, \text{Co.1 and Co.2 are both satisfied.}\}$$

$$\bigcup \{\mathbf{a}_i = 0 \,|\, \text{User } i \text{ violates a constraint.}\} \tag{36}$$

is the set of feasible spectrum access strategies of user i, and K_i is the cardinality of $\mathcal{S}_i$.

As OP.1 is a discrete optimization problem, finding its globally optimal solution usually needs the exhaustive search algorithm that has to search for $\mathcal{O}(\prod_{i \in \mathcal{N}} K_i)$ combinations. However, exhaustive search

Table 1. The definition of constraints

Constraints	Definition
Co.1	The users from different transmission pairs must not select the same channel.
Co.2	$c_i \geq c_i^{min}, \forall i \in \mathcal{N}$

algorithm is often impractical due to the following two reasons: 1) the computational complexity of exhaustive search algorithm will progressively increase with usually increases exponentially the number of the users and the strategies, which is prohibitive in practical systems; 2) it is very difficult to pick out the strategy profiles that violate the stated constraints in advance, which makes exhaustive search even more in efficient. Considering the fact that searching the globally optimal solution is highly complex and probably inefficient in practice, we would like to propose more efficient and practical approaches to find a solution (probably suboptimal) to formulated discrete optimization problem, with the performance close to the global optimum. For this purpose, we would invoke game theory in the next section.

3. Joint Resource Management Game

Since it is difficult to solve the discrete optimization problem (35) especially in large scale networks, we would like to introduce a game theoretical framework, namely potential game. Potential games [1] are a class of games where the maximizer of the potential functions constitute the pure strategy Nash equilibrium. In other words, if the potential function is carefully designed so that it admits a maximizer and accommodates itself to the objective of the optimization problem, a local search algorithm can be proposed to obtain optimal or near optimal solutions. Therefore, we will study the joint resource management from the perspective of a potential game in this section.

3.1 Game Theoretic Reformulation

In order to optimize a certain performance metric, the decision made by each user should depend on the decision made by the other one. This leads to a joint resource management game among the users as follows.

$$\mathcal{G} = [\mathcal{N}, \{\mathcal{A}_i\}_{i\in\mathcal{N}}, \{u_i\}_{i\in\mathcal{N}}] \tag{37}$$

where $\mathcal{N}$ is the set of users, $\mathcal{A}_i$ which is the set of all possible resource management strategies is the set of pure strategies (actions) and u_i is the utility function of user *i*.

To single out the pure strategy of user *i*, let $\mathbf{a}_{-i} = (m_{-i}, P_{-i})$ denote the set of pure strategies of user *j* $\neq i$, i.e., $\mathbf{a}_{-i} = (\{\mathbf{a}_j\}_{j\in\mathcal{N}\setminus i})$. Furthermore, the pure strategy profile is denoted as

$$(\mathbf{a}_i, \mathbf{a}_{-i}) = (\{\mathbf{a}_i\}_{i\in\mathcal{N}}) = (\mathbf{a}_1, \cdots, \mathbf{a}_N). \tag{38}$$

The utility is a measure of the relative satisfaction of the decision maker and may not be unique. Any function that includes the elements of $\{\mathcal{A}_i\}_{i\in\mathcal{N}}$ in the desired order is a candidate utility function [14]. Normally, different utility functions may result in different analysis and conclusions. How to design the utility function depends on the preference requirement of the terminals in the system.

Note that, in order to guarantee the feasibility of the Nash equilibrium, $\forall i \in \mathcal{N}$, u_i is defined in (39)

$$u_i(\mathbf{a}_i,\mathbf{a}_{-i}) = \begin{cases} \sum_{k\in\mathcal{N}} c_k(\mathbf{a}_i,\mathbf{a}_{-i}) - \varepsilon_i P_i + \sum_{k\in\mathcal{N}} \beta_k^c \Theta\left(c_k(\mathbf{a}_i,\mathbf{a}_{-i}) - c_k^{min}\right), \forall j \in \mathcal{N}, \hat{m}_j(\mathbf{a}_i,\mathbf{a}_{-i}) = 0 \\ -\varepsilon_i P_i - \beta^m, \exists j \in \mathcal{N}, \hat{m}_j(\mathbf{a}_i,\mathbf{a}_{-i}) = -1. \end{cases} \tag{39}$$

where $\mathbf{a}_i = (m_i, P_i) \in \mathcal{A}_i$, $\beta^m \geq 0$ is a penalty factor of Co.1 and $\beta_i^c \geq 0$ is a penalty factor of Co.2 of user *i*, $\Theta(s)$ is an *penalty function* defined as

$$\Theta(s) = \begin{cases} -1, s < 0 \\ 0, otherwise \end{cases} \tag{40}$$

and $\hat{m}_j(\mathbf{a}_i,\mathbf{a}_{-i})$ is defined in (41).

$\hat{m}_j(\mathbf{a}_i,\mathbf{a}_{-i}) = 0$, when $m_j = m_{j+1}$ if *j* is odd, or $m_j = m_{j-1}$, if *j* is even, or $m_j \neq m_{j'}, \forall j' \in \mathcal{N} \setminus j$, or $P_j =$ 0.

$\hat{m}_j(\mathbf{a}_i,\mathbf{a}_{-i}) = -1$, when $m_j = m_{j'}$, where $j' \neq j+1$, if j is odd, and $j' \neq j-1$, if j is even, $\forall j' \in \mathcal{N} \setminus j$, and $P_j > 0$ (41)

Note that, the physical meaning of (41) is: once two users from different communication pairs select the same channel, then $\hat{m}_j(\mathbf{a}_i,\mathbf{a}_{-i}) = -1$ and every users will be severely punished, if this situation does happen, then $\hat{m}_j(\mathbf{a}_i,\mathbf{a}_{-i}) = 0$ and no punishment will be imposed.

The equation (40) is an indication function of the Co.1 and the term $\sum_{i\in\mathcal{N}} \beta_i^c \Theta\left(c_i(\mathbf{a}_i,\mathbf{a}_{-i}) - c_i^{min}\right)$ in (39) represents the minimum transmission rates constraint (i.e., Co.2).

Given the utility function defined in (39), the user who selects a strategy that violates Co.1 or Co.2 will be severely punished.

If the Co.2 is violated, the punishment is implemented by each user itself. As a global constraint that depends on all uses, if Co.1 is violated, the relay node shall penalize all users to guarantee the global requirement.

We will prove in the next subsection that, under certain conditions, our proposed utility function can guarantee the feasibility of the system with pure strategy set $\mathcal{A}_i$ for each user.

Note that, the penalty factors reflect the penalty intensity when users violate the stated constraints, and the penalty functions indicate whether users violate the stated constraints or not. Thus β^m can be considered to have a unit *bits/s/Hz* and β^c is considered to have no unit so that u_i is consistent with the unit of c_i in *bits/s/Hz*, $\forall i \in \mathcal{N}$.

In the resource management game, each user selects a strategy, such that $\mathbf{a}_i$ maximizes his utility on the strategies of all the other terminals in the system. Formally, it can be expressed as:

$$\mathcal{G} : \max_{\{\mathbf{a}_i \in \mathcal{A}_i\}_{i\in\mathcal{N}}} u_i(\mathbf{a}_i,\mathbf{a}_{-i}), \forall i \in \mathcal{N}. \tag{41}$$

3.2 Properties of the Proposed Game

In this subsection, we will show the relation between the game $\mathcal{G}$ and the network utility optimization problem (35). To build a link between $\mathcal{G}$ and the network utility optimization problem (35), we analyze the property of $\mathcal{G}$ and its solution, i.e., Nash equilibrium.

For this purpose, let us define a function $U1$ in (42). Then, we have the following result showing that $\mathcal{G}$ is a potential game.

$$U1(\mathbf{a}_i,\mathbf{a}_{-i}) = \begin{cases} U + \sum_{k\in\mathcal{N}} \beta_k^c \Theta\left(c_k(\mathbf{a}_i,\mathbf{a}_{-i}) - c_k^{min}\right), & \forall j \in \mathcal{N}, \hat{m}_j(\mathbf{a}_i,\mathbf{a}_{-i}) = 0 \\ -\sum_{k\in\mathcal{N}} \varepsilon_k P_k - \beta^m, & \exists j \in \mathcal{N}, \hat{m}_j(\mathbf{a}_i,\mathbf{a}_{-i}) = -1. \end{cases} \tag{42}$$

Lemma 1 Given the utility function defined in (25), the $\mathcal{G}$ is a potential game with potential function U1.

Lemma 1 can be proved by using the definition of potential games. Based on Lemma 1, we can provide the following result on the Nash equilibrium of $\mathcal{G}$ and the relation between $\mathcal{G}$ and the global network optimization problem (35).

Theorem 9 Given the utility function defined in (35), if $\beta_i^c > U_{max} \triangleq \max_{\{\mathbf{a}_i\in\mathcal{A}_i\}_{i\in\mathcal{N}}}(U)$ when $P_i >$ 0, and $\beta_i^c = 0$ when $P_i = 0$, $\forall i \in \mathcal{N}$, then $\mathcal{G}$ has at least one feasible pure strategy Nash equilibrium, and the optimal solution to OP.1 constitutes a maximizer of U1 that is also a feasible pure strategy Nash equilibrium of $\mathcal{G}$.

Proof: Without loss of generality (w.l.o.g.), we assume that $(\mathbf{a}_1^o, \mathbf{a}_2^o, \cdots, \mathbf{a}_N^o)$ is a maximizer for U1. Then we can obtain

$$U1(\mathbf{a}_i^o,\mathbf{a}_{-i}^o) \geq U1(\mathbf{a}_i,\mathbf{a}_{-i}^o), \forall i \in \mathcal{N}, \mathbf{a}_i \neq \mathbf{a}_i^o, and \mathbf{a}_i \in \mathcal{A}_i. \tag{43}$$

Thus, based on the definition of the pure strategy Nash equilibrium, we know that $(\mathbf{a}_1^o, \mathbf{a}_2^o, \cdots, \mathbf{a}_N^o)$ is a pure strategy Nash equilibrium of $\mathcal{G}^P$ which is defined as

$$\mathcal{G}^P = [\mathcal{N}, \{\mathcal{A}_i\}_{i\in\mathcal{N}}, \{u_i = U1\}_{i\in\mathcal{N}}]. \tag{44}$$

It is not difficult to see that the set of pure strategy Nash equilibrium of $\mathcal{G}^P$ coincides with the set of the pure strategy Nash equilibrium of $\mathcal{G}$. Thus, $(\mathbf{a}_1^o, \mathbf{a}_2^o, \cdots, \mathbf{a}_N^o)$ is a pure strategy Nash equilibrium of $\mathcal{G}$. When the conditions in Theorem 9, it is straight to observe that if $U1(\mathbf{a}_i,\mathbf{a}_{-i}) \geq 0$ with $\mathbf{a}_i \in \mathcal{A}_i, \forall i \in \mathcal{N}$, then $(\mathbf{a}_i, \mathbf{a}_{-i})$ should be feasible. Since $U1(\mathbf{a}_1^o, \mathbf{a}_2^o, \cdots, \mathbf{a}_N^o) \geq U1(0,0,\cdots,0) = 0$, $(\mathbf{a}_1^o, \mathbf{a}_2^o, \cdots, \mathbf{a}_N^o)$ is a feasible strategy profile.

Furthermore, if $(\mathbf{a}_i, \mathbf{a}_{-i})$ is a feasible strategy profile, then from (29), we can obtain $U1(\mathbf{a}_i,\mathbf{a}_{-i}) = U(\mathbf{a}_i,\mathbf{a}_{-i})$. Thus, $U1(\mathbf{a}_1^o, \mathbf{a}_2^o, \cdots, \mathbf{a}_N^o) = U(\mathbf{a}_1^o, \mathbf{a}_2^o, \cdots, \mathbf{a}_N^o)$. That is, $(\mathbf{a}_1^o, \mathbf{a}_2^o, \cdots, \mathbf{a}_N^o)$ is also an optimal solution to OP.1. Thus, the optimal solution to OP.1 constitutes a feasible pure strategy Nash equilibrium of $\mathcal{G}$. Hence the theorem is proved.

Theorem 9 indicates that the globally optimal solution to OP.1 must be an pure strategy Nash equilibrium of $\mathcal{G}$, although the inverse may not hold. This implies that we shall find a pure strategy Nash equilibrium of $\mathcal{G}$ to solve OP.1. The existence of feasible pure strategy Nash equilibrium of $\mathcal{G}$ is guaranteed under a mild condition. Furthermore, if the pure strategy Nash equilibrium of $\mathcal{G}$ is unique, then it must be feasible and also is the globally optimal solution to OP.1. Yet, it is still unclear whether a pure strategy Nash equilibrium can satisfy the specified constraints or not. This problem, i.e., the feasibility of Nash equilibrium, is investigated in the following theorem.

4. Joint Resource Allocation Algorithms

Based on the theoretic analysis in the previous section, in this section, we propose two joint channel, relay mode and discrete power selection algorithms to obtain the feasible and efficient resource management strategies. In the proposed algorithms, according to Theorem 9, we choose

$$\begin{cases} \beta_i^c = \beta > \max(\beta_{max}, U_{max}), if \quad P_i > 0, \forall i \in \mathcal{N} \\ \beta_i^c = 0, if \quad P_i = 0, \forall i \in \mathcal{N} \\ \beta^m = N\beta \end{cases} \tag{45}$$

to guarantee feasible strategies. In practice, one can predefine a sufficiently large β for all users and the relay node.

4.1 Algorithm 1

First, we propose an asynchronous update algorithm based on the concept of random better reply dynamic [3]. The algorithm is formally described in Algorithm 1 (A1) where $\mathbf{a}_{i,t}$ denotes the strategy of user i at iteration t. The convergence of A1 is studied in the following Theorem 10.

Asynchronous Updating Algorithm Based on Random Better Reply Dynamic (A1)

Step 1: The users send training sequences and informs c_i^{min} to the relay node, the relay node estimates the channel state information (CSI) of all users.

Step 2: Set the initial resource management strategy $\mathbf{a}_{i,0} = 0$, the utility $u_{i,0} = 0, \forall i \in \mathcal{N}$, and $t = 0$.

Step 3: Repeat

Step 4: Update $t = t + 1$.

Step 5: Update $\mathbf{a}_{i,t} = \mathbf{a}_{i,t-1}$, $u_{i,t} = u_{i,t-1}$, $\forall i \in \mathcal{N}$.

Step 6: All users inform the relay node of their currently strategies $\mathbf{a}_{i,t}, \forall i \in \mathcal{N}$.

Step 7: FOR$\{i \in N\}$

Step 8: Active user i randomly chooses a strategy $\mathbf{a}_{i,t}^r = (m_{i,t}^r, P_{i,t}^r)$ with uniform probability and then informs the relay node of the chosen strategy. The relay node evaluates $\hat{m}_j(\mathbf{a}_{i,t}, \mathbf{a}_{-i,t}), \forall j \in \mathcal{N}$.

Step 9: IF $\exists j \in \mathcal{N}, \hat{m}_j(\mathbf{a}_{i,t}, \mathbf{a}_{-i,t}) = -1$

Step 10: The relay node feedbacks an indication to active user i.

Step 11: **ELSE**

Step 12: The relay node evaluates $u_i(\mathbf{a}^r_{i,t}, \mathbf{a}_{-i,t})$ and feedbacks it to active user *i*.

Step 13: **EDNIF**

Step 14: Active user *i* updates his strategy following: if $u_i(\mathbf{a}^r_{i,t}, \mathbf{a}_{-i,t}) > u_{i,t}$, then

$$\mathbf{a}_{i,t} = \mathbf{a}^r_{i,t}, u_{i,t} = u_i(\mathbf{a}^r_{i,t}, \mathbf{a}_{-i,t}).$$

Step 15: Otherwise, the strategy of user *i* remain unchanged.

Step 16: **ENDFOR**

Step 17: UNTIL the algorithm converges.

Step 18: RETURN $\{\mathbf{a}_i(t-1)\}$

Theorem 10 If (45) holds, A1 converges to a feasible pure strategy Nash equilibrium of $\mathcal{G}$ in finite steps. with probability 1, i.e., let *T* be a random variable such that $(\mathbf{a}_{i,t}, \mathbf{a}_{-i,t}) \in \mathcal{A}^*$ where $\mathcal{A}^*$ is pure strategy Nash equilibrium set of $\mathcal{G}$, for all $t > T$, then $Prob(T < \infty) = 1$.

Proof: The better reply dynamic generates a Markov process on the finite state space $\{\mathcal{A}_i\}_{i\in\mathcal{N}}$ of action profiles. In A1, once a user changes his strategy, we regard it as a strategy change event. Let t^j denote the time when the *j*th strategy change event happens. Then, once the active user *i* changes his resource management strategy at instant t^j, we have

$$u_i(\mathbf{a}_{i,t^j}, \mathbf{a}_{-i,t^j}) - u_i(\mathbf{a}_{i,t^j-1}, \mathbf{a}_{-i,t^j}) > 0. \tag{46}$$

Since $\mathcal{G}$ is a potential game with potential function U1, then

$$U1(\mathbf{a}_{i,t^j}, \mathbf{a}_{-i,t^j}) - U1(\mathbf{a}_{i,t^j-1}, \mathbf{a}_{-i,t^j}) = u_i(\mathbf{a}_{i,t^j}, \mathbf{a}_{-i,t^j}) - u_i(\mathbf{a}_{i,t^j-1}, \mathbf{a}_{-i,t^j}) > 0. \tag{47}$$

Therefore, the utility of each user and the network utility are strictly increasing.

Since $\mathcal{G}$ is a finite game and the optimal value of U1 is finite, i.e., $U1 < +\infty$, thus A1 must ends in finite steps.

When A1 converges to a resource management strategy profile $(\mathbf{a}^*_i, \mathbf{a}^*_{-i})$, for an arbitrary user *i* and an alternative pure strategy $\mathbf{a}_i \neq \mathbf{a}^*_i$ and $\mathbf{a}_i \in \mathcal{A}_i$, we have

$$u_i(\mathbf{a}^*_i, \mathbf{a}^*_{-i}) - u_i(\mathbf{a}_i, \mathbf{a}^*_{-i}) \geq 0. \tag{48}$$

According to the definition of pure strategy Nash equilibrium in [2], the resource management strategy profile $(\mathbf{a}^*_i, \mathbf{a}^*_{-i})$ is a pure strategy Nash equilibrium of $\mathcal{G}$. This means that if the initial point of A1, $(\mathbf{a}_i, \mathbf{a}_{-i})$, is not a pure strategy Nash equilibrium of our proposed game, then there is a positive probability of moving from $(\mathbf{a}_i, \mathbf{a}_{-i})$ to $(\mathbf{a}^*_i, \mathbf{a}^*_{-i})$ in a finite number of periods where $(\mathbf{a}^*_i, \mathbf{a}^*_{-i}) \neq (\mathbf{a}_i, \mathbf{a}_{-i})$ since our proposed game is a potential game which has finite improvement property (FIP) [1].

In contrast, if the initial point of A1 is a pure strategy Nash equilibrium of our proposed game (denoted as $(\mathbf{a}_i^*,\mathbf{a}_{-i}^*)$), then the probability of moving from $(\mathbf{a}_i^*,\mathbf{a}_{-i}^*)$ to $(\mathbf{a}_i, \mathbf{a}_{-i})$ in a finite number of periods is zero where $(\mathbf{a}_i^*,\mathbf{a}_{-i}^*) \neq (\mathbf{a}_i,\mathbf{a}_{-i})$.

Given the parameters in (45) and the conditions in Theorem 2 are satisfied, the pure strategy Nash equilibrium $(\mathbf{a}_i^*,\mathbf{a}_{-i}^*)$ must be feasible. Hence the theorem is proved.

At each iteration of A1, the relay node only needs to compute the utility of each user once and each user only needs to make a comparison based on the feedback from the rely node. The amount of feedback is small, since it is just a real number of user's utility or an indication information. Thus, the signaling overhead and computation complexity of A1 are low.

From Theorems 9 and 10, A1 converges to a feasible pure strategy Nash equilibrium efficiently and even can *converge the optimal solution to* OP.1 when $\mathcal{G}$ possesses unique pure strategy Nash equilibrium. Later we show that A1 works well in practice. In some situation, $\mathcal{G}$ may have multiple pure strategy Nash equilibriums in practice. In this case, a pure strategy Nash equilibrium may be a locally optimal solution. Nevertheless, we would like to propose another algorithm that can achieve faster convergence and better global performance in the next subsection.

5.2 Algorithm 2

In this subsection, we propose an iterative algorithm based on the best response dynamic algorithm which is formally described in Algorithm 2 (A2), where $\mathbf{a}_i^k$ is the *k*th strategy of user *i*, $\mathbf{a}_{i,t}$ denotes the strategy of user *i* at iteration *t*, $K_{rep} \in \mathbb{Z}^+$ and $K_{rep} \geq 1$. The convergence property of A2 is investigated in Theorem 11.

Iterative Algorithm Based on Best Response Dynamic (A2)

Step 1: Let $k = 1$.
Step 2: While $k \leq K_{rep}$
Step 3: Let the initial strategy $\mathbf{a}_{i,0} = \mathbf{a}_i^k, \forall i \in \mathcal{N}$ and $t = 0$.
Step 4: **REPEAT**
Step 5: Update $t = t + 1$.
Step 6: Update $\mathbf{a}_{i,t} = \mathbf{a}_{i,t-1}$.
Step 7: FOR $i \in N$
Step 8: $\mathbf{a}_{i,t} = \arg\max_{\mathbf{a}_i \in \mathcal{A}_i} U1(\mathbf{a}_i,\mathbf{a}_{-i,t})$.
Step 9: **ENDFOR**
Step 10: UNTIL $U1(\mathbf{a}_{i,t},\mathbf{a}_{-i,t}) = U1(\mathbf{a}_{i,t-1},\mathbf{a}_{-i,t-1}), \forall i \in \mathcal{N}$.
Step 11: Save the converged strategy profile and its corresponding value of U1.
Step 12: $k = k + 1$;
Step 13: **END**
Step 14: RETURN the strategy profile with largest value of U1.

Theorem 11 Given that (45) holds, A2 converges to a pure strategy Nash equilibrium of $\mathcal{G}$. Furthermore, A2 converges to an optimal solution to OP.1 when $N = 2$ and $K_{rep} = \min(\{K_i\}_{i\in\mathcal{N}})$.

Proof: To prove the first part, consider, consider step 4 to step 10 in A2. Once an update of user *i*'s strategy occurs at iteration $t > 0$, we have

$$U1(\mathbf{a}_{i,t+1}, \mathbf{a}_{-i,t}) > U1(\mathbf{a}_{i,t}, \mathbf{a}_{-i,t}), \forall i \in \mathcal{N}. \tag{49}$$

For any user $i \in \mathcal{N}$, once

$$U1(\mathbf{a}_{i,t+1}, \mathbf{a}_{-i,t}) = U1(\mathbf{a}_{i,t}, \mathbf{a}_{-i,t}) \tag{50}$$

A2 will go to step 11. Thus, A2 has a strictly increasing procedure in step 4 to step 10. Since $U1 < +\infty$ and all users have finite discrete strategies, A2 must converge. Furthermore, if A2 converges to a strategy profile $(\mathbf{a}_i^*, \mathbf{a}_{-i}^*)$, we can have

$$U1(\mathbf{a}_i^*, \mathbf{a}_{-i}^*) \geq U1(\mathbf{a}_i, \mathbf{a}_{-i}^*) \tag{51}$$

where $\mathbf{a}_i \neq \mathbf{a}_i^*$ and $\mathbf{a}_i \in \mathcal{A}_i$.

Obviously, (51) coincides with the definition of pure strategy Nash equilibrium in game theory [2]. Thus, $(\mathbf{a}_i^*, \mathbf{a}_{-i}^*)$ is a pure strategy Nash equilibrium of $\mathcal{G}^P$. From Lemma 1, we know that $\mathcal{G}$ is a potential game with potential function U1. Therefore, we can conclude that $(\mathbf{a}_i^*, \mathbf{a}_{-i}^*)$ is also a pure strategy Nash equilibrium of $\mathcal{G}$. Moreover, given the parameters in (45) and based on Theorem 11, if (45) holds, $(\mathbf{a}_i^*, \mathbf{a}_{-i}^*)$ must be feasible. Hence the first part is proved.

Now we prove the second part. When $N = 2$ and $K_{rep} = \min(\{K_i\}_{i\in\mathcal{N}})$, A2 can search all possible strategy profiles in $\mathcal{G}$ and find out all pure strategy Nash equilibrium of $\mathcal{G}$. Thus, the output strategy profile must be a maximizer of U1. From Theorem 10, we can further conclude that the output strategy profile of A2 is an optimal solution to OP.1. Hence the second part is proved.

From the above analysis, one can see that A2 is with linear complexity, since it only need to search $\mathcal{O}(T\sum_{i\in\mathcal{N}} K_i)$ combinations. Thus, the complexity of A2 is significantly reduced compared to the exhaustive search algorithm. From Theorems 1,2,11, A2 can converge to an optimal solution to OP.1 when $\mathcal{G}$ has just one pure strategy Nash equilibrium if (34) holds. However, even if A2 may not converge to the optimal solution to OP.1 when $N > 2$ and $\mathcal{G}$ possesses multiple pure strategy Nash equilibrium, it also can help us find a feasible pure strategy Nash equilibrium with higher Pareto efficiency than A1 when $K_{rep} = \min(\{K_i\}_{i\in\mathcal{N}})$ which will be shown in next section. Actually, in the next section, it is shown that the performance of A2 usually is very close to that of exhaustive search.

5. Numerical Results

In this section, we present numerical results via computer simulations to illustrate the performance of the proposed algorithms. Here, it is assumed that the channel gains of the users include the path loss and fading effects. Mathematically, the channel gain from user *i* to channel *m* is $|h_{im}|^2 = \frac{|h'_{im}|^2}{d_{im}^{\alpha}}$, where α is the path loss exponent, h'_{im} follows the circularly symmetric complex Gaussian (CSCG) distribution with zero mean and unit variance, and d_{im} is the distance between the user *i* and the channel *m*.

$\forall i \in \mathcal{N}, \forall m \in \mathcal{M}, d_{im}$ is assumed to be distributed on the interval $[\underline{d}, \overline{d}]$ with uniform distribution. In this chapter, we set $\alpha = 4$.

In the figures and tables in this section, the term ``RS" means that the users randomly select resource management strategies and all of them will be shut down if the selected strategy profile is infeasible. The term ES means the exhaustive search algorithm that can find the optimal solution but with high complexity.

The first case (case 1) is an one-user-pair-multiple-channel system, i.e., $N = 2$, $M = 2$. The bandwidth is $B = 2MHz$, the noise power is $\sigma^2 = 5 \times 10^{-15} Watts$, $P_R = 0.1 Watts$, $\underline{d} = 200m$, $\overline{d} = 500m$, and the transmission power set is $\mathcal{P}_1 = \mathcal{P}_2 = \{0, 0.005, 0.05, 0.5\} Watts$. Correspondingly, we have $K_1 = K_2 = 13$ and $\mathcal{M} = 1, 2$. Furthermore, we set $c_1^{min} = c_2^{min} = 2Mbps$ and $\varepsilon1 = \varepsilon2 = \varepsilon_1 = \varepsilon_2 = 5 \times 10^5$.

In the second case (case 2), we focus on a multiple-user-pair-multiple-channels system where $M = 3$, $N = 2, 4, 6, 8, 10$. Here, the bandwidth is $B = 2Mbps$, the noise power is $\sigma^2 = 1 \times 10^{-15} Watts$, $P_R = 0.1 Watts$, $\underline{d} = 200m$, $\overline{d} = 500m$, and the transmission power set is $\mathcal{P}_i = \{0, 0.005, 0.05, 0.5\} Watts, \forall i \in \mathcal{N}$. We set $c_i^{min} = 2Mbps$, $\varepsilon_i = 5 \times 10^5$, $\forall i \in \mathcal{N}$.

We also consider another case (case 3) where $N = 4$, $M = 2, 3, 4, 5, 6, 7, 8$ while the other parameters are the same as those in case 2.

Figure 3. plots the average network utilities of different algorithms in case 1 over 1000 different channel realizations, where A2' denotes the algorithm A2 with $K_{rep} = 1$ and A2 denotes the algorithm A2 with $K_{rep} = \min(\{K_i\}_{i \in \mathcal{N}})$. From Fig.3, we can see that A1, A2 and ES significantly outperform

Figure 3. The average network utilities of case 1

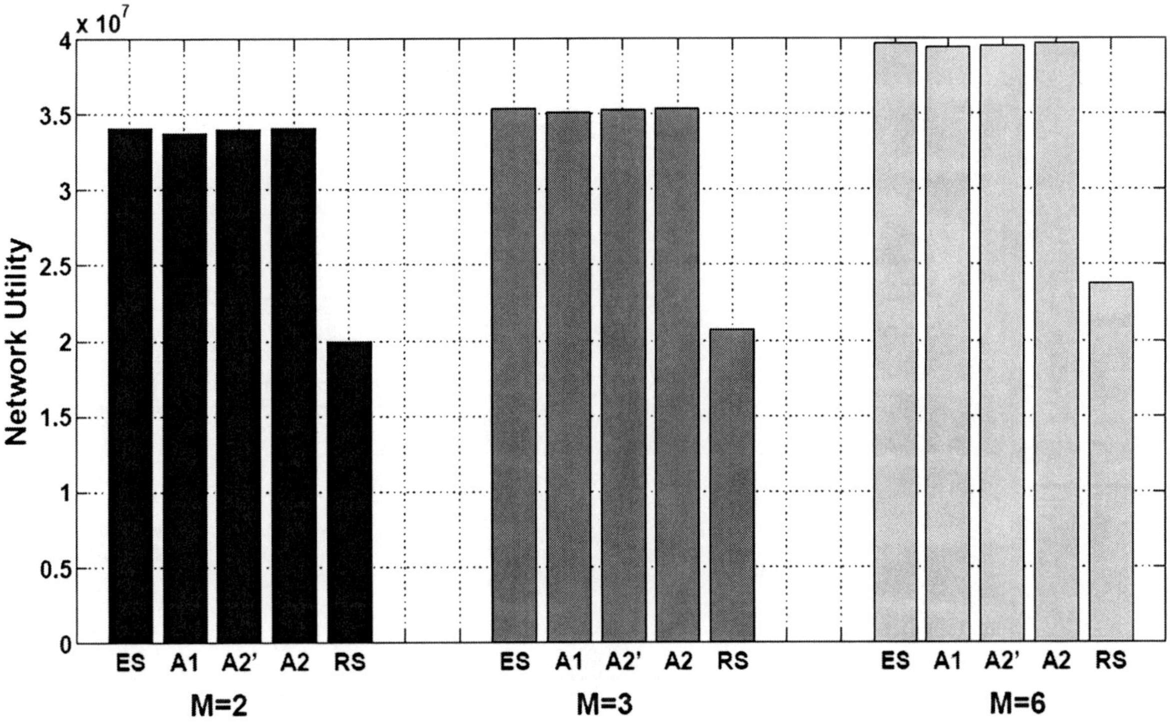

Figure 4. The average network utility with different number of users (case 2)

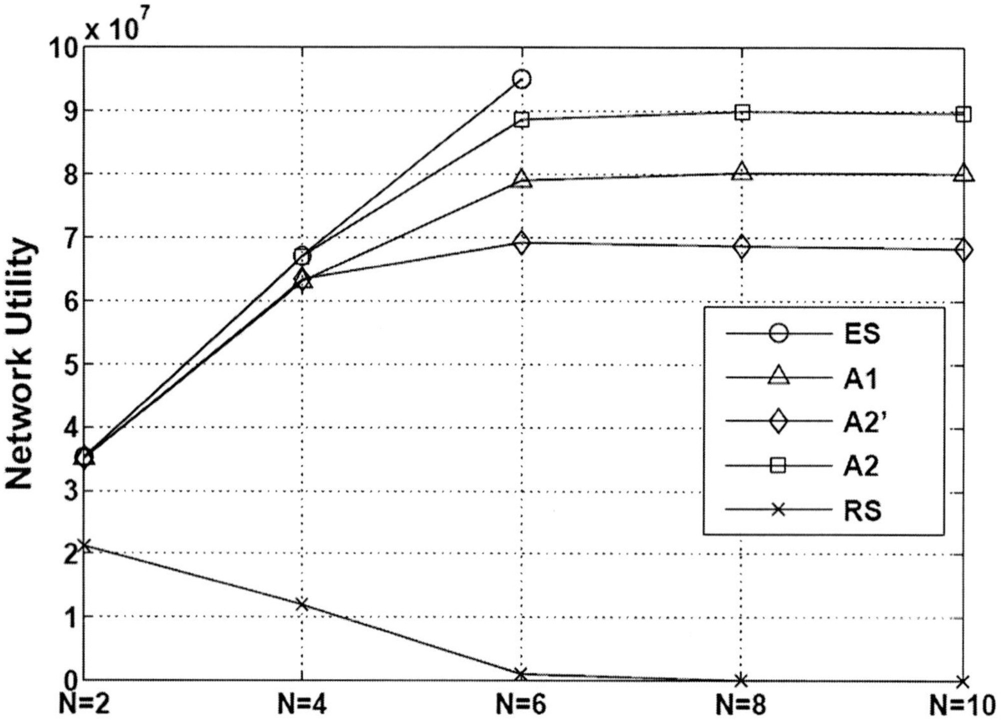

Figure 5. The average network utility with different number of channels (case 3)

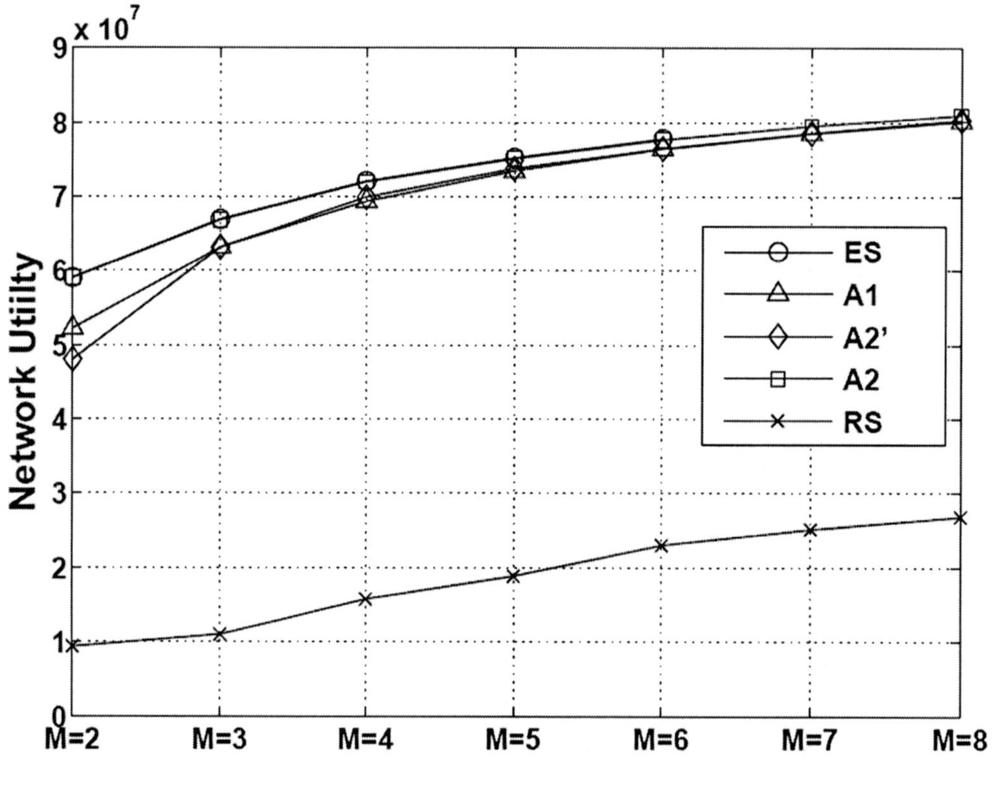

RS. Both A1 and A2 with $K_{rep} = 1$ can achieve near the optimal performance as ES. A2 with $K_{rep} = \min(\{K_i\}_{i \in \mathcal{N}}$ even can reach the optimal performance (i.e., the performance obtained by ES). Note that the average network utilities of all algorithms increase with the number of channels. This may follow that fact that more available channels would provide more diversity gains.

In Fig.4, we plot the average network utility of case 2 over 1000 channel realizations, and in Fig.5, we plot the average network utility of case 3. The performance of ES is lacking in Fig.4 when $N > 6$ and in Fig.5 when $M > 6$ due to the high complexity.

It is shown in Fig.4 and 5 that A1, A2 and ES all outperform RS considerably. In Fig.4, when $N \leq 4$, A2 has nearly the same average network utility as that obtained by ES, and A1 works as well as A2 with $K_{rep} = 1$. When the number of users becomes large ($N \geq 6$), A2 with $K_{rep} = \min(\{K_i\}_{i \in \mathcal{N}})$ will achieve much higher average network utility than A1 and A2 with $K_{rep} = 1$. Furthermore, A1 outperforms A2 with $K_{rep} = 1$ when $N \geq 6$. In Fig.5, one can see that A2 always achieve the optimal performance, while A1 performs as well as A2 with $K_{rep} = 1$ when $M > 2$. The gap between ES and A1 (A2 with $K_{rep} = 1$) becomes smaller as more channels are available.

CONCLUSION

This chapter has introduced the concepts and properties of potential games. The game theoretic algorithms which can be used to find the Nash equilibrium of the potential games are given. Then, the authors showed two examples of the application of potential games to wireless networks. It is shown in the examples that potential game is a power tool to analyze resource allocation problem in wireless networks.

ACKNOWLEDGMENT

The authors would like to thank Prof. Youyun Xu and Prof. Yueming Cai for their useful comments.

REFERENCES

Ai, X., Srinivasan, V., & Tham, C. (2008). Optimality and complexity of pure Nash equilibria in the coverage game. *IEEE Journal on Selected Areas in Communications*, *26*(7), 117–1182. doi:10.1109/JSAC.2008.080914

Friedman, J. W., & Mezzetti, C. (2001). Learning in games by random sampling. *Journal of Economic Theory*, *98*(1), 55–84. doi:10.1006/jeth.2000.2694

Gunduz, D., Yener, A., Goldsmith, A., & Poor, V. (2013). The multi-way relay channel. *IEEE Transactions on Information Theory*, *59*(1), 51–63. doi:10.1109/TIT.2012.2219156

Hossain, E., Niyato, D., & Han, Z. (2009). *Dynamic spectrum access and management in cognitive radio networks*. Cambrige University Press. doi:10.1017/CBO9780511609909

Katti, S., Gollakota, S., & Katabi, D. (2007). Embracing wireless interference: analog network coding. In *Proceedings of ACM SIGCOMM*. ACM. doi:10.1145/1282380.1282425

Liang, C., & Dandekar, K. R. (2007). Power management in MIMO ad hoc networks: A game-theoretic approach. *IEEE Transactions on Wireless Communications*, *6*(4), 1164–1170. doi:10.1109/TWC.2007.348307

MacKenzie, A. B., & DaSilva, L. A. (2007). *Game theory for wireless engineers*. Morgan & Claypool Publisher.

Menon, R., MacKenzie, A., Buehrer, R., & Reed, J. (2009). Interference avoidance in networks with distributed receivers. *IEEE Transactions on Communications*, *8*(10), 5225–5237.

Monderer, D., & Shapley, L. S. (1996a). Potential games. *Games and Economic Behavior*, *14*(1), 124–143. doi:10.1006/game.1996.0044

Monderer, D., & Shapley, L. S. (1996b). Fictitious play property for games with identical interests. *Journal of Economic Theory*, *1*(1), 258–265. doi:10.1006/jeth.1996.0014

Myerson, R. (1991). *Game theory: analysis of conflict*. Cambridge, UK: Harvard University Press.

Rosenthal, R. W. (1973). A Class of Games Possessing Pure-Strategy Nash Equilibria. *International Journal of Game Theory*, *2*(1), 65–67. doi:10.1007/BF01737559

Saraydar, C. U., Mandayam, N. B., & Goodman, D. J. (2002). Efficient power control via pricing in wireless data networks. *IEEE Transactions on Communications*, *50*(2), 291–303. doi:10.1109/26.983324

Scutari, G., Barbarossa, S., & Palomar, D. P. (2006). Potential games: A framework for vector power control problems with coupled constraints. In *Proceedings of IEEE ICASSP*. IEEE. doi:10.1109/ICASSP.2006.1660950

Uykan, Z., & Jantti, R. (2014). Joint optimization of transmission-order selection and channel allocation for bidirectional wireless links—part I: Game theoretic analysis. *IEEE Transactions on Wireless Communications*, *13*(7), 4003–4013. doi:10.1109/TWC.2014.2314633

Xiao, M. B., & Shroff, N. B. (2001). Utility-Based power control in cellular wireless systems. In Proceedings of IEEE INFOCOM (pp. 412-421). IEEE.

Xing, Y., & Chandramouli, R. (2008). Stochastic learning solution for distributed discrete power control game in wireless data networks. *IEEE/ACM Transactions on Networking*, *16*(4), 932–944. doi:10.1109/TNET.2007.911424

Xu, Y., Wang, J., Wu, Q., Anpalagan, A., & Yao, Y. (2012). Opportunistic spectrum access in cognitive radio networks: Global optimization using local interaction games. *IEEE Journal of Selected Topics in Signal Processing*, *6*(2), 180–194. doi:10.1109/JSTSP.2011.2176916

Yilmaz, E., Zakhour, R., Gesbert, D., & Knopp, R. (2010). Multi-pair two-way relay channel with multiple antenna relay station. In *Proceedings of IEEE International Conference on Communications (ICC)*. IEEE. doi:10.1109/ICC.2010.5502396

Zhang, S., Liew, S., & Lam, P. P. (2006). Hot topic: physical-layer network coding. In *Proceedings of ACM MobiCom* (pp. 358-365). ACM.

Zhong, W., Chen, G., Jin, S., & Wong, K. K. (2014). Relay selection and discrete power control for cognitive relay networks via potential game. *IEEE Transactions on Signal Processing*, *62*(20), 5411–5424. doi:10.1109/TSP.2014.2347261

Zhong, W., Xu, Y., Tao, M., & Cai, Y. (2010). Game theoretic multimode precoding strategy selection for MIMO multiple access channels. *IEEE Signal Processing Letters*, *17*(6), 563–566. doi:10.1109/LSP.2010.2047315

Zhong, W., Xu, Y., & Tianfield, H. (2011). Game-theoretic opportunistic spectrum sharing strategy selection for cognitive MIMO multiple access channels. *IEEE Transactions on Signal Processing*, *59*(6), 2745–2759. doi:10.1109/TSP.2011.2121063

Chapter 2
Coalition Formation Game for Wireless Communications

Dan Wu
PLA University of Science and Technology, China

Yueming Cai
PLA University of Science and Technology, China

ABSTRACT

With the emergence of cooperation as a new communication paradigm, and the need for self-organizing, distributed networks, it is necessary to seek suitable game theoretical tools that allow to analyze and study the behavior and interactions of the nodes in future communication networks. In this context, this chapter introduces the coalition formation game theory, and its potential applications in communication and wireless networks. Specifically, it presents the fundamental components, the key properties, the mathematical techniques, the solution concepts, and describes the methodologies for applying these games in several applications drawn from the state-of-the-art research in communications.

INTRODUCTION

As we know, there has been significant growth in research activities, which exploit game theory to model and analyze the communication networks, and help the network devices to make independent and rational strategic decisions by low complexity distributed algorithms. In general, game theory can be divided into two branches, namely, non-cooperative and cooperative game theory. The former provides effective tools for studying the competitive behavior of rational players, and its potential applications in communication and wireless networks become more perfect. The latter focuses on collaborative scenarios. In particular, the coalition formation games are the main branch of cooperative games, which describes the formation of cooperating groups of players, referred to as coalitions, for the purpose of strengthening the players' positions in a game.

Note that, it is a trend that cooperation emerges as a new networking paradigm, which is capable of improving the performance from the physical layer up to the network layer. However, implementing cooperation in large scale, self-organizing, and distributed networks needs to face several obstacles, e.g.,

DOI: 10.4018/978-1-4666-8642-7.ch002

modeling, flexibility, scalability, optimality, and complexity. In this context, this chapter focuses on the coalition formation game theory to deal with such obstacles. Unfortunately, there are only several works which are restricted to study very limited aspects of cooperation in networks.

This chapter aims to overview the state-of-the-art research advances, from coalition formation games to their communication applications. Firstly, the fundamental components are presented, including the key properties, the mathematical techniques of coalition formation games. Then, two applications that are drawn from the state-of-the-art research in communications are introduced to describe the methodologies for applying these games. In particular, they address the major challenges in applying coalition formation games to the understanding and designing of communication systems, with emphasis on both new analytical techniques and novel application scenarios (Saad, Han, Debbah, Hjørungnes, & Baar, 2009a). Finally, the future application-oriented approaches to coalition formation games are discussed, and the available research directions for using this strong analytical tool of coalitional games are further pointed out.

INTRODUCTION TO COALITION FORMATION GAMES

In essence, a coalition formation game is defined as a triplet $(\mathcal{N}, v, \mathcal{S})$. Specifically, $\mathcal{N} = \{1, \ldots N\}$ denotes a set of players, which try to form coalitions to strengthen their positions in the game. Any coalition represents an agreement among the players in the same coalition to act as a single entity. v denotes the coalition value, which quantifies the worth of a coalition in a game. The definition of the coalition value determines the form and type of the game. In particular, $\mathcal{S}$ is a coalitional structure or a partition of $\mathcal{N}$, i.e., a collection of coalitions $\mathcal{S} = \{S_1, \ldots, S_l\}$, such that $\forall i \neq k$, $S_i \cap S_k = \varnothing$, and $\bigcup_{i=1}^{l} S_i = \mathcal{N}$. It is imposed by an external factor on the game, such as physical restrictions in the problem.

The most common form of a coalition formation game is the characteristic form, where the value of a coalition just depends on the members of this coalition. As such, the coalition value v can be treated as a characteristic function, and further, the characteristic function can be either transferable utility (TU) or nontransferable utility (NTU). More precisely, the value of a game in characteristic form with TU is a function over the real line defined as $v: 2^{\mathcal{N}} \to \mathbb{R}$. It associates with each coalition $S_i \subseteq \mathcal{N}$ a real number quantifying the values of S_i. In other words, the coalition values in TU games are treated as monetary values which are distributed among the members in a coalition by using an appropriate fairness rule. The amount of utility that a player $j \in S_i$ obtains from the division of $v(S_i)$ constitutes the utility of player and is denoted by x_j. The vector $\mathbf{x} \in \mathbb{R}^{|S_i|}$, whose element x_j is the utility of player $j \in S_i$, shows a utility allocation. Here, $|\bullet|$ represents the cardinality of a set. In addition, the coalition values in NTU games can not be assigned a single real number, or rigid restrictions exist on the distribution of the utility. Taking coalition S_i in an NTU game for example, its value, $v(S_i)$ is no longer a function over the real line, but a set of utility vectors, $v(S_i) \subseteq \mathbb{R}^{|S_i|}$, where each element x_j of a vector $\mathrm{x} \in v(S_i)$ represents a utility that player $j \in S_i$ can obtain within coalition S_i given a certain strategy selected by j while being a member of S_i. According to this definition, a TU game can be regarded as a particular case of the NTU framework. Coalition formation games in characteristic form with TU or NTU are one of the most important types of games.

From the formulation of the coalition formation game, coalitional structure and cost for cooperation play a major role in such a game. In addition, the coalition formation game is robust to environmental changes incurred by the factors which may affect the network's topology, e.g., the new players joining or the current players leaving. The important characteristics which classify a game as a coalition formation game are as follows (Saad et al., 2009a): i) The game can support both TU and NTU frameworks in characteristic form. However, the presence of a cost for forming coalitions limit the gains which forming a coalition brings to its members. Hence, the game is generally not superadditive, and its optimal structure is seldom the grand coalition. ii) The focus is on investigating the coalitional structure to track, for example, which coalitions will form, what is the optimal coalition size, and how can we assess the structure's characteristics.

Furthermore, the solution concepts are important for solving formation coalition games, which are not only related to the stability of the coalition structure to guarantee that no group of players has an incentive to leave the current coalitions, but also assess the utilities that the coalition structure achieves as well as the fairness criteria which is used for distributing these utilities. Note that, in the presence of $\mathcal{S}$, the total value available for coalition S_i is divided among its members, rather than that the value of the grand coalition $v(\mathcal{N})$ is distributed among all players. Then, the concept of group rationality should be substituted by relative efficiency. To be more precise, given an allocation vector $\mathbf{x} \in \mathbb{R}^{|\mathcal{N}|}$, relative efficiency implies that, for each coalition $S_i \in \mathcal{S}$, $\sum_{j \in S_i} x_j = v(S_i)$ (Aumann & Drèze, 1974). Moreover, the allocation vector x is feasible for coalition S_i, if $\sum_{j \in S_i} x_j \leq v(S_i)$, and it is blocked by coalition S_i, if $\sum_{j \in S_i} x_j < v(S_i)$. Accordingly, many solution concepts, such as the core, the Shapley value, and the nucleolus, have been defined or can be defined not only with reference to the all-player coalition, but also with reference to an arbitrary coalition structure. The former complies with the assumption that the grand coalition would form due to the superadditivity property, and the latter is applied to appropriately defined games on each of the coalitions in the coalition structure mainly by replacing group rationality with relative efficiency. With regards to solutions, the first concern is the Shapley value for TU games.

Definition 1 (Aumann & Drèze, 1974): Given $(\mathcal{N}, v, \mathcal{S})$, the Shapley value is characterized by a unique mapping ϕ which satisfies the following axioms (ϕ_j is the utility given to player j by the Shapley value ϕ, i.e., $\phi(v) = (\phi_1, \ldots \phi_{|\mathcal{N}|})$ in $\mathbb{R}^{|\mathcal{N}|}$.): i) Relative efficiency: For all $S_i \in \mathcal{S}$, $\sum_{j \in S_i} \phi_j(v) = v(S_i)$. ii) Symmetry: For all permutations of $\mathcal{N}$ under which $\mathcal{S}$ is invariant. iii) Additivity: $\phi(v + w) = \phi(v) + \phi(w)$, where the game $(v + w)$ is defined by $(v + w)(S_i) = v(S_i) + w(S_i)$ for all S_i. iv) Dummy: If player j is a dummy player, i.e., $v(S_i \cup \{j\}) - v(S_i) = 0$ for every $S_i \subset \mathcal{N}$, then $\phi_j(v) = 0$.

Definition 1 implies that the restriction to S_i of the Shapley value for the game $(\mathcal{N}, v, \mathcal{S})$ is the value for the game $(S_i, v \mid S_i)$. In other words, the Shapley value ϕ can be determined by computing separately $\phi(v \mid S_i)$, $\forall S_i \in \mathcal{S}$. We refer the readers to the work of Aumann and Drèze (1974) for extensive textbook treatment of the redefinitions of the rest of solution concepts. In general, the problem of finding the core and the nucleolus in a coalition formation game is more complex than the Shapley value. In the work of Aumann and Drèze (1974), the basic idea for finding them is to model a game which is equivalent to v, and accordingly, the core and nucleolus in this equivalent game will be considered. However, in the work of Aumann and Drèze (1974), the definitions of such solutions are restricted to static coalition formation games with TU and in characteristic form. In fact, once NTU games or dynamic coalition formation games are considered, the difficulty of such solutions further increases, especially for getting the solutions in a distributed fashion. To this end, the most common way is to redefine the solution concepts or present alternatives which are specific to the investigated game, such as the works of Sandholm et al. (1999), Ray (2007), Apt and Witzel (2007), and Arnold and Schwalbe (2002). It is worth mentioning that Arnold and Schwalbe (2002) propose the definition of the core in dynamic coalition formation games with TU, which is more practical for varying communication environments.

Assumption 1 The utility of each player is restricted to a finite set $\mathcal{D}_j$ of all integral multiples of ϱ in the closed interval $\left[v(\{j\}), \max_{S \in \mathcal{CS}} \sum_{S_i \in S} v(S_i)\right]$, where $\mathcal{CS}$ is the set of all the coalition structures, and the smallest money unit ϱ is constructed according to the work of Newton (2010, pp. 1223-1235).

Definition 2 (Arnold & Schwalbe, 2002): For any TU game, an allocation $\mathbf{x} \in \mathbb{R}^{|\mathcal{N}|}$ is a core allocation if, i) $x_j \in \mathcal{D}_j$, $\forall j \in \mathcal{N}$, $\sum_{j \in \mathcal{N}} x_j \leq \max_{S \in \mathcal{CS}} \sum_{S_i \in S} v(S_i)$, and if the inequality is strict, then $\sum_{j \in \mathcal{N}} x_j + \varrho \geq \max_{S \in \mathcal{CS}} \sum_{S_i \in S} v(S_i)$, ii) $\sum_{j \in S_i} x_j \geq v(S_i)$, $\forall S_i \subset \mathcal{N}$.

Note that, by Definition 2, if a group of players can form a coalition which provides its members higher utility values, this coalition will block the proposal. Moreover, Definition 2 means that no coalition $S_i \subset \mathcal{N}$ has an incentive to reject the proposed utility allocation, to deviate from the current coalition, to form another coalition, and to divide the surplus. In this regard, the set of these allocations is called the core, which is directly related to the stability of coalition structure.

DYNAMIC COALITION FORMATION ALGORITHM

Considering the external or internal factors, such as, the changes in the coalition structure driven by that some players join in or leave the game, the changes to the game nature has potential to occur. Hence, in a coalition formation game, the key lies in the formation of the coalition structure and the evolution of this structure. In many applications, coalition formation involves finding a coalitional structure that either maximizes the total utility (social welfare) if the game is TU, or finding a structure with Pareto optimal utility distribution for the players if the game is NTU (Saad et al., 2009a). To this end, there exist two types of approaches to form coalitions, i.e., from centralized and distributed perspectives.

For the former, the corn solution is to find an optimal partition by iterating over all the partitions of the player set $\mathcal{N}$. Obviously, it is computationally complex and impractical, that is, the computational complexity grows exponentially with the number of players. In particular, for some specific cases, such as the works of Sandholm et al. (1999), Ray (2007), Apt and Witzel (2007), and Arnold and Schwalbe (2002), it is available to utilize the properties of the game, especially of the coalition value v to reduce the centralized complexity.

For the latter, the key is that the players have an autonomy on deciding whether or not they join in a coalition. It is more useful and general for many practical applications. In fact, such a type of approaches is rich, which involves heuristic approaches, Markov chain-based methods, theory-based methods, and approaches based on bargaining theory or other negotiation techniques from economics.

More precisely, Gharehshiran and Krishnamurthy (2010) develop a distributed dynamic coalition formation algorithm based on Markov chain, which can determine the optimal coalition structure and measurement allocations by reaching the core of the defined coalition formation game. It involves the following steps, i.e., Revision Strategy, Evaluating the Best Strategy for the Next Period, Experimentation, Best-reply Process, and Recursion. In particular, with regard to experimentation, every player takes a random draw from the Bernoulli trial to revise its strategy. However, incompatibility of players' choices may arise, which may cause a player to lose the opportunity of improving its performance. Moreover, the basic idea of the best-reply rule is that a player switches coalitions only if its expected utility in the new coalition exceeds its current utility, and it demands the most it can get. It should be noted that the player can switch only between existing coalitions, or form a singleton coalition. Maybe there exists a blocking coalition which can not be formed directly, according to the current coalition structure. As a result, every member of the blocking coalition has an incentive to destabilize the prevailing structure. It implies that the player is willing to try with suboptimal strategies whenever there is a chance that this might lead to a preferred coalition structure. For such an algorithm, it is important to detect the sensor nodes in the potential blocking coalitions. Gharehshiran and Krishnamurthy (2010) propose two detection methods, i.e., exhaustive and randomized search methods. The former one requires to check $\left(2^{|\mathcal{N}|}-1\right)$ different combinations of players, and the computational cost at each iteration is $O\left(2^{|\mathcal{N}|}\right)$. The latter one constructs a sample set by taking ζ ($1 \leq \zeta \leq |\mathcal{N}|$) random samples from all possible coalitions, and the examination is made only for this sample set. Then, the computational cost at each iteration is $O(\zeta)$. However, this method may misjudge the players which should have been in the blocking coalitions, thus it leads to slower convergence. This misclassification possibility is inversely proportional to the cardinality of the sample set.

Furthermore, Apt and Witzel (2009) provide the mathematical framework which can be used to develop application-specific coalition formation algorithms. That is, such an algorithm entails three major elements:

- well-defined preference relations for comparing collections of coalitions,
- simple rules for forming or breaking coalitions,
- logical conceptions for assessing the stability of a coalitional structure.

For comparing collections of coalitions, there exist a number of preference relations, which are defined in many literature, e.g., the works of Saad et al. (2009a) and Apt and Witzel (2009). In addition, they are referred to as a unified representation $\rhd$. Two of the more common preference relations are introduced as follows:

Definition 3 Given two collections $\mathcal{A} = \{A_1, \ldots, A_m\}$ and $\mathcal{B} = \{B_1, \ldots, B_n\}$ to the same players, $\mathcal{A}$ is preferred over $\mathcal{B}$ by the Nash preference relation, denoted as $\mathcal{A} \rhd_{\text{Nash}} \mathcal{B}$, if and only if $\prod_{i=1}^{m} v(A_i) > \prod_{k=1}^{n} v(B_k)$ holds.

Definition 4 Given two allocations x and y that are allotted by $\mathcal{A} = \{A_1, \ldots, A_m\}$ and $\mathcal{B} = \{B_1, \ldots, B_n\}$, respectively, to the same players, $\mathcal{A}$ is preferred over $\mathcal{B}$ by the Pareto preference relation, denoted as $\mathcal{A} \rhd_{\text{Pareto}} \mathcal{B}$, if and only if $\mathbf{x} \succ_{\text{Pareto}} \mathbf{y}$ holds. Here, $\mathbf{x} \succ_{\text{Pareto}} \mathbf{y}$ iff $\forall j \in \mathcal{N}$, $x_j \geq y_j$, and $\exists j \in \mathcal{N}$, $x_j > y_j$.

Note that, the Nash preference relation is generally suitable for TU games. The intuition behind it is that when the sum $\sum_{i=1}^{m} v(A_i)$ is fixed, the product $\prod_{i=1}^{m} v(A_i)$ is largest when all $v(A_i)$ are equal. Hence, the Nash preference relation favors an equal distribution. Moreover, the Pareto preference relation is suitable for both TU and NTU games. It is defined by comparing the individual utility of each player which is derived from its coalition. In this way, it is guaranteed that the Pareto preference relation is anonymous in the sense that the names of the players do not play a role.

Using such orders, Apt and Witzel (2009) present two main rules for forming or breaking coalitions, referred to as merge and split rules. The two rules allow to modify a partition of $\mathcal{N}$. The basic idea behind them is detailed as follows:

- Merge rule. Merge any collection of disjoint coalitions $\{A_1,...,A_m\}$, where $\left\{\bigcup_{i=1}^{m} A_i\right\} \triangleright \left\{A_1,\ldots,A_m\right\}$, thus $\left\{A_1,\ldots,A_m\right\} \rightarrow \left\{\bigcup_{i=1}^{m} A_i\right\}$.
- Split rule. Split any coalition $\left\{\bigcup_{i=1}^{m} A_i\right\}$, where $\left\{A_1,\ldots,A_m\right\} \triangleright \left\{\bigcup_{i=1}^{m} A_i\right\}$, thus $\left\{\bigcup_{i=1}^{m} A_i\right\} \rightarrow \left\{A_1,\ldots,A_m\right\}$.

Theorem 1 Independent of the selected order, any arbitrary sequence of the two rules is shown to converge to a final partition of $\mathcal{N}$.

Stability is one of the most important concepts in a coalition formation game. For assessing the stability of the final partition, Apt and Witzel (2007, 2009) propose the concept of a defection function $\mathbb{D}$, which is a function that associates with each partition, a group of other partitions, or a family of collections in $\mathcal{N}$. Intuitively, $\mathbb{D}$ can be interpreted as follows: For a partition $\mathcal{S}$, the family $\mathcal{D}(\mathcal{S})$ consists of all the collections whose players can leave the partition $\mathcal{S}$ by forming a new, separate, group of players which are divided by the collection. In what follows three natural types of such a function will be considered to yield three different notions of a $\mathbb{D}$-stable partition. Each can assess whether there is an incentive for the players in partition $\mathcal{S}$ to deviate and form other partitions or collections.

The first notion of stability is $\mathbb{D}_{hp}$ stability, and the defection function $\mathbb{D}_{hp}$ allows any group of players to leave the partition $\mathcal{S}$ of $\mathcal{N}$ through merge-and-split operations to create another partition in $\mathcal{N}$. A partition $\mathcal{S}$ is $\mathbb{D}_{hp}$-stable, if no players in $\mathcal{S}$ are interested in leaving $\mathcal{S}$ through merge-and-split to form other partitions in $\mathcal{N}$. It can be treated as merge-and-split process of a partition, or a kind of equilibrium with respect to merge-and-split. However, one drawback is that coalitions within such partitions may be prone to deviations due to the partial reversibility of the merge-and-split algorithm.

Theorem 2 (Apt & Witzel, 2009) A partition $\mathcal{S} = \{S_1,\ldots,S_l\}$ is $\mathbb{D}_{hp}$-stable, if and only if the following two conditions are satisfied: i) For each $i \in \{1, ..., l\}$ and for each partition $\{B_1, ..., B_n\}$ of S_i, $\left\{B_1,\ldots,B_n\right\} \ntriangleright S_i$ holds. ii) For each $\Phi \subseteq \{1,\ldots,l\}$, $\bigcup_{i\in\Phi} S_i \ntriangleright \left\{S_i \middle| i \in \Phi\right\}$ holds. Here, $\ntriangleright$ is the opposite rule of the preference operator $\triangleright$.

Definition 5 (Apt & Radzik, 2006) Given a partition $\mathcal{S} = \{S_1,\ldots,S_l\}$, a partition $\hat{\mathcal{S}} = \left\{\hat{S}_1,\ldots,\hat{S}_m\right\}$ is called $\mathcal{S}$-homogeneous if, for each $k \in \{1, ..., m\}$, some $i \in \{1, ..., l\}$ exist such that either $\hat{S}_k \subseteq S_i$ or

...

Theorem 3 (Apt & Radzik, 2006) A partition $\mathcal{S}$ is $\mathbb{D}_{hp}$-stable if and only if $\mathcal{S} \triangleright_{\text{Pareto}} \hat{\mathcal{S}}$ holds for all $\mathcal{S}$-homogeneous partitions $\hat{\mathcal{S}}$ in $\mathcal{N}$.

The second notion of stability is $\mathbb{D}_p$ stability, and the defection function $\mathbb{D}_p$ allows a group of players to leave $\mathcal{S}$ only as the group of all players. Note that, every $\mathbb{D}_p$-stable partition is also $\mathbb{D}_{hp}$-stable. The $\mathbb{D}_p$-stable partitions are exactly those that maximize the social welfare in the set of all partitions. However, in specific situations, the notion of a $\mathbb{D}_p$-stable partition may be inadequate.

Theorem 4 (Apt & Witzel, 2007) A partition $\mathcal{S}$ of $\mathcal{N}$ is $\mathbb{D}_p$-stable if and only if it is $\triangleright_{\text{Pareto}}$-maximal, where we call a partition $\triangleright_{\text{Pareto}}$-maximal if, for any partition $\hat{\mathcal{S}}$ of $\mathcal{N}$, which is different from $\mathcal{S}$, $\mathcal{S} \triangleright_{\text{Pareto}} \hat{\mathcal{S}}$ holds.

The third notion of stability is $\mathbb{D}_c$ stability, and the defection function $\mathbb{D}_c$ allows any group of players to leave $\mathcal{S}$ and create an arbitrary collection in $\mathcal{N}$. In particular, $\mathbb{D}_c$ stability is stronger than the former two, and the $\mathbb{D}_c$-stable partition has numerous attractive properties:

- If it exists, a $\mathbb{D}_c$-stable partition is the unique outcome of any arbitrary iteration of merge-and-split and is a unique $\mathbb{D}_{hp}$-stable partition.
- A $\mathbb{D}_c$-stable partition $\mathcal{S}$ is a unique $\triangleright$-maximal partition, that is for all partitions $\hat{\mathcal{S}} \neq \mathcal{S}$ of $\mathcal{N}$, $\mathcal{S} \triangleright \hat{\mathcal{S}}$. When $\triangleright$ corresponds to $\triangleright_{\text{Pareto}}$, the $\mathbb{D}_c$-stable partition $\mathcal{S}$ is the partition that presents a Pareto optimal utility distribution for all the players.
- A $\mathbb{D}_c$-stable partition does not contain any coalitions prone to deviations.

Theorem 5 (Apt & Witzel, 2007) A partition $\mathcal{S} = \{S_1, \ldots, S_l\}$ of $\mathcal{N}$ is $\mathbb{D}_c$-stable, if and only if the following conditions are satisfied:

- For each $i \in \{1, \ldots, l\}$ and each pair of disjoint coalitions A and B such that $\{A \cup B\} \subseteq S_i$, $\{A \cup B\} \triangleright \{A, B\}$ holds.
- A coalition $A \subset \mathcal{N}$ formed of players belonging to different $S_i \in \mathcal{S}$ is $\mathcal{S}$-incompatible, if for no $i \in \{1, \ldots, l\}$ $A \subset T_i$ holds. $\mathbb{D}_c$-stability requires that for all $\mathcal{S}$-incompatible coalitions $\{A\}[\mathcal{S}] \triangleright \{A\}$ where $\{A\}[\mathcal{S}] = \{A \cap S_i, \forall i \in \{1, \ldots l\}\}$ is the projection of coalition A in partition $\mathcal{S}$.

APPLICATION OF COALITION FORMATION GAMES

1. Cooperative Communication Scheme in Clustered Wireless Sensor Networks

As we known, how to improve energy efficiency is one of the most dominating concerns for wireless sensor networks (WSNs). In particular, cooperative communication techniques can offer significant energy saving improvements in WSNs. The essential step towards practically exploiting this energy saving potential is to design an effective cooperative communication scheme. Note that, there is an inherent tradeoff between the energy consumption and the QoS provisioning. Such a fact inspires us to design a cooperative communication scheme from a new perspective, which improves both the energy efficiency and the desired QoS provisioning. Furthermore, the key element of such a scheme is the selection and coordination of cooperative nodes. Specifically, since single cooperative node selection strategies can achieve the same diversity-multiplexing tradeoff as that of multiple cooperative node selection strategies, they are practically appealing. Combining with power control, the energy saving potential of cooperative communication can be improved dramatically. However, for some special cases, where the sensor nodes have almost equally low energy, and the long-haul transmission distance is large, the consumed energy is only concentrated in the "best" cooperative node, and thus the resulting energy efficiency may be degraded. Hence, it is better to determine the optimal cooperative node set according to the actual scenario. In comparison, it becomes more complicated for joint cooperative node set selection and power allocation. To solve this problem with lower complexity, the solutions are derived i) in a quasi-closed form, ii) by using stochastic optimization tools to approach the optimal strategy, and iii) by transforming into some sub-optimal problems.

In practice, the sensor nodes in the same cluster i) prefer to cooperate with one another to achieve a certain joint work, and ii) care about their own energy consumption. In other words, the friendliness and the selfishness coexist for each sensor node. The relationship between them can be quantified by the balance between the benefit and the cost reaped by each node. Therefore, we can refer to coalition formation game theory to study the complex interactions among sensor nodes while accounting for the inherent benefit-cost tradeoff involved in the formations. In what follows, we take the work of Wu et al. (2012) for example, which exploits coalition formation game theory to design a cooperative communication scheme to achieve the optimal solution of a random tradeoff between the outage performance and the network lifetime in a clustered WSN.

1.1 System Model and Problem Formulation

Consider a clustered WSN, where there are one cluster head (CH) and $N - 1$ CNs. The set of sensor nodes is denoted by $\mathcal{N} = \{\mathrm{CH}, \mathrm{CN}_1, \ldots, \mathrm{CN}_{N-1}\}$. As shown in Figure 1, the cooperative communication scheme is carried out by two phases:

Phase 1: The local broadcasting transmission. The CH chooses $n_t - 1$ CNs to form the set of assisting CNs, and broadcasts the data to them.

Phase 2: The long-haul cooperative MISO transmission. The CH and the assisting CNs jointly transmit to the sink node, based on distributed space time codes (DSTC).

During both phases, the channels are modeled by a path loss exponent δ and frequency flat Rayleigh fading. More precisely, $h_{\mathrm{CH,CN}_j} \sim \mathcal{CN}(0,1)$, $h_{\mathrm{CH,sink}} \sim \mathcal{CN}(0,1)$ and $h_{\mathrm{CN}_j,\mathrm{sink}} \sim \mathcal{CN}(0,1)$ are unitary power, Rayleigh fading coefficients. Moreover, the CH knows its channel knowledge with the CNs in the same cluster, and the long-haul distance is the same for each one, denoted by d.

In order to guarantee the QoS requirement of the WSN, the outage probability P_{out} should not be larger than the threshold value $P_{\mathrm{out}}^{\mathrm{thr}}$, under the fixed outage capacity C_{out}. Accordingly, setting the long-haul cooperative MISO communication rate to the outage capacity C_{out}, the capacity region of the local broadcasting transmission is constrained to:

Figure 1. An illustrative example of the cooperative communication scheme

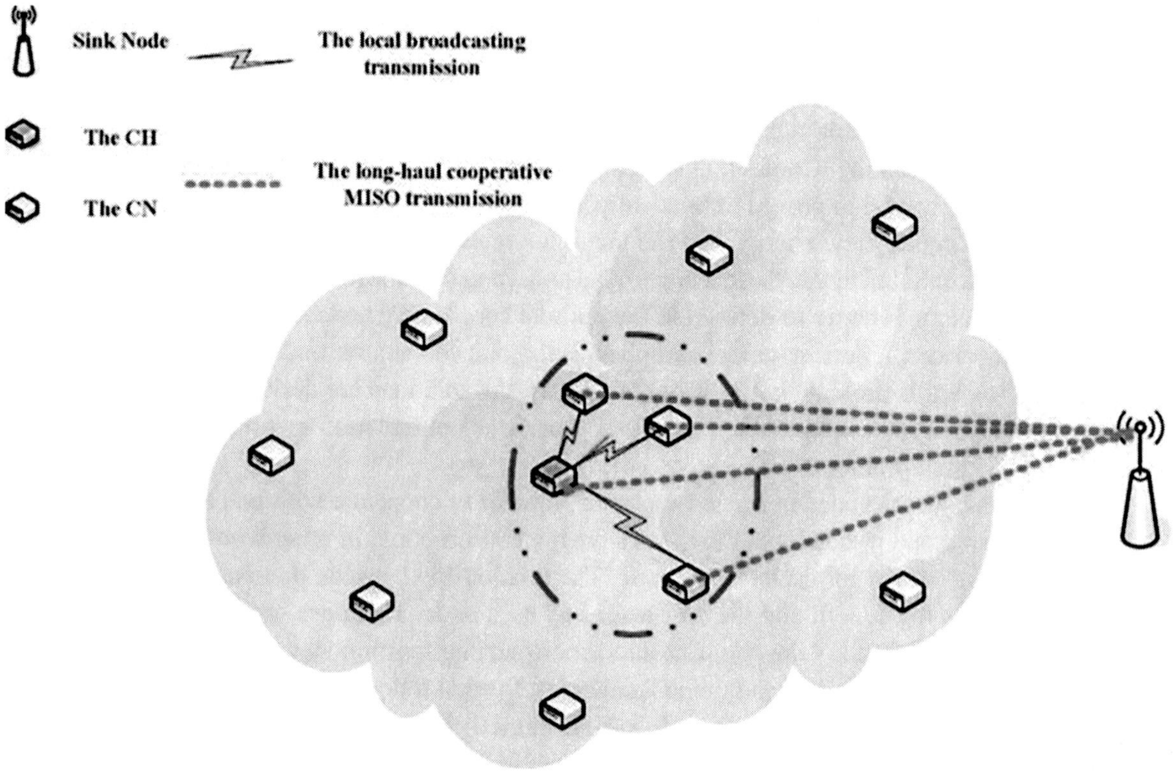

$$\min_{\mathrm{CN}_j \in \mathcal{J}} \left\{ \frac{1}{2} \log_2 \left(1 + \frac{p_{\mathrm{t,CH}}^{\mathrm{local}}}{\sigma^2} \kappa d_j^{-\delta} \left| h_{\mathrm{CH,CN}_j} \right|^2 \right) \right\} \geq C_{out} \tag{1}$$

where $\mathcal{J}$ is the set of assisting CNs, and $|\mathcal{J}| = n_t - 1$, $p_{\mathrm{t,CH}}^{\mathrm{local}}$ is the power needed for broadcasting, σ^2 is the Gaussian noise variance, d_j is the local distance between the CH and the CN_j, and κ is a constant which depends on the propagation environment.

Then, during the long-haul cooperative MISO transmission, each transmitting member has the same transmit power, namely, $p_{\mathrm{t,CH}}^{\mathrm{MISO}} = p_{\mathrm{t,CN}_j}^{\mathrm{MISO}} = p_{\mathrm{t}}^{\mathrm{MISO}} / n_t$, $\forall \mathrm{CN}_j \in \mathcal{J}$. Then, the outage probability is

$$P_{\mathrm{out}} = \Gamma \left(n_t, \frac{\left(2^{2C_{out}} - 1 \right) \sigma^2 \kappa^{-1} d^{\delta}}{p_{\mathrm{t}}^{\mathrm{MISO}} / n_t} \right) \tag{2}$$

where $\Gamma(n_t, b) = \frac{1}{(n_t - 1)!} \int_0^b x^{n_t - 1} e^{-x} dx$.

Furthermore, the inherent tradeoff between the QoS provisioning and the energy consumption is characterized as the tradeoff between the outage performance and the network lifetime. In particular, maximizing the network lifetime is translated into maximizing the minimum residual energy among all the nodes during each transmission. Hence, such a tradeoff is described as a multi-variable optimization problem with a system design parameter $\beta \in [0,1]$ that controls the desired tradeoff between them. The optimization problem can be expressed as

$$\max_{\mathcal{J}, p_{t,CH}^{local}, p_t^{MISO}} J_1 \tag{3}$$

subject to

$$E_{CH} - \frac{1}{2}\left[p_{ct} + (1+\alpha) p_{t,CH}^{local}\right] - \frac{1}{2}\left[p_{ct} + \frac{(1+\alpha) p_t^{MISO}}{n_t}\right] \geq V \tag{3.1}$$

$$E_{CN_j} - \frac{1}{2} p_{cr} - \frac{1}{2}\left[p_{ct} + \frac{(1+\alpha) p_t^{MISO}}{n_t}\right] \geq V, \ \forall CN_j \in \mathcal{J} \tag{3.2}$$

$$V \geq 0 \tag{3.3}$$

$$P_{out}^{thr} \geq \Gamma\left(n_t, \frac{\left(2^{2C_{out}} - 1\right)\sigma^2 \kappa^{-1} d^{\delta}}{p_t^{MISO} / n_t}\right) \tag{3.4}$$

$$p_{t,CH}^{local} \geq \frac{\left(2^{2C_{out}} - 1\right)\sigma^2 \kappa^{-1}}{\min_{CN_j \in \mathcal{J}} \left\{ d_j^{-\delta} \left| h_{CH,CN_j} \right|^2 \right\}} \tag{3.5}$$

where E_{CH} and E_{CN_j} is the current energy supply in the CH and the CN_j, respectively. p_{ct} is the transmitter circuit power, and p_{cr} is the receiver circuit power. α is a constant parameter which is dependent on the modulation scheme and the associated constellation size. Moreover, J_1 is defined as

$$J_1 = \beta\left(P_{out}^{thr} - \Gamma\left(n_t, \frac{\left(2^{2C_{out}} - 1\right)\sigma^2 \kappa^{-1} d^{\delta}}{p_t^{MISO} / n_t}\right)\right) + (1-\beta) V \tag{4}$$

Note that, for fixed $\mathcal{J}$, the maximum of the second term is obtained at the point where $p_{t,CH}^{local}$ is minimum, according to constraint (3.5). Then, the objective function (3) can be horizontally decomposed into the concatenation of two subproblems: i) the long-haul transmit power per sensor node, and ii) the set of assisting CNs, which is given by $\max_{\mathcal{J}, p_{t,CH}^{local}} \left(\max_{p_t^{MISO}} J_1 \right)$. Here, the first subploblem is solved by the convex optimization theory and the Lambert W function. For the detailed analysis, the reader can be referred to [11, Section III]. Then, a coalition formation game is modeled to solve the second subproblem, where the characteristic function is designed based on the results of the former one.

1.2 Solution to The Set of Assisting CNs Subproblem

The set of assisting CNs subproblem is modeled as a $(\mathcal{N}, \mathcal{S}, v)$ coalition formation game in a characteristic form, where $\mathcal{S} = \{S_1, \ldots, S_l\}$ shows a partition of $\mathcal{N}$, and the utility of every coalition $v(S_i)$ is evaluated by using a characteristic function as follows

$$v(S_i) = \begin{cases} J_1\left(p_{\mathrm{t,opt}}^{\mathrm{MISO}}\right), & \mathrm{CH} \in S_i, S_i \in \mathcal{S}, \text{and} \left|S_i\right| > 1, \\ J_2\left(p_{\mathrm{t,opt}}^{\mathrm{SISO}}\right), & S_i = \{\mathrm{CH}\}, S_i \in \mathcal{S}, \\ 0, & \text{others.} \end{cases} \tag{5}$$

where J_2 is given by

$$J_2 = \beta\left(P_{out}^{thr} - \Gamma\left(1, \frac{\left(2^{C_{out}} - 1\right)\sigma^2\kappa^{-1}d^{\delta}}{p_{\mathrm{t}}^{\mathrm{SISO}}}\right)\right) + (1-\beta)\left(E_{\mathrm{CH}} - p_{ct} - (1+\alpha)\, p_{\mathrm{t}}^{\mathrm{SISO}}\right) \tag{6}$$

Similarly, $p_{\mathrm{t,min}}^{\mathrm{SISO}}$ can be obtained by solving the equation $P_{out}^{thr} = \Gamma\left(1, \left(2^{C_{out}} - 1\right)\sigma^2\kappa^{-1}d^{\delta} / p_{\mathrm{t}}^{\mathrm{SISO}}\right)$. Then, $J_1\left(p_{\mathrm{t,opt}}^{\mathrm{MISO}}\right)$ and $J_2\left(p_{\mathrm{t,opt}}^{\mathrm{SISO}}\right)$ can be determined by substituting $p_{\mathrm{t,opt}}^{\mathrm{MISO}}$ and $p_{\mathrm{t,opt}}^{\mathrm{SISO}}$ derived from the first subproblem into (4) and (6) given $\mathcal{J} = S_i \setminus \{\mathrm{CH}\}$, respectively. From the resulting coalition structure, the set of assisting CNs, i.e., $\mathcal{J} = S(\mathrm{CH}) \setminus \{\mathrm{CH}\}$, can be obtained, where $S(\mathrm{CH})$ is the coalition which has the CH. Note that, the utility in (5) quantifies the degree of performance improvement given coalition structure and optimal long-haul transmit power, that is, represents the amount of net revenue received by a coalition. Hence this amount can be arbitrarily apportioned among the coalition members, which justifies the TU nature of the game. In addition, we denote a utility allocation as $\mathbf{x} = \left(x_{\mathrm{CH}}, x_{\mathrm{CN}_1}, \ldots, x_{\mathrm{CN}_{N-1}}\right)$. Similar to Assumption 1, we assume that the utility of each sensor node is restricted to a finite set $\mathcal{D}_j$ of all integral multiples of ϱ in the closed interval $\left[v(\{j\}), \max_{\mathcal{S} \in \mathcal{CS}} \sum_{S_i \in \mathcal{S}} v(S_i)\right]$.

Proposition 1 The core of the proposed coalition formation game is nonempty.

Proof: An allocation $\mathbf{x}$ is constructed as follows. Assume that the coalition structure $\mathcal{S} = \{S_1, \ldots S_l\}$ has been formed, where a coalition S_1 has the CH, $v(S_1) \geq v(S_i \cup \{\mathrm{CH}\})$, $\forall S_i \subseteq S_1 \setminus \{\mathrm{CH}\}$, and $v(S_1) \geq v(S_i \cup \{\mathrm{CH}\})$, $\forall S_i \subseteq \mathcal{N} \setminus S_1$. Moreover, each sensor node obtains a utility, i.e., $x_{\mathrm{CH}} = v(S_1)$, $x_{\mathrm{CN}_j} = 0$, $\forall \mathrm{CN}_j \in \mathcal{N}$. Then, the allocation $\mathbf{x}$ is feasible, and is not blocked by any coalition. According to Definition 2, the resulting allocation $\mathbf{x}$ is in the core. It shows that the core of the proposed game is nonempty. ■

Theorem 6 From the resulting coalition structure $\mathcal{S}$ with a core allocation **x**, the optimal set of assisting CNs $\mathcal{J}_{\text{opt}}$ can be derived.

Proof: Suppose that the set of assisting CNs is not optimal, and then, there exists a coalition $S_i \subset \mathcal{N}$ which has the CH and $v(S_i) > \sum_{j \in S_i} x_j$. It can be shown that coalition S_i blocks allocation **x**, and this contradicts the case that **x** is in the core. Hence, coalition *S*(CH) in the resulting coalition structure $\mathcal{S}$ can coincide with the optimal set of assisting CNs $\mathcal{J}_{\text{opt}}$. ■

1.3 Description and Analysis of Algorithm

An algorithm is presented to obtain the optimal solution of the tradeoff between the outage performance and the network lifetime. It exploits a dynamic coalition formation process based on the best-reply process with trial opportunity, which is derived from the distributed dynamic coalition formation algorithm in the work of Gharehshiran and Krishnamurthy (2010), and consists of the coalition structure generation process and the negotiation process. The algorithm is described as follows:

Step 1: At $t = 0$, initialize the coalition structure, i.e., $\mathcal{S}^0 = \{\{\text{CH}\},\{\text{CN}_1\},\ldots,\{\text{CN}_{N-1}\}\}$. For any coalition $S^0 \in \mathcal{S}^0$, calculate the optimal long-haul transmit power $p_{\text{t,opt}}^{\text{SISO}}$ according to the unique closed-form solution in [11, Section III]. Then, calculate the utility of every coalition according to (5), and initialize the allocation, i.e., $\mathbf{x}^0 = \left(x_{\text{CH}}^0, x_{\text{CN}_1}^0, \ldots, x_{\text{CN}_{N-1}}^0\right)$, where $x_{\text{CH}}^0 = v(\{\text{CH}\})$, $x_{\text{CN}_1}^0 = \ldots = x_{\text{CN}_{N-1}}^0 = 0$. For expression simplicity, the initial state of the proposed game can be characterized by a tuple $\omega^0 = \left(\mathcal{S}^0, \mathbf{x}^0\right)$.

Step 2: At $t > 0$, the state of the proposed game is $\omega^t = \left(\mathcal{S}^t, \mathbf{x}^t\right)$. Choose a sensor node $j \in \mathcal{N}$ with probability $1/N$ to do the following steps. Other sensor nodes keep their strategies, i.e., for a sensor node $k \neq j$, $S^{t+1}(k) = S^t(k)$, and $x_k^{t+1} = x_k^t$, where $S(k)$ represents the coalition in which node k exists. If there exists a coalition $B(j) \subseteq \mathcal{N}$, which can block the allocation $\mathbf{x}^t$, i.e., $v(B(j)) > \sum_{k \in B(j)} x_k^t$, then sensor node j can take Strategy 1 with positive probability ϵ_j, and take Strategy 2 with positive probability $1 - \epsilon_j$. Otherwise, sensor node j only takes Strategy 1.

- **Strategy 1: Best-reply Process.** For any coalition $S^t \cup \{j\}$, $S^t \in \mathcal{S}^t \cup \varnothing$, compute the optimal long-haul transmit power $p_{\text{t,opt}}^{\text{MISO}}$ (or $p_{\text{t,opt}}^{\text{SISO}}$) according to [11,Section III]. Then, calculate the utility of every coalition, .., according to (5). Determine the best strategy for the next time $t + 1$, i.e.,

$$ST_j^{t+1}(\omega^t) = \left\{S^{t+1}(j) = S^t \cup \{j\} \middle| \arg\max_{S^t \in \mathcal{S}^t \cup \varnothing} \left(v(S^t \cup \{j\}) - \sum_{k \neq j, k \in S^t} x_k^t\right)\right\} \tag{7}$$

$$x_j^{t+1}(\omega^t) = \max_{S^t \in \mathcal{S}^t \cup \varnothing} \left(v(S^t \cup \{j\}) - \sum_{k \neq j, k \in S^t} x_k^t\right) \tag{8}$$

where $\mathcal{ST}_j^{t+1}(\omega^t)$ denotes the chosen collection in which sensor node j can obtain its maximum expected utility, given the others' utility values. In this way, the utility of node j at the next time $t+1$ is $x_j^{t+1} = x_j^{t+1}(\omega^t)$. If $\left|\mathcal{ST}_j^{t+1}(\omega^t)\right| \neq 1$, randomly choose $\mathcal{S}^{t+1}(j) \in \mathcal{ST}_j^{t+1}(\omega^t)$ with equal probability $1/\left|\mathcal{ST}_j^{t+1}(\omega^t)\right|$. Then, go to Step 3.

- **Strategy 2: Trial Action.** Choose $\mathcal{S}^{t+1}(j) = S^t \cup \{j\},\ \forall S^t \in \mathcal{S}^t \cup \varnothing$ with equal probability. Choose $x_j^{t+1} \in \mathcal{D}_j$ with equal probability $1/\left|\mathcal{D}_j\right|$. Then, go to Step 3

Step 3: Replace $t+1$ with t. If the state of $t+1$ is equal to that of t, stop the algorithm. Otherwise, go to Step 2.

In the proposed algorithm, the CH chooses a sensor node (including itself) with equal probability $1/N$ to revise its strategy at each iteration, and judges whether this sensor node is in a blocking coalition or not. As a result, the coalition formation process avoids the additional managing center and coordination. In order to detect the sensor nodes in the potential blocking coalitions, the proposed detection method is described as follows: The process only chooses a sensor node to revise its strategy at each iteration. Once this sensor node finds a blocking coalition, the search process will stop. At worst, the computational cost is $O(2^{N-1})$, which happens when there is no blocking coalition for this sensor node, or this blocking coalition can not be found until all possible coalitions are searched. In fact, the search process requires a significantly lower number of attempts, however, the convergence is not affected. Similar to the work of Gharehshiran and Krishnamurthy (2010), the CH represents the role of the base station for each cluster, and is in charge for determining the stable coalition structure with a core allocation. Due to the proposed detection method and the special characteristic function, the CH does not have heavy computational burden. Moreover, there exists little control information exchange, except the overhead coming from each CN reports its current energy to the CH (From this overhead, the CH can obtain its CSI with the CNs in the same cluster.), and the one coming from the CH sends a message to inform the assisting CNs. Hence, the energy consumption which comes from the data exchange penalty is referred to as the cooperation costs, and the overhead caused by the control information exchange is ignored.

Proposition 2 The proposed algorithm results in the process which can be defined as a finite Markov chain with state space $\Omega = \left\{\omega = (\mathcal{S}, \mathbf{x}) \middle| \mathcal{S} \in \mathcal{CS}, \mathbf{x} \in_{\times_{j\in\mathcal{N}}} \mathcal{D}_j\right\}$.

Proof: The proposed algorithm gives rise to the process which only has a finite number of states. For any state $\omega^t \in \Omega$ at the time t, the next state ω^{t+1} just depends on the current state and not on the past. Hence, the process is a finite Markov chain. The Markov chain is completely characterized by the set of all states and transition probabilities. More precisely, the transition probability from state ω^t to ω^{t+1} can be expressed as

$$P_{\omega^t\omega^{t+1}} = \begin{cases} \frac{1}{N}\gamma_j\left(\omega^{t+1}\middle|\omega^t\right), & \omega^{t+1} \neq \omega^t, \\ 1 - \sum_{\omega^{t+1}\in\Omega,\omega^{t+1}\neq\omega^t} P_{\omega^t\omega^{t+1}}, & \omega^{t+1} = \omega^t \end{cases} \tag{9}$$

where sensor node j is the one which is chosen to change its strategy in order to transit from state ω^t to ω^{t+1}. Let $S^t(j)$ and $S^{t+1}(j)$ denote the coalitions, which node j belongs to, in state ω^t and ω^{t+1}, respectively. Then, $\gamma_j\left(\omega^{t+1}\middle|\omega^t\right)$ is the probability of node j choosing $S^{t+1}(j)$ if it is a member of $S^t(j)$ in state ω^t, that is

$$\gamma_j\left(\omega^{t+1}\middle|\omega^t\right)=\begin{cases}\dfrac{\epsilon_j}{\left|\mathcal{ST}_j^{t+1}\left(\omega^t\right)\right|}+\dfrac{1-\epsilon_j}{\left|\Xi_j^{t+1}\left(\omega^t\right)\right|}, & x_j^{t+1}=x_j^{t+1}\left(\omega^t\right)\text{ and }S^{t+1}(j)\in\mathcal{ST}_j^{t+1}\left(\omega^t\right),\\ \dfrac{1-\epsilon_j}{\left|\Xi_j^{t+1}\left(\omega^t\right)\right|}, & x_j^{t+1}\neq x_j^{t+1}\left(\omega^t\right)\text{ or }S^{t+1}(j)\notin\mathcal{ST}_j^{t+1}\left(\omega^t\right),\end{cases} \tag{10}$$

where $\epsilon_j\in\left(0,1\right]$ is the probability of node j taking Strategy 1, and we set $\epsilon_j=1$ if there is no $B(j)$. Moreover, $\Xi_j^{t+1}\left(\omega^t\right)$ is defined as

$$\Xi_j^{t+1}\left(\omega^t\right)=\left\{\left(S^{t+1}(j),x_j^{t+1}\right)\middle|S^{t+1}(j)=S_i^t\cup\{j\},\forall S_i^t\in\mathcal{S}^t\cup\varnothing,x_j^{t+1}\in\mathcal{D}_j\right\} \tag{11}$$

which represents the set of strategies available to node j at the next time $t+1$, given the specific state ω^t at the time t. ■

Definition 6 (Arnold & Schwalbe, 2002) A set $\Psi\subset\Omega$ is ergodic if for any $\omega\in\Psi$, $\omega'\notin\Psi$, $p_{\omega\omega'}=0$, and no nonempty proper subset of Ψ has this property. Singleton ergodic sets are called absorbing states, i.e., $\omega\in\Omega$ is absorbing if $p_{\omega\omega}=1$.

Definition 7 (Arnold & Schwalbe, 2002) A state $\omega=\left(\mathcal{S},\mathbf{x}\right)$ is an absorbing state of the best-reply process if and only if the following three conditions are all met: i) $x_j\geq v\left(\{j\}\right)$, $\forall j\in\mathcal{N}$, ii) $\forall S_i\in\mathcal{S}$: $\sum_{j\in S_i}x_j\leq v\left(S_i\right)$, and if the inequality is strict, then, $v(S_i)<\sum_{j\in S_i}x_j+\varrho$, iii) $\forall j\in\mathcal{N}$: $\nexists S_i\in\mathcal{S}$: $x_j<v\left(S_i\cup\{j\}\right)-\sum_{k\neq j,k\in S_i}x_k$.

Proposition 3 The set of allocations associated with an absorbing state of the proposed algorithm coincides with the set of core allocations of the proposed game.

Proof: On the one hand, if an allocation **x** is in the core of the proposed game, the allocation **x** will never be blocked by any coalition. Thus, each sensor node is not motivated to revise its current coalition and utility. For this case, the state is absorbing. On the other hand, assume that a state $\omega=(\mathcal{S},\mathbf{x})$ is absorbing, but **x** is not in the core. Then, there at least exists a coalition $S_i\subseteq\mathcal{N}$, which blocks the allocation **x**. Each sensor node $j\in S_i$ may obtain a lower utility than it could achieve. Once it can be chosen with probability $1/N$, it will switch coalitions or change its utility by operating the best-reply process or taking the trial action. It means that the state is not stable, which contradicts the assumption that the state $\omega=(\mathcal{S},\mathbf{x})$ is absorbing. ■

Proposition 4 The proposed algorithm will converge to a stable coalition structure with a core allocation **x**, whose probability is one as time tends to infinity.

Proof: For a Markov chain, no matter where the process starts, the probability that the process reaches an ergodic set after t steps goes to one as t tends to infinity (Kemeny & Snell, 1976). Hence, the proposed algorithm can be guaranteed to converge to an ergodic set Ψ. If $|\Psi|>1$, each state $\omega=\left(\mathcal{S},\mathbf{x}\right)$

is not absorbing according to Definition 6, and each allocation **x** is not in the core according to Proposition 3. In other words, there at least exists a coalition which blocks **x**. Accordingly, the sensor nodes are divided into two categories: i) The sensor nodes in the blocking coalitions. For example, the sensor node j firstly forms a singleton coalition with a positive probability and obtains the utility $v(\{j\})$. Then, the singleton coalition $\{j\}$ repeats to adjust its strategy to approach a potentially better coalition with a positive probability. ii) The sensor nodes which are not in the blocking coalitions. They form some non-singleton coalitions and get the opportunities to adjust their own utilities. Due to the fact that the core in the proposed game is nonempty, an absorbing state can finally be reached. The readers can be referred to the work of Arnold and Schwalbe (2002) for more detailed proof. Note that, lots of states are generated in turn during the above process, and all the states belong to Ψ. Therefore, starting from any initial state ω, there is a positive probability to reach an absorbing state. This contradicts the assumption that ω is an element of an ergodic, and it follows that all the ergodic sets are singletons. That is, the proposed algorithm will converge to an absorbing state with probability one as time tends to infinity. Combined with Proposition 3, the proposed algorithm will converge to a coalition structure with a core allocation **x**. Since the core is a subset of the stable set of v (Branzei, Dimitrov, & Tijs, 2005), the resulting coalition structure is stable. ■

1.4 Simulation Results and Analysis

Assume that some CNs are randomly located within a circle with a 30m-radius around the CH. Here, we only take LEACH for example to organize the clusters. Note that many other clustering protocols can also be applied. For both the local broadcasting channels and the long-haul cooperative MISO channels, Rayleigh fading coefficients are modeled as unitary power, complex Gaussian random variables. The constant κ is set to 1, and the path loss exponent δ is set to 3. The Gaussian noise variance σ^2 is 10^{-12}W, and the outage capacity $\mathcal{C}_{out}$ is 1.4bps/Hz. In addition, the model introduced in the work of Cui et al. (2004) is used to calculate the circuit power.

In Figure 2(a), a snapshot of the coalition structure $\mathcal{S}=\{\{\text{CH},\text{CN}_1,\text{CN}_5\},\{\text{CN}_2\},\{\text{CN}_3\},\{\text{CN}_4\}\}$. The retaining energy of CN_2 is 0.5J. In (b), a snapshot of the coalition structure $\mathcal{S}=\{\{\text{CH},\text{CN}_1,\text{CN}_2,\text{CN}_5\},\{\text{CN}_3\},\{\text{CN}_4\}\}$. The retaining energy of CN_2 is 1.5J. The others are the same with those of Figure 2(a).

Consider that the cluster has $N = 6$ sensor nodes, the long-haul transmission distance is 200m, and the threshold value of the outage probability is $P_{out}^{thr} = 10^{-4}$. The current energy of the CH is 2J, and the current energy of all the CNs is less than that of the CH. Figure 2 indicates the final coalition structures through the proposed scheme. Specifically, for Figure 2(a), since CN_1 and CN_5 are closer to the CH, and both of them have more energy, they become the candidates to help the CH. From the resulting coalition structure, the optimal set of assisting CNs, i.e., $\mathcal{J} = \{\text{CN}_1,\text{CN}_5\}$, is derived. Moreover, for Figure 2(b), its setting stays the same, except that CN_2 has more retaining energy. For this reason, the optimal set of assisting CNs becomes $\mathcal{J} = \{\text{CN}_1,\text{CN}_2,\text{CN}_5\}$. Hence, it can be concluded that the relative configuration of the network and the current energy of the sensor nodes has visible effect on the coalition structure and the allocations in the core.

Figure 3 indicates the sensor nodes positions and the final coalition structures through the proposed scheme, when the cluster comprises N = 10, 15, 20 and 25 sensor nodes, respectively. From the resulting coalition structures, we can determine which CNs are chosen to collaborate with the CH. Moreover, it

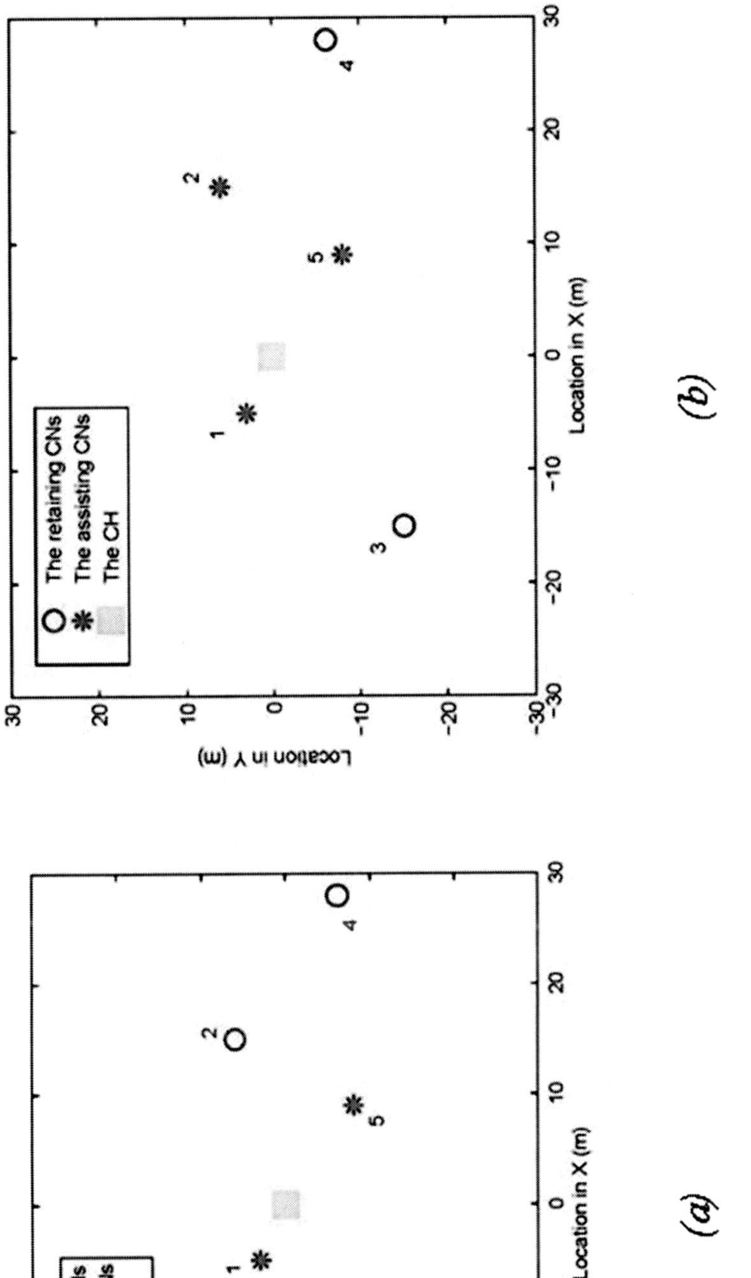

Figure 2. Sensor nodes positions and final coalition structures resulting from the proposed scheme

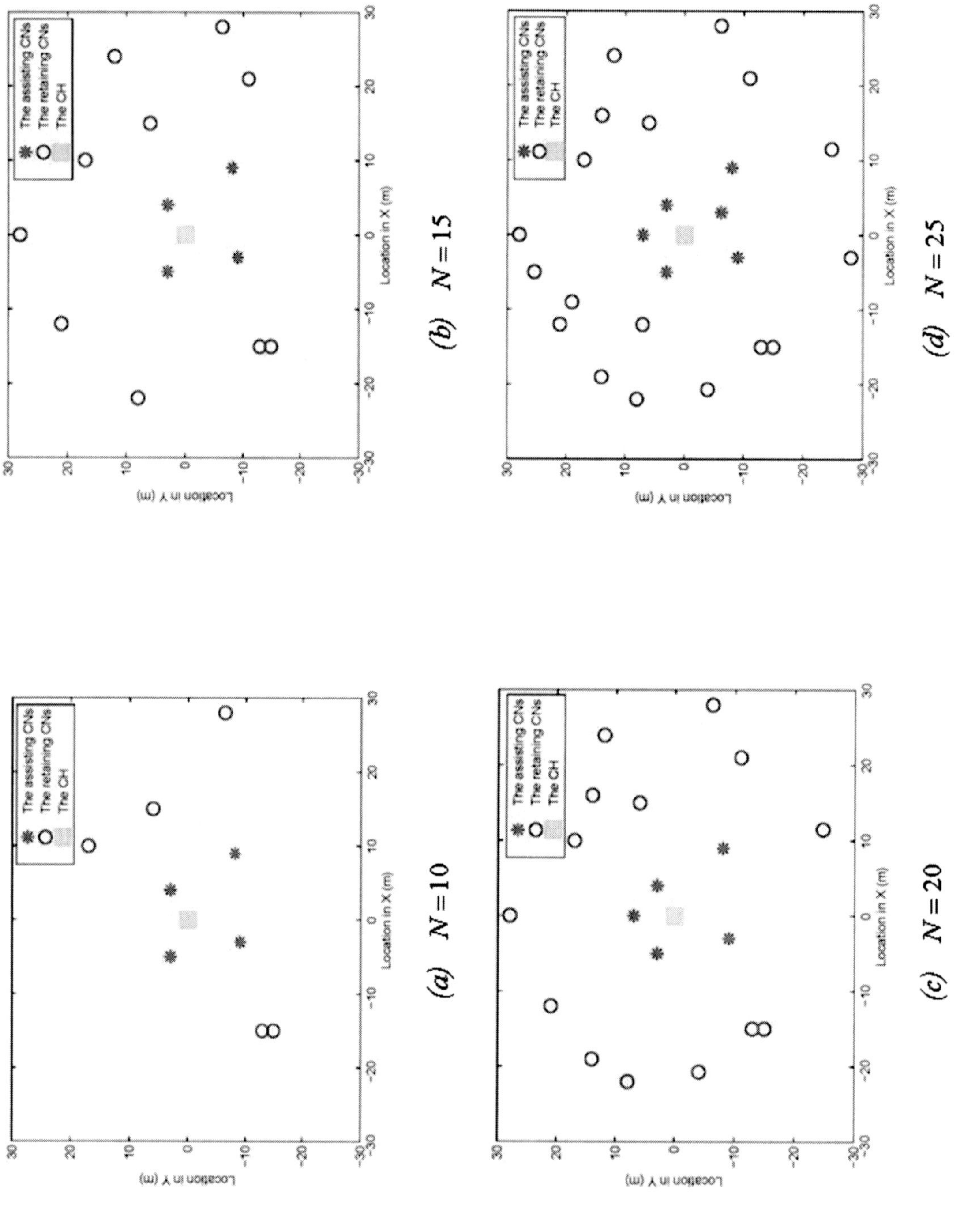

Figure 3. Sensor nodes positions and final coalition structures for the different number of the sensor nodes

Figure 4. The outage performance and the minimal retaining energy as a function of β

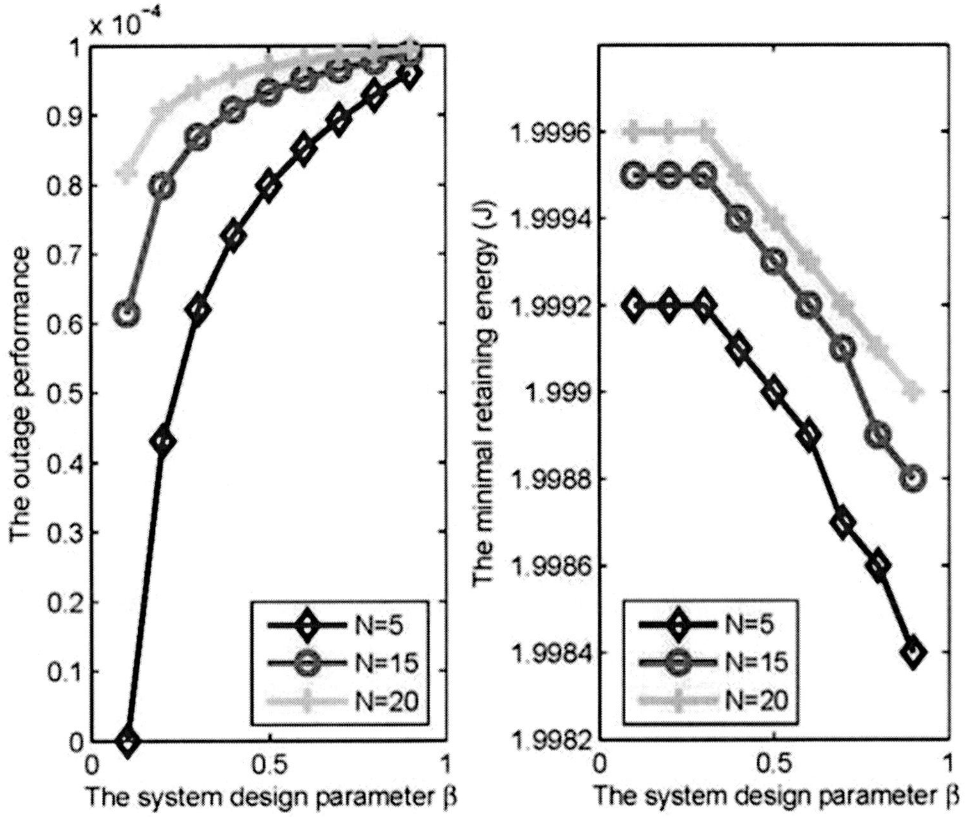

can be observed that: i) more sensor nodes do not result in more CNs which could be chosen to help the CH, but ii) more sensor nodes afford more opportunities to find more good assisting CNs. It should be noted that similar simulation results can be achieved when more nodes are employed.

Figure 4 shows the outage performance and the network lifetime tradeoff curve as the system design parameter β is varied from 0.1 to 0.9. β represents the different attention paid to them. The more β is, the more attention the outage performance is paid to, and the more the energy may consume. In addition, according to the resulting coalition structures which have the different number of sensor nodes, once there are more CNs to cooperate with the CH, the outage performance and the minimal retaining energy can both be improved.

Figure 5 presents the tradeoff with different distances between the cluster and the sink node. In Figure 5(a), the non-cooperative SISO communication scheme is plotted as a reference. With the increasing of the long-haul transmission distance, the tradeoff value drops off sharply. However, in cooperative communication scheme, due to the optimal set of assisting CNs, any increase in energy assumption is averaged among them, and the tradeoff value decreases tinily with the distance. Hence, the proposed scheme can jointly maximize the outage performance and the network lifetime. In addition, Figure 5(b) shows the tradeoff value for the resulting coalition structures which have the different number of sensor nodes. Particularly, the tradeoff in the case of $N = 10$ is the same as that of $N = 15$, since the additional good CNs are not found.

Figure 5. The tradeoff value vs. the long-haul transmission distance in different algorithms

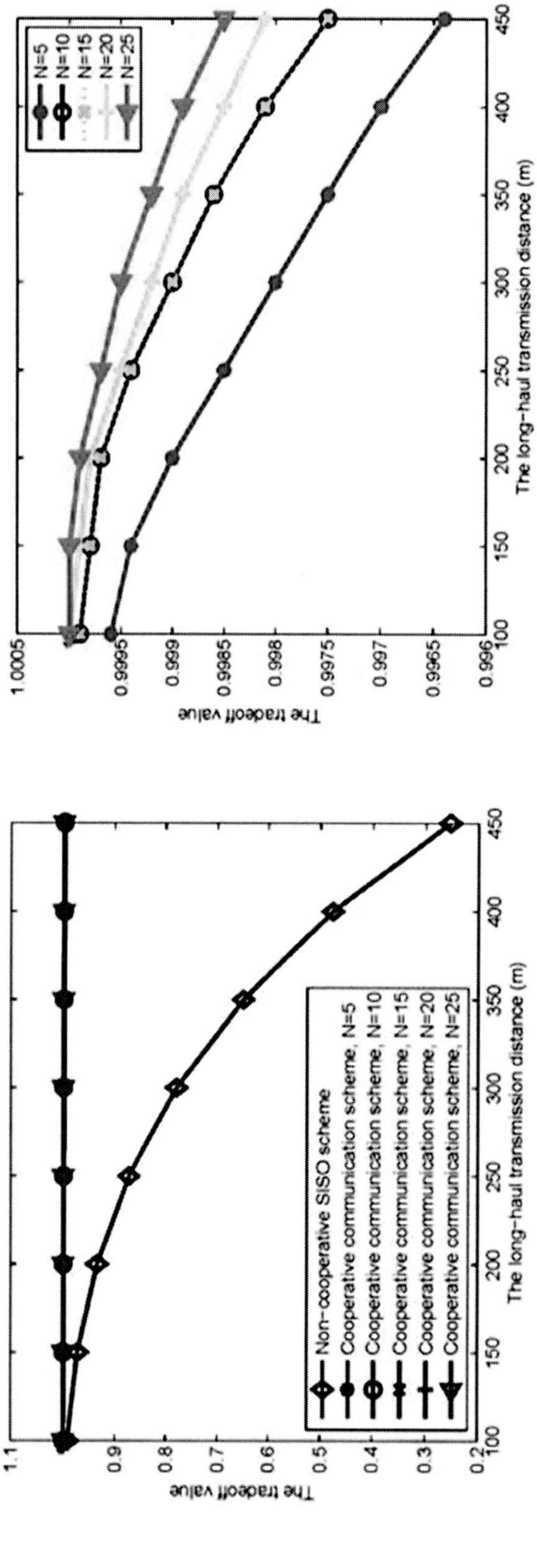

(a) The comparison of the different schemes

(b) The tradeoff value vs. the different number of the sensor nodes

2. Energy-Efficient Resource Sharing for D2D Communications

Recently, device-to-device (D2D) communication has received significant attention due to the advantages in enhancing network throughput, saving energy consumption, improving spectrum utilization. However, since the link resources of cellular users are reused, one of the crucial issues is the intelligent resource sharing between potential D2D pairs and cellular users to guarantee excellent quality and services.

Note that, the design of resource sharing scheme is a systematic project. On one hand, the D2D communication flexibly operates in multiple modes, e.g., the non-orthogonal and orthogonal D2D communication sharing modes, the cellular mode, etc. Thus, an efficient mode selection method is necessary to build up a good platform for sharing of resources. On the other hand, the resource scheduling also belongs to the scope of the resource sharing strategies, which involves how the potential D2D pair chooses resource sharers and splits its transmit power among its sharers, and how the cellular users control their transmit power. The objective is to optimize the network performance over the shared resources under practical constraints on services. Due to lack of convexity, globally optimal scheduling is difficult to derive. To reduce the complexity, most existing works exploit the heuristic or suboptimal methods.

Furthermore, it is an important research topic to achieve the mode selection and resource scheduling together in a unified optimization framework from the energy-efficient perspective. Nonetheless, mathematically speaking, such an issue is typically modeled as a non-convex multi-objective optimal problem based on the discrete space and the continuous space. Due to the combinatorial characteristics, this joint optimization turns out to be a NP-hard problem. Moreover, most of the existing literature assumes that the transmission environment is static or quasi-static. In practical terms, since the mobility of nodes results in mutable channel quality and traffic loads, it is desirable to incorporate the dynamics of the environment and strategy decision into this process, especially in a distributed manner.

Intuitively, the potential D2D pair and their resource sharers form a group, in which each member enjoys the cooperation gain (e.g., the improvement in energy efficiency), however, it subjects to the cooperation cost (e.g., the interference between the D2D and cellular users). Hence, investigating the resource sharing is equivalent to studying how users can form collaborative groups and exploit resources in the same group to improve the profit of each member. As such, it is available to refer to coalition formation game theory to design the resource sharing scheme. In what follows, we take the work of Wu et al. (2014) for example, which formulates the uplink resource sharing in mobile D2D communication as a non-transferable coalition formation game, with the characteristic function which accounts for the gains in terms of energy efficiency and the costs in terms of mutual interference, and develops a distributed coalition formation algorithm based on the merge-and-split rule and the Pareto order. Then, the resulting coalition structure shows the energy-efficient sharing strategy on the joint mode selection, uplink link reusing allocation, and power management.

2.1 Network Model and Problem Formulation

As Figure 6, a single cell scenario of multi-user cellular network is considered, where there exist N users which together form a set $\mathcal{N}$. Each user has access to a frequency band with a bandwidth normalized to be 1. Moreover, some of them intend to employ D2D communication as an underlay of the cellular network. These users come in pairs, which are called potential D2D pairs, and together constitute a set $\mathcal{N}_D$. For each potential D2D pair, its transmitter sends the information to the corresponding receiver

Figure 6. Network model of D2D communication underlaying cellular networks

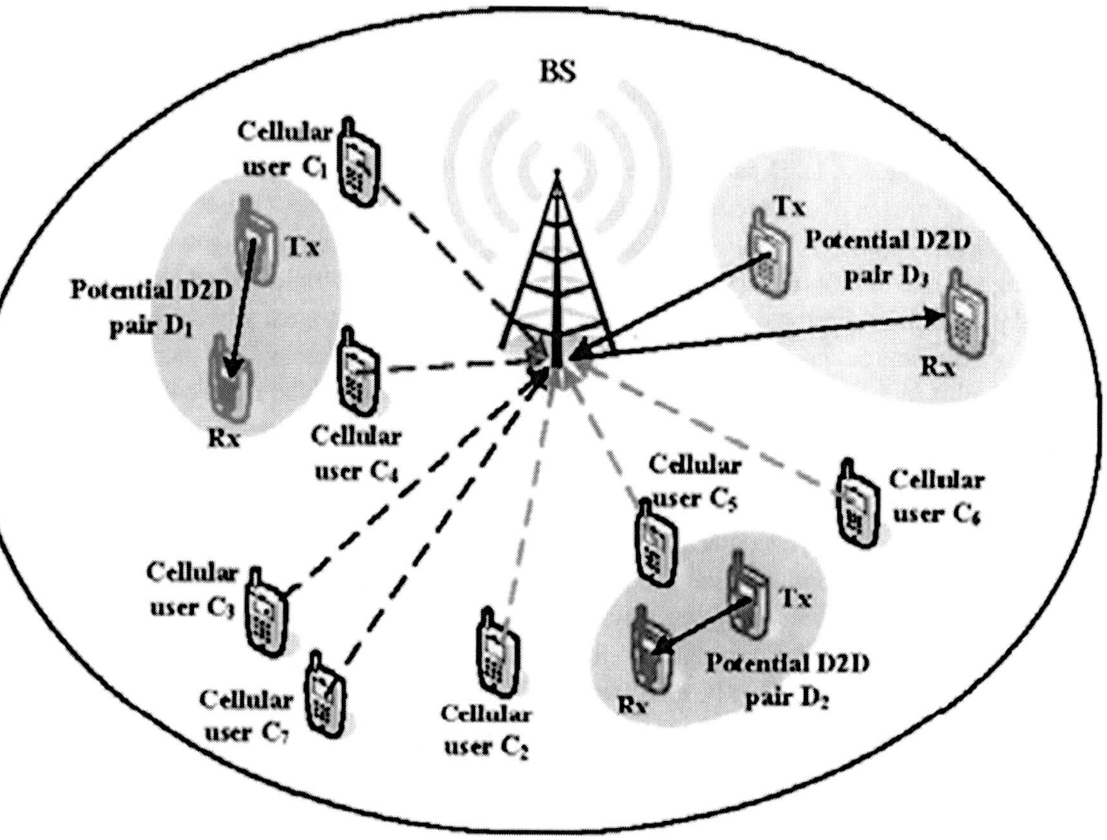

over a link by reusing cellular resources. The rest of users communicate with the nearest BS directly, which are treated as the cellular users, and comprise a set $\mathcal{N}_C$. Apparently, $\mathcal{N} = \mathcal{N}_D \bigcup \mathcal{N}_C$. In addition, the propagation channel model contains path loss and frequency flat Rayleigh fading, and σ^2 is the Gaussian noise variance.

Once there exist the potential D2D pairs to enable D2D operation, the non-orthogonal D2D communication sharing mode is employed, in which the D2D pair shares the same uplink resources with the ones selected out of cellular users. Then, these cellular users form a group. With potential D2D pair D_i for example, if cellular user C_j devotes its uplink link to D_i, their throughputs can be obtained respectively by

$$R_{D_i} = W_{D_i} \sum\nolimits_{C_j \in \mathcal{N}_{D_i}} \log_2\left(1 + \gamma_{D_i}^{C_j}\right) \tag{12}$$

$$R_{C_j} = W_{D_i} \log_2\left(1 + \gamma_{C_j}\right),\ \forall C_j \in \mathcal{N}_{D_i} \tag{13}$$

where $W_{D_i} = \left(\left|\mathcal{N}_{D_i}\right| + 2\right) / \left|\mathcal{N}_{D_i}\right|$, $\mathcal{N}_{D_i}$ is a set which is composed of the cellular users helping D_i, and $|\bullet|$ represents the cardinality of a set. The received SINR of the receiver of D_i on the uplink link of C_j is given by $\gamma_{D_i}^{C_j} = \frac{p_{D_i}^{C_j} H_{D_i}^{C_j}}{p_{C_j} H_{C_j,D_i} + W_{D_i} \cdot \sigma^2}$, and the received SINR for cellular user C_j at the BS is given respectively by $\gamma_{C_j} = \frac{p_{C_j} H_{C_j,B}}{p_{D_i}^{C_j} H_{D_i,B}^{C_j} + W_{D_i} \cdot \sigma^2}$, where $p_{D_i}^{C_j}$ is the transmit power of the transmitter of D_i on the uplink link of C_j, and p_{C_j} is the transmit power of C_j. Also, the channel gains from C_j to the BS and from C_j to the receiver of D_i are denoted as $H_{C_j,B}$ and H_{C_j,D_i}, respectively. The channel gains from the transmitter of D_i to its receiver and from the transmitter of D_i to the BS by reusing the uplink link of C_j are expressed as $H_{D_i}^{C_j}$ and $H_{D_i,B}^{C_j}$, respectively.

Since D_i no longer occupies the uplink and downlink links between itself and the BS, it is assumed that the bandwidth of the two links can be allocated equally among the cellular users in $\mathcal{N}_{D_i}$. Accordingly, it characterizes the advantages in spectrum utilization quantitatively, and is treated as the benefit which the cellular users get from sharing the resource with D_i. Once potential D2D pair D_i determines the cellular communication, the cellular mode is employed, in which the transmitter and receiver of D_i receive the help from the BS in the relaying manner. Consequently, its throughput is given by

$$R_{D_i} = \log_2\left(1 + \min\left\{\gamma_{D_i}^{\mathrm{Tx}}, \gamma_{D_i}^{\mathrm{Rx}}\right\}\right) \tag{14}$$

where $\gamma_{D_i}^{\mathrm{Tx}}$ and $\gamma_{D_i}^{\mathrm{Rx}}$ are the received SINRs for the transmitter of D_i at the BS and the receiver of D_i, respectively. Actually, the BS could coordinate its transmit power to guarantee $\gamma_{D_i}^{\mathrm{Rx}}$ is not smaller than $\gamma_{D_i}^{\mathrm{Tx}}$, which means that Eq. (14) can be simplified into $R_{D_i} = \log_2\left(1+\gamma_{D_i}^{\mathrm{Tx}}\right)$.

In practical, the number of cellular users is larger than that of D2D pairs, and the cellular users access the BS by using standard control signaling. If the uplink link of C_j is not reused by any D2D pair, its throughput is

$$R_{C_j} = \log_2\left(1 + \frac{p_{C_j} H_{C_j,B}}{\sigma^2}\right), \ \forall C_j \in \mathcal{N}_C \setminus \left\{\bigcup_{D_i \in \mathcal{N}_D} \mathcal{N}_{D_i}\right\} \tag{15}$$

Note that, the resource sharing between D2D pairs and cellular users should not incur the loss of communication quality of the D2D and cellular links. In other words, the QoS requirements of both the potential D2D pairs and the cellular users, especially in terms of the throughput targets of users, should always be satisfied, namely,

$$R_{D_i} \geq R_{D_i}^{thr}, \ \forall D_i \in \mathcal{N}_D \tag{16}$$

$$R_{C_j} \geq R_{C_j}^{thr}, \ \forall C_j \in \mathcal{N}_C \tag{17}$$

where $R_{D_i}^{thr}$ and $R_{C_j}^{thr}$ are the throughput threshold of potential D2D pair D_i and cellular user C_j, respectively.

Furthermore, the energy efficiency is treated as a performance metric. Here, the energy efficiency of each user is defined as the ratio of the throughput to the total power consumption. For D_i and C_j, the energy efficiency is formally expressed as $EE_{D_i} = R_{D_i} / p_{D_i}^{total}$, $\forall D_i \in \mathcal{N}_D$, and $EE_{C_j} = R_{C_j} / p_{C_j}^{total}$, $\forall C_j \in \mathcal{N}_C$, where $p_{D_i}^{total} = (1+\alpha)\sum_{C_j \in \mathcal{N}_{D_i}} p_{D_i}^{C_j} + p_{ct} + p_{cr}$ and $p_{C_j}^{total} = (1+\alpha) p_{C_j} + p_{ct}$. The total average power consumption results from the power amplifiers and the other circuit blocks. Specifically, α is dependent on the modulation scheme and the associated constellation size. p_{ct} and p_{cr} are the power consumed by circuits at the transmitter side and at the receiver side, respectively. Without loss of generality, α, p_{ct} and p_{cr} are the same for each user.

2.2 Energy-Efficient Uplink Resource Sharing as a Coalition Formation Game

Mathematically, the energy-efficient uplink resource sharing problem is modeled as a coalition formation game with a NTU. According to the process of coalition formation and the resulting partition, the solutions to the joint problem of mode selection, uplink reusing allocation and power management is derived, while accounting for the tradeoff between the benefits in terms of energy efficiency and the costs in terms of the mutual interference. To be precise, the proposed coalition formation game can be defined by the triplet $(\mathcal{N}, v, \mathcal{S})$ in a characteristic form. Here, $\mathcal{N}$ denotes the set of players, i.e., the users in the single-cell cellular network, v is a characteristic function, and $\mathcal{S} = \{S_1, \ldots, S_k, \ldots, S_l\}$ shows a coalition structure of $\mathcal{N}$.

Since the users are energy-aware, they prefer to consume their power as little as possible, subject to the constraints on the desired QoS performance, especially in terms of the throughput targets of users. To this end, we can obtain the transmit power of each user by achieving their throughput threshold, that is by transforming the constraints in (16) and (17) into equality constraints. The reader can be referred to [15, Proposition 1], which points out how the potential D2D pair splits its transmit power among its sharers, and how the cellular users control their transmit power.

Furthermore, in the proposed game, the characteristic function is dependent on the energy efficiency. Specifically, given a coalition $S_k \in \mathcal{S}$, the coalition value set is defined as

$$v(S_k) = \left\{\phi(S_k) \in \mathbb{R}^{|S_k|} \middle| \phi_j(S_k), \forall j \in S_k\right\} \tag{18}$$

where $\phi(S_k)$ is a vector, whose element $\phi_j(S_k)$ represents a utility that player $j \in S_k$ can obtain within coalition S given a certain strategy selected by j. Taking coalition S_k for example, the setting of $\phi_j(S_k)$ is classified and discussed in the following:

- $|S_k \cap \mathcal{N}_D| \le 1$ and $|S_k| = 1$. In this case, where coalition S_k has only a member, either a potential D2D pair or a cellular user, we have $\phi_j(S_k) = EE_j$.
- $|S_k \cap \mathcal{N}_D| = 1$ and $|S_k| > 1$. If player $j \in S_k$ is a potential D2D pair, it may improve its energy efficiency by sharing multiple uplinks with different cellular users. However, the energy efficiency may be limited by the potential interference between D2D and cellular users. In this regard, it is sufficient to use the energy efficiency to quantify a tradeoff between the achievable gains and the costs, namely, $\phi_j(S_k) = EE_j$. If player $j \in S_k$ is a cellular user, its energy efficiency also plays a

leading role in performance evaluation. Potential D2D pair D_i will free up the occupied frequency bands to it, so that it can experience an increase in its energy efficiency. However, as the number of such cellular users increases, the splitting of the released frequency bands occurs more, and the mutual interference becomes more serious. Then, these constraints make its energy efficiency worse off. Moreover, D_i should reward the cellular user for donating its own resource. Intuitively, the reward may be defined as the improvement of D_i on its utility, compared with the non-shared case. In this regard, we have $\phi_j(S_k) = \lambda_j \cdot (\phi_{D_i}(S_k) - \phi_{D_i}) + EE_j$, where λ_j is a positive constant. As a tunable parameter, it can be adjusted to satisfy heterogeneous service requirements.

- $|S_k \cap \mathcal{N}_D| > 1$. There exist more than one D2D pair to reuse the same resources, the utility of each player is set to $-\infty$. That is because the harmful interference to the cellular network is generated, and thus, the player has clearly no interest in cooperating.

From the above analysis, it shows that the utility obtained by each player in S_k is related to the joint strategies that these players in S_k select. Hence, the coalition value can not be arbitrarily apportioned among them, and then,

Proposition 5 Given the characteristic function in (18), the proposed energy-efficient uplink resource sharing coalition formation game has a NTU.

In such a case, by well performing uplink resource sharing, the energy efficiency of potential D2D pairs and cellular users can be improved. However, this advantage steadily weakens as the number of cellular users denoting their own resources increases, and the cost in terms of mutual interference limits this advantage further. The increase in this cost depends on many factors, such as the transmit power of the users in the same coalition, the coalition size, etc. Hence, the proposed game has the following property:

Proposition 6 The proposed energy-efficient uplink resource sharing coalition formation game is non-superadditive. Instead, disjoint independent coalitions will form in the cellular network.

Proof: Given a number of users positioned at different locations in the cellular network, cooperation for improving the energy efficiency entails costs, as a consequence of Proposition 5. Hence, in a practical scenario, it is highly likely that, when they cooperate for forming the grand coalition $\mathcal{N}$ of all users, either: i) there exist more than one D2D pair to reuse the same resource, so that $\phi_j \rightarrow -\infty$ and hence they have no incentive to join the grand coalition, or ii) there are too many cellular users with which a D2D pair shares the resource, so that $(|\mathcal{N}_{D_i}|+2)/|\mathcal{N}_{D_i}|$ becomes small, however, the mutual interference increases, and thus, these cellular users have incentive to deviate from the grand coalition. In a word, the presence of the cost for cooperation renders the game non-superadditive. That is, the grand coalition of all the users will seldom form (it only forms if there are only one D2D pair and multiple cellular users, and all of them can coordinate interference well, which is unrealistic in a large scale cellular network). Accordingly, in the most general form, the partition consists of disjoint independent coalitions. ■

For the proposed coalition formation game with the resulting coalition structure $\mathcal{S} = \{S_1, \ldots, S_k, \ldots, S_l\}$, a disjoint coalition $S_k \in \mathcal{S}$ has the following types: It has i) one cellular user, ii) one potential D2D pair, and iii) one potential D2D pair and some cellular users. Actually, the proposed game provides a theoretical foundation for solving the uplink resource sharing problem to optimize the energy efficiency of users. The resulting coalition structure shows how to choose between the non-orthogonal D2D communication sharing mode and cellular mode for the potential D2D pairs, and how to allocate the uplink reusing among them. That is, if coalition S_k is the second type, the D2D pair in coalition S_k gets back to the cellular communication. If coalition S_k is the third type, the D2D user reuses the uplink resources of other members of coalition S_k.

2.3 Distributed Coalition Formation Algorithm for Energy-Efficient Uplink Resource Sharing

In many practical applications, it is desirable that the coalition formation process takes place in a low-complexity and distributed manner. Here, a distributed coalition formation algorithm based on the merge-and-split rule and the Pareto order is developed, whereby the users in the proposed game have an autonomy to decide whether to join a coalition or not. In particular, due to the non-transferable nature of the proposed game, the Pareto order is an adequate preference relationship for comparing collections of coalitions. Then, the energy-efficient uplink resource sharing algorithm is described as follows.

Step 1: The users of $\mathcal{N}$ are organized into an initial partition $\mathcal{S}_0 = \mathcal{N}$, that is, each user solely forms a coalition.

Step 2: In $\mathcal{S}_0 = \mathcal{N}$, the coalitions, each of which has a potential D2D pair and tends to merge, form a D2D collection $\mathcal{A}_{0,D}$. The other coalitions, each of which is a singleton set, and has only one cellular user, form a cellular collection $\mathcal{A}_{0,C}$. According to Proposition 1 in the work of Wu et al. (2014) and Eq. (18), calculate the coalition value set of each coalition.

Step 3: An iteration consisting of a number of successive merge operations is repeated until all the coalitions have made their local merge decisions, resulting in a partition $\tilde{\mathcal{S}}$. Table I shows the merge process of the proposed algorithm.

Step 4: The coalitions in the resulting partition $\tilde{\mathcal{S}}$ accepts some split operations, if any is possible, until it convergence to a final partition $\mathcal{S}$.

Through users periodically repeating, the proposed algorithm can provide self-adaptation to the dynamic network environment incurred by the user mobility. The key point is to change the frequency of merge-and-split operations according to various movement speeds. Note that, with the increase of the speed, the number of merge-and-split operations increases.

Theorem 7 Starting from any initial coalitional structure $\mathcal{S}_0$, the energy-efficient uplink resource sharing algorithm will always converge to a final partition $\mathcal{S}$ which is composed of a number of disjoint independent coalitions.

Proof: Beginning with any initial partition $\mathcal{S}_0$, the merge process of the proposed algorithm is firstly mapped to a sequence of merge operations. Every merge operation yields the transformation, i.e., $\mathcal{S}_{\tau}^{n_{\tau}} \rightarrow \mathcal{S}_{\tau+1}^{n_{\tau+1}}$, which shows that during iteration index τ after the occurrence of n_{τ} merge operations, a user makes a single merge operation to form a new partition $\mathcal{S}_{\tau+1}^{n_{\tau+1}}$ at iteration $\tau + 1$. Combined with Definition 4 and the characteristic function in (18), it follows that every single merge operation results in a partition which either is a new partition or remains unchanged from the previous partition. Obviously, Bueif the current partition always stays the same, the merge operations will terminate. If every single merge operation leads to a new partition, the above transformations will also terminate. That is because the number of partitions of a set is finite, i.e., the Bell number, and then, the number of transformations is finite. Further, a sequence of split operations will occur, and similarly, they always stop. Hence, the proposed algorithm can converge to a final partition $\mathcal{S}$ which is composed of a number of disjoint coalitions. ■

Table 1. Merge process of energy-efficient uplink resource sharing algorithm

Begin
At iteration $\tau \geq 0$, the partition $\mathcal{S}_\tau$ is formed, where the D2D collection and the cellular collection are get by $\mathcal{A}_{\tau,D}$ and $\mathcal{A}_{\tau,C}$, respectively;
Calculate the coalition value set of each coalition $S_{\tau,k} \in \mathcal{S}_\tau$ by Eq. (18);
while ($\tau \geq 0$)
if (there exists a coalition in $\mathcal{A}_{\tau,D}$ which has more than one member)
State with this coalition;
else
Start with any other coalition in $\mathcal{A}_{\tau,D}$;
end if
The selected coalition $S_{\tau,k}$ determines coalition $S_{\tau,m}$, which can provide the best uplink resource among the coalitions in $\mathcal{A}_{\tau,C}$ which never attempts to merge with $S_{\tau,k}$;
Calculate the coalition value set of coalition $\{S_{\tau,k} \cup S_{\tau,m}\}$ by Eq. (18);
if ($\{S_{\tau,k} \cup S_{\tau,m}\} \triangleright_{\text{Pareto}} \{S_{\tau,k}, S_{\tau,m}\}$)
The merging occurs to form the new coalition $S_{\tau+1,k} = \{S_{\tau,k} \cup S_{\tau,m}\}$;
$\mathcal{A}_{\tau+1,C} = \mathcal{A}_{\tau,C} \setminus S_{\tau,m}$; $\mathcal{S}_{\tau+1} = \{\mathcal{S}_\tau \setminus S_{\tau,k}\} \cup S_{\tau+1,k}$;
else
$\mathcal{A}_{\tau+1,C} = \mathcal{A}_{\tau,C}$; $\mathcal{S}_{\tau+1} = \mathcal{S}_\tau$;
end if
if ($\mathcal{A}_{\tau+1,C}$ has a coalition which does not attempt to merge with $S_{\tau,k}$)
$\mathcal{A}_{\tau+1,D} = \{\mathcal{A}_{\tau,D} \setminus S_{\tau,k}\} \cup S_{\tau+1,k}$;
else
$\mathcal{A}_{\tau+1,D} = \mathcal{A}_{\tau,D} \setminus S_{\tau,k}$;
end if
if ($\mathcal{A}_{\tau+1,D} = \varnothing$ or $\mathcal{A}_{\tau+1,C} = \varnothing$)
$\tau = -1$;
else
$\tau = \tau + 1$;
end if
end while

Furthermore, the stability of the resulting partition $\mathcal{S}$ can be investigated based on the concept of defection function $\mathbb{D}$. In the following, it is verified that the resulting partition $\mathcal{S}$ is not only $\mathbb{D}_{hp}$-stable, but also stronger $\mathbb{D}_p$-stable.

Proposition 7 The partition $\mathcal{S}$ that results from the proposed energy-efficient uplink resource sharing algorithm is $\mathbb{D}_{hp}$-stable.

Proof: The resulting partition $\mathcal{S}$ can not be subject to any further merge or split. For instance, for any coalition S_k and any set $\mathcal{K} \subseteq \{1,\ldots,l\}$, it is assumed that $\cup_{k\in\mathcal{K}} S_k \triangleright_{\text{Pareto}} \{S_k | k \in \mathcal{K}\}$, and then, the partition $\mathcal{S}$ has a motivation to be changed through application of the merge operations on $\{S_k | k \in \mathcal{K}\}$. Obviously, it contradicts with the fact that $\mathcal{S}$ results from a termination of merge-and-split iteration, and hence, $\bigcup_{k\in\mathcal{K}} S_k \ntriangleright_{\text{Pareto}} \{S_k | k \in \mathcal{K}\}$. Similarly, for any $k \in \mathcal{K}$, any partition of S_k is not preferred over S_k by the Pareto order. That is, a split rule no longer works for $\mathcal{S}$. ■

Proposition 8 The resulting partition $\mathcal{S}$ from the proposed energy-efficient uplink resource sharing algorithm is $\mathbb{D}_p$-stable.

Proof: According to the characteristics of the resulting partition, any possible partition in $\mathcal{N}$ which is different from $\mathcal{S}$, $\hat{\mathcal{S}}$, is $\mathcal{S}$-homogeneous. As Proposition 7 and Theorem 3, since the resulting partition $\mathcal{S}$ is $\mathbb{D}_{hp}$-stable, $\mathcal{S} \triangleright_{\text{Pareto}} \hat{\mathcal{S}}$ holds for any $\hat{\mathcal{S}}$ in $\mathcal{N}$. That is, $\mathcal{S}$ is $\triangleright_{\text{Pareto}}$-maximal. To this end, the resulting partition $\mathcal{S}$ is $\mathbb{D}_p$-stable. ■

In general, the complexity of the proposed algorithm is closely related to the number of merge-and-split operations. According to the characteristics of the partition in the proposed game, only the coalition which has a potential D2D pair is subject to the merge operations. The most extreme case occurs when each potential D2D pair attempts to merge with all the others, however, these operations all fail. The total number of merge attempts will be $|\mathcal{N}_D| \cdot |\mathcal{N}_C|$. In practice, the merge process requires a significantly lower number of operations, due to the following facts: i) By running the proposed algorithm, the initial N users in independent states will form larger coalitions. Then, the subsequent runs of the proposed algorithm aim at reducing the number of merge attempts per coalition. ii) Due to the costs for cooperation and QoS requirements, not all cellular users are suitable for participating in D2D communications, and the merge possibilities for the remaining users will decrease for the partition with the larger coalition. Moreover, since splitting involves all the possible partitions of the set formed by the members in a coalition, it is treated as a complex operation. However, the split operations are also only applied to the coalition which has a potential D2D pair and cellular users. For such a coalition, the number of all possible partitions is $\sum_{n=0}^{|S_k|-1} \frac{(|S_k|-1)!}{n!(|S_k|-n-1)!}$. Similarly, a coalition is not usually required to search for all the split forms, due to the practical aspects, e.g., the increasing cost, the lower complexity for sequential search, etc.

2.4 Simulation Results and Analysis

The simulation setup is as follow: Consider a hexagonal cell with a radius of 500m. The distance between the transmitter and receiver of a potential D2D pair is no further than 50m. Taking an example of D_i and C_j, $H_{D_i}^{C_j} = (d_{D_i})^{-\kappa} |h_{D_i}^{C_j}|^2$, where d_{D_i} is the distance between potential D2D pair D_i, $h_{D_i}^{C_j}$ is the unitary power, Rayleigh fading channel coefficient, and κ is the path loss exponent and is set to 3. All the settings are similar for H_{C_j,D_i}, $H_{C_j,B}$, and $H_{D_i,B}^{C_j}$. The Gaussian noise variance σ^2 is 10^{-12}W. For comparison, two cases are considered as benchmarks: i) Cellular: All users are in cellular mode. ii) Exhaustive optimal: The optimal resource sharing strategy is found via exhaustive search among all possible combinations.

Figure 7 shows the user positions and the final partitions through the proposed algorithm with different numbers of users. The solid diamond and the empty one indicate the transmitter and the receiver of potential D2D pair, respectively. The solid circle represents the cellular user. If the potential D2D pair and cellular users have the same color, they share the same uplink resources. In Figure 7(a), the resulting partition is $\mathcal{S} = \{\{D_1, C_1\}, \{D_2, C_3\}, \{C_2\}, \{C_4\}\}$. As such, the uplink resource sharing is derived, that is, both D_1 and D_2 choose the non-orthogonal D2D communication sharing mode, and reuse the uplinks of C_1 and C_3, respectively, and C_2 and C_4 achieve traditional cellular communications. Moreover, Figure 7(b) demonstrates that the proposed algorithm is still effective for large-scale scenarios. By the proposed algorithm, the potential D2D pairs reuse the resources of one or more than one cellular user on a case by case basis, and in particular, they prefer to cellular users, being away from themselves, to avoid interference.

For different values of λ, Figure 8 plots normalized average individual utility versus the number of users. Regardless of the number of users, the increase in the value of λ incurs the increase on the normalized average individual utility value. That is because each cellular user receives more reward for sharing its own resources with increasing λ. Then, the more cellular users the potential D2D pairs can attract, the more likely they find the suitable sharers.

Take potential D2D pair D_1 in Figure 7(a) for example, Figure 9 shows the normalized energy efficiency with different distance between potential D2D pair D_1 and the BS. Obviously, in cellular case, the increase of the distance results in more energy consumption to guarantee the QoS requirements, and then, the energy efficiency of D_1 drops. In the proposed algorithm, D_1 leads to less interference to the cellular users when it moves away from the BS. Accordingly, it can use power very efficiently on the uplink links of cellular users.

Figure 10 plots the comparison of normalized sum of energy efficiency in cellular case, the exhaustive optimal case and the proposed algorithm. Once there exists a potential D2D pair which can share the uplink resources with the cellular users, the energy efficiency of the D2D pair is raised substantially. Obviously, the proposed algorithm has a better performance than that of the cellular case. Although the exhaustive optimal case can achieve the best resource sharing strategies, its complexity is so high that it can not be applied to practical scenarios. Note that, when the number of users is small ($N \leq 9$), the performance of the proposed algorithm well coincides with that of the exhaustive optimal case. When the number of users is large, the result of the exhaustive optimal case only has a slightly better performance than that of the proposed algorithm. For example, when $N = 10$, their performance difference is about 0.12%, however, when $N = 14$, their performance difference is about 0.56%. This is because the competition becomes drastic and complex with the increase in the number of players.

3. Future Applications of Coalition Formation Games

Potential applications of coalition formation games in communication networks are numerous and diverse. Beyond the applications presented above, coalition formation games have already been applied in the work of Saad et al. (2009b) to improve the network capacity by attracting the transmitters to form cooperative MIMO, in the work of Saad et al. (2009c) to improve the sensing performance for the unlicensed secondary users in cognitive radio networks, and in the work of Saad et al. (2009d) to improve the physical layer security of wireless nodes through cooperation among the transmitters, etc. By far, potential applications of coalition formation games have involved cooperative networks, wireless sensor

Figure 7. Users' positions and final coalition structures with different numbers of users

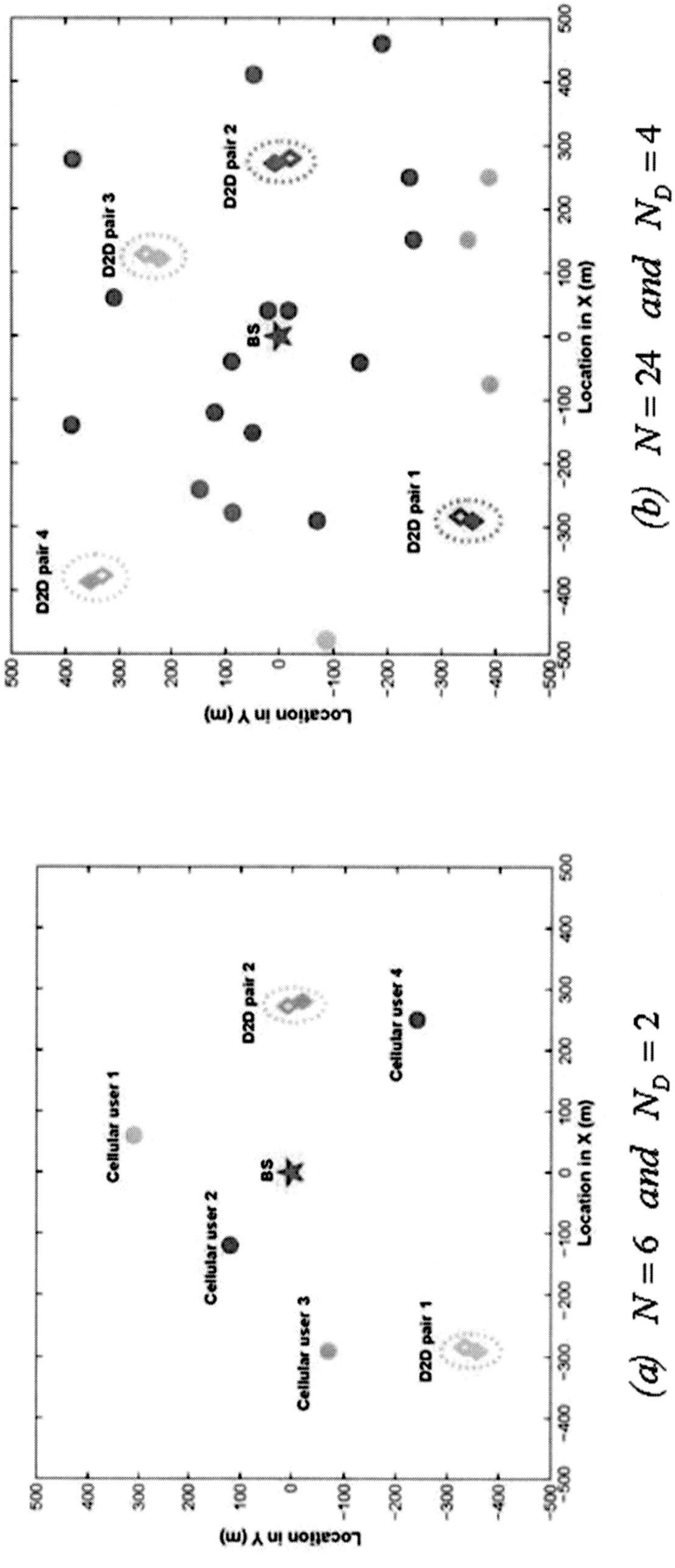

Figure 8. Comparison of normalized average individual utility values for different values of λ

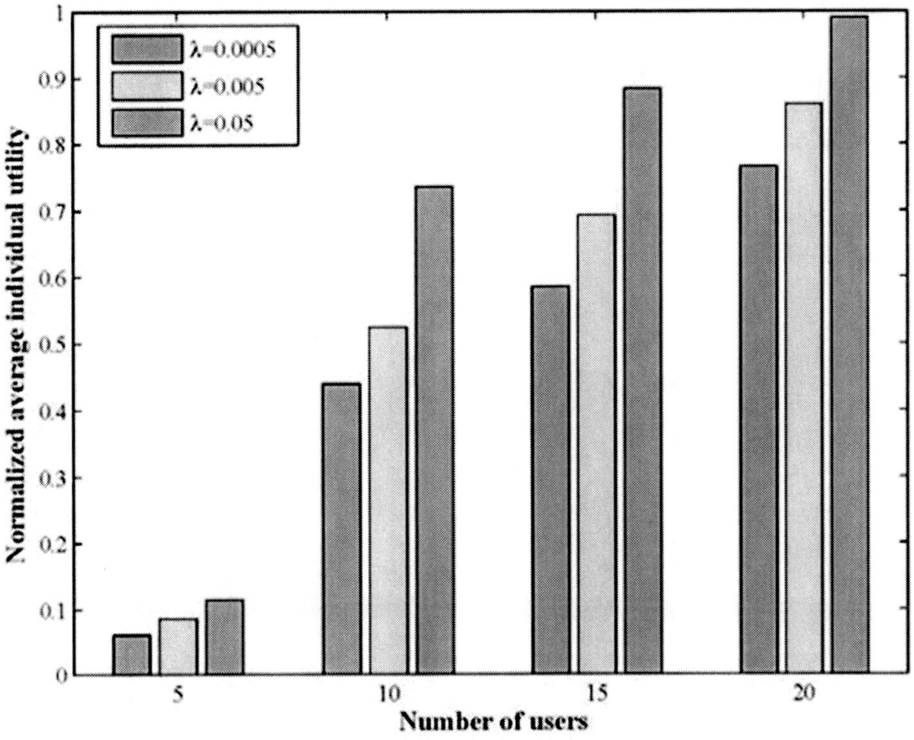

Figure 9. Normalized energy efficiency vs. distance between the potential D2D pair and the BS

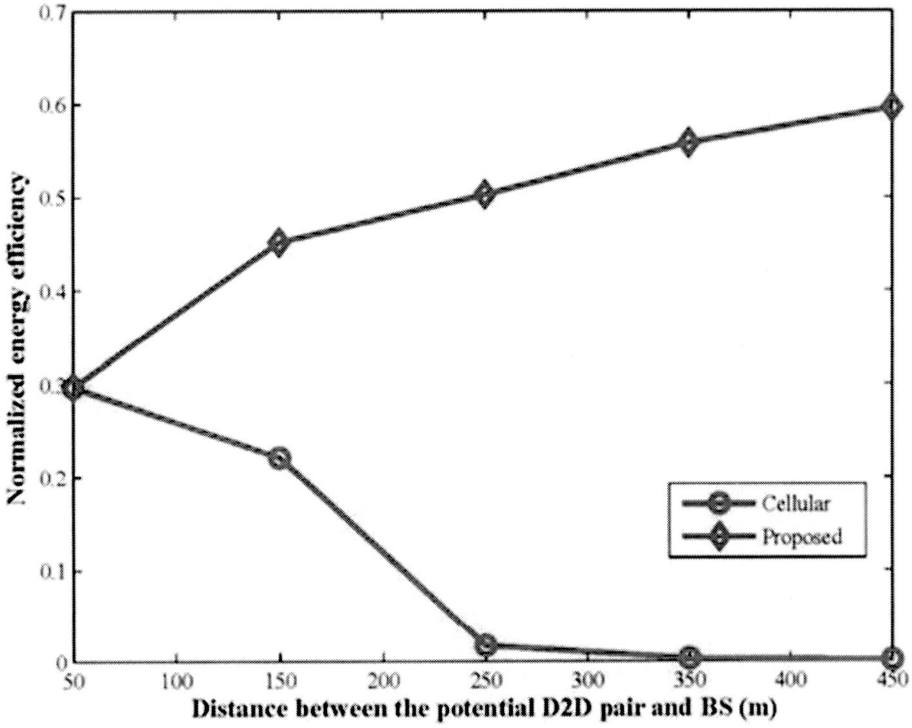

Figure 10. Normalized sum of energy efficiency vs. number of users

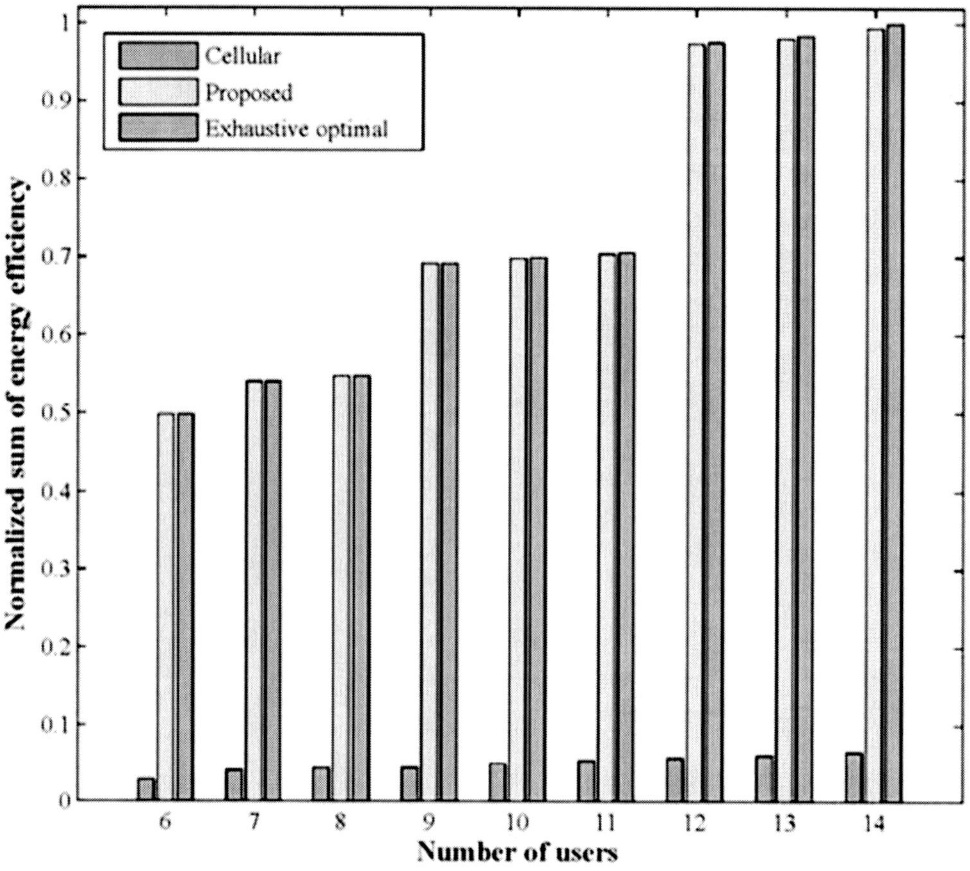

networks, ad hoc self-configuring networks, and many others. A comprehensive survey is studied in the work of Saad et al. (2009a). Indeed, coalition formation game theory is a strong tool for modeling the problems which involves the study of cooperative wireless nodes behavior when a cost is present, and for presenting solid theoretical basis to develop distributed algorithms. In particular, the communication networks, which are self-configuring, self-organizing, self-optimizing, and self-protecting, are becoming more mainstream. Not only that, due to the changes in topology, service requirements, application scenarios, etc, the practical communication environments are always dynamic. In this regard, coalition formation games can not only present an adequate framework for the modeling and analysis of these networks, but also help to characterize the adaptation of behavioral responses to the environment for the users, especially with flexibility and functional scalability.

Note that, in the presence of a cooperative cost, this partition form game falls in the class of coalition formation games. Although the main applications in this chapter require a characteristic form, coalition formation games in partition form are of major interest and have potential applications in communication networks, e.g., the work of Mathur et al. (2008). To this end, coalition formation games in partition form are ripe for many future applications.

In addition, a novel framework from coalition formation game theory, known as hedonic coalition formation games, has been widely studied. In particular, this class of games entails several interesting properties that can be applied, not only in economics, e.g., the works of Bogomonlaia and Jackson (2002), and Dreze and Greenberg (1980), but also in wireless networks, e.g., the works of Saad et al. (2009a, 2011), and Hao et al. (2012). Importantly, it provides a suitable analytical tool for studying the formation of coalitions, among a number of players, in which the utility of any player depends solely on the members of the coalition to which the player belongs, and the coalitions are formed as a result of the preferences of the players over their possible coalitions' set. These characteristics facilitate to obtain the solutions not only in the local and distributed way but also in the low-complexity and low-overhead manner for the dynamic environments. Moreover, these games admit different stability concepts which are extensions to well-known concepts such as the core or the Nash equilibrium. To this end, hedonic games are a very useful analytical tool that has a very strong potential to model problems in wireless communication networks.

CONCLUSION

This book chapter provides comprehensive theory and technical details of coalition formation games, which involve the fundamental properties, the main solution concepts, and the in-depth analysis of the methodologies and approaches for using these games. Moreover, several practical examples are provided to indicate the potential applications of coalition formation games in wireless communication networks. The future opportunities for using such a strong analytical tool in diverse applications are also pointed out. In a word, whenever a cooperative scheme that brings gains, but the gains are limited by a cost imposed by external or internal factors with environmental changes, one can utilize coalition formation games to assess the stability of the coalition structure, identify fairness criteria in allocating the utilities derived from cooperation, and analyze the interaction behaviors among the user.

REFERENCES

Apt, K., & Radzik, T. (2006). *Stable partitions in coalitional games*. Retrieved from arXiv:cs/0605132v1 [cs.GT].

Apt, K., & Witzel, A. (2007). *A generic approach to coalition formation*. Retrieved from arXiv preprint arXiv:0709.0435.

Apt, K., & Witzel, A. (2009). A generic approach to coalition formation (extended version). *International Journal of Game Theory*, *11*(3), 347–367. doi:10.1142/S0219198909002352

Arnold, T., & Schwalbe, U. (2002). Dynamic coalition formation and the core. *Journal of Economic Behavior & Organization*, *49*(3), 363–380. doi:10.1016/S0167-2681(02)00015-X

Aumann, R., & Drèze, J. (1974). Cooperative games with coalition structures. *International Journal of Game Theory*, *3*(4), 217–237. doi:10.1007/BF01766876

Bogomonlaia, A., & Jackson, M. (2002). The stability of hedonic coalition structures. *Games and Economic Behavior*, *38*(2), 201–230. doi:10.1006/game.2001.0877

Branzei, R., Dimitrov, D., & Tijs, S. (2005). *Models in Cooperative Game Theory: Crisp, Fuzzy and Multichoice Games*. New York, NY: Springer.

Cui, S., Goldsmith, A. J., & Bahai, A. (2004). Energy-efficiency of MIMO and cooperative MIMO techniques in sensor networks. *IEEE Journal on Selected Areas in Communications*, *22*(6), 1089–1098. doi:10.1109/JSAC.2004.830916

Dreze, J., & Greenberg, J. (1980). Hedonic coalitions: Optimality and stability. *Econoemetrica*, *48*(4), 987–1003. doi:10.2307/1912943

Gharehshiran, O. N., & Krishnamurthy, V. (2010). Coalition formation for bearings-only localization in sensor networks—a cooperative game approach. *IEEE Transactions on Signal Processing*, *58*(8), 4322–4338. doi:10.1109/TSP.2010.2049201

Hao, X., Cheung, M. H., Wong, V. W. S., & Leung, V. C. M. (2012). Hedonic coalition formation game for cooperative spectrum sensing and channel access in cognitive radio networks. *IEEE Transactions on Wireless Communications*, *11*(11), 3968–3979. doi:10.1109/TWC.2012.092412.111833

Kemeny, J. G., & Snell, J. L. (1976). *Finite Markov Chains*. New York, NY: Springer.

Mathur, S., Sankaranarayanan, L., & Mandayam, N. (2008). Coalitions in cooperative wireless networks. *IEEE Journal on Selected Areas in Communications*, *26*(9), 1104–1115. doi:10.1109/JSAC.2008.080908

Newton, J. (2010). Non-cooperative convergence to the core in Nash demand games without random errors or convexity assumptions. *Journal of Economic Theory*, *24*(6), 1223–1235.

Ray, D. (2007). *A Game-Theoretic Perspective on Coalition Formation*. New York, NY: Oxford University Press. doi:10.1093/acprof:oso/9780199207954.001.0001

Saad, W., Han, Z., Baar, T., Debbah, M., & Hjørungnes, A. (2011). Hedonic coalition formation for distributed task allocation among wireless agents. *IEEE Transactions on Mobile Computing*, *10*(9), 1327–1344. doi:10.1109/TMC.2010.242

Saad, W., Han, Z., Debbah, M., & Hjørungnes, A. (2009). A distributed coalition formation framework for fair user cooperation in wireless networks. *IEEE Transactions on Wireless Communications*, *8*(9), 4580–4593. doi:10.1109/TWC.2009.080522

Saad, W., Han, Z., Debbah, M., & Hjørungnes, A. (2009). Coalitional games for distributed collaborative spectrum sensing in cognitive radio networks. In *Proceddings of the 28th Conference on Computer Communications (IEEE INFOCOM '09)*. Rio de Janeiro, Brazil: IEEE. doi:10.1109/INFCOM.2009.5062135

Saad, W., Han, Z., Debbah, M., & Hjørungnes, A. (2009). Physical layer security: Coalitional games for distributed cooperation. In *Proceedings of the 7th Intl. Symposium on Modeling and Optimization in Mobile, Ad Hoc, and Wireless Networks (WiOpt '09)*. Seoul, Korea: IEEE.

Saad, W., Han, Z., Debbah, M., Hjørungnes, A., & Baar, T. (2009). Coalitional game theory for communication networks. *IEEE Signal Processing Magazine*, *26*(5), 77–97. doi:10.1109/MSP.2009.000000

Sandholm, T., Larson, K., Anderson, M., Shehory, O., & Tohme, F. (1999). Coalition structure generation with worst case guarantees. *Artificial Intelligence*, *111*(1-2), 209–238. doi:10.1016/S0004-3702(99)00036-3

Wu, D., Cai, Y., Zhou, L., & Wang, J. (2012). A cooperative communication scheme based on coalition formation game in clustered wireless sensor networks. *IEEE Transactions on Wireless Communications*, *1*(3), 1190–1200. doi:10.1109/TWC.2012.012712.111049

Wu, D., Wang, J., Hu, R. Q., Cai, Y., & Zhou, L. (2014). Energy-efficient resource sharing for mobile device-to-device multimedia communications. *IEEE Transactions on Vehicular Technology*, *63*(5), 2093–2103. doi:10.1109/TVT.2014.2311580

KEY TERMS AND DEFINITIONS

Coalition Formation Game: A coalition formation game is defined as a triplet $(\mathcal{N}, v, \mathcal{S})$. Specifically, $\mathcal{N} = \{1, \ldots N\}$ denotes a set of players, which try to form coalitions to strengthen their positions in the game. Any coalition represents an agreement among the players in the same coalition to act as a single entity. v denotes the coalition value, which quantifies the worth of a coalition in a game. The definition of the coalition value determines the form and type of the game. In particular, $\mathcal{S}$ is a coalitional structure or a partition of $\mathcal{N}$, i.e., a collection of coalitions $\mathcal{S} = \{S_1, \ldots, S_l\}$, such that $\forall i \neq k$, $S_i \cap S_k = \varnothing$, and $\bigcup_{i=1}^{l} S_i = \mathcal{N}$.

Transferable Utility Game: For any TU game, an allocation $\mathbf{x} \in \mathbb{R}^{|\mathcal{N}|}$ is a core allocation if, i) $x_j \in \mathcal{D}_j$, $\forall j \in \mathcal{N}$, $\sum_{j \in \mathcal{N}} x_j \leq \max_{\mathcal{S} \in \mathcal{CS}} \sum_{S_i \in \mathcal{S}} v(S_i)$, and if the inequality is strict, then $\sum_{j \in \mathcal{N}} x_j + \varrho \geq \max_{\mathcal{S} \in \mathcal{CS}} \sum_{S_i \in \mathcal{S}} v(S_i)$, ii) $\sum_{j \in S_i} x_j \geq v(S_i)$, $\forall S_i \subset \mathcal{N}$.

Shapley Value: Given $(\mathcal{N}, v, \mathcal{S})$, the Shapley value is characterized by a unique mapping ϕ which satisfies the following axioms (ϕ_j is the utility given to player j by the Shapley value ϕ, i.e., $\phi(v) = (\phi_1, \ldots \phi_{|\mathcal{N}|})$ in $\mathbb{R}^{|\mathcal{N}|}$.): i) Relative Efficiency, ii) Symmetry, iii) Additivity, and iv) Dummy.

Nash Preference Relation: Given two collections $\mathcal{A} = \{A_1, \ldots, A_m\}$ and $\mathcal{B} = \{B_1, \ldots, B_n\}$ to the same players, $\mathcal{A}$ is preferred over $\mathcal{B}$ by the Nash preference relation, denoted as $\mathcal{A} \rhd_{\text{Nash}} \mathcal{B}$, if and only if $\prod_{i=1}^{m} v(A_i) > \prod_{k=1}^{n} v(B_k)$ holds.

Pareto Preference Relation: Given two allocations **x** and **y** that are allotted by $\mathcal{A} = \{A_1, \ldots, A_m\}$ and $\mathcal{B} = \{B_1, \ldots, B_n\}$, respectively, to the same players, $\mathcal{A}$ is preferred over $\mathcal{B}$ by the Pareto preference relation, denoted as $\mathcal{A} \rhd_{\text{Pareto}} \mathcal{B}$, if and only if $\mathbf{x} \succ_{\text{Pareto}} \mathbf{y}$ holds. Here, $\mathbf{x} \succ_{\text{Pareto}} \mathbf{y}$ iff $\forall j \in \mathcal{N}$, $x_j \geq y_j$, and $\exists j \in \mathcal{N}$, $x_j > y_j$.

$\mathbb{D}_{hp}$ **Stability:** A partition $\mathcal{S} = \{S_1, \ldots, S_l\}$ is $\mathbb{D}_{hp}$-stable, if and only if the following two conditions are satisfied: i) For each $i \in \{1, \ldots, l\}$ and for each partition $\{B_1, \ldots, B_n\}$ of S_i, $\{B_1, \ldots, B_n\} \ntriangleright S_i$ holds. ii) For each $\Phi \subseteq \{1, \ldots, l\}$, $\bigcup_{i \in \Phi} S_i \ntriangleright \{S_i | i \in \Phi\}$ holds. Here, $\ntriangleright$ is the opposite rule of the preference operator $\rhd$.

$\mathbb{D}_p$ **Stability:** A partition $\mathcal{S}$ of $\mathcal{N}$ is $\mathbb{D}_p$-stable if and only if it is $\triangleright_{\text{Pareto}}$-maximal, where we call a partition $\triangleright_{\text{Pareto}}$-maximal if, for any partition $\hat{\mathcal{S}}$ of $\mathcal{N}$, which is different from $\mathcal{S}$, $\mathcal{S} \triangleright_{\text{Pareto}} \hat{\mathcal{S}}$ holds.

$\mathbb{D}_c$ **Stability:** A partition $\mathcal{S} = \{S_1, \ldots, S_l\}$ of $\mathcal{N}$ is $\mathbb{D}_c$-stable, if and only if the following conditions are satisfied: i) For each $i \in \{1, \ldots, l\}$ and each pair of disjoint coalitions *A* and *B* such that $\{A \cup B\} \subseteq S_i$, $\{A \cup B\} \triangleright \{A, B\}$ holds. ii) A coalition $A \subset \mathcal{N}$ formed of players belonging to different $S_i \in \mathcal{S}$ is $\mathcal{S}$-incompatible, if for no $i \in \{1, \ldots, l\}$ $A \subset T_i$ holds. $\mathbb{D}_c$-stability requires that for all $\mathcal{S}$-incompatible coalitions $\{A\}[\mathcal{S}] \triangleright \{A\}$ where $\{A\}[\mathcal{S}] = \{A \cap S_i, \forall i \in \{1, \ldots l\}\}$ is the projection of coalition *A* in partition $\mathcal{S}$.

Chapter 3
Distributed Learning of Equilibria with Incomplete, Dynamic, and Uncertain Information in Wireless Communication Networks

Yuhua Xu
PLA University of Science and Technology, China

Jinlong Wang
PLA University of Science and Technology, China

Qihui Wu
PLA University of Science and Technology, China

ABSTRACT

New decision-making paradigms addressing the requirements of flexibility, adaptability and intelligence are needed for future wireless networks. Moreover, mutual interactions should be captured when all the devices are autonomous and smart. Game theory is a powerful tool to study such interactions. However, since it is a branch of applied mathematic and mainly studied in economic, some featured challenges should be addressed when applied in wireless networks. This chapter bridges game theory and practical wireless applications, by focusing on the incomplete, dynamic and uncertain information constraints. Four kinds of distributed learning algorithms including stochastic learning automata, payoff-based log-linear learning, learning by trial and error, and no-regret learning are discussed. The learning procedures and basic theoretical results are presented, and their applications in wireless networks are reviewed. Contrastive analysis on environment dynamics, solution concepts, synchrony, convergence, and convergent results is discussed, and some future research directions are given.

DOI: 10.4018/978-1-4666-8642-7.ch003

1. INTRODUCTION AND BACKGROUND

With the explosive increase in wireless transmission demands, traditional static and pre-determined decision approaches can not meet the requirements of flexibility, adaptability and intelligence for the terminals (users). Cognitive radio has established an innovative framework for smart communication systems which adapt to dynamic environment by observing the environment, making intelligent decisions and reconfiguring the hardware (Haykin, 2005; Mitola & Maguire Jr, 1999). With the great success of cognitive radio, autonomous decision-making has been regarded as a promising approach for future wireless communication networks, in which the devices are required to be autonomous and smart. Basically, mutual interaction should be well addressed when all the devices are autonomous and smart. In particular, the interactions may include competition, collision, interference and coordination. Game theory is a powerful tool to study the interactions among multiple autonomous decision-makers and has been extensively applied in wireless communication networks, e.g., power control (Saraydar, Mandayam, & Goodman, 2002), medium access control (Cui, Chen, & Low, 2008), channel assignment (Nie & Comaniciu, 2006), rooting selection (Kannan & Iyengar, 2004) and many other wireless applications. In game-theoretic solutions, equilibria are the desirable and stable outcomes, in which no decision-maker is willing to change its strategy unilaterally.

For game-theoretic solutions, there are three key steps: designing the game model, identifying the existence of equilibria and finding them. Since game theory is a branch of applied mathematic and was mainly studied in the field of economic, some new challenges with regard to information constraints should be addressed when it is applied to wireless communication networks. However, the most existing studies in the literature did not consider the featured information constraints in wireless communication networks, which may not be suitable for practical applications. The purpose of this chapter is to bridge game theory and practical wireless applications, by taking into account the un-neglectable information constraints in wireless communication networks, which are mainly classified into the following three categories:

- **Incomplete**: a decision-maker only has partial information about the environment; in addition, it only knows the individual information but knows nothing about other users.
- **Dynamic**: the system states/ variables are time-varying.
- **Uncertain**: the observations are not equal to the real values, e.g., the observed values are corrupted by noise.

For presentation, we denote the above as IDU information constraints. IDU information constraints are common in wireless communications, some examples are: 1) in a multi-channel access system, a user only has information of a small part of channels rather than that of all channels due to hardware limitation, 2) in a distributed spectrum access system, a user does not know the strategies of other users, 3) in cognitive radio networks, the channel occupation of the primary users are always time-varying, 4) the observations may be corrupted by noise.

Learning is a powerful approach to cope with the above information constraints. Specifically, it is characterized by having the ability to adjust current strategy based on the history information. Therefore, learning and game models can be naturally integrated to provide efficient solutions for practical wireless decision problems.

In this chapter, a comprehensive survey of learning algorithms of equilibria with IDU information constraints is presented. For each kind of learning algorithm, the learning procedure is firstly presented. Then, its properties in terms of convergence and optimality are analyzed, and the existing studies on distributed learning of equilibria in wireless communication networks are surveyed and compared.

There are some survey papers on learning in wireless networks, e.g., learning and reasoning approaches for cognitive radio networks were surveyed in (Gavrilovska, Atanasovski, Macaluso, & DaSilva, 2013), machine learning algorithms for cognitive radio networks were reviewed in (Bkassiny, Li, & Jayaweera, 2013; Clancy, Hecker, Stuntebeck, & O'Shea, 2007), decision-theoretic approaches for distributed channel section were surveyed and discussed in (Xu et al., 2013), and learning equilibria with partial information was reported in (Rose, Lasaulce, Perlaza, & Debbah, 2011). The main differences between this chapter and existing studies are: 1) we focus on incomplete, dynamic and uncertain information constraints, and 2) each kind of learning technology is formally and mathematically investigated.

The rest of this chapter is organized as follows. In Section 2, a general framework of distributed learning of equilibria in wireless communication networks is presented. We review and compare four kinds of distributed technologies for practical wireless communications in Sections 3--6 respectively. Specifically, stochastic learning automata was presented and analyzed in Section 3, the log-linear learning in Section 4, learning by trial and error in Section 5 and no-regret learning in Section 6. Finally, contrastive analysis for the four kinds learning technologies and future research directions are presented in Section 7, and conclusion is provided in Section 8.

2. A GENERAL FRAMEWORK OF DISTRIBUTED LEARNING OF EQUILIBRIA

In this section, we present a general framework of distributed learning of equilibria in wireless communication networks. Firstly, the preliminary of non-cooperative game is described. Secondly, the principles for formulating game models are discussed. Finally, the general procedure of distributed learning with incomplete, dynamic and uncertain information is discussed.

2.1 Preliminary of Non-Cooperative Game

Since the decisions of the users in wireless communications are interactive, the optimization problem can be formulated as a non-cooperative game. Specifically, a non-cooperative game is denoted as $\mathcal{G} = (\mathcal{N}, \mathcal{A}_n, u_n)$, where $\mathcal{N}$ is the player (user) set, $\mathcal{A}_n$ is the action set of player n and u_n is the utility function of player n. Denote $a_n \in \mathcal{A}_n$ as the chosen action of player n and a_{-n} as the action profile of all other players except n. The achieved utility function is jointly determined by the action profile of the all the users, which means that it is usually expressed as $u_n(a_n, a_{-n})$.

In non-cooperative games, the players selfishly maximize their individual utility functions. Therefore, stable solutions are desirable for the multiuser systems. In the following, two stable concepts of game models are presented and analyzed.

Definition 1 (Nash equilibrium): An action profile $a^* = (a_1^*, \ldots, a_N^*)$ is a pure strategy Nash equilibrium if and only if no player can improve its utility function be deviating unilaterally, i.e.,

$$u_n(a_n^*, a_{-n}^*) \geq u_n(a_n, a_{-n}^*), \forall n \in \mathcal{N}, \forall a_n \in \mathcal{A}_n.$$

The concept of NE was firstly coined by John. Nash (Nash, 1950), who was awarded by the 1994 Nobel Prize in economics. It is noted that a distinct feature of NE is that all the players make their decisions selfishly and independently. Essentially, the players do not necessarily make their decisions independently. In contrast, as shown below, correlated-decision may improve the performance significantly. To achieve correlated decisions, correlated equilibrium (CE) is a promising solution.

The concept of CE was firstly established by Robert J. Aumann (Aumann, 1987), who was awarded the 2005 Nobel Prize in economics. Compared with NE, CE admits the following advantages: (i) the set of correlated equilibria is structurally simpler than that of Nash equilibria. The reason is that CE is a convex set, whereas Nash equilibria are isolated points at the extrema of this set, and (ii) since the set of correlated equilibria is convex, fairness between players can be well addressed.

Formally, CE is defined in the following, which can be regarded as an important generation of Nash equilibrium. Suppose that there is a third party sending a recommendation signal to all the players[1]. These recommendation signals allow players to coordinate their actions and to perform joint randomization over their strategies according to a certain probability distribution. Mathematically, it follows that:

Definition 2 (Correlated equilibrium): A probability distribution π over $\mathcal{A}_1 \times \cdots \times \mathcal{A}_N$ is a correlated equilibrium if and only if for all $n \in \mathcal{N}$, for all action $a_n \in \mathcal{A}_n$ and all alternative action $a_n' \in \mathcal{A}_n$,

$$\sum_{a_{-n} \in \mathcal{A}_{-n}} \pi(a_n, a_{-n}) \left(u_n(a_n, a_{-n}) - u_n(a_n', a_{-n}) \right) \geq 0,$$

where $\pi(a_n, a_{-n}) \geq 0$ represents the probability that player n takes action a_n while all other players take action profile a_{-n}. The inequality indicates that when the recommendation to player n is to choose action a_n, then choosing any other action a_n' rather than a_n can not yield a higher expected throughput for player n. The set of correlated equilibria is nonempty, closed and convex for every finite game. In fact, every NE is also a CE and Nash equilibria correspond to the special case where $\pi(a_n, a_{-n})$ is a product of each individual user's probability for different actions. Moreover, CE may include the distribution that is not in the convex hull of the NE distributions.

Since game theory is a branch of applied mathematic, which means that some designing principles should be addressed when they are applied to wireless communication networks. In general, the action of the players in wireless communication systems will be repeated in several times. In particular, denote the received payoff of player n after each play as r_n. Due to the open, dynamic and uncertain nature of wireless environment, r_n is not deterministic. For wireless communications, the following are two basic principles for designing game models:

1). **Guaranteeing the equilibria close to the global optima of the system.** Ideally, the perfect design is that the equilibria are also the global optima. Since the players in game models are self-interested, it may cause dilemma and inefficiency, which is known as the tragedy of common (Kameda & Altman, 2008). Therefore, attentions should be paid to improving the performance of the equilibria.

2). **Determining the optimization metrics according to the traffic demand.** For example, some applications care about the expected value of the optimization metrics, whereas some applications care about not only the expected values but also the variance. Moreover, some applications care about the outage probability in which the metric is greater than a particular threshold, i.e., the probability that the received signal-to-noise-ratio (SNR) is greater than the required threshold for correctly demodulating. In methodology, different optimization metrics lead to different solutions.

2.2 General Procedure of Distributed Learning of Equilibria with Incomplete, Dynamic, and Uncertain Information

In this subsection, we present a general procedure of distributed learning of equilibria with incomplete, dynamic and uncertain information constraints. As stated before, the actions chosen by the users would be repeatedly played. For analysis, we assume that the decision epoch is discrete and is divided into slots with equal length. Denote $a(k) = \{a_1(k),\dots,a_N(k)\}$ as the action profile and $r(k) = \{r_1(k),\dots,r_N(k)\}$ as the received payoffs of the users in slot k respectively. Note that the received payoffs may be random or corrupted by noise. After receiving the payoffs, the users employ a learning rule to update their actions in the next decision epoch. Generally, this procedure can be expressed as follows

$$a_n(k+1) = F[a_n(1),\dots,a_n(k); r_n(1),\dots,r_n(k)],$$

where F is the learning rule. The learning procedure is completely distributed, since each user is only relying on its individual information and does not require information of other players. Note that the above procedure explicitly provides with an efficient solution for the incomplete information constraint in wireless communication networks. For different learning algorithms, their properties are different. In the following, four kinds of distributed learning algorithm for achieving equilibria with incomplete, dynamic and uncertain information constraints will be analyzed and surveyed.

3. STOCHASTIC LEARNING AUTOMATA

In this section, we first present the basic stochastic learning automata (SLA) algorithm (Obaidat, Papadimitriou, & Pomportsis, 2002; Sastry, Phansalkar, & Thathachar, 1994; Thathachar & Sastry, 2002; Verbeeck & Nowe, 2002), analyze its theoretical properties, and finally survey and compare existing studies with regard to the application of SLA in wireless communication networks.

3.1 Algorithm Description and Basic Theoretical Results

The SLA algorithm iterates in mixed strategy forms. Specifically, let the mixed strategy for player n at the k th slot be denoted by a probability distribution $p_n(k) \in \Delta(\mathcal{A}_n)$, where $\Delta(\mathcal{A}_n)$ is the set of probability distributions over the available action set $\mathcal{A}_n$. In the SLA algorithm, the game is played once in a slot according to the mixed strategy profile of the players. After each play, each player receives a payoff $r_n(k)$ and updates its mixed strategy based on the received payoff. The update rule is simple and easy to implement in practice. In particular, if an action is selected and a positive payoff is received, the probability of choosing this action in the next iteration increases. The commonly used update rule in the literature is the linear reward-inaction (L_{R-I}) (Verbeeck & Nowe, 2002), which can be expressed as follows:

$$q_{nm}(k+1) = q_{nm}(k) + b\tilde{r}_n(k)(1 - q_{nm}(k)), m = a_n(k)$$
$$q_{nm}(k+1) = q_{nm}(k) - b\tilde{r}_n(k)q_{nm}(k), m \neq a_n(k)$$

where $\tilde{r}_n(k)$ is the normalized received payoff in slot k, and b is the step size. It is emphasized that other forms of update rules can be used, e.g., linear reward-penalty and linear reward- ε -penalty (Verbeeck & Nowe, 2002). The reason for using L_{R-I} is that it has analytical tractability when being incorporated with game models, which will be analyzed below.

It is noted that the linear reward-inaction is completely distributed since the strategy update of each player is only relying on its individual information. Also, the received payoffs can be random and dynamic (Sastry et al., 1994). In this sense, the SLA algorithm is an efficient solution for the incomplete, dynamic and uncertain information in wireless communication networks. Although the convergence of the SLA algorithm for generic games is hard to guarantee, it asymptotically converges to the Nash equilibria of potential games.

Definition 3 ((Monderer & Shapley, 1996)). A game is an (exact) potential game if there exists a potential function $\phi : A_1 \times \cdots \times A_N \rightarrow R$ such that for all $n \in \mathcal{N}$, all $a_n \in \mathcal{A}_n$, and $a_n' \in \mathcal{A}_n$,

$$u_n(a_n, a_{-n}) - u_n(a_n', a_{-n}) = \phi(a_n, a_{-n}) - \phi(a_n', a_{-n})$$

In other words, the change in the utility function caused by an arbitrary unilateral action change of every player is the same with that in the exact potential function.

Theorem 1: When the learning step size of SLA is sufficiently small, i.e., $b \rightarrow 0$, the linear reward-inaction algorithm asymptotically converges to the Nash equilibria of potential games with random payoffs.

The above theorem can be proved by the method of stochastic approximation. Specifically, it was firstly proved in (Sastry et al., 1994) that it asymptotically converges to the Nash equilibria of coordination games with random payoffs. In coordination games, all players have the same utility function. Very recently, it was proved in (Xu, Wang, Wu, Anpalagan, & Yao, 2012b) that it also asymptotically converges to the Nash equilibria of potential games with random payoffs. With this promising convergence property, the linear reward-inaction algorithm can be widely applied in wireless communication networks.

3.2 The Application of Stochastic Learning Automata in Wireless Networks

In this subsection, the application of the SLA algorithm in distributed wireless games are surveyed and compared. Specifically, the summary and comparison are shown in Table 1.

In a series of recent studies (Wu et al., 2013; Xu et al., 2012b; Y. Xu, Wu, Wang, Shen, & Anpalagan, in press), we investigated SLA-based game-theoretic optimization approaches for cognitive radio networks. Specifically, we formulated the problem of distributed channel selection in cognitive radio

Table 1. Summary and comparison of SLA algorithms in wireless games

Resource Optimization Problem	Game Model	Convergence	Distinct Features	Ref.
Distributed channel selection in cognitive radio networks	Potential game	Converge to a pure strategy NE	The spectrum occupation are random and the corresponding idea probabilities are unknown.	(Xu et al., 2012b)
Distributed channel selection for interference mitigation	Potential game	Converge to a pure strategy NE	The channel gains between two arbitrary users are random.	(Wu et al., 2013)
Channel sensing order selection in cognitive radio networks	Potential game	Converge to a pure strategy NE	The active player set is time-varying	(Xu et al., in press)
Power control for cognitive radio networks with QoS and interference temperature constraints	Potential game	Converge to a pure strategy NE	----	(Xing et al., 2007)
Distributed discrete power control	Generic game	Converge in two-player games	The power levels of the users are discrete	(Xing & Chandramouli, 2008)
Precoding Strategy Selection for MIMO	Potential game	Converge to a pure strategy NE	The payoffs of all the players in each play are the same	(Zhong et al., 2010)
Joint relay selection and discrete power control for cognitive relay networks	Potential game	Converge to a pure strategy NE	----	(Wei et al., 2014)
Finding the optimal voting rule for decision fusion in cooperative spectrum sensing	A two-stage dynamic game	Converge to a pure strategy NE	----	(Yuan et al., 2011)

networks as a potential game and proposed a SLA-based algorithm to converge to efficient pure strategy Nash equilibria in (Xu et al., 2012b). It is noted that the considered spectrum environment therein is random and unknown; more specifically, the spectrum occupation are random and the corresponding idle probabilities are unknown. The optimization objective in (Xu et al., 2012b) is to maximize the network throughput, while the optimization objective in (Wu et al., 2013) is to minimize the expected aggregate interference. The considered environment in (Wu et al., 2013) is practical as the channel gains between arbitrary two users are random. In the third work (Xu et al., in press), we considered the problem of channel sensing order optimization, which was also formulated as a potential game. A distinct feature in (Xu et al., in press) is that the set of active players is time-varying.

The SLA algorithm has also been applied for distributed power control. In (Xing, Mathur, Haleem, Chandramouli, & Subbalakshmi, 2007), the problem of distributed power control for cognitive radio networks with QoS and interference temperature constraints was formulated as a potential game and a SLA-based algorithm was proposed to achieve desirable results. In addition, a discrete power control game was studied in (Xing & Chandramouli, 2008). Since the formulated game in (Xing & Chandramouli, 2008) is generic, the convergence of SLA algorithm needs to be re-investigated. It is proved that the SLA algorithm converges in a two-player game therein.

Besides the above-mentioned channel access and power control problems, the SLA algorithms can be used in other research issues in wireless communications. In (Zhong, Xu, Tao, & Cai, 2010), the problem of precoding strategy selection for multiple-input multiple-output (MIMO) multiple access channels was formulated as a potential game and a SLA-based algorithm was also proposed. It should be noted that the payoffs of all the players after each play are the same. In (Wei, Gang, Shi, & Kai-Kit,

2014), the problem of joint relay selection and discrete power control for cognitive relay networks was studied using a SLA-based potential-game theoretic approach. Also, the SLA algorithm is applied to find the optimal voting rule for decision fusion in cooperative spectrum sensing (Yuan, Leung, Cheng, & Chen, 2011), in which a two-state game model was formulated.

Due to the promising feature of adaptive decision-making feature, there are also a large numbers of studies using the SLA algorithms for wireless communications, which were not analyzed under the framework of game model. Examples are given by, e.g., rate adaptation in IEEE 802.11 networks (Joshi, Ahuja, Singh, & Agrawal, 2008), congestion avoidance for healthcare wireless sensor networks (Misra, Tiwari, & Obaidat, 2009), polling protocols for wireless LANs(Nicopolitidis, Papadimitriou, & Pomportsis, 2003, 2004), routing optimization for ad hoc networks (Akbari Torkestani & Meybodi, 2010a, 2010b; Torkestani & Meybodi, 2011), stochastic traffic engineering problem in multihop cognitive wireless mesh networks (Yang, Chi, & Yuguang, 2010), and wireless data broadcasting with dynamic and a priori unknown client demands (Nicopolitidis, Papadimitriou, & Pomportsis, 2002). It has been shown in these studies that the SLA algorithms converge to stable and desirable solutions.

To summarize, the SLA algorithm is a powerful tool for wireless networks, and can be applied into various wireless optimization problems. It performs well in both dynamic wireless environment. In particular, when incorporating with game models, the convergent solutions can be theoretically analyzed.

4. PAYOFF-BASED LOG-LINEAR LEARNING

It is noted that the SLA algorithm presented in Section 3 asymptotically converges to pure strategy Nash equilibria of potential games. When there are multiple equilibria, it randomly converges to an arbitrary one. In this section, we introduce another distributed learning algorithm that asymptotically converges to the best Nash equilibria of potential games.

4.1. Algorithm Description and Basic Theoretical Results

The update rule of the payoff-based log-linear learning algorithm is based on the well known spatial adaptive play (Young, 2001), in which the player needs to know the information about other players. In the following, we first present the update procedure of the binary log-linear linear algorithm (Marden, Arslan, & Shamma, 2009), which is the basic of payoff-based log-linear learning algorithms.

Suppose that there are N players. At the k th iteration, denote $\{a_1(k),\ldots,a_N(k)\}$ as the chosen action profile of all the players, and $\{u_1(k),\ldots,u_N(k)\}$ as their received payoffs. An arbitrary player, say n, is randomly chosen to perform exploration and learning. In particular, player n randomly chooses an action $\hat{a}_n(k) \in \mathcal{A}_n$ with equal probability $1/|\mathcal{A}_n|$, where $|\mathcal{A}_n|$ is the number of available actions of player n. With the new action profile, the received payoff of player n is denoted as $\hat{u}_n(k)$. In the $(k+1)$th iteration, the chosen player n updates its strategy according to the following rule:

$$\Pr[a_n(k+1) = \hat{a}_n(k)] = \frac{e^{\hat{u}_n(k)\beta}}{e^{\hat{u}_n(k)\beta} + e^{u_n(k)\beta}}$$

$$\Pr[a_n(k+1) = a_n(k)] = \frac{e^{u_n(k)\beta}}{e^{\hat{u}_n(k)\beta} + e^{u_n(k)\beta}},$$

where β is the learning parameter. Meanwhile, all other players keep their actions unchanged, i.e., $a_{-n}(k+1) = a_{-n}(k)$. The above procedure is repeated until some stop criterion is met, e.g., the maximum iteration number is achieved, or the probability of choosing a particular action of every player is sufficiently approaching one.

Denote $\mathcal{A}$ as the set of available channel profiles of all nodes, i.e., i.e., $\mathcal{A} = \mathcal{A}_1 \otimes \cdots \otimes \mathcal{A}_N$, and $\mu(a)$ as the stationary distribution of an action profile a. Although the convergence of the binary log-linear learning algorithm for general game models is hard to guarantee, it admits promising properties in potential games.

Theorem 2 (Young, 2001): If all the players in a potential game adhere to the binary log-linear learning algorithm, the unique stationary distribution $\mu(a) \in \Delta(\mathcal{A})$ of any action selection profile $a \in \mathcal{A}$, $\forall \beta > 0$, is given by:

$$\mu(a) = \frac{e^{\beta\Phi(a)}}{\sum_{s \in \mathcal{A}} e^{\beta\Phi(s)}},$$

where Φ is the potential function. Theorem 2 can be proved by studying the stationary probability distribution of the Markov process induced by the binary log-linear learning algorithm (Young, 2001). Based on Theorem 2, the following theorem can be straightforwardly obtained.

Theorem 3 (Xu, Wu, Shen, Wang, & Anpalagan, 2013): With a sufficiently large β, the binary log-linear learning algorithm asymptotically converges to an action profile that maximizes the potential function.

According to Theorem 3, if the maximizer of the potential function is also the global optimal solution of the original problem, then the binary log-linear learning algorithm asymptotically converges to the global optima. This is interesting since the payoff based learning algorithm is only relying on the individual information and does not need information about other players.

There are some improved variants of the binary log-linear learning algorithm, e.g., the Max-logit algorithm (Song, Wong, & Lee, 2011) and the synchronous log-linear learning algorithm (Marden & Shamma, 2012). Specifically, the update rule of the Max-logit algorithm is similar to the binary log-linear algorithm, which is characterized by, and the difference is that the denominator is replaced by $\max\{e^{\hat{u}_n(k)\beta}, e^{u_n(k)\beta}\}$. It is shown in (Song et al., 2011) that the Max-logit algorithm converges faster than the binary log-linear learning algorithm. In addition, it is noted that only one player updates its action in each iteration in the binary log-linear algorithm. To overcome this limitation, the synchronous log-linear learning algorithm in which all the players can update simultaneously was proposed and analyzed in (Marden & Shamma, 2012).

4.2 The Application of Payoff-Based Logit-Linear Learning Algorithms in Wireless Networks

The payoff-based logit-linear learning algorithms have began to draw attention in wireless communications. In the following, the applications of the payoff-based log-linear learning algorithm in distributed wireless networks are listed and compared. For presentation, B-logit stands for the binary logit-linear learning algorithm, Max-logit for the Max-logit learning algorithm, and S-logit for the synchronous log-linear learning algorithm. Specifically, the summary and comparison are shown in Table 2.

Table 2. Summary and comparison of payoff-based learning algorithms in wireless networks

Resource Optimization Problem	Learning Algorithm	Ref.
Distributed channel selection for MAC-layer interference minimization	B-logit	(Xu, Wu, Shen, et al., 2013)
MAC-layer interference minimization with partially overlapping channels	S-logit	(Xu, Wu, Wang, et al., 2013)
Gateway selection for multi-domain wireless networks	Max-logit	(Song et al., 2011)
Joint power allocation and user scheduling in multicell networks	B-logit	(Zheng et al., 2014)

Currently, the payoff-based learning algorithms were mainly applied into wireless optimization problems which can be formulated as potential games. In (Xu, Wu, Shen, Wang & Anpalagan, 2013), the problem of distributed channel selection for MAC-layer minimization was formulated as a potential game and the B-logit learning algorithm was successfully applied. In (Xu, Wu, Wang, Shen, & Anpalagan, 2013), the MAC-layer interference mitigation problem was further extended from to orthogonal channels to partially overlapping channels (Mishra, Banerjee, & Arbaugh, 2005; Mishra, Rozner, Banerjee, & Arbaugh, 2005), and the S-logit learning algorithm was used to achieve the global optima. In (Song et al., 2011), the problem of gateway selection for multi-domain wireless networks was formulated as a potential game and the Max-logit learning algorithm was proposed to achieve the global optimal solution with faster convergence speed. Following the idea of local altruistic game which was originally proposed in (Xu, Wang, Wu, Anpalagan, & Yao, 2012a), the problem of joint power control and user scheduling in multicell networks was formulated as a potential game and the B-logit algorithm was applied to achieve desirable solutions (Zheng et al., 2014).

5. LEARNING BY TRIAL AND ERROR

It is noted that both the stochastic learning automata and the payoff-based log-linear learning algorithm presented in Section 3 and Section 4 converge to Nash equilibria of potential games, which are interesting and useful. However, they can not converge for generic games, which are more common in wireless communications. In this section, we introduce the learning by trial and error (TE) algorithm (Young, 2009), which converges to efficient NE of generic games.

5.1. Algorithm Description and Basic Theoretical Results

The TE algorithm is characterized by the mood-driven action selection of each player. Specifically, each player has four moods: content (c), discontent (d), hopeful (h) and watchful (w). The players take different strategies in different moods, and their moods are changed based on the received payoffs. Specifically, the state of player n in the kth iteration is a tuple $z_n(k) = (m_n(k), \bar{a}_n(k), \bar{u}_n(k))$, where $m_n(k)$ is the current mood of player n, $\bar{a}_n(k)$ is the benchmark action, and $\bar{u}_n(k)$ is the benchmark utility (payoff), respectively. The TE learning algorithm starts from a random action for each player and iterates by updating the mood, benchmark action and benchmark utility of the players according to the following rules (Young, 2009):

1) When **content** ($m_n(k) = c$), the player chooses an action in the $k+1$ th iteration according to the following rule: with probability $1-\varepsilon$, it takes the current benchmark action, i.e., $a_n(k+1) = \bar{a}_n(k)$, and with probability ε, it randomly tries new actions, i.e., $a_n(k+1) = s, s \neq \bar{a}_n(k)$. Based on the chosen action $a_n(k+1)$ and the actions profiles of other users, the received payoff of the player is given by $u_n(k+1)$.

If a content player takes a new action and receives a higher payoff, i.e., $u_n(k+1) > \bar{u}_n(k)$, it keeps the content mood and updates the benchmark action and benchmark utility to the new action and the new utility, i.e., $z_n(k+1) = (c, s, u_n(k+1))$; otherwise, it keeps the content mood, the benchmark action and utility unchanged in the next iteration, i.e., $z_n(k+1) = (c,, \bar{a}_n(k), \bar{u}_n(k))$. If a content player takes the benchmark action, and the received payoff is higher than the benchmark utility, it keeps the benchmark utility and benchmark action, but changes its mood to hopeful, i.e., $z_n(k+1) = (h,, \bar{a}_n(k), \bar{u}_n(k))$. On the other hand, playing the benchmark action and receiving a lower payoff than the benchmark utility, it still keeps the benchmark utility and benchmark action, but changes its mood to watchful, i.e., $z_n(k+1) = (w,, \bar{a}_n(k), \bar{u}_n(k))$.

2) When **hopeful**, it always takes the benchmark action, i.e., $a_n(k+1) = \bar{a}_n(k)$. If the received payoff is higher than the benchmark utility, i.e., $u_n(k+1) > \bar{u}_n(k)$, it becomes content, maintains the benchmark action, and updates the benchmark utility to the new utility, i.e., ... Otherwise, it becomes watchful and maintains the benchmark action and benchmark utility, i.e., $z_n(k+1) = (w,, \bar{a}_n(k), \bar{u}_n(k))$.
3) When **watchful**, it always takes the benchmark action, i.e., $a_n(k+1) = \bar{a}_n(k)$. If the received payoff is higher than the benchmark utility, i.e., $u_n(k+1) > \bar{u}_n(k)$, it returns to content, maintains the benchmark action, and updates the benchmark utility to the new utility, i.e., $z_n(k+1) = (c,, \bar{a}_n(k), u_n(k+1))$. Otherwise, it becomes discontent and maintains the benchmark action and benchmark utility, i.e., $z_n(k+1) = (w,, \bar{a}_n(k), \bar{u}_n(k))$.
4) When **discontent**, the player takes an action $a_n(k+1)$, which is randomly chosen from its action set A_n. With probability $p = \phi(u_n(k+1), \bar{u}(k))$, it updates its state as $z_n(k+1) = (c,, a_n(k+1), u_n(k+1))$. With probability $1-p$, it updates its state as $z_n(k+1) = (d,, \bar{a}_n(k), \bar{u}_n(k))$. Note that the function $\phi(x; y)$ is monotone increasing in x and monotone decreasing in y.

Definition 3 (Young, 2009): A game G is said to be *interdependent* if any subset $K \subset N$ of players can influence the payoff of at least one player not in K, or equivalently,

$$\exists i \notin K, \exists a_i^* \neq a_i, \text{such that } u_i(a_i^*, a_{-i}) \neq u_i(a_i, a_{-i})$$

Interestingly, interdependence holds for generic games, and the asymptotical convergence behavior of the TE learning algorithm is characterized by the following theorem.

Theorem 4 (Young, 2009): Let G be an interdependent n-player game which has at least one pure Nash equilibrium. If all the players use the TE learning algorithm with experimentation probability ε and response function ϕ, then for sufficiently small ε, the TE learning algorithm converges to pure Nash equilibria with frequency $1-\varepsilon$.

It is promising to see that the TE learning algorithm converges to pure strategy NE with higher probability. Furthermore, there are two slightly modified variants which converge to more efficient results. The algorithm proposed in (Pradelski & Young, 2012) converges to one pure strategy Nash equilibrium that maximizes the welfare of game, while the algorithm proposed in (Marden, Young, & Pao, 2012) converges to an action profile (not necessarily be a pure strategy NE) that maximizes the welfare of the game. Specifically, the update rules in (Pradelski & Young, 2012) was slightly changed while the mood set was reduced to two moods (content and discontent) in (Marden et al., 2012).

5.2 The Application of Learning by Trial and Error in Wireless Networks

Due to the fully distributed implementation and efficient convergent solutions, the learning by trial and error algorithms have been applied in wireless communications. In the following, the application of the learning by trial and error algorithms in distributed wireless networks are briefly surveyed. For presentation, TE-1 stands for the original algorithm coined in (Young, 2009), while TE-2 stands for the the variant proposed in (Marden et al., 2012). Specifically, the summary and comparison are shown in Table 3.

Rose Perlaza et al. studied the problem of distributed power minimization in a self-configuring network with SINR constraints in (Rose, Perlaza, Debbah, & Le Martret, 2012), and investigated the problem of channel selection and power control in clustered ad-hoc networks in (Rose, Perlaza, Le Martret, & Debbah, 2014), using the TE-1 algorithms. In addition, they further proposed an enhanced TE-1 algorithm for joint channel and power allocation in tactical cognitive networks in (Rose, Le Martret, & Debbah, 2013). Very recently, the problem of distributed carrier aggregation was formulated as a repeated game and solved by the TE-1 algorithm (Ahmadi, Macaluso, & Dasilva, 2013).

As stated before, the TE-2 algorithm converges to an action profile that maximizes the welfare of the game. This is promising for wireless optimization problems. In (Singh & Chaporkar, 2013), the decentralized user association problem for multiple technology networks was solved by the TE-2 algorithm. A

Table 3. Summary and comparison of learning by trial and error algorithms in wireless networks

Resource Optimization Problem	Learning Algorithm	Distinct Feature	Ref.
Distributed power minimization in a self-configuring network with SINR constraints	TE-1	----	(Rose et al., 2012)
Joint channel selection and power control in clustered ad hoc networks	TE-1	----	(Rose et al., 2014)
Joint channel and power allocation in tactical cognitive networks	Enhanced TE-1	----	(Rose, Le Martret, et al., 2013)
Distributed carrier aggregation	TE-1	----	(Ahmadi et al., 2013)
Decentralized user association for multiple technology networks	TE-2	The constraints on user requirements are considered	(Singh & Chaporkar, 2013)
Power control for clustered ad hoc networks	TE-2	----	(Rose, Perlaza, et al., 2013)
Utility-based resource allocation scheme for multi-channel decentralized networks	TE-2	----	(Sheng et al., 2014)

distinct feature in (Singh & Chaporkar, 2013) is that the constraints on user requirements are considered. In (Rose, Perlaza, Le Martret, & Debbah, 2013), the power control problem, which is based on the TE-2 algorithm, is analyzed for clustered ad hoc networks. Also, based on TE-2 algorithm, a utility-based resource allocation scheme for multi-channel decentralized networks was studied in (Sheng et al., 2014).

6. NO-REGRET LEARNING

The above three kinds of learning algorithms converge to Nash equilibria. In this section, we present the no-regent learning algorithm which converges to correlated equilibria (CE) of generic games. Note that the pioneer work on distributed learning algorithms converging to CE was investigated by Hart Mas-Colell. They first coined the regret-matching algorithm (Hart & Mas-Colell, 2000), and proposed a significant variant, which is fully distributed and named as modified regret-matching algorithm (Hart & Mas-Colell, 2001).

6.1. Algorithm Description and Basic Theoretical Results

No-regret learning is an uncoupled algorithm which is based on the minimization of internal regret (Cesa-Bianchi & Lugosi, 2006). The rule is simple and can be described as follows (Hart & Mas-Colell, 2000): at each iteration, each player establishes a vector of average internal regrets for not having played other actions with respect to its current action; at the next iteration, it switches to an action with a probability that is proportional to the average internal regret of that action. Specifically, for two distinct actions j and $s \in A$, the regent of player n from s to j is given by:

$$R_n^k(s,j) = [D_n^k(s,j)]^+ = \max\{D_n^k(s,j), 0\},$$

where

$$D_n^k(s,j) = \frac{1}{k} \sum_{i \le k : a_n(i) = s} \left(u_n(j, a_{-n}(i)) - u_n(a(i)) \right),$$

which is the payoff loss had it chosen action j every time it played action s. However, it is noted that the first item in the bracket can not be computed by player n, since player n knows neither the action profiles of other players, i.e., $a_{-n}(i)$, nor its received payoff, i.e., $u_n(j, a_{-n}(i))$. To cope with this problem, an estimation for the internet-regret was proposed in (Hart & Mas-Colell, 2001). Specifically, the estimation is given by:

$$C_n^k(s,j) = \frac{1}{k} \sum_{i \le k : a_n(i) = s} \left[\frac{p_n^j(i)}{p_n^s(i)} u_n(a_n(i)) - u_n(a(i)) \right]$$

where $p_n^s(i)$ is the probability at the i th iteration and will be illustrated below. Thus, the modified regret can be expressed as:

$$Q_n^k(s,j) = [C_n^k(s,j)]^+ = \max\{C_n^k(s,j),0\}.$$

Suppose that the action of player n in the k th iteration is s, then it updates the action in the $(k+1)$ th iteration according to the following probabilities:

$$p_n^j(k+1) = (1-\delta)\min\left\{\frac{1}{\mu}Q_n^k(s,j), \frac{1}{m_n-1}\right\} + \delta\frac{1}{m_n}, j \neq s$$
$$p_n^s(k+1) = 1 - \sum_{j\in A_n, j\neq s}(1 - p_n^j(k+1)),$$

where $m_n = |A_n|$ is the number of available action of player n, $0 < \delta < 1$ and $\mu > 2M_n(m_n - 1)$. Note that M_n is an upper bound of $|u_n(a)|$ for all possible action profile a.

Theorem 5 (Hart & Mas-Colell, 2001): If every player adheres to the no-regret learning algorithm, then the empirical distributions of play converge almost surely as $t \to \infty$ to the set of correlated equilibria of the game.

Note that the above presented algorithm is for games with finite action set. For games with infinite action set (e.g., power control), an efficient no-regret learning can be found in (Stoltz & Lugosi, 2007).

6.2 The Application of No-Regret Learning Algorithms in Wireless Networks

Due to the attractive features of simple implementation and efficient convergent solutions, the no-regret learning algorithm has been extensively applied in wireless networks. The applied algorithms include the regret-matching algorithm (Hart & Mas-Colell, 2000), the fully distributed modified regret-matching algorithm (Hart & Mas-Colell, 2001) and the continuous regret-matching algorithm (Stoltz & Lugosi, 2007). Specifically, the applications are surveyed and compared in Table 4.

There are large numbers of studies on channel access/assignment using no-regret learning algorithms. In (Han, Pandana, & Liu, 2007), the problem of distributed channel selection for cognitive radio networks using CE and the regret-matching algorithm was studied. In (Lu, He, Wang, & Li, 2009), an energy-efficient distributed channel access algorithm was proposed using the modified regret-matching learning. In (Tan, Sim, & Chuah, 2008), the problem of joint channel assignment and power control for wireless ad hoc networks was solved by the continuous regret-matching algorithm. In (Yu, Chen, Sun, Fan, & Shen, 2010), the problem of distributed channel assignment for wireless sensor networks was studied using the modified regret-matching algorithm, and this problem is further studied in a subsequent work (Chen et al., 2015). In (Maskery, Krishnamurthy, & Zhao, 2009), the problem of distributed channel access for cognitive radio networks was studied and an adaptive regret tracking algorithm was proposed. Note that the proposed adaptive regret tracking is based on the regret-matching algorithm.

The application of no-regret learning algorithms in wireless networks is across-the-board. In a series work (Krishnamurthy, Maskery, & Yin, 2008; Maskery & Krishnamurthy, 2007a, 2007b), the regret-matching algorithms were used to study the problem of distributed sensor activation in sensor networks. In (Wu, Zhou, & Cai, 2012), the regret-matching algorithm was used to study energy-efficient resource allocation for orthogonal frequency division multiple access (OFDMA) systems. In (Wu, Cai, Zhou, Zheng, & Zheng, 2013), the regret-matching algorithm was also used to study joint relay selection and power control for cooperative ad hoc networks. In addition, based on the modified regret-matching

Table 4. Summary and comparison of no-regret learning algorithms in wireless networks

Resource Optimization Problem	Learning Algorithm	Ref.
Distributed channel selection for cognitive radio networks	Regret-matching	(Han et al., 2007)
Energy-efficient distributed channel access	Modified regret-matching	(Lu et al., 2009)
Joint channel assignment and power control for wireless ad hoc networks	Continuous regret-matching	(Tan et al., 2008)
Distributed channel assignment for wireless sensor networks	Modified regret-matching	(Chen et al., 2015; Yu et al., 2010)
Distributed channel access for cognitive radio networks	Adaptive regret tracking	(Maskery et al., 2009)
Distributed sensor activation in sensor networks	Regret-matching	(Krishnamurthy et al., 2008; Maskery & Krishnamurthy, 2007a, 2007b)
Energy-efficient resource allocation for OFDMA systems	Regret-matching	(Wu,Cai, et al., 2012)
Joint relay selection and transmit power control for for cooperative ad hoc networks	Regret-matching	(Wu, Zhou, et al., 2013)
Discrete power allocation for fast fading MIMO access channel	Modified regret-matching	(Belmega et al., 2010)
Downlink spectrum strategy for LTE/3GPP femtocells	Modified regret-matching	(Huang & Krishnamurthy, 2011)

algorithm, a discrete power allocation problem for fast fading MIMO access channel was studied in (Belmega, Lasaulce, Debbah, & Hjorungnes, 2010), and an efficient downlink spectrum strategy for LTE/3GPP femtocells was investigated in (Huang & Krishnamurthy, 2011).

7. CONTRASTIVE ANALYSIS AND FUTURE RESEARCH

In this section, we discuss the above presented learning algorithms for wireless communications, from a global and contrastive perspective. Moreover, we outline some future research directions.

7.1 Contrastive Analysis

In this subsection, some important concerns of the above presented learning algorithms are discussed. Specifically, the concerns including environment dynamics, solution concepts, synchrony, convergence, and convergent results are briefly discussed in Table 5.

1) Environment requirements. The stochastic learning automata (SLA) algorithm can be applied in both static and dynamic environment; on the contrary, all other three learning algorithms are only suitable for static environment. For wireless communications, the environment is always dynamic, e.g., the time-varying spectrum occupation and channel rate fluctuation due to fading.
2) Solution concepts. The no-regret learning algorithm converges to CE while all other learning algorithms converge to NE. As discussed before, CE is structurally simpler than NE and can achieve higher fairness.

Table 5. Comparison results for the four learning algorithms

Learning Algorithms	Environment Dynamics	Solution Concepts	Synchrony	Convergence	Convergent Results
Stochastic learning automata	Static or dynamic	NE	Synchronous	Asymptotically converges in potential games.	Converges to an arbitrary NE.
Payoff-based log-linear learning	Static	NE	B-logit and Max-logit are asynchronous, while S-logit is synchronous	Asymptotically converges in potential games	Converges to an pure strategy NE that maximizes the potential function.
Learning by trial and error	Static	NE	Synchronous	Asymptotically converges in generic games.	The algorithm originally proposed in (Young, 2009) converges to a pure strategy NE, the algorithm in (Pradelski & Young, 2012) converges to one pure strategy Nash equilibrium that maximizes the welfare of game, while that in (Marden et al., 2012) converges to an action profile that maximizes the welfare of the game.
No-regret learning	Static	CE	Synchronous	Asymptotically converges in generic games.	Converges to an arbitrary CE.

3) Synchrony. In the context of learning in games, synchrony means that all the players update their actions simultaneously, while asynchrony means that the players update the actions in sequence, i.e., only one players updates its action at a time. For payoff-based log-linear learning algorithms, B-logit and Max-logit are asynchronous while S-logit is synchronous. All other three learning algorithms are synchronous. For wireless communications, it should be noted that an asynchronous learning algorithm requires a coordination mechanism to schedule the learning.
4) Convergence. Both stochastic learning automata and payoff-based log-linear learning asymptotically converge in potential games. In contrast, both learning by trial and error algorithm and no-regret learning algorithm asymptotically converge in generic games.
5) Convergent results. The improved learning by trial and error algorithms converges to desirable results that maximize the welfare, while other three learning algorithms converges to NE or CE. Specifically, the stochastic learning automata algorithm converges to an arbitrary NE, and the no-regret algorithm converges to an arbitrary CE. The payoff-based log-linear algorithm converges to an pure strategy NE that maximizes the potential function. Thus, for the latter three algorithms, the game modes need to be carefully designed to make the stable solutions (NE or CE) close to the global optima.

From the above analysis and comparison, it is noted that different learning algorithms admit different advantages and shortcomings. Thus, the choice of learning algorithms for wireless communications should be scenario-dependent and application-dependent.

7.2 Future Research

It is seen that learning for wireless communications has definitely drawn a beautiful and exciting future. In particular, with the rapid development of wireless technologies, e.g., 5G and Internet of Things (IoT), some new requirements and challenges with respect to learning algorithms should be investigated. We list some future research problems below:

1) Design game models and learning algorithms for large-scale wireless networks. In most existing studies, the player set is small or moderate. However, the players are expected in large number in further wireless networks, e.g., the small cells are densely deployed in 5G (Xu et al., 2014) and the the number of terminals in IoT may be huge (Wu et al., 2014). Thus, it is important to design game models and learning algorithms which are suitable for large scale wireless networks. In particular, the learning algorithm should scale up with the increase in the number of players. Some efficient approaches are graphical games and hierarchical games (Xu et al., 2015).
2) Design heterogeneous learning algorithms. In existing studies, it is noted that all the players employed the same learning algorithm, in terms of the update rules and learning parameters. However, this assumption is for theoretic analysis but not true in practice. In future wireless networks, the players may belong to different holders or operators, and hence may prefer to different learning algorithms; in addition, even belonging to the same holder, they may have different processing ability and individual preference, and may choose heterogeneous update rules and learning parameters. Technically, introducing heterogeneity into the learning algorithms will affect the convergence properties and asymptotic behavior significantly (Xu et al., 2015), which needs to be carefully studied.

8. CONCLUSION

Game theory is a powerful tool to study the interactions among multiple autonomous decision-makers and has been extensively used in wireless communication networks, and is expected to proceed to play an important rule for future wireless networks. However, since game theory is a branch of applied mathematic and is mainly studied in the field of economic, some new challenges with regard to information constraints should be addressed when it is applied to wireless communication networks. The purpose of this chapter is to bridge game theory and practical wireless applications, by taking into account the incomplete, dynamic and uncertain information constraints in wireless communication networks. Four important uncoupled learning algorithms including stochastic learning automata, payoff-based log-linear learning, learning by trial and error, and no-regret learning were presented and discussed. Specifically, the learning procedures and basic theoretical results were presented, and their applications in wireless networks were surveyed and compared. In addition, contrastive analysis for the four learning algorithms in terms of environment dynamics, solution concepts, synchrony, convergence, and convergent results were briefly discussed, and some future research directions were given.

ACKNOWLEDGMENT

This work was supported by the National Science Foundation of China under Grant No. 61401508 and 61172062.

REFERENCES

Ahmadi, H., Macaluso, I., & Dasilva, L. A. (2013). Carrier aggregation as a repeated game: Learning algorithms for efficient convergence to a Nash equilibrium. In *Proceedings of the 2013 IEEE Global Communications Conference (GLOBECOM)*, Atlanta, GA: IEEE. doi:10.1109/GLOCOM.2013.6831243

Akbari Torkestani, J., & Meybodi, M. R. (2010a). An intelligent backbone formation algorithm for wireless ad hoc networks based on distributed learning automata. *Computer Networks*, *54*(5), 826–843. doi:10.1016/j.comnet.2009.10.007

Akbari Torkestani, J., & Meybodi, M. R. (2010b). Mobility-based multicast routing algorithm for wireless mobile Ad-hoc networks: A learning automata approach. *Computer Communications*, *33*(6), 721–735. doi:10.1016/j.comcom.2009.11.019

Aumann, R. J. (1987). Correlated equilibrium as an expression of Bayesian rationality. *Econometrica*, *55*(1), 1–18. doi:10.2307/1911154

Belmega, E. V., Lasaulce, S., Debbah, M., & Hjorungnes, A. (2010). Learning distributed power allocation policies in MIMO channels. In *Proceedings of the European Signal Processing Conference (EUSIPCO-2010)*. Aalborg, Denmark: EURASIP.

Bkassiny, M., Li, Y., & Jayaweera, S. K. (2013). A survey on machine-learning techniques in cognitive radios. *IEEE Communications Surveys and Tutorials*, *15*(3), 1136–1159. doi:10.1109/SURV.2012.100412.00017

Cesa-Bianchi, N., & Lugosi, G. (2006). *Prediction, learning, and games*. Cambridge, UK: Cambridge University Press. doi:10.1017/CBO9780511546921

Chen, J., Yu, Q., Chai, B., Sun, Y., Fan, Y., & Shen, X. S. (2015). Dynamic channel assignment for wireless sensor networks: A regret matching based approach. *IEEE Transactions on Parallel and Distributed Systems*, *26*(1), 95–106. doi:10.1109/TPDS.2014.2307868

Clancy, C., Hecker, J., Stuntebeck, E., & O'Shea, T. (2007). Applications of machine learning to cognitive radio networks. *IEEE Wireless Communications*, *14*(4), 47–52. doi:10.1109/MWC.2007.4300983

Cui, T., Chen, L., & Low, S. H. (2008). A game-theoretic framework for medium access control. *IEEE Journal on Selected Areas in Communications*, *26*(7), 1116–1127. doi:10.1109/JSAC.2008.080909

Gavrilovska, L., Atanasovski, V., Macaluso, I., & DaSilva, L. (2013). Learning and reasoning in cognitive radio networks. *IEEE Communications Surveys and Tutorials*, *15*(4), 1761–1777. doi:10.1109/SURV.2013.030713.00113

Han, Z., Pandana, C., & Liu, K. J. R. (2007). Distributive opportunistic spectrum access for cognitive radio using correlated equilibrium and no-regret learning. In *Proceedings of the 2007 IEEE Wireless Communications and Networking Conference (WCNC)*. Hong Kong, China: IEEE. doi:10.1109/WCNC.2007.8

Hart, S., & Mas-Colell, A. (2000). A simple adaptive procedure leading to correlated equilibrium. *Econometrica*, *68*(5), 1127–1150. doi:10.1111/1468-0262.00153

Hart, S., & Mas-Colell, A. (2001). *A reinforcement procedure leading to correlated equilibrium*. New York: Springer. doi:10.1007/978-3-662-04623-4_12

Haykin, S. (2005). Cognitive radio: Brain-empowered wireless communications. *IEEE Journal on Selected Areas in Communications*, *23*(2), 201–220. doi:10.1109/JSAC.2004.839380

Huang, J. W., & Krishnamurthy, V. (2011). Cognitive base stations in LTE/3GPP femtocells: A correlated equilibrium game-theoretic approach. *IEEE Transactions on Communications*, *59*(12), 3485–3493. doi:10.1109/TCOMM.2011.093011.100693

Joshi, T., Ahuja, D., Singh, D., & Agrawal, D. P. (2008). SARA: Stochastic automata rate adaptation for IEEE 802.11 networks. *IEEE Transactions on Parallel and Distributed Systems*, *19*(11), 1579–1590. doi:10.1109/TPDS.2007.70814

Kameda, H., & Altman, E. (2008). Inefficient noncooperation in networking games of common-pool resources. *IEEE Journal on Selected Areas in Communications*, *26*(7), 1260–1268. doi:10.1109/JSAC.2008.080922

Kannan, R., & Iyengar, S. S. (2004). Game-theoretic models for reliable path-length and energy-constrained routing with data aggregation in wireless sensor networks. *IEEE Journal on Selected Areas in Communications*, *22*(6), 1141–1150. doi:10.1109/JSAC.2004.830937

Krishnamurthy, V., Maskery, M., & Yin, G. (2008). Decentralized adaptive filtering algorithms for sensor activation in an unattended ground sensor network. *IEEE Transactions on Signal Processing*, *56*(12), 6086–6101. doi:10.1109/TSP.2008.929664

Lu, Y., He, H., Wang, J., & Li, S. (2009). Energy-efficient dynamic spectrum access using no-regret learning. In *Proceedings of the 7th International Conference on Information, Communications and Signal Processing* (ICICS 2009). Beijing, China: Springer.

Marden, J. R., Arslan, G., & Shamma, J. S. (2009). Cooperative control and potential games. *IEEE Transactions on Systems, Man, and Cybernetics. Part B, Cybernetics*, *39*(6), 1393–1407. doi:10.1109/TSMCB.2009.2017273 PMID:19369160

Marden, J. R., & Shamma, J. S. (2012). Revisiting log-linear learning: Asynchrony, completeness and payoff-based implementation. *Games and Economic Behavior*, *75*(2), 788–808. doi:10.1016/j.geb.2012.03.006

Marden, J. R., Young, H. P., & Pao, L. Y. (2012). Achieving pareto optimality through distributed learning. In *Proceedings of the 51th IEEE Conference on Decision and Control (CDC2012)*. Hawaii, HI: IEEE.

Maskery, M., & Krishnamurthy, V. (2007a). Decentralized activation in a ZigBee-enabled unattended ground sensor network: a correlated equilibrium game theoretic analysis. In *Proceedings of the 2007 IEEE International Conference on Communications (ICC),* Glasgow, UK: IEEE. doi:10.1109/ICC.2007.645

Maskery, M., & Krishnamurthy, V. (2007b). Decentralized adaptation in sensor networks: Analysis and application of regret-based algorithms. In *Proceedings of the 46th IEEE Conference on Decision and Control (CDC).* New Orleans, LA: IEEE. doi:10.1109/CDC.2007.4434388

Maskery, M., Krishnamurthy, V., & Zhao, Q. (2009). Decentralized dynamic spectrum access for cognitive radios: Cooperative design of a non-cooperative game. *IEEE Transactions on Communications*, *57*(2), 459–469. doi:10.1109/TCOMM.2009.02.070158

Mishra, A., Banerjee, S., & Arbaugh, W. (2005). Weighted coloring based channel assignment for WLANs. *Mobile Computing and Communications Review*, *9*(3), 19–31. doi:10.1145/1094549.1094554

Mishra, A., Rozner, E., Banerjee, S., & Arbaugh, W. (2005). Exploiting partially overlapping channels in wireless networks: Turning a peril into an advantage. In *Proceedings of the 5th ACM SIGCOMM conference on Internet Measurement.* Philadelphia, PA: ACM. doi:10.1145/1330107.1330144

Misra, S., Tiwari, V., & Obaidat, M. S. (2009). Lacas: Learning automata-based congestion avoidance scheme for healthcare wireless sensor networks. *IEEE Journal on Selected Areas in Communications*, *27*(4), 466–479. doi:10.1109/JSAC.2009.090510

Mitola, J., & Maguire, G. Q. Jr. (1999). Cognitive radio: Making software radios more personal. *IEEE Personal Communications*, *6*(4), 13–18. doi:10.1109/98.788210

Monderer, D., & Shapley, L. S. (1996). Potential games. *Games and Economic Behavior*, *14*(1), 124–143. doi:10.1006/game.1996.0044

Nash, J. F. (1950). Equilibrium points in n-person games. (1950). *Proceedings of the National Academy of Sciences of the United States of America*, *36*(1), 48–49. doi:10.1073/pnas.36.1.48 PMID:16588946

Nicopolitidis, P., Papadimitriou, G. I., & Pomportsis, A. S. (2002). Using learning automata for adaptive push-based data broadcasting in asymmetric wireless environments. *IEEE Transactions on Vehicular Technology*, *51*(6), 1652–1660. doi:10.1109/TVT.2002.802978

Nicopolitidis, P., Papadimitriou, G. I., & Pomportsis, A. S. (2003). Learning automata-based polling protocols for wireless LANs. *IEEE Transactions on Communications*, *51*(3), 453–463. doi:10.1109/TCOMM.2003.809788

Nicopolitidis, P., Papadimitriou, G. I., & Pomportsis, A. S. (2004). Distributed protocols for ad hoc wireless LANs: A learning-automata-based approach. *Ad Hoc Networks*, *2*(4), 419–431. doi:10.1016/j.adhoc.2003.09.004

Nie, N., & Comaniciu, C. (2006). Adaptive channel allocation spectrum etiquette for cognitive radio networks. *Mobile Networks and Applications*, *11*(6), 779–797. doi:10.1007/s11036-006-0049-y

Obaidat, M. S., Papadimitriou, G. I., & Pomportsis, A. S. (2002). Guest editorial learning automata: Theory, paradigms, and applications. *IEEE Transactions on Systems, Man, and Cybernetics. Part B, Cybernetics*, *32*(6), 706–709. doi:10.1109/TSMCB.2002.1049604 PMID:18244877

Pradelski, B. S., & Young, H. P. (2012). Learning efficient Nash equilibria in distributed systems. *Games and Economic Behavior*, *75*(2), 882–897. doi:10.1016/j.geb.2012.02.017

Rose, L., Lasaulce, S., Perlaza, S. M., & Debbah, M. (2011). Learning equilibria with partial information in decentralized wireless networks. *IEEE Communications Magazine*, *49*(8), 136–142. doi:10.1109/MCOM.2011.5978427

Rose, L., Le Martret, C. J., & Debbah, M. (2013). Joint channel and power allocation in tactical cognitive networks: Enhanced trial and error. In *Proceedings of the 2013 Military Communications and Information Systems Conference (MilCIS)*. Canberra, Australia: IEEE.

Rose, L., Perlaza, S. M., Debbah, M., & Le Martret, C. J. (2012). Distributed power allocation with SINR constraints using trial and error learning. In *Proceedings of the 2012 Wireless Communications and Networking Conference (WCNC)*. Paris, France: IEEE doi:10.1109/WCNC.2012.6214083

Rose, L., Perlaza, S. M., Le Martret, C. J., & Debbah, M. (2013). Achieving Pareto optimal equilibria in energy efficient clustered ad hoc networks. In *Proceedings of the 2013 IEEE International Conference on Communications (ICC2013)*. Budapest, Hungary: IEEE. doi:10.1109/ICC.2013.6654723

Rose, L., Perlaza, S. M., Le Martret, C. J., & Debbah, M. (2014). Self-organization in decentralized networks: A trial and error learning approach. *IEEE Transactions on Wireless Communications*, *13*(1), 268–279. doi:10.1109/TWC.2013.112613.130405

Saraydar, C. U., Mandayam, N. B., & Goodman, D. (2002). Efficient power control via pricing in wireless data networks. *IEEE Transactions on Communications*, *50*(2), 291–303. doi:10.1109/26.983324

Sastry, P. S. S., Phansalkar, V. V., & Thathachar, M. (1994). Decentralized learning of Nash equilibria in multi-person stochastic games with incomplete information. *IEEE Transactions on Systems, Man, and Cybernetics*, *24*(5), 769–777. doi:10.1109/21.293490

Sheng, M., Xu, C., Wang, X., Zhang, Y., Han, W., & Li, J. (2014). Utility-Based Resource Allocation for Multi-Channel Decentralized Networks. *IEEE Transactions on Communications*, *62*(10), 3610–3620. doi:10.1109/TCOMM.2014.2357028

Singh, M., & Chaporkar, P. (2013). An efficient and decentralised user association scheme for multiple technology networks. In *Proceedings of the 2013 11th International Symposium on Modeling & Optimization in Mobile, Ad Hoc & Wireless Networks (WiOpt)*. Tsukuba Science City, Japan: IEEE.

Song, Y., Wong, S. H., & Lee, K.-W. (2011). Optimal gateway selection in multi-domain wireless networks: a potential game perspective. In *Proceedings of the* 17th annual international conference on Mobile computing and networking (MOBICOM). Beijing, China: ACM.

Stoltz, G., & Lugosi, G. (2007). Learning correlated equilibria in games with compact sets of strategies. *Games and Economic Behavior*, *59*(1), 187–208. doi:10.1016/j.geb.2006.04.007

Tan, C. K., Sim, M. L., & Chuah, T. C. (2008). Game theoretic approach for channel assignment and power control with no-internal-regret learning in wireless ad hoc networks. *IET Communications*, *2*(9), 1159–1169. doi:10.1049/iet-com:20070547

Thathachar, M., & Sastry, P. S. (2002). Varieties of learning automata: An overview. *IEEE Transactions on Systems, Man, and Cybernetics. Part B, Cybernetics*, *32*(6), 711–722. doi:10.1109/TSMCB.2002.1049606 PMID:18244878

Torkestani, J. A., & Meybodi, M. R. (2011). Weighted Steiner connected dominating set and its application to multicast routing in wireless MANETs. *Wireless Personal Communications*, *60*(2), 145–169. doi:10.1007/s11277-010-9936-4

Verbeeck, K., & Nowe, A. (2002). Colonies of learning automata. *IEEE Transactions on Systems, Man, and Cybernetics. Part B, Cybernetics*, *32*(6), 772–780. doi:10.1109/TSMCB.2002.1049611 PMID:18244883

Wei, Z., Gang, C., Shi, J., & Kai-Kit, W. (2014). Relay selection and discrete power control for cognitive relay networks via potential game. *IEEE Transactions on Signal Processing*, *62*(20), 5411–5424. doi:10.1109/TSP.2014.2347261

Wu, D., Cai, Y., Zhou, L., Zheng, Z., & Zheng, B. (2013). Cooperative strategies for energy-aware ad hoc networks: A correlated-equilibrium game-theoretical approach. *IEEE Transactions on Vehicular Technology*, *62*(5), 2303–2314. doi:10.1109/TVT.2012.2235189

Wu, D., Zhou, L., & Cai, Y. (2012). Energy-efficient resource allocation for uplink orthogonal frequency division multiple access systems using correlated equilibrium. *IET Communications*, *6*(6), 659–667. doi:10.1049/iet-com.2011.0625

Wu, Q., Ding, G., Xu, Y., Feng, S., Du, Z., Wang, J., & Long, K. (2014). Cognitive Internet of Things: A New Paradigm Beyond Connection. *IEEE Internet of Things Journal*, *1*(2), 129–143. doi:10.1109/JIOT.2014.2311513

Wu, Q., Xu, Y., Wang, J., Shen, L., Zheng, J., & Anpalagan, A. (2013). Distributed channel selection in time-varying radio environment: Interference mitigation game with uncoupled stochastic learning. *IEEE Transactions on Vehicular Technology*, *62*(9), 4524–4538. doi:10.1109/TVT.2013.2269152

Xing, Y., & Chandramouli, R. (2008). Stochastic learning solution for distributed discrete power control game in wireless data networks. *IEEE/ACM Transactions on Networking*, *16*(4), 932–944. doi:10.1109/TNET.2007.911424

Xing, Y., Mathur, C. N., Haleem, M. A., Chandramouli, R., & Subbalakshmi, K. (2007). Dynamic spectrum access with QoS and interference temperature constraints. *IEEE Transactions on Mobile Computing*, *6*(4), 423–433. doi:10.1109/TMC.2007.50

Xu, J., Wang, J., Zhu, Y., Yang, Y., & Zheng, X. (2014). Cooperative distributed optimization for the hyper-dense small cell deployment. *IEEE Communications Magazine*, *52*(5), 61–67. doi:10.1109/MCOM.2014.6815894

Xu, Y., Anpalagan, A., Wu, Q., Shen, L., Gao, Z., & Wang, J. (2013). Decision-theoretic distributed channel selection for opportunistic spectrum access: Strategies, challenges and solutions. *IEEE Communications Surveys and Tutorials*, *15*(4), 1689–1713. doi:10.1109/SURV.2013.030713.00189

Xu, Y., Wang, J., Wu, Q., Anpalagan, A., & Yao, Y.-D. (2012a). Opportunistic spectrum access in cognitive radio networks: Global optimization using local interaction games. *IEEE Journal of Selected Topics in Signal Processing*, *6*(2), 180–194. doi:10.1109/JSTSP.2011.2176916

Xu, Y., Wang, J., Wu, Q., Anpalagan, A., & Yao, Y.-D. (2012b). Opportunistic spectrum access in unknown dynamic environment: A game-theoretic stochastic learning solution. *IEEE Transactions on Wireless Communications*, *11*(4), 1380–1391. doi:10.1109/TWC.2012.020812.110025

Xu, Y., Wang, J., Wu, Q., Du, Z., Shen, L., & Anpalagan, A. (2015). A game-theoretic perspective on self-organizing optimization for cognitive small cells. *IEEE Communications Magazine*, *53*(7), 100–108. doi:10.1109/MCOM.2015.7158272

Xu, Y., Wu, Q., Shen, L., Wang, J., & Anpalagan, A. (2013). Opportunistic spectrum access with spatial reuse: Graphical game and uncoupled learning solutions. *IEEE Transactions on Wireless Communications*, *12*(10), 4814–4826. doi:10.1109/TWC.2013.092013.120862

Xu, Y., Wu, Q., Wang, J., Shen, L., & Anpalagan, A. (2013). Opportunistic spectrum access using partially overlapping channels: Graphical game and uncoupled learning. *IEEE Transactions on Communications*, *61*(9), 3906–3918. doi:10.1109/TCOMM.2013.072913.120881

Xu, Y., Wu, Q., Wang, J., Shen, L., & Anpalagan, A. (in press). Robust multiuser sequential channel sensing and access in dynamic cognitive radio networks: Potential games and stochastic learning. *IEEE Transactions on Vehicular Technology*.

Yang, S., Chi, Z., & Yuguang, F. (2010). Stochastic Traffic Engineering in Multihop Cognitive Wireless Mesh Networks. *IEEE Transactions on Mobile Computing*, *9*(3), 305–316. doi:10.1109/TMC.2009.111

Young, H. P. (2001). *Individual strategy and social structure: An evolutionary theory of institutions*. Princeton, NJ: Princeton University Press.

Young, H. P. (2009). Learning by trial and error. *Games and Economic Behavior*, *65*(2), 626–643. doi:10.1016/j.geb.2008.02.011

Yu, Q., Chen, J., Sun, Y., Fan, Y., & Shen, X. (2010). Regret matching based channel assignment for wireless sensor networks. In *Proceedings of the 2010 IEEE International Conference on Communications (ICC)*. Cape Town, South Africa: IEEE. doi:10.1109/ICC.2010.5502547

Yuan, W., Leung, H., Cheng, W., & Chen, S. (2011). Optimizing voting rule for cooperative spectrum sensing through learning automata. *IEEE Transactions on Vehicular Technology*, *60*(7), 3253–3264. doi:10.1109/TVT.2011.2161001

Zheng, J., Cai, Y., Liu, Y., Xu, Y., Duan, B., & Shen, X. (2014). Optimal power allocation and user scheduling in multicell networks: Base station cooperation using a game-theoretic approach. *IEEE Transactions on Wireless Communications*, *13*(12), 6928–6942. doi:10.1109/TWC.2014.2334673

Zhong, W., Xu, Y., Tao, M., & Cai, Y. (2010). Game theoretic multimode precoding strategy selection for mimo multiple access channels. *IEEE Signal Processing Letters*, *17*(6), 563–566. doi:10.1109/LSP.2010.2047315

KEY TERMS AND DEFINITIONS

Cognitive Radio: A radio can adapt to dynamic environment by observing the environment, making intelligent decisions and reconfiguring the hardware.

Dynamic Information: The system states/variables are time-varying.

Incomplete Information: A decision-maker only has partial information about the environment; in addition, it only knows its individual information and knows nothing about other users.

Learning: A decision-maker updates its current action based on its historical action-payoff information.

Uncertain Information: The observations are not equal to the real values.

ENDNOTES

1 In fact, the third party can fictitiously exist and the CE can be achieved in a distributed and autonomous manner.

Section 2

Game Theory in Emerging Heterogeneous Small Cell Networks

Chapter 4
Game-Theoretic Approaches in Heterogeneous Networks

Chih-Yu Wang
National Taiwan University, Taiwan & Academia Sinica, Taiwan

Hung-Yu Wei
National Taiwan University, Taiwan

Mehdi Bennis
University of Oulu, Finland

Athanasios V. Vasilakos
Lulea University of Technology, Sweden

ABSTRACT

Improving capacity and coverage is one of the main issues in next-generation wireless communication. Heterogeneous networks (HetNets), which is currently investigated in LTE-Advanced standard, is a promising solution to enhance capacity and eliminate coverage holes in a cost-efficient manner. A HetNet is composed of existing macrocells and various types of small cells. By deploying small cells into the existing network, operators enhance the users' quality of service which are suffering from severe signal degradation at cell edges or coverage holes. Nevertheless, there are numerous challenges in integrating small cells into the existing cellular network due to the characteristics: unplanned deployment, intercell interference, economic potential, etc. Recently, game theory has been shown to be a powerful tool for investigating the challenges in HetNets. Several game-theoretic approaches have been proposed to model the distributed deployment and self-organization feature of HetNets. In this chapter, the authors first give an overview of the challenges in HetNets. Subsequently, the authors illustrate how game theory can be applied to solve issues related to HetNets.

INTRODUCTION

High system capacity is one of the fundamental requirements to access current wireless communication. While most advanced signal and transmission techniques potentially enhance the performance of wireless systems (Parkvall, Furuskar, & Dahlman, 2011), they eventually reach the theoretical limitation due to the physical laws: the signal quality. Most of next generation wireless networks are planned to operate in high frequency spectrum. In such spectrum, the signals will degrade significantly in long distance and indoor environments. This suggests that more areas will experience weak signal receptions unless the network deployment is densified.

DOI: 10.4018/978-1-4666-8642-7.ch004

In order to boost network capacity in a flexible and cost-efficient manner, the concept of Heterogeneous Networks (HetNets) has been introduced in LTE-A standard (Lopez-Perez et al., 2011). A heterogeneous network consists of macrocells, which are deployed for serving large coverage areas, and low-power and low-cost nodes such as picocells, femtocells, relay nodes, or remote radio heads (RRHs), which provide services in areas with dedicated capacity. The wireless signal quality can be greatly enhanced through the assistance from the low-power nodes when they are properly deployed in the coverage holes in the macrocells. Additionally, these newly deployed small cells can be served for offloading purposes for help reducing the heavy loading in macrocells. The extra capacity offered by these cells can be used to handle more demands in the cellular network, or even redirect them directly to home and company's intranet or the Internet. Lastly, these low-cost nodes are more economically attractive as they usually require lower-cost infrastructure (pico and relay nodes) and lower requirements in terms of backhaul connections (femtocell).

The deployment of HetNets, however, is a serious challenge to the service operators. The deployment scheme depends on the purpose of the service provider. In case of traffic offloading, the deployment should be directly handled by the service provider according to the statistic or predicted demands in certain areas. In such a case, the problem is similar to the traditional cell planning except that the small cell deployment could be dynamic according to the current demands.

On the other hand, when it comes to the cell quality enhancement, the traditional cell planning and deployment solution is impractical to HetNets, especially for the femtocells. This is due to the significant larger number of cell sites and uncertainty in coverage holes. Additionally, the coverage holes are also difficult to be found without the assistance and demands from end-users. This poses a strong link between the deployment of femtocells and the demands from users. These users, who should be considered rational, determine the deployment of femtocells by either sending requests to the service provider or installing low-power nodes by themselves. Unlike the service provider, who concerns the overall system performance, these rational users care about their own benefits only. Additionally, heterogeneous components (pico/femtocells, macrocells, UEs) in HetNet may have different objectives (Khan, Tembine, & Vasilakos, 2012) and preferences on the network configurations. This increases the difficulty to have a proper organization among all these components. The conflict of interests between the service provider and end users may lead to inefficient deployment of HetNets if not carefully addressed. Therefore, it is important to understand how these rational users interact with the service provider in the deployment and configuration of HetNets. Game theory, a powerful tool for analyzing a distributed system with rational users, is a natural choice for studying above scenarios in HetNets.

BACKGROUND

Game theory is a study on the mathematical models of the strategic interactions between individual players in a game. The outcome of the game depends on the interactions among participated players. The players are considered rational and have certain valuations on the outcome of the game. Since the game outcome depends on the interaction, a player will be aware of the (expected) actions of other player and will make her decisions accordingly in order to reach her most desired outcome. Most researchers are interested in predicting the final outcome of a game, in which various types of equilibrium concept, such as Nash equilibrium, correlated equilibrium, core, ...etc, can be applied. Nevertheless, it is possible that the expected outcome is an inefficient one since players are involved in multiple individual interests on the outcome. In such cases, some researchers study on tuning the final outcome by refining the game rule or structure in order to improve the system performance.

The deployment and configuration of HetNets is a highly dynamic and distributed process within the cellular system, as the authors previous mentioned. Additionally, the involvement of rational users introduces the conflict of interests into the HetNet systems. In such systems, the authors must apply a game-theoretic approach to the problems in HetNets in order to predict how rational users and service provider cooperate or compete in such a dynamic and distributed system. Based on the results, the authors may then propose game-theoretic solutions to improve the system performance if the outcome with original designs is undesired.

In this chapter, the authors would like to provide a comprehensive overview on the game-theoretic approaches to different type of issues in HetNets. The authors first briefly introduce the components and characteristics of HetNets in Section 2. Then, the authors give a simple tutorial on the basic elements in game theory for readers to understand the solution concepts provided by game theory in Section 3. Next in Section 4, the authors provide a survey on different game-theoretic approaches which have been proposed in the literature. The authors will discuss how these approaches and corresponding solution concepts work in the targeting issues in HetNets. Finally, the authors come up with a list of open problems in HetNets which are potentially suitable for game-theoretic approaches in Section 5. The authors draw the conclusions in Section 6.

A typical HetNet in LTE-A is composed of lower-power base stations (BSs) underlying in the existing macrocell system. These small BSs are intended to increase the signal strength, offload the macrocells, and enhance the spectrum utilization. The deployment of HetNets can be planned and conducted by the service provider in advance, or requested and deployed by users themselves. The service area and operating spectrum of the small cells is usually partly or fully overlapping with the macrocell. Heterogeneous small cell base stations have been introduced in HetNets (Lopez-Perez et al., 2011). The authors briefly state as follows:

- **Picocells** are low-power (23 to 30dBm) cell towers providing similar features as macrocells except smaller coverage (hundreds of meters) and user load (tens of users). They use the same backhaul as the macrocells and are deployed by the service provider.
- **Relays** are small stations that deliver the data between macrocells and MSs in a multi-hop over-the-air scheme. They are mostly deployed by the service provider in order to extend the coverage of existing networks. A relay requires over-the-air backhaul capacity between macrocell BS and uses a similar transmit power as picocells.
- **RRHs** are radio control units that are connected directly to the macrocells through fibers but deployed with a distance from the macrocell BS. The macrocell has full control on the RRHs and operate them as its own wireless interface.
- **Femtocells** are also known as Home eNode Bs (HeNBs) in LTE systems. A femtocell BS can be regarded as a simple, low-transmission power (i.e. 23 dBm or less) base station installed by users in an unplanned manner. Through the deployment of femtocells, subscribers are able to access to networks via broadband backhaul. That is, femtocells may utilize Internet protocol (IP) and flat base station architectures. Femtocells may operate in open-access, closed-subscribed group (CSG), or hybrid-access scheme, depending on the choice of the cell owner.

In these possible choices of small cells, the femtocell has the following advantages: It increases indoor signal coverage and system capacity on demands, providing higher link quality with lower transmission power, and utilizes the existing broadband connection as its backhaul. Nevertheless, the femtocell system faces several challenges. In what follows, the authors will focus on the challenges in HetNets using femtocells.

CHALLENGES IN FEMTOCELLS

There are numerous challenges that have been identified in femtocells. In this chapter, the authors focus on self-organization, intercell interference, and economic potentials. These challenges cover two fundamental but different perspectives, technologic and economic, of the femtocells.

Self-Organization

Self-organization is a key component of heterogeneous networks. As the authors mentioned previously, the deployment of femtocells in HetNets is likely to be conducted by users themselves without planning. Additionally, the ability for the service provider to organize and control the deployed cells may be limited by 1) the scalability in network size, and 2) the low QoS provided by the backhaul connection. An improperly deployed and configured femtocell may eventually lead to poor performance due to severe intercell interference and load unbalance among cells.

Intercell Interference

Intercell interference takes place when femtocells operate in a licensed spectrum and their coverage overlaps with other base stations in the same spectrum, as shown in Fig. 1. Intercell interference poses a great technical challenge to the deployment of co-channel femtocells. This issue becomes more challenging than in traditional cellular networks because 1) unplanned deployment and limited control from the service provider makes the centralized-optimization solution in traditional networks impractical, 2) self-organization feature provided by the femtocells may tend to optimize their own performance instead of the overall network performance, such as always operating in CSG mode even it still has unused resource for other MSs, and 3) dynamic deployment from the users make the environment and network deployment changes more frequently and therefore a fast-response mechanism is required to dynamically conduct the interference mitigation.

Economic Potentials

One of the main reasons to introduce femtocell is to reduce the deployment cost of the service provider when implementing the next-generation wireless networks. Therefore, it is necessary to analyze if femtocells indeed are profitable to the service provider by considering both the deployment cost and the added value to the service. The value addition of femtocells mainly comes from the enhanced service quality, such as higher throughput, lower delay, and less congestions and blocking probability. These performance improvements under the femtocell technique should be evaluated in advance before introducing such

Figure 1. An illustration of various interference scenarios in heterogeneous networks

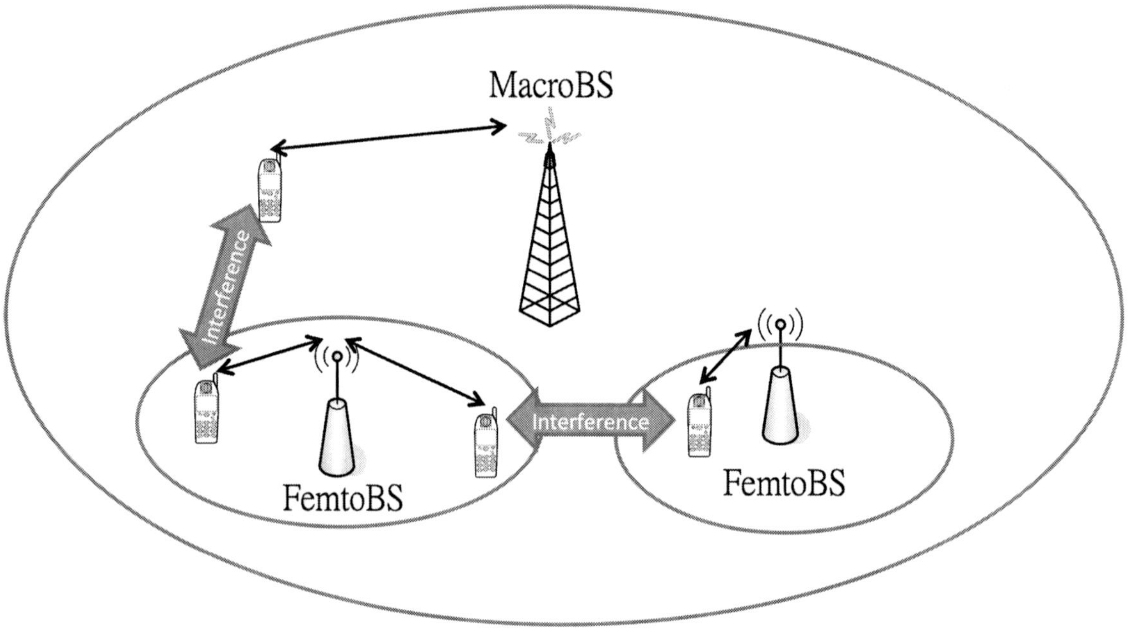

techniques. Additionally, these objective metrics need to be translated into the subjective valuation from users. Users may be heterogeneous with different environments, preferences, and requirements for the wireless service. How these users will behave is an important topic to investigate.

GAME THEORY

Game theory is a powerful tool applied to model and to analyze the outcome of interactions among multiple decision-makers. A traditional game consists of three basic elements: **players**, **strategies**, and **utilities**. Players are the participants and decision makers in the game. They can take some predefined **actions** to affect the interaction with other players and make influence on the final game result. Players are individual decision makers - given specified information (game rules, state of the system, applied actions of other players, etc) of the game, they apply strategies to decide the actions (reactions) taken in the game. Given a strategy profile which describes the strategies applied by each player, the game will produce a corresponding outcome.

Utility functions, as quantified evaluations to the outcomes of a game, map the outcomes into real-value spaces. A player's evaluation of the outcome is given by a utility function. Since different game strategies may bring out different outcomes, the authors can see the element of utility as a function of a strategy evaluation.

In most cases, the authors assume that all players are rational. Thus, they tend to adopt strategy that can maximize their utility. In such games, every player is trying to maximize their own utility. If the players refuse to collude with each other, the game can be modeled as a non-cooperative game, and players compete with each others directly. For the case that players may cooperate with each other, the authors model it as a cooperative game, where players may form groups (coalitions) in order to gain advantages in the competition. A cooperative game is suitable for describing specific application such as admission control and cluster formation, where coalitions naturally forms in such systems. Nevertheless, due to space limitation, the authors focus on non-cooperative game approaches in this chapter as it is popularly applied in HetNet issues.

In most game models, the purpose of the theoretic analysis is to find out the equilibrium, namely, the most likely produced outcome of a steady state in the system. Under the framework of a game theory, the behavior and interactions of players can be properly modeled. For researchers in communication areas, game theory is very useful to analyze problems involving interactions among elements in the system, the resource allocation problem particularly.

As discussed in Section 1, game theory is suitable for analyzing and resolving general communication problems with interactions, like femtocell systems. With the involvement of femtocells, the network structure becomes much complicated and centralized optimization costs much, especially when the network changes dynamically. Game theory encourages each player to do self-optimization. By applying an appropriate mechanism design, distributed self-optimization behavior for all players (i.e. femtocell) may lead to high system performance and at the same time reduce the computation costs.

In the following sections, the authors provide a review on the game-theoretic approaches for various topics in HetNets.These issues are suitable for game theory as they involve interactions among players/agents/decision makers and distributed self-optimization. For each issue, the authors first briefly introduce the challenge of the issue. Then, the authors discuss how the game-theoretic approaches can be applied in this issue in HetNets.

INTERFERENCE MITIGATION

Intercell interference is one of the most challenging issue in HetNets. Without a proper interference mitigation technique, the service quality of macrocells in HetNets will be harmed by the interference from pico/femtocells, as illustrated in Fig. 1. Due to the limitation on the control ability of the service provider on the femtocells and the scalability issue, self-organized and distributed interference mitigation techniques are required. The interference mitigation is a classic issue and has been studied by researchers using various game-theoretic approaches, such as strategic game (Chandrasekhar, Andrews, Muharemovic, Shen & Gatherer, 2009), potential game (Mustika,Yamamoto, Murata, & Yoshida, 2011), Bayesian game (He, Debbah & Altman, 2010), and Stackelberg game (Guruacharya, Niyato, Kim & Hossain, 2013). The authors will introduce each of them in the following paragraphs.

Strategic Game

Strategic game is a type of games that all players behave simultaneously with perfect knowledge on other players' possible actions. Specifically, suppose that there is a game that involves two or more players, in which each player is assumed to know the actions of the other players. All players choose their actions simultaneously, and then the outcome of the game is also settled. In such a case, a rational player should predict what actions other players will choose *before* she chooses her action.

The expected outcome of the game can be found through finding the Nash equilibrium. Let's assume that there exists an action profile that after each player has chosen her action accordingly, no player can increase her utility from changing her action when other players' actions remain unchanged. If such an action profile is applied, no players have the incentive to deviate from the applied action since the deviation gives her equal or less utility. When the above conditions are met, the action profile constitutes a Nash equilibrium.

Femtocells can be considered as the players in the strategic game, while their actions are their applied transmission power, occupied resources, or other operations that will potentially influence the service quality of other cells. Then, a femtocell's utility can be defined as the service quality, such as the throughput or delay time, experienced by UEs in the cell. An example is illustrated in Fig. 2, where multiple femtoBSs are determining their transmission power. Given other femtocell's transmission power, a femtocell may have her optimal transmission power that maximizes her utility. In such an approach, the authors would like to identify the stable outcome of the game, that is, the Nash equilibrium in the HetNet. Strategic game approach is straight forwarding, but the results may not be appealing: the Nash equilibrium can be an inefficient outcome comparing to the optimal solution due to the competitive effect in strategic game. Some regulation designs, such as penalties on the femtocells, may be necessary to improve the system performance.

Chandrasekhar et al. (2009) is one of the first group to discuss the problem of uplink interference in overlay macrocell-femtocell systems with game-theoretic analysis. They consider a strategic game model in this work. They first provide a detailed analysis on how the upper bound of femtocell signal-interference to noise ratio (SINR) target affect the feasible range of cellular (macrocell) SINR, and discover a product relation between them. Then, they formulated the problem into a power control game with penalty in caused interference in the utility of femtocells. In their game model, a femtocell should determined her power considering the positive valuation from his experienced SINR and the penalty from the interference caused to other cells. The outcome of the game is determined by the transmission powers each femtocell selected. This forms a strategic game. They then proposed a distributed iterative power adaption algorithm to find the desired Nash equilibrium. Additionally, the macrocell may feedback

Figure 2. A strategic game approach to heterogeneous networks

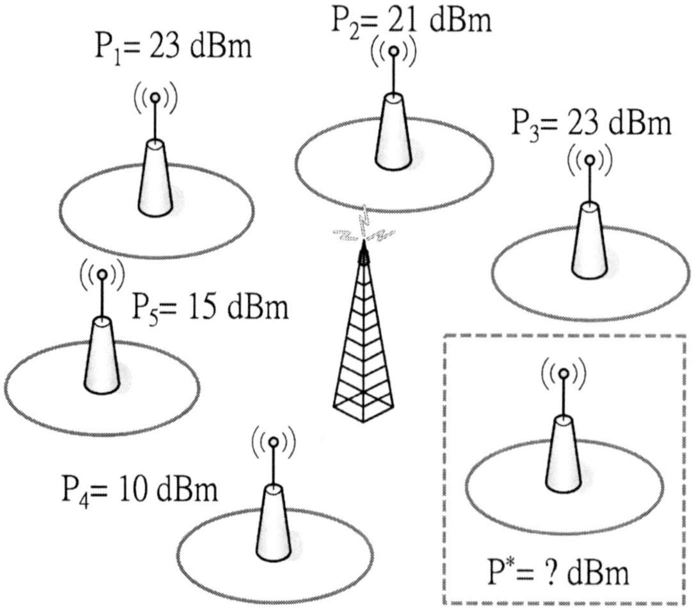

to the femtocells if its target SINR is satisfied. If not, all femtocells are required to lower their target SINR and therefore lower their transmission power. This hierarchical control technique is very useful for service providers to regulate the rational users/cells in the system.

Potential Games

Potential game is another popular game model, especially in power control problem. It is a special type of strategic game: In such a game, the incentive of any player in the game to change their actions can be expressed in a global function called the potential function. Specifically, there exists a function taking the applied actions of all players in the game as inputs. Whenever a player changes her actions and has some utility increase/decrease, the output of the function increase/decreases in the same direction or with the same amount. It has been proved that the set of pure Nash equilibrium in a potential game can be found by simply locating the local optima of the potential function. Additionally, a distributed iterative update algorithm in a potential game always converges to a Nash equilibrium. This reduces the complexity to implement the Nash equilibrium.

Potential game is one of the ideal approaches to HetNets since it guarantees a distributed iterative update algorithm will always converge to a stable Nash equilibrium, which is desirable for self-organized femtocells. Nevertheless, it requires a global potential function, which may not exist in a general HetNets. The convergence speed is also a concern when the deployment is dynamic.

Mustika et al. (2011) consider the interference mitigation as a resource block allocation problem. They formulate the resource allocation problem as a potential game, in which femtocell users select the desired subset of resource blocks considering both the interference from others choosing the same resource blocks and the caused interference to others. Since it is a potential game, the convergence of the best response dynamic algorithm is guaranteed.

Bayesian Game

Bayesian game is a game model with uncertainty in the type of players. In a typical Bayesian game, at least one player is unsure of the type (and so the payoff function) of another player. Players have initial beliefs about the type of each player and can update their beliefs according to Bayes' Rule in the game. When the information of the system is incomplete, i.e., the channel gains of femtocell users are unknown, an appropriate choice of game models is Bayesian game, in which a belief on the unknown information is introduced. Notice that the performance of the system does not necessary degrade in Bayesian game comparing to complete information game. Incomplete information sometimes results in less fierce competition and therefore prevent performance degradation. Nevertheless, the challenges in Bayesian game is correctly identifying and learning the distribution of the unknown information, such as the channel quality distributions. An incorrectly constructed distribution usually leads to the sub-optimal operations in the system.

Take the work of He et al. (2010) as an representative example. They develop a general model for resource allocation problem in multi-user fading multiple access channels (MAC). In their system, each users do not have the knowledge of other users' fading channel gains. Therefore they design and analyze the problem in a Bayesian game-theoretic approach. They observe that by introducing the symmetric strategy restriction into the system, the performance of the network under the optimal strategy can be close to the Pareto optimal one.

Stackelberg Game

Stackelberg game is a sequential game specifically for the systems with hierarchical structure. In a Stackelberg game, two types of players, leaders and followers, are defined. In the game process, the leader should apply or announce her action first. Then, the followers response to the leader's action accordingly. Since all players are rational, the followers should choose their actions that maximize their own utility. By using this insight, the leader can predict the rational responses of the followers if she chooses certain actions. The leader then can choose her action that maximize her own utility based on her analysis on the rational response of the followers.

Stackelberg game is ideal for HetNets consists of both macrocell and femtocells. The leaders, which should be macrocells in HetNets, have the advantages to apply their actions wisely before the followers, which are the femtocells in HetNets (Fig. 3). By strategically determining their applied action, the macrocells can lead the game to their desired outcome when they have enough information to predict the response of the femtocells in HetNets. This reflects the control ability of macrocells on the operations of femtcoells, such as the penalty on undesired interference. Nevertheless, the requirements to fully understand the response of femtocells given any possible action of macrocell in Stackelberg game may be impractical when the HetNet is complex. In such a case, learning techniques, such as reinforcement learning, could be applied to help macrocell find her best strategy in the Stackelberg game.

Guruacharya et al. (2013) consider the downlink power allocation game in the overlay system as a Stackelberg game. Macrocells, which are the leaders, compete with each other in the power control game. Femtocells, on the other hand, are defined as followers. The femtocells apply the water-filling power control algorithm as their responses in this work. The simulation results shows that under such a design, on average the users under macrocells have lower throughput than the femtocells.

Figure 3. Stackelberg game approach to heterogeneous networks

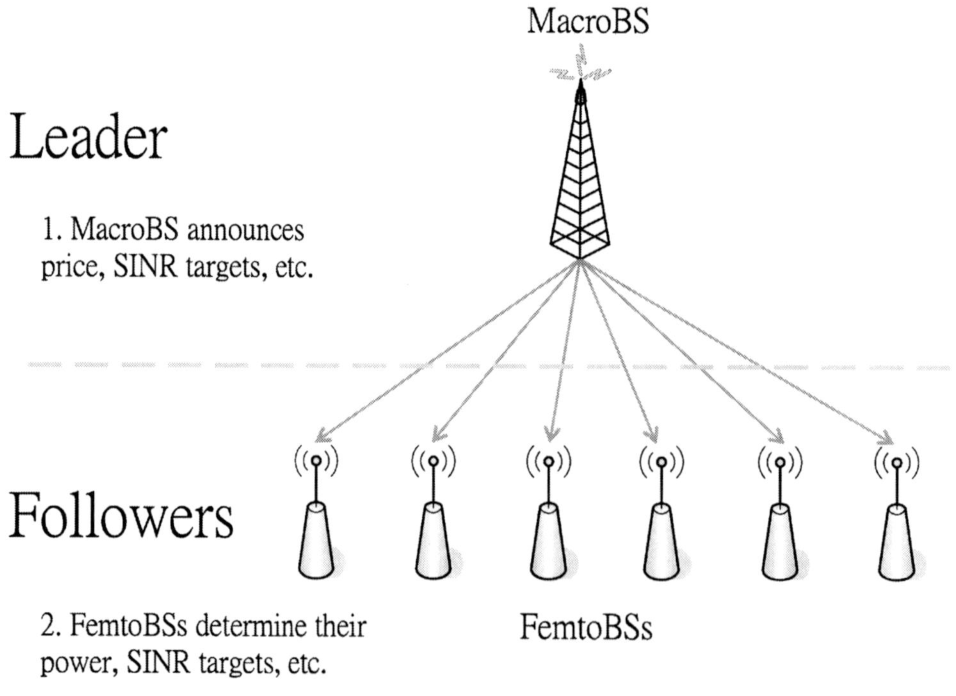

Kang et al. (2012) introduce the pricing process into the resource allocation problem. Macrocells impose a unit-price on the interference to charge the femtocells for their interference. The Stackelberg game model is applied here with macrocells as the leaders to predict the response of femtocells under the given price. Their simulation results show that the pricing process effectively reduce the interference to the desired level.

HYBRID-ACCESS DESIGN

Femtocell can operate in three different access modes: OSG, CSG, and hybrid modes. When femtocell is operating in OSG or hybrid mode, macrocell users may connect to the femtocell when required. Several studies have demonstrated that either static CSG or OSG modes may result in system inefficiency in interference management and bandwidth sharing. Hybrid mode as a design to balance interference and efficiency is suggested by Choi et al. (2008). However, as the authors mentioned earlier, the femtocells may not be wiling to do so. Therefore, it is necessary to provide some incentive to stimulate the femtocells operating in OSG or hybrid mode. Stackelberg game is a proper approach to this issue. From a game-theoretic view, the service provider or the macrocell, which are the leaders, should act first by announcing some femtocell-preferred policy or benefits in order to attract the rational femtocells, which are the followers, to provide hybrid-access. In such an issue, Stackelberg game and mechanism design approaches are proper choices for stimulating rational femtocells to operate in hybrid mode.

Stackelberg Game

Stackelber game is popularly applied in hybrid-access system due to the natural leader–follower relation between macrocell and femtocell services. Yi et al. (2012) consider the case that two service providers construct a HetNet: one holds the traditional macrocell service, while the other is operating pico/femtocell service. They assume that the system is in split-spectrum mode and the macrocell service provider has the license of all operating spectrum. Therefore, the femtocell service provider needs to rent from the macrocell service provider in order to provide the service. They formulate the system as a three-stage Stackelberg game and show that under such a game structure, the equilibrium will be a win-win situation: the macrocell service provider will lease the spectrum to the femtocell service, while the femtocell service providers will run its service in hybrid-access mode.

Chen et al. (2012) propose a refund-based mechanism to provide some compensations to femtocells who are willing to partly allow macrocell users to access them. A Stackelberg game model is proposed in which a macrocell is the leader and femtocells are followers. The macrocell first determines the total amount of refund it is willing to give to the femtocells, and then the femtocells determine the open time given the macrocell's decision. By properly adjust the refund to compensate the cost of femtocells, the macrocell can stimulate the femtocells to operate in hybrid-access mode.

Mechanism Design

Mechanism design is a branch of game theory. System designers apply the techniques to design an algorithm or a procedure to achieve some good properties in the system. By using the techniques of mechanism design, the system can achieve several desirable properties. For example, the property of social welfare maximizes the utility summation of all players. Note that such rules are not against the players' nature of rationality and selfishness. That is, the players' decisions are still based on their own profits under the setup. Since mechanism design provides a set of elegant tools for a general system design, it is fairly suitable to apply the concept to wireless communication protocols or algorithm designs, especially when the element of players' interactions involves. In HetNets, mechanism design can be a powerful tool when the service provider would like to improve the efficiency of a system with rational players or prevent undesired cheating behaviors from the players. In hybrid-access stimulating issue, the service provider may use mechanism design to provide more incentive for the femtocells to provide hybrid-access service.

Ko and Wei (2011) propose a resource allocation mechanism in an overlay macrocell-femtocell network. The resource is allocated in a two-tier structure: the macrocell determines the resource allocation for macrocell users and femtocells, and then the femtocells then determine the allocation of the granted resource to their own users. By inducing a game-theoretic analysis, they prove that their design can induce the users to truthfully report their traffic demands. Additionally, their proposed mechanism also induce the femtocells to allow access from the MSs whenever the MSs have higher data rates under the femtocells comparing to under the macrocells.

Chen et al. (2012) also consider an auction-based approach to deal with the hybrid access problem. They propose a reverse auction, which contains multiple seller (femtocells) and single buyer (macro servicer provider), for determining the hybrid service area of each femtocell. The femtocells who win the auction are required to provide the hybrid access service in certain areas with a payment from the macro service provider in return. By properly design the price and winner determination rule, the proposed reserver auction is proved to be cheat-proof.

REVENUE MAXIMIZATION

Except the technical issues in femtocells, the potential economic value of HetNets is a popular topic in recent years. As mentioned in previous sections, the service providers will not deploy HetNets unless it has been proved to be potentially beneficial to the revenue and profit of the service provider. Game theory is a natural choice for studying such a topic since it originally serves for studying the rational behaviors and interactions of human beings in societies, especially in markets.

Mechanism Design

Shetty et al. (2009) discuss the effects of femtocells on the revenue of service providers. They study a monopolist service provider's optimal pricing strategy considering the subjective evaluations of the heterogeneous users. A Stackelberg game is constructed with the service provider, who decides the servicer price, as the leader, and the users, who make the purchases of macro/femto service, are the followers. Their framework includes two spectrum deployment models: common spectrum and split-spectrum. Their analysis shows that when the intercell interference can be resolved efficiently, the common spectrum model generates higher revenue than the split-spectrum one. Results also show that femtocells increases the profit of the service provider by decreasing the loading of macrocells while charging higher for the femtocell services.

Wang and Wei (2013) show that the differentiated contract, which can generate more profits when user's valuation on the service is varied, has great potential to improve the overall profit of HetNets. They propose a wireless service differentiation framework, which is based on mechanism design in game theory, to investigate the profit service providers make under a variety of differentiated contracts in the overlay macrocell-femtocell network. In their model, the price and maximum data rate are the differentiating factors in the contract structure. They draw a comparison between the shared-spectrum and split-spectrum systems and find out that it is difficult to further extract profit from MSs in a split-spectrum system. In the shared-spectrum system, nevertheless, there indeed exist differentiated contracts which can generate more profits by raising service prices for the MSs with good service qualities in femtocells, while providing cheaper prices to other MSs with poor service qualities.

Stackelberg Game

In the case of multiple markets, the effect of competitions on the revenue and the choice of deployment model is of interest. Ren et al. (2013) illustrate a model with one incumbent service provider and one entry service provider. Both providers compete for the market with their selected femtocell spectrum deployment scheme, i.e., split-spectrum or shared-spectrum schemes. Their analysis, which is based on the Stackelberg game model, shows that the existence of an incumbent will lower the incentive of the entry service provider to join the competition as the expected profits is harmed by the competition among them.

FUTURE RESEARCH DIRECTIONS

As illustrated so far, game theory is one of the most powerful tools to resolve the problems involving multiple decision makers, especially in decentralized systems. Up to now, the authors have given many contexts in discussing the possibilities in applying game theory to challenges in HetNets. Nevertheless, it is still worthwhile to give more potential research topics which are suitable for game-theoretic approaches here.

- Resource allocation in HetNets
    - In time domain, how do the authors determine the scheduling or queueing rules?
    - How do the authors manage the subcarriers in an OFDMA based system?
    - How do the authors adaptively or statically allocating the uplink and downlink channel in an overlaying scheme?
- Power control in uplink/downlink scenario
    - When the authors consider the interactions between the neighboring base stations, what is the optimized solution when bargaining involved?
    - How to determine the strategy of the transmission power when considering a tradeoff between quality and battery power consumption?
- Self-organization Networks (SON)
    - Is there any mechanism that can give incentive for small cells to provide open or hybrid-access, without requiring monetary transfer process?
- Potential economic value of HetNets

- What is the optimal pricing scheme that maximizes the revenue/profit?
- What resource allocation scheme is more profitable and cost-efficient?
- How the service providers introduced HetNet considering the competition effects?

In additional to distributed solutions to HetNets, game-theoretic modeling is also useful to determine the role and the controlling policies of the service provider in a centralized system, as the authors illustrated in Stackelberg game and Mechanism designs. If the service providers can gain the insights of the users' behavior and their interactions by modeling the system as a game, they can simply optimize the system performance by controlling the key factors or set essential rules for the players. Here the authors would like to provide some open research problems which can apply the centralized techniques:

- How do the authors determine the position of small cells in a general geographical topology, considering their rational demands?
- How do the authors determine the service price of HetNet service, considering their potentially malicious behaviors?

Finally, the implementability of the proposed game theoretic approaches in practical systems, such as current and future LTE-Advanced releases, is still an open challenge and deserves further studies. Generally, the key issues to implement a game-theoretic approach to a communication system is to 1) linking action in game to operations in the targeting system, 2) establish required information exchange process among players, and 3) evaluate the utility functions of users/devices in the targeting system. A well-designed game-theoretic approach should carefully address all these issues in order to guarantee the implementability of the proposed solution based on the approach. For practical protocols such as LTE-Advanced systems, most operations and information exchange process are already predefined and cannot be altered easily, which may become some constraints on the proposed approach. Additionally, it also requires some prior works like customer survey in order to choose the proper utility function for users. Given that game-theoretic approaches are on the rise in this area, the authors expect to see more interesting related studies in the near future.

CONCLUSION

In this chapter, the authors introduce HetNets as a promising solution to increase capacity and coverage in next-generation wireless communications. Nevertheless, due to the unplanned deployment and distributed operating characteristics, several challenges exist in the deployment and configuration of HetNets. Game theory, which is a pothe authorsrful tool for studying distributed systems with self-interest players, has the potential to resolve most difficult challenges in HetNets. The authors illustrate how game theory can be applied in HetNet through reviewing the existing literature. According to their results, the authors confirm that game theory indeed can resolve most of the key challenges in HetNets. Finally, the authors pin down the remaining problems in HetNets and briefly mention the potential of game theory in addressing these open challenges.

REFERENCES

Chandrasekhar, V., Andrews, J. G., Muharemovic, T., Shen, Z., & Gatherer, A. (2009). Power control in two-tier femtocell networks. *IEEE Transactions on Wireless Communications*, *8*(7), 4316–4328. doi:10.1109/TWC.2009.081386

Chen, Y., Zhang, J., & Zhang, Q. (2012). Utility-Aware Refunding Framework for Hybrid Access Femtocell Network. *IEEE Transactions on Wireless Communications*, *11*(5), 1688–1697. doi:10.1109/TWC.2012.031212.110002

Chen, Y., Zhang, J., Zhang, Q., & Jia, J. (2012). A reverse auction framework for access permission transaction to promote hybrid access in femtocell network. In *Proceedings of IEEE International Conference on Computer Communications (INFOCOM)*. IEEE.

Choi, D., Monajemi, P., Shinjae, K., & Villasenor, J. (2008). Dealing with Loud Neighbors: The Benefits and Tradeoffs of Adaptive Femtocell Access. In *Proceedings of IEEE Global Communications Conference (GLOBECOM)*. IEEE. doi:10.1109/GLOCOM.2008.ECP.540

Guruacharya, S., Niyato, D., Kim, D. I., & Hossain, E. (2013). Hierarchical competition for downlink power allocation in ofdma femtocell networks. *IEEE Transactions on Wireless Communications*, *12*(4), 1543–1553. doi:10.1109/TWC.2013.022213.120016

He, G., Debbah, M., & Altman, E. (2010). A Bayesian Game-Theoretic Approach for Distributed Resource Allocation in Fading Multiple Access Channels. *EURASIP Journal on Wireless Communications and Networking*, *2010*(1), 8:1–8:12.

Kang, X., Zhang, R., & Motani, M. (2012). Price-based resource allocation for spectrum-sharing femtocell networks: A stackelberg game approach. *IEEE Journal on Selected Areas in Communications*, *30*(3), 538–549. doi:10.1109/JSAC.2012.120404

Khan, M. A., Tembine, H., & Vasilakos, A. V. (2012). Game Dynamics and Cost of Learning in Heterogeneous 4G Networks. *IEEE Journal on Selected Areas in Communications*, *30*(1), 198–213. doi:10.1109/JSAC.2012.120118

Ko, C.-H., & Wei, H.-Y. (2011). On-Demand Resource-Sharing Mechanism Design in Two-Tier OFDMA Femtocell Networks. *IEEE Transactions on Vehicular Technology*, *60*(3), 1059–1071. doi:10.1109/TVT.2011.2106171

Lopez-Perez, D., Guvenc, I., de la Roche, G., Kountouris, M., Quek, T. Q. S., & Zhang, J. (2011). Enhanced intercell interference coordination challenges in heterogeneous networks. *IEEE Wireless Communications*, *18*(3), 22–30. doi:10.1109/MWC.2011.5876497

Mustika, I. W., Yamamoto, K., Murata, H., & Yoshida, S. (2011) Potential Game Approach for Self-Organized Interference Management in Closed Access Femtocell Networks. In *Proceedings of IEEE 73rd Vehicular Technology Conference (VTC Spring)*. IEEE.

Parkvall, S., Furuskar, A., & Dahlman, E. (2011). Evolution of LTE toward IMT-advanced. *IEEE Communications Magazine*, *49*(2), 84–91. doi:10.1109/MCOM.2011.5706315

Ren, S., Park, J., & van der Schaar, M. (2013). Entry and Spectrum Sharing Scheme Selection in Femtocell Communications Markets. *IEEE/ACM Transactions on Networking*, *21*(1), 218–232. doi:10.1109/TNET.2012.2198073

Shetty, N., Parekh, S., & Walrand, J. (2009). Economics of Femtocells. In *Proceedings of IEEE Global Telecommunications Conference (Globecom)*. IEEE.

Wang, C.-Y., & Wei, H.-Y. (2013). Profit maximization in femtocell service with contract design. *IEEE Transactions on Wireless Communications*, *12*(5), 1978–1988. doi:10.1109/TWC.2013.032513.111789

Yi, Y., Zhang, J., Zhang, Q., & Jiang, T. (2012). Spectrum leasing to femto service provider with hybrid access. In *Proceedings of IEEE International Conference on Computer Communications (INFOCOM)*. IEEE.

ADDITIONAL READING

Nisan, N., Roughgarden, T., Tardos, E., & Vazirani, V. V. (Eds.). (2007). *Algorithmic Game Theory*. Cambridge, UK: Cambridge University Press. doi:10.1017/CBO9780511800481

KEY TERMS AND DEFINITIONS

Femtocell: A simple, low-transmission power (i.e. 23 dBm or less) base station installed by users in an unplanned manner.

Game Theory: A mathematical tool to model and to analyze the outcome of interactions among multiple rational decision-makers.

Heterogeneous Networks (Cellular): A cellular network consists of macrocells, which are deployed for serving large coverage areas, and low-power and low-cost nodes such as picocells, femtocells, relay nodes, or remote radio heads (RRHs), which provide services in areas with dedicated capacity.

Interference: An undesired signal, noise, or anything else that may alter or disrupt the message in the process of transmissions.

Mechanism Design: A branch of game theory for system designers to design an algorithm or a procedure to achieve some good properties in the system.

Nash Equilibrium: A strategy profile of a game where all players in the game do not have the incentive to alter their strategy given other players have applied the strategies described in the profile.

Stackelberg Game: A two-stage sequential game with leaders and followers, where the leaders apply their actions before the followers.

Chapter 5
Interference Mitigation with Power Control and Allocation in the Heterogeneous Small Cell Networks

Xiaodong Xu
Beijing University of Posts and Telecommunications, China

Rui Gao
Beijing University of Posts and Telecommunications, China

Mingyang Li
Beijing University of Posts and Telecommunications, China

Yunting Wang
Beijing University of Posts and Telecommunications, China

ABSTRACT

With the introduction of Small Cell into current cellular structure, the ever-growing demand for mobile traffic gets the opportunity to be fulfilled. But the overlapped dense deployment caused by the Small Cell Heterogeneous Network (HetNet) also arouses the interference problems. In order to solve those problems, this chapter focuses on Game Theory based uplink power control and downlink power allocation strategies for the interference mitigation of the heterogeneous Small Cell Network (SCN). For the uplink scenario, this chapter proposes the non-cooperative game model based power control algorithm, which can optimize the initial transmission power of both Macro Cell and Small Cell users through the Nash Equilibrium solution. For the downlink scenario with multiple service types in the SCN, the non-cooperative game model based scheme is proposed to optimize the transmission power allocation with constraints of different Quality of Service (QoS) requirements. The simulation results show the merits of the proposed strategies over current works.

INTRODUCTION

Nowadays, the mobile Internet applications and corresponding versatile mobile services are influencing every aspects of our daily life, forcing the mobile network to face the challenges for revolutionary development to cope with the significant increase of the mobile traffic. Besides the spectrum expanding, the radio transmission efficiency improving, the mobile network architecture is also considered as

DOI: 10.4018/978-1-4666-8642-7.ch005

the most important potential for further empowering the capacity increase of the 5th Generation mobile systems (5G) (Andrews, 2013), (Hanzo et al., 2012). Fueled by the requirements of the capacity increase and improved coverage for home and office use, the Heterogeneous Network (HetNet) with Small Cell Network (SCN) deployments have attracted significant interests in the mobile communication industry and standard organizations (3GPP, 2012). The heterogeneous SCN will be a radically different network design that could provide a cost- and energy-efficient solution to cope with the forecast traffic growth. SCN is founded by the idea of dense deployments of self-organizing, low-cost, low-power base stations that are substantially smaller than the traditional Macro Cell equipment, which is also introduced into the standard framework of the 3rd Generation Partnership Project (3GPP) (3GPP, 2013) and so on. The small Cell evolved Node B (eNB) is a low power access node, working in an authorized or unauthorized spectrum. The Small Cell integrates Femto Cell, Pico Cell, Micro Cell and distributed Remote Radio Head (RRH) technologies. The covering ranges of Small Cell are much smaller than the Macro Cell, which can cover from 10 meters of indoor environment to the fields of 2km according to different transmission power settings (3GPP, 2013). The Small Cells could be deployed on the building walls or urban furniture, in the streets or inside the hotspot buildings, by which, the transmission range of Small Cells can offer better signal reception at indoor users. With the combination of the already deployed Coordinated Multi-Point Transmission/Reception (CoMP) and Relay techniques (3GPP, 2013), the heterogeneous SCN could bring more potential gain over current network architecture. Therefore, the heterogeneous SCN are expected to offload part of capacity from the Macro Cell, thereby reducing the Capital Expenses (CAPEXs) and Operational Expenses (OPEXs) for the network operators.

Although SCNs can provide significant benefits, they may also suffer severe interferences with the high density heterogeneous deployments, emanating from other nearby Small Cells and Macro Cells with the shared spectrum manner. The dynamic scheduling modes result in the random variation of inter-cell interferences, such as the Small Cell dynamically switching ON/OFF. The delayed exchange of load situations will further lead more complicated and unpredictable inter-cell interferences. Especially, cell edge users are far from the eNB and the relatively higher signal attenuation usually needs larger transmission power, which will produce a strong interference that could seriously deteriorate the adjacent cell performances. In these scenarios, the inter-cell interference will be one of the main obstacles to limit the performance in the heterogeneous SCN.

Furthermore, the heterogeneous SCN consists of both Macro Cells and a variety of Small Cells. In the heterogeneous SCN, the existing several eNB types, such as the Macro Cell eNBs, Micro Cell eNBs, Pico Cell eNBs and Femto Cell eNBs, have different transmission power limitations. For example, the Macro Cell eNBs are with high transmission power limitation and Small Cell eNBs are with low transmission power. According to the standards of 3GPP, the typical transmission power of Macro Cell eNBs and Small Cell eNBs are 46 dBm and 30 dBm respectively (3GPP, 2013).

The introduction of the Small Cell into the Macro Cell deployments could enhance the system capacity and user throughput, but this layout will also affect the user association. For the Long Term Evolution (LTE)/LTE-Advanced based systems, the Reference Signal Received Power-based (RSRP-based) user association is widely used (3GPP, 2013). In this scheme, users are associated with the cells of strongest downlink transmission power. As the Macro Cell eNBs have higher transmission power than the Small Cell eNBs, users may connect with the Macro Cell even though the path loss of the adjacent Small Cell is much lower than the Macro Cell. This will also cause more severe interferences. Moreover, because the coverage area of Small Cells is much smaller than the Macro Cells, the potential offloading gains from the Macro Cells to the Small Cells are also limited.

Therefore, in order to reduce the interferences and increase the offloading gain of the Macro Cells by Small Cells, it is worthy to increase the coverage areas of Small Cells for attracting more users for association. This solution can be achieved by the technique of Cell Range Expansion (CRE) (3GPP, 2013). The CRE is defined as expanding the coverage area of Small Cells via adding an offset for the RSRP of the Small Cell eNBs. According to the CRE, the user association algorithm can be biased so that users can be connected with Small Cells even if the actual RSRP of the Small Cells are lower than those of the Macro Cells.

The CRE guarantees the traffic offloading and increases the hot spot capacity in the HetNet. But there is also a disadvantage of deploying the CRE, which lies when user communicates with the Small Cell may experience strong interferences from the Macro Cells in the heterogeneous SCN scenario, where Macro Cells and Small Cells are operated on the shared spectrum or intra-frequency manner.

A solution can be implemented to maintain the system performances through user connected with both the Small Cell eNB and the Macro Cell eNB simultaneously due to the different cell coverage ranges in the uplink (UL) and downlink (DL) respectively. That is, the user associations of the uplink and downlink are separated among different eNBs. The UL/DL separated users can communicate with two cells, one of which offers highest uplink throughput through adjacent Small Cells, one of which offers highest downlink received signal strength through high transmission power of Macro Cells. The user communicates with the Macro Cell from which the highest RSRP is received in the downlink. In the uplink, the user communicates with the Small Cell, where the user's path loss is the lowest (3GPP, 2013). In general, UL/DL separation could be used for both intra-frequency deployments and inter-frequency deployments. But when Macro Cells and Small Cells are deployed with the intra-frequency, it will be more beneficial with the UL/DL separation. Moreover, the UL/DL separation can also mitigate the interferences due to the power imbalance problem (3GPP, 2013).

For the LTE/LTE-Advanced based systems, the uplink and downlink transmission technology are different because of the different multiple access schemes adopted. The Single Carrier-Frequency Division Multiple Access (SC-FDMA) scheme is applied in the uplink and the Orthogonal Frequency Division Multiple Access (OFDMA) scheme is implemented in the downlink, which also affect the interference situation and interference mitigation solution for the uplink and downlink scenarios. In this chapter, the interference mitigation researches are correspondingly divided into two parts for the uplink and downlink, which will be fit for the UL/DL separation scenario discussed above.

Uplink Power Control

For the uplink of the heterogeneous SCN, the existing LTE uplink power control scheme adopts the method of combining the open-loop and closed-loop power control approaches. The open-loop power control approach is based on the path loss between the users and the eNB to set up an initial uplink transmission power. The close-loop power control approach is designed with the constantly information exchange with eNBs to fine-tune the power for adaptation to the varying channel conditions.

But for the dual-link connection condition in the UL/DL separation mode, there still exists the following problems in the LTE uplink power control scenario. According to current 3GPP TR36.932 specifications, the backhaul between the Macro Cell eNB and Small Cell eNB is non-ideal. Thus, the connection by the users between the Macro Cell eNB in downlink and Small Cell eNB in uplink is via the non-ideal backhaul link. When the serving Macro Cell eNB in the downlink need to exchange the information through the serving Small Cell eNB in the uplink, there will be no prompt path loss information in the uplink that could be used for supporting the uplink closed-loop power control to adjust the parameters. Therefore, the current power control scheme in the LTE uplink will be no longer accurate for the UL/DL separation scenario.

Aiming at the above problem, the uplink power control strategy for the UL/DL separation scenario will be researched in this chapter.

Downlink Power Allocation

For the downlink of the heterogeneous SCN, since the LTE downlink adopts the OFDMA technology, the downlink transmission signals for different users in each cell are orthogonal to each other, so there will be no necessity of power control for overcoming the distance effect. In order to cope with the path loss and shadow fading of different users in the network, LTE can flexibly schedule the radio resources through the frequency domain to avoid the potential interferences. In addition, the adoption of the power control in downlink would disrupt the measurement of Channel Quality Indication (CQI) in the downlink, thus affecting the accuracy of the downlink scheduling. Therefore, LTE network did not adopt the power control approach in the downlink, but only adopting a static or semi-static power allocation (Wang, 2010). The purpose of the power allocation in downlink is to minimize the transmission power on the premise of meeting the requirements of the user reception quality and reducing the interferences. In the heterogeneous SCN, because the different cell eNBs have different transmission power limitations and different user QoS requirements, the efficiency of the downlink power allocation is very important to the system performances of the throughput and Energy Efficiency. This will also be the research focus in this chapter.

Main Focuses of this Chapter

Therefore, in order to solve above problems, this chapter focuses mainly on researches of Game Theory based uplink power control and downlink power allocation strategy for the interference mitigation of the heterogeneous SCN. The research targets are improving the system throughput and Energy Efficiency. For the uplink power control scenario, this chapter proposes the non-cooperative game model based power control algorithm with the convex pricing function, which can optimize the initial transmission power of both Macro Cell and Small Cell users through the Nash Equilibrium solution. The existence and uniqueness of the Nash Equilibrium for the proposed game model are verified, which provide the feasible solution for the cross-tier power control in the heterogeneous SCN. The proposed scheme can bring a significant increase in the system throughput, reduce the outage probability as well as enhance the energy efficiency over current works.

For the downlink power allocation with multiple services in the SCN, the non-cooperative game based scheme is proposed to optimize the transmission power with constraints of different QoS requirements. The corresponding Nash Equilibrium is solved with the verifications of its existence and uniqueness. The system Energy Efficiency performances are improved with constraints of guaranteeing different users QoS requirements.

LITERATURE REVIEW

In order to cope with the above uplink power control and downlink power allocation problems, which are the key challenges for the heterogeneous SCN deployment, there have been several researches focused on these areas so far.

Uplink Power Control

About the uplink power control strategy, a class of energy-efficient sleep mode algorithms for Small Cells is introduced in (Ashraf, Boccardi, & Ho, 2011) and (Claussen, Ashraf, & Ho, 2010). However, there are no priorities in interference handling. 3GPP proposes a utility-based distributed algorithm to allocate resources (3GPP 2013). The interferences are managed by a decentralized approach based on multi-armed bandit solutions.

The game theory has also been applied in the uplink power control scenario of wireless networks in (Chandrasekhar, Andrews, Muharemovict, Shen, & Gatherer, 2009), (Ahmed, Peng, Ahmad, Li, & Abana, 2013), (Varma, Mhiri, Treust, Lasaulce, & Samet), (Hkimi, Langar, Secci, Boutaba, & Pujolle, 2013) and (Li, Zheng, Su, & Wen, 2013). In (Chandrasekhar et al., 2009), a power control scheme based on game theory is studied within the HetNet deployments. This algorithm makes each femto cell maximize its individual utility with a SINR based reward less than an incurred cost. In (Ahmed et al, 2013), two-tier Macro and Femto cell networks are modeled via the Stackelberg game, where the macro cell imposes price on each Femto cell who causes interference to protect users. An algorithm is proposed based on the Stackelberg game through the convex optimization method. In (Varma, Mhiri, Treust, Lasaulce, & Samet, 2013), a solution is proposed for a relevant game with a new energy efficiency metric which considers cross-layer approaches. The two-tier networks are deployed with the Macro cell with co-channel femto cells. The proposed algorithm takes into account of the effects about the presence of a queue with a finite length at the transmitter. For the cooperative femto cell network scenario, a joint resource and power allocation strategy is proposed for the interference managements. The resource and power allocation problem is modeled as an operations research game. In this game, imputations are inferred form cooperative game theory, using utility components results of partial optimization (Hkimi et al., 2013).

In (Li et al, 2013), a distributed power control scheme for the uplink scenario is proposed in two-tier macro and femto cell networks. Single-channel and multiple-channel femto cell networks are considered in this scenario. In single-channel scenario, each user has a price that reflects other users' compensations. Power levels and prices can be updated with an asynchronous distributed power control algorithm. Furthermore, by increasing the users' prices, a cellular link protection scheme is also proposed for protecting the users. For the multiple-channel scenario, a Stackelberg game model is proposed, in which each user prices its limited power then sells it to subchannels.

In (Tsiropoulou, Katsinis, & Papavassiliou, 2013), a convex pricing model of users' uplink transmission power of mobile networks supporting multiple services is established for the efficient distributed power control. A utility function is established for each user and this utility function represents the expected trade-off between QoS-aware actual uplink throughput and the corresponding uplink power consumption. This scheme firstly formulates a multi-service uplink power control game model, where each user maximizes his utility-based performance in some imposed physical limitations and it is verified that there has a unique Nash equilibrium. Then, this scheme introduces a usage-based convex pricing policy of the transmission power. This scheme offers a more effective method compared to the linear pricing schemes. Finally, a multiservice uplink power control game with convex pricing is formulated and its Pareto optimal Nash equilibrium is determined.

In (Tsiropoulou et al., 2013), a problem of efficient power control scheme in the uplink of two-tier macro and femto cell networks is addressed. Two different dimensions, the corresponding constraints and properties are considered. The network utility maximization theory is used for a common framework. Each user is associated with a defined QoS-aware utility function. Macro cell users are given higher priority to mitigate the inter-cell/cross-tier interferences and the femto cell users will be penalized through a convex cost function. This work formulates the problem as a non-cooperative game, and solves it through the supermodel theory towards determining the Nash Equilibrium. In order to obtain the game's Nash Equilibrium solution, a distributed iterative algorithm is proposed.

Downlink Power Allocation

For the downlink power allocation problem, a novel power allocation algorithm based on non-cooperative game model is proposed (Tian, Liu, & Li, 2012). The downlink power allocation scheme in OFDMA cellular systems uses the pricing mechanism of the non-cooperative game theory. The data rate is set as the pricing factor, and the objective function is the variance of transmission power. The genetic algorithm is applied to obtain the optimal solution. In (Fu et al., 2014), a distributed transmission scheduling algorithm and a centralized transmission scheduling algorithm are proposed to choose cell-edge users and cooperative eNBs for each physical resource block. This work further proposes a non-cooperative power allocation game in order to adjust the initial power allocation to reduce interferences. In (Zhang & Zhang, 2008), a power allocation scheme is proposed for downlink OFDMA mobile network based on cooperative game theory. The users' data throughput is firstly defined as a revenue function. Then, in order to maximize the aggregate payoff of the users, the source allocation is modeled as a cooperative game. By assuming an equal power allocation, a suboptimal subcarrier allocation is proposed to achieve the Nash bargaining solution. An optimal power allocation algorithm can ensure QoS requirements of users by maximizing the sum payoff.

In (Pan, Nix, & Beach, 2008), a distributed resources (spectrum and power) sharing algorithm is proposed based on cooperative game for downlink transmission in OFDMA-based system. Both the system capacity and user fairness are improved. By considering resource allocation for both intra- and inter-coalitions, an optimal resource allocation algorithm is proposed. A non-cooperative game model and a cooperative game model are established for the inter-coalition resource allocation, where the cooperative game is solved by a nonsymmetrical Nash bargaining solution.

Game Theory

The Game theory is now developed into a sophisticated theoretical system, which can be applied to many fields of economics, political sciences and public choices. Game theory has its own basic model, where a complete game model requires at least three elements, namely Player, Strategy Set and Utility Function, as the game processes have been expressed in (Liu et al., 2013) and so on.

Player, is the game's main decision-maker, developing the game strategies. Players are different in different game models, they can be either an individual or a group, but these groups must participate in the game for a common goal or interest. For the uplink power control problem addressed in this chapter, players are the UL/DL separation users locating on the small cell edges. For the downlink power allocation strategies, players, to offer services for the same user, are the one or several small cell eNBs.

Strategy Set, is the strategy set of the players. It is also the method or the value that each player can choose during the process of decision-making. In different game models, the Strategy Set is also different. In this chapter, it is a condition to set transmission power of users in UL or small cell eNBs in DL.

Utility Function, is the optimization goal of the players. Utility function is generally composed of two parts, namely, revenue function and cost function, to represent the gain and loss for the Strategy Sets. The former represents the interests or revenue of the players, while the latter represents the punishment of the players. The cost function could be modeled as linear or non-linear, the specific choice is determined according to the actual different situations. In this chapter, in the uplink power control scenario, the non-linear pricing function is explored. While in the downlink power allocation scenario, the linear pricing function is implemented.

In addition, the game model category can be divided into cooperative game and non-cooperative game in general case. The differences between the two lie in whether the players have a binding agreement, if the agreement is possessed, it is the cooperative game. Otherwise, it will be a non-cooperative game. In this chapter, both the uplink power control strategy and downlink power allocation strategy are considered with the non-cooperative game model.

Nash Equilibrium, namely non-cooperative game equilibrium, is an important concept in the Game Theory named after John Nash. Nash published two important articles about non-cooperative games in 1950 and in 1951. These two articles define the non-cooperative game process and its equilibrium, proving the existence of equilibrium, therefore laying the foundation for non-cooperative game theory.

Definition of Nash Equilibrium (Ma, Lv, & Lu, 2013):

If the vector $\boldsymbol{a}^* = (a_1^*, a_2^*, ..., a_n^*)$ in the strategy space is a Nash equilibrium, then no decision-makers can unilaterally improve its own utility, namely $u_i(a_i^*, \boldsymbol{a}_{-i}^*) \geq u_i(a_i, \boldsymbol{a}_{-i}^*)$, $\forall i \in N, \forall a_i \in S_i$, $\boldsymbol{a}_{-i}^* = (a_1^*, ..., a_{i-1}^*, a_{i+1}^*, ..., a_N^*)$.

Nash equilibrium is a stable solution of a game, that is, no decision-makers can improve their utility through their own decisions. In other words, if the decision-makers change the strategies at this time, its own utility will also be reduced. In Nash equilibrium, each rational decision-maker does not individually change strategies. Nash equilibrium provides a very important analytical method, making game theory research get meaningful results.

GAME THEORY BASED UPLINK POWER CONTROL AND DOWNLINK POWER ALLOCATION STRATEGY

As mentioned in the Introduction, the dual connection supported by the UL/DL separation mode could solve the interference problem caused by the power imbalances of high Macro Cell transmission power to the Small Cell edge users, especially in the case of intra-frequency heterogeneous deployment scenarios. The typical deployment scenario for the user dual connectivity with UL/DL separation mode in the heterogeneous SCN is shown as Figure 1.

Based on the above deployment scenario, the strategies of uplink power control and downlink power allocation for the heterogeneous SCN will be described as follows with the system model, Game Theory approaches implementation, power control/power allocation algorithm designs and the performance evaluations.

Figure 1. The deployment of UL/DL separated small cell edge users

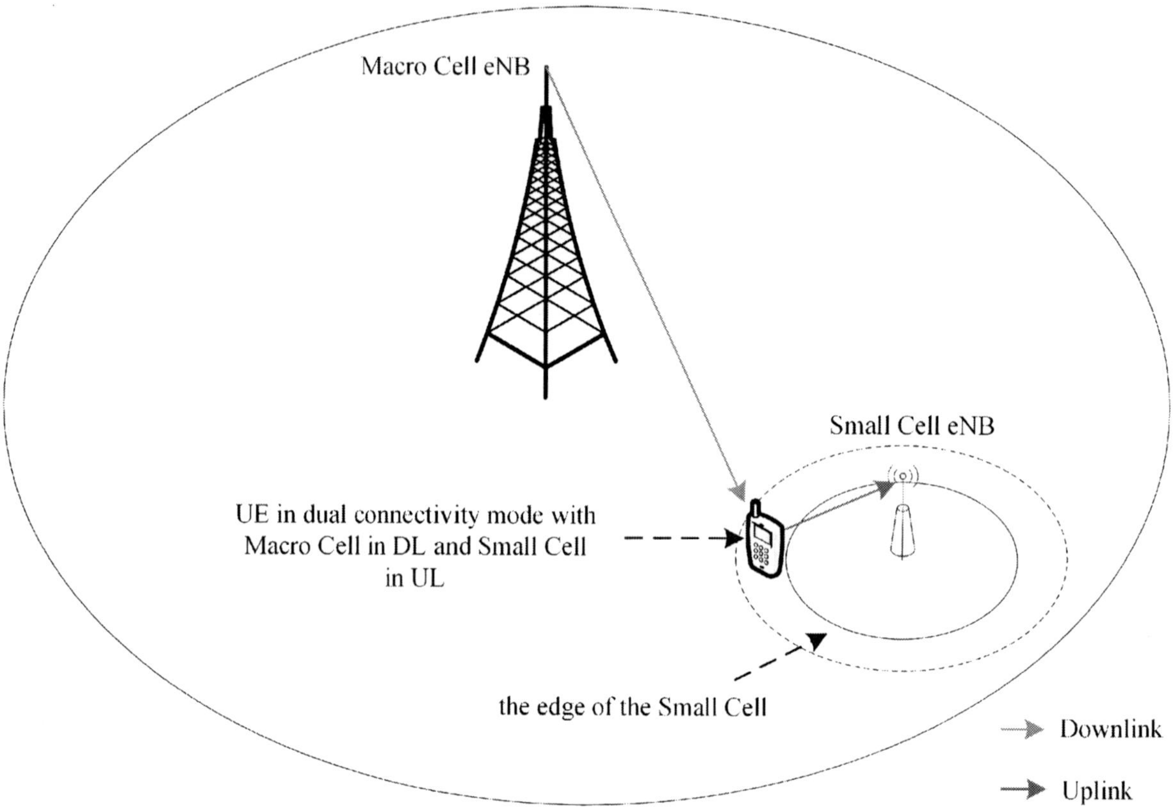

System Model

Assuming the Macro Cells and Small Cells are deployed with the intra-frequency mode, the cross-tier interferences are considered with the deployment scenario of several Small Cells located within the coverage of each Macro Cell. The coverage radius of Macro Cell is R_m, the coverage radius of Small Cell is R_s.

Uplink Scenario

According to the 3GPP specifications, assuming that there is Inter Cell Interference Coordination (ICIC) scheme implemented in the Macro Cell tier, such as Fractional Frequency Reuse (FFR), the interferences between different Macro Cells can be sufficiently avoided (Xu, Li, Yang, & Zhu, 2012). Without loss of generality, this chapter will concern one typical Macro Cell with multiple Small Cell clusters inside of this Macro Cell coverage. As the Small Cells will share the spectrum with the co-located Macro Cell, if the cross-tier interferences inside one Macro Cell could be successfully solved, the interferences in the whole network will be mitigated with the aid of ICIC schemes applied in the Macro Cell tier.

For the uplink scenario, the users are deployed in the Macro Cell coverage, while the users within the UL/DL separation mode are deployed in the Small Cell coverage with CRE region enabled, as shown in Figure 2.

Figure 2. Uplink system model for the UL/DL separation mode users in the small cell edge with CRE

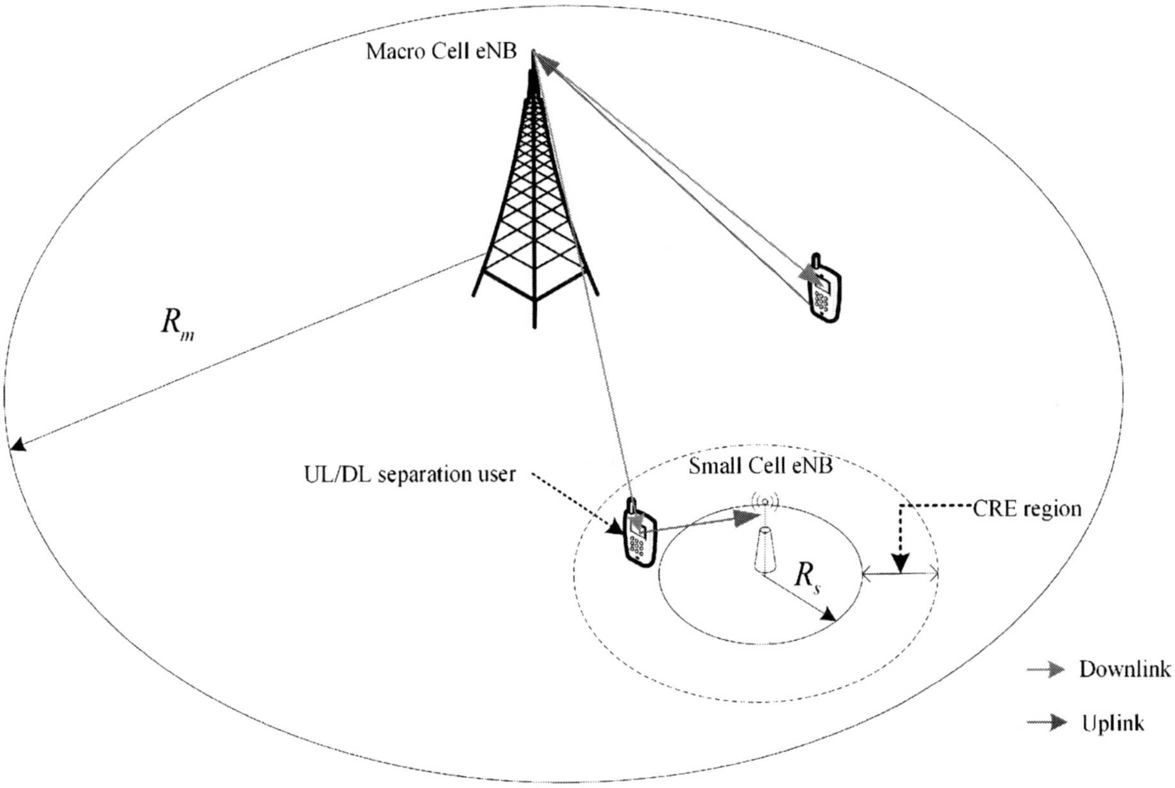

The users are denoted by $i \in \{0, 1, 2, ..., N\}$, the user No.0 is especially reserved for the representation of the user served by the Macro Cell, the remaining N represent the users located in the Small Cell coverage. Each user's transmission power is denoted by p_i, Additive Gaussian white noise variance yields to the users is denoted by σ^2. The channel gain factor is denoted by $g_{i,j}$, indicating the channel gain from the user i to the eNB j. For simplicity, this chapter does not consider the Rayleigh Fading and Shadow Fading. Then the Signal to Interference and Noise Ratio (SINR) that each eNB receives from the user in the uplink is described as:

$$\gamma_i = \frac{p_i g_{i,i}}{I_i + \sigma^2} = \frac{p_i g_{i,i}}{\sum_{j \neq i} p_j g_{i,j} + \sigma^2} \tag{1}$$

where $I_i = \sum_{j \neq i} p_j g_{i,j}$ is the interference measured by the Small Cell associated with the user i.

Downlink Scenario

Besides the same basic cellular deployment scenario with the uplink scenario, the downlink users distributed in the coverage of the Macro Cells and Small Cells are shown in Figure 3.

Figure 3. Downlink system model of the heterogeneous SCN

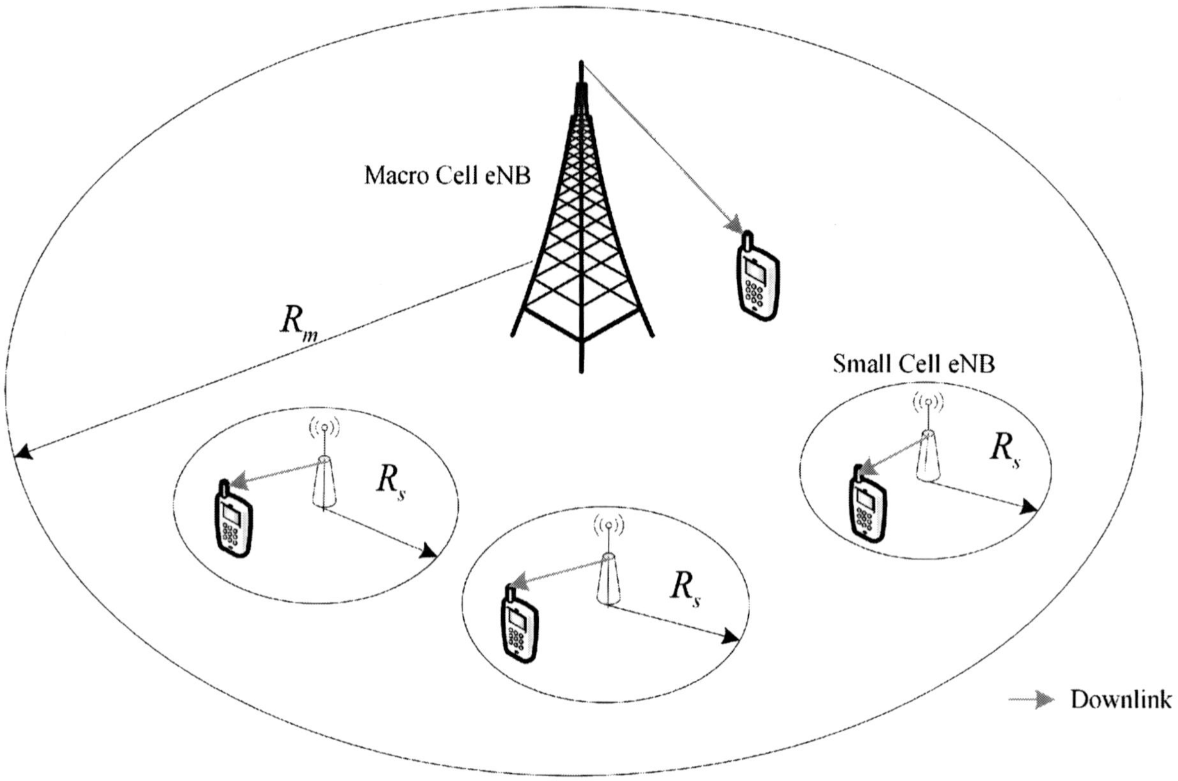

In the downlink scenario, the Macro Cell and Small Cell eNBs are denoted as $k \in \{1, 2, ..., N\}$, the former N_1 represent the Macro Cell eNBs, the remaining $(N - N_1)$ represent the Small Cell eNBs. Correspondingly, the users deployed by each eNB are represented respectively by $j \in \{1, 2, ..., N\}$, the former N_1 represent the users of the Macro Cell eNBs, and the remaining $(N - N_1)$ represent the users with association of the Small Cell eNBs. The transmission power of the eNB for the user k is expressed by p_k, the variance of the additive Gaussian white noise suffered by users is expressed by σ^2. The channel gain factor is $g_{k,j}$, referring to the channel gain from the eNB j to the user k. As the same assumptions with the uplink system model, the Rayleigh Fading and Shadow Fading are not included in this chapter. Then each user receives the downlink SINR sent by its' service eNB k is

$$\gamma_k = \frac{p_k g_{k,k}}{I_k + \sigma^2} = \frac{p_k g_{k,k}}{\sum_{j \neq k} p_j g_{k,j} + \sigma^2} \tag{2}$$

where $I_k = \sum_{j \neq k} p_j g_{k,j}$ is the interference measured by the user k in downlink.

Game Theory Based Solution for Uplink and Downlink Scenarios

Uplink Power Control with Non-cooperative Game Model

According to above system models, the uplink power control strategy for interference mitigation will be researched via the Game Theory approach with the transmission power limitations. The convex pricing based non-cooperative game model is applied for the Macro Cell and Small Cell overlaid heterogeneous deployment downlink power control. The convex pricing is designed as the exponential pricing function of each cells' transmission power, which can also reflect the cross-tier interference level.

Furthermore, in the overlaid deployment of the Macro Cell and Small Cell, the main responsibility for the Small Cells will be maximally offload the higher-QoS-requirement services while the Macro Cell mainly handles the high-mobility users or the guaranteed lower-QoS-requirement services. Therefore, this chapter has the assumption that the users served by the Macro Cell will have more priority in the power control based interference mitigation via the non-cooperative game model.

For there is only one Macro Cell eNB and *N* Small Cell eNBs in the coverages area, the uplink users could form a *N+1* non-cooperative power control game model composed of one user in the Macro Cell eNB and *N* users in Small Cell coverages.

$$G = [N, \{p_i\}, \{u_i(p_i, \gamma_i | \mathbf{p}_{-i})\}] \tag{3}$$

where:

- $N = \{0, 1, 2, ..., N\}$represents the set of Macro Cell user and Small Cell users;
- $\{p_i\} = \{p_i \in [0, p_{max}]\}$ represents the transmission power limitations of the user i, $p_{max} > 0$ is the maximum transmission power;
- $u_i(p_i, \gamma_i | \mathbf{p}_{-i})$ represents the utility function of the user i, where γ_i is the SINR of the user i, $\mathbf{p}_{-i}$ is the transmission power of other users in addition to the user i.

The user's utility function is composed of the revenue function and the cost function (Hkimi et al., 2013) (Li et al., 2013), which is defined as:

$$u_i(p_i, \gamma_i | \mathbf{p}_{-i}) = R(\gamma_i, \Gamma_i) - C(p_i, \mathbf{p}_{-i}) \tag{4}$$

where $R(\gamma_i, \Gamma_i)$ represents the revenue function of the user i, $C(p_i, \mathbf{p}_{-i})$ represents the corresponding pricing function.

$R(\gamma_i, \Gamma_i)$ is defined as (Li et al., 2013):

$$R(\gamma_i, \Gamma_i) = 1 - e^{-a_i(\gamma_i - \Gamma_i)} \tag{5}$$

The revenue function means when $\gamma_i \leq \Gamma_i$ declines rapidly with the decrease of γ_i. When $\gamma_i > \Gamma_i$, the value of the revenue function will rise slowly with the increase of γ_i. Γ_i represents the user's target SINR. a_i is a fixed parameter that has a positive value.

The pricing function $C(p_i, \mathbf{p}_{-i})$ is defined as (Liu et al., 2013):

$$C(p_i, \mathbf{p}_{-i}) = \frac{b_i g_{0,i}}{I_i}(e^{p_i} - 1) \tag{6}$$

The pricing function means the larger the transmission power of the user i, the greater the interference to other users will be, the more punishments the user utility function will be suffered. The pricing function is designed as the non-linear function of the transmission power p_i for the user i. The parameter b_i is a fixed parameter that has the positive value. $\frac{g_{0,i}}{I_i}$ reflects the interferences of the user i to the Macro Cell users.

The goal of the game process is to maximize the utility function of every user when the game reaches the Nash Equilibrium:

$$\max_{0 \le p_i \le p_{\max}} u_i(p_i, \gamma_i \mid \mathbf{p}_{-i}), \ \forall i = 0,1,2,\ldots,N \tag{7}$$

When the game finally reaches the Nash Equilibrium, each user's transmission power will be defined as $\mathbf{p}^* = (p_1^*, p_2^*, \ldots, p_N^*)$, where

$$u_i(p_i^*, \gamma_i^* \mid \mathbf{p}_{-i}^*) \ge u_i(p_i, \gamma_i \mid \mathbf{p}_{-i}), \ \forall i = 0,1,2,\ldots,N \tag{8}$$

Then, $\mathbf{p}^* = (p_1^*, p_2^*, \ldots, p_N^*)$ is the Nash Equilibrium solution of the proposed game model. Here will be the proofs for the existence and uniqueness of the corresponding Nash Equilibrium.

The Existence of the Nash Equilibrium

Theorem 1 (Liu et al., 2013), (Yates, 1995)**:**

For the propose game model $G = [N, \{p_i\}, \{u_i(p_i, \gamma_i \mid \mathbf{p}_{-i})\}]$, $\forall i = 0,1,2,\ldots,\mathrm{N}$, if the Nash Equilibrium of the game exists, only the following two conditions will need to be proved that the game model, which satisfies:

(1) All of the strategy sets p_i is the compact convex set within the Euclid space $\mathbb{R}^N$.
(2) All of the utility function $u_i(p_i)$ is continuous and is the p_i quasi-concave function.

Proof:

(1) Due to the strategy set $p_i = [0, p_{\max}]$, apparently strategy p_i is a compact set. In the one-dimensional space, convex set is a single point or a continuous line, and therefore p_i is a convex set.
Then, the condition 1 is satisfied.
(2) Quasi-concave function's definition:

Function $f(x)$ is defined on a subset D of $\mathbb{R}^N$, if and only if $f(x)$ satisfies the following properties when $f(x)$ is quasi-concave:

$$f(\lambda x_1 + (1-\lambda)x_2) \geq \min(f(x_1), f(x_2)), \qquad \lambda \in [0,1] \tag{9}$$

In other words, the so-called quasi-concave function refers to the curve without the lower convex phenomenon with respect to the horizontal axis. Thus, the upper convex function is quasi-concave.

Due to

$$\frac{\partial u_i}{\partial p_i} = \frac{a_i g_{i,i}}{I_i} e^{-a_i(\gamma_i - \Gamma_i)} - \frac{b_i g_{0,i}}{I_i} e^{p_i} \tag{10}$$

When the fixed parameter is not zero,

$$\frac{\partial^2 u_i}{\partial^2 p_i} = -(\frac{a_i g_{i,i}}{I_i})^2 e^{-a_i(\gamma_i - \Gamma_i)} - \frac{b_i g_{0,i}}{I_i} e^{p_i} < 0 \tag{11}$$

So, $u_i(p_i)$ is continuous and is the quasi-concave function of p_i.

Thus the existence of the Nash Equilibrium of the proposed game model is proved.

The Uniqueness of the Nash Equilibrium

In the Equation (10), let $\frac{\partial u_i}{\partial p_i} = 0$, namely

$$\frac{\partial u_i}{\partial p_i} = \frac{a_i g_{i,i}}{I_i} e^{-a_i(\gamma_i - \Gamma_i)} - \frac{b_i g_{0,i}}{I_i} e^{p_i} = 0 \tag{12}$$

Getting

$$p_i = \min \begin{Bmatrix} [\frac{I_i}{I_i + a_i g_{i,i}}(a_i \Gamma_i + \ln \frac{a_i g_{i,i}}{b_i g_{0,i}})]^+ \\ p_{\max} \end{Bmatrix} \tag{13}$$

Then the iterative formula will be:

$$p_i^{(k+1)} = \min \begin{Bmatrix} [\frac{p_i^{(k)}}{p_i^{(k)} + a_i \gamma_i^{(k)}}(a_i \Gamma_i + \ln \frac{a_i g_{i,i}}{b_i g_{0,i}})]^+ \\ p_{\max} \end{Bmatrix} \tag{14}$$

In order to provide a general framework for showing the convergence of the power control process, the standard interference function $p = f(p)$ will be introduced. One function is proved to be standard if it satisfies the following properties:

(1) Non-negativity: $f(p) > 0$.
(2) Monotony: if $p \geq p'$, then $f(p) \geq f(p')$.
(3) Scalability: For all of the $\alpha > 1$, then $\alpha f(p)\ f(\alpha \mathrm{p})$.

Theorem 2 (Liu et al., 2013), (Yates, 1995)**:**

If the function $f(p)$ is standard, and if the iteration function $p^{(k+1)} = f(p^{(k)})$ is convergent, it will converge to the unique fixed point. The proof are listed as follows.

(1) Assuming

$$f_i(p) = \min \begin{cases} [\dfrac{p_i^{(k)}}{p_i^{(k)} + a_i \gamma_i^{(k)}}(a_i \Gamma_i + \ln \dfrac{a_i g_{i,i}}{b_i g_{0,i}})]^+ \\ p_{\max} \end{cases} \tag{15}$$

Then obviously, $p_i^* > 0$ so $f(p) > 0$ is satisfied.

(2) Equation (15) could be converted to

$$f_i(p) = \min \begin{cases} [\dfrac{1}{1 + a_i \dfrac{\gamma_i^{(k)}}{p_i^{(k)}}}(a_i \Gamma_i + \ln \dfrac{a_i g_{i,i}}{b_i g_{0,i}})]^+ \\ p_{\max} \end{cases} \tag{16}$$

Because if $p \geq p'$, then $\dfrac{\gamma}{p} \leq \dfrac{\gamma}{p'}$, $\dfrac{1}{1 + a\dfrac{\gamma}{p}} \geq \dfrac{1}{1 + a\dfrac{\gamma}{p'}}$. Therefore, $f_i(p) \geq f_i(p')$ and the monotonicity is satisfied.

(3) In order to prove the third feature, a function $F(\alpha) = \alpha f_i(p) - f_i(\alpha p)$ is defined, whose first-order partial derivative with respect to α is:

$$\frac{\partial F(\alpha)}{\partial \alpha} = \frac{\alpha(\alpha - 1)[(\alpha + 1)p + 2\gamma]p\gamma}{(p + \gamma)^2 (\alpha p + \gamma)^2}(a_i \Gamma_i + \ln \frac{a_i g_{i,i}}{b_i g_{0,i}}) \tag{17}$$

To all of the $\alpha > 1$, $\frac{\partial F(\alpha)}{\partial \alpha} > 0$ and $F(1) = 0$, $F(\alpha) > F(1)$ can be obtained. This means to all of the $\alpha > 1$, $\alpha f_i(p) > f_i(\alpha p)$.Therefore, the scalability property is satisfied.

The proofs of the Theorem 2 are completed.

Thus the uniqueness of the Nash Equilibrium solution of the proposed non-cooperative game model has been proved.

Initial Transmission Power Adjustment: Dynamic Power Adjust Algorithm

Based on the above Nash Equilibrium solution obtained for the proposed non-cooperative game model based uplink power control process, this chapter proposes the Dynamic Power Adjust (DPA) algorithm to adjust the transmission power for each user with the power limitation p_{max}, which could further optimize the performance of the users.

The steps of the DPA algorithm are described as follows:

Step1: After the dual connectivity users associate with the uplink Small Cell eNB, the Small Cell eNB assigns the resource information blocks to the user and interacts with the corresponding co-located Macro Cell. The following steps need to be completed before the uplink Small Cell eNBs of the users allocate the resource blocks to the user.

Step1-1: User-centered deployment scenario with the radius of the focused area *d*, which is shown as the Figure 4.

Step1-2: The downlink Macro Cell eNB locates the users within its coverage area, then the Macro Cell eNB will find several users of the adjacent Macro Cell eNBs who are using the same resource blocks with the focused center-user, represented by M-UE shown in Figure 4. The downlink Macro Cell eNBs measure the interferences that multiple M-UE suffered and then evaluate averaging interferences through $I_{M\_UE\_before}$.

Step1-3: Each adjacent Small Cell eNB can locate its users within the coverage area. Therefore, each of the adjacent Small Cell eNB can identify several users of other Small Cells within the focused area who are using the same resource blocks with the focused center-user, represented by S-UEs. Each adjacent Small Cell eNB measures the interferences that suffer from multiple S-UEs and sends it to the co-located Macro Cell eNB. The Macro Cell eNB evaluates the average interferences through $I_{S\_UE\_before}$ of multiple S-UEs.

Step2: The Small Cell eNB connected with the user in the uplink allocates the resource blocks to the user. The following steps needs to be completed after the allocation.

Step2-1: The Macro Cell eNB measures the downlink interferences that multiple M-UE suffered and evaluate the averaging interference through $I_{M\_UE\_after}$;

Step2-2: Each adjacent Small Cell eNB measures the interferences that suffered from the multiple S-UEs and sends it to the Macro Cell eNB, the Macro Cell eNBs calculate the average interferences through $I_{S\_UE\_after}$ of multiple S-UEs.

Step3: The Macro Cell eNB monitors the downlink access of the user before and after the allocation process, within the defined area of the user-centered with the radius *d*. The changes of the interferences that suffered from the multiple M-UEs are calculated through $\Delta I_{M\_UE} = I_{S\_UE\_after} - I_{S\_UE\_before}$.

Figure 4. Deployment scenario with the user as the center

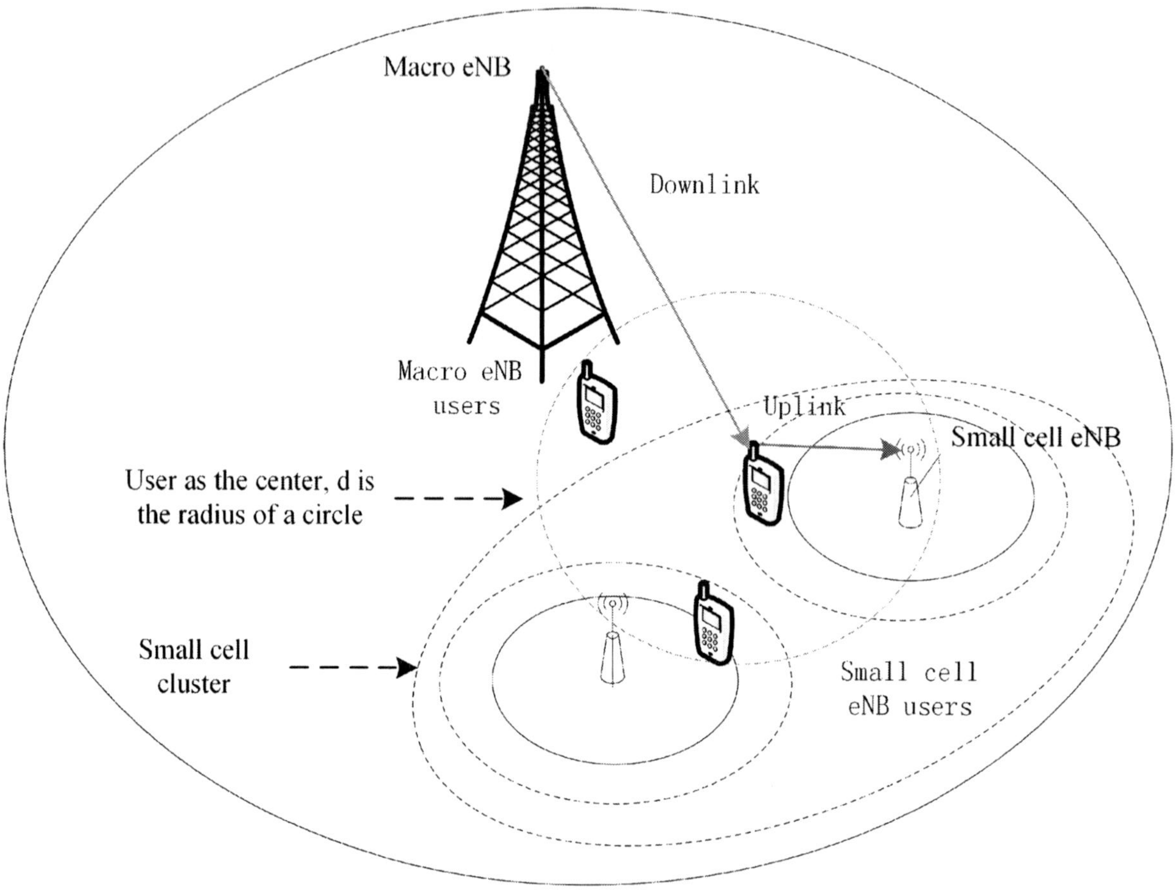

Step4: The Macro Cell eNB monitors the downlink access of the user before and after the allocation process, within the defined area of the user-centered with the radius d, the changes of the interferences that suffered from the multiple S-UEs are calculated through $\Delta I_{M\_UE} = I_{S\_UE\_after} - I_{S\_UE\_before}$.

Step5: The Macro Cell eNB compare the size of ΔI_{M_UE}, ΔI_{S_UE} and thresholds $\Delta I_{threshold}$:

If $\Delta I_{M\_UE} > \Delta I_{threshold}$ and $\Delta I_{S\_UE} > \Delta I_{threshold}$, then it indicates that the transmission power of the user is reduced by one step, the step length is Δ_1;

If either ΔI_{M_UE} or ΔI_{S_UE} is less than $\Delta I_{threshold}$, then it indicates that the transmission power of the user is reduced by one step, the step length is Δ_2;

If $\Delta I_{M\_UE} < \Delta I_{threshold}$ and $\Delta I_{S\_UE} < \Delta I_{threshold}$, then it indicates that the transmission power of the user is reduced by one step, the step length is Δ_3; And $\Delta_1 > \Delta_2 > \Delta_3$.

Step6: The Macro Cell eNB sends the power adjustment command to the user. Then, the user adjusts the transmission power accordingly. The uplink power control strategy is completed.

Downlink Power Allocation with Non-Cooperative Game Model

As described in above system model for the downlink scenario, there are N_1 Macro Cell eNBs and $(N-N_1)$ Small Cell eNBs, which can form non-cooperative game model based power allocation game composed with N_1 Macro Cell eNBs and $(N-N_1)$ Small Cell eNBs in the downlink scenario:

$$G=[N,\{p_k\},\{u_k(p_k,\gamma_k|\,\mathbf{p_{-k}})\}] \tag{18}$$

where:

- $N=\{0, 1, 2, ..., N\}$ represents a set of Macro Cell eNBs and Small Cell eNBs
- $\{p_k\}=\{p_k\,|\,p_k\in[0, p_{max}]\}$ represents the power set of eNB k, $p_{max}>0$ is the maximum transmission power of the eNB.
- $u_k(p_k, \gamma_k\,|\,\mathbf{p}_{-k})$ represents the utility function of eNB k, γ_k is the SINR of user k, $\mathbf{p}_{-k}$ is a set of all other eNBs' transmission power except eNB k.

The utility function of the eNB is composed of revenue function and cost function, which is defined as:

$$u_k(p_k,\gamma_k|\,\mathbf{p_{-k}})=a_k\frac{f(\gamma_k)}{p_k}-b_kp_k \tag{19}$$

where $\frac{f(\gamma_k)}{p_k}$ is the revenue function of eNB, the meaning of which is the Energy Efficiency of the system.

Assuming that each eNB only schedule one user at a moment, the Energy Efficiency of the system is defined as the Energy Efficiency calculated by the scheduled user in downlink. The b_kp_k stands for pricing function of the eNB, he meaning of which is the punishment to the eNB with the increase of eNB's transmission power. The parameters a_k and b_k are both positive constant parameters.

The $f(\gamma_k)$ is defined as:

$$f(\gamma_k)=1-e^{-\frac{\gamma_k}{2}} \tag{20}$$

where the Energy Efficiency of users will increase with the increment of γ_k slowly.

In conclusion, the target of the designed non-cooperative game process is to make the utility function of each eNB reaching the maximum utility:

$$\max_{0\le p_k\le p_{\max}} u_k(p_k,\gamma_k|\,\mathbf{p}_{-k}),\ \forall k=1,2,...,N \tag{21}$$

When each eNB reaches the Nash Equilibrium, the transmission power of each eNB is defined as $\mathbf{p}^*=(p_1^*,p_2^*,...,p_N^*)$, making

$$u_k(p_k^*,\gamma_k^* \mid \mathbf{p}_{-k}^*) \geq u_k(p_k,\gamma_k \mid \mathbf{p}_{-k}),\ \forall k=1,2,...,N \tag{22}$$

where $\mathbf{p}^* = (p_1^*, p_2^*,..., p_N^*)$ is the Nash Equilibrium solution of this game process. The existence and uniqueness of the above Nash Equilibrium will be proved as follows.

The Existence of the Nash Equilibrium Solution

According to the definition of the utility function and pricing function of the proposed game model, this model is a non-cooperative supermodel game. The strategy space S_k is the transmission power of eNBs, i.e. P_k. There exits the theorem about the existence of Nash Equilibrium solution of the supermodel game.

Lemma 1 (Topkks, 1979):

For each player k, P_k is tight and u_k is upper semi-continuity to p_k for each $\mathbf{p}_{-k}$. If a game is a supermodel game, its pure strategy Nash Equilibrium set E is non-empty, and it has minimum equilibrium $\mathbf{p}_S$ and maximum equilibrium $\mathbf{p}_L$, i.e., $\mathbf{p}_S < \mathbf{p} < \mathbf{p}_L$.

This chapter introduces a algorithm of complete nonsynchronous to form a power sequence converged to minimum equilibrium solution $\mathbf{p}_S$.

Assume that eNB k updates its transmission power in real-time according to the time slot $T_k = \{t_{k1}, t_{k2}, t_{k3}...\}$, of which $t_{kl} < t_{k(l+1)}$ and $t_{k0} = 0$ for all eNBs. Let all point-in-time of $T_1 \cup T_2 \cup ... \cup T_N$ by increasing time to form set $T = \{\tau_1, \tau_2, \tau_3...\}$. Assume that there are not two complete coincident point-in-time.

Power Allocation Algorithm (Zhang, Chu, & Wen, 2013)**:**

The flow of the power sequence is:

1) set the initial power value when $t = 0$: $\mathbf{p} = \mathbf{p}(0)$. Let $l = 1$.
2) for all l and $\tau_l \in T$

for all eNBs k and $\tau_l \in T_k$

i. According to $\mathbf{p}(\tau_{l-1})$, calculate $p_k^* = \arg\max_{p_k \in \mathbf{p}_k} u_k(p_k, \gamma_k \mid \mathbf{p}_{-k})$
ii. if $\gamma_k^* \geq \gamma_k^{threshold}$

so $p_k(t_l) = \min(p_k^*, p_{\max})$

else

remove eNB k in this iteration, continue to calculate in next iteration.Theorem 3:

The proposed Power Allocation Algorithm converging to the Nash Equilibrium solution and the minimum equilibrium solution $\mathbf{p}_S$ is the Nash Equilibrium solution.

Reference (Zhou, Liu, Yuan, & Cheng, 2008) proves this theorem. This theorem shows that Nash Equilibrium solution of supermodel game exists and can be reached through the minimum value or maximum value of strategy set of the proposed *Power Allocation Algorithm*. But it's not sure that the minimum equilibrium solution is unique. This chapter will further judge whether there exists a unique solution through comparing solutions of Nash Equilibrium set. The following section will also prove that the $\mathbf{p}_S$ is the optimal solution of Nash Equilibrium set.

The Uniqueness of the Nash Equilibrium Solution

Theorem 4:

If **x**, **y** ∈ E are both Nash Equilibrium of the proposed game model, and **x** ≥ **y**, then $u_k(\mathbf{x}) \leq u_k(\mathbf{y})$ to all k.

Proof:

To all player eNB k in the game, when p_k is a fixed constant, the utility function $u_k = a_k \frac{f(\gamma_k)}{p_k} - b_k p_k$ will decrease with the increment of $\mathbf{p}_{-k}$.

Therefore, because of $\mathbf{x}_{-k} \geq \mathbf{y}_{-k}$, then

$$u_k(x_k, x_{-k}) \leq u_k(x_k, y_{-k}) \tag{23}$$

And **y** is a Nash Equilibrium solution of the proposed game as

$$u_k(x_k, y_{-k}) \leq u_k(y_k, y_{-k}) \tag{24}$$

According to (23), (24),

$$u_k(x) \leq u_k(y) \tag{25}$$

The proof is completed.

Therefore, the existence and uniqueness of Nash Equilibrium solutions of the proposed game model are proved.

According to Theorem 4, the minimum Nash Equilibrium solution brings the highest utility to the system than any other Nash Equilibrium solutions.

Because of $\mathbf{p}_s < \mathbf{p}$ and $\mathbf{p} \in E$, we can get

$$\text{To all } k,\ u_k(\mathbf{p}_S) \geq u_k(\mathbf{p}) \tag{26}$$

So, the non-cooperative game model based approach proposed by this chapter has Nash Equilibrium solutions, and the highest utility solution is the minimum Nash Equilibrium solution.

Performance Evaluation

In this section, the system-level simulation is implemented to evaluate the performance of the proposed Game Theory based uplink power control and downlink power allocation strategies. In order to evaluate the most severe case for interference mitigations of the heterogeneous SCN, a full load and full buffer network environments are established. The detailed simulation assumptions are largely based on 3GPP LTE-Advanced evaluation methodology (3GPP, 2013) for getting more practical results that could be comparable with other evaluations.

Simulation Environment and Parameter Setting

According to the system model, the Macro Cell eNBs and the Small Cell eNBs are deployed as the intra-frequency mode. There are several Macro Cell eNBs exist in the system. Furthermore, there are *M* Small Cell clusters deployed within the coverage of each Macro Cell eNB and *N* Small Cells distributed in each Small Cell cluster where *N* is a variable in the simulation. A 19-hexagonal Macro Cell model with 3 sectors per Macro Cell is employed in the simulation. The Small Cell eNB supports the UL/DL separation users within the area enabled by the Cell Range Extension.

The channel gain factor is denoted as $g_{i,j}$, which refers to the channel gain from the user *j* to the eNB *i*, as shown in the following expression (Hkimi et al., 2013). Where K_{f_i} is the uplink path-loss, K_{f_0} and K_c are the downlink path-loss for the users of Macro Cell eNBs and Small Cell eNBs. The α refers to the path loss factor and the $D_{i,j}$ refers to the distance from the user *j* to the eNB *i*.

$$g_{i,j} = \begin{cases} K_c^{-1}\min(D^{-\alpha_c},1) & i=j=0 \\ K_{f_i}^{-1}\min(D_{i,j}^{-\alpha_{f_i}},1) & i,j>0 \\ K_{f_0}^{-1}\min(D_{0,j}^{-\alpha_{f_0}},1) & i=0, j=0 \\ K_{f_i}^{-1}\min(D_{i,0}^{-\alpha_c},1) & i>0, j=0 \end{cases} \tag{27}$$

In addition, the resource blocks are used as three types of services, including the Signaling Radio Bearers (SRB), Guaranteed Bit Rate (GBR) and non-GBR services. For the specific radio bearer resource, SRB is only used to transmit the Radio Resource Control (RRC) and Non Access Stratum (NAS) messages. GBR means minimum bit rate system needs to guarantee even in the situation of lack of network resources. Non-GBR means services need to bear the demand of reducing rate in the situation of network congestion. And non-GBR bearing can be established for a long time because it doesn't need to occupy fixed network resources. GBR bearing is established when it is need and its priority higher than non-GBR.

This chapter considers the influence of the proportion of different service types for the system performance evaluation. Assume there are *K* users located in each cell randomly, of which K_1 users belong to GBR service, K_2 users belong to non-GBR service. The proportion of GBR service and non-GBR service is $\eta = K_1/K_2$. Therefore, the Energy Efficiency of the system will change with the proportion η. In addition, without loss of the generality, the Macro Cell eNBs will mainly handle GBR service, and Small Cell eNBs mainly support non-GBR service for improving the hot-spot throughput performance.

More simulation parameters are shown in Table 1.

Comparison Algorithms

For the uplink scenario, this chapter proposes the non-corporate game model based uplink power control strategy with non-linear pricing function and corresponding DPA algorithm. The game model proposed in the reference (Zappone, Alfano, Buzzi, & Meo, 2011) adopts the non-corporate power control scheme with linear pricing (NPC-LP), but the game model proposed in this chapter adopts non-linear pricing (NPC-NLP). This section will compare three kinds of algorithms: the non-linear game model based on the proposed DPA algorithm (denoted by NPC-NLP-DPA), non-linear pricing game model (denoted by NPC-NLP) and the linear pricing game model (denoted by NPC-LP).

Table 1. Simulation Parameters

Parameters	Setting
Numbers of Macro Cell eNBs	19
Inter-site distance between the Macro Cell eNB	750m
The number of Small Cell clusters in Macro Cell eNB	10
The number of Small Cells in the clusters	10
The radius of Small Cells in the clusters	150m
CRE Bias	14dB
Maximum transmission power of Macro Cell eNB p_{max}	46dBm
Maximum transmission power of Small Cell eNB p_{max}	30dBm
Minimum SINR threshold value of user $\gamma^{threshold}$	2dB
η	1~10
Γ_i	5dB
Uplink path loss K_{f_i}	37dB
Downlink path loss $K_{f_0} = K_c$	28dB
α_{f_i}	3
$\alpha_c = \alpha_{f_0}$	4
Carrier frequency	2.0GHz
The standard deviation of shadow fading	8dB
The autocorrelation distance of shadow fading	50 m
Noise density	-174dBm/Hz

For the downlink scenario, the non-cooperative game model based power allocation strategy (denoted by NPGP) proposed in this chapter will be compared with Equal Power Allocation (denoted by EPA) as the performance baseline to evaluate the performance.

Simulation Results

The simulation results are presented with 1000 random network instantiations, generated according to the aforementioned simulation environment and parameters. The performance metrics include the user percentages below the target SINR, the user average SINR at the Nash Equilibrium solution, the iterations needed for the solution convergence, the system Energy Efficiency in the situation of different services scale and different numbers of Small Cell deployed when the system is finally balanced.

Figure 5 shows the percentages of users below the target SINR when the UL/DL separation mode applied within the Small Cell CRE region at the Nash Equilibrium solution. From the simulation results comparison of the three algorithms, as the increase of the Small Cells number within the Macro Cell

Figure 5. The percentage of the UL/DL separation users below the target SINR

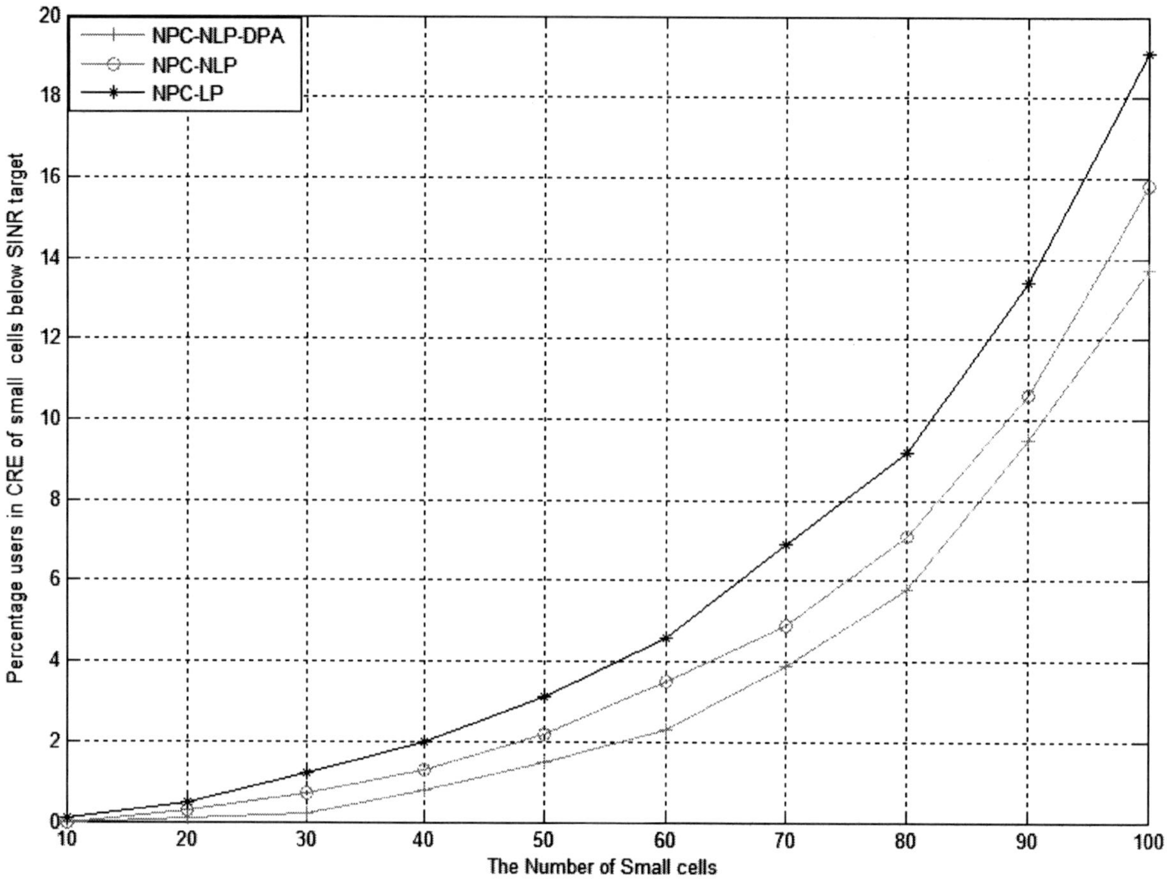

eNB coverages, the percentages of users below the target SINR also increases. This is due to the Small Cell eNBs are at the intra-frequency mode and densely deployed, the interferences between the users in the adjacent Small Cell coverages are increasingly stronger, particularly for the Small Cell edge users. According to the comparison results, it shows that the performance of the proposed non-cooperative game model based power control strategy with non-linear pricing function of DPA algorithm is the best. The percentage of users below the target SINR is the smallest, which indicates that adopting the non-cooperative game model with the non-linear pricing function has the better performance than adopting the linear pricing function. As the non-linear pricing function adopted in this chapter is the exponential function of the user's transmission power, when the user transmission power increases, the interferences to other users will also increase. Thus the penalizations to the users will grow exponentially, which will make a tradeoff of the user transmission power growth.

Figure 6 shows the average SINR when the UL/DL separation users in the Small Cell CRE region are at the Nash Equilibrium solution. From the simulation results of the three comparison algorithms, as the increase of the Small Cell number within the coverage of the co-located Macro Cells, the user average SINR decreases. The reason is mentioned above that the interferences between the users are more and more consistent under the situation of intra-frequency mode and dense Small Cell deployment. Similarly, according to the comparisons, it shows that the performance of the proposed non-cooperative

Figure 6. SINR of the UL/DL separation users at the Nash equilibrium solution

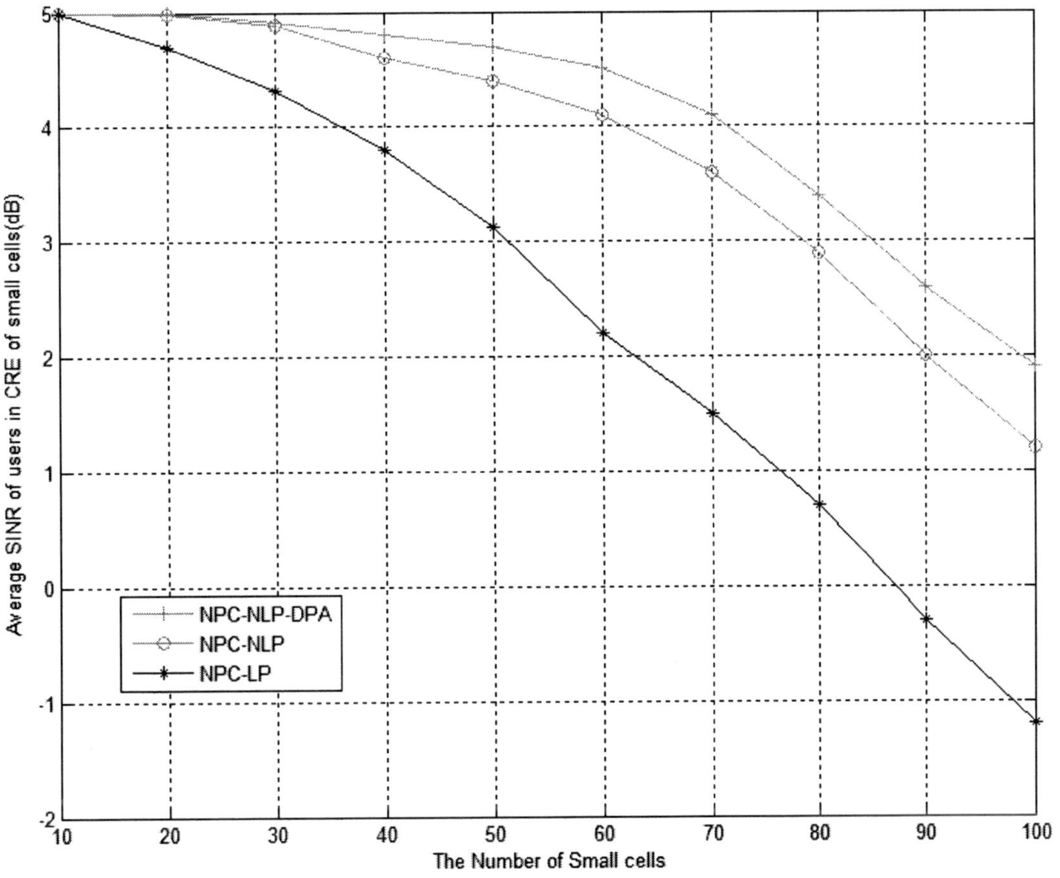

game model based power control strategy with the non-linear pricing function is the best. The performance of the linear pricing game model is the worst. For example, within the coverage of the Macro Cell eNBs, the number of the Small Cells reaches 100. The DPA algorithm makes the user average SINR as 2dB, and the NPC-NLP makes the average SINR of the users as about 1dB, but the NPC-LP makes the user average SINR below 0dB. The implementation of the DPA algorithm doubles the performance improvement of the proposed game model.

Furthermore, Figure 7 describes the impact of the implementation complexity of the DPA algorithm from the iteration of the game solution mentioned in this chapter. It can be seen from the simulation result, as the Small Cell numbers within the Macro Cell coverage increases, the required number of the iterations increases rapidly. The proposed DPA algorithm could adjust the initial value of the user transmission power, thus effectively reducing the iterations number. For instance, when the number of the Small Cell eNBs within the Macro Cell coverage reaches 50, the iteration number required by the DPA algorithm reduces by about one tenth.

Figure 7. The iterations of the algorithm

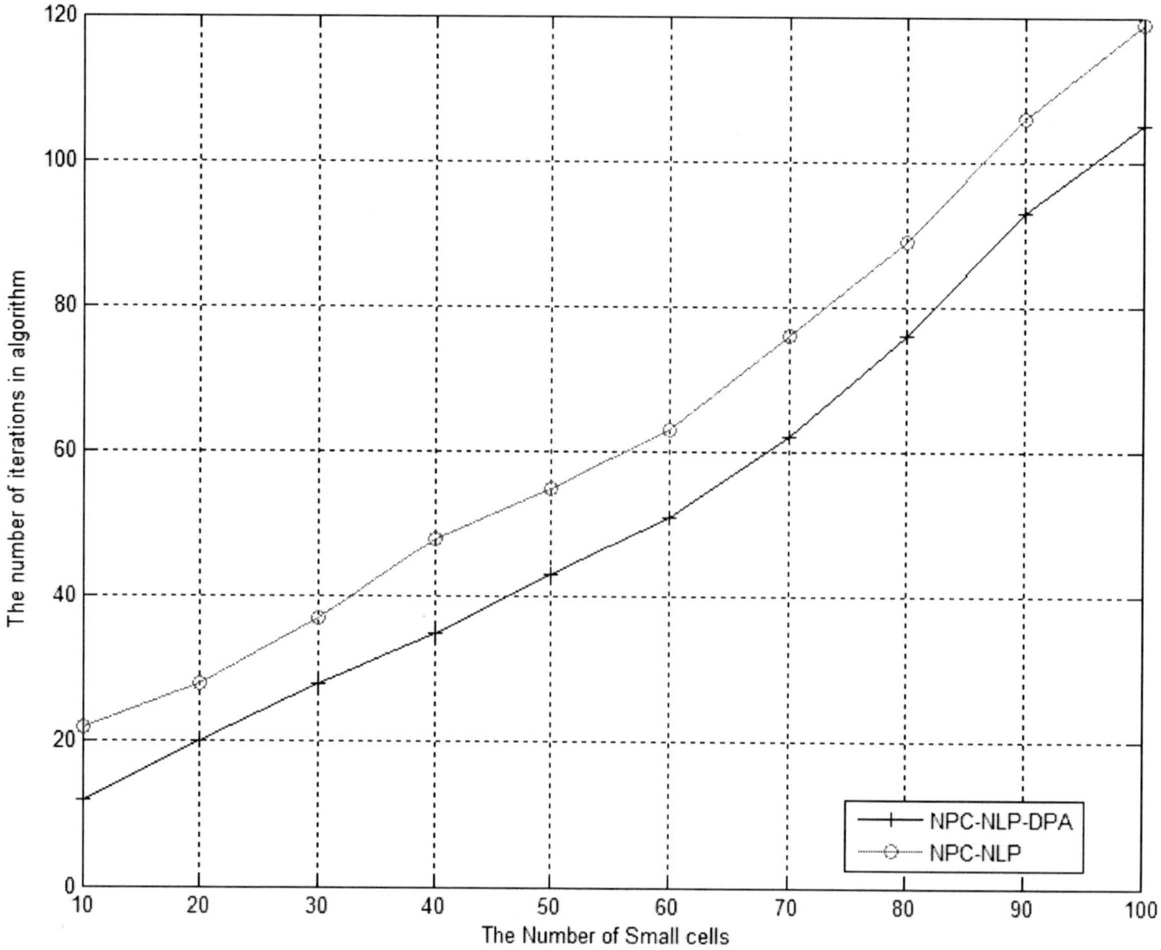

For the downlink scenario, the simulations results about the system Energy Efficiency and the average transmission power are provided. Among the simulation results, Figure 8 and Figure 9 show the changes of system Energy Efficiency and average transmission power of eNB when the independent variable is set as the user service scale η in the fixed Small Cell number situation.

Figure 8 shows the system energy efficiency and different user services ratio when the Small Cell number is set as 200. It means that there are 10 Small Cell clusters in each Macro Cell and 10 Small Cells in each cluster. The ratio of GBR service user and non-GBR service user in the deployed area changes from 1 to 10. Regardless of NPGP algorithm or EPA algorithm, both the system energy efficiency decrease with the increase of η. As a case study of NPGP algorithm, when η changes from 1 to 10, the system energy efficiency reduces from 10^8 to 10^7. As shown previously, GBR service need the system ensures bearing minimum bit ratio, even in the situation that lacks of network resources, corresponding bit ratio will be maintained. On the contrary, non-GBR service doesn't need to occupy fixed network resources, which can be established for a long time. So, the increase of GBR service user will result in the decline of system energy efficiency. In addition, it can be seen that NPGP algorithm have a higher energy efficiency than EPA algorithm with all the value of η.

Figure 8. The system energy efficiency vs. η when the small cell number as 200

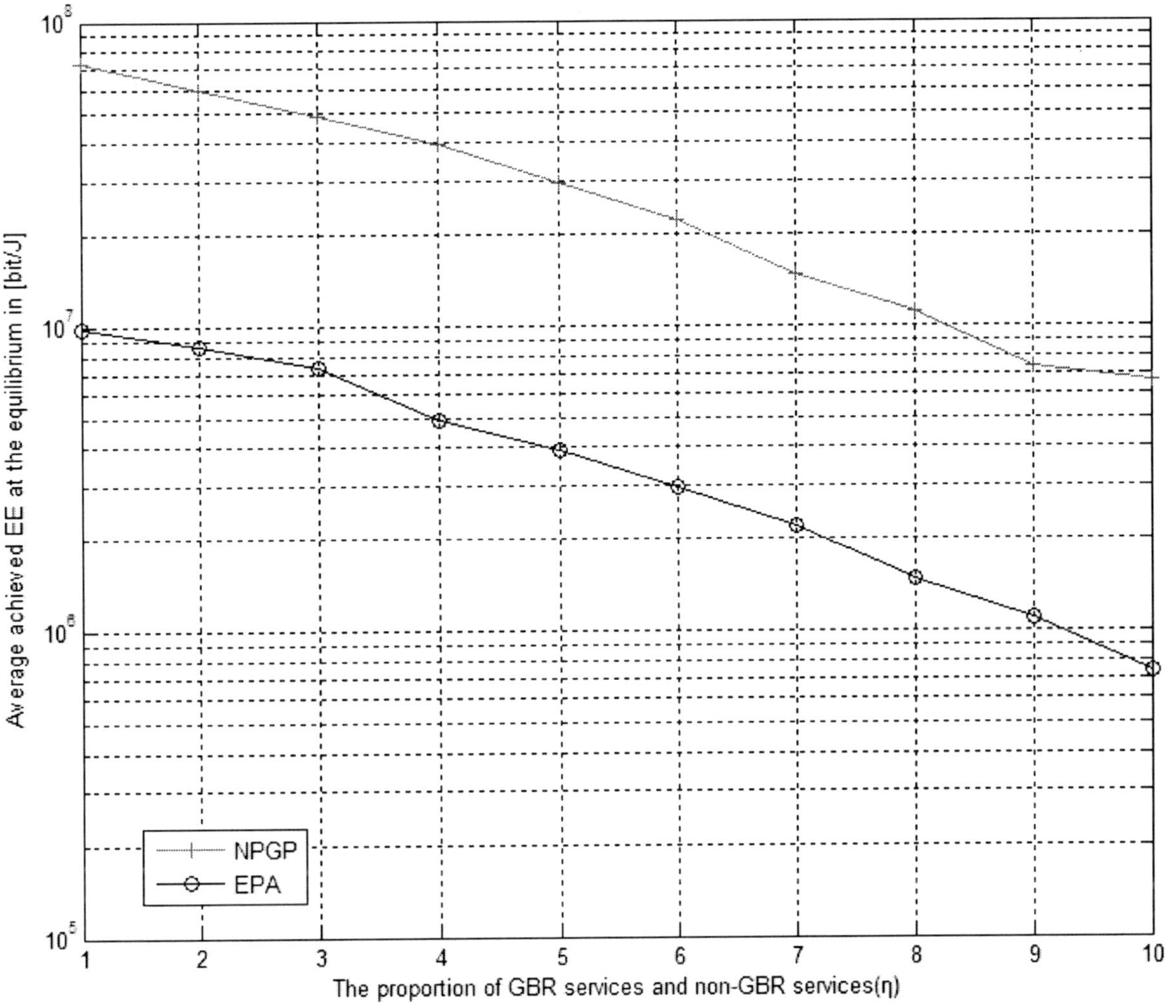

Figure 9 shows the average transmission power of the Macro Cell eNBs, Small Cell eNBs and service ratio when the Small Cell number is set as 200. Comparing with the EPA algorithm, with the increasing of service ratio η, NPGP algorithm makes average transmission power of Macro Cell eNBs increase, and average transmission power of Small Cell eNBs decrease. Specially, when η changes from 1 to 10, average transmission power of Macro Cell eNBs increases about 5.7dBm. The average transmission power of Small Cell eNBs decreases about 5.6dBm. The flexible adjusting of the average transmission power of Macro Cell and Small Cell eNBs can fit for users' different service requirements which is preferably to improve the system Energy Efficiency performance.

Figure 10 and Figure 11 show the changes of the system Energy Efficiency and average transmission power vs. the Small Cell numbers when the user service ratio η is fixed.

Figure 10 shows the system Energy Efficiency with the Small Cell numbers when the user service ratio η is set as 1. As the number of Small Cell clusters in each Macro Cell coverage increase from 1 to 10, there are also 10 Small Cells deployed in each Small Cell cluster. Whether it is NPGP algorithm or

Figure 9. Average transmission power vs. η with of small cell numbers as 200

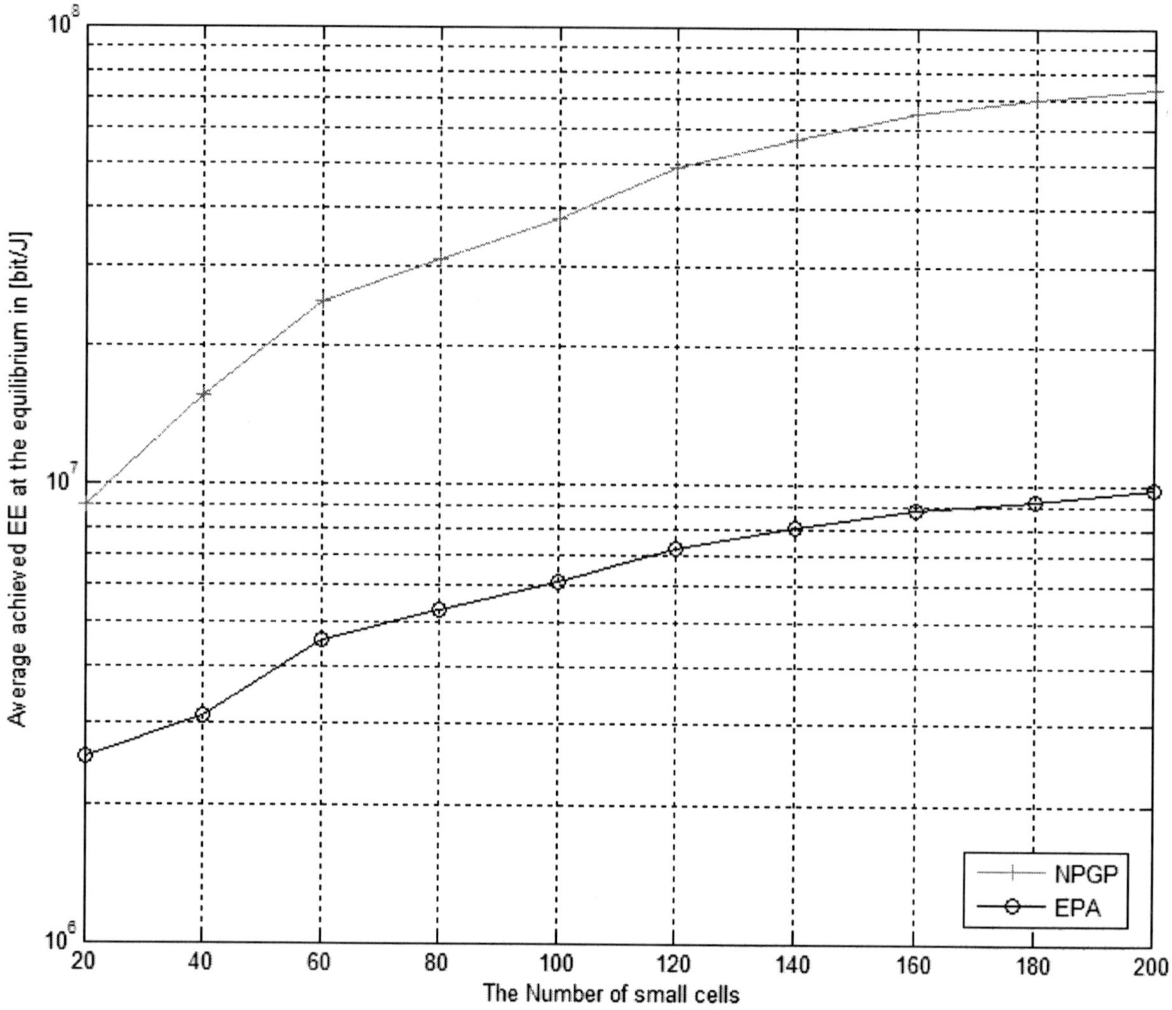

EPA algorithm, the system energy efficiency will increase with the increment of the Small Cell numbers. Taking NPGP algorithm for example, when the Small Cell numbers increase from 20 to 200, the system energy efficiency increases from 10^7 to 10^8 . With the increasing of Small Cells numbers, the coverage areas of the heterogeneous SCN are bigger and more comprehensive. For the overlapping coverage, Small Cell eNBs can adjust transmission power flexibly to further improve the system energy efficiency. In addition, it can be seen from Figure 10, in the situation of different Small Cells numbers, the NPGP algorithm achieves higher system energy efficiency than the EPA algorithm.

Figure 11 shows the average transmission power of Small Cell eNB and Macro Cell eNBs vs. the Small Cell numbers when the user service ratio η is 1. With the increasing of user service ratio, the NGPG algorithm obtains the best consistent performances of the average transmission power of Macro Cell eNBs, but the average transmission power of Small Cell eNBs is decreasing. Specially, when the Small Cell numbers increase from 20 to 200, the average transmission power of Macro Cell eNBs remains 46dBm. The average transmission power of Small Cell eNBs decreases about 6dBm. This is because the ensuring GBR service users in Macro Cell eNB coverage go with a swing, whether the change of Small

Figure 10. The system energy efficiency vs. small cells numbers when η as 1

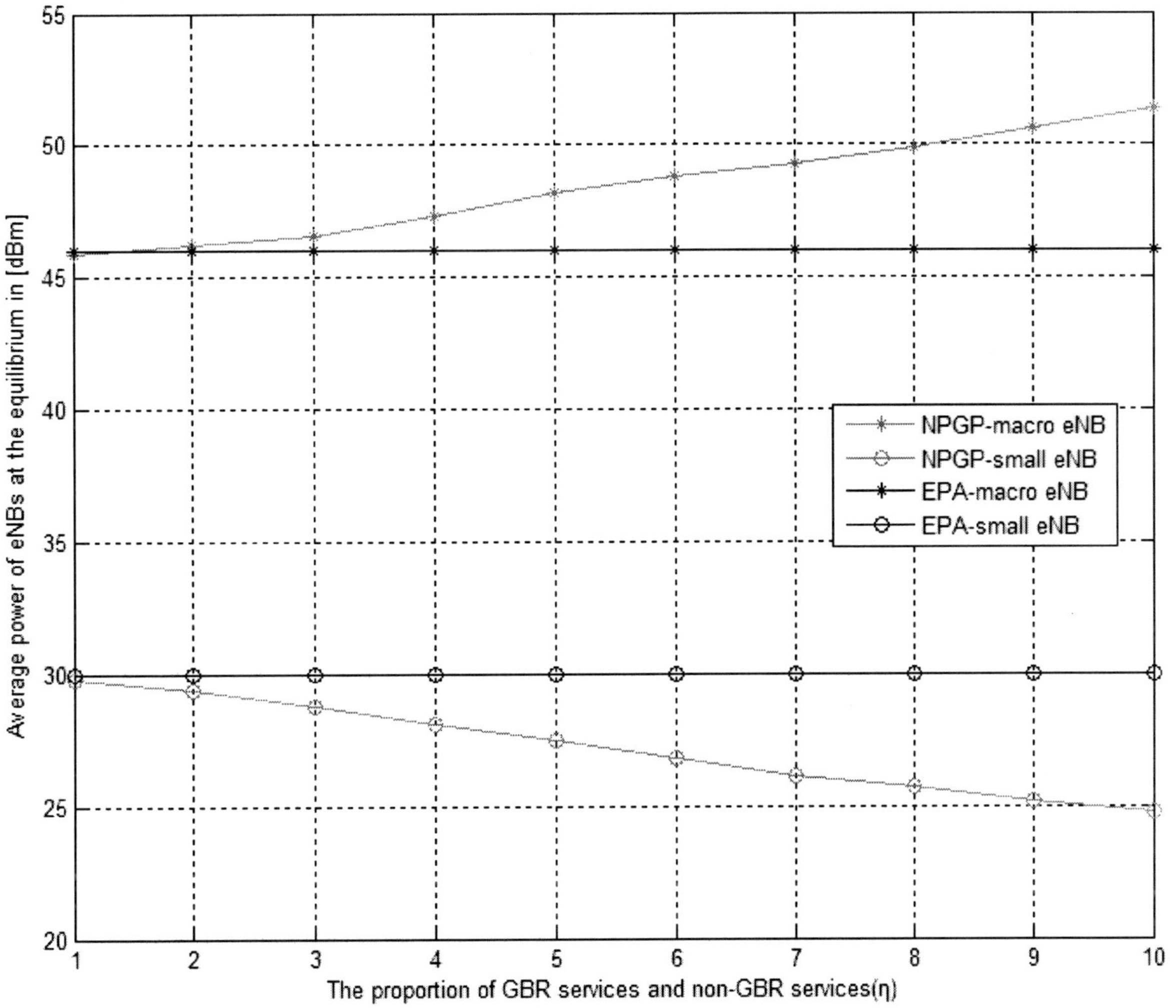

Cell numbers, the transmission power of Macro Cell eNBs will remain constant. But for the Small Cells, in order to serve the non-GBR services, the more numbers of Small Cells, the less transmission power of Small Cell eNBs transmission need.

FUTURE RESEARCH DIRECTIONS

The heterogeneous Small Cell Networks are attracting more and more focuses both from academia and industry for its potentials of improving the system capacity and user experiences. But the interference mitigation and energy efficiency are also facing new challenges within the densely deployed multi-tier network environment.

This chapter focuses on above problems and researches on the cross-tier uplink interferences mitigation and energy efficient downlink power allocation for different service requirements through the Game Theory approaches. Two kinds of game model are implemented and two kinds of the pricing function

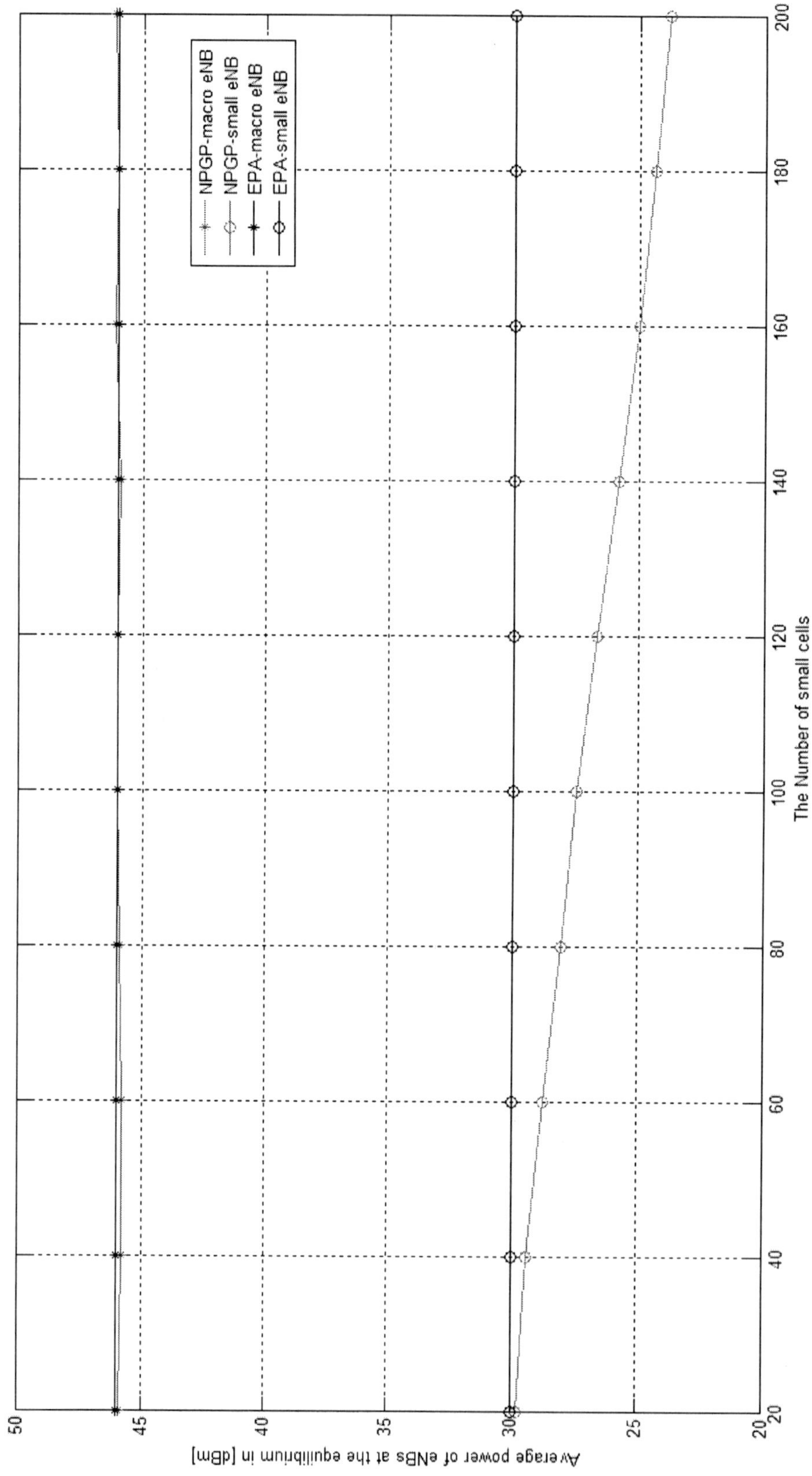

Figure 11. Average transmission power vs. small cells numbers when η as 1

designs are considered in the research for fitting the different scenarios of the system model. The optimization solutions of the proposed game model based schemes are deduced with Nash Equilibriums respectively. Although the system performances of the proposed schemes are further improved over the current works, there still exit open questions for future researches in the HetNet deployments with multi-tier cellular network environments.

The future research directions include:

1. **Combination of Interference Coordination and Power Control for Cross-tier Interference Mitigation:** The research on the interference mitigation in this chapter mainly focuses on the Small Cell tier, while the interference mitigation for the Macro Cell tier is also important to further improve the performance of heterogeneous deployments. According to the 3GPP, the Fractional Frequency Reuse based Inter-cell Interference Coordination (ICIC) for the LTE and LTE-Advanced networks are the basic schemes to perform the inter-cell interference mitigation. But for the heterogeneous deployment scenario, the cross-tier interference situations are more complex than single-tier network, which is currently hard to accurately model the interferences from multi-Macro Cells with their own Small Cell clusters into the non-cooperative game model proposed in this chapter. This research direction is worth to be addressed in future with improved game models.
2. **User Plane and Control Plane Separation for the Dense Deployment of Heterogeneous SCNs:** Dense deployment of the Small Cells in the HetNet will cause large increase of system signaling overheads. Furthermore, it is hard to improve the throughput of each user through optimizing the allocated multi-dimensional resources of many eNBs. A solution to solve this questions is to separate the network User Plane and Control Plane. Under the above circumstances, each user could be served by more than one eNB coordinately. Then, one eNB may take charge in handling the GBR service, while another eNB could support the non-GBR services as many as possible. The role the eNB playing is not decided by its transmission power, but the QoS requirements of its serving users. Based on this new kind of dynamic network framework, the network infrastructures could get more efficiently usage and the energy efficiency could be further improved. Moreover, the current cellular networking strategies also need to be updated with more researches on the user-centric resource allocation and scheduling in the future for the User Plane and Control Plane separations. The Game Theory is believed to find more applications in the above research scenarios.
3. **Complexity Reduction of Game Theory based Solutions:** Game Theory based solutions implemented by this chapter have good results comparing to current works, but the required iterations that reach the optimization are relatively large, which may have limitations for the strategies realizing in the actual network. If the proposed Game Theory based solutions are taken into actual system in future, the utility function and pricing function design still need to be improved in order to reduce the complexity via decrease of the iteration times.

CONCLUSION

Addressing on the interference mitigation of heterogeneous Small Cell Network deployment, with the application of Game Theory approaches, this chapter researches on the uplink power control and downlink power allocation strategies for the LTE/LTE-Advanced based system.

For the uplink scenario, especially for the cell edge user within the UL/DL separation mode, this chapter proposes the non-cooperative game model based power control scheme with convex pricing function. The system model for the UL/DL separation is established with the utility function and non-linear pricing function design for this scenario. The Nash Equilibrium of the proposed non-cooperative game is deduced with proving of its existence and uniqueness. The feasibility and rationality of the proposed game model are verified and the non-linear pricing function design is evaluated with better performance than linear pricing function via the system-level simulations. The corresponding Dynamic Power Adjust algorithm successfully adjusts the initial transmission power of both the Macro Cell and the Small Cell users with optimal power control effects. The system throughput and inter-cell interference mitigation performances are improved over current works.

As to the downlink scenario, this chapter proposes the non-cooperative game model based downlink power allocation in the heterogeneous deployments with over-laid Macro Cell and Small Cells. Based on the established downlink system model with multiple service types and different user QoS demands, the transmission powers with constraints of different QoS requirements are optimized. The corresponding Nash Equilibrium is solved with the verifications of its existence and uniqueness. The system energy efficiency performances are improved via the system-level simulation results with constraints of guaranteeing different user service QoS requirements.

The Game Theory approach will have more and more applications in the research area of wireless networking optimization in the heterogeneous Small Cell Networks. The intra-tier and cross-tier networking strategies, such as the power control, power allocation, offloading and resource management, could be benefited a lot from the appropriate game models.

REFERENCES

3GPP R2-130416. (2013). *Small cell challenges and benefits of dual connectivity*. Ericsson.

3GPP R2-131678. (2013). *Further discussions on UL/DL split*. Ericsson.

3GPP R2-131902. (2013). *On the potential UL/DL imbalance problem*. Nokia Corporation.

3GPP R2-131958. (2013). *Intra-frequency small cell enhancements*. Alcatel Lucent.

3GPP R2-132042. (2013). *Study of Small Cell Enhancement Gains for Dual Connectivity Scenario*. NEC Corporation.

3GPP TR36.872. (2013). *Small cell enhancements for E-UTRA and E-UTRAN Physical layer aspects (Release 12)*. Academic Press.

3GPP TR 36.814. (2013). *E-UTRA; Further Advancements for E-UTRA Physical Layer Aspects, v. 11.1.0*. Academic Press.

3GPP TR 36.839 V 11.1.0. (2012). *Mobility Enhancements in Heterogeneous Networks*. Academic Press.

3GPP TR 36.842. (2013). *Small Cell enhancements for E-UTRA and E-UTRAN-Higher layer aspects, v12.0.0*. Academic Press.

3GPP TR 36.842. (2013). *Study on Small Cell Enhancements for E-UTRA and E-UTRAN - Higher layer aspects, v0.4.1*. Academic Press.

3GPP TR 36.872. (2013). *Small cell enhancements for E-UTRA and E-UTRAN-Physical layer aspects, v12.1.0*. Academic Press.

3GPP TR 36.932. (2013). *Scenarios and requirements for small cell enhancements for E-UTRA and E-UTRAN, v12.1.0*. Academic Press.

3GPP TS 36.101. (2014). *Evolved Universal Terrestrial Radio Access (E-UTRA); User Equipment (UE) radio transmission and reception, v12.4.0*. Academic Press.

3GPP TS 36.104. (2014). *Evolved Universal Terrestrial Radio Access (E-UTRA); Base Station (BS) radio transmission and reception, v12.4.0*. Academic Press.

3GPP TS 36.133. (2014). *Evolved Universal Terrestrial Radio Access (E-UTRA); Requirements for support of radio resource management, v12.4.0*. Academic Press.

3GPP TS 36.213. (2014). *Evolved Universal Terrestrial Radio Access (E-UTRA); Physical layer procedures, v12.2.0*. Academic Press.

Ahmed, M., Peng, M., Ahmad, I., Li, Y., & Abana, M. (2013, October). Stackelberg game based optimized power allocation scheme for two-tier femtocell network. In *Proceedings of Wireless Communications & Signal Processing (WCSP)* (pp. 1-6). IEEE. doi:10.1109/WCSP.2013.6677160

Andrews, J. G. (2013). Seven ways that HetNets are a cellular paradigm shift. *Communications Magazine, IEEE, 51*(3), 136–144. doi:10.1109/MCOM.2013.6476878

Buzzi, S., Colavolpe, G., Saturnino, D., & Zappone, A. (2012). Potential games for energy-efficient power control and subcarrier allocation in uplink multicell OFDMA systems. *IEEE Journal of Selected Topics in Signal Processing*, *6*(2), 89–103.

Chai, B., Deng, R., Cheng, P., & Chen, J. (2012, December). Energy-efficient power allocation in cognitive sensor networks: A game theoretic approach. In *Proceedings of Global Communications Conference (GLOBECOM)* (pp. 416-421). IEEE.

Chandrasekhar, V., Andrews, J. G., Muharemovict, T., Shen, Z., & Gatherer, A. (2009). Power control in two-tier femtocell networks. *IEEE Transactions on Wireless Communications*, *8*(8), 4316–4328.

Charalambous, C. D., Djouadi, S. M., & Denic, S. Z. (2005). Stochastic power control for wireless networks via SDEs: Probabilistic QoS measures. *IEEE Transactions on Information Theory*, *51*(12), 4396–4401.

Chaves, F. D. S., de Oliveira Neto, R. A., & Cavalcanti, F. R. P. (2006, September). Energy efficiency improved game-theoretic distributed power control algorithm for best effort services in wireless communications systems. In *Proceedings of Telecommunications Symposium, 2006 International* (pp. 874-879). IEEE. doi:10.1109/ITS.2006.4433394

Chen, X., Zhang, H., Chen, T., & Lasanen, M. (2013, June). Improving energy efficiency in green femtocell networks: a hierarchical reinforcement learning framework. In *Proceedings of Communications (ICC)* (pp. 2241-2245). IEEE. doi:10.1109/ICC.2013.6654861

Claussen, H., Ashraf, I., & Ho, L. T. (2010). Dynamic idle mode procedures for femtocells. *Bell Labs Technical Journal, 15*(2), 95–116. doi:10.1002/bltj.20443

Dong, L. (2007). Supermodel game Nash equilibrium stability. Wuhan University of Science and Technology, 20(1), 55-57.

Dou yinan, the research of the classification of network users according to the multi-dimensional characteristics. (2010). Beijing: Beijing University of Posts and Telecommunications.

Fu, S., Wu, B., Wen, H., Ho, P.-H., & Feng, G. (2014). Transmission Scheduling and Game Theoretical Power Allocation for Interference Coordination in CoMP. *IEEE Transactions on Wireless Communications, 13*(1), 112–123. doi:10.1109/TWC.2013.120113.121646

Grandhi, S. A., Vijayan, R., Goodman, D. J., & Zander, J. (1993). Centralized power control in cellular radio systems. *IEEE Transactions on Vehicular Technology, 42*(4), 466–468. doi:10.1109/25.260766

Hanzo, L., Haas, H., Imre, S., O'Brien, D., Rupp, M., & Gyongyosi, L. (2012). Wireless myths, realities, and futures: from 3G/4G to optical and quantum wireless. *Proceedings of the IEEE, 100*(Special Centennial Issue), 1853-1888.

Helmy, A., Musavian, L., & Le-Ngoc, T. (2013). *Energy-efficient power adaptation over a frequency-selective fading channel with delay and power constraints.* Academic Press.

Hkimi, M., Langar, R., Secci, S., Boutaba, R., & Pujolle, G. (2013, June). Demands rescaling for resource and power allocation in cooperative femtocell networks. In *Proceedings of Communications (ICC)* (pp. 6235-6239). IEEE. doi:10.1109/ICC.2013.6655605

Li, W., Zhang, H., Zheng, W., Su, T., & Wen, X. (2012, December). Energy-efficient power allocation with dual-utility in two-tier OFDMA femtocell networks. In Proceedings of Globecom Workshops (GC Wkshps) (pp. 535-540). IEEE. doi:10.1109/GLOCOMW.2012.6477630

Liu, J., Zheng, W., Li, W., Wang, X., Xie, Y., & Wen, X. (2013, April). Distributed uplink power control for two-tier femtocell networks via convex pricing. In *Proceedings of Wireless Communications and Networking Conference (WCNC)* (pp. 458-463). IEEE.

Ma, Y., Lv, T., & Lu, Y. (2013, April). Efficient power control in heterogeneous Femto-Macro cell networks. In *Proceedings of Wireless Communications and Networking Conference (WCNC)* (pp. 4215-4219). IEEE.

Mériaux, F., Valentin, S., Lasaulce, S., & Kieffer, M. (2012, October). An energy-efficient power allocation game with selfish channel state reporting in cellular networks. In *Proceedings of Performance Evaluation Methodologies and Tools (VALUETOOLS)* (pp. 290-297). IEEE. doi:10.4108/valuetools.2012.250288

Obaid, N., & Czylwik, A. (2013, September). The impact of deploying pico base stations on capacity and energy efficiency of heterogeneous cellular networks. In *Proceedings of Personal Indoor and Mobile Radio Communications (PIMRC)* (pp. 1904-1908). IEEE. doi:10.1109/PIMRC.2013.6666454

Pan, Y., Nix, A., & Beach, M. (2008, September). A game theoretic approach to distributed resource allocation for OFDMA-based relaying networks. In *Proceedings of Personal, Indoor and Mobile Radio Communications, 2008* (pp. 1-5). IEEE.

Rasti, M., Sharafat, A. R., & Zander, J. (2011). Pareto and energy-efficient distributed power control with feasibility check in wireless networks. *IEEE Transactions on Information Theory*, *57*(1), 245–255.

Saraydar, C. U., Mandayam, N. B., & Goodman, D. (2002). Efficient power control via pricing in wireless data networks. *IEEE Transactions on Communications*, *50*(2), 291–303. doi:10.1109/26.983324

The 25th China Internet Development Statistics Report. (2012). Beijing: China.

Tian, Y., Liu, H., & Li, Z. (2012, December). Power Allocation Algorithm Based on Non-cooperative Game Theory for LTE Downlink. In *Proceedings of Web Intelligence and Intelligent Agent Technology (WI-IAT)* (Vol. 1, pp. 643-647). IEEE. doi:10.1109/WI-IAT.2012.81

Tsiropoulou, E. E., Katsinis, G. K., Vamvakas, P., & Papavassiliou, S. (2013, September). Efficient uplink power control in multi-service two-tier femtocell networks via a game theoretic approach. In *Proceedings of Computer Aided Modeling and Design of Communication Links and Networks (CAMAD)* (pp. 104-108). IEEE. doi:10.1109/CAMAD.2013.6708098

Varma, V. S., Elayoubi, S. E., Lasaulce, S., & Debbah, M. (2012, October). A flow level perspective on base station power allocation in green networks. In *Proceedings of Performance Evaluation Methodologies and Tools (VALUETOOLS)* (pp. 83-88). IEEE. doi:10.4108/valuetools.2012.250249

Varma, V. S., Mhiri, M., Treust, M. L., Lasaulce, S., & Samet, A. (2013, July). On the benefits of repeated game models for green cross-layer power control in small cells. In *Proceedings of Communications and Networking (BlackSeaCom)* (pp. 137-141). IEEE. doi:10.1109/BlackSeaCom.2013.6623397

Xu, Z., Li, G. Y., Yang, C., & Zhu, X. (2012). Throughput and optimal threshold for FFR schemes in OFDMA cellular networks. *IEEE Transactions on Wireless Communications*, *11*(8), 2776–2785.

Yates, R. D. (1995). A framework for uplink power control in cellular radio systems. *IEEE Journal on Selected Areas in Communications*, *13*(7), 1341–1347.

Yingmin, W. (2010). *TD-LTE technology principle and system design*. Beijing: People's Posts and Telecommunications Press.

Zander, J. (1992). Performance of optimum transmitter power control in cellular radio systems. *IEEE Transactions on Vehicular Technology*, *41*(1), 57–62.

Zappone, A., Alfano, G., Buzzi, S., & Meo, M. (2011, September). Energy-efficient non-cooperative resource allocation in multi-cell OFDMA systems with multiple base station antennas. In *Proceedings of Online Conference on Green Communications (GreenCom)* (pp. 82-87). IEEE. doi:10.1109/GreenCom.2011.6082512

Zhang, G., & Zhang, H. (2008, November). Adapative resource allocation for downlink OFDMA networks using cooperative game theory. In *Proceedings of Communication Systems, 2008* (pp. 98-103). IEEE.

Zhang, H., Chu, X., & Wen, X. (2013). *Energy Efficient Power Control in Femtocells with Interference Pricing. In 4G Femtocells* (pp. 51–61). Springer New York.

Zhou, P., Liu, W., Yuan, W., & Cheng, W. (2008, March). Energy-efficient joint power and rate control via pricing in wireless data networks. In *Proceedings of Wireless Communications and Networking Conference, 2008* (pp. 1091-1096). IEEE. doi:10.1109/WCNC.2008.197

KEY TERMS AND DEFINITIONS

Cell Range Extension: The Cell Range Extension is defined as expanding the coverage area of Small Cells via adding an offset for the Reference Signal Received Power of the Small Cell evolved NodeBs. By which, the user association can be biased so that users can be connected with Small Cells even if the actual Reference Signal Received Power of the Small Cells are lower than those of the Macro Cells for offloading system throughput purposes.

Heterogeneous Network: A heterogeneous network often indicates the scenario of multiple types of access nodes co-located in the same coverage area, typically as Macro Cells and Small Cells overlaid deployments.

Interference Mitigation: The interference mitigation technologies are implemented in LTE and LTE-Advanced systems for alleviating the inter-cell interferences and other kinds of interferences existing in the network.

Macro Cell: The Macro Cells are radio access nodes with higher Radio Frequency transmission power in the cellular communication system, which are often deployed with low density or in suburb regions to serve wider coverage areas.

Power Allocation: In LTE and LTE-Advanced systems, power allocation is applied in downlink to dynamically allocate the transmission powers for better system performances.

Power Control: Power control is the intelligent adjustment of transmitter power output in a communication system to achieve better performance.

Small Cell: Small Cells are low transmission power radio access nodes that operate in licensed and unlicensed spectrum in the cellular communication system. They typically have a coverage range from 10m to 1km.

Chapter 6
Game Theory-Based Radio Resource Optimization in Heterogeneous Small Cell Networks (HSCNs)

Mugen Peng
Beijing University of Posts and Telecommunications (BUPT), China

Chengdan Sun
Beijing University of Posts and Telecommunications (BUPT), China

Yaohua Sun
Beijing University of Posts and Telecommunications (BUPT), China

Manzoor Ahmed
Beijing University of Posts and Telecommunications (BUPT), China

ABSTRACT

To optimize radio resource allocation, the game theory is utilized as a powerful tool because its characteristic can be adaptive to the distribution characteristics of in heterogeneous small cell networks (HSCNs). This chapter summarizes the recent achievements for the game theory based radio resource allocation in HSCNs, where macro base stations (MBSs) and dense small cell base stations (SBSs) share the same frequency spectrum and interfere with each other. Two kinds of game models are introduced to optimize the radio resource allocation, namely the non-cooperative Stackelberg and the cooperative coalition. System models, optimization problem formulation, problem solution, and simulation results for these two kinds of game models are presented. Particularly, the Stackelberg models for HSCNs are presented with the Stackelberg equilibrium and the closed-form expressions. The coalition formations for traditional HCSNs, cloud small cell networks, and heterogeneous cloud small cell networks are introduced. Simulation results are shown to demonstrate the proposed game theory based radio resource optimization strategies converged and efficient.

DOI: 10.4018/978-1-4666-8642-7.ch006

I. BACKGROUND OVERVIEW

Mobile data applications such as high-quality wireless video streaming, social networking and machine-to-machine communication are growing exponentially and it is envisioned that asymmetric digital subscriber line (ADSL)-like user experiences would be supported in mobile networks. This implies the wireless networks emerge a major transformation shifting from primarily delivering voice and text services to transporting data and connecting to the Internet and internet of things (IoTs). It has been revealed that data traffic already constitutes more than 97% of the total bits transmitted in many mobile networks (IMT.Vision, 2013). Unfortunately, the current code division multiple access (CDMA) based third generation (3G), and the orthogonal frequency division multiplexing (OFDM) based long term evolution (LTE) cellular networks wouldn't maintain candidates to meet requirements for the significant increasing traffic with a low energy consumption. To improve the transmission rate, the traditional cellular network can increase the power transmitted from a base station (BS), creating serious interference to other serving user equipments (UEs) with high energy consumption. Therefore, the traditional cellular networks are reaching their breaking points, and conventional cellular architectures that are devised to cater to large coverage areas and optimized for homogeneous traffic are facing unprecedented challenges. Since these bursting packet traffics are mainly generated in indoors or hot spots, there is an increasing interest to deploy relay stations (RSs), distributed antennas, and access points (APs) of small cellular cells (such as picocells, femtocells, small cells, etc.) in residential homes, subways, and offices. These new network entities, which may be either operator-deployed and/or consumer-deployed, form a mix of low-power cells underlying the macrocell network. Consequently, the heterogeneous small cell network (HSCN) is presented as a new network paradigm evolution to the fifth generation (5G) wireless systems.

HSCN is a new technology that can cost-efficiently improve system coverage and capacity (Chandrasekhar, Andrews, & Gatherer, 2008). High power nodes (HPNs), such as macro base stations (MBSs), are deployed for the extended coverage and the high mobility support, while low power nodes(LPNs), such as femto access points (FAPs), small cell base stations (SBSs), pico base stations (PBSs), and RSs, help to supply the high achievable data rates in the local coverage of some hot spots. By deploying additional LPNs within the local-area range and bringing LPNs closer to the desired UEs, HSCNs can potentially improve spatial resource reuse and extend the coverage, thus allowing future cellular systems to achieve higher data rates and consume lower energy, while retaining the uninterrupted connectivity and seamless mobility of cellular networks.

To provide seamless coverage and high serving quality in multiple radio access networks (RANs), heterogeneous wireless access systems are firstly proposed to complete interworking and cooperative functionalities. Subsequently, inter-HetNets leverage the frequency spectrum and facilitate the flexible utilizations of alternate frequency bands (e.g., free unlicensed bands) across different radio access technologies (RATs), which adds an extra degree of freedom in terms of spectral efficiency, energy efficiency (EE), and transmission reliability. Lately, the single wireless access network with a huge number of LPNs in multi-tier layers is proposed to enable the dense reuse of spectrum and reduce power consumption via the flexible deployment of LPNs in the coverage holes and hot spots. In a HSCN system, the frequency spectrum can be orthogonally used by HPNs and LPNs, namely overlay mechanism, and the cross-tier interference wherein could be avoided with the expense of occupying with additional frequency bandwidth. In comparison, the frequency spectrum for HPNs and LPNs is non-orthogonally and fully re-used for the underlay mechanism, which could provide efficient utilization of the available resources under mitigating the resultant co-channel interference to the receivers (Quek, Roche, Guvenc, & Kountouris,

2013). Due to the attractive features and potential advantages to reuse the scarce spectral frequency, the HSCN is a promising alternative to improve spectral and energy efficiency performances, which has already attracted significant attentions in the wireless industry and research community during the last few years (Ghosh et al., 2005).

There are numbers of technical studies and researches associated with various aspects of HSCNs, including deployments (Lim & Cimini, 2012), system models (Ghimire & Rosenberg, 2013) and performance metrics (Mukherjee, 2012). Damnjanovic et al. (2011) gave a high-level overview of the topology and the deployment options for HSCNs including the 3rd generation partnership project (3GPP) LTE air interface, network nodes, spectrum allocation options was provided, along with the state-of-the-art enabling mechanisms for heterogeneous deployments. A comprehensive survey related to stochastic geometry models for single-tier, multi-tier, and cognitive cellular scenarios was discussed in (ElSawy, Hossain, & Haenggi, 2013), where a taxonomy model was presented. Meanwhile, due to the significant theoretical performance gains, HSCNs have attracted the attention of standardization bodies, such as 3GPP LTE-Advanced and IEEE 802.16 wireless metropolitan area networks (WMANs). The 3GPP has made substantial progress in related fields including network architecture, backhaul, security, radio frequency (RF) requirements, and new features to enable the deployment of home eNodeBs (HeNBs) in femtocells. In 3GPP Release 10, working items for HSCNs have been formed to study the non-carrier-aggregation based interference control, mobility enhancements of HeNBs, and wireless LAN offload. In 3GPP Releases 10 and 11, the self-organizing network (SON) for HSCNs was recommended into the technical requirements, and the corresponding working items have been stated recently (3GPP TS 32.500, 2011). Currently, there are working items in 3GPP Release 12 focusing on further enhancements of interference control and joint operation with carrier aggregation in HSCNs. There are also activities regarding coordinated multipoint (CoMP) transmission and reception with non-ideal backhaul and network-assisted interference cancelation and suppression (3GPP TR 36.902, 2011). In addition, a new study group on hierarchical networks has been already formed in IEEE 802.16 WMAN, where the usage cases, network architectures, key features and requirements, backhaul enhancements as well as the simulation modeling methodologies are contained (IEEE 802.16ppc-10/0008r3, 2010).

For successful rollouts and commercial operations, the underlay HSCNs come with their own challenges, and there are significant technical issues that still need to be addressed to improve spectral and energy efficiency, including the interference coordination and interference cancelation (ICIC), and cooperative radio resource management (CRRM). In the co-channel deployment based underlay scenario, UEs severed by LPNs will suffer both the strong inter-tier interferences from HPNs, and the intra-tier interference from a huge number of random adjacent LPNs. The inter-tier and intra-tier interferences need to be mitigated with helps of ICIC techniques. Also, to coordinate the interference of UEs in different tiers and satisfy the wide variety of UEs/service demands in an efficient manner by exploiting their varying characteristics and properties, CRRM jointly allocates and manages the scarce radio resources to mitigate the interference and achieve good quality of service (QoS) for UEs. Due to the selfishness of LPNs and complexity of ICIC and CRRM in HSCNs, the game theory is proposed to improve spectral and energy efficiency in HSCNs.

A. Game Theory

Game theory, a branch of applied mathematics, is used to study how rational entities, referred to as players, make decisions in a situation of conflict. It provides a rich set of mathematical tools to model and analyze interactions among the players. Economists have long used it as a tool for examining the actions of economic agents such as firms in a market. However, game theory has also been applied to wireless network in recent years.

Generally speaking, the application is motivated by the following three reasons. First, it is reasonable to assume that the nodes like UEs in wireless network are programmed to maximize the expected value of some utility function, and thus if the nodes are taken as players, the strong rationality assumption in game theory can be met. Second, the optimization objectives of the nodes are sometimes conflict. For instance, every distributed mobile user tends to transmit at its maximal power level to improve its performance, regardless of how this action affects the other users in the network, subsequently giving rise to conflict scenarios. Third, many of the problems which must be solved in wireless network are known to be NP-hard, but fortunately game theory deals primarily with distributed optimization—individual users, who are selfish, make their own decisions instead of being controlled by a central authority, making network highly scalable.

By now, much research focuses on applying non-cooperative game and cooperative game to wireless network. A non-cooperative game is a game reflecting a competitive situation where each player needs to take its decision independently of the other players, given the possible choices of the other players and their effect on the player's objectives or utilities. In (Bacci, Bulzomato, & Luise, 2011), non-cooperative transmission power control and cooperative subcarrier allocation are jointly performed for energy efficiency maximization in an Orthogonal Frequency Division Multiple Access (OFDMA) system. Miao and Himayat and Li and Talwar (2011) investigated energy efficiency power optimization schemes for interference limited communications using non-cooperative game theory. In (Niyato & Hossain, 2008), a non-cooperative game-theoretic framework for the radio resource management (that is, bandwidth allocation and admission control) was proposed in 4G heterogeneous wireless networks. However, these games in (Bacci et al., 2011; Miao et al., 1983; Niyato & Hossain, 2008) only consider the recourse management efficiency without taking QoS requirement into account. Therefore, a non-cooperative joint power and rate control game with pricing scheme was proposed in (Mehbodniya & Aiffissa, 2012) where each FU seeks to choose its possible transmission power and rate in order to maximize its own utility when the network guarantees its target Signal-to-Interference plus-Noise Ratio (SINR) as QoS requirement is introduced. Moreover, in (Pantisano, Bennis, Saad, Debbah, & Latva-aho, 2013), the authors tackled the intra-tier femtocell coexistence using a cooperative game based theoretical approach in order to manage inference better.

As one classical non-cooperative game model, Stackelberg game is formulated with MBSs as the leaders and HBSs as the followers (MUs as the leaders and FUs as the followers). The leaders can make the decision and commit the strategy on the price before other followers. In (He, Lasaulce, & Hayel, 2011), a Stackelberg game was formed between a number of macrocells and femtocells to suppress interference between the cells of the two types. The game requires the macrocells to calculate interference and send it back to each femtocell instantly. Significant overhead occurs. In (Wang, Han, & Liu, 2009), a two-level Stackelberg game was formulated for distributed power control in cooperative communication networks. The source and the relays need to instantly and frequently exchange the price and the information about how much power to buy, whenever the power is adjusted.

For cooperative game theory, it provides analytical tools to study the behavior of rational players when they cooperate. A main branch of cooperative games is coalition formation game which has attracted much attention. In (Pantisano et al., 2013), the authors modeled interference management problem as a coalitional game, where interference alignment (IA) technique was used and recursive core was introduced as the solution. However, IA has practical limitations such as synchronization, low signal-to-noise ratio (SNR) performance and channel state information acquisition overhead. The authors in (Pantisano, Bennis, Saad, Verdone, & Latva-Aho, 2011) proposed a cooperative model for femtocell resource sharing to avoid interference using a coalitional game with transferable utility (TU) in partition form. However, the computational complexity of the game solution is high.

B. Structure of This Chapter

In HSCNs, under improper or uncoordinated resource allocation paradigms, a small cell can cause severe interference to macro cells as well as to other small cells. Such interference will degrade the spectrum efficiency gain brought by deploying small cells. Therefore, developing efficient schemes to allocate resource to suppress interference in HSCNs is essential. Here, the resource includes transmitted power on the shared frequency band, the time to do transmission, and so on.

Motivated by the high efficiency of the game model, we take it as the tool to optimize resource allocation to suppress interferences. More specifically, we use Stackelberg game to allocate the transmitted power of small cell user equipments (SUEs) on the shared frequency band to suppress their interference to an MBS. And to suppress the interference among small cells, we use coalition formation game to make SBSs organized into different coalitions within each of which the downlink transmission is scheduled based on time division multiple access (TDMA).

The rest of this chapter is organized as follows. In Section II, Stackelberg game is used to allocate the transmitted power of SUEs on the shared frequency band to suppress their interference to an MBS, in which the MBS is modeled as the leader and SUEs are modeled as followers. Moreover, the QoS constraints of the leader and followers are both considered. Then a distributed algorithm is proposed, and the convergence condition is derived. In Section III, the coalition formation game is used to allocate the transmission time among SUEs on the shared frequency band to coordinate co-tier interference. Within each coalition, TDMA based transmission mode is used. Also, a distributed algorithm is proposed, and its convergence is proved. In Section IV, the coalition formation game for cloud small cell networks are presented. Further, the coalition formation game for heterogeneous cloud small cell networks are discussed in Section V, followed by the summary in Section VI.

II. NON-COOPERATIVE GAME BASED RESOURCE OPTIMIZATION

Non-cooperative game provides an analytical framework characterizing the interactions and decision-making process involving several players with partially or totally conflicting interests, where each player needs to take its decision independently of the other players, given the possible choices of the other players and their effect on the player's objectives or utilities. A non- cooperative game is usually presented in strategic form, which is actually a triplet $\mathcal{G} = \left\{\mathcal{K}, \left\{S_k\right\}_{k=1}^{K}, \left\{u_k\right\}_{k=1}^{K}\right\}$ where $\mathcal{K} = \{1,..,k,..,K\}$ is the set of game players, S_k is the set of available strategies for player *k*, and $u_k : S \rightarrow \mathbb{R}$ is the utility (payoff) function for player *k* with $S = S_1 \times \ldots \times S_k \times \ldots \times S_K$ being Cartesian product of the strategy sets. The most accepted solution concept for a non-cooperative game is Nash equilibrium (NE), introduced by John F. Nash in his seminal work (Nash, 1950), which is a game state where no players can improve its utility by changing its strategy unilaterally. When studying the Nash equilibrium of a game, the key points of interest are existence and multiplicity, considering a non-cooperative game can admit zero, one, or multiple Nash equilibria. Leveraging the use of the uniqueness theorem in (Yates,1995), the non-cooperative game is guaranteed to have a unique NE if the best-response functions (Han, Niyato, & Saad, 2012) of a non-cooperative game $\mathcal{G}$ are standard functions (Jayaweera & Tinaming, 2009) for all players.

However, the appropriate application of non-cooperative game theory to the resource optimization in heterogeneous small cell networks (HSCNs) is a challenging task. On one hand, the resource types in HSCNs cover a wide range, such as power, frequency, time-slots, code-words, space resource, *et.al.* On the other hand, such a large number of game models are available, making it tougher to match. Moreover, it is desirable to design low-cost, low-complexity algorithms to enable users to operate efficiently, optimize their performance metrics, and meet QoS demands. Motivated by the above contents, in this section, Stackelberg game based power optimization is taken as an example providing insights into how to pick appropriate game models according to the actual situation.

A. Stackelberg Game Based Power Allocation with Stackelberg Equilibrium

In HSCNs, because of the limitation of frequency spectrum, users are usually assumed to operate on the same frequency band, making it essential to optimize their transmit power so as to ensure good performance at the same time minimize interference. In HSCNs, MBSs are deployed to provide seamless coverage while SBSs are used to provide hotspots enhancements, making MBSs prior to SBSs. This priority means MBSs can impose requirements on SBSs, and SBSs are only allowed to operate on the condition that cross-tier interference to MBSs is tolerable. Hence, Stackelberg game, as a hierarchical model, is suitable to HSCNs. By leveraging the practice in (Sun, Peng, Sun, & Li, 2014), we provide an insight into optimizing power of SUEs from the perspective of Stackelberg game, where the benefits of an MBS and SUEs are jointly considered. The formulated hierarchical game is divided into two levels: the MBS, as the leader, plays a seller-level game, and SUEs, as followers, play a buyer-level game. The MBS sells the quota of tolerable interference to SUEs at a certain price. Moreover, the QoS constraints of both the leader and followers are considered simultaneously in the price update by using a discount factor. Each player is selfish and wants to maximize his own benefit. Specifically, the MBS can be viewed as a seller who aims to maximize the utility which consists of its data rates and the payment of followers. Each SUE can be seen as a buyer aiming at maximizing its benefit at the least cost.

In the following, our focus is mainly on the analysis of game solution and algorithm implementation.

1) System Model

Consider the uplink scenario in an HSCN where an MBS is placed and SBSs are randomly deployed under the coverage of the MBS. As shown in Figure 1, both MBS and SBSs occupy with the same frequency, and each base station serves one user. The SINR at SBS k and the MBS can be respectively expressed as:

$$SINR_k(p_k, \mathbf{p}_{-k}, p_0) = \frac{h_{k,k} p_k}{h_{k,0} p_0 + \sum_{l \neq k}^{l \in \kappa} h_{k,l} p_l + \sigma_k^2},$$

Figure 1. Non-cooperative heterogeneous small cell

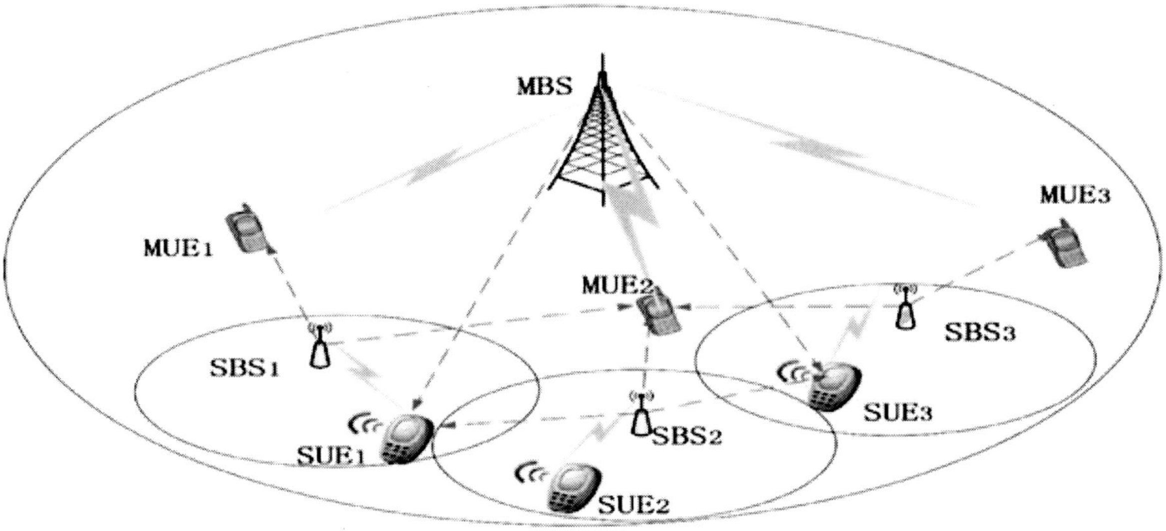

$$SINR_0(\mathbf{p}, p_0) = \frac{h_{0,0} p_0}{\sum_{k=1}^{K} h_{0,k} p_k + \sigma_0^2},$$

where the MBS is labeled as 0, $\mathbf{p}=[p_1, p_2 \dots p_K]^T$ denotes the transmit power profile vector with p_k being the transmit power of SUE k, and ..; $\mathbf{p}_{-k}=[p_1, \dots, p_{k-1}, p_{k+1}, \dots p_K]^T$ denotes the transmit power profile vector except p_k; p_0 denotes the transmit power of the MUE. In addition, SINR threshold θ is used to guarantee the MUE's QoS requirement, and to meet the requirement, the MBS has to control the uplink interference from SUEs. This can be implemented by imposing a price on received interference from SUEs. SUEs then update power allocation strategies to maximize their individual utility functions in a non-cooperative way, based on the announced price from the MBS. Given the announced price, a non-cooperative game exists among SUEs. Generally, every individual SUE has its own QoS requirement. To satisfy these requirements, SUEs need to turn back different upper bounds for the price announced in next scheduled slot, referred to as recommended prices. After that, the announced price is updated according to the instant uplink interference and the recommended prices. Furthermore, it should be addressed that this system works in a distributed way, and no central controller is needed.

2) Problem Formulation

For MUE, the optimization problem is expressed as:

Problem 2.1:

$\max U_{MBS}(\alpha, \mathbf{p})$,

subject to

$$\frac{h_{0,0}p_0}{\sum_{k=1}^{K} h_{0,k}p_k + \sigma_0^2} \geq \theta,$$

where $U_{MBS}(\alpha, \mathbf{p}) = \lambda \times \log\left(1 + \frac{h_{0,0}p_0}{\sum_{k=1}^{K} h_{0,k}p_k + \sigma_0^2}\right) + \sum_{i=1}^{K} \alpha h_{0,k}p_k$; λ is the utility gain of per unit transmission rate for the MBS; α is the price imposed on per unit interference from SUEs; θ is the lower bound of the MBS's SINR. For each SUE, the objective optimization function can be expressed as:

Problem 2.2:

$\max U_k(p_k, \mathbf{p}_{-k}, \boldsymbol{\alpha})$,

subject to

$0 \leq p_k \leq p_{\max}$

$SINR_k(p_k, \mathbf{p}_{-k}, \alpha) \geq \beta_k \ \forall k \in \kappa$,

where $U_k(p_k, \mathbf{p}_{-k}, \alpha) = \mu \times \log\left(1 + \frac{h_{k,k}p_k}{h_{k,0}p_0 + \sigma_k^2 + \sum_{l \neq k}^{l \in \kappa} h_{k,l}p_l}\right) - \alpha h_{0,k}p_k$, μ is the utility gain of per unit transmission rate for SUE .., and β_k is SUE k's QoS requirement.

Definition 2.1: Let α^* and $\mathbf{p}^*$ be the optimal solution for the leader and followers respectively. Strategy profile $(\alpha^*, \mathbf{p}^*)$ is called Stackelberg equilibrium if:

$$U_{MBS}(\alpha^*, \mathbf{p}^*) \geq U_{MBS}(\alpha, \mathbf{p}^*),$$
$$U_k(\alpha^*, \mathbf{p}^*) \geq U_k(\alpha^*, \mathbf{p}).$$

3) Problem Solutions

Based on above formulation, the backward induction method is introduced to analyze SE, which is defined in **Definition 2.1**. Firstly, the solutions for SUEs are derived. Based on the assumptions about followers' action strategies, the solution for the MBS is analyzed afterwards. For every specific price, the NE of the non-cooperative game among SUEs can be reached iteratively. By applying the iterative water-filling method (Yu, Ginis, & Cioffi, 2002), the solution for **Problem 2.2** can be obtained:

$$p_k^* = \frac{\mu}{\alpha h_{0,k} \ln 2} - \frac{h_{k,0}p_0 + \sigma_k^2}{h_{k,k}} - \frac{\sum_{l \neq k}^{l \in \kappa} h_{k,l}p_l}{h_{k,k}}.$$

Define $p_k^* \triangleq \Phi_k(\alpha, p_0, \mathbf{p}_{-i})$. Substituting (9) into (7), the upper bound of the price for SUE k is denoted as:

$$\alpha_k^U \le \frac{\mu h_{k,k}}{h_{0,k}\ln 2} \times \frac{1}{(\beta_k + 1)\left(h_{k,0}p_0 + \sigma_k^2 + \sum_{l \neq k}^{l \in \kappa} h_{k,l}p_l\right)},$$

which is carried back to the MBS as the recommended price.

4) Convergence Proof

Definition 2.2: Let (**X**,d) be a metric space with metric d, and g be a function mapping **X** into itself. If there is a α<1 such that $d(g(\mathbf{x}), g(\mathbf{x}')) \le \alpha d(\mathbf{x}, \mathbf{x}')$, for all $\mathbf{x}, \mathbf{x}' \in \mathbf{X}$, then g is called a contraction map or a contraction.

Theorem 2.1(Banach Contraction Theorem): If (**X**,d) is a complete metric space and g:**X**→**X** is a contraction, then g has a unique fixed point $\mathbf{x}^*$ that is globally asymptotically stable.

Next we prove the iteration formulation (9) converges to an unique stable value by leveraging the concepts of contraction and Banach Contraction Theorem (Shum, Leung, & Sung, 2007). Here, a $K \times K$ matrix is introduced:

$$\mathbf{W}_{m,n} = \begin{cases} \dfrac{h_{m,n}}{h_{m,m}} & m \neq n \\ 0 & m = n \end{cases} \quad m, n \in \kappa.$$

Define $\Phi(\alpha, p_0, \mathbf{p}) \triangleq [\Phi_1(\alpha, p_0, \mathbf{p}_{-1}), \ldots\ldots, \Phi_K(\alpha, p_0, \mathbf{p}_{-K})]^T$. As mentioned above, every SUE updates the power according to $\Phi_k(\alpha, p_0, \mathbf{p}_{-k})$. Finally, by applying **Definition 2.2**, the following theorem is got.

Theorem 2.2: The iteration procedure (11) converges if $\|\mathbf{W}\| < 1$, where $\|\bullet\|$ is any kinds of matrix norm.

Proof:

For simplicity, $\Phi(\alpha, p_0, \mathbf{p})$ can be rewritten as

$$\theta(\alpha, p_0, \mathbf{p}) = \left[\theta(\alpha) - \theta - \mathbf{W}\mathbf{p}\right],$$

where

$$\Gamma(\alpha) \triangleq \left[\frac{\mu}{\alpha h_{0,1}\ln 2}, \frac{\mu}{\alpha h_{0,2}\ln 2} \ldots\ldots \frac{\mu}{\alpha h_{0,K}\ln 2}\right]^T; \theta \triangleq \left[\frac{h_{1,0}p_0 + \sigma_1^2}{h_{1,1}}, \frac{h_{2,0}p_0 + \sigma_2^2}{h_{2,2}} \ldots\ldots \frac{h_{K,0}p_0 + \sigma_K^2}{h_{K,K}}\right]^T.$$

Distance function $d(\mathbf{x},\mathbf{y}) \triangleq \|\mathbf{x}-\mathbf{y}\|$ is selected as the metric. According to **Definition 2.2**, the following is obtained:

$$\|\Phi(\alpha, p_0, \mathbf{p}_1) - \Phi(\alpha, p_0, \mathbf{p}_2)\| = \|\mathbf{W}(\mathbf{p}_1 - \mathbf{p}_2)\| \le \|\mathbf{W}\|\|\mathbf{p}_1 - \mathbf{p}_2\| \le \|\mathbf{p}_1 - \mathbf{p}_2\|$$

As observed, $\Phi(\alpha, p_0, \mathbf{p})$ is a contraction map. From Banach Contraction Theorem, the convergence is verified.

Then we are dedicated to solving **Problem 2.1** which is equivalent to a multi-criteria optimization problem and can be formulated as follows:

$$\max \sum_{l=1}^{K} \alpha h_{0,k} p_k,$$

$$\max \lambda \times \log\left(1 + \frac{h_{0,0} p_0}{\sum_{k=1}^{K} h_{0,k} p_k + \sigma_0^2}\right).$$

The MBS's utility consists of two aspects: the profit from Shannon rates and the profit from selling the interference quota. Constraint (4) guarantees the basic QoS requirement for the MUE. Intuitively, considering the QoS requirements of SUEs, the MBS isn't allowed to raise the Shannon rates blindly if basic communication need is satisfied. Consequently, the profit from selling interference quota becomes the major revenue. Furthermore, optimization problem (13) gains much larger influence than optimization problem (14) if the interference from SUEs exceeds some threshold mathematically. Therefore, we come to (13) for suboptimal solution. Replacing p_k with equation (9), the optimization problem is rewritten as:

$$\max \sum_{k=1}^{K} \left(\frac{\mu}{\ln 2} - \frac{\alpha h_{0,k}\left(h_{k,0} p_0 + \sigma_k^2\right)}{h_{k,k}} - \frac{\alpha h_{0,k}\left(\sum_{l\neq k}^{l\in\kappa} h_{k,l} p_l\right)}{h_{k,k}} \right),$$

subject to:

$$\sum_{k=1}^{K} \left(\frac{\mu}{\alpha \ln 2} - \frac{h_{0,k}\left(h_{k,0} p_0 + \sigma_k^2\right)}{h_{k,k}} - \frac{h_{0,k}\left(\sum_{l\neq k}^{l\in\kappa} h_{k,l} p_l\right)}{h_{k,k}} \right) \le \frac{h_{0,0} p_0}{\theta} - \sigma_0^2.$$

Based on KKT conditions and the complementary slackness theory, the optimal solution for (15) is derived as the optimal price for the MBS, which is expressed as:

$$\alpha^{*}=\frac{K\mu}{\ln 2\left(\left(\frac{h_{0,0}p_{0}}{\theta}-\sigma_{0}^{2}\right)+\sum_{k=1}^{K}\frac{h_{0,k}\left(h_{k,0}p_{0}+\sigma_{k}^{2}\right)}{h_{k,k}}+\sum_{k=1}^{K}\frac{h_{0,k}\left(\sum_{l\neq k}^{l\in\kappa}h_{k,l}p_{l}\right)}{h_{k,k}}\right)}.$$

Considering the recommended prices from SUEs, the MUE announces the updated price according to $\alpha=(1-\varepsilon)\alpha^{*}+\varepsilon\alpha^{U}$, where α^{U} is the minimum value of recommended prices, i.e., $\alpha^{U}=\min_{k}\alpha_{k}^{U}$, and $\varepsilon\in 0,1]$ is the discount factor, reflecting the emphasis on followers' performance relative to the leader's performance. $\varepsilon=0$ indicates the MUE's QoS must be satisfied fully and SUEs' QoS are neglected. $\varepsilon=1$ means the opposite. Other values mean a tradeoff between MUE's QoS and followers' QoS.

5) Distributed Optimization Algorithm

A distributed power optimization algorithm related to the price update is proposed, as detailed in **Algorithm 1**. This algorithm is periodically performed to overcome the channel variation because of the mobility. And restrictions can also be imposed on the upper-bound number of iterations required to obtain the solution, where a sub-optimal solution is reached.

6) Simulation Results

Here we present the evolutions of optimal price and recommended price along with MBS's θ. All points in the graph correspond to convergent prices at the specific θ. To guarantee the higher θ, the higher price is expected by the MBS link. In intuition, the recommended price decreases when followers require higher QoS β selfishly. Therefore, the β =0.5 line evolves below the β =0.25 line all along. There is no doubt that there exists a threshold θ, below which both the leader's QoS and followers' QoS can be satisfied simultaneously. Intersection A and B correspond to the critical values of θ for β=0.25 and β =0.25 respectively. It's worthy noting that a tradeoff exists between the optimal price and the recommended price, which essentially reflects the tradeoff between the MUE's QoS and SUE's QoS.

The impact of MUE's QoS requirements on the sum-rates of SUEs and MUE's rates is depicted in this figure. Intuitively, larger rates are obtained by the leader when higher QoS is satisfied, as shown in the bottom sub-figure. To meet higher QoS requirements of the MUE, the interference from followers must be suppressed by charging larger prices. In such a case, SUEs are forced to transmit at lower power, leading to lower data rates consequently. Surely there exists a θ referred as the threshold, below which SUEs' QoS can also be satisfied. The impact of the discount factor ε on data rates is also presented in our simulations. Below the threshold θ=0.63, recommended prices are greater than optimal prices of the MUE. The larger ε is, the higher the announced price becomes, leading to higher rates of the MUE and lower rates of SUEs. It means the opposite when ε is greater than 0.63. Therefore, a tradeoff exists between the MUE's QoS and SUEs' QoS by changing ε.

Algorithm 1. Algorithm for price and power update

MBS link:
4) Initialize parameters $\Delta=1\times10^{-4}$; t=0; ε; $\alpha_k^U(0)=0$, $\forall k$. Then choose a reasonable price randomly, and broadcast it.
5) At every scheduled slot, **repeat:**
2-1) Collect recommended prices from SBS links, and find $\alpha^U(t)=\min_k \alpha_k^U(t)$. Compute optimal price according to (17).
2-2) Compute announced price according to $\alpha(t+1)=(1-\varepsilon)\alpha^*(t)+\varepsilon\alpha^U(t)$.
2-3) **if** $\lvert\alpha(t+1)-\alpha(t)\rvert>\Delta$ broadcast $\alpha(t+1)=(1-\varepsilon)\alpha^*(t)+\varepsilon\alpha^U(t)$
else
broadcast $\alpha(t+1)=\alpha(t)$.
end if
2-4) $t=t+1$.
6) Stop if some criteria are met.
SBS links:
4) Initialize $\Delta_1=1\times10^{-5}$, **p**=0.
5) repeat:
2-1) Detect $\alpha(t+1)$, from leader, let n=0.
2-2) Update $\mathbf{p}(n+1)$ according to (9).
while $\lvert\mathbf{p}(n+1)-\mathbf{p}(n)\rvert>\Delta=1$
$n=n+1$; update $\mathbf{p}(n+1)$ according to (9).
end while
2-3) transmit at $\mathbf{p}(n+1)$.
6) Stop if some criteria are met.

B. Stackelberg Game Based Power Allocation with Closed-Form Expressions

While we dealt with SE via iteration in the previous subsection due to no close form solution, the closed-form solution of Stackelberg game can still be obtained under some assumptions (Ahmed, Peng, Ahmad, Li, & Abana, 2013), where the upper optimization of the leader and lower optimization of followers are respectively formulated as:

$$\max_{\lambda\succ 0}\sum_i\sum_n \lambda_i^n p_i^n g_i^n,$$

subject to

Figure 2. Announced price and recommended price vs. θ

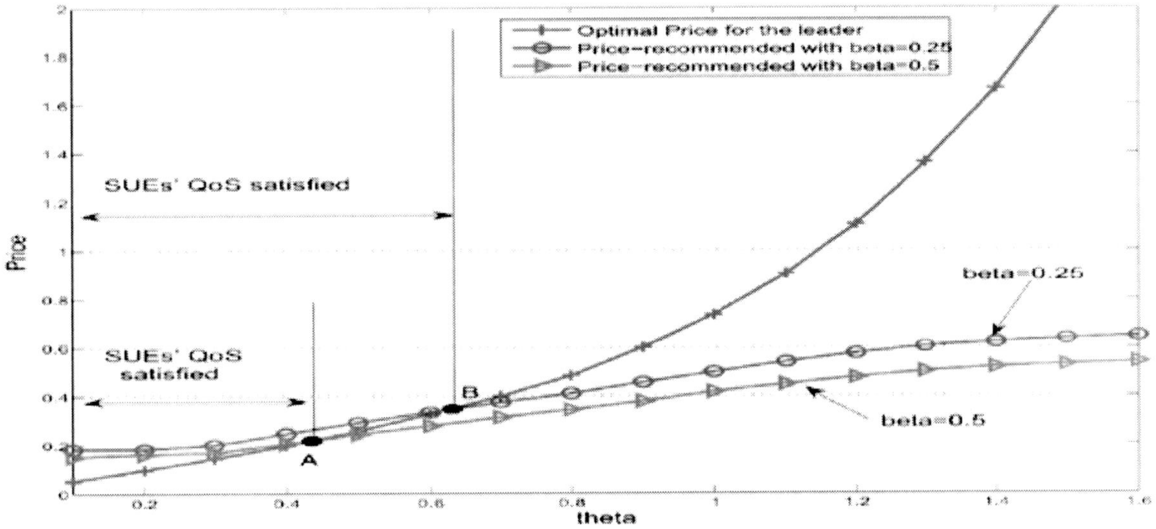

Figure 3. Sum-rates of SUEs and the rate of MBS vs. θ at ε =0, ε =0.2 and ε =0.7. Top: Sum-rates of overall SBS links vs. θ. Bottom: Rate of MBS link vs. θ

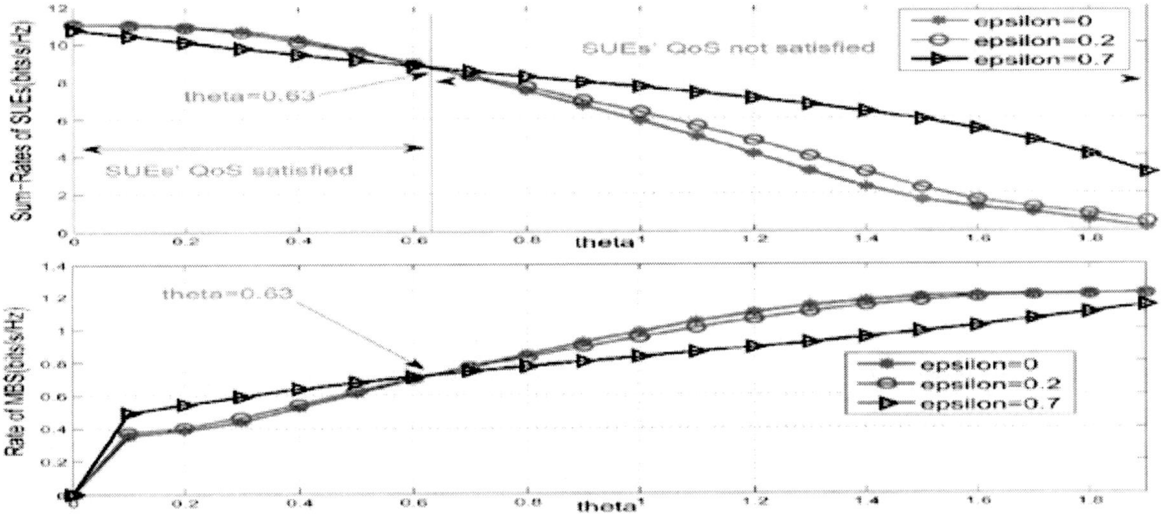

$$\sum_i \sum_n p_i^n g_i^n \geq I_{\max},$$

and

$$\max_{p_i^n \succ 0} \sum_n \left(R_i^n \left(p_i^n, p_{-i}^n \right) - \lambda_i^n p_i^n g_i^n \right),$$

subject to

$\sum_n p_i^n \leq P_f^{\max}, p_i^n \geq 0.$

The upper optimization can be solved by the Lagrangian method:

$$L(p,\beta,\mu) = \sum_n \left(\log\left(1+\gamma_i^n\left(p_i^n, p_{-i}^n\right)\right) - \lambda_i^n p_i^n g_i^n\right) + \beta_i\left(\sum_n p_i^n - P_f^{\max}\right) - \sum_n p_i^n \mu_i^n,$$

where β_i and μ_i are the Lagrangian multipliers associated with inequality constraints. A typical water filling solution can be obtained as:

$$p_i^{n*} = \left[\frac{1}{\beta_i + \lambda_i^n g_i^n} - \frac{I_F^n + \sigma^2}{h_i^n}\right]^+,$$

where $\beta_i = \dfrac{N}{P_f^{\max} + \sum_n \dfrac{I_F^n + \sigma^2}{h_i^n}}$ and $I_F^n = \sum_{j \neq i} p_j h_{i,j}$.

To find closed-form solution, we assume that aggregated interference I_F^n is upper bounded as $I_F^n \leq w$. For the worst case condition, when small cells are densely deployed, I_F^n is assumed equal to w. On the other hand, for the ideal case when small cells are sparsely deployed, the aggregated interference I_F^n is assumed to be zero. Under this assumption, the utility function of all SUEs can de decoupled in term of their p_i's. Now the optimization in (18) can be rewritten as:

$$\max_{\lambda^n \succ 0} \sum_i \sum_n \left[\frac{\lambda_i^n g_i^n}{\beta_i + \lambda_i^n g_i^n} - \frac{\lambda_i^n g_i^n \left(I_F^n + \sigma^2\right)}{h_i^n}\right]^+,$$

which can be maximized if we maximize the first term and (or) minimize the second term. So the original is converted to a minimization problem, as follows:

$$\min_{\lambda^n \succ 0} \sum_i \sum_n \frac{\lambda_i^n g_i^n \left(I_F^n + \sigma^2\right)}{h_i^n},$$

subject to

$$\sum_i \sum_n \frac{1}{\beta_i / g_i^n + \lambda_i^n} \leq I_{\max} + \sum_i \sum_n \frac{g_i^n \left(I_F^n + \sigma^2\right)}{h_i^n},$$

which can be solved by standard Lagrangian method. For MUE, the optimal price is calculated as:

$$\lambda_i^{n*} = \frac{\sum_i \sum_n \sqrt{\frac{g_i^n \left(I_F^n + \sigma^2\right)}{h_i^n}}}{I_{\max} + \sum_i \sum_n \frac{g_i^n \left(I_F^n + \sigma^2\right)}{h_i^n}} \sqrt{\frac{h_i^n}{g_i^n \left(I_F^n + \sigma^2\right)}} - \beta_i / g_i^n.$$

The equation (23) and (27) give the optimal solution p_i^{n*} and λ_i^{n*} for the simplified problem, respectively. The Stackelberg equilibrium is reached as long as the MBS and SUEs update their price and power according to the closed-form solution.

III. COALITION FORMATION FOR INTERFERENCE MITIGATION IN HSCNS

Coalitional game is a growing field of cooperative game theory for studying interactions among selfish entities. It is essentially an aggregation of analytic tools that enables centralized and distributed decision process. In coalitional games, the main focus is on group of players (or coalitions) instead of individual players. In order to provide application oriented approach to such games, Saad et al. (2012) grouped the coalitional games into following categories, i.e., i) canonical games, ii) coalition formation games, and iii) coalitional graph games. The main focus of this chapter is on coalition formation games, which can be either in characteristic or partition form. Furthermore, each of the forms is classified as transferable utility (TU) and non-transferable utility (NTU) games. In a nutshell, this part comprises a unified treatment of coalition formation game theory along with the applications to the wireless communication networks.

The salient features of coalition formation games includes: i) In general, the game is non-superadditive; ii) The coalition formation gains are constrained by the cooperation cost incurred in forming coalitions; iii) The target is to examine the network partition to get the answers of the following questions, such as the optimal coalition size, which coalitions will form, and how we can evaluate the structure's characteristics; iv) The game is also affected due to the internal and external factors, i.e., the change in the number of players and strength of each player.

As there are no specific rules for coalition formation, but in (Apt & Witzel, 2009), general rules are provided for designing distributed coalition formation algorithms to cope with application specific needs. The main elements are (a) well-defined orders appropriate for comparing two coalition structures, (b) simple rules for making and breaking the coalition, and (c) adequate impressions for evaluating the stability of a coalition structure. Coalitional games exhibit to be a very effective tool for modeling efficient, robust, and fair cooperation schemes in communication networks. In the following section, we highlight some examples on the use of coalition formation theory in 4G heterogeneous networks.

In these days, the factors like incessant trend of multimedia application usage through small cells in heterogeneous networks, coupled with further densification of SBSs and impeded by the limited functionality of SBSs, has made the interference management problem among SBSs more challenging. In this regard, ensuring different QoS requirements for distinct real time applications are really challenging, and requires an autonomous, distributed, low-complexity and cooperative interference management solution.

The coalitional game theory was utilized in an attempt to study the interference management problem in (Ahmed et al., 2015). Specifically, an HSCN with small cells densely deployed was considered. Since the SUE served by a certain SBS is usually within the coverage of neighboring SBSs, it will definitely suffers severe co-tier interference from them due to spectrum reuse, resulting in an extremely low quality of service and a low spectrum efficiency.

To handle such co-tier interference, a cooperative model was studied, which includes the formulation of the interference management problem in an HSCN as a coalition formation game in partition form with the objective of maximizing the individual SBS throughput and the system utility. Within the framework of the proposed game, SBSs transmissions in a coalition are coordinated by TDMA, so that the interference among them can be avoided. Moreover, a simplified distributed merge-only coalition formation algorithm was proposed, which runs on each SBS to search autonomously for the best coalition. All the SBSs can finally form a stable partition, which lies in the recursive core. Also, performance analysis of the proposed algorithm, in terms of stability, convergence, and complexity, was studied. In addition, for readers' better understanding, the dynamics of coalition formation was captured via Markov chain. Finally, through Monte Carlo simulations, effectiveness of the proposed scheme was verified by comparing the results with the classical case (Pantisano et al., 2013) and the non-cooperative case.

A. System Model

As is said above, we consider an HSCN with SBSs densely deployed, and all the SBSs are within the coverage of an MBS. Then Shannon capacity expression was used to derive the link rate R_i achieved by the SUE-i in the non-cooperative case, as follows:

$$R_i = W \log_2\left(1 + \frac{\left|h_{i,i}\right|^2 d_{i,i}^{-\alpha} P_i}{I_{i,M} + \sum_{j=1, j\neq i}^{N} \left|h_{i,j}\right|^2 d_{i,j}^{-\alpha} P_j + \sigma_i^2}\right),$$

where W is the allocated bandwidth to the link operating between SBS-i and SUE-i. Note that among three terms in the denominator of Equation (28), the first one is the interference from the MBS which is symbolized by $I_{i,M} = \left|h_{i,o}\right|^2 d_{i,o}^{-\alpha} P_o$, and the second and third ones are interferences from all the other links and variance of AWGN respectively. Here, we can see that in a non-cooperative network, where small cells are densely deployed, the transmission rate of each link is mainly constrained by the co-tier interference, i.e., mutual interference among small cell links, especially when multimedia applications are running.

In order to mitigate the co-tier interference among active links, SBSs can cooperate by forming coalitions with one another as shown in Figure 4. Within each coalition, transmissions of SBSs are coordinated via TDMA, that is, each SBS is allocated with a period of contention free transmission time, so that interference among SBSs in the coalition can be effectively avoided. Note that interference from SBSs outside the coalition still exists due to channel reuse.

SBSs' decisions on how to form coalitions are partially based on the reports from their respective SUE, which indicates their interference level on the allocated bandwidth got from the received signal strength indicators (RSSIs) of the neighboring SBSs. Another factor that affects SBSs' decisions is the

Figure 4. An illustrative example of the coalition model

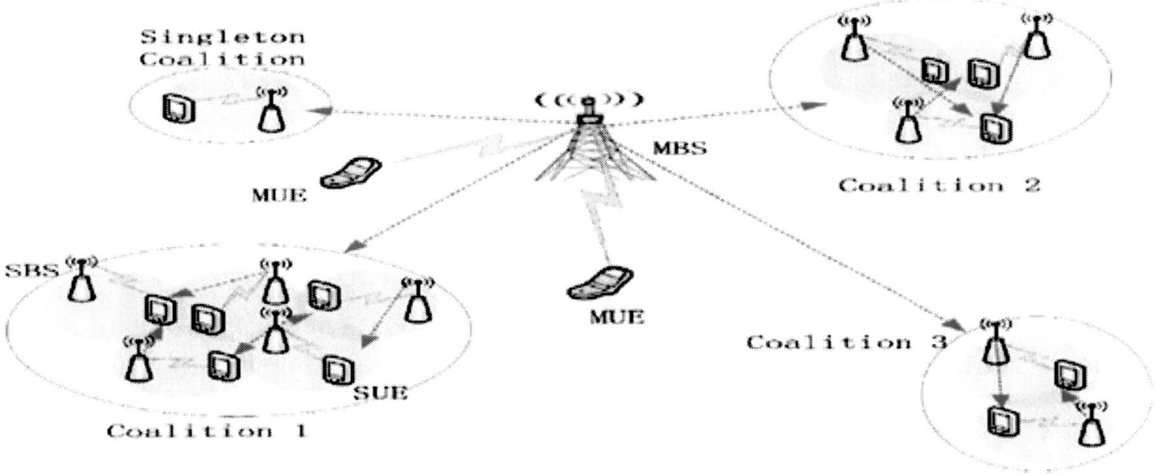

inherent coordination cost incurred in coalition formation. We use transmission power needed for information exchange to capture such a cost. Due to the nature of the wireless signal, if the information broadcast by a coalition member can be received by the farthest member, it can also be received by all the other members in the coalition. Therefore, the cost for forming a coalition C_k, i.e., P_{C_k} , is considered as the sum of power required by each SBS when broadcasting information to its farthest SBS inside the coalition, that is

$$P_{C_k} = \sum_{i \in C_k}^{|C_k|} P_{\hat{i},i},$$

where $P_{\hat{i},i} = \frac{\zeta_o \sigma_i^2}{|h_{\hat{i}i}|^2 d_{\hat{i}i}^{-\alpha}}$ is the power spent by SBS-i to broadcast the information to its farthest SBS-$\hat{i}$ in coalition C_k. Here, ζ_o denotes the minimum SNR required at the SBS-$\hat{i}$. From above expression we can see that the power cost is not only dependent on the coalition size but also on the spatial distribution of SBSs. Taking the cost into consideration, SBSs will have incentive to form a coalition only when the cost does not exceed the maximum tolerable limit of cost $\hat{P}_{\max}$, i.e.,

$$P_{C_k} \leq \hat{P}_{\max}$$

Considering the fact that all SBSs are selfish and rational, their decision making process is modeled as a coalitional game in the following subsections.

B. Proposed Game Model

In this subsection, the interactions of M SBSs are modeled as a coalition formation game, where SBSs have a strong incentive to cooperate and form coalitions to mitigate interference and improve their performances.

Definition 3.1 (Coalition Game): A coalition game $\mathcal{G} = (\mathcal{M}, \mathcal{V})$ with non-transferable utility (NTU) is defined by a finite set of players $\mathcal{M}$, a partition function $\mathcal{V}$: $\Pi(\mathcal{M}) \rightarrow \mathbb{R}^M$ that maps for every partition $\pi_{\mathcal{M}}$ and coalition $C_k \subseteq \pi_{\mathcal{M}}$ to a closed convex subset of $\mathbb{R}^{|C_k|}$ which contains the payoff vectors that players (i.e., SBSs) in a coalition C_k can achieve (Saad, Han, Debbah, Hjungnes, & Baser, 2009).

In the proposed game model, the SBS $i \in C_k$ is allocated a fraction of time ψ_i for transmission using TDMA. Assuming that the time slot is normalized, we have $\sum_{i \in C_k} \psi_i = 1$. In such a case, the utility of coalition $C_k \subseteq \pi_{\mathcal{M}}$ can be defined as

$$\mathcal{U}(C_k, \pi_{\mathcal{M}}) = \sum_{i \in C_k} \psi_i W \log_2(1 + \frac{|h_{i,i}|^2 d_{i,i}^{-\alpha} P_i}{I_{i,M} + I_{i,J} + \sigma_i^2}).$$

Here, $I_{i,J}$ is the overall interference that the link-i suffers, i.e.,

$$I_{i,J} = \sum_{j \in \widehat{C_k}, j \neq i} |h_{i,j}|^2 d_{i,j}^{-\alpha} P_j,$$

where $\widehat{C_k}$ is the set of SBSs which generates interference on link-i. From (31) we can see that the interference within coalition C_k is completely suppressed, and the utility of each SBS in the coalition is increased. Furthermore, since the utility of C_k is affected by $\widehat{C_k}$ which is decided by how the SBSs outside C_k form, therefore, the proposed game was a coalitional game in partition form.

Taking the power cost of forming a coalition into consideration, the value of coalition C_k can be expressed as

$$V(C_k, \pi_M) = \begin{cases} U(C_k, \pi_M), & if\ P_{C_k} \leq \hat{P}_{\max} \\ 0, & otherwise, \end{cases}$$

The above formula indicates whenever the power cost of coalition exceeds the limit, the coalition can not be formed and the value of the coalition must be zero. Within the coalition C_k, the individual payoff of SBS-i, which is defined as the achieved rate of its active link, is

$$x_i(C_k, \pi_M) = \xi_i V(C_k, \pi_M),$$

where $\xi_i = \frac{R_i}{\sum_{j \in C_k} R_j}$ represents the ratio of achieved data rates when operating non-cooperatively. Then for $\pi_{\mathcal{M}} = \{C_1 \ldots\ldots C_k\}$, the partition value can be given as

$$V(\pi_M) = \sum_{j=1}^{k} V(C_k, \pi_M).$$

In order to solve this problem, recursive core was exploited as the solution concept for the proposed coalition formation game in partition form. The recursive core is a desirable outcome of a coalition formation process when externalities, i.e., mutual interference between coalitions of SBSs is taken into account. As the payoff of all the SBSs in the game depends on the way the network is partitioned, the recursive core of the game can be seen as a set of partitions that allows the SBSs to organize in such a way that maximizes their payoff. Interested readers are referred to (Koczy, 2006) for detail description of recursive core.

C. Coalition Formation Algorithm

In this subsection, it is analyzed how coalitions in the HSCN can be formed in a distributed way and whether the final partition lies in the recursive core.

Merge-split scheme is mostly used to help players find the best coalition (Apt & Witzel, 2009). However, such schemes are apt to lead to cycles, causing complexity and consequently the system instability, because SBSs can split from one coalition and join another to get higher payoff. Therefore, in (Saad, Han, Debbah, & Hjorungnes, 2009), the authors of the paper proposed simple merge-only method for coalition formation, and a comparison relation $\triangleright$ was defined for comparing two collections based on Pareto order, which only cares about individual value.

Generally, the optimal solution for a coalitional game is obtained through exhaustive search, which is impractical in our scenario, because the possible partitions computed by the Bell number increases exponentially with the number of SBSs. Since SBSs in the HSCN are rational and autonomous, a distributed algorithm to form a coalition structure that can improve individual user-rate and network sum-rate is highly desirable.

The aforementioned **Algorithm 2** is self-explanatory and requires no detailing.

D. Performance Analysis

In this subsection, the stability, convergence, and complexity properties of the proposed algorithm are summarized in Proposition 3.1, Theorem 3.1, and Proposition 3.2, respectively.

Proposition 3.1: The final outcome of Algorithm 2 belongs to the recursive core and is stable.

Proof: In the proposed algorithm, each SBS makes decision independently and merges with the coalition partner, which can improve its payoff via Pareto order. Since each SBS selects the coalition partners within its interference list, the remaining SBSs in the networks are unaffected and their coalition values remain the same. After every step of the proposed algorithm, the sum of payoff of the coalitions increases without decreasing individual SBS payoff. So, the overall network utility increases on each

Algorithm 2. Distributed Coalition Formation Algorithm in HSCN for Interference Management

Initial Stage:
At the beginning of all time at τ=0 with the network being partitioned as $\Pi^0 = \{\{C_1\}, \{C_2\}, \ldots, \{C_M\}\}$
Phase 1- Interferers (aggressors) Discovery:
(c) Each SBS determines its potential coalition partners through their active SUEs by sensing the neighborhood SBSs (aggressors) RSSI.
(d) Each concerned SBS maintains the list by sorting the interference of aggressors in descending order i.e. from the highest to the lowest interference level.
Phase 2- Coalition Formation of SBSs:
Coalition formation using merge only rule occurs.
repeat
(f) Update time-index: $\tau=\tau+1$.
(g) Each SBS-*i* calculates the cooperation cost starting from the leading aggressor up to the last member of the maintained list, which ensures the power constraint in (30)
(h) The potential payoff is calculated as in (34).
If payoff is improved **then**
(iii) SBS-*i* sends cooperation proposal towards aggressor in order to merge.
(iv) The transmission power in (29) is updated, which accounts for the cooperation cost.
end if
(i) Each SBS merges with the aggressor which maximizes payoff (which is an un-dominated outcome).
(j) The resulting coalition is included in the recursive core.
until: Convergence to a stable coalition partition in the recursive core.
Phase 3- Inner Coalition Scheduling:
Within each coalition, TDMA is performed by allocating separate time slot to every member of the coalition.

step by Pareto dominance. Finally a stage comes when there is no SBS with whom coalition can be formed, meaning that highest payoff has been achieved. Therefore, such final outcome characterizes an un-dominated coalition that is part of the recursive core of the considered game and is stable.

Theorem 3.1: The coalition formation process with **Algorithm2** converges to a final stable partition in a finite number of steps.

Proof: Since the number of SBSs are finite, the coalitions that they can form are also finite. Moreover, because an SBS is allowed to merge with another coalition only when the merge process leads to Pareto improvement in the new coalition, no loop appears during the merge process, implying that the number of steps is bounded. In such a case, the coalition formation process based on merge-only method must converge to a final stable partition in a finite number of steps.

Proposition 3.2: The complexity of Algorithm 2 is less than $M(M\text{-}1)/2$.

Proof: The complexity of our algorithm is related to the number of merge operations performed by each SBS, which is directly related to the total number of cooperation proposals sent by each of the M SBSs. We consider two extreme cases, (i) when all the cooperation proposals are rejected, and (ii) when all the cooperation proposals are accepted. Considering the first case, each SBS-*i* sends at most $|\omega_i|$ proposals to the neighborhood aggressors where $|\omega_i| \leq M\text{-}1$. Now, if the initial SBS sends (M-1) proposals and the next SBS sends (M-2) proposals and so forth, the maximum number of proposals is given by the series $(M\text{-}1)+(M\text{-}2)+\ldots+2+1=M(M\text{-}1)/2$ Thus, the complexity order of the proposed algorithm in the worst case is $O(M^2)$. Similarly, in the second case, where all the proposals are accepted, now the total number of proposals is only M-1. Therefore, in the best case complexity order reduces to $O(M)$. Since in our coalitional game, each SBS, being distributed, only tends to merge with neighbor interferers, the complexity of the proposed algorithm is much smaller than $M(M\text{-}1)/2$.

E. Dynamics of the Coalition Formation

To seek and capture a better insight into how the presented coalition formation game reaches a final stable partition or coalition structure, dynamics of the process was studied i.e., a time-evolving sequence of steps experienced by SBSs before attaining a stable partition. The temporal partitions formed in the steps are modeled as a sequence of random variables describing the states of the system, the transitions of which can be represented by a Markov chain (Arnold & Schwalbe, 2002). For the detail insight, readers are referred to (Ahmed et al., 2015).

The performance of the proposed scheme was compared with that of classic case (Pantisano et al., 2013) and the non-cooperative case. In the proposed scheme, frequency reuse of 1 was implemented, i.e., there is no orthogonal frequency resource allocation among SBSs; which implies more bandwidth for every SBS and consequently higher spectral efficiency. Moreover, it involves no spectrum sensing power cost, which means more opportunities for cooperation. Also, the proposed simplified algorithm is based on merge-only rule instead of merge and split method, therefore, it achieved fast convergence, stability and improvement in performance. The results of system payoff, in terms of the total rate achieved, versus the number of SBSs are shown in Figure 5. From the results we can see that when $M \leq 3$, the system payoff of all the schemes are almost the same. This is because in a sparsely deployed network, SBSs are more likely to work non-cooperatively. However, as the number of SBSs in a certain area increases, i.e., the deployment of SBSs is getting dense, the proposed scheme performs much better than the classical case and non-cooperative case, because by cooperation, interference among densely deployed SBSs

Figure 5. System payoff versus the number of SBSs

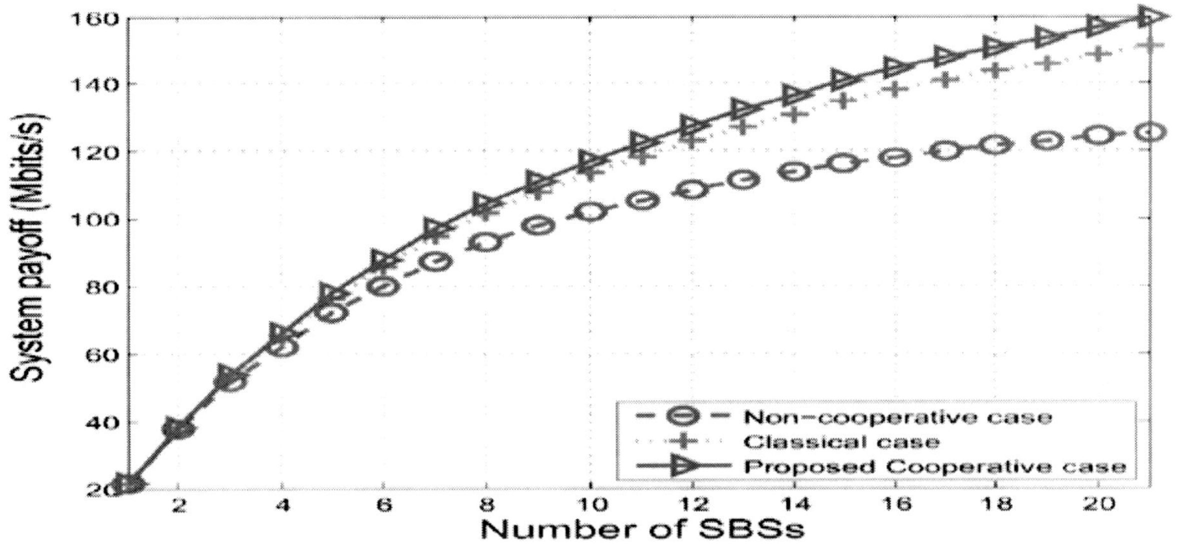

has been successfully eliminated. For M=21, the system payoff of our scheme has been improved up to 5.6% and 27.3% respectively, in comparison to the classical case and the non-cooperative case. We then examine SBS's average satisfaction level with the proposed scheme. Since various types of real-time services have different QoS requirements, e.g., data rate requirement, which means different achievable rate will bring subscribers different service experiences.

Then, we exploit a monotonically non-decreasing function, i.e., sigmoid function, to approximately capture the satisfaction level of a user. The simulation results are presented in Figure 6. It can be observed that for a fewer SBSs in the network, all the aforementioned schemes have almost the same performance. Nonetheless, with the increase in the number of SBSs, our proposed scheme performs much better than the classical case and non-cooperative case. This is because in our scheme, the co-tier interference within a coalition is effectively mitigated by cooperation and thereby SBS's achievable data rates increases.

IV. COALITION FORMATION FOR INTERFERENCE MITIGATION IN CLOUD SMALL CELL NETWORKS

Cloud small cell networks(C-SCNs) are promising technological paradigms to improve capacity and extend coverage. However, the severe interference resulting from dense deployment of remote radio heads (RRHs) prevents to achieve high spectrum efficiency. This section investigated the cooperative interference management among RRHs in C-SCNs. To successfully suppress the intra-coalition interference, RRHs can form coalitions within each of which non-coherent joint transmission is performed. Here, such cooperation behavior of RRHs is modeled as a coalition formation game. And to find the solution for the game, an enhanced concept, called defection order, is proposed. Based on this concept, a self-optimal Coalition Formation Algorithm (SOCFA) is proposed in (Sun et al., 2014).

Figure 6. Average satisfaction per SBS versus the number of SBSs

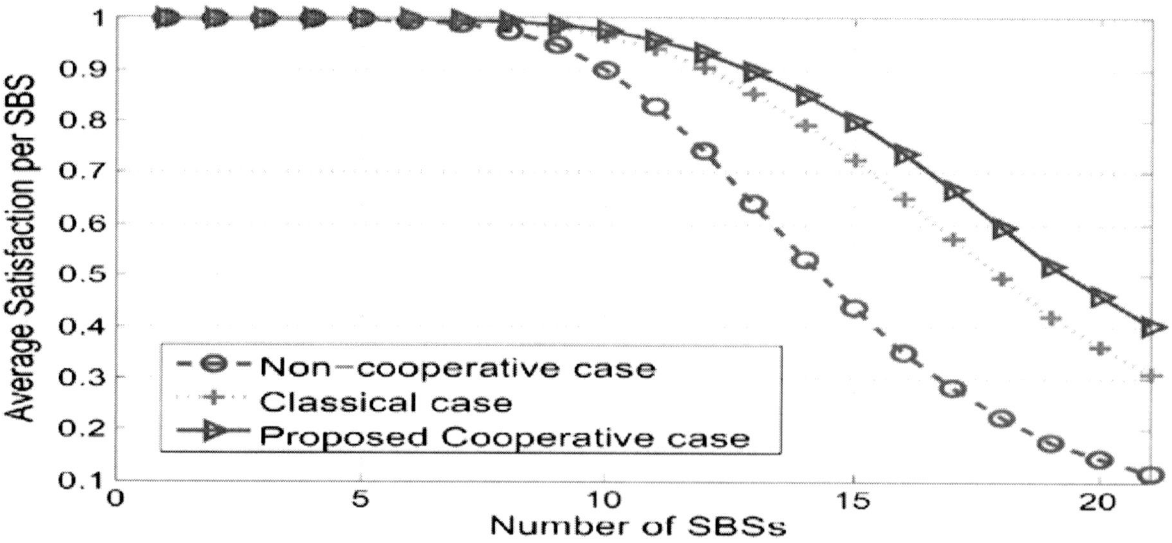

A. System Model for C-SCNs

Owing to the advantages of low latency, high infrastructure utilization, and low expenditure, C-SCNs stand out to become a promising technology in recent years, making it necessary to research the cooperative transmission/reception in C-SCNs. This section considers the downlink transmission scenario in a C-SCN with a single MBS and N RRHs, and the MBS is responsible for the basic coverage while the RRHs are used to improve the hotspots' performance. In addition, all the RRHs are connected to the cloud BBU pool via the low-latency optical fiber fronthaul. Assume that N RRHs share the same frequency, and each of them has one antenna serving only one UE. The UE is identified by the same label as its corresponding RRH. The perfect channel state information (CSI) is assumed to be obtained by the UE and then fed back to the corresponding RRH.

Let $\mathcal{N} = \{1,\ldots,N\}$ be the set of RRHs, $g_{i,j}$ be the channel gain from RRH j to UE i, and A_i^T and .. be the transmitting antenna gain of RRH i and the receiving antenna gain of UE i, respectively. Given the transmission power p_i of RRH i, the received signal to interference plus noise ratio (SINR) of UE i can be written as:

$$SINR_i = \frac{p_i A_i^T g_{i,i} A_i^R}{\sigma^2 + I_{i,\mathcal{N}} + I_0},$$

where σ^2 is the variance of additive Gaussian white noise; $I_{i,\mathcal{N}} = \sum_{j \neq i}^{j \in \mathcal{N}} p_j A_j^T g_{i,j} A_i^R$ and I_0 are the interference from other RRHs and the MBS, respectively. Obviously, I_0=0 if the overlaying RRHs and the MBS use different frequency bands. The achievable Shannon rates of UE i is expressed as:

$$R_i = B_i \cdot \log_2\left(1 + \frac{p_i A_i^T g_{i,i} A_i^R}{\sigma^2 + I_{i,\mathcal{N}} + I_0}\right),$$

where B_i demotes the bandwidth allocated to the UEi.

As for each RRH, the obtained utility is closely related to the performance of its UE. Thus, for each RRH $i \in \mathcal{N}$, its utility function is defined as the spectrum efficiency of its corresponding UE i, which describes the level of satisfaction,

$$u_i\left(\mathcal{N}\right) = \frac{R_i}{B_i} = \log_2\left(1 + \frac{p_i A_i^T g_{i,i} A_i^R}{\sigma^2 + I_{i,N} + I_0}\right).$$

Regarding to (38), the utility of a RRH is degraded by the co-channel interference. In order to reduce such interference, RRHs tend to cooperate by forming coalitions.

B. Coalitional Game Formulation

Since RRHs are selfish and rational, cooperation among them can be formulated into the framework of coalitional game. In this section, the cooperation among RRHs in C-SCNs is first modeled as a coalition formation game, and then the necessary conditions for RRHs to cooperate are analyzed.

A coalition $S_k \subseteq N$ is a non-empty subset of $\mathcal{N}$, and N RRHs are divided into disjoint coalitions. Thus, $\forall i, j \in \{1,\ldots,K\}$ and $i \neq j$ where K is the total number of total coalitions, it's obtained that $\cup S_i = \mathcal{N}$ and $S_i \cap S_j = \varnothing$. The set of K coalitions is called the *coalition structure*. Obviously, each coalition structure forms a partition on $\mathcal{N}$.

Within a coalition, transmission of RRHs is coordinated exploiting TDMA, that is, $\forall i \in S_k$, its transmission time should be a fraction of the normalized time slot α_i, which satisfies:

$$\sum_{i \in S_k} \alpha_i = 1, 0 < \alpha_i \leq 1.$$

Consequently, there will not be more than one user occupying the same channel at the same time-slot within a coalition, which efficiently mitigates the interference inside the coalition. The cooperation among RRHs can be formulated as a coalitional game in simplified-partition form with transferable utility (TU). To better understand the model, the following two definitions are made:

Definition 4.1: A coalitional game in simplified-partition form is defined as a pair $G = (\mathcal{N}, v)$, where $\mathcal{N}$ is the set of players (i.e. RRHs) and v is a mapping that determines the payoffs of these players. In simplified-partition form, the value of a coalition S depends on both the members of S and the players in $\mathcal{N} \setminus S$, but this value has no dependence on the specific way of partitioning $\mathcal{N} \setminus S$.

Definition 4.2: A transferable utility function is a value function v that assigns a real number $v(S, \mathcal{N})$ to each coalition $S \subseteq \mathcal{N}$, which represents the total utility players in S can achieve, satisfying ...

Note that for the game in simplified-partition form, the utility of each RRH in a coalition, say S_k, dose not depend on how RRHs in $\mathcal{N} \setminus S_k$ are partitioned, which makes it different from the game in partition form, where the utility of each RRH in S_k will be influenced by how RRHs in $\mathcal{N} \setminus S_k$ are partitioned.

It's assumed that, for the game in simplified partition form, all RRHs in a coalition serve the same UE simultaneously at a specific instant, which means non-coherent joint transmission is used within a coalition. Consequently, for every UE of RRHs in S_k, all RRHs in $\mathcal{N} \setminus S_k$ cause interference regardless of the specific way that they are partitioned. In such a case, the number of interferers for $i \in S_k$ is reduced from N-1 to N-$|S_k|$, such that the interference can be efficiently mitigated. The signal and interference of UE i when RRH i joins a coalition can be respectively written as:

$$P_i^{received} = p_i A_i^T g_{i,i} A_i^R + \sum_{l \neq i}^{l \in S_k} p_l A_l^T g_{i,l} A_i^R,$$

$$I_{i,\mathcal{N} \setminus S_k} = \sum_{j \in \mathcal{N} \setminus S_k} p_j A_j^T g_{i,j} A_i^R,$$

which shows that part of the interference links are converted into useful message links compared with (36). The utility of RRH $i \in S_k$ is defined as:

$$u_i\left(S_k, \mathcal{N}\right) = \alpha_i \log_2\left(1 + \frac{P_i^{received}}{\sigma^2 + I_{i,\mathcal{N} \setminus S_k} + I_0}\right),$$

where α_i is the fraction of the time duration which has been introduced in the previous part.

The value of each coalition S_k is defined as the sum of the utility of all RRHs in the coalition, which is given by:

$$u(S_k, \mathcal{N}) = \sum_{i \in S_k} u_i\left(S_k, \mathcal{N}\right).$$

With $\mathcal{N}$ and $u(S_k, \mathcal{N})$ defined in (43), the cooperative interference management can be modeled as a coalitional game in simplified-partition form according to the **Definition 4.1**. The necessary conditions for cooperation in C-SCNs are summarized into **Theorem 4.1**.

$$\sum_{i \in S_j} \alpha_i \log_2\left(1 + \frac{g_{i,i} p + \sum_{j \in S_k}^{j \neq i} g_{i,j} p}{\sigma^2 + \sum_{l \in \mathcal{N} \setminus S_k} g_{il} p}\right) > \sum_{i \in S_k} \log_2\left(1 + \frac{g_{i,i} p}{\sigma^2 + \sum_{l \in \mathcal{N}}^{l \neq i} g_{il} p}\right),$$

$$\frac{1}{K+1} \cdot \log_2\left(1 + \frac{g_{i,i} p + \sum_{k \in S_k} g_{i,k} p}{\sigma^2 + \sum_{l \in \mathcal{N} \setminus S_K}^{l \neq i} g_{i,l} p}\right) > \log_2\left(1 + \frac{g_{i,i} p}{\sigma^2 + \sum_{j \in \mathcal{N}}^{j \neq i} g_{i,j} p}\right).$$

Theorem 4.1: The necessary conditions for RRH i to join a coalition S_k which consists of K RRHs is:

$$g_{i,i} < \frac{\sigma^2 + I_{S_k} + I_{\mathcal{N} \setminus S_k}}{p} \cdot \left[\sqrt[K]{1 + \frac{I_{S_k}}{\sigma^2 + I_{\mathcal{N} \setminus S_k}}} - 1\right],$$

where p and σ^2 are the transmitting power and background noise, respectively; I_{S_k} and $I_{\mathcal{N} \setminus S_k}$ represent the total signals of RRHs in coalition S_k and in $\mathcal{N} \setminus S_k$, respectively.

Proof: When forming a coalition, the basic super-additivity within the coalition must be satisfied. Moreover, one RRH will participate in a coalition only when its utility increases after forming a new coalition. Assume that the transmission time within a coalition is equally shared among RRHs, *i.e.*,

$\alpha_i = \frac{1}{K+1}$, (44) can be simplified as (45). Denote $\sum_{j\in\mathcal{N}}^{j\neq i} g_{i,j}p = I_{S_k} + I_{\mathcal{N}\setminus S_k}$, where $I_{S_k} = \sum_{k\in S_k} g_{i,k}p$ means the total signals from RRHs in coalition S_k, and $I_{\mathcal{N}\setminus S_k} = \sum_{l\in\mathcal{N}\setminus S_K}^{l\neq i} g_{i,l}p$ means the interference from RRHs in $\mathcal{N}\setminus S_k$. Then (45) is further simplified as:

$$\left(1+\frac{g_{i,i}p}{\sigma^2 + I_{S_k} + I_{\mathcal{N}\setminus S_k}}\right)^K < 1+\frac{I_{S_k}}{\sigma^2 + I_{\mathcal{N}\setminus S_k}}.$$

According to (46), we get the necessary channel gain condition if RRH i tends to join S_k:

$$g_{i,i} < \frac{\sigma^2 + I_{S_k} + I_{\mathcal{N}\setminus S_k}}{p}\cdot\left[\sqrt[K]{1+\frac{I_{S_k}}{\sigma^2 + I_{\mathcal{N}\setminus S_k}}}-1\right].$$

The formula (47) indicates that when the RRH i is suffering from bad channel quality, *i.e.*, $g_{i,i}$ is small, it prefers to compromise by joining a coalition. On the other hand, the possibility of forming a coalition increases if I_{S_k} is large, which means that RRH i is adjacent to coalition members in S_k. In fact, the interference from the neighborhood is severe when two RRHs are close enough to each other, these RRHs therefore are inclined to form coalitions to mitigate the interference. Otherwise, they tend to act alone due to the low interference. Although I_{S_k} can be increased by including more RRHs in S_k, the increase of coalition members K reduces the fraction of time-slots, which consequently degrades the performance per RRH. There exists a tradeoff between the increase of I_{S_k} and the reduction of time fraction. Thus, the number of RRHs in a coalition should be balanced.

C. Coalition Formation Algorithm

Generally, the optimal solution for a coalitional game is obtained through exhaustive search. However, the number of all possible partitions, which is calculated by the Bell Number, increases exponentially with the number of game players (*i.e.*, RRH), making exhaustive search impractical when RRHs are densely deployed. Under such circumstances, most of the existing works focus on finding suboptimal solutions in a distributed way. In this section, a centralized practical algorithm for the coalitional game is proposed, and some properties of the proposed algorithm such as convergence, complexity as well as fairness are analyzed.

1) Defection Order

Considering the nature of selfishness, RRHs aim to maximize their utilities when choosing the coalition. To better understand the nature of coalition formation, a criteria, *i.e., defection order*, is first described in **Definition 4.3**.

Definition 4.3: For a given coalition composing of l players $S_k=\{k_1,k_2,\ldots,k_l\}$, and a newly formed coalition $S_k^{new}=S_k\cup\{i\}$, S_k^{new} is said to defect order S_k, *i.e.*, $S_k^{new}\rhd_D S_k$ if the following conditions are satisfied:

$$S_k^{new}\rhd_D S_k \Leftrightarrow \begin{cases} u\left(S_k^{new},\mathcal{N}\right)>\sum_{j\in S_k^{new}} u_j\left(\mathcal{N}\right) \\ \dfrac{u\left(S_k^{new},\mathcal{N}\right)}{\left|S_k^{new}\right|}>\dfrac{u\left(S_k,\mathcal{N}\right)}{\left|S_k\right|} \\ u_i\left(S_k^{new},\mathcal{N}\right)>u_i\left(S^{old},\mathcal{N}\right) \end{cases},$$

where S^{old} is the coalition in which RRH i stays before joining coalition S_k; and $\left|S_k\right|=l\$,\$\left|S_k^{new}\right|=l+1$; and $u_i\left(S_k^{new},\mathcal{N}\right)$ is the utility of RRH i after joining coalition S_k^{new}, while $u_i\left(S^{old},\mathcal{N}\right)$ is RRH i's utility staying in the original coalition S^{old}; and $\dfrac{u\left(S_k^{new},\mathcal{N}\right)}{\left|S_k^{new}\right|}$ and $\dfrac{u\left(S_k,\mathcal{N}\right)}{\left|S_k\right|}$ are the average utility of members in S_k^{new} and S_k respectively.

The *defection order* $\rhd_D$ lists three necessary conditions under which a RRH can deviate from one coalition to another. These conditions are: (i) the utility of the new coalition should be greater than the sum of the individual utility of coalition members when they do not cooperate, which is also referred to as utilitarian order; (ii) compared to the average utility per RRH in the coalition without the new comer, the average utility per RRH in the newly formed coalition should be increased after absorbing the new comer, which is referred to as average order; (iii) the utility of the defected RRH i should be increased after deviating from the original coalition, which is referred to as individual order.

2) The Proposed Algorithm

With the *defection order*, the central node, *i.e.*, BBU, is able to partition selfish RRHs into different coalitions, taking into account both the individual utility and coalition utility. The corresponding *defection order* based algorithm, which BBU uses to find the best coalition formation solution, is proposed in **Algorithm 3**. The goal of proposed algorithm is to reach a stable coalition partition, where no RRHs will deviate from their coalitions.

The proposed self-optimal coalition formation algorithm (SOCFA) is composed of three stages: initial stage, coalition formation stage and transmission stage. Initially, the network is partitioned in a non-cooperative way. The coalition formation procedure is performed by the BBU in the second stage, where the BBU continually traverses all coalitions in the partition for each individual RRH and divides each RRH into the coalition ensuring the maximum $u_i\left(S_k^{new},\mathcal{N}\right)$ and $\dfrac{u\left(S_k^{new},\mathcal{N}\right)}{\left|S_k^{new}\right|}$, until a stable coalition partition is reached. Here, the maximum $u_i\left(S_k^{new},\mathcal{N}\right)$ ensures the utility of the current RRH

Algorithm 3. Self-optimal coalition formation algorithm

1: **Initial Stage:**
The network is partitioned by $\mathcal{H}=\{1,...i,...,N\}$ in non-cooperative way.
2: **Coalition Formation Stage:**
The BBU repeats:
For RRH i=1:N
4) Discover the existing coalitions $\mathcal{H}$ and collect the related information.
5) Traverse every coalition S_k. If the *defection order* is satisfied and the newly formed coalition doesn't exist in the history *H*(*i*), record $u_i\left(S_k^{new},\mathcal{N}\right)$ and $\frac{u\left(S_k^{new},\mathcal{N}\right)}{\left\|S_k^{new}\right\|}$.
6) Transfer the RRH to the coalition which ensures the maximum $u_i\left(S_k^{new},\mathcal{N}\right)$ and $\frac{u\left(S_k^{new},\mathcal{N}\right)}{\left\|S_k^{new}\right\|}$, and add the new coalition into History *H*(*i*), and update $\mathcal{H}$.
End For
Until: Converge to a stable coalition partition.
3: **Transmission Stage:**
Within each coalition, TDMA based non-coherent joint transmission is performed.

is maximized, while the maximum $\frac{u\left(S_k^{new},\mathcal{N}\right)}{\left|S_k^{new}\right|}$ guarantees the utilities of other members in the selected coalition are maximized. These two constraints ensure that, the partition of selfish RRH doesn't impair the performance of original members in the selected coalition, and original members of the selected coalition can benefit from allowing the participation of other RRHs as well. Finally, RRHs in the same coalition access the channel by TDMA in the transmission stage. Note that **Algorithm 3** is periodically performed in the BBU pool, such that the variation of channel conditions can be overcome.

3) Performance Evaluation

The proposed SOCFA algorithm has some advantages, as described in the following theorem and corollaries.

Theorem 4.2: The complexity of the proposed SOCFA is $\mathcal{O}\left(\frac{B_{N+1}}{B_N}-1\right)$, where B_N is the Bell number.

Proof: The number of ways to partition a N-elements set into *r* non-empty subsets (*i.e.*, $|\mathcal{H}|=r$) is denoted by the Stirling number of second kind as:

$$S(N,r)=\frac{r^N}{r!}\sum_{k=0}^{r}(-1)^k\binom{r}{k}\left(1-\frac{k}{r}\right)^N,$$

where $1\le r\le N$. Obviously, $S(N,1)=1$, $S(N,N)=1$, and $S(N,r)=0$, $\forall r>N$ or $r<1$.

On the other hand, the possible ways to partition an N-elements set are given by the famous Bell number B_N. In such a case, the relationship between Bell number and the Stirling number of second kind should be:

$$B_N=\sum_{r=1}^{N}S(N,r).$$

From the statistics viewpoint, the possibility of dividing *N* RRHs into *r* non-empty subsets (*i.e.*, coalition number is *r*) is given by:

$$P(|\mathcal{H}|=r)=\frac{S(N,r)}{B_N}.$$

The expected number of coalitions is:

$$\mathbb{E}\left[|\mathcal{H}|\right]=\sum_{r=1}^{N}\frac{rS(N,r)}{B_N}.$$

Since the Stirling number of second kind can be generated by the recurrence relation $S(N+1,r) = S(N,r-1) + rS(N,r)$, it's obtained that:

$$\sum_{r=1}^{N}rS(N,r)=\sum_{r=1}^{N}S(N+1,r)-\sum_{r=1}^{N}S(N,r-1),$$

where:

$$\sum_{r=1}^{N}S(N+1,r)=\sum_{r=1}^{N+1}S(N+1,r)-S(N+1,N+1)=\sum_{r=1}^{N+1}S(N+1,r)-1,$$

and

$$\sum_{r=1}^{N} S(N,r-1) = \sum_{r=1}^{N+1} S(N,r-1) - S(N,N) = \sum_{r=1}^{N+1} S(N,r-1) - 1.$$

Therefore, (53) can be rewritten as:

$$\sum_{r=1}^{N} rS(N,r) = \sum_{r=1}^{N+1} S(N+1,r) - 1 - \left\{\sum_{r=1}^{N+1} S(N,r-1) - 1\right\}$$
$$= \sum_{r=1}^{N+1} S(N+1,r) - \sum_{r=1}^{N+1} S(N,r-1) = B_{N+1} - B_N.$$

Substituting (56) into (53):

$$\mathbb{E}\left[\left|\mathcal{H}\right|\right] = \frac{B_{N+1}}{B_N} - 1.$$

Therefore, the complexity of SOCFA is linear with the coalition number, *i.e.*, the complexity is $\mathcal{O}\left(\frac{B_{N+1}}{B_N} - 1\right)$. Note that with the introduction of variable $H(i)$, the complexity of SOCFA can be further reduced.

Lemma 4.1: The average number of members in a coalition is $\frac{NB_N}{B_{N+1} - B_N}$.

Proof: This conclusion can be derived directly from **Theorem 4.2**. According to (57), the expectation of number of coalitions in the system is $\mathbb{E}\left[\left|\mathcal{H}\right|\right] = \frac{B_{N+1}}{B_N} - 1$. Therefore, the average number of members in a coalition should be $\frac{NB_N}{B_{N+1} - B_N}$.

Corollary 4.1: The proposed algorithm SOCFA is converged.

Proof: Considering that the total number of coalitions is finite according to **Theorem 4.2**, and thus the number of such satisfying coalitions is limited. Although such a merge-and-split algorithm may lead to cycles among these satisfying coalitions and hence the system may not converge, our introduced *history* variable $H(i)$ prevents the RRH from revisiting a previously joined coalition again. Therefore, the partition structure finally converges and no deviation or defection occurs.

Corollary 4.2: The fairness of RRHs can be guaranteed by SOCFA.

Proof: According to the SOCFA, RRH i may maximize its utility by joining coalition where it can obtain the maximum $u_i\left(S_k^{new}, \mathcal{N}\right)$. However, the participation of RRH i may degrades the performance of RRHs that are already in S_k, since the new comer i may reduce the occupied time fraction of the original coalition members. In such a case, incorporating the new comer is unfair for original members. Nevertheless, such unfairness can be efficiently eradicated by the *defection order* which requires

$\frac{u\left(S_k^{new},\mathcal{N}\right)}{\left|S_k^{new}\right|} > \frac{u\left(S_k,\mathcal{N}\right)}{\left|S_k\right|}$, meaning that the new comer is accepted only if it can also benefit the original members in the jointed coalition. The fairness of RRHs in the system is thus guaranteed.

D. Simulation Results and Analysis

In this part, the characteristic and performance of the proposed game scheme are evaluated. The effectiveness and fairness of the proposed game scheme are evaluated, and the results are compared with the ones in the non-cooperative scheme and the conventional broadcasting based coalitional scheme (referred to as broadcasting scheme). It's considered that, *N* RRHs are randomly placed in a circular region with radius *R*. The coverage range of each RRH is a circle with radius of 20*m*. Only one UE is currently engaged in the downlink communication with each individual RRH. For simplicity, all RRHs in a coalition, say S_k, have an equal share of the transmission time, namely $\alpha_i = \frac{1}{\left|S_k\right|}$, and they all transmit with the same power p_i=30*dBm*. Here all antenna gains are normalized to the unit $A_i^T = A_i^R = 1, \forall i \in \mathcal{N}$. The sum of noise power and interference power from the macro BS is set to -30*dBm*. The pathloss model is set according to BeFemto specifications as:

$$PL(dB) = 18.7 \times \log_{10}(d) + 46.8 + 20 \times \log_{10}(2.47/5)$$

Note that all results shown in the analysis are computed by averaging 10000 runs.

The average and maximum sizes of the coalition vs. the number of RRHs are shown in Figure 7. Given the specific coverage distance, both the average and maximum coalition sizes increase as *N* increases. The influence of coverage distances on the size of coalition is also illustrated in Figure 7. It's observed that, given specific number of RRHs in the system, the coalition sizes, both in average and maximum,

Figure 7. Coalition size vs. number of RRHs N

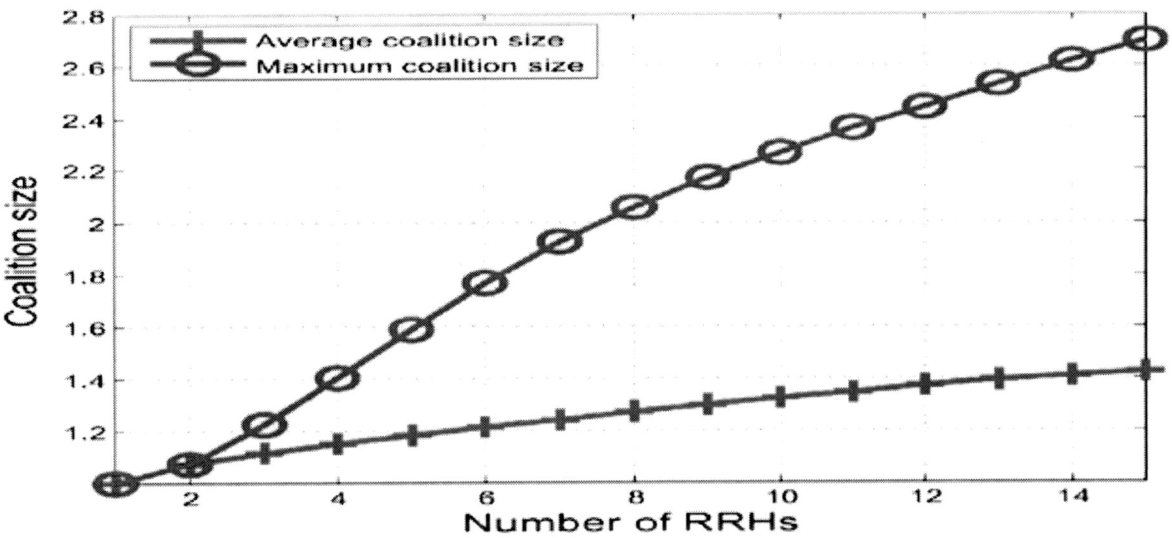

increase as the coverage distance decreases. Such phenomenons are due to the fact that, in low-density network where the interference among RRHs is low, RRHs choose to act non-cooperatively, resulting in singleton coalitions mainly. As the density of RRHs increases, the interference increases. Since forming coalitions with nearby players can efficiently suppress the severe interference, a cooperative strategy becomes more rewarding. Under such circumstance, RRHs tend to cooperate, which leads to the increase of coalition size. It should be pointed out that, own to the *defection order*, a coalition keeps to be in medium-size and will not expand to a large one.

Figure 8 exhibits how the RRH numbers and coverage distances influence the complexity of our proposed SOCFA, in terms of iterations. Figure 8 shows that the iteration number increases linearly to the number of RRHs. This is due to the fact that, given specific number of RRHs, smaller coverage distances generally mean larger possibility of cooperation and more optional coalitions. However, the iteration number of our proposed SOCFA doesn't exceed 3 even if the density of RRHs is very high, which means our SOCFA converge quickly.

The performances of the proposed game are verified, in terms of aggregate spectrum efficiency, average suffering interference per RRH, average utility per RRH and the system fairness. To this aim, our algorithm is compared with two other schemes: the non-cooperative scheme and the broadcasting scheme. In the non-cooperative scheme, all RRHs transmit with the maximum power, without cooperation with others. While in the broadcasting scheme, RRHs cooperate to transmit, and part of the power is consumed to broadcast information so as to form coalitions, leading to the reduction of the transmitting power.

Figure 9 shows the overall system utility in terms of the aggregate spectrum efficiency, as a function of the number of RRHs N. Figure 9 shows that the non-cooperative scheme has the worst performance. This is because the interference in the non-cooperative scheme is much greater than the ones in the other two schemes. With cooperation, the broadcasting scheme can suppress the interference and thus improves the system performance efficiently. Compared with the broadcasting scheme, our SOCFA algorithm owns more transmitting power. Thus, the SOCFA algorithm performs better than the broadcasting scheme. Figure 9 images that, as the number of RRHs increases, the performance gap among

Figure 8. Iterations vs. RRH number

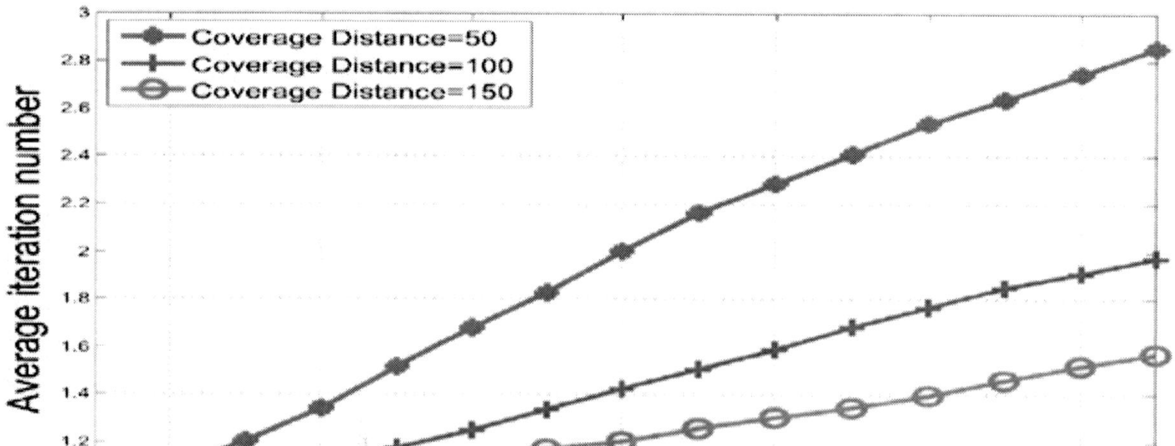

Figure 9. System utility vs. number of RRHs N

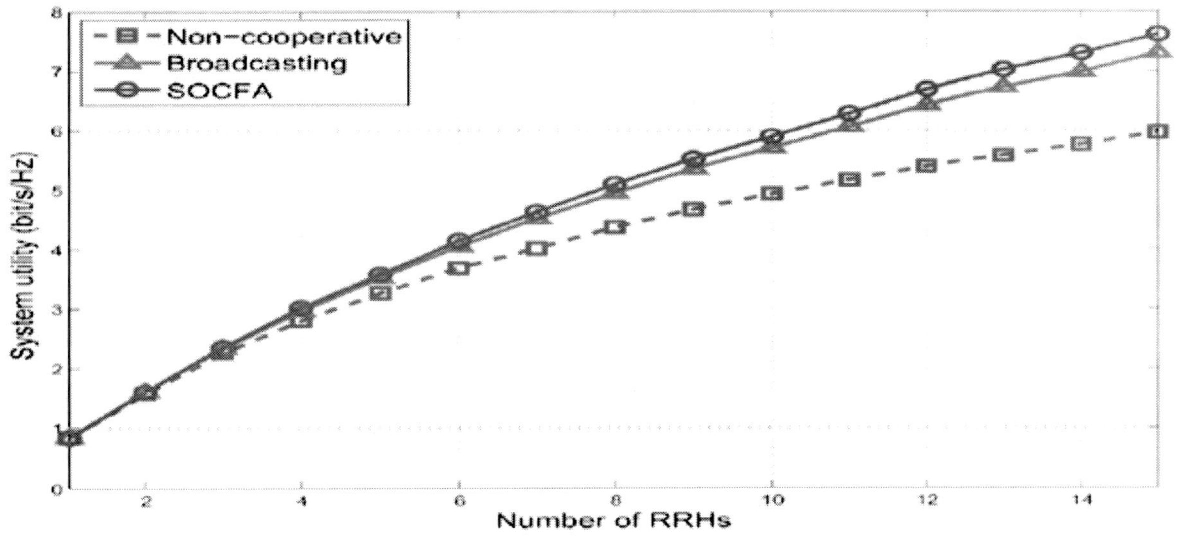

these schemes increases. At N=15, the aggregate spectrum efficiency is improved 4.52% and 28% by SOCFA algorithm compared to the broadcasting scheme and the non-cooperative scheme, respectively, which means cooperation becomes more important when the interference becomes more severe. Similar conclusions can be drawn by analyzing Figure 10, where the results of overall system performance vs. the coverage range are shown.

On the other hand, as the number of RRHs increases, the suffering interference per RRH is getting more severe in three schemes, as shown in Figure 11. As a consequence, the average utility of each individual RRH decreases, as shown in Figure 12. Figure 12 and Figure 11 show that, through coopera-

Figure 10. System utility vs. coverage distances for N=15

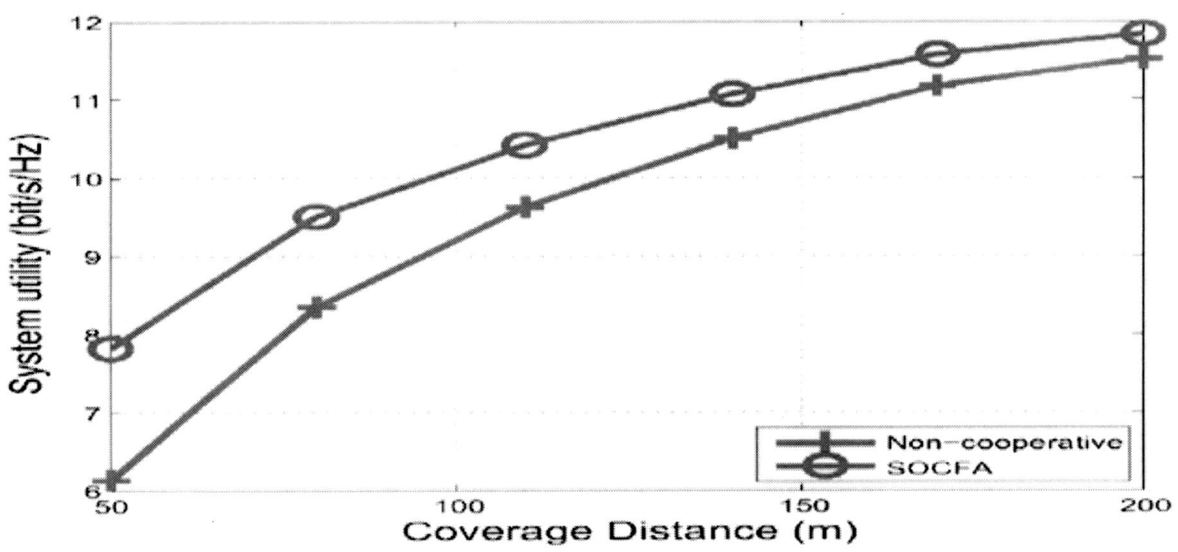

Figure 11. Average suffering interference per RRH vs. number of RRHs N

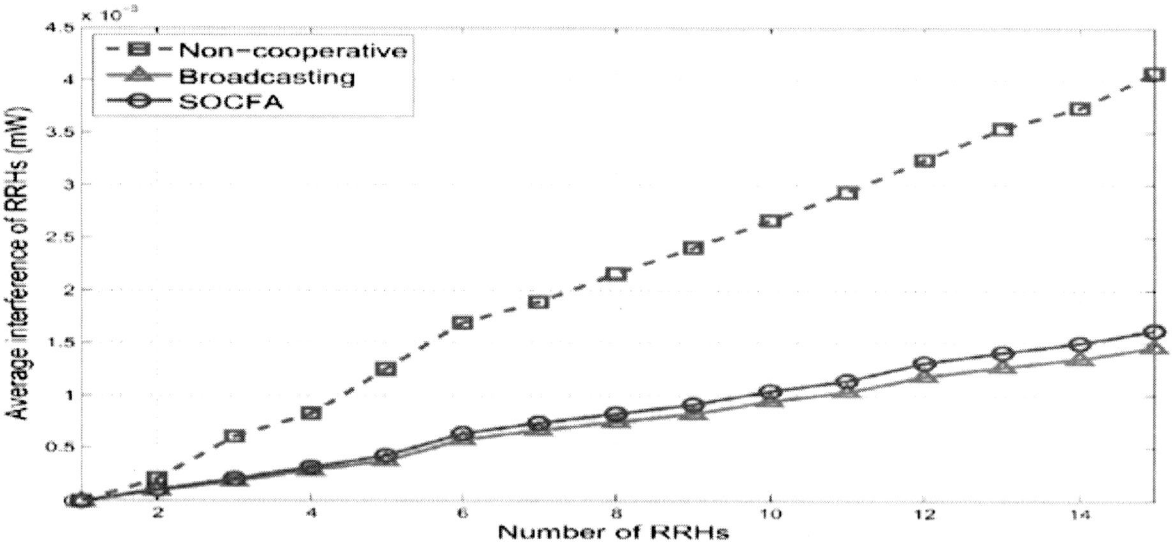

Figure 12. Average utility per RRH vs. number of RRHs N

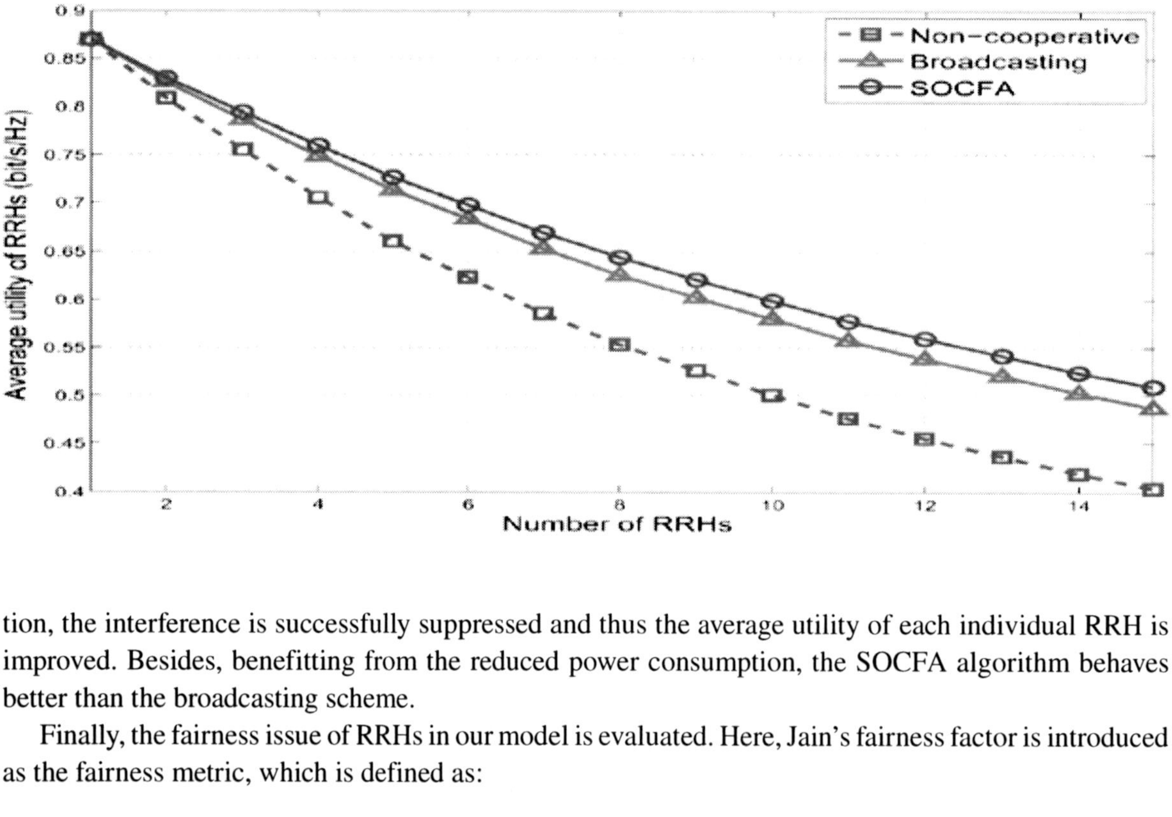

tion, the interference is successfully suppressed and thus the average utility of each individual RRH is improved. Besides, benefitting from the reduced power consumption, the SOCFA algorithm behaves better than the broadcasting scheme.

Finally, the fairness issue of RRHs in our model is evaluated. Here, Jain's fairness factor is introduced as the fairness metric, which is defined as:

$$\mathcal{F}_J = \frac{\left(\sum_{i=1}^{N} x_i\right)^2}{N \cdot \sum_{i=1}^{N} x_i^2},$$

Note that the fairness index varies between 0 and 1, with a larger value meaning more fairness.

Figure 13 illustrates that, the fairness factors of both our SOCFA algorithm and the non-cooperative scheme approximate 1 closely. This is due to the fact that, RRHs are deployed randomly, and thus each individual RRH statistically shares the same position in C-SCNs. With the increase in RRH numbers, the system fairness decreases, fluctuating closely to 1. This bias is caused by the insufficiency of simulation rounds and the deficient sample space. However, our SOCFA algorithm is more fair to RRHs than the non-cooperative one, since the impact of deviating RRHs on the original coalition members has been taken into consideration. Therefore, our proposed algorithm can significantly improve the system utility and guarantee the fairness of RRHs at the same time.

V. COALITION FORMATION FOR INTERFERENCE MITIGATION IN HETEROGENEOUS CLOUD SMALL CELL NETWORKS

In the heterogeneous cloud small cell networks (H-CSCNs), the small cells are underlaied within a macro cell. The intra-tier interference coordination problem in SCNs was modelled as a coalition formation game in characteristic form (Sun, Zhang, Peng, & Sun, 2014), where SBSs as game players can form coalitions and do their transmissions by using TDMA based non-coherent joint transmission mode to coordinate the interference within the coalition. To find the solution for the coalition formation game, a condition called Filter Condition that is used to find coalitions available to join is firstly introduced. Based on Filter Condition, the coalition formation algorithm for centralized implementation is proposed. Beyond that, the proposed algorithm is carefully designed such that the partition loop problem that may occur is successfully avoided.

Figure 13. Fairness vs. RRH number

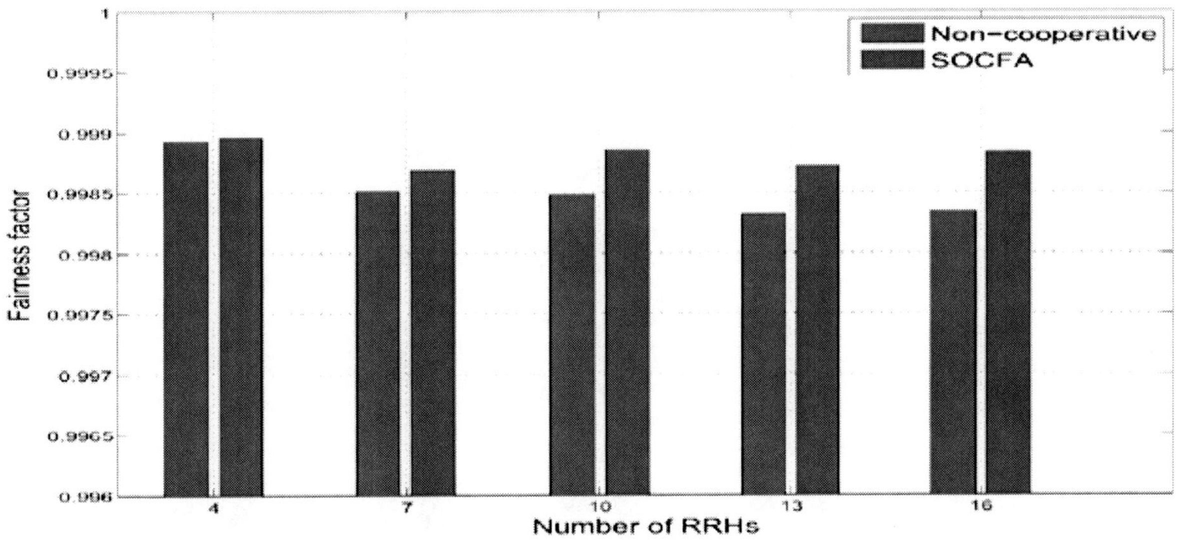

A. System Model for H-CSCNs

As shown in Figure 14, the downlink transmission scenario in an H-CSCNs is considered, where the MBS is responsible for basic coverage and SBSs are connected to the centralized BBU pool through fiber fronthauls. Let $\mathcal{N} = \{1,...,N\}$ be the set of N SBSs. For each SBS $i \in \mathcal{N}$, it can serve multiple UEs by means of OFDMA with only one UE being served on one sub-channel. Assuming that the sub-channels used by SBSs are independent of one another, in the following we only focus on one specific sub-channel. In such a case, UEs operating on that sub-channel can be identified by the same label as their SBSs. Nevertheless, we will show how to extend single sub-channel case to multiple sub-channel case in our proposed algorithm.

Since more than one SBS can operate on a certain sub-channel, when SBSs are densely deployed, one UE can suffer severe interference from SBSs in different small cells. To coordinate the interference, SBSs can form coalitions each of which is a non-empty subset of $\mathcal{N}$. Within each coalition, TDMA mode is used, and in each time-slot only one UE in the coalition is served by all SBSs in the same coalition through simultaneous data transmission. With such TDMA based non-coherent joint transmission, a UE only suffers interference from SBSs in other coalitions. Suppose SBS i joins coalition S_k, and then the useful signal and total interference from other coalitions received by its UE can be respectively written as:

$$P_i^{received} = p_i g_{i,i} + \sum_{l \neq i}^{l \in S_k} p_l g_{i,l},$$

and

$$I_{i,\mathcal{N} \setminus S_k} = \sum\nolimits_{j \in \mathcal{N} \setminus S_k} p_j g_{i,j},$$

Figure 14. System Model for H-CSCNs

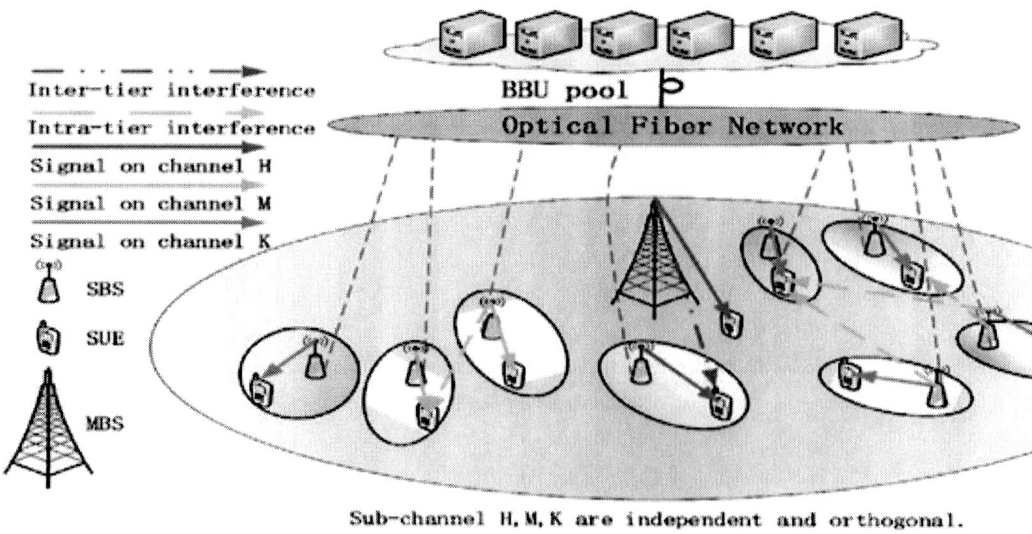

where p_i is the transmission power of SBS i, and $g_{i,j}$ is the channel gain from SBS j to UE i.

Assuming all the UEs in the same coalition equally share the channel in terms of time, the utility of SBS $i \in S_k$ defined as spectrum efficiency can be given as:

$$u_i(S_k, N) = \frac{1}{|S_k|} \log_2 \left(1 + \frac{P_i^{received}}{\sigma^2 + I_{i,N \setminus S_k}} \right),$$

where $|S_k|$ is the number of members of coalition S_k, $\frac{1}{|S_k|}$ is the time duration used by SBSs in coalition S_k to serve a UE in a normalized slot, and σ^2 is the variance of the Gaussian noise.

Then, the total utility of coalition S_k is defined as the sum of all its members' utilities as follow:

$$u(S_k, \mathcal{N}) = \sum_{i \in S_k} u_i(S_k, \mathcal{N}).$$

From (63), we can see the utility of an SBS only depends on the coalition it joins. Considering that SBSs are selfish and rational, the interactions of SBSs to form coalitions can be formulated as a coalition formation game.

B. The Coalition Formation Game

To give a better understanding of our model, two definitions are listed here.

Definition 5.1 (Coalitional Game in Characteristic Form) (Saad et al., 2009): A coalitional game in characteristic form is a coalitional game where the value of a coalition S depends solely on the members of that coalition, with no dependence on how the players in $\mathcal{N} \setminus S$ are structured.

Definition 5.2 (Transferable Utility): A transferable utility is a value function v that assigns a real number to each coalition $S \subseteq \mathcal{N}$, i.e., $v(S, \mathcal{N})$, which represents the total utility players in S can achieve, satisfying $v(\phi, \mathcal{N}) = 0$.

According to the two definitions, our coalition formation game is in characteristic form with TU.

C. Coalition Formation Algorithm

In this section, criteria for SBSs to find coalitions available to join are introduced. Later on, the process of forming coalitions for SBSs is formulated as a coalition formation game in characteristic form with transferable utility (TU).

1) Filter Condition

To help SBSs identify coalitions available to join, a condition called Filter Condition which contains two sub-conditions is introduced.

Definition 5.3 (Join Condition): Suppose SBS i is originally in coalition S and the candidate coalition SBS i considers to join is S'. Then the criteria for SBS i to deviate from S and join S' is called Join Condition as follows:

$$u_i(\tilde{S},\mathcal{N}) > u_i(S,\mathcal{N}),$$
$$u(\tilde{S},N) > u_i\left(\{i\},N\right) + \sum_{j\in S'} u_j\left(\{j\},N\right),$$
$$\frac{u(\tilde{S},\mathcal{N})}{|\tilde{S}|} > \frac{u(S',\mathcal{N})}{|S'|},$$

where $\tilde{S}$ is the newly formed coalition after SBS i joins S', i.e., $\tilde{S} = S' \bigcup \{i\}$.

From (64) we can see, the first condition guarantees that SBS i has enough incentive to deviate from the old coalition and join a new one, since its utility can be improved in the newly formed coalition. The second condition promises that the total utility of the newly formed coalition is larger than the sum of its members' non-cooperative utility, only based on which the SBSs are willing to cooperate. Then, the third condition makes sure that the average utility per SBS in the newly formed coalition increases, which interprets why the new member will be accepted.

Definition 5.4 (Split Condition): Suppose SBS i is originally in coalition S. Then the criteria for SBS i to deviate from S and act alone is called Split Condition as follows:

$$u_i(\{i\},\mathcal{N}) > u_i(S,\mathcal{N}),$$
$$u(S',\mathcal{N}) > \sum_{j\in S'} u_j(\{j\},\mathcal{N}),$$

where S' denotes the coalition formed by the remaining players in coalition S after SBS i deviates, i.e., $S' = S / \{i\}$.

From (65) we can see that the first condition is the basic motivation for an SBS to deviate, and the second condition is the requirement for the remaining SBSs in S' to cooperate.

However, it should be declared that when coalition S just has two players, only if $u_i(\{i\},\mathcal{N}) > u_i(S,\mathcal{N})$, will SBS i deviate from coalition S to act alone.

2) The Proposed Algorithm

With the criteria introduced, an algorithm to find the solution for the coalition formation game is able to be proposed. The coalition formation algorithm is given in **Algorithm 4**, and the algorithm can be executed in the centralized BBU. According to **Algorithm 4**, when it is the turn of SBS i to make decision it will find all the available coalitions which meet Filter Condition but are not recorded in variable H_i. After that, SBS i will choose to join the coalition which ensures the maximal individual utility.

Algorithm 4. Algorithm for coalition formation

The BBU pool first finds which SBSs are using the same sub-channel based on channel state information (CSI) fed back by SBSs, and then it executes the following process for each sub-channel:
3. Initial Stage:
A variable H is initialized for each SBS to record coalitions which it has ever joined.
4. Coalition Formation Stage:
Do iteration:
For each SBS using the chosen sub-channel:
4) According to the Filter Condition, the chosen player SBS i filters the existing coalitions based on Join Condition. Moreover, if coalition S_i, the current coalition SBS i belongs to, has more than one player, the singleton of SBS i also needs to be filtered based on Split Condition.
5) Record the coalition set $\overline{S}_i$ which contains coalitions meeting the Filter Condition.
6) If $\overline{S}_i$ is not an empty set, and then find the coalition $\hat{S}_i$ that guarantees the highest individual utility from $\overline{S}_i$.
If $\hat{S}_i \in H_i$, SBS i won't move.
Else if $\hat{S}_i$ is $\{i\}$, SBS i will deviate from the current coalition to form singleton $\{i\}$, and $\{i\}$ is added to H_i. For each $SBS{=}{\in}S_i$, the individual utility is updated.
Else SBS i will join the existing coalition $\hat{S}_i$ that will be added to H_i afterwards. For each $SBS \in S_i \cup \hat{S}_i$, the individual utility is updated.
End if.
End if.
Else SBS i won't move.
End if.
End for.
End iteration if no SBS deviates from the current coalition.
3. Transmission Stage:
Under the schedule of the BBU pool, TDMA based non-coherent joint transmission is performed within each coalition.

D. Performance Analysis

In this subsection, performance of the proposed algorithm in terms of convergence and stability is investigated.

1) Convergence

First of all, the convergence of **Algorithm 4** is discussed in the following proposition.

Proposition 5.1: Algorithm 4 guarantees to converge.

Proof: Considering the amount of different partitions of SBS set $\mathcal{N}$ is limited which is given by Bell number (Rota, 1964), the number of coalitions available for SBS i to join is limited. Furthermore, with variable H_i, SBS i is prevented from revisiting previously joined coalitions again. Therefore, convergence of the algorithm is guaranteed.

2) Stability

The outcome stability is an issue of crucial importance in a coalition formation game. In the following, we first give the definitions of stable partition and partition loop, and after that we discuss the stability of the proposed algorithm.

Definition 5.5 (Stable Partition): A partition of a coalition formation game is said to be stable if no player wants to deviate from the current coalition under this partition.

It is desirable to have a stable partition for the proposed coalition formation game. However, if the coalition formation algorithm is not properly designed, partition loop may occur.

Definition 5.6 (Partition Loop): For a coalition formation game with player set $\mathcal{N} = \{1,..., N\}$, if coalition formation algorithm is not properly designed, players will repeatedly join existing coalitions they have ever joined or form singleton, making partition updating process enter a loop which can be described by transition probability given as follows:

$$P(X^{(kN+i+1)} = \pi_{i+1} \mid X^{(kN+i)} = \pi_i) = 1, i \neq N,$$
$$P(X^{(kN+i+1)} = \pi_1 \mid X^{(kN+i)} = \pi_i) = 1, i = N,$$

where $X^{(n)}$ represents the partition of $\mathcal{N}$ at time n with initial state $X^{(0)}=\{\{1\},\ldots,\{N\}\}$ composed of singletons of players, $N>=2$, $i\in\{1,2,3,\ldots,N\}$, $k\in\{0,1,2,\ldots\}$, and π_i is one of the partitions in the loop. Such loop is called partition loop.

From **Definition 5.5** and **Definition 5.6**, we can see that if partition loop $\pi_1 \to \pi_2 \to \cdots \to \pi_N \to \pi_1 \to \pi_2 \to \cdots \to \pi_N \to \pi_1 \to \cdots$ occurs, a final and stable partition can't be achieved. To give an intuitive understanding about partition loop, here, we use our proposed algorithm without variable H as an example.

Consider a system composed of 3 SBSs, i.e., N=3 and $\mathcal{N} = \{1,2,3\}$. The utilities of these SBSs meet the following relationship:

$$u_2\left(\{2\},\mathcal{N}\right) > u_3\left(\{3\},\mathcal{N}\right) > u_1\left(\{1\},\mathcal{N}\right),$$
$$u_1\left(\{1,2\},\mathcal{N}\right) > u_1\left(\{1,3\},\mathcal{N}\right) > u_2\left(\{2\},\mathcal{N}\right) > u_1\left(\{1\},\mathcal{N}\right),$$
$$u_2\left(\{2,3\},\mathcal{N}\right) > u_2\left(\{1,2\},\mathcal{N}\right) > u_2\left(\{2\},\mathcal{N}\right),$$
$$u_3\left(\{1,3\},\mathcal{N}\right) > u_3\left(\{2,3\},\mathcal{N}\right) > u_3\left(\{3\},\mathcal{N}\right).$$

Without variable H, we can verify that the following partition loop will come out when SBSs acts non-cooperatively at the beginning:

{{1,2}, {3}} → {{1}, {2,3}} → {{1,3}, {2}} → {{1,2}, {3}} → {{1}, {2,3}} → {{1,3}, {2}} →, which means SBS 1 first forms coalition with SBS 2 then SBS 2 chooses to form coalition with SBS 3. After that SBS 3 chooses to form coalition with SBS 1 and later on SBS 1 decides to form coalition with SBS 2 again.

The partition loop also exists in other coalition formation schemes such as (Saad et al., 2009). To avoid the loop and give a stable partition, merge-only algorithm is proposed in (Guruacharya, Niyato, Bennis, & Kim, 2013), where split, one of the basic actions of coalitions, is forbidden. Different from the aforementioned algorithms, with variable H, our proposed algorithm is able to incorporate the split process, meanwhile, lead to a stable partition as stated in **Proposition 5.2**, which means that partition loop is eliminated.

Proposition 5.2: With variable H, the final outcome generated by our algorithm is stable.

Proof: We prove this proposition by contradiction. According to **Proposition 5.1**, there must be a final partition for our algorithm. Suppose that the final partition is not stable, then from **Definition 5.5**, the current partition can still be updated by the algorithm and a new partition will be generated, which contradicts the fact that the algorithm is converged. Hence, the final partition is stable and this stable partition, together with the corresponding individual utility vector, constitutes the final stable outcome.

E. SIMULATION RESULTS AND ANALYSIS

In this subsection, we evaluate the performance of our proposed algorithm by simulation. Consider that N SBSs are randomly deployed in a circular region with radius of R, while the coverage of each SBS is a circle with radius of 20m. The pathloss model is set according to BeFemto specifications as (Zhang, Song, Han, Saad, & Lu, 2013):

$$PL(dB) = 18.7 \times \log_{10}(d) + 46.8 + 20 \times \log_{10}(2.7/5)$$

Since it is assumed that all the sub-channels used by an SBS are orthogonal and independent, here we only focus on the performance of a single sub-channel where the SBS only serves one UE with transmission power $30dBm$.

First, the convergence of the proposed algorithm is verified and the results are shown in Figure 15. From Figure 15, we can see that the proposed algorithm always converges. On the other hand, given the radius of a circular region, the average iteration of the proposed scheme increases as the number of SBSs grows. This is because dense deployment of SBSs causes severe intra-tier interference, and hence an SBS has more incentive to cooperate, which makes the interactions among SBSs more complicated. This can also explain why the average iteration of the proposed scheme increases as the radius of a circular region gets smaller for a fixed number of SBSs.

Next, we evaluate the system performance of our proposed scheme by comparing it with that of the non-cooperative scheme in which SBSs don't form coalitions and just act alone.

In Figure 16, the overall system utility in terms of aggregate spectrum efficiency VS. the number of SBSs N is shown. From Figure 16, we can see that for the same number of SBSs, the overall system utility achieved by our algorithm is larger than that obtained by the non-cooperative one. This is because by cooperation one UE suffers less interference from other SBSs and thus the individual utility of the corresponding SBS is improved. From Figure 16, we can also see that as the number of SBSs in a region increases, the gap between the aggregate spectrum efficiency gained by our proposed algorithm and the non-cooperative algorithm widens. Specifically, when the system has 13 SBSs the system performance

Figure 15. Iteration number for convergence vs. Number of SBSs N

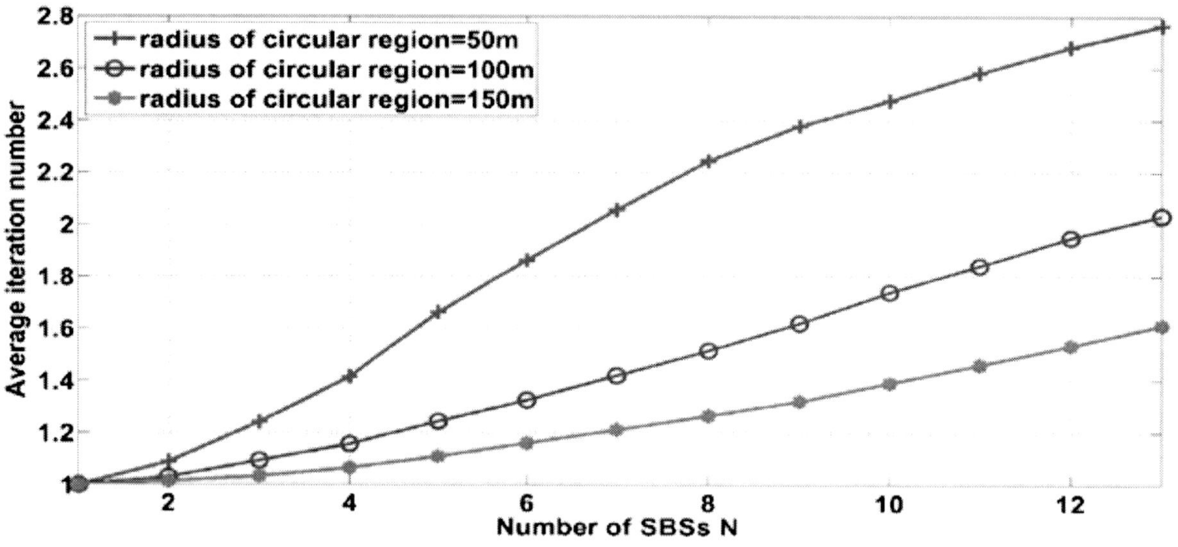

Figure 16. System utility vs. Number of SBSs N

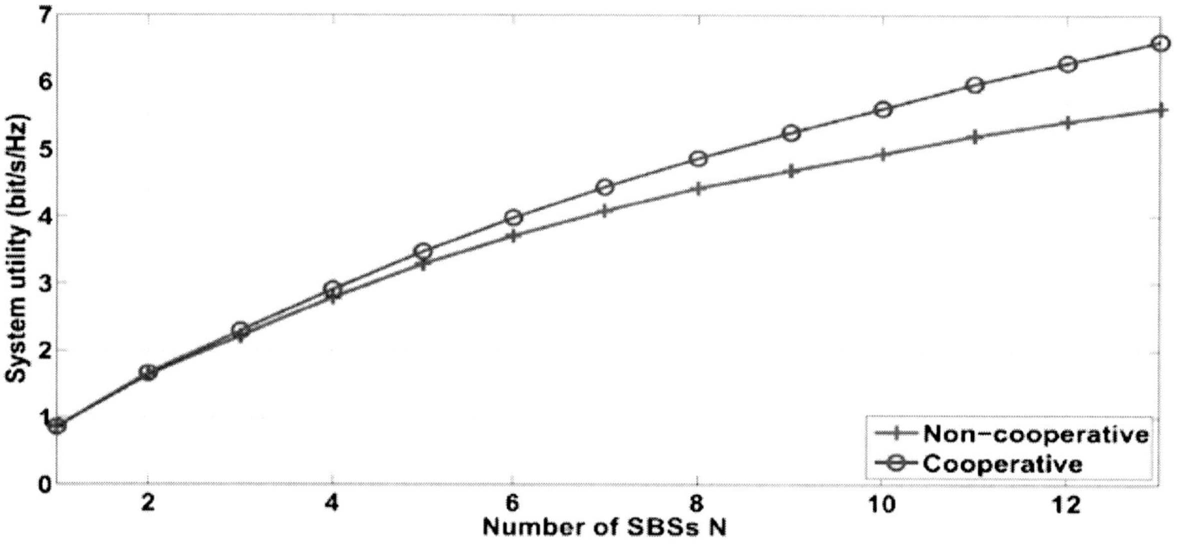

is improved by 21.91% with our scheme. This means that when the SBSs are deployed more densely and thus intra-tier interference becomes more severe, cooperation is more necessary. Such a conclusion can also be verified by Figure 17 and Figure 18, which indicate that with the density of SBSs increases, the average interference per UE suffers is significantly suppressed, and the average individual utility per SBS is greatly improved, respectively.

Figure 17. Average interference per UE suffers vs. Number of SBSs N

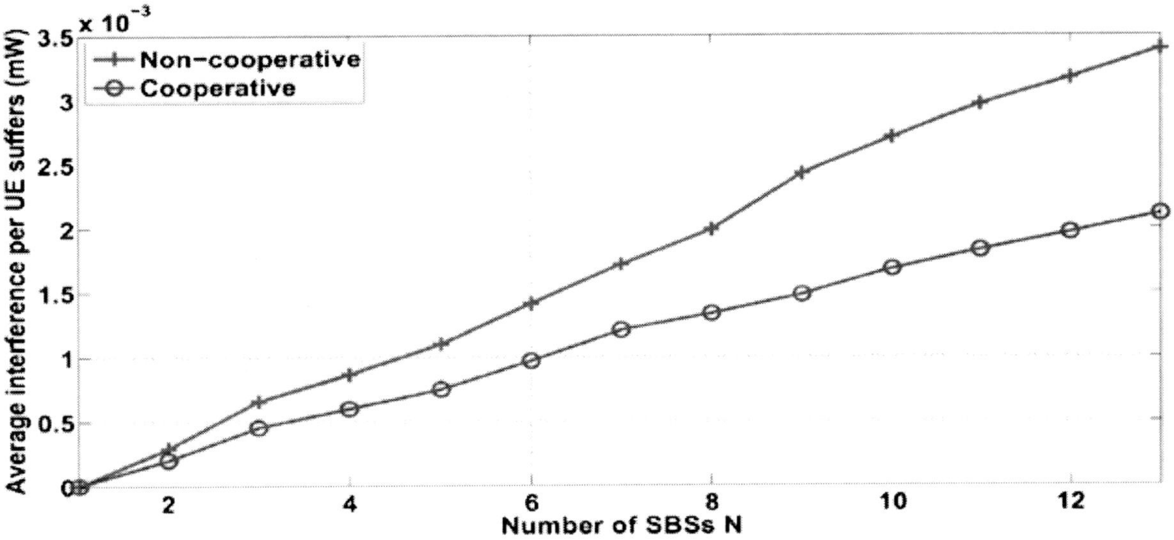

Figure 18. Average utility per SBS vs. Number of SBSs N

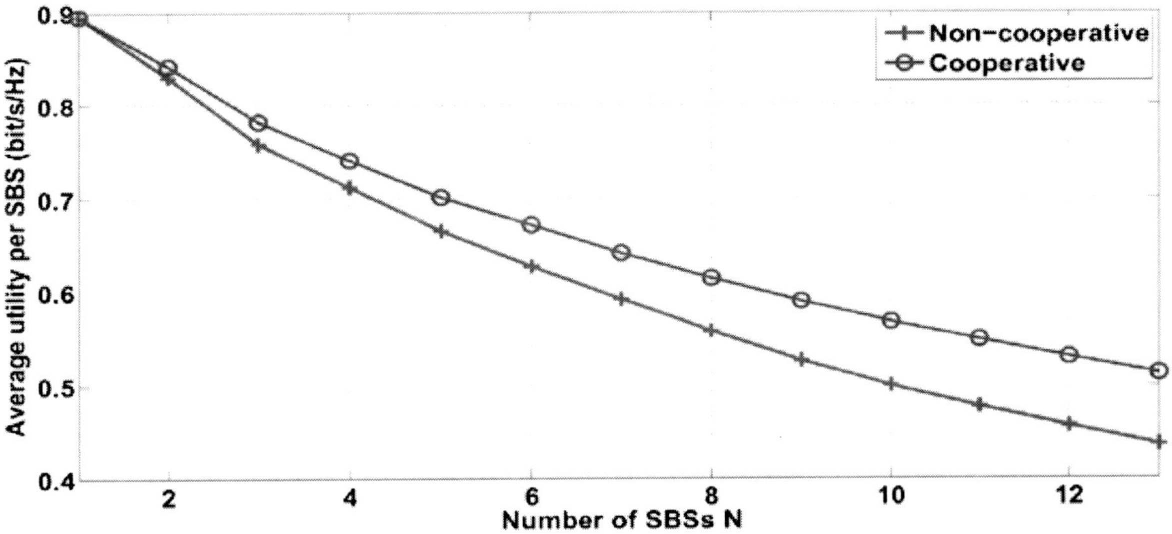

VI. SUMMARY

In this chapter, we have applied game theory to the resource optimization in HSCNs. Specifically, Stackelberg game is used to optimize the uplink transmitted power of SUEs to suppress their interference to an MBS, and to coordinate the interference among small cells, coalition formation game is used to make SBSs organized into different coalitions within each of which the downlink transmission is scheduled based on TDMA.

Many published works have shown that the game theory based resource allocation optimization model is an appropriate tool for analyzing problems existing in HSCNs. In the future, with the emergence of novel technologies such as green communication, IA, and cloud radio access, it is expected that game-theory-based models will become more abundant, notably in the context of communication problems involving user interaction, resource allocation, base station association, and distributed optimization.

REFERENCES

Ahmed, M., Peng, M., Ahmad, I., Li, Y., & Abana, M. (2013). Stackelberg game based optimized power allocation scheme for two-tier femtocell network. *Wireless Communications and Signal Processing (WCSP), 2013 International Conference on* (Vol.411, pp.1-6). IEEE.

Ahmed, M., Wang, W., Zhang, B., Sun, C., Ahmad, I., Rehman, W. U., & Yan, S. et al. (2015). A dynamic coalition formation framework for interference management in dense small cell networks. *Transactions on Emerging Telecommunications Technologies*, *26*(4), 718–731. doi:10.1002/ett.2901

Apt, K., & Witzel, A. (2009). A generic approach to coalition formation (extended version). *International Game Theory Review*, *11*(3), 347–367. doi:10.1142/S0219198909002352

Arnold, T., & Schwalbe, U. (2002). Dynamic coalition formation and the core. *Journal of Economic Behavior & Organization*, *49*(3), 363–380. doi:10.1016/S0167-2681(02)00015-X

Bacci, G., Bulzomato, A., & Luise, M. (2011). Uplink power control and subcarrier assignment for an OFDMA multicellular network based on game theory. *International Conference Perfect Evaluation methodology Tools (Value-Tools)*. doi:10.4108/icst.valuetools.2011.245691

Chandrasekhar, V., Andrews, J. G., & Gatherer, A. (2008). Femtocell networks: A survey. *IEEE Communications Magazine*, *46*(9), 59–67. doi:10.1109/MCOM.2008.4623708

Damnjanovic, A., Montojo, J., Wei, Y., Ji, T., Luo, T., Vajapeyam, M., & Malladi, D. et al. (2011). A survey on 3GPP heterogeneous networks. *IEEE Wireless Communications*, *18*(3), 10–21. doi:10.1109/MWC.2011.5876496

ElSawy, H., Hossain, E., & Haenggi, M. (2013). Stochastic geometry for modeling, analysis, and design of multi-tier and cognitive cellular wireless networks: A survey. *IEEE Communications Surveys and Tutorials*, *13*(3), 996–1019. doi:10.1109/SURV.2013.052213.00000

Ghimire, J., & Rosenberg, C. (2013). Resource allocation, transmission coordination and user association in heterogeneous networks: A flow-based unified approach. *IEEE Transactions on Wireless Communications*, *12*(3), 1340–1351. doi:10.1109/TWC.2013.013013.120940

Ghosh, A., Mangalvedhe, N., Ratasuk, R., Mondal, B., Cudak, M., Visotsky, E., & Novlan, T. D. et al. (2012). Heterogeneous cellular networks: From theory to practice. *IEEE Communications Magazine*, *50*(6), 54–64. doi:10.1109/MCOM.2012.6211486

3. GPP TR 36.902. (2011). *Self-configuring and self-optimizing network (SON) use cases and solutions. v9.3.1*. Academic Press.

3. GPP TS 32.500. (2011). *Self-organizing networks (SON); concepts and requirements. v11.1.0.* Academic Press.

Guruacharya, S., Niyato, D., Bennis, M., & Kim, D. I. (2013). Dynamic coalition formation for network MIMO in small cell networks. *IEEE Transactions on Wireless Communications, 12*(10), 5360–5372. doi:10.1109/TWC.2013.090513.130516

Han, Z., Niyato, D., & Saad, W. (2012). *Game theory in wireless and communication networks.* Cambridge, United Kingdom: Cambridge University Press.

He, G., Lasaulce, S., & Hayel, Y. (2011). Stackelberg games for energy-efficient power control in wireless networks. *2011. Proceedings - IEEE INFOCOM*, 2(3), 591–595.

IEEE 802.16ppc-10/0008r3. (2010). *Study report on hierarchical networks*. IEEE.

Jayaweera, S. K., & Tinaming, L. (2009). Dynamic spectrum leasing in cognitive radio networks via primary-secondary user power control games. *IEEE Transactions on Wireless Communications, 8*(6), 3300–3310. doi:10.1109/TWC.2009.081230

Koczy, L. A. (2006). A recursive core for partition function form games. *Meteor Research Memorandum, 63*(1), 41-51.

Lim, G., & Cimini, L. J. (2012). Energy-efficient cooperative relaying in heterogeneous radio access networks. *IEEE Communications Letters, 1*(5), 476–479. doi:10.1109/WCL.2012.070312.120366

Mehbodniya, A., & Aiffissa, S. (2012). Game-theoretic approach for interference management in heterogeneous multimedia wireless personal area networks. *IET Communications, 6*(15), 2278–2286. doi:10.1049/iet-com.2012.0297

Miao, G., Himayat, N., Li, G., & Talwar, S. (2011). Distributed interference-aware energy-efficient power optimization. *IEEE Transactions on Wireless Communications, 10*(4), 1323–1333. doi:10.1109/TWC.2011.021611.101376

Mukherjee, S. (2012). Distribution of downlink SINR in heterogeneous cellular networks. *IEEE Journal on Selected Areas in Communications, 30*(3), 575–585. doi:10.1109/JSAC.2012.120407

Nash, J. (1950). Equilibrium points in n-person games. *Proceedings of the National Academy of Sciences of the United States of America, 36*(1), 48–49. doi:10.1073/pnas.36.1.48 PMID:16588946

Niyato, D., & Hossain, E. (2008). A Non-cooperative game-theoretic framework for radio resource management in 4G heterogeneous wireless access networks. *IEEE Transactions on Mobile Computing, 7*(3), 332–345. doi:10.1109/TMC.2007.70727

Pantisano, F., Bennis, M., Saad, W., Debbah, M., & Latva-aho, M. (2013). Interference alignment for cooperative femtocell networks: A game-theoretic approach. *IEEE Transactions on Mobile Computing, 12*(11), 2233–2246. doi:10.1109/TMC.2012.196

Pantisano, F., Bennis, M., Saad, W., Verdone, R., & Latva-Aho, M. (2011). Coalition formation games for femtocell interference management: A recursive core approach. *Wireless Communications and Networking Conference (WCNC)*, 2011 IEEE (Vol.34, pp.1161-1166). IEEE. doi:10.1109/WCNC.2011.5779295

Quek, T., Roche, G., Guvenc, I., & Kountouris, M. (2013). *Small cell networks: Deployment, PHY techniques, and resource Allocation*. Cambridge, UK: Cambridge University Press. doi:10.1017/CBO9781139061421

Rota, G. C. (1964). The number of partitions of a set. *The American Mathematical Monthly*, *71*(5), 498–504. doi:10.2307/2312585

Saad, W., Han, Z., Debbah, M., & Hjorungnes, A. (2009). A distributed coalition formation framework for fair user cooperation in wireless networks. *IEEE Transactions on Wireless Communications*, *8*(9), 4580–4593. doi:10.1109/TWC.2009.080522

Saad, W., Han, Z., Debbah, M., Hjungnes, A., & Baser, T. (2009). Coalition game theory for communication networks: A tutorial. *IEEE Signal Processing Magazine*, *26*(5), 77–97. doi:10.1109/MSP.2009.000000

Shum, K. W., Leung, K.-K., & Sung, C. W. (2007). Convergence of iterative water-filling algorithm for Gaussian interference channels. *IEEE Journal on Selected Areas in Communications*, *25*(6), 1091–1100. doi:10.1109/JSAC.2007.070804

Sun, C., Peng, M., Sun, Y., & Li, Y. (2014). Distributed power control for device-to-device network using stackelberg game. *Wireless Communications and Networking Conference (WCNC)*, 2014 IEEE (pp.1344-1349). IEEE. doi:10.1109/WCNC.2014.6952365

Sun, C., Peng, M., Zhang, B., Sun, Y., Li, Y., & Wang, C. (2014). A coalitional game scheme for cooperative interference management in cloud radio access networks. *Transactions on Emerging Telecommunications Technologies*, *25*(9), 954–964. doi:10.1002/ett.2838

Sun, Y., Zhang, B., Peng, M., & Sun, C. (2014). Interference coordination in small cell networks using coalition formation game. *Communications in China (ICCC), 2014 IEEE/CIC International Conference on* (pp.642-646). IEEE.

IMT Vision. (2013). Framework and overall objectives of the future development of IMT for 2020 and beyond. *Working document toward preliminary draft new recommendation ITU-R M.[IMT.Vision]*, Working Party 5D/TEMP/126 Rev. 1.

Wang, B., Han, Z., & Liu, K. (2009). Distributed relay selection and power control for multiuser cooperative communication networks using stackelberg game. *IEEE Transactions on Mobile Computing*, *8*(7), 975–990. doi:10.1109/TMC.2008.153

Yates, R. D. (1995). A framework for uplink power control in cellular radio systems. *IEEE Journal on Selected Areas in Communications*, *13*(7), 1341–1347. doi:10.1109/49.414651

Yu, W., Ginis, G., & Cioffi, J. M. (2002). Distributed multiuser power control for digital subscriber lines. *IEEE Journal on Selected Areas in Communications*, *20*(5), 1105–1115. doi:10.1109/JSAC.2002.1007390

Zhang, Z., Song, L., Han, Z., Saad, W., & Lu, Z. (2013). Overlapping coalition formation games for cooperative interference management in small cell networks. *Wireless Communications and Networking Conference (WCNC)* (pp.643-648). IEEE. doi:10.1109/WCNC.2013.6554639

KEY TERMS AND DEFINITIONS

Cloud Small Cell Network: A network architecture where cloud computing is embedded into small cell networks and small cell base stations are connected to the cloud via low-latency fronthaul links.

Coalition Formation Game: A cooperative game that partitions a set of players into different cooperative groups according to a given coalition formation rule.

Game Theory: A methodology that is suitable to study the interactions among players when conflicts of interest exist.

Heterogeneous Small Cell Networks: A network architecture where low power nodes are deployed to supply high achievable data rates in hot spot areas while macro base stations are deployed to extend coverage and support mobility.

Interference Management: A way to improve the performance of communication systems by properly allocating radio resources among communication nodes to reduce mutual interference.

Power Control: A way to mitigate the interference caused by a node to other nodes by controlling the transmit power of the node while meeting the lowest quality of service requirement.

Stackelberg Game: A hierarchical game that contains leaders and followers. Leaders move first, and then followers take actions in respond to the decisions of leaders.

Chapter 7
Game Theory-Based Coverage Optimization for Small Cell Networks

Yiqing Zhou
Beijing Key Laboratory of Mobile Computing and Pervasive Device, Institute of Computing Technology, Chinese Academy of Sciences, China

Lin Tian
Beijing Key Laboratory of Mobile Computing and Pervasive Device, Institute of Computing Technology, Chinese Academy of Sciences, China

Liang Huang
National Computer Network Emergency Response Technical Team, Coordination Center of China, China

Jinglin Shi
Beijing Key Laboratory of Mobile Computing and Pervasive Device, Institute of Computing Technology, Chinese Academy of Sciences, China

ABSTRACT

Focusing on the coverage optimization of small cell networks (SCN), this chapter starts with a detailed analysis on various coverage problems, based on which the coverage optimization problem is formulated. Then centralized and distributed coverage optimization methods based on game theory are described. Firstly, considering the coverage optimization with a control center, a modified particle swarm optimization (MPSO) is presented for the self-optimization of SCN, which employs a heuristic power control scheme to search for the global optimum solution. Secondly, distributed optimization using game theory (DGT) without a control center is concerned. Considering both throughput and interference, a utility function is formulated. Then a power control scheme is proposed to find the Nash Equilibrium (NE). Simulation results show that MPSO and DGT significantly outperform conventional schemes. Moreover, compared with MPSO, DGT uses much less overhead. Finally, further research directions are discussed and conclusions are drawn.

DOI: 10.4018/978-1-4666-8642-7.ch007

INTRODUCTION

Recently, small cell networks (SCN) have been widely adopted to provide seamless cellular coverage for large indoor or outdoor areas. However, coverage optimization is a big challenge for SCN because these small cells are usually omnidirectional and ad-hoc deployed plug-and-play devices without precise network planning as traditional cellular networks. Game theory is promising to provide self coverage optimization for SCN. Aiming to minimize coverage problems such as coverage holes, loud neighbor overlap and cell overload, this chapter presents how to carry out self-optimization for the coverage in SCN based on game theory in centralized and distributed ways. Two algorithms are introduced, respectively, i.e., modified particle swarm optimization (MPSO) and distributed optimization using game theory (DGT). Their performances are evaluated via simulations.

BACKGROUND

With the development of smart phones, tablet computers and mobile application, the requirement for mobile data has increased exponentially these years. The cliffy increase has imposed great pressure on cellular networks. Small cells are regarded as a promising solution to the ever growing wireless capacity needs (Chandrasekhar, Andrews & Gatherer, 2008), with low-power (e.g. less than 20dBm) and low-cost (e.g. target cost 100$) wireless access points (WAP) that operate in licensed spectrum and are operator-managed and featured with edge-based intelligence ("Small Cell Forum", 2011; Hoydis, Kobayashi & Debbah, 2011). There are various types of access points, such as Femtocell, Picocell and Metrocell. Small cells have already been successfully used in residential environment to offload traffic from macrocells, improve user experience and lower cost per bit. Operators are now exploiting the potential of small cells to provide high capacity cellular service in both indoor and outdoor environments (Hoydis et al., 2011) through densely deployed small cell networks (SCN). SCN, also known as small cell grid, enterprise small cells and small cell clusters, is a group of densely deployed small cells together to extend the usage of small cells from residential environment to enterprise, urban and rural environments, i.e. high-traffic business environments, including office, retail malls, metro and municipal buildings, and so on ("Small Cell Forum", 2011). Thanks to Self-Organizing Network (SON) capabilities of small cells, SCN can outperform picocells or distributed antenna systems (DAS) by increasing system capacity and reducing radio planning and network deployment cost. However, new challenges arise for the coverage optimization because small cells are plug-and-play devices and deployed without precise network planning as macrocells.

Although self-configuration and self-optimization enabled small cells can automatically adjust their radio parameters (Claussen, 2007; Claussen, Ho &Samuel, 2008; Jo, Mun, Moon & Yook, 2010; Li, Qian & Kataria, 2009; Li, Macuha, Sousa, Sato & Nanri, 2009), most existing studies consider a heterogeneous network consisting of small cells and macrocells but not small cell clusters, focusing on maximizing indoor coverage while minimizing interference to public space without neighboring small cell collaboration. Downlink power control is studied for macrocell small cell overlay to guarantee quality of services (QoS) of users in both femtocells and macrocells (X. Li et al., 2009). A pilot and data power control method is also proposed, which can ensure a constant femtocell coverage radius and mitigate uplink interference to macrocells (Claussen, 2007). In addition, a cognitive femtocell scheme is introduced in the sense that channel reuse pattern is cognitively determined according to each femtocell's channel environment to avoid interference (Y. Li et al., 2009). Moreover, different auto-configuration and self-optimization methods are studied and a mobility event based self-optimization is proposed to optimize the coverage of femtocells in house environment in (Claussen et al., 2008). Finally, for two-tier femtocell networks, a self-optimized coverage coordination scheme is presented in (Jo et al., 2010), based on statistics of the signal and the interference power measured in the downlink of a femtocell. The proposed method can provide sufficient indoor coverage and reduce coverage leakage to outdoor areas.

It can be seen that most of the existing research focus on auto-configuration and self-optimization of a single small cell or with the overlap of a macrocell. If these traditional self-optimization methods are employed in SCN, each small cell will independently set the transmission power according to the requirements of its users or the interference constraints (Hong, Yun & Cho, 2009), which may result in coverage holes or strong interference between neighboring small cells. Moreover, the multiple home NodeBs constructing the small cell cluster are deployed jointly without precise network planning. Therefore, to seamlessly cover a specific area with small cell clusters, it is necessary to study the coverage optimization. This problem concerns jointly tuning pilot power of multiple home NodeBs in an optimized way, which is multi-dimensional and needs tight neighboring cell collaboration.

Studies on coverage optimization in SCN have been carried out (Ashraf, Claussen & Ho, 2010; Ho, Ashraf & Claussen, 2009; Hong, Yun & Cho, 2009; Treust, Tembine, Lassaulce & Debbah, 2010). Imran proposed distributed load balancing algorithm through updating small cell pilot transmit power in (Ashraf et al., 2010). Coverage holes can be minimized by adjusting the minimum pilot power threshold. In addition, a genetic algorithm is introduced for enterprise small cell optimization in (Ho et al., 2009), where a power control strategy tree is created for each small cell in an offline simulated way. However, the optimization does not converge to a static state but keeps adjusting power in one direction until incurring new problems. Moreover, distributed coverage optimization using cooperative power control in SCN with game theory is studied in (Treust et al., 2010). The algorithm uses a long-run repeated game which does not converge. Furthermore, distributed power control scheme for small cells is studied in (Hong et al., 2009) considering a central macrocell and several small cells. A non-cooperative game is formulated and proved to be a supermodular game. However, the assumption that every user in the cell has the same channel gains and equal capacity cannot hold in practice.

Due to the plug-and-play feature of small cell devices and the unplanned nature of SCN, the coverage optimization of SCN is expected to be carried out in a self-organized way. Moreover, since game theory is a promising methodology for self-optimization, it is desirable to investigate the coverage self-optimization of SCN based on game theory. The chapter firstly analyzes various coverage problems in SCN, such as coverage holes and overloading, based on which the coverage optimization problem is formulated. Then game theory is introduced to solve the problem in a centralized and distributed way. Two example algorithms, modified particle swarm optimization (MPSO) and distributed optimization using game theory (DGT), are introduced, respectively. Aiming to optimize the coverage of SCN, i.e., minimize coverage hole, avoid loud neighbor overlap and overload, MPSO is proposed, which works in a centralized SON architecture, automatically detects coverage problems and finds optimized transmit power using game theory. Firstly, associations are set up between the small cells and particles, transmit power and particle position, and power adjustment and particle velocity. In MPSO, received pilot signal strength is measured by user entities (UEs) and reported to small cell access points (SAP). This information as well as small cell load are shared among SAPs through backhaul connection (i.e. extended Home NodeB Application Protocol) and are used to evaluate coverage performance and detect coverage problems. Then MPSO can be employed to iteratively update SAP's pilot transmit power and find the best pilot transmit power for each SAP. On the other hand, distributed SON architecture is considered in DGT, which requires less on backhaul connection and is more suitable for network to scale up. Game theory is introduced to model the action of small cells. Throughput is used as payoff of each small cell which encourages each small cell to increase transmit power and remove coverage holes. Interference value is used as punishment to small cells to avoid selfish behavior leading the network to an undesirable equilibrium. A power adjust scheme is proposed which maximizes net utility of each cell considering both throughput and interference of each small cell. The DGT algorithm converges to the Nash Equilibrium (NE) which improves both network throughput and coverage ratio of SCN. The performance of MPSO and DGT is evaluated extensively with simulations. It is shown that the MPSO significantly outperforms single particle PSO (SPSO) and fixed power scheme with low complexity. Moreover, the performance of MPSO and DGT is compared to that of the simulated annealing algorithm (SA). It is shown that both MPSO and DGT outperform SA with higher coverage ratio, higher throughput and less iterations. In addition, MPSO and DGT converge quickly and are suitable for online optimization.

SYSTEM DESCRIPTION

A SCN is a wireless network consisting of multiple small base stations (SBS), working together in open access mode to cover a non-residential area such as a huge office or outside field. The small cells in the cluster can be organized in a centralized SON architecture or distributed SON architecture (3th Generation Partnership Project [3GPP], 2010) as shown in Fig.1 and Fig.2, respectively. Note that the OAM&SON (OAM: Operation Administration and Maintenance) entity in Fig.1 is the control center responsible for the coverage optimization of the SCN, while there is no such entity in distributed SON.

Consider a small cell cluster network where UEs and small cells are indexed as $i \in I$, $I=\{1,\ldots,M\}$ and $j \in J$, $J=\{1,\ldots,N\}$, respectively. The pilot transmit power of SBS j is denoted as p_j, which is usually a proportion β (e.g. 10%) of the total transmit power of small cell, P_{tx}, that is: $p_j=\beta P_{tx}$. The data transmit power of SBS j is denoted as p_j^d. The channel impact between SBS j and UE i is expressed as g_{ij} and the background noise is σ^2. Thus, the received pilot signal strength at UE i from cell j can be described as:

$$P_{rx}(i,j) = p_j g_{ij} \tag{1}$$

Thus the SINR of the data signal at UE i from cell j is given by

$$SINR_i = \frac{g_{ij} p_j^d}{\sigma^2 + \sum_{k \neq j} g_{ik} p_j^d} \tag{2}$$

which is limited by the background noise and also the interference from other cells.

Figure 1. Small cell networks with centralized SON

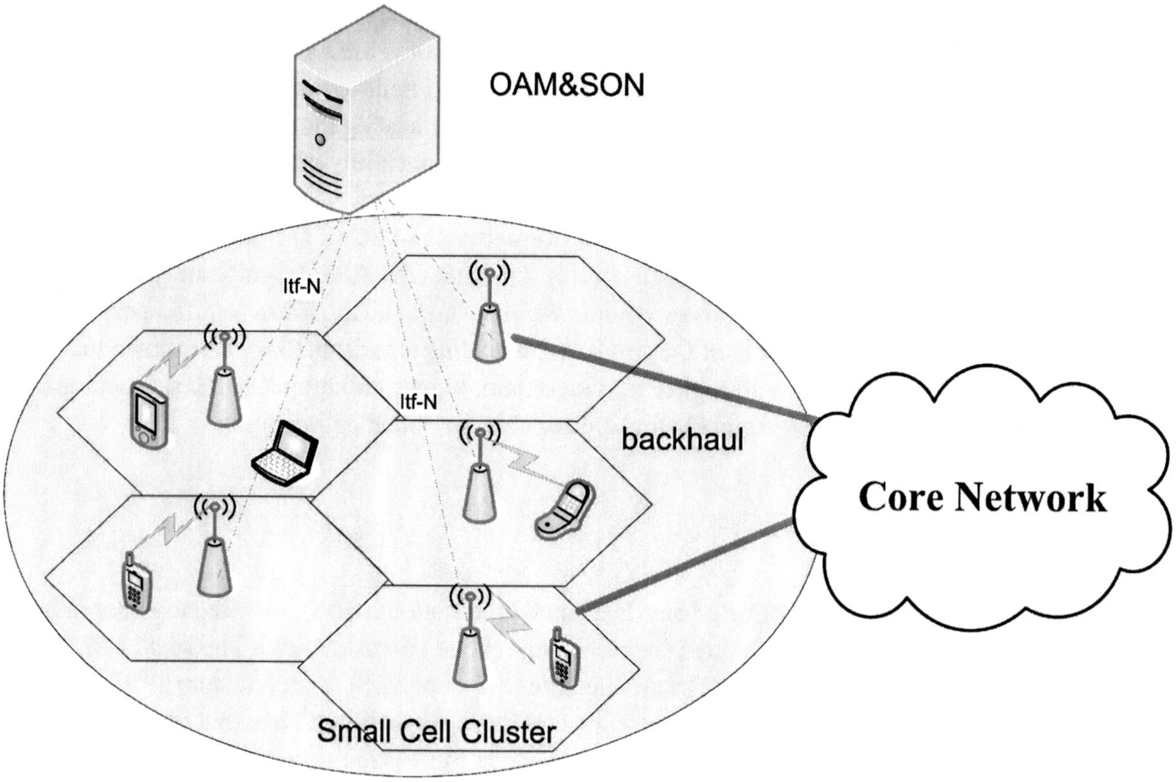

Figure 2. Small cell networks with distributed SON

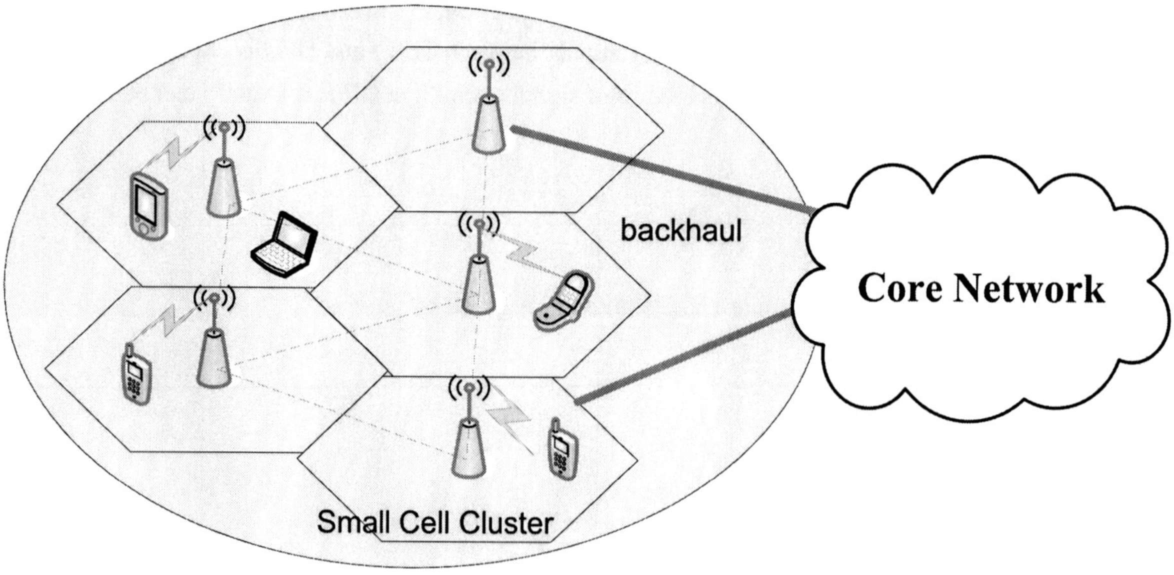

It is well known that poor coverage can easily degrade the user experience, which results in out-of-service, call setup failure, calls drop, poor voice quality, low download speed, and so on. The coverage problems can be described as follows.

Coverage Holes

Coverage holes are areas where the received pilot signal strength P_{rx} is below a threshold P_{cover} which is required by a UE to access the network, which can be defined as

$$\max P_{rx}(i,j) < P_{\text{cover}}, j \in J \tag{3}$$

For example, Fig.3(a) illustrates the situation when a UE *U* is in a coverage hole, i.e., it is not covered by SBS *A*, *B* or *C*. Note that when UEs are in coverage holes, it is impossible to measure their P_{rx}. One solution is to detect weakly covered users and reduce coverage holes by improving these weakly covered users. As shown in Fig.3(a), a weakly covered user UE *E* can be defined as:

$$P_{cover} < \max P_{rx}(i,j) < P_{weak}, j \in J \tag{4}$$

where P_{weak} is the threshold setting to detect the weak coverage.

Loud Neighbor Overlap

To seamlessly cover an area, cells should overlap with each other to avoid coverage holes and ensure that UEs can handover from one cell to another. However, if there is too much overlap, UEs in the overlapped area may receive multiple strong signals, which results in ping pong handovers. UEs may even fail to initiate service due to high interference level. As shown in Fig.3(b), a UE *V* is in loud neighbor overlap if there is more than one strong signals, which can be described as

$$\begin{cases} J_{overlap} = \{j \mid P_{rx}(i,j) > P_{strong}\} \\ |J_{overlap}| \geq 2 \end{cases} \tag{5}$$

where $J_{overlap}$ is the set of the small cells from which the UE *V* receives a strong pilot signal and P_{strong} is the threshold used to judge if a received pilot signal is strong or not. If a UE is detected to be in a loud neighbor overlap area, all the cells with a strong pilot should decrease transmission power to make the received pilot power in the overlapped area to a reasonable level.

Cell Overload

The third problem is cell overload mainly caused by the limited capacity of a small cell. Although small cell capacity has been increased from the previous 4 channels to 16, 32 and even 64 channels, load balance is still needed when serving UEs in a cell reaches the cell capacity to avoid further call drops. Cell overload is defined as:

Figure 3. Illustrations of coverage problems

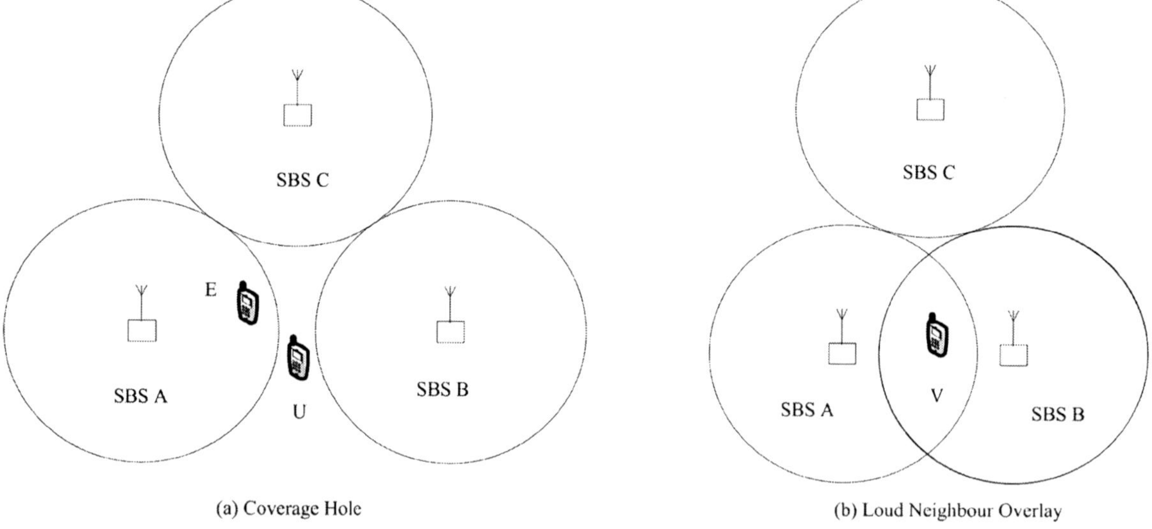

$$|U(j)| \geq \Omega \tag{6}$$

where $U(j)$ is the served user set in the small cell j and Ω is the cell capacity measured by the number of UEs. In practice, to prevent cell overload, load balance is carried out when the cell is highly loaded, i.e., when $|U(j)| \geq \Omega - c_o$ (c_o is the reserved resource measured by the number of UEs). Therefore, once high loading is detected by the small cell, it should shrink the coverage area to offload some users at the cell edge to neighbor cells, which on the contrary need to increase the transmission power to enlarge the coverage.

Low Throughput

Another coverage problem is low throughput, which also has a strong impact on user experience. The throughput of the cell can directly reflect coverage condition. Given the received signal SINR (see (2)), the normalized throughput of UE i can be described as:

$$t_i = \log_2(1 + SINR_i) = \log_2(1 + \frac{g_{ij} p_j^d}{\sigma^2 + \sum_{k \neq j} g_{ik} p_k^d}) \tag{7}$$

Thus the total normalized throughput of the SBS j is:

$$T_j = \sum_{i \in U(j)} \log_2(1 + \frac{g_{ij} p_j^d}{\sigma^2 + \sum_{k \neq j} g_{ik} p_k^d}) \tag{8}$$

Obviously UE with stronger received power and less interference will get higher throughput while either weak received power or strong interference may generate low throughput. Thus low throughput reflects either weak coverage or strong interference.

SCN Coverage Optimization

As described above, coverage problems in small cell clusters mainly include coverage holes, loud neighbor overlap, cell overload and low throughput. The target of coverage optimization is to minimize these problems. It should be noted that to reduce coverage holes, loud neighbor overlap and cell overload, a control center could be deployed to make the small cells collaborate with each other. Frequent information exchange is needed between SBS and central controller which needs sufficient backhaul connection. If there is no control center in the system or the backhaul connection is limited, distributed coverage optimization should be employed which aims to maximize the throughput and only needs little information exchange between neighbors.

For centralized method, the target is to reduce coverage hole, avoid loud neighbor overlap and cell overload. Thus the objective function can be formulated as an area coverage ratio, given by

$$f = \frac{\sum S_{cover}}{\sum S_{cover} + \sum S_{hole} + \sum S_{overlap}} \tag{9}$$

where ΣS_{cover} is the area properly covered without coverage holes and overlapping, ΣS_{hole} and $\Sigma S_{overlap}$ are the area of coverage holes and overlapped area with loud neighbors, respectively, and $\Sigma S_{cover}+\Sigma S_{hole}+\Sigma S_{overlap}$ is the target area.

However, in real networks, it is difficult to measure the area of these spaces. Therefore, instead of the area coverage ratio shown in (9), a user coverage ratio (see (10) below) is employed to evaluate the coverage quality. The optimization problem is thus formulated as follows:

$$\max \frac{N_{served}}{N_{served} + N_{hole} + N_{loud} + N_{overload}} \tag{10}$$

s.t.

$$P_{\min} \le P_{tx}(j) \le P_{\max}, j \in J \tag{11}$$

$$| U(j) | \le \Omega \tag{12}$$

where N_{served}, N_{hole}, N_{loud}, and $N_{overload}$ are the number of normal users, users in holes, users in loud neighbor overlap area and users rejected by cell overload, respectively; $P_{\min}$ and $P_{\max}$ are the minimum and maximum transmission powers of the SBSs, respectively. As described before, N_{served}, N_{hole}, N_{loud} and $N_{overload}$

can be obtained from measurement reports of UEs. Obviously, when there are more users in normal state and less users in holes, loud neighbor overlap area and overload, the user coverage ratio (10) increases, demonstrating a better coverage quality.

For distributed method, the target is to maximize each SBS's throughput with the same capacity and power limit in Equ.(11)(12). The object function can be described as:

$$\forall j \in J, \max \sum_{i \in U_j} \log_2 (1 + \frac{g_{ij} p_j^d}{\sigma^2 + \sum_{k \neq j} g_{ik} p_k^d}) \tag{13}$$

The game theory based solutions of the two methods will be described in detail in the following two parts.

CENTRALIZED COVERAGE OPTIMIZATION USING MPSO

As shown in Figure 1, in centralized SON architectures, each SBS communicates to OAM entity through *Itf-N* interface via backhaul connection. The OAM entity gets the global view of the network, and coverage problems can be detected using the method described in the previous part. Then centralized coverage optimization is carried out in the entity to minimize coverage holes, loud neighbor overlap and cell overload. The objective function can be formulated as the maximization of coverage ratio, given by (10)-(12). It can be seen that the problem (10) is multi-dimensional and NP-hard (Siomina, Varbrand, &Yuan, 2006). A modified PSO is proposed to solve this multi-dimension problem.

Overview of Particle Swarm Optimization

PSO is a population based stochastic optimization technique, inspired by social behavior of bird flocking or fish schooling (Kennedy & Eberhart, 1995). It has been successfully applied in various research areas to solve multidimensional problems. In PSO, each single solution is a "bird" in the search space, which is called "*particle*". The optimization objective is the fitness values of all the particles. A group of particles fly through the problem space by following the current optimum particles and finally the global optimum solution can be obtained ("Swarm Intelligence").

In small cell clusters, the behavior of small cells is quite swarm like: they cannot be too close or too far away; change of one entity (small cell) may affect neighboring entities; neighboring entities need to collaborate together to get a global best solution. Therefore, PSO can be employed to solve the coverage optimization problem of small cell clusters. First of all, associations are set up between the small cell cluster and the particles in the swarm, pilot transmit power of small cells and particle position, and power adjustment and the velocity of the particle. Moreover, (10) can be taken as the fitness function in PSO.

In standard PSO (Kennedy & Eberhart, 1995; Shi & Eberhart, 1998), each particle position P_i represents a possible solution, where $i \in [1,N]$ is the index of the particle and N is the population of the particle group. Accordingly, in coverage optimization for SCN, the pilot transmit power of the SCN i can be associated to the particle position, i.e., $\overrightarrow{P_i} = \left[Ptx_1^i, Ptx_2^i, \ldots, Ptx_n^i\right]$, where n is the total number of SAPs in the cluster. Assuming that there are N SCNs, the pilot transmit power of SCN i is updated according to PSO algorithm as follows

$$\overline{P}_i(t+1) = \overline{P}_i(t) + \overline{\delta}_i(t+1) \tag{14}$$

$$\vec{\delta}_i(t+1) = \omega\vec{\delta}_i(t) + c_1 r_1 \left[\vec{P}^{i}_{sbest}(t) - \vec{P}_i(t)\right] + c_2 r_2 \left[\vec{P}_{gbest}(t) - \vec{P}_i(t)\right] \tag{15}$$

where $\overline{P}_i(t+1)$ and $\vec{P}_i(t)$ are the updated and current and pilot transmit powers of SCN i, respectively, and $\overline{\delta}_i(t+1)$ and $\vec{\delta}_i(t)$ are the updated and current power adjustment, respectively. Moreover, ω is the inertia weight which is usually set to a positive value decreasing from 1.4 to 0.5 to improve the performance of PSO (Shi & Eberhart, 1998). $\overline{P}^{i}_{sbest}(t)$ is the best solution found by particle i in the previous iterations, i.e., the pilot transmit power vector for the SCN that can achieve the best coverage . $\vec{P}_{gbest}(t)$ is the best pilot transmit power vector found by all N particles in the previous t iterations. r_1 and r_2 are random values uniformly distributed in [0, 1]. The second term in (15) $c_1 r_1 \left[\vec{P}^{i}_{sbest}(t) - \vec{P}_i(t)\right]$ stands for *personal influence* and the third term $c_2 r_2 \left[\vec{P}_{gbest}(t) - \vec{P}_i(t)\right]$ is called *social influence*. c_1 and $c_2$are weights for personal and social terms, respectively, and both can be set to 2 (Kennedy & Eberhart, 1995).

To find the global optimum solution, the standard PSO needs multiple particles (multiple SCNs) to search through the solution space. However, in practice, a more general case is to find the optimum coverage of one SCN, where the standard PSO degrades to single particle swarm optimization (SPSO), Accordingly, (15) becomes

$$\delta_i(t+1) = \omega\delta_i(t) + c_1 r_1 \left[P^{i}_{sbest}(t) - P_i(t)\right] \tag{16}$$

where there is no social influence term since only one SCN is concerned. As a result, SPSO has limited capability to find the global optimum solution and the performance is not so good as the standard PSO.

Modified Particle Swarm Optimization

A modified PSO (MPSO) has been proposed according to small cell cluster environment, which can enhance the performance of SPSO, reduce the calculation and is more powerful in searching the global optimum solution for coverage optimization of small cell clusters (Huang, Zhou, Hu, Han & Shi, 2012). With the centralized SON architecture, the SON entity can get the knowledge of the whole network. Thus it is easy to use the current coverage information to guide the PSO search process: if the signals from two SBSs are both too strong, then either one may need to reduce the transmission power; if there is a coverage hole or weak coverage, then the nearest SBS needs to increase the transmission power so that the received signal power of weakly covered users can be improved. Also, to avoid cell overload, the highly loaded cell can decrease power to offload edge users to the loudest neighbor cells while the neighbor cells may need to increase power so that the UEs can handover to them without call drops. Based on these rules, a heuristic reactive power adjustment scheme can be introduced

$$\lambda_j(t) = \omega_1 \Delta^j_{hole} + \omega_2 \Delta^j_{loud} + \omega_3 \Delta^j_{load} \tag{17}$$

where $\Delta^j_{hole} > 0$ is the power increment to be applied to SBS j at iteration t according to weak coverage detection and coverage hole avoidance, $\Delta^j_{loud} < 0$ is the power decrement of SBS j to reduce loud neighbor overlap, and Δ^j_{load} is the power modification needed in the cell overload avoidance. ω1, ω2, ω3 are the weights of the three different adjustment which can be set to a positive value in [0,1]. Denoting $\vec{\lambda}(t) = [\lambda_1(t), \lambda_2(t), .., \lambda_N(t)]$ and taking (15) by omitting index i, PSO is modified by introducing $\vec{\lambda}(t)$ as the social influence (MPSO):

$$\vec{P}(t+1) = \vec{P}(t) + \vec{\delta}(t+1) \tag{18}$$

$$\vec{\delta}(t+1) = \omega\vec{\delta}(t) + c_1 r_1 [\vec{P}_{sbest}(t) - \vec{P}(t)] + c_2 r_2 \cdot \vec{\lambda}(t) \tag{19}$$

$$\vec{P}(t+1) = \begin{cases} \min(\vec{P}(t+1), P_{\max}) \\ \max(\vec{P}(t+1), P_{\min}) \end{cases} \tag{20}$$

where $\vec{P}(t) = [Ptx_1(t), Ptx_2(t), ..., Ptx_n(t)]$ is the pilot transmit power of the concerned SCN at iteration t and $\vec{\delta}(t+1)$ is the power adjustment.

In detail, MPSO consists of three algorithms. Algorithm1 is the framework of MPSO. Algorithms 2 and 3 are coverage check and power adjustment algorithms, respectively.

The MPSO combines swarm intelligence with coverage characters of cellular network. The calculated velocity through power control mechanism accelerates the search speed of PSO even when there is only one particle in the swarm. Simulation results will be discussed later.

DISTRIBUTED COVERAGE OPTIMIZATION USING GAME THEORY

Although MPSO can provide a solution for centralized coverage optimization, a centralized controller should be employed and sufficient backhaul connection is needed to support frequent information change between SBS and the controller. When small cells are deployed in distributed architectures without a centralized controller, different optimization algorithms should be designed, such as the distributed coverage optimization using game theory (DGT) proposed in (Huang et al., 2013). As discussed before, low throughput can also reflect coverage problems such as coverage hole or loud neighbor overlap. Therefore, in distributed SCNs, the target of coverage optimization could be to maximize the throughput of each SBS.

It is noted that with higher transmission power, users under the SBS can get higher SINR and thus the SBS gets higher throughput. However, higher transmission power also produces higher interference to neighbors and reduces the throughput of neighboring SBSs. Thus, the neighboring SBS may increase

Algorithm 1. Pseudo-code of the SCN coverage optimization using MPSO

Initialize n small cell position evenly in the office
Initialize m user position randomly
Initialize small cell pilot transmit power $P_i(t)$ randomly
Initialize power change value $\delta_i(t)$ randomly within maximum velocity
set periodically measurement report to all the UEs in the cluster
while coverage < target AND iteration < target_times
do
collect measurement information form UEs
coverage_check
if cover_ratio > best_ratio
best_ratio = cover_ratio
$\overrightarrow{P}_{sbest}(t) = \overrightarrow{P}(t)$ //save the best solution
end
update_power
end
set femtocell pilot power with P_{gbest}

its transmit power to get higher throughput. This is a non-cooperative game and can easily get P_{max} as the Nash Equilibrium. However, this Nash Equilibrium is obviously inefficient and should be avoided (Treust et al., 2010).

To avoid the selfish behavior of SBS leading the SCN to an undesired equilibrium with high interference, a punishment cost is introduced to limit the interference between neighboring cells, which is described as the function of interference to neighboring SBS normalized by the variance of the noise, given by

$$c_j = \sum_n \frac{p_j^d g_{nj}}{\sigma^2} \tag{21}$$

where c_j is the punishment cost of SBS j, n is the index of the UEs in the neighboring cells, g_{nj} is the channel impact from SBS j to the user n. Note that different punishment has been introduced in previous research. For example, the linear price function of power cp_j^d is used as price part in uplink power control game in (Douros, Toumpis & Polyzos, 2012). However, this cannot reflect the influence to other players. Equ.(21) shows that higher transmission power will get more punishment which reflects higher interference. Moreover, SBS may get more punishment if there are many neighboring edge users interfered by this cell. Considering both the throughput of the own cell and the interference to other cells, the net utility function of SBS j is given by

Algorithm 2. Pseudo-code of coverage_check

1: for each connected UE i in the SCN
2: do
3: $S = sort(P_{rx}(i,J),'descend')$ // sort and store to array S by descend order, $P_{rx}(i,J)$ is the pilot power set received by UE i from all cells in the cluster
4: if $S(1)<P_{cover}$
5: inc_coverage_hole
6: $N_{hole}=N_{hole}+1$
7: elseif $S(1)<P_{weak}$
8: select the serving cell j
9: $si_j=\max(si_j, P_{weak} - S(1))$;//self_increase
10: $k=find(P_{rx}(i,J)=S(2))$ //max neighbor
11: **if** $S(2)<P_{cover}$
12: $ni_k=\max(ni_k, P_{cover} - S(2))$ //neighbor increase
13: **end**
14: elseif $S(2)>P_{strong}$//loud neighbor overlap
15: $N_{loud}=N_{loud}+1$;
16: $sd_j = \max(sd_j, (S(1) - P_{strong})*0.8)$;//self_decrease
17: **for** $t=2$; $S(t)>P_{strong}$ AND $t<M$; t++
18: **do**
19: $K=find(P_{rx}(i,J)==S(t))$ //loud neighbor
20: $nd_k=\max(nd_k, S(t) - P_{strong})$ //neighbor decrease
21: **end**
22: end
23: if $Load(j)>\Omega-c_o$ //High load
24: $sd_j=\max(sd_j, S(\Omega-c_o) - P_{cover})$;
25: **for** $t=\mathbf{\Omega}$-co: $Load(j)$
26: **do**
27: $S=sort(P_{rx}(t,J)$, 'descend')
28: $k=find(P_{rx}(t,J)==S(2))$
29: $ni_k=\max(ni_k, P_{cover} - S(2))$
30: **end**
31: end
32: if $Load(j)>U$
33: $N_{overload}=N_{overload}+Load(j)-\Omega$;
34: $Load(j)=\Omega$;
35: end
36: end
37: $cover\_ratio=N_{served}/(N_{served}+N_{hole}+N_{loud}+N_{overload})$

Algorithm 3. Pseudo-code of update_power of small cells in the SCN

1: **for** each small cell i in the SCN
2: do
3: λ_i=max(si_i+ni_i) – min(sd_i+nd_i)
4: $\delta_i(t+1) = \omega\delta_i(t) + r_1c_1\left[P^i_{sbest} - P_i(t)\right] + r_2c_2\lambda_i$
5: $\delta_i(t+1)$ = max(-5, $\delta_i(t+1)$) //set velocity max to 5
6: $\delta_i(t+1)$ = min(5, $\delta_i(t+1)$)
7: $P_i(t+1)$ = min($P_i(t)$ + $\delta_i(t+1)$, P_{max})
8: $P_i(t+1)$ = max($P_i(t)$ + $\delta_i(t+1)$, P_{min})
9: end

$$Nu_j(p_j^d, p_{-j}^d) = \sum_{i\in U_j} \log_2\Big(1 + \frac{g_{ij}p_j^d}{\sigma^2 + \sum_{k\neq j} g_{ik}p_k^d}\Big) - \alpha\sum_n \frac{p_j^d g_{nj}}{\sigma^2} \tag{22}$$

where p_{-j}^d is the power of the rest small cells; the parameter α is used to control the weight of the cost. A small α means that the performance of SBS j is more important and gives a rise to high transmission power so that the throughput and coverage of SBS j could be improved. On the other hand, a large α means more punishment to interference and SBS j will use low transmission power to avoid strong interference. Therefore, power control for the SBS cluster is that each SBS tries to maximize its net utilities with the limited transmission power. The coverage problem can be formulated as game G

$$G = (J, \{p_j^d\}, \{Nu_j\}) \tag{23}$$

where J is the player set (i.e., SBSs in the cluster), p_j^d is the strategy played by player j which is the data transmission power of SBS j, and Nu_j is the utility of player j. Thus, the strategy of each SBS is to maximize utility within the power limit. That is:

$$\begin{aligned} &\max\ Nu(p_j^d, p_j^d) \\ &s.t.\quad p_j^d = (1-\beta)P_{tx} \\ &\qquad P_{\min} \le P_{tx} \le P_{\max} \end{aligned} \tag{24}$$

Take the first order derivative of Nu w.r.t. p_j^d and it can be obtained

$$\frac{dNu}{dp_j^d} = \sum_{i\in U_j} \frac{\frac{g_{ij}}{\sigma^2 + \sum_{k\neq j} g_{ik}p_k^d}}{1 + \frac{g_{ij}p_j^d}{\sigma^2 + \sum_{k\neq j} g_{ik}p_k^d}} - \alpha\sum_n \frac{g_{nj}}{\sigma^2} \tag{25}$$

Take the second order derivation of Nu w.r.t. p_j^d and it can be obtained

$$\frac{d^2 Nu}{d(p_j^d)^2} = -\sum_{i \in U_j} \frac{(\frac{g_{ij}}{\sigma^2 + \sum_{k \neq j} g_{ik} p_k^d})^2}{(1 + \frac{g_{ij} p_j^d}{\sigma^2 + \sum_{k \neq j} g_{ik} p_k^d})^2} < 0 \tag{26}$$

It can be seen from Equ.(22-26) that game (23) is non-cooperative and the objective function (22) is continuous in p_j^d and quasi-concave in p_j^d (Equ. (26)). Following the *"Nash equilibrium existence theorem"* in Debreu(1952), Glicksberg(1952) and Rosen(1965) for each SBS *j*, the optimization problem in game *G* has a Nash Equilibrium p^* satisfies:

$$\forall j \in J, \sum_{i \in U_j} \frac{\frac{g_{ij}}{\sigma^2 + \sum_{k \neq j} g_{ik} p_k^*}}{1 + \frac{g_{ij} p_j^*}{\sigma^2 + \sum_{k \neq j} g_{ik} p_k^*}} = \alpha \sum_n \frac{g_{nj}}{\sigma^2} \tag{27}$$

In the NE, no SBS can improve its throughput without affecting other SBS's throughput. To reach NE of the game, each SBS uses best response strategy based on the information of iteration *t*

$$p_j^d(t+1) = sol(F_{j,-j}(t)) \tag{28}$$

where $p_j^d(t+1)$ is the solution to equation $F_{j,-j}(t)$, which is given by

$$F_{j,-j}(t): \sum_{i \in U_j} \frac{\frac{g_{ij}}{\sigma^2 + \sum_{k \neq j} g_{ik} p_k^d(t)}}{1 + \frac{g_{ij} p_j^d(t+1)}{\sigma^2 + \sum_{k \neq j} g_{ik} p_k^d(t)}} = \alpha \sum_{n \in U_{-j}} \frac{g_{nj}}{\sigma^2} \tag{29}$$

Given $P_{rx}^d(i,j) = p_j^d g_{ij}$, Equ.(28) and (29) can be rewritten as

$$p_j^d(t+1) = sol\left\{ \sum_{i \in U_j} \frac{\frac{P_{rx}^d(i,j)(t)}{(N_o + I_i(t)) p_j^d(t)}}{1 + \frac{P_{rx}^d(i,j)(t) p_j^d(t+1)}{(N_o + I_i(t)) p_j^d(t)}} = \alpha \sum_{n \in U_{-j}} \frac{P_{rx}^d(n,j)(t)}{N_o p_j^d(t)} \right\} \tag{30}$$

where $I_i(t) = \sum_{k \neq j} g_{ik} p_k^d(t)$ is the interference from neighboring cells received by user *i* at iteration *t*. Note that all variables other than $p_j^d(t+1)$ can be easily obtained by SBS, so the equation can be solved and $p_j^d(t+1)$ is obtained. Also discrete search methods can also be employed to find p_j^* with the maximum *Nu* based on the measured path loss and interference level according to Equ. (22). Each SBS plays the game with strategy in Equ(30) until a Nash Equilibrium reaches.

The DGT algorithm is described as follows:

Step 1: Assume all the SBSs in the cluster are synchronized and adjust their power at the same predefined time period.

Step 2: Each SBS monitors its own performance and invokes power control when performance metric drops and notifies neighbors to performance distributed power control together. Then the whole network performs coverage optimization.

Step 3: SBS set measurement report request to UEs in the cell, including measurement of the serving cell and the neighboring cells.

Step 4: SBS get UE received signal power from serving SBS and neighboring SBSs through measurement report messages.

Step 5: SBS feeds back the received interference to SBS *j* according to Equ.(21).

Step 6: Each SBS calculates target $p_j^d(t+1)$ according to Equ.(30) or search the p_j^d that maximize the net profit according to Equ.(22).

Step 7: Each SBS applies the new transmit power.

Step 8: The algorithm stops if every SBS get the same transmit power as the current one which means NE reaches.

The efficiency and convergence of the DGT algorithm will be described in next part.

SIMULATION RESULTS

Performance of MPSO and DGT is evaluated through simulation using Matlab2011b. Consider an open office environment with a length of 300 meters and a width of 200 meters. A small cell network with M=6 small cells is shown in Figure 4 with N=60 users uniformly deployed.

The parameters of the system are listed in Table 1. The initial transmission power of SBS is set to a random value from [P_{min}, P_{max}], pilot transmission power takes up 10% of the total power and the rest is for data power. Performance of MPSO and DGT is evaluated through coverage ratio, throughput, converge steps and signaling overhead. The performance of MPSO and DGT is also compared with traditional coverage optimization algorithm using simulated annealing (SA) (Siomina et al, 2006).

Performance of MPSO

First of all, the average coverage performances of fixed and random pilot power are shown in Figure 5, which are obtained from 1000 simulations. The blue bars show the average user coverage ratio of the cluster when each SBS has the same fixed pilot transmit power, varying from -20dBm to 10dBm. It can be seen that with low fixed pilot transmit power, the small cell cluster only covers a small fraction of the area. Then coverage ratio increases with higher pilot transmit power as coverage holes are reduced. However, as the pilot power is increased further, the coverage ratio drops because more loud neighbor overlaps are generated. The coverage ratio can reach a peak of 82% at the power of -10dBm and -5dBm.

Figure 4. Small cell network deployment

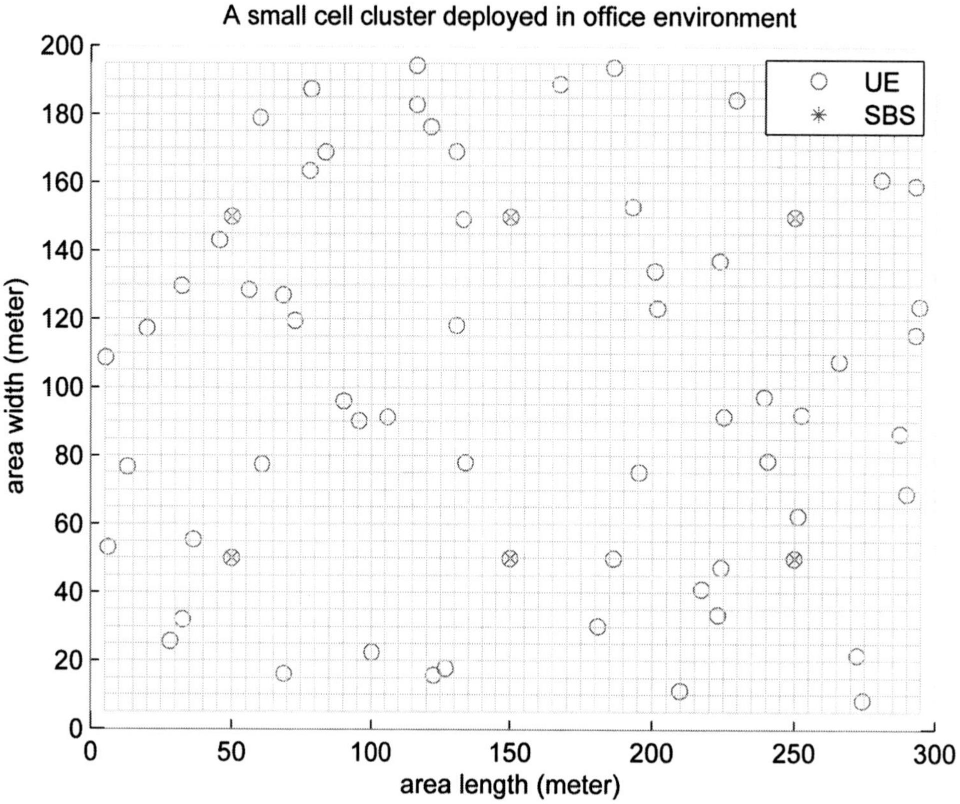

Table 1. Simulation Parameters

Parameters	Value
Path loss model:	Short distance:38.5+20*log10(*d*) dB *d* (meter); other cases:28+35log10(*d*)
Frequence Bandwidth	5MHz
Small cell capability	16 users/cell, high load=14 users, c_o=2
P_{min}, P_{max}	-20 dbm,10dbm
P_{strong}, P_{weak}, P_{cover}	-95 dbm, -100 dbm,-96dbm
PSO Parameters	c_1=2, c_2=2, ω decreases lineally from 1.4 to 0.5 (Shi & Eberhart, 1998), max PSO velocity: 5
Noise Power	-109 dbm
α,β	α=0.1, β=10%

Moreover, the coverage performances are also shown when the pilot transmit power of each SBS is randomly chosen from [P_{min}, P_{max}], where the red bar shows the average user coverage ratio, and the green bar stands for the area coverage ratio. It can be seen that the values of area and user coverage ratio are close to each other and both are below 60%. Obviously, both fixed and random pilot transmit power cannot provide good coverage performance for SCN.

Figure 5. Coverage performance with fixed pilot transmit power

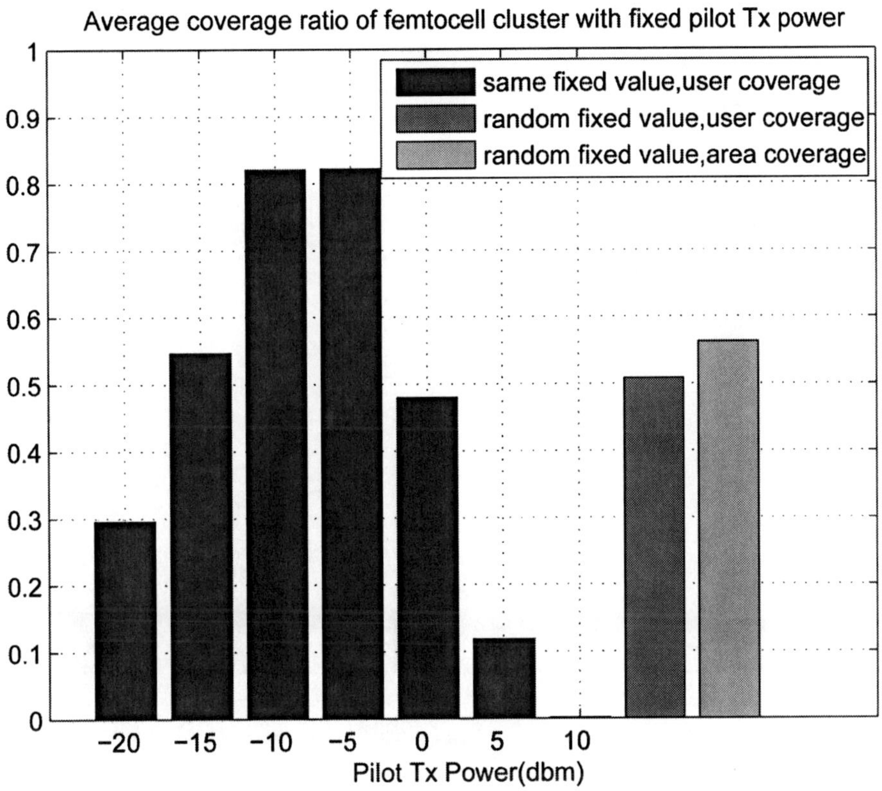

Before demonstrating the performance of MPSO, the effect of system parameter P_{weak} is evaluated, which is used to define edge users, which are used to avoid or eliminate coverage holes in MPSO. Figure 6 shows the performance of MPSO as a function of P_{weak} in terms of coverage ratio and iteration times. It can be seen that when P_{weak} varies from -99dBm to -95dBm, the iteration number decreases considerably while the coverage ratios change slightly. As a whole, when P_{weak} is set to -96dBm, MPSO reaches the highest user and area coverage ratios with the least iteration times. Therefore, P_{weak} is set to 96dBm in the following evaluations.

Next, the convergence performance of MPSO is shown in Figure 7 and Figure 8. The algorithm is initialized with a set of random pilot power, i.e., [-10.53 -6.96 -2.14 -15.75 5.74 0.83] dBm, which could only cover a small fraction of the users and the whole area. As shown in Figure 7, both the user and area coverage ratios increase rapidly within the first few iterations. The algorithm reaches 100% user coverage after 257 iterations with an optimum power allocation of [-6.85 0.22 -8.36 -3.85 -7.09 0.48] dBm. The area coverage ratio grows slower than the user coverage ratio and reaches 98.8% when user coverage reaches 100%. Meanwhile, Figure 8 shows the number of users in coverage holes, loud neighbor overlap and cell overload as a function of iteration times. It can be seen that loud neighbor overlap is serious at first, but mitigated rapidly as iterations go on. On the other hand, although there is no coverage hole and cell overload at the beginning of the algorithm, some holes and overload are detected when the algorithm tries to reduce loud neighbor overlap area. When the algorithm converges, there is no coverage hole, or cell overload or loud neighbor overlap. As shown in Figure 7 and Figure 8, MPSO can converge rapidly and is suitable for online coverage optimization.

Figure 6. Coverage performance with different P_{weak}

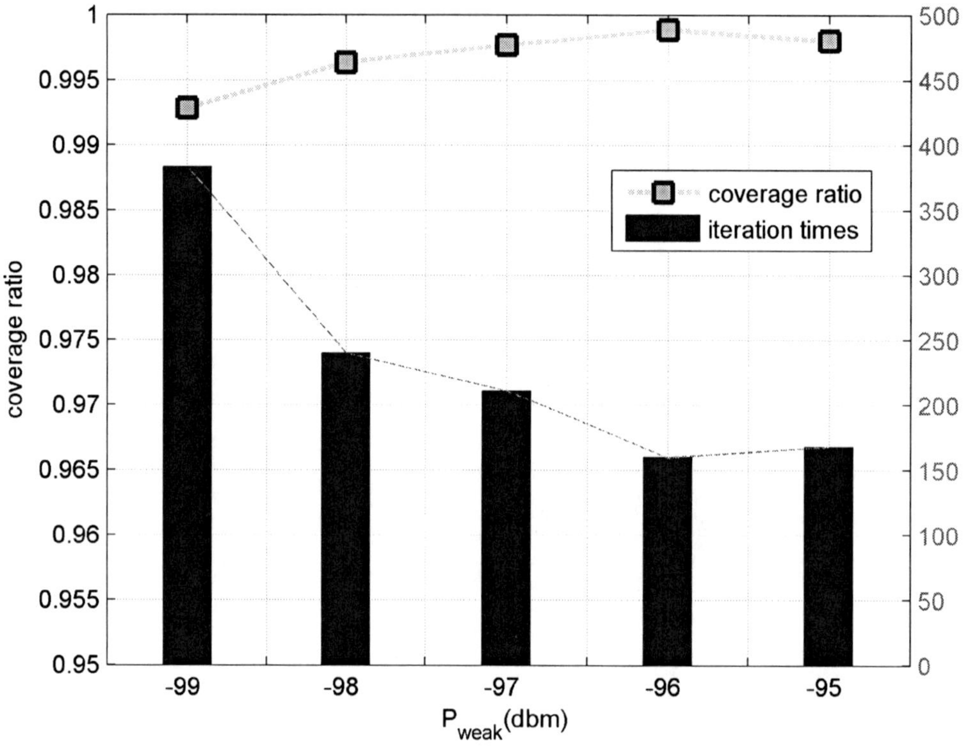

Figure 7. Coverage optimization process of MPSO

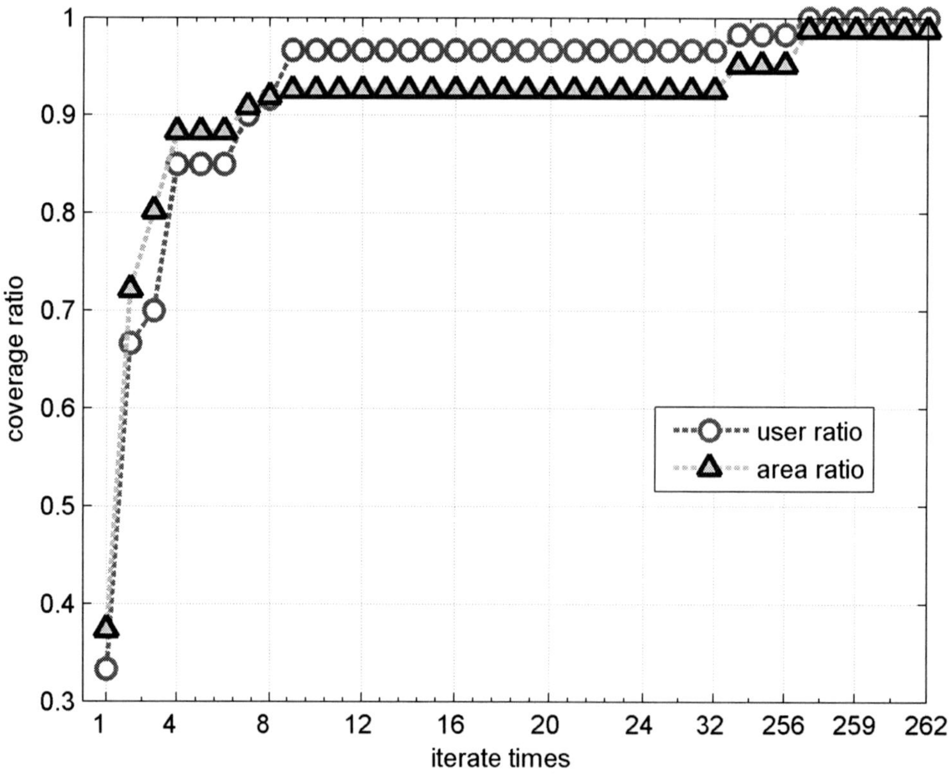

Figure 8. Coverage details in SCN

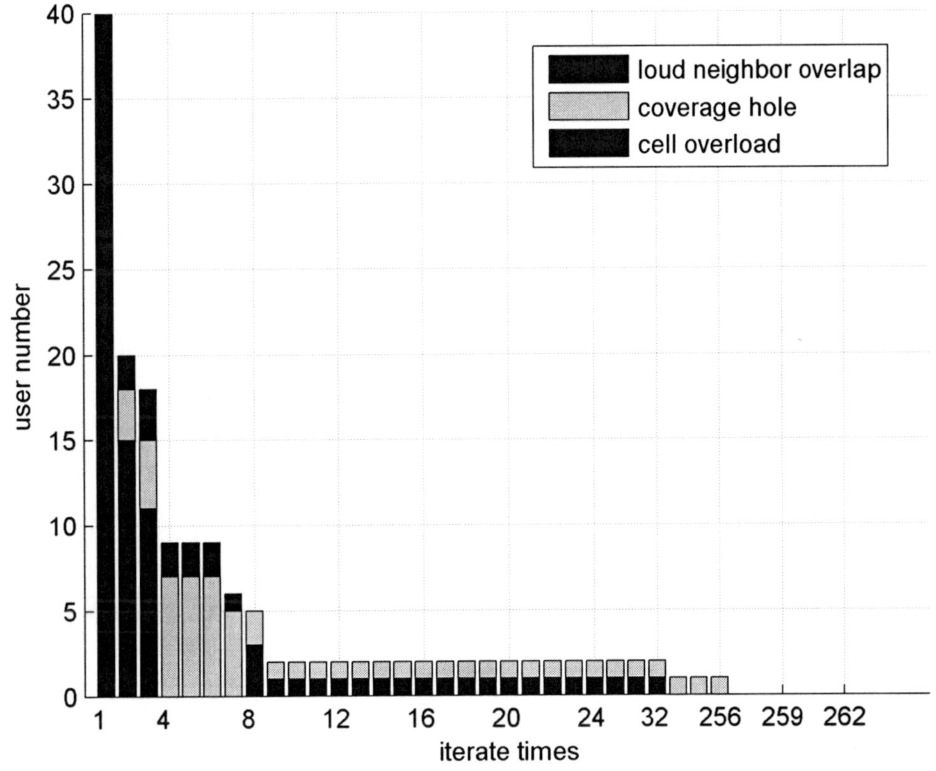

Finally, Figure 9 illustrates the average coverage performance of SPSO and MPSO. As stated before, when only one SCN (or small cell cluster) is concerned, the standard PSO reduces to SPSO, which fails to find a global best solution and needs up to 1000 iterations to get a user coverage ratio of 54.196% and an area coverage ratio of 58.470%. Using heuristic power control scheme as the social influence term, MPSO is designed for the coverage optimization of single small cell cluster, which reaches a user coverage ratio of 99.786% and an area coverage ratio of 95.778%with 12 iterations. It can be seen that MPSO significantly outperforms SPSO and fixed pilot power mode with low complexity. Moreover, from Figure 5 to Figure 7, it can be seen that in various scenarios, the area coverage ratio and the user coverage ratio present similar trends and their values are close to each other. Therefore, both area and user coverage ratios can be used to demonstrate the coverage quality. User coverage ratio is more suitable for practical systems.

Performance of DGT

For distributed coverage optimization, the total power of each SBS is initialized to P_{min}, and coverage optimization is carried out at each SBS simultaneously. Each SBS calculates its own best power based on the last state of the SCN and set the total transmit power to the new power. SBS plays the game repeatedly until equilibrium is reached.

Before demonstrating the performance of DGT, the effect of control parameter α is investigated. As described before, α is used to control the weight of interference punishment. Figure 10 demonstrates the

Figure 9. Coverage performance of SPSO and MPSO

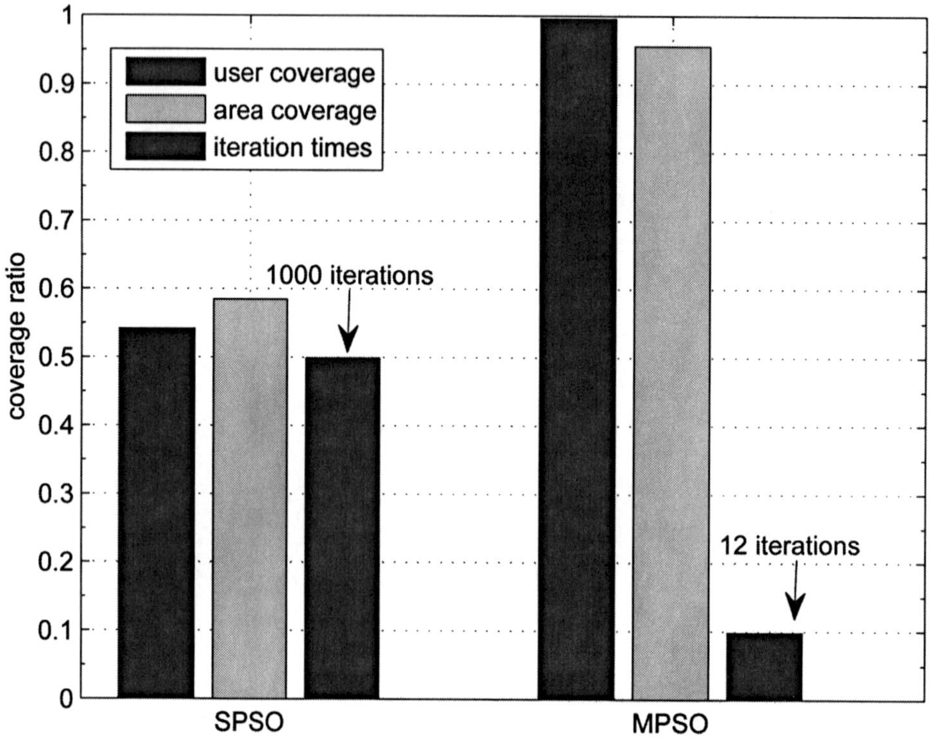

Figure 10. Coverage performance with different α

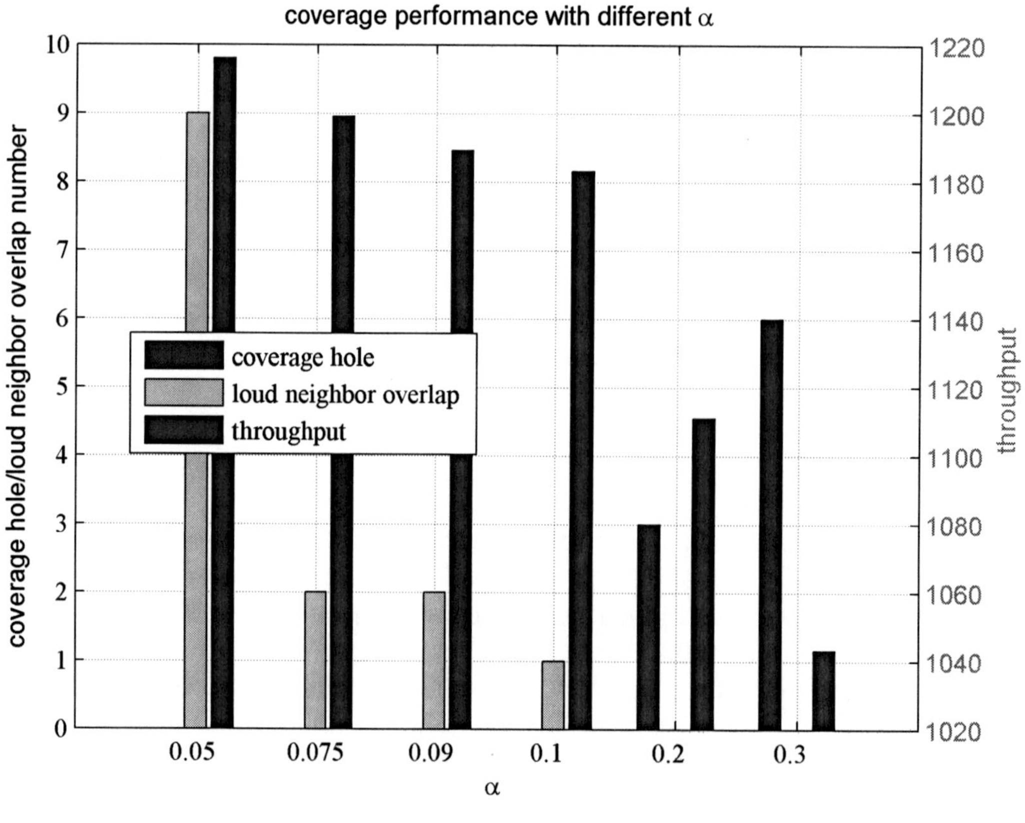

performance of DGT with different α. Coverage hole and loud neighbor overlap numbers are compared together with network throughput. It can be shown from the figure that with smaller value, SBS can get higher transmit power and thus more loud neighbor overlap although the network throughput is also higher. With the increase of α, transmit power of SBS will be limited to suppress interference and the number of neighbor overlap decreases together with slightly decrease of network throughput. When α continues to increase, loud neighbor eliminated and coverage hole occurs with a sharp decrease of network throughput. It is demonstrated that increasing the value α can remove loud neighbor overlap and decreasing the value α can remove coverage hole in the cluster. DGT can get the highest coverage ratio with α set to 0.1 while at the same time keeps a reasonable throughput which is a good balance between coverage hole, loud neighbor overlap and throughput. Thus α is set to 0.1 in the following experiments.

Figure 11 demonstrates the coverage ratio and throughput of the cluster during the power control game. It is shown that the SCN is initialized with very small coverage ratio and low throughput due to lots of coverage holes. However, the coverage holes are quickly removed after the first round of game. Both coverage ratio and throughput increase sharply with only one slot shot game. Coverage ratio even

Figure 11. Coverage performance of DGT

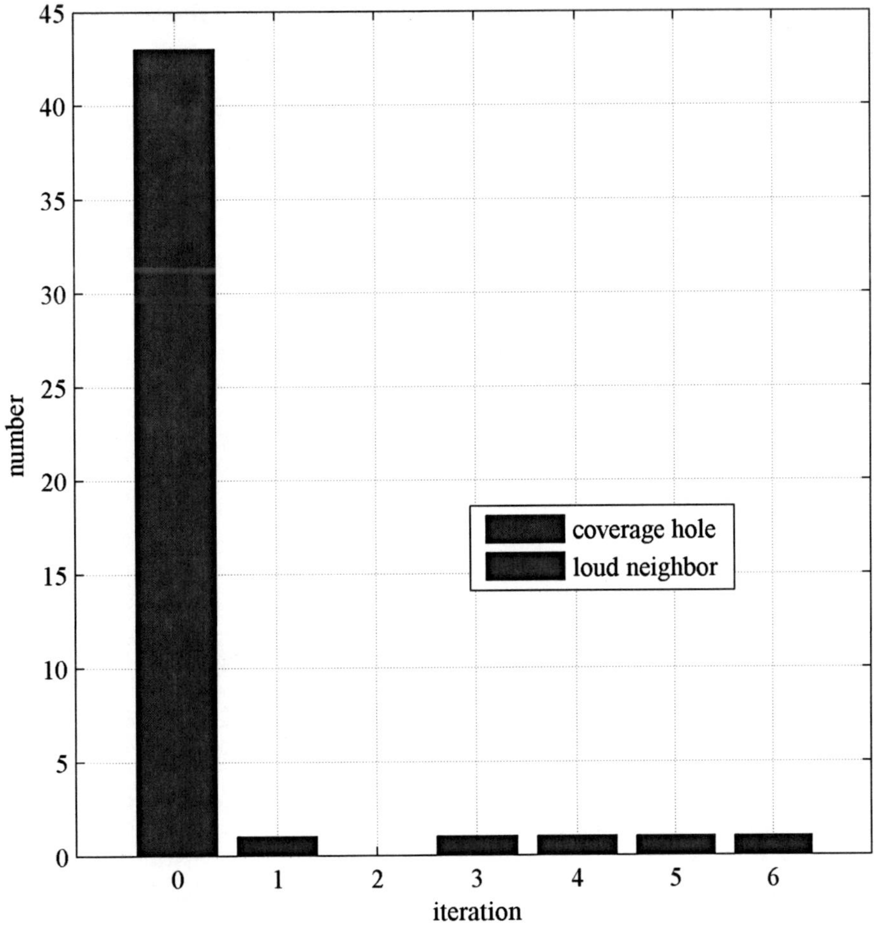

reaches 100% at the second iteration. However, it can also be seen that a loud neighbor occurs at later iterations, although the system throughput is still improved slightly. The game goes on with little variations on coverage ratio and throughput and reaches the equilibrium at the fourth iteration.

Figure 12 shows the change in SBS power during the optimization. Each SBS is initialized with the minimum -20 dBm power, then increases sharply at the first iteration. The transmit powers of some SBSs continue to increase, while others decreases at later iterations according to the game theory. Finally, the power became stable after the fourth iteration. It can be seen from Figure 11 and Figure 12 that DGT is effective even at the first iteration and the algorithm converges quickly with only 4 iterations.

Comparison of MPSO and DGT

The average coverage ratio, throughput and convergence steps performance of MPSO and DGT are shown in Figure 13, compared to those of SA. It can be seen that MPSO improves the coverage ratio to 100% with 12 iterations and reaches a normalized throughput of 1147.87 while DGT reaches a coverage of 98.33% with only 4 iterations, and provides the highest normalized throughput of 1183.13. SA

Figure 12. Adjustment of SBS transmit power during the game

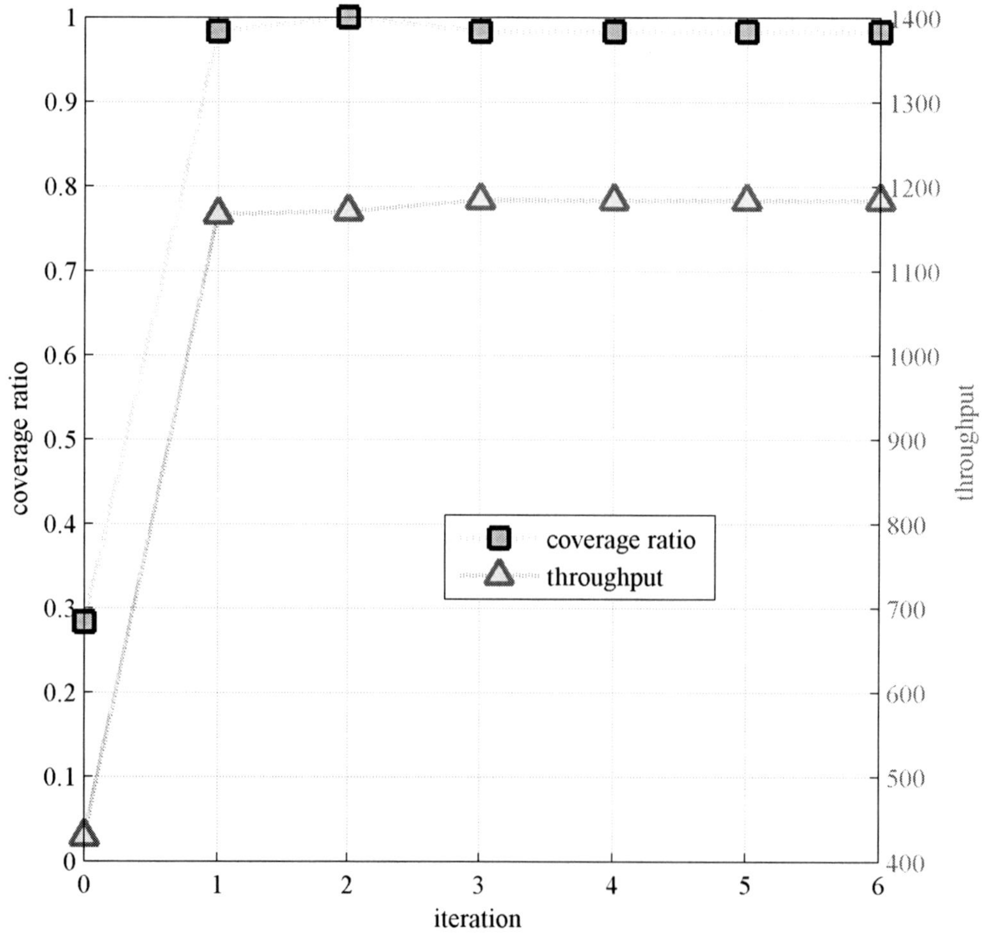

Figure 13. Performance comparison of different coverage optimization algorithms

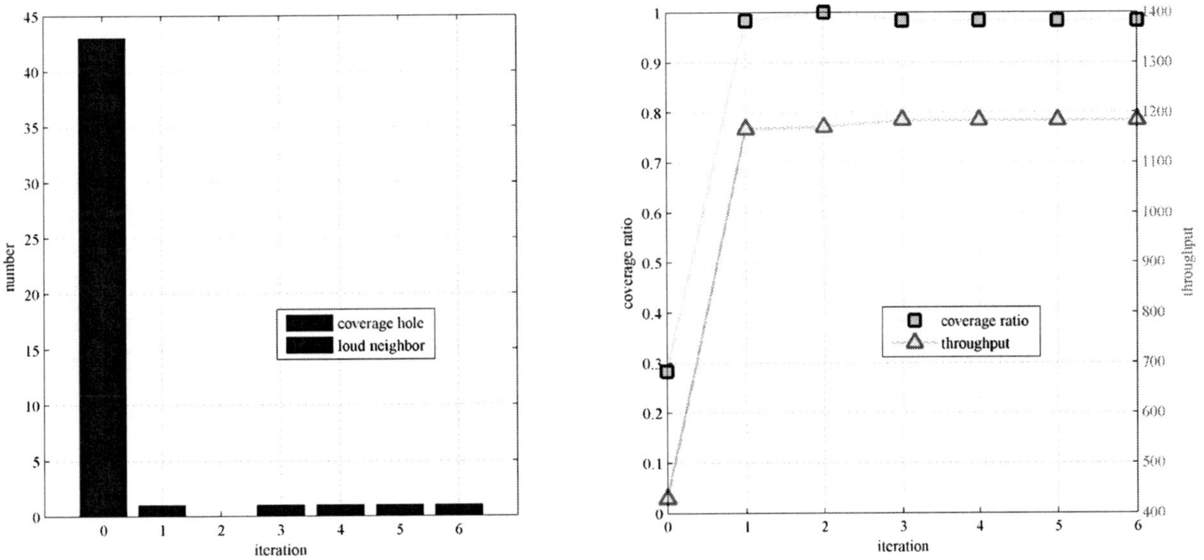

reaches a coverage of 95% with a throughput of 1022.9 but uses 215 iterations. Both MPSO and DGT outperform SA with higher coverage ratio and higher throughput using less iterations. MPSO shows an excellent coverage ratio while DGT performs best in throughput and convergence.

Furthermore, the signal overhead of the three algorithms are also compared in Table 2. For centralized algorithms, such as MPSO and SA, measurement information is collected and sent to centralized SON entity which then notifies each SBS the optimized power in each iteration. For DGT, only interference value is sent to neighboring cells at each step. In our system, the measurement information sent from SBS to OAM entity contains a fixed 4 byte head part and a list of received signal strength $P_{rx}(i,j)$, which is encoded to 8 bytes each. Thus, the average uplink overhead for SA and MPSO at each iteration is 4+(M*N*8)/M. And the power contol command from OAM to SBS contains a new power value which is encoded to 4 bytes. For DGT, SBS only needs to exchange interference value,which is encoded to 4 bytes, to neiboring SBSs, usually 3 neighbors. Thus the costs for each SBS in DGT is 4*3 bytes in each iteration. It can be seen that DGT uses much less signaling overhead than SA or MPSO, especially when the network size is large, i.e. large *M* and *N*.

As a whole, it can be concluded that both MPSO and DGT outperform SA with higher coverage ratio, higher throughput and less iterations. Also both algorithms converge quickly and are suitable for online optimization.

Table 2. Signaling Overhead

Algorithms	Iterations	Data(Bytes/iteration)	Control(Bytes/iteration)
SA	215	4+(*M**N**8)/*M*	4
MPSO	12	4+(*M**N**8)/*M*	4
DGT	4	4*3=12	0

FUTURE RESEARCH DIRECTIONS

In this chapter, the coverage of pure small cell networks in indoor environments is concerned and the conventional regular cell deployment is assumed (see Figure 4). For future researches, this work should be firstly extended from the pure SCN to a more general and practical network model, such as heterogeneous networks with both macro cells and small cells. Secondly, this work should also be extended to outdoor environments in future mobile cellular networks with a large amount of small cells, which is known as ultra dense cellular networks (UDN). Note that as the density of small cells increases, the cellular topology becomes more and more irregular and the regular deployment assumption used in this chapter does not stand any more (Garcia, Zhou & Shi, 2014). Therefore, it worths effort to study the coverage performance in UDN with irregular network topologies. Since smart technologies based on game theory have the ability of self-adjustment, they should be promising solutions to UDN coverage optimizations.

CONCLUSION

SCN or small cell cluster is a promising solution to cover large indoor/outdoor areas. However, the coverage optimization is a big challenge when using SCN in practice. This chapter focuses on the coverage problems in SCN and introduces centralized and distributed self-optimization algorithms based on game theory. Modified PSO with heuristic power control scheme as social influence is introduced in centralized optimization, while game theory is adopted in distributed optimization with a net utility function using cell throughput as game profit and interference to neighboring cells as cost. Simulation results show that both MPSO and DGT algorithm outperforms SA algorithm with higher coverage ratio and higher throughput while uses less iterations. Also, DGT uses less signaling cost compared with MPSO and SA. It can be seen from the research on MPSO and DGT that game theory can be well exploited to carry out self-optimization for SCN, in either a centralized and distributed way. It is necessary to further explore game theory which is promising to provide more advanced solutions for cooperative problems in wireless communication networks.

REFERENCES

Ashraf, I., Claussen, H., & Ho, L. T. W. (2010). Distributed radio coverage optimization in enterprise femtocell networks. In *Proceeding of 2010 IEEE International Conference in Communication (ICC)*. Cape Town: IEEE. doi:10.1109/ICC.2010.5502072

Chandrasekhar, V., Andrews, J. G., & Gatherer, A. (2008). Femtocell networks: A survey. *IEEE Communications Magazine*, *46*(9), 59–67. doi:10.1109/MCOM.2008.4623708

Claussen, H. (2007). Performance of macro-and co-channel femtocells in a hierarchical cell structure. In *Proceeding of IEEE 18th Annual International Symposium on Personal, Indoor and Mobile Radio Communications (PIMRC2007)*. Athens: IEEE. doi:10.1109/PIMRC.2007.4394515

Claussen, H., Ho, L. T. W., & Samuel, L. G. (2008). Self-optimization of coverage for femtocell deployments. In *Proceeding of Wireless Telecommunications Symposium 2008 (WTS2008)*, Pomona, CA: IEEE doi:10.1109/WTS.2008.4547576

Debreu, G. (1952). A social equilibrium existence theorem. National Academy of Sciences, 38, 886–893.

Douros, V. G., Toumpis, S., & Polyzos, G. C. (2012). Power control under best response dynamics for interference mitigation in a two-tier femtocell network. In *Proceeding of 2012 10th International Symposium on Modeling and Optimization in Mobile, Ad Hoc and Wireless Networks (WiOpt)*. Paderborn: IEEE.

Garcia, V., Zhou, Y. Q., & Shi, J. L. (2014). Coordinated Multipoint Transmission in Dense Cellular Networks with User-Centric Adaptive Clustering. *IEEE Transactions on Wireless Communications*, *13*(8), 4297–4308. doi:10.1109/TWC.2014.2316500

Glicksberg, L. (1952). A further generalization of the Kakutani fixed point theorem with application to Nash equilibrium points. *American Mathematical Society, 3*(1), 170-174.

Ho, L. T. W., Ashraf, I., & Claussen, H. (2009). Evolving femtocell coverage optimization algorithms using genetic programming. In *Proceeding of IEEE 20th Annual International Symposium on Personal, Indoor and Mobile Radio Communications (PIMRC2009)*. Tokyo: IEEE. doi:10.1109/PIMRC.2009.5450062

Hong, E. J., Yun, S. Y., & Cho, D. H. (2009). Decentralized power control scheme in femtocell networks: A game theoretic approach. In *Proceeding of IEEE 20th Annual International Symposium on Personal, Indoor and Mobile Radio Communications (PIMRC2009)*. Tokyo: IEEE. doi:10.1109/PIMRC.2009.5449782

Hoydis, J., Kobayashi, M., & Debbah, M. (2011). Green small-cell networks—a cost- and energy-efficient way of meeting the future traffic demands. *IEEE Vehicular Technology Magazine*, *6*(1), 37–43. doi:10.1109/MVT.2010.939904

Huang, L., Zhou, Y. Q., Han, X., Wang, Y. Y., Qian, M. L., & Shi, J. L. (2013). Coverage optimization for small cell clusters using game theory. In *Proceeding of 2013 IEEE Wireless Communication and Networking Conference (WCNC)*. Shanghai: IEEE. doi:10.1109/WCNC.2013.6554917

Huang, L., Zhou, Y. Q., Hu, J. L., Han, X., & Shi, J. L. (2012). *Coverage optimization for femtocell clusters using modified particle swarm optimization*. In *Proceeding of 2012 IEEE International Conference in Communication (ICC)*. Ottawa: IEEE. doi:10.1109/ICC.2012.6364494

Jo, H. S., Mun, C., Moon, J., & Yook, J. G. (2010). Self-optimized coverage coordination in femtocell networks. *IEEE Transactions on Wireless Communications*, *9*(10), 2977–2982. doi:10.1109/TWC.2010.090210.081313

Kennedy, J., & Eberhart, R. (1995). *Particle swarm optimization*. Paper presented at the IEEE International Conference on Neural Networks, Australia.

Le Treust, M., Tembine, H., Lassaulce, S., & Debbah, M. (2010). Coverage games in small cell networks. In Proceeding of 2010 Future Network and Mobile Summit. Florence: IEEE.

Li, X. F., Qian, L. J., & Kataria, D. (2009). Downlink power control in co-channel macrocell femtocell overlay. In *Proceeding of 43rd Annual Conference on Information Sciences and Systems, 2009 (CISS 2009)*. Baltimore, MD: IEEE. doi:10.1109/CISS.2009.5054750

Li, Y. Y., Macuha, M., Sousa, E. S., Sato, T., & Nanri, M. (2009). Cognitive interference management in 3G femtocells. In *Proceeding of IEEE 20th Annual International Symposium on Personal, Indoor and Mobile Radio Communications (PIMRC2009)*. Tokyo: IEEE. doi:10.1109/PIMRC.2009.5450030

Rosen, J. B. (1965). Existence and uniqueness of equilibrium points for concave n-person games. *Econometrica*, *33*(3), 520–534. doi:10.2307/1911749

Shi, Y., & Eberhart, R. (1998). A modified particle swarm optimizer. In *Evolutionary Computation Proceedings, 1998. The 1998 IEEE International Conference on IEEE World Congress on Computational Intelligence*. Anchorage: IEEE doi:10.1109/ICEC.1998.699146

Siomina, I., Varbrand, P., & Yuan, D. (2006). Automated optimization of services coverage and base station antenna configuration in UMTS networks. *IEEE Wireless Communications*, *13*(6), 16–25. doi:10.1109/MWC.2006.275194

Small Cell Forum. (2011). Retrieved June 15, 2013, from http://www.smallcells.org

Swarm Intelligence. (n.d.). Retrieved June 15, 2013, from http:// www.swarmintelligence.org

The 3rd Generation Partnership Project. (2010). *Telecommunication Management; Self-Organizing Networks (SON); Concepts and requirements* (3GPP TS 32.500 V10.0.0). Retrieved June 15, 2013, from www.qtc.jp/3GPP/Specs/32500-a00.pdf

KEY TERMS AND DEFINITIONS

DGT: Distributed optimization using game theory, a distributed SON technology proposed by the chapter authors for SCN coverage optimization, which requires less on backhaul connection and is more suitable for network to scale up, compared to MPSO.

MPSO: Modified particle swarm optimization, a centralized SON technology proposed by the chapter authors which could automatically detects coverage problems and finds optimized transmit power using game theory

NE: Nash equilibrium, named after John Nash, is an important concept in game theory. Consider a non-cooperative game with no less than two players. Each player is supposed to know the equilibrium strategies of the others. If each one has chosen a strategy and no one can benefit by only changing one's own strategy, then the current set of strategies and the corresponding payoffs are known as a Nash equilibrium.

PSO: Particle swarm optimization, is a population based stochastic optimization technique, inspired by social behavior of bird flocking or fish schooling. It has been successfully applied in various research areas to solve multidimensional problems.

SA: Simulated annealing, inspired by annealing in metallurgy, which using heating and controlled cooling to improve the quality of materials. Consider the global optimization problem with a large search space. Since it may be highly complicated to find the global optimum, SA provides a probabilistic and metaheuristic method to find a good approximation with a fixed complexity.

SCN: Small cell networks, also known as small cell grid, enterprise small cells and small cell clusters, is a group of densely deployed small cells together to extend the usage of small cells from residential environment to enterprise, urban and rural environments, i.e. high-traffic business environments, including office, retail malls, metro and municipal buildings, and so on

SON: Self-Organizing Network, designed to reduce the operating and maintenance cost of mobile cellular networks. With SON technologies, the network can be automatically configured, managed, optimized and healed.

Chapter 8
Resource Allocation for Device-to-Device Communications in LTE-A Network:
A Stackelberg Game Theory Approach

Shaoyi Xu
Beijing Jiatong University, China

Chunmei Xia
Beijing Jiaotong University, China

ABSTRACT

In the Long Term Evolution-Advanced (LTE-A) system, Device-to-device (D2D) communication underlying cellular networks can bring some advantages such as high data rates, low delay, low power consumption by reusing frequency bands with cellular user equipments (UEs). While at the same time, inter-cell and intra-cell interference is inevitable. The D2D users and cellular UEs will compete or cooperate to coordinate interference and share resources which incurs game theory an effective mathematical tool. This chapter proposed a Stackelberg game based algorithm to jointly allocate power and resources when the uplink frequency is shared with LTE-A users. In the game, the evolved NodeB (eNB) and D2D UEs are grouped to form the seller-buyer pair and the eNB sets prices to reduce the interference that it suffers meanwhile maximizes its revenue. For given specified prices, the D2D users compete for the resources to communicate with each other and reach their individual utility maximization. Simulation results prove that satisfying performance can be achieved by using the proposed mechanisms.

INTRODUCTION

With the development of mobile communications, the requirements on data rates are increasing sharply, nevertheless the scarcity of spectrum is becoming increasingly serious. To get higher total capacity, Device-to-device (D2D) communication underlaying the cellular network has brought much attention as a promising component to improve spectral efficiency. Different from the infrastructure based cellular

DOI: 10.4018/978-1-4666-8642-7.ch008

network, D2D users do not communicate via the central coordination (base station, NodeB or evolved NodeB) but operate as an underlay and communicate directly with each other. Excluding the unnecessary core network involvement, D2D communication can also provide several advantages such as high bit rates, low delay, low power consumption, low cost, plug-and-play convenience, and flexibility. It has been integrated into the Third Generation Partnership Project (3GPP) Long Term Evolution-Advanced (LTE-A) system as an underlay to a cellular network. The LTE-A system, in which the total bandwidth of the transmitted signal can be as large as 100MHz by aggregating multiple carriers, is of particular interest as it is the major technology approved by the International Telecommunications Union (ITU) as fulfilling the International Mobile Telecommunications-Advanced (IMT-A) requirements. Note that, the D2D communication can flexibly operate in multiple modes, e.g., the non-orthogonal and orthogonal D2D communication sharing modes, the cellular mode, etc. While by reusing the LTE-A cellular resources, D2D communication enhances spectrum utilization and improves cellular coverage. However, its utilizing cellular spectrum poses new challenges because the system must cope with new interference between the cellular and D2D communications. Consequently, intra-cell interference is no longer negligible and how to allocate physical resource blocks (PRBs) between cellular user equipments (UEs) and D2D UEs and control transmission power of all transmitters is a key problem. If there is no interaction among cellular users and D2D users, all of them will not control their transmit power, so that serious interference may lead to an inefficient outcome.

Modern game theory was born in 1928 when John von Neumann published his minimax theorem. As a branch of applied mathematics, game theory plays exceedingly important roles. On the other hand, as an important tool, game theory has pervaded almost all fields of sciences and social sciences (Nie & Zhang, 2008). The games studied in game theory are well-defined mathematical objects. To be fully defined, a game must specify the following elements: the players of the game, the information and actions available to each player at each decision point, and the payoffs for each outcome. A game theorist typically uses these elements, along with a solution concept of their choosing to deduce a set of equilibrium strategies for each player. As a result, when these strategies are employed, no player can profit by unilaterally deviating from their strategy. These equilibrium strategies determine an equilibrium to the game, namely, a stable state in which either one outcome occurs or a set of outcomes occur with known probability. So far, the game theory has been applied for the cognitive radio networks, heterogeneous network, small cell, and so on.

As D2D communication is an underlay to the cellular networks, the concept of the game is well suited for the cell system to decrease the interference. In this chapter, a Stackelberg game based scheme will be described to coordinate the intra-cell and inter-cell interference in the hybrid network. In the Stackelberg game based work, the eNB and D2D UE make up a seller-to-buyer pair to improve the sum throughput, where the eNB is viewed as the seller and D2D UE is the buyer. The seller acts first, then the buyer observes the seller's behavior and decides its strategy. In detail, the seller charges some fees to increase its performance of every link, whereas the buyer improves its rates by paying for some fees. Finally, the game will lead to equilibrium, making the cellular link and D2D link communicate reliably. By comparing with traditional methods, simulations prove the effectiveness and efficiency of our proposed scheme.

BACKGROUND

In this section, we first give a brief overview of D2D communications and Stackelberg game and then give the main idea of the proposed Stackelberg game based scheme.

There are some new trends in current mobile smart terminals, context-aware applications for example, are regarded as one of the most important value added services. Besides, social networks and machine-to-machine (M2M) applications are growing fast and changing people's life greatly. In another side, the explosive growth of the mobile user population and multimedia services are faster than the spectrums that service providers can obtain and thus cause a severe overload problem in the current cellular networks. Driven by the new services and to cater for the higher data rate and system capacity required by LTE-A, the 3GPP defined D2D communications underlaying LTE networks as a critical solution which has received much attention recently. D2D users communicate directly with each other or more hops without involving the forwarding by the aid of the central coordinator. Excluding the unnecessary core network involvement, D2D communication is an appealing concept and has become integral to the LTE-A world to form a hybrid network (3rd Generation Partnership Project, 2012).

Figure 1 illustrates the basic and extended concept of D2D communication and hybrid networks. In Figure 1, a user equipment (UE) can communicate with another UE directly or by the aid of a base station. Moreover, multiple D2D subsystems can share resources in a cell.

Actually, the trait of D2D communication is not new in the cellular system and it has been identified in some specifications such as Private Radio Communication System (PRCS), Japanese Handy Phone System (HPS), Digital Enhanced Cordless Telecommunications (DECT), etc. Many companies devote actively much in this research. Qualcomm, for example, is heading and has been very active recently. A prototype of "FlashLinQ" modem has been developed since 2006 to enable direct D2D communication over a licensed spectrum. The prototype effort covers the protocol stack from the host software interface to the physical layer algorithm. "FlashLinQ" is even a new trademark registered in 2008 by Qualcomm (Xu, Wang, Chen, Peng, & Kwak, 2012).

The key motivation for utilizing D2D communication underlaying a cellular network is to keep local communication local. D2D communication shares the same resources with the cellular system whereas

Figure 1. Illustration of D2D communication as an underlay to a cellular system

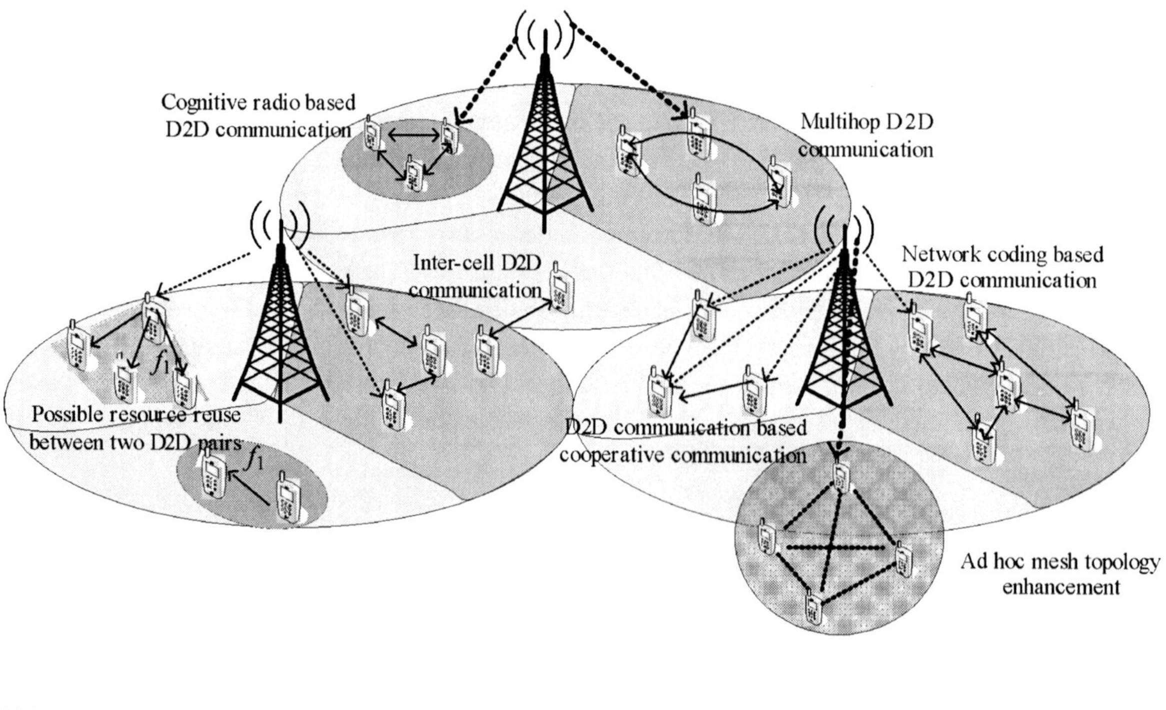

under the control of the evolved NodeB (eNB). In such a case, an eNB can still control the resource and power assigned for D2D transmission to limit the interference to the primary cellular networks. Although the concept of D2D communication as an underlay to a cellular network is not new in a cellular system, it is new in LTE systems. As a result, some new challenges are incurred. Among them interference management in a hybrid network is a critical issue in that the interference from D2D links can be expected to reduce the cellular capacity and efficiency and vice versa.

For the D2D communication underlaying the cellular system, both uplink (UL) and downlink (DL) resources can be shared between D2D subsystems and cellular networks. Due to heavier traffic and fast resource scheduling in LTE DL transmission, most works emphasized that D2D devices make use of the cellular UL spectrum leading to the intra-cell and inter-cell interference which must be controlled efficiently. As Figure 2 shows that the eNB suffers the interference from the D2D transmitters, while D2D receivers suffer the interference from the cellular users if cellular users share channels with D2D users in the uplink transmission.

In such a hybrid system, D2D users competing for channels and transmission power with the cellular users and the eNB can be viewed as the selfish players as well, the process of competition can be modeled as a Stackelberg game. Stackelberg games, which are also called leader-follower games, are initially proposed by Stackelberg in 1952 (von Stackelberg, 1952) based on some economic monopolization phenomena. In a Stackelberg game, one player acts as a leader and the rest as followers. The problem is then to find an optimal strategy for the leader, assuming that the followers react in such a rational way that they optimize their objective functions given the leader's actions (Nie & Zhang, 2008).

Under a Stackelberg game based model, D2D users must pay some cost for reusing channels with the cellular users and at the same time improve their communication performance. According to some novel algorithms, the equilibrium of coordinating the interference and choosing the channels is reached among the players. Finally, through the coordination of the game, two objectives are achieved, one is to

Figure 2. Illustration of interference when reusing UL spectrums

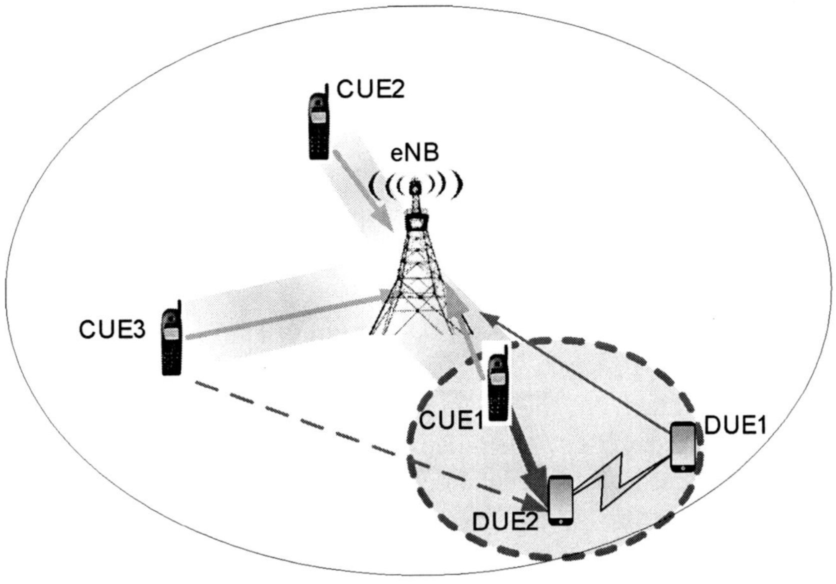

make the D2D communication reliable, the other is to eliminate the interference efficiently. Therefore, Stackelberg game theory has been widely recognized as a useful mathematical tool and designed for resource allocation, interference management and energy saving, and so on.

Aiming to guarantee the performance of licensed cellular users or maximize the sum throughput of the D2D subsystem, power control and resource allocation schemes are developed in (Wen, Zhu, Lin,Zhang & Yang, 2013; Le 2012; Wang, Zhu, Zhao, Li & Lei, 2013; Xu, Song & Han, 2013). Under different constraints and resource sharing methods, the original optimization qutstion is formulated and further solved using Lagrangian Duality Optimization (LDO) with Karush–Kuhn–Tucker (KKT) conditions (Wen, 2013; Le 2012; Wang, 2013) or Game theory (Wen, 2013; Xu 2013). To manage the interference between D2D and cellular networks, in (Fu, Yao & Gao, 2013), a transmit beamforming approach is designed at the D2D transmitter to maximize the throughput of D2D communication based on the estimated channel state information (CSI). Multiple-Input and Multiple-Output (MIMO) based methods are suggested in (Tang, Zhu & Ding, 2013) to deal with the intra-cell interference by designing appropriate precoding or utilizing a column generation method (Phunchongharn, Hossain & Kim, 2013). Interference control schemes were presented in (Min, Lee, Park, & Hong, 2011) based on a predefined interference limited area. Authors in (Xu et al., 2013) adopted a combinatorial auction approach to assign the cellular users channels to the D2D users for sum throughput maximization. An algorithm for joint scheduling and power allocation was proposed in (Wang, Song, Han, Zhao, & Wang, 2013), which developed a Stackelberg game framework, grouping a cellular UE and a D2D UE to form a leader-to- follower pair.

Different from prior works which have considered little about jointing power control and resource allocation, in our work, we add PRBs allocation based on the power control. In our based on Stackelberg game work, the eNB and D2D UE make up a seller-to-buyer pair to improve the sum throughput, where the eNB is viewed as the seller and D2D UE is the buyer. The eNB takes part in the communication of all cellular links, so we can consider that the PRBs are owned by the eNB, namely the seller. The seller acts first, then the buyer observes the seller's behavior and decides its strategy. In detail, the seller charges some fees to increase its performance of every link, whereas the buyer improves its rates by paying for some fees. Finally, the game will lead to an equilibrium, making the cellular link and D2D link communicate reliably. By comparing with traditional methods, simulations prove the effectiveness and efficiency of our proposed scheme. Simulation results prove that satisfying performance can be achieved by using the proposed mechanisms.

The main contributions of this chapter can be summarized as follows.

1) A Stackelberg game based work is proposed firstly. The eNB and D2D UE make up a seller-to-buyer pair to improve the sum throughput, where the eNB is viewed as the seller and D2D UE is the buyer. The proposed approach leads to a good trade-off between the sum throughput and the rate of D2D pairs.
2) Different from prior works which have considered little about joint power control and resource allocation, PRBs allocation are considered in this chapter based on the power control.
3) During the PRBs allocation, this chapter proposed the scheme to deal with the confliction when D2D UEs choose the same PRBs. The proposed mechanism is realized by announcing updated prices by the eNB and this leads to the suboptimal PRB selection results to the DUEs but satisfying interference level to the eNB.

By comparing with the traditional methods, extensive numerical results are provided to demonstrate the effectiveness of the approach in this chapter. The rest of this chapter is organized as follows. After presenting the D2D communication system model as an underlay to the cellular network, the Stackelberg game based algorithm and some theoretical analyses will be discussed in detail. Then the simulation results and performance evaluation are presented. Finally, conclusions are drawn.

SYSTEM MODEL

Consider a single cell scenario with multiple UEs and one eNB located at the center of the cell, including two kinds of UEs, cellular UEs (CUEs) and D2D UEs (DUEs).The DUEs are in pairs, consisting of one transmitter (D2D-Tx) and one receiver (D2D-Rx). The system model of D2D communication underlaying the cellular network by reusing UL resources of the cellular network is illustrated in Figure 3. The single cell has N CUEs and M D2D pairs, labeled with $i = 1,\ldots, N$, $J = 1,\ldots, M$, respectively, and $N \geq M$. We assume that non-overlapping PRBs are utilized by CUEs, and one PRB is only allocated to a CUE, so there is no interference between CUEs. Another, since the PRBs allocated to the CUEs have been decided by the eNB, M D2D pairs are selected to share M PRBs from N PRBs with the CUEs. Therefore there is also no interference between D2D pairs. As shown in Figure 3, in UL transmission, CUEs transmit data to the eNB, while D2D-Tx transmits to its corresponding D2D-Rx.The two DUEs in a pair is close enough to satisfy the reliable communication. The D2D pair 1 is selected to share the same PRB with CUE1, while the D2D pair 2 is chosen to share the same PRB with CUE2. Therefore, the eNB suffers the interference from D2D-Tx1 and D2D-Tx2. Also D2D-Rx1 and D2D-Rx2 are exposed to the interference from CUE1 and CUE2.

For simplicity, we assume that the eNB knows the accurate channel gains of all links. The channel gain between eNB and CUEi is defined as $L_{ci,\ b}$. The channel gain between eNB and DUEj is viewed as $L_{di,\ b}$. $L_{ci,\ dj}$, L_{jj} denote the channel gains between CUEi and DUEj, D2D-Txj and D2D-Rxj respectively. The transmit power of CUEi is determined by using a similar but simpler rule as defined in 3GPP (Teng, Guo, Honig, Xiao, & Liu, 2012)

$$10\lg P_i^{ue} = \min\left\{P - 10\lg L_{ci,b}, 10\lg \frac{IoT_{th}^d BN}{L_{ci,dj}}\right\}, \tag{1}$$

where P_i^{ue} represents the transmit power of the CUEi, and P_i is the target power received at the eNB. N denotes the power spectral density of additive white Gaussian noise (AWGN). B is the bandwidth of one PRB. IoT_{th}^d represents the threshold of the interference over thermal (IoT) of D2D-Rx and is used to restrict the interference from CUE to D2D-Rx according to $\frac{P_i^{ue} L_{ci,dj}}{BN} \leq IoT_{th}^d$. We define one set of binary variable β_{ji} ($j = 1, \ldots, M$; $i = 1, \ldots, N$) to denote the situation of resource allocation. $\beta_{ji} = 1$ if the jth D2D pair is chosen to employ the PRB with the ith CUE, otherwise $\beta_{ji} = 0$. We derive the received SINR at the D2D-Rxj from the Shannon theory as

Figure 3. The system model of D2D communication underlaying cellular network by reusing the UL resources of cellular network

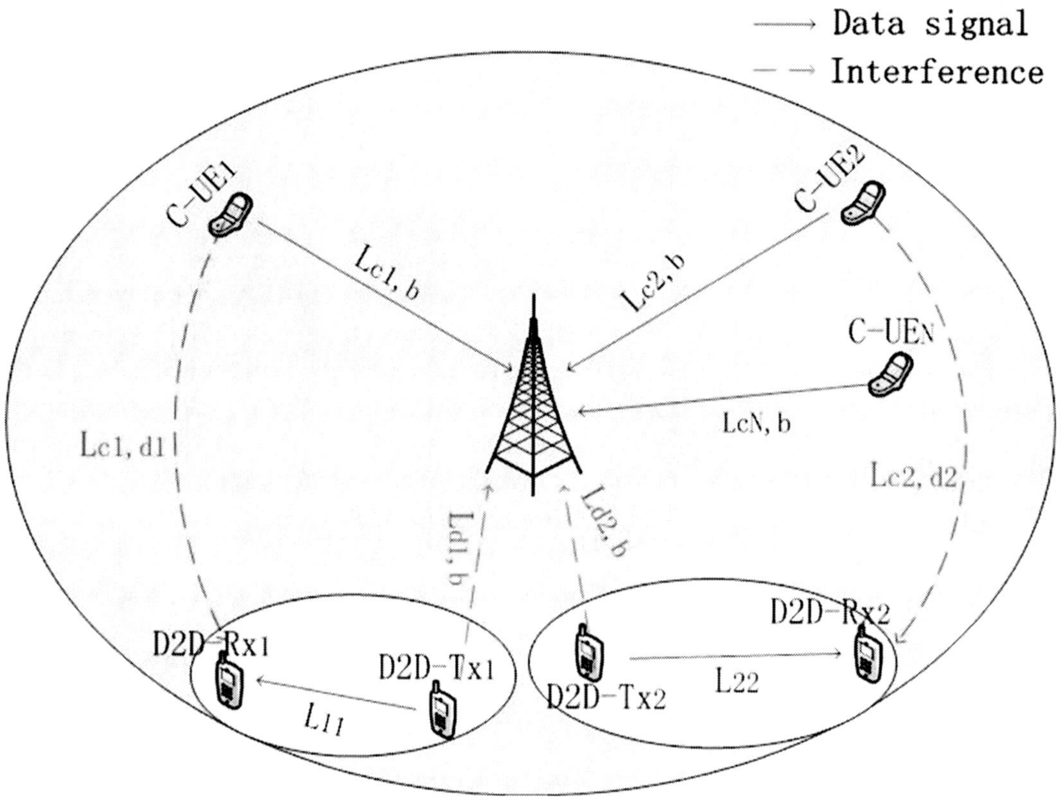

$$\gamma_j^d = \frac{P_{ji}^d L_{jj} \beta_{ji}}{BN + P_i^{ue} L_{ci,dj}}. \tag{2}$$

The SINR at the eNB corresponding to CUEi is

$$\gamma_i^c = \frac{P_i^{ue} L_{ci,b}}{BN + P_{ji}^d L_{dj,b} \beta_{ji}}, \tag{3}$$

where P_{ji}^d is the transmit power of D2D-Txj. The subscript *ji* or *ij* means the *j*th DUE sharing the same PRB with the *i*th CUE. The system sum rate during the UL transmission can be denoted as

$$R = \sum_{i=1}^{i=N} B \log_2 \left(1 + \gamma_i^c\right) + \sum_{j=1}^{j=M} B \log_2 \left(1 + \gamma_j^d\right). \tag{4}$$

The objective of this chapter is to appropriately allocate PRBs and transmission power of DUEs to maximize the utility function when UL spectrums are reused with LTE-A systems based on the Stackelberg games.

STACKELBERG GAME BASED RESOURCE ALLOCATION SCHEME

As illustrated in Figure 3, the interference is no longer negligible during the UL phase of the cellular network. If there is no interaction among eNB and DUEs, both of them will not control their transmit power, so that serious interference may lead to an inefficient outcome. Inevitably, we employ the Stackelberg game to coordinate the interference between eNB and DUEs, in order to guarantee the high quality of communication in two kinds of modes, namely, cellular mode and D2D mode.

In the game, the eNB is the Stackelberg seller, it firstly measures the interference temperature and then sets the interference price to increase the rates of every cellular link. The DUEs are buyers and choose their actions of power to maximize their individual payoffs, according to the prices announced by the eNB. As a seller, the eNB charges the DUEs some fees for sharing the PRBs. Here the fees are fictitious and defined as a part of transmit power of DUEs to complete the coordination. The buyer contends PRBs according to the price, which is decided by the eNB. Finally, through the coordination in this chapter, two objectives are achieved, one is to make the D2D communication reliable, the other is to improve the total data rates of every cellular link. In this way, equilibrium can be reached.

FORMULATION OF PAYOFF FUNCTIONS

Here we analyze the payoff functions of a one-seller-one-buyer pair, namely, the functions of the eNB and the DUEs. The utility of the eNB is defined as

$$U_{ij}(P_{ji}^{d},\beta_{ji}) = B\log_2(1+\frac{P_i^{ue}L_{ci,b}}{BN+P_{ji}^{d}L_{dj,b}\beta_{ji}}) + B\log_2(1+\frac{P_{ji}^{d}L_{jj}\beta_{ji}}{BN+P_i^{ue}L_{ci,dj}}) - B\log_2(1+\frac{P_i^{ue}L_{ci,b}}{BN}), \tag{5}$$

where U_{ij} represents the utility function of the eNB corresponding to the *i*th CUE sharing the same PRB with the *j*th D2D pair. The optimization problem for the eNB is to set proper transmit power and PRBs to maximize its utility, i.e.,

$$\max U_{ij}(P_{ji}^{d},\beta_{ji}), i=1,\ldots,N, j=1,\ldots,M. s.t. \beta_{ji} \in (0,1). \tag{6}$$

For the *j*th DUE, the payoff function can be written as

$$\Pi_d^j(P_{ji}^{d},\beta_{ji}) = B\log_2(1+\frac{P_{ji}^{d}L_{jj}\beta_{ji}}{BN+P_i^{ue}L_{ci,dj}}) - \lambda P_{ji}^{d}L_{dj,b}. \tag{7}$$

In the formula, λ is the weight factor to guarantee the two parts in the same order of the magnitude. The optimization problem of the DUE is to find suitable transmit power resulting in maximizing its own payoff, namely,

$$\max \Pi_j^d(P_{ji}^d, \beta_{ji})$$

s.t.

$$\begin{aligned} & P_{\min} \leq p_j^d \leq P_{\max}, \qquad C1 \\ & P_j^d \leq \frac{IoT_{th}^b BN}{L_{dj,b}} \qquad C2 \end{aligned} \tag{8}$$

where $P_{\min}$ and $P_{\max}$ stand for the minimum and maximum transmit power of D2D-Tx, respectively. P_j^d represents the transmit power of D2D-Txj, while IoT_{th}^b denotes the threshold of IoT of the eNB. The constraint C2 is the charging price, decided by the eNB, based on $\frac{P_j^d L_{dj,b}}{BN} \leq IoT_{th}^b$ to decrease the interference from DUEs.

THE BEHAVIOR OF THE BUYER

Although the DUE surrenders some fees for buying the PRB from the eNB, the fees can not be too high, otherwise the interference that the eNB suffers will be too large. Similarly, a low fee may lead to inefficient D2D communication. We suppose that both of the transmit power and the channel of the CUE have been known by the eNB. Given β_{ji}, we can derive the relationship between the payoff function and the transmit power of D2D-Txj. When $P_{ji}^d = 0, \Pi_j^d = 0$. As P_{ji}^d increases, Π_j^d also increases. However, Π_j^d will begin to decrease with P_{ji}^d continuing to grow since the logarithmic function increases slower than the linear function. To maximize the function by choosing proper transmit power, the first derivative of the payoff function is calculated and is expressed as

$$\frac{\partial \Pi_j^d}{\partial P_{ji}^d} = \frac{BL_{jj}\beta_{ji}}{\ln 2(BN + P_i^{ue}L_{ci,dj} + P_{ji}^d L_{jj}\beta_{ji})} - \lambda L_{dj,b} = 0\,. \tag{9}$$

The solution is further expressed as

$$\hat{P}_{ji}^d = \frac{B}{\lambda L_{dj,b}(\ln 2)} - \frac{BN + P_i^{ue} L_{ci,dj}}{L_{jj}\beta_{ji}}. \tag{10}$$

Substitute the formula (10) into the second order derivative, we obtain

$$\frac{\partial^2 \Pi_j^d}{\partial P_{ji}^{d2}} = -\frac{B}{\ln 2}\left(\frac{L_{jj}\beta_{ji}}{BN + P_i^{ue}L_{ci,dj} + P_{ji}^d L_{jj}\beta_{ji}}\right)^2 < 0. \tag{11}$$

Hence, the solution in (10) is a maximum point. According to the constraints (C1) and (C2), the best response is searched as follows

$$P_{ji}^d \in \{P_{\min}, \min\{\hat{P}_{ji}^d, \frac{IoT_{th}^b BN}{L_{dj,b}}, P_{\max}\}\}. \tag{12}$$

From the analysis above, the buyer will choose its optimal solution P^{d^*} in the condition of given β_{ji}, written as

$$\Pi_j^d(P^{d^*}, \beta_{ji}) > \Pi_j^d(P_{ji}^d, \beta_{ji}). \tag{13}$$

For all D2D pairs, the Nash Equilibrium (NE) can be described as $P^{d^*} = \{P_1^{d^*}, P_2^{d^*}, ..., P_M^{d^*}\}$.

THE BEHAVIOR OF THE SELLER

Based on the analysis above, we have achieved the best solution of the DUEs, next we will analyze the eNB's behavior.

First, the value of D2D-Tx's transmit power derived from (12) must ensure $U_{ij} \geq 0$, meaning that the rates should be increased after DUEs getting into the cell. Note that in this chapter, we calculate all the values of $P_{ji}^{d^*}$ on every PRB, then substitute it into U_{ij}, where i=1,…,N, j=1,…,M. As the value of U_{ij} increases, the performance of the ith cellular link becomes better. The maximum also means that the ith cellular link chooses the jth D2D pair to share the same resource with itself to increase its rates. We select the maximum from all values to ensure that not only the payoff function of DUE reaches the maximum but also the CUE's rates are maximized. Finally, the sum throughput of the cell can reach to its maximum. Simultaneously, both of the two utility functions achieve the balance. Based on the above description, we obtain Theorem 1.

Theorem 1: When satisfying $U_{ij}(P_{ji}^{d^*}, \beta_{ji}^*) > U_{ij'}(P_{j'i}^d, \beta_{j'i})$, for $\forall j' \neq j$, the players' NE point is ($P_{ji}^{d^*}, \beta_{ji}^*$).

Then the whole cell's optimal solution can be expressed as

$$\{P^{d^*}, \beta^*\} = \{(P_{1i}^{d^*}, \beta_{1i}^*), (P_{2i'}^{d^*}, \beta_{2i'}^*), ..., (P_{Mi''}^{d^*}, \beta_{Mi''}^*)\}.$$

Proof: For every optimal transmit power of D-UE, namely, $P_{ji}^{d^*}$, is different for different j. Given i, substituting (10) into the eNB's utility function (5), we derive M values of the seller's function. And we can certainly find the maximum from the M values, described as

$$U_{ij} = (P_{ji}^{d^*}, \beta_{ji}) > U_{ij'}(P_{j'i}^{d*}, \beta_{j'i}); j \neq j', \forall j'. \tag{14}$$

Therefore, the β_{ji} is the NE point on the premise of $P_{ji}^{d^*}$, in that the *i*th cellular link can obtain the most increment of rates when selecting the *j*th D2D pair to share the same PRB or the *i*th D2D pair has the highest priority of using the PRB. Thereby the NE point between the eNB and the *i*th D-UE is $(P_{ji}^{d*}, \beta_{ji}^{*})$, in turn, we can derive the whole cell's NE point, based on that any player has no incentive to change its strategy.

PRBS ALLOCATION

Based on the analysis of both the DUEs and the eNB, we can derive the proper transmit power of the D2D-Tx. Then we take the PRB allocation into account. According to the behavior of the eNB, the D-UEs sort the values of U_{ij} in the descending order for ∀*j*, *i*, and select the PRB to maximize U_{ij}. At the same time, the maximum also means that the priority the D2D pair reusing the PRB is the highest. Then the D2D pair will communicate on this PRB, meaning $\beta_{ji} = 1$. To efficiently allocate PRBs, three policies should be satisfied.

If there is no D2D pairs that maximize U_{ij} and $U_{i'j}$ at the same time or both the D2D pair *j* and *j'* maximize $U_{ij'}$, it means that there is no conflict between D2D pairs using PRBs. D-UEs choose the maximum U_{ij} and define $\beta_{ji} = 1$.Since one D2D pair is allocated only one PRB, written as, $\sum_{i=1}^{N} \beta_{ji} = 1$, the process ends when $\sum_{j=1}^{M} \sum_{i=1}^{N} \beta_{ji} = M$.

The repeated situation turns up as follows.

(1) If $U_{ij}(P_{ji}^{d*}, \beta_{ji}) = U_{i'j}(P_{ji'}^{d*}, \beta_{ji'})$, or described as one D2D pair reusing two PRBs with CUEi and CUEi', then the DUE chooses to make its payoff function lager. In other words, if the former is lager, it will share the PRB with CUEi and define $\beta_{ji} = 1$. According to this method, the process of PRBs allocation ends when all of the D2D pairs are allocated to suitable PRBs.

(2) If there exists multiple D2D pairs to maximize U_{ij} which means that different D2D pairs select the same PRB to use. Then they will transmit data on this PRB simultaneously but the confliction is unknown. However, the interference that eNB suffers from such conflictions will grows until to an intolerable threshold, then the eNB announces a new price to the DUEs. This new defined price leads to decreased transmit power of DUEs, consequently leading to the decline of the DUEs' performance. Finally the DUEs will select the suboptimal PRB to communicate instead of the optimal one.

In this way, the performance may not be good enough for every player but it is perfect for the whole cell, since there is no player breaking the state after reaching to the NE. The whole process can be expressed in Algorithm 1.

Algorithm 1. Joint D2D power control and PRBs allocation

Step 1: Given Channel State Information (CSI) and the weight factor λ.

Step 2: eNB announces the price, $P_j^d \leq \frac{IoT_{th}^b BN}{L_{dj,b}}$.

Step 3: **if** the D-UE affords the price, then go to 4.

Step 4: Calculate the optimal solution $P_{ji}^{d^*}$ on all PRBs as

$$P_{ji}^{d*} = \frac{B}{\lambda L_{dj,b}(\ln 2)} - \frac{BN + P_i^{ue} L_{ci,dj}}{L_{jj}}, \forall i.$$

if $U_{ij}(P_{ji}^{d^*}, \beta_{ji}) > U_{i^{'}j}(P_{ji^{'}}^{d^*}, \beta_{ji^{'}}); i \neq i^{'}, \forall i^{'}, j$

then define $\beta_{ji} = 1$ and $\beta_{ji^{'}} = 0$

Step 5: Begin to communicate.

if $\beta_{ji} = 1$ and $\beta_{ji^{'}} = 0$

then go back to 2

end if

else $U_{ij}(P_{ji}^{d*}, \beta_{ji}) = U_{i^{'}j}(P_{ji^{'}}^{d*}, \beta_{ji^{'}})$

then define $\beta_{ji} = 1$ making $\prod_j^d$ larger

Go to 5.

end if

else reduce its power $\hat{P}_{ji}^{d*}$, noting that $\prod_j^d$ decreases.

then choose the suboptimal U_{ij}, go back to 5

end if

Step 6: Finish the procedure until $\sum_{j=1}^{M} \sum_{i=1}^{N} \beta_{ji} = M$.

SIMULATION RESULTS AND ANALYSIS

In this session, we will present some simulation results to evaluate the performance of the proposed scheme. During the simulation, the proposed scheme is compared with the formal scheme, which only adjusts the transmit power of UEs with PRBs random allocation. In the simulation, the number of D2D pairs in the cell changes from 0 to 10. And other simulation parameters and values used are listed in Table1.

To study the average rate of UE with different distance between C-UE and D-UE which share the same PRB in our proposed scheme, we plot the curve of the rate and show it in Figure 4. Compared with the rate of C-UEs without sharing resources, the rate of C-UE plus D-UE both grows in the proposed and the formal scheme. However, due to the fact that the rate of D-UE is strictly increasing with the distance in this chapter, the performance of the rate is improved 50% than the formal scheme. In addition, the rate increases less with the distance increasing because the interference has been small due to the large distance.

Figure 5 illustrates the changing of the sum throughput of cellular communication and D2D communication vs. not only the different number of D2D pairs but also the different distance between the D-UE and C-UE which share the same PRB. *M* represents the number of D2D pairs in this cell. In this condition, the distance between D-UE and eNB is randomly fixed from the value smaller than 500 m.

Table 1. Simulation parameters and values

Parameter	Values
Radius of the cell	500m
Distance between D2D-Tx and D2D-Rx	20-30m
Maximum transmit power of C-UE	23dBm
Transmit power of D2D-Tx	20-22.5dBm
Number of CUEs	30
Bandwidth	180KHz
Thermal noise power density	-174dBm/Hz
Path loss model of cellular subsystem	128.1+37.6logd(Km)
Path loss model of D2D subsystem	148.1+40logd(Km)
Sensitivity of eNB	-147dBm/Hz
weight factor λ	13.578dB

Figure 4. The average rate of UEs with different distance between D-UE and C-UE

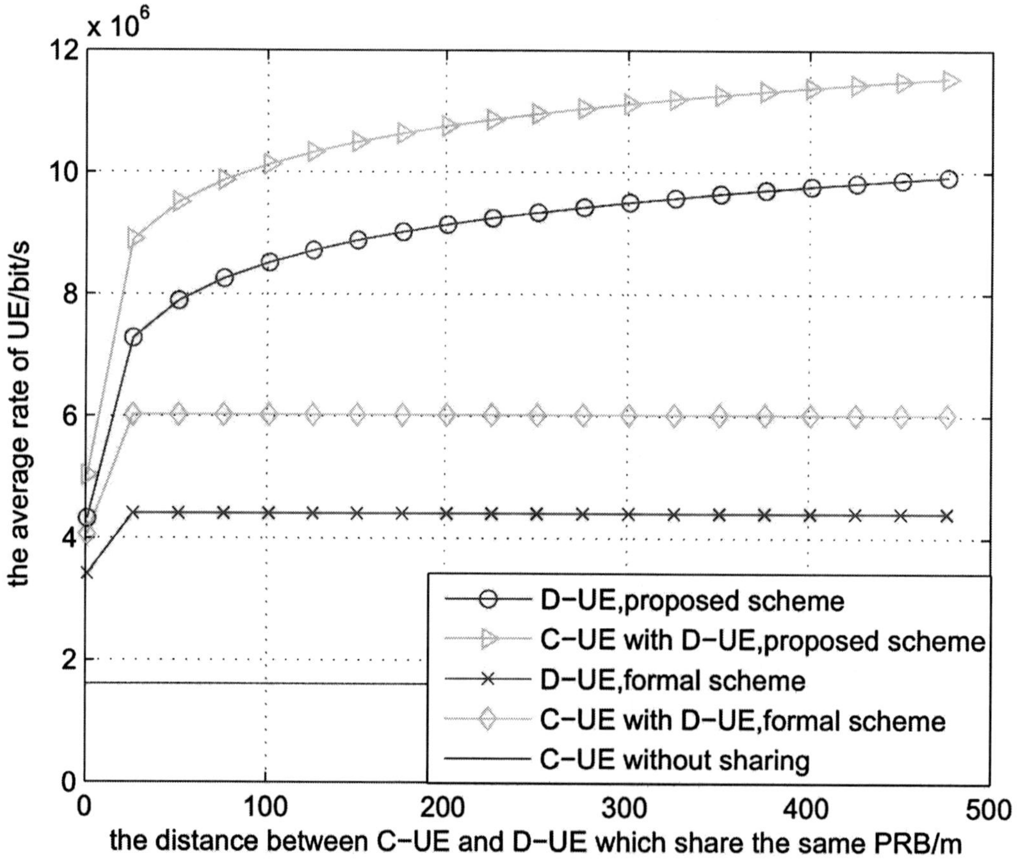

From the result, D2D communication can improve the sum throughput obviously. As Figure 5 shows, the sum throughput basically increases with the growth of the number of D2D pairs sharing the same resource. Whereas it should be pointed out that the throughput is smaller with ten D2D pairs than that with five when the distance between C-UE and D-UE is small in the proposed approach due to the fact that the interference is larger with smaller distance. However, the performance is still better than the traditional scheme. Moreover it also can be seen that blindly increasing the number of D2D pairs can't continue to grow the sum throughput of the cell because of too large interference. Apparently, the large distance between UEs using the same PRB means smaller interference and can also improve the sum throughput.

Meanwhile, according to the Figure 6, we analyze the sum throughput of cellular and D2D links with different numbers of D2D pairs and different distance between the D-UE and eNB. Similar to the Figure 5, the sum throughput both of the proposed approach and the formal scheme increases with the rising of the distance between the eNB and D-UE and the number of D2D pairs can not be blindly increased. At the beginning, the throughput is smaller than that in the traditional scheme when there are ten D2D pairs because the transmission power of a DUE is controlled to a small value in the formal scheme. Consequently, its throughput increases less than ours with the distance growing. Under the circumstances, the throughput increases at least 13% when there are 10 D2D pairs in this cell, comparing with the other two schemes.

To evaluate the relationship between the sum rates and the D-UE's transmission power, we plot the changing curve and show it in Figure 7. The number of D2D pairs is set to 5. From the figure we can conclude that the value of the throughput tends to smooth due to the interference as the value of transmit power is about 150mW.

Figure 5. The sum throughput of cellular communication and D2D communication with different numbers of D2D pairs and different distance between the D-UE and C-UE which share the same PRB

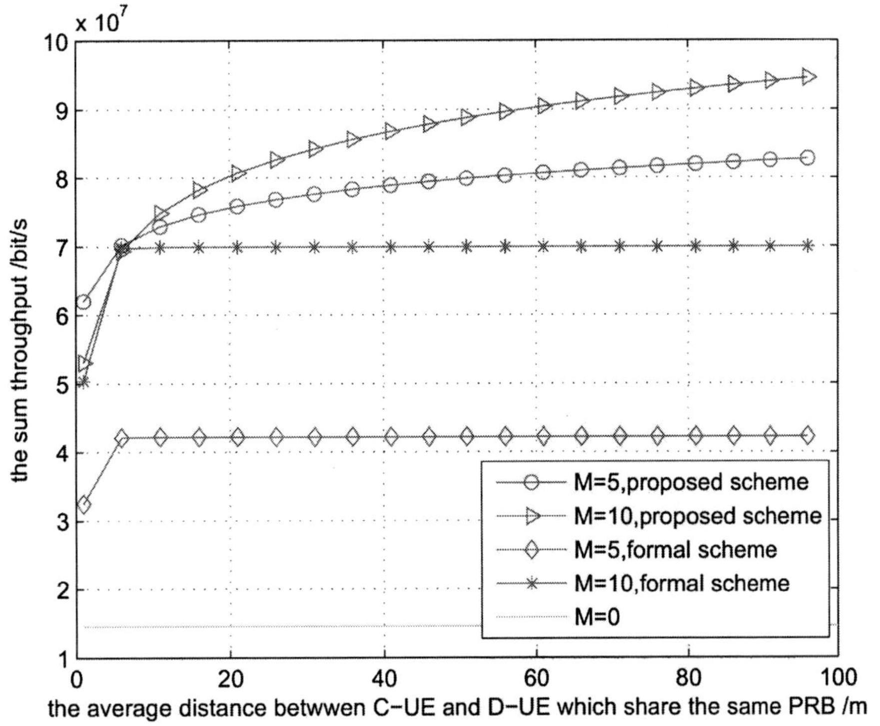

Figure 6. The sum throughput of cellular communication and D2D communication with different numbers of D2D pairs and different distance between the D-UE and eNB

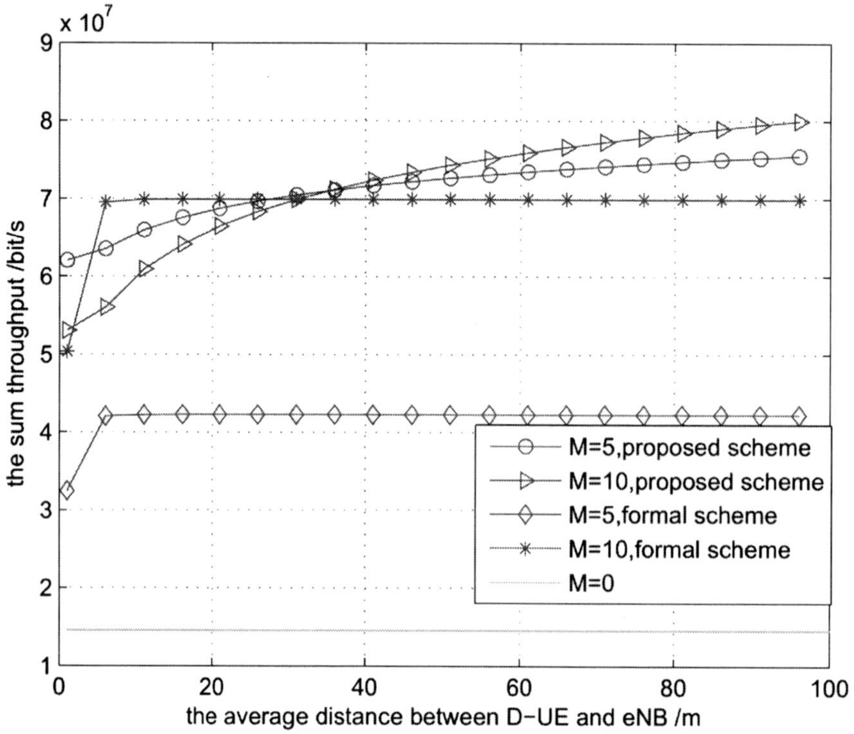

Figure 7. The sum throughput of cellular communication and D2D communication with different transmit power of D2D-Tx

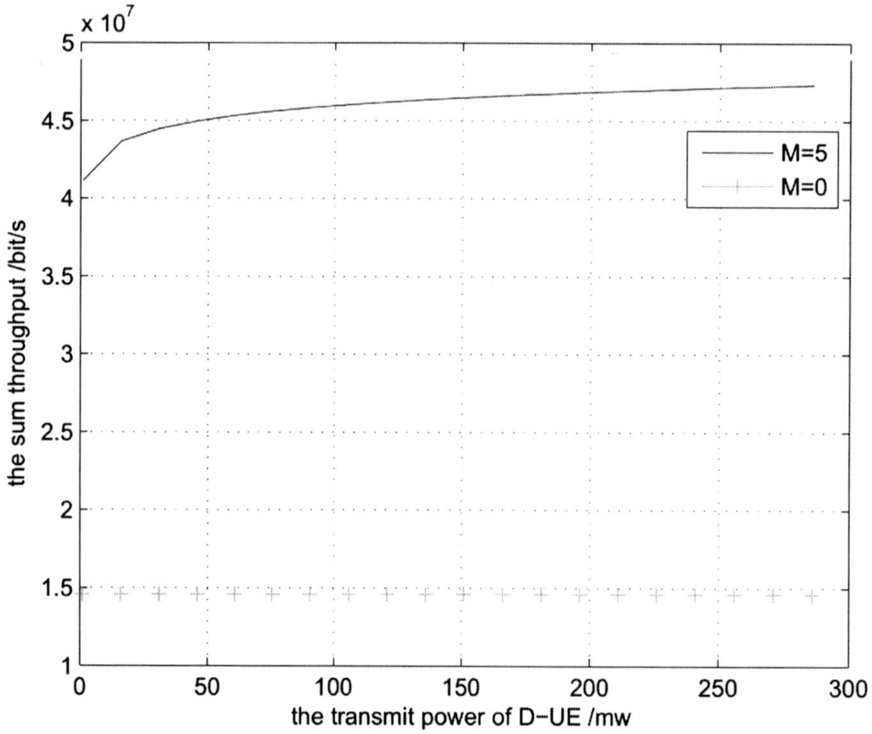

CONCLUSION

This chapter proposed an algorithm which joins power control and resource allocation and concentrates on reusing the uplink resources. Specially, a Stackelberg game based algorithm is developed in which the eNB and DUE are grouped to form the seller-buyer pair. By adopting the proposed method, the interference can be well controlled under the constraints, and better throughput performance can be obtained. Also, the resource allocation and power control proposed can lead to a good trade-off between the sum throughput and the rate of D2D pairs. Finally, the simulation results show that the proposed method not only controls the interference to the eNB, but also decreases the interference to D2D-Rx. Meanwhile, the throughput and the rate can be improved significantly.

REFERENCES

Doppler, K., Rinne, M., Wijting, C., Ribeiro, C., & Hugl, K. (2009). Device-to-device communication as an underlay to LTE-advanced networks. *IEEE Communications Magazine*, *47*(12), 42–49. doi:10.1109/MCOM.2009.5350367

Fu, W., Yao, R., Gao, F., Li, J. C., & Lei, M. (2013). *Robust null-space based interference avoiding scheme for D2D communication underlaying cellular networks*. Paper presented at IEEE Wireless Communications and Networking Conference (WCNC), Shanghai, China. doi:10.1109/WCNC.2013.6555244

Janis, P., Koivunen, V., Ribeiro, C. B., Doppler, K., & Hugl, K. (2009). *Interference-Avoiding MIMO Schemes for Device-to-Device Radio Underlaying Cellular Networks*. Paper presented at IEEE Personal, Indoor and Mobile Radio Communications (PIMRC), Tokyo, Japan. doi:10.1109/PIMRC.2009.5450284

Le, L. B. (2012). *Fair resource allocation for device-to-device communications in wireless cellular networks*. Paper presented at IEEE Global Communications Conference (GLOBECOM), Anaheim, California, USA.

Min, H., Lee, J., Park, S., & Hong, D. (2011). Capacity enhancement using an interference limited area for device-to-device uplink underlaying cellular networks. *IEEE Transactions on Wireless Communications*, *10*(12), 3995–4000. doi:10.1109/TWC.2011.100611.101684

Nie, P., & Zhang, P. (2008,). *A Note on Stackelberg Games*. Paper presented at IEEE Control and Decision Conference, Yantai, China.

Phunchongharn, P., Hossain, E., & Kim, D. I. (2013). Resource allocation for device-to-device communications underlaying LTE-advanced networks. *IEEE Wireless Communications*, *20*(4), 91–100. doi:10.1109/MWC.2013.6590055

3. rd Generation Partnership Project. (2012). *TR 22.803 V0. 3.0, 3rd Generation Partnership Project; Technical Specification Group SA; Feasibility Study for Proximity Services (ProSe) (Release 12), May, 2012*. Academic Press.

Tang, H., Zhu, C., & Ding, Z. (2013). *Cooperative MIMO precoding for D2D underlay in cellular networks*. Paper presented at IEEE International Conference on Communications (ICC), Budapest, Hungary. doi:10.1109/ICC.2013.6655469

Teng, F., Guo, D., Honig, M. L., Xiao, W., & Liu, J. (2012). *Power Control Based on Interference Pricing in Hybrid D2D and Cellular Networks*. Paper presented at IEEE Global Communications Conference (GLOBECOM), Anaheim, CA. doi:10.1109/GLOCOMW.2012.6477655

Von Stackelberg, H. (1952). *The Theory of the Market Economy*. Oxford, UK: Oxford University Press.

Wang, F., Song, L., Han, Z., Zhao, Q., & Wang, X. (May, 2013). *Joint scheduling and resource allocation for Device-to-device underlay communication*. Paper presented at IEEE Wireless Communications and Networking Conference (WCNC), Shanghai, China. doi:10.1109/WCNC.2013.6554552

Wang, J., Zhu, D., Zhao, C., Li, J. C., & Lei, M. (2013). Resource sharing of underlaying device-to-device and uplink cellular communications. *IEEE Communications Letters*, *17*(6), 1148–1151. doi:10.1109/LCOMM.2013.042313.130239

Wen, S., Zhu, X., Lin, Z., Zhang, X., & Yang, D. (2013). *Distributed resource management for Device-to-Device (D2D) communication underlay cellular networks*. Paper presented at IEEE Personal Indoor and Mobile Radio Communications (PIMRC), London, UK. doi:10.1109/PIMRC.2013.6666402

WINNER+/CELTIC Deliverable CELTIC/CP5-026 D1.3. (2008). *Innovative Concepts in Peer-to-Peer and Network Coding*.

Xu, C., Song, L., Han, Z., Zhao, Q., Wang, X., Cheng, X., & Jiao, B. (2013). Efficiency resource allocation for device-to-device underlay communication systems: A reverse iterative combinatorial auction based approach. *IEEE Journal on Selected Areas in Communications*, *31*(9), 348–358. doi:10.1109/JSAC.2013.SUP.0513031

Xu, S. Y., Wang, H. M., Chen, T., Peng, T., & Kwak, K. S. (2012). Device-to-Device Communication Underlaying Cellular Networks: Connection Establishment and Interference Avoidance. *Transactions on Internet and Information Systems (Seoul)*, *6*(1), 203–228.

Yu, C., Doppler, K., Ribeiro, C. B., & Tirkkonen, O. (2011). Resource sharing optimization for device-to-device communication underlaying cellular networks. *IEEE Transactions on Wireless Communications*, *10*(8), 2752–2763. doi:10.1109/TWC.2011.060811.102120

Zulhasnine, M., Huang, C., & Srinivasan, A. (2010). *Efficient resource allocation for device-to-device communication underlaying LTE network*. Paper presented at IEEE Wireless and Mobile Computing, Networking and Communications (WiMob), Niagara, Canada. doi:10.1109/WIMOB.2010.5645039

KEY TERMS AND DEFINITIONS

Long Term Evolution-Advanced (LTE-A): LTE Advanced is a mobile communication standard and a major enhancement of the Long Term Evolution (LTE) standard. It was formally submitted as a candidate 4G system to ITU-T in late 2009 as meeting the requirements of the IMT-Advanced standard, and was standardized by the 3rd Generation Partnership Project (3GPP) in March 2011 as 3GPP Release 10.

Device-to-Device (D2D) Communication: D2D users do not communicate via the central coordination (base station, NodeB or evolved NodeB) but operate as an underlay and communicate directly with each other. Excluding the unnecessary core network involvement, D2D communication can also provide several advantages such as high bit rates, low delay, low power consumption, low cost, plug-and-play convenience, and flexibility.

Inter-Cell Interference: In D2D underlaying cellular networks, there is two subsystems, namely, D2D subsystems and cellular subsystems. Reusing resources, D2D subsystems will interfere with cellular subsystems and vice versa. Such interference is called Inter-Cell interference.

Intra-Cell Interference: In D2D underlaying cellular networks, there is two subsystems, namely, D2D subsystems and cellular subsystems. Within D2D subsystems, D2D pairs reuse resources so that there is interference among D2D pairs. Such interference is called Intra-Cell Interference.

Resource Allocation: Resource allocation is the assignment of available resources (including transmission power, frequencies, scheduling periods, space, etc.) for data and signaling transmission in wireless communications.

Interference Coordination: A solution by applying restrictions to the radio resource management block, improving favorable channel conditions across subsets of users that are severely impacted by the interference, and thus attaining high spectral efficiency.

Stackelberg Game: A strategic game in economics in which the leader firm moves first and then the follower firms move sequentially. It is named after the German economist Heinrich Freiherr von Stackelberg who published Market Structure and Equilibrium (Marktform und Gleichgewicht) in 1934 which described the model.

Chapter 9
Distributed Dynamic Resource Allocation for OFDMA-Based Cognitive Small Cell Networks Using a Regret-Matching Game Approach

Wei-Sheng Lai
National Chiao Tung University, Taiwan

Tsung-Hui Chang
National Taiwan University of Science and Technology, Taiwan

Ta-Sung Lee
National Chiao Tung University, Taiwan

ABSTRACT

Game theoretical approaches have been used to develop distributed resource allocation technologies for cognitive heterogeneous networks. In this chapter, we present a novel distributed resource allocation strategy for cognitive small cell networks based on orthogonal frequency-division multiple access. In particular, we consider a heterogeneous network consisting of macrocell networks overlaid with cognitive small cells that opportunistically access the available spectrum. We focus on a regret-matching game approach, aiming at maximizing the total throughput of the small cell network subject to cross-tier interference and quality of service (QoS) constraints. The regret-matching game approach exploits a regret procedure to learn the optimal resource allocation strategy from the regrets of the actions of cognitive users. Furthermore, the regret-matching game approach is extended to the joint resource allocation and user admission control problem. Numerical results are presented to demonstrate the effectiveness of the proposed regre-matching approaches.

DOI: 10.4018/978-1-4666-8642-7.ch009

INTRODUCTION

The provision of a guaranteed quality of service (QoS) and high data rate to a large number of users is one of the most prominent challenges for system operators in future beyond-4G (B4G) and 5G networks. A small cell network (sometimes also called a heterogeneous network) has been recognized as a potential solution to offer better service coverage and higher spectral efficiency. In addition, a combination of a small cell network with cognitive techniques to opportunistically exploit unlicensed bands [e.g., the 2.4-GHz industrial, scientific, and medical(ISM) and 5-GHz Unlicensed National Information Infrastructure (U-NII) bands] and bands not occupied by the macrocell system can further improve the system throughput by offloading the traffic demands from mobile users. This idea has motivated the extension of the modern Long-Term Evolution (LTE)-Advanced system to an unlicensed spectrum, leading to the so-called LTE-U system. It is expected that advanced dynamic resource allocation strategies would serve one of the most important roles in future cognitive small cell networks in order to make the best usage of available resources while effectively controlling co-tier and cross-tier interference.

In this chapter, a heterogeneous network consisting of macrocell networks overlaid with cognitive small cell networks that opportunistically access the available spectrum is considered. In particular, the aim is to develop distributed dynamic resource allocation strategies, in which cognitive small cells manage to maximize their individual utility independently, subject to cross-tier interference and QoS constraints. To this end, a game-theoretic framework is proposed for distributed dynamic resource allocation for cognitive small cell networks.

This chapter contains two parts. In the first part, a distributed dynamic resource allocation problem for the interference management of cognitive small cell networks is studied. Specifically, the design problem is formulated as a regret-matching game problem that accounts for the sensing information gathered by the small cells. The regret-matching game approach is a type of adaptive learning algorithm in which the small cells learn the best resource allocation strategies (actions) based on certain "regret" values that are predictable from current and past actions. It is shown that the small cells can independently and selfishly maximize their individual utility and efficiently mitigate co-tier interference by integrating the regret-matching game approach with environmental sensing information. More importantly, it is provable that the approach can converge to a set of so-called coarse correlated equilibrium.

In the second part of this chapter, the regret-matching game approach is further extended to a joint resource allocation and user admission control problem. The considered joint design problem is important because mobile users in densely deployed small cell networks are subject to complex co-tier and cross-tier interference constraints. To ensure users that have accessed the network experience a uniform and guaranteed QoS, it is necessary to optimize both the resource allocation and admission control strategies jointly. Extensive simulation results will be provided to demonstrate the effectiveness of the proposed game-theoretic approaches. In addition, detailed discussions of the challenges and promising future research directions will be presented as well.

The rest of this chapter is organized as follows. In the next section, the background of the cognitive small cell network is briefly described. The system model and problem statements are then presented in the following section. Afterwards, the distributed dynamic resource allocation problem is studied, and it is shown how the regret-matching procedure in game theory can be used to solve it. Next, our work is extended to the joint resource allocation and user admission control problem. Finally, some numerical results of the proposed algorithm are presented with the conclusions.

BACKGROUND

A small cell network embedded with cognitive radio (CR) technology has been recognized as a promising approach for the deployment of B4G/5G mobile communication systems(Wang & Liu,2011).In CR, secondary users (SUs) are allowed to use unused bands opportunistically under the premise of not affecting primary users (PUs). The concept of CR has been introduced into small cell networks in order to improve the reuse factor and to provide a uniform experience for indoor mobile users. The motivation behind the use of small cells is that their coverage and capacity are relatively poor, despite the large portion of data traffic originating from homes and offices. Because the deployment of small cell networks is usually achieved by sharing the resources of conventional cellular networks, efficient dynamic resource utilization is required for improving spectrum usage. In practice, the mitigation of interference in the absence of coordination between cellular users and small cells becomes a critical issue because small cells are undefined user plug-in systems.

In cognitive small cell networks, small cells are required to sense the spectrum status before accessing the spectrum (Kang *et al.*, 2008).Therefore, the dynamic resource allocation problem considered in this chapter is for a sensing-based two-tier network. Small cell access points (SAPs) can sense spectra as resource blocks (RBs) and detect whether and which RBs are used by the macrocell user equipment (MUE). In particular, the modified maximum–minimum eigenvalue(MME)-based sensing scheme (Zeng & Liang, 2009; Lai *et al.*, 2012) is adopted, in which the SAPs can perform eigenvalue decomposition of the sample covariance matrix and cooperatively obtain and share these eigenvalues with SAPs in the same cluster (Lai *et al.*, 2012). On the basis of this sensing information, small cell user equipments (SUEs) can strategically use the spectrum and execute resource allocation algorithms. However, proper simultaneous spectrum access management for all SUEs remains challenging, even when the sensing results are perfect. Cross-tier (small-cell-to-macrocell) and co-tier (small-cell-to-small cell) interference can still occur.

Considerable research has been conducted to address this resource allocation problem. The work of Kang *et al.*(2008) decomposed the problem into a two-level optimization problem and used a subgradient algorithm to solve it. The problem formulation in Khoshkholgh *et al.* (2010) considered the distribution of channel gains between the MUEs and the SUEs. Game-theoretic approaches are also considered for this problem. The small cells act as players and act to select available spectrum to access and perform suitable power allocation. A useful algorithm called the regret-matching algorithm in game theory can provide an effective way to solve such a competition problem in a distributed system, which has been proposed in Kang *et al.* (2008), Hung *et al.* (2011), Maskeryet *et al.* (2009), Bennis *et al.* (2012), and Nie and Comaniciu (2005). A regret-matching algorithm is a type of learning algorithm, and its basic idea is to learn regret from the historical actions that occur in previous time instants and adjust to minimize the regret value (Hart & Mas-Colell, 2000; Hart, 2004).

In the work of Hung *et al.*(2011), the authors consider the downlink spectrum allocation problem with a pure overlay spectrum access strategy and design a local utility function incorporating self-interest, fairness, and power consumption. Furthermore, Bennis *et al.* (2012) designed actions with joint spectrum and power allocation; therefore, the action space became larger. Nieand and Comaniciu (2005) showed that the regret-matching learning algorithm has similar performance as the potential game (a potential game needs higher communication overhead for information exchange; in contrast, the regret-matching algorithm does not exchange any information with other players).In practice, the pure overlay strategy is inefficient because the demand for spectrum has dramatically increased in cellular systems. Similar to the requirements mentioned in the cooperative spectrum sensing phase, it is also desirable that only a small amount of data be exchanged, and the computational complexity of the spectrum allocation algorithm remains low. However, these studies did not consider the avoidance of interference from small cells to other small cells and MUEs. Moreover, these small cells need to be no cooperative and competitive. This is because a processing center, which can fuse the sensing information and dynamically and efficiently allocate the spectrum and power for all SAPs, is not always feasible(Al-Rubaye *et al.*, 2011).To this end, a new distributed approach is presented for game-theoretic resource allocation based on spectrum sensing information.

SYSTEM MODEL AND PROBLEM STATEMENTS

Network Architecture

An uplink orthogonal frequency-division multiple access (OFDMA) system is shown in Fig. 1(a). Here, only one macrocell base station (MBS) is considered. In this macrocell, it is assumed that there are N_{SAP} SAPs, N_{SUE} SUEs, and N_{MUE} MUEs that coexist and share N_{RB} RBs. It is assumed that the SAPs have multiple antennas, and both the SUEs and MUEs have a single antenna. Other system parameters for resource allocation are listed in Table 1. The MUEs and small cells are chosen and uniformly distributed in the macrocell service region (Maskeryet *et al.*, 2009), and the small cells operate in a closed subscriber group (CSG). This means that each SUE is served by one SAP only. Here, it is assumed that each SAP has only one SUE.

The spectrum access strategy is illustrated in Fig. 1(b).In this subsection, it is presumed that the MUEs already occupy some RBs and that the sensing procedure of the SUEs could identify the occupied RBs. Because sensing is not always perfect, the spectrum access status may occupy two scenarios. The misdetection scenario occurs when the sensing detects occupied RBs as unoccupied. The other scenario is the occurrence of a false alarm when the sensing incorrectly detects available RBs as occupied. These two scenarios cause the spectrum reuse factor to fall.

Depending on the SUE requirements, there are primarily two types of access strategies (Khoshkholgh *et al.*, 2010).The overlay strategy is employed to only allow SUEs to access the spectrum when no MUE signals are detected to ensure no interference to the MUE. In addition, the underlay strategy allows SUEs to always access the spectrum on a specific constraint of cross-interference to the MUEs. However, the strategy in Fig. 1(b) is implemented in this study, in which three RBs are sensed as unoccupied and allowed access by all small cells. One RB still causes cross-tier interference when a misdetection occurs. Therefore, the consideration here is not a pure overlay access strategy but a mixed strategy.

Problem Statements

It is assumed that all small cells can access the unoccupied licensed or unlicensed spectrum. Owing to the issue of security and the limited availability of backhaul bandwidth (Zemlianov & Veciana, 2005), interference mitigation in small cell networks faces practical challenges due to the lack of coordination between SAPs. Mathematically, the SAPs are to independently and selfishly/rationally maximize the individual rate, which is defined as follows:

Figure 1. Network architecture and spectrum access strategy: (a) coexistence of small cell and macrocell systems and (b) detection of available RBs

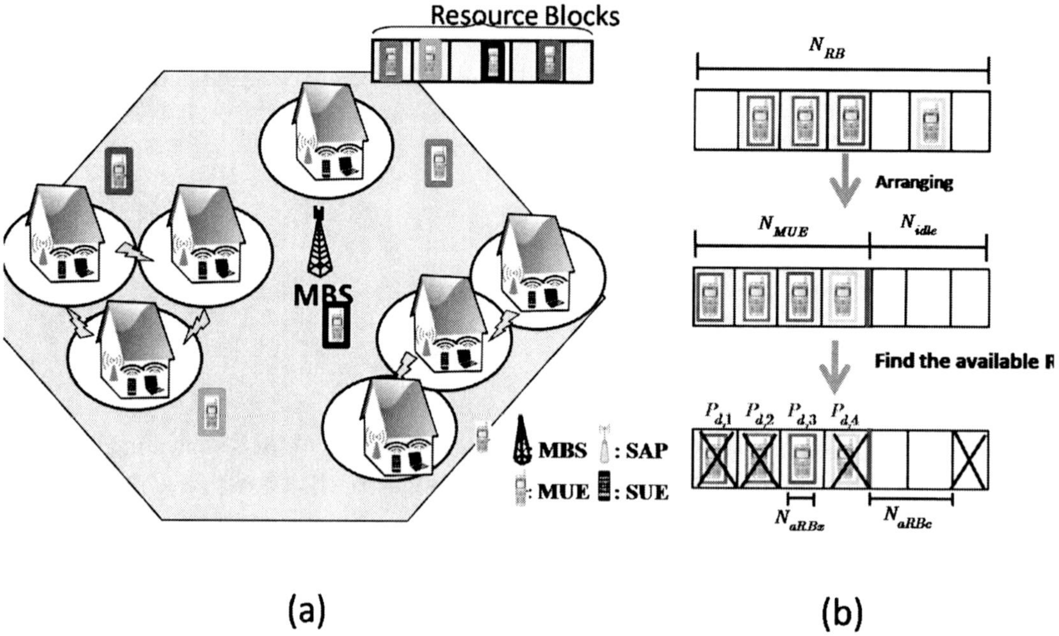

Table 1. Dynamic resource allocation parameters

N_{SAP}	Total number of SAPs in the same cluster
N_{SUE}	Total number of SUEs in the same cluster
N_{MUE}	Total number of MUEs
N_{RB}	Total number of RBs
N_{idle}	Total number of unoccupied RBs ($N_{idle}=N_{RB}$-N_{MUE})
N_{aRB}	Total number of available RBs by sensing $N_{aRB}=N_{aRBx}+N_{aRBc}$
N_{aRBx}	Number of available RBs with cross-tier interference
N_{aRBc}	Number of available RBs that may encounter co-tier interference
N_{iter}	Total number of iterations needed to approach an exhaustive search
N_{pre}	Total number of pre-assignments for interference avoidance
N_{acc}	Total number of accessed RBs by each SUE

$$R_k = \sum_{f=1}^{N_{acc}} \log_2 \left\{ 1 + \frac{a_k(f) \cdot h_k(f) p_k(f)}{\left[I_{ss,k}(f) + I_{ms,k}(f) + \sigma_k^2(f) \right]} \right\}, \forall k, \quad (1)$$

where $h_k(f)$ is the path-loss and fast-fading channel gain between SAP k and the served SUE, and $p_k(f)$ is the power allocation of SUE k at RB f. $\sigma_k^2(f)$ is the noise term at RB f, and N_{acc} is the total number of accessed RBs by each SUE. The interference terms $I_{ss,k}(f)$ and $I_{ms,k}(f)$ are the co-tier interference and cross-tier interference at SAP k, which is defined as follows:

$$I_{ss,k}(f) = \sum_{\substack{i=1 \\ i\neq k}}^{N_{SUE}} w_{ss,k,i}(f)\, a_k(f),\ \forall k, \tag{2}$$

$$I_{ms,k}(f) = \sum_{j=1}^{N_{MUE}} w_{ms,k,j}(f)\, a_k(f),\ \forall k, \tag{3}$$

where $w_{ss,k,i}(f)$ is the co-tier interference between the kth SAP and the ith SUE on the fth RB. $w_{ms,k,j}(f)$ is the cross-tier interference between the jth MUE and the kth SAP on the fth RB. $a_k(f)\in\{0,1\}$ represents whether the fth RB is used. If $a_k(f)=1$, the fth RB is used; otherwise, it is not. Here, $\sum_{f=1}^{N_{aRB}} a_k(f) = N_{acc}$, where N_{acc} is the total number of accessed RBs by each SUE, and N_{aRB} is the number of available RBs.

Because the SAPs belong to a secondary subscriber, they cannot seriously affect the operation of the system (i.e., the MBS). Let $P_I^{\max}(f)$ be the maximum tolerable interference of the MBS over RB f. Here, a tolerable interference power is set to constrain the interference from the SAPs to the MBS as follows:

$$I_{\text{tol}}(f) = \sum_{k=1}^{K} w_{sm,k}(f)\, a_k(f) \le P_I^{\max}(f), \tag{4}$$

where $w_{sm,k}(f)$ is the cross-tier interference from SAP k to the MBS.

SENSING-BASED REGRET-MATCHING GAME FOR DISTRIBUTED DYNAMIC RESOURCE ALLOCATION

The global resource allocation optimization problem is to maximize the sum of each local utility function subject to an interference constraint. Hence, the global resource allocation optimization objective function can be represented as follows:

$$\arg\max_{\mathcal{A}} \sum_{k} U_k^t(\mathcal{A}), \tag{5}$$

where U_k^t is the general utility function of each SUE, $\mathcal{A}$ is the joint action space defined as $\mathcal{A} \triangleq \prod_{k\in K} \mathcal{A}_k$, and $\mathcal{A}_k$ is the individual action space. If the action spaces of every SUE are the same, the size of joint action space becomes $|\mathcal{A}| = |\mathcal{A}_k|^K$.

To reach the global optimum solution with the decentralized approach, an idea is adopted from the regret-matching algorithm in game theory (Hung & Krishnamurthy, 2011; Hart & Mas-Colell, 2000). The basic idea of the regret-matching algorithm is to learn the "regret" of its own actions that had been taken at every time instant and attempt to minimize the regret value. In the work of Hart and Mas-Colell (2000), if all players in the game follow the regret-matching algorithm to switch their action, the distribution of the joint action space converges to a set of correlated equilibria. The definition of a correlated equilibrium is presented below (Hung *et al.*, 2011):

Definition 1: Correlated Equilibrium

In a correlated equilibrium, no player can increase his or her own local utility by changing his or her own action without changing the actions of other players (i.e., a player is not better off with any arbitrary action alone):

$$\sum_{\mathbf{a}_{-k}^{t}\in\mathcal{A}_{-k}} \pi_{CE}\left(\mathbf{a}_{k}^{t},\mathbf{a}_{-k}^{t}\right)\cdot U_{k}\left(\mathbf{a}_{k}^{t}\right) \geq \sum_{\mathbf{a}_{-k}^{t}\in\mathcal{A}_{-k}} \pi_{CE}\left(\hat{\mathbf{a}}_{k}^{t},\mathbf{a}_{-k}^{t}\right)\cdot U_{k}\left(\hat{\mathbf{a}}_{k}^{t}\right),\ k=1,2,\ldots,K, \tag{6}$$

where $\left(\mathbf{a}_{k}^{t},\mathbf{a}_{-k}^{t}\right)=\mathbf{a}^{t}$ is defined as the instantaneous action of the *k*th player at time instant *t*, i.e. $\left(\mathbf{a}_{k}^{t}\right)$, and the action of all players except the *k*th player, i.e., $\left(\mathbf{a}_{-k}^{t}\right)$. $\pi_{CE}\left(\mathbf{a}_{k}^{t},\mathbf{a}_{-k}^{t}\right)$ is the joint probability distribution of the correlated equilibrium in the joint action space $\mathcal{A}$. The correlated equilibrium is a solution concept that is more general than the well-known Nash equilibrium (Hung *et al.*, 2011). A strategy assigns an action to every possible observation a player can make. If no player would want to deviate from the recommended strategy, the distribution is called a correlated equilibrium. A detailed discussion and explanation will be presented after our proposed algorithm. Here, the problem of concern is first described, which can be formulated in a game form in the following.

Game-Theoretic Problem Formulation

Our problem is formulated in a game formant solved by the regret-matching algorithm in this subsection. Here, each SUE in the same cluster acts as a player performing an action $\mathbf{a}_{k}^{t}$ to acquire certain RBs at each discrete instant *t*. The objective function of the resource allocation problem is to independently and selfishly/rationally maximize the individual utility U_{k}^{t} at each time *t*. To describe a game, let us introduce the following four elements (Hung *et al.*, 2011):

- **Game**: The spectrum (RB) allocation problem is formulated among SUEs in the same cluster in the uplink OFDMA system as a non-cooperative game. In game theory, a non-cooperative game means that all players make their own decisions independently. A non-cooperative game has a distributed nature, as each SAP does not require the information of others. Consequently, the system is scalable with the number of SUEs.

- **Player**: The N_{SUE} SUEs coexisting in the same cluster are modeled as players competing to acquire the available RBs.
- **Actions**: The access indicator set is defined as the action spaces of the SUE. The joint and individual action spaces $\mathbf{a}^t$ and $\mathbf{a}_k^t$ are defined as follows:

$$\begin{cases} \mathbf{a}_k^t = \left[a_k^t(1),\ldots,a_k^t(N_{aRB})\right] \in \mathcal{A}_k \\ \mathbf{a}^t = \left[\left(\mathbf{a}_1^t\right)^T,\ldots,\left(\mathbf{a}_K^t\right)^T\right]^T \in \mathcal{A}, \end{cases} \tag{7}$$

where $a_k^t(f) \in \{0,1\}$ represents whether the fth RB is used by SUE k at time t, for all $f \in \{1,\ldots,N_{aRB}\}$. $\mathcal{A}_k$ is the individual action space of size $|\mathcal{A}_k|$, and $\mathbf{a}^t$ is the joint action, whereby the joint action space is $\mathcal{A} \triangleq \prod_{k\in K} \mathcal{A}_k$. As mentioned before, all players make their own decisions and carry out their actions independently in a non-cooperative game.

- **Utility function**: Let $U_k^{t'}$ denote a new local utility function of SUE k at time instant t as follows:

$$U_k^{t'} = \sum_{f=1}^{N_{aRB}} \max\left(U_k^t(f) + P_k^t(f), 0\right), \tag{8}$$

where U_k^t is the local utility function of SAP k at time instant t, which is defined as follows:

$$U_k^t(f) = \sum_{f=1}^{N_{acc}} \log_2\left\{1 + \frac{a_k^t(f)\cdot h_k(f)\,p_k(f)}{\left[I_{ss,k}^t(f) + I_{sm,k}^t(f) + \sigma_k^2(f)\right]}\right\}, \tag{9}$$

where the interference terms $I_{ss,k}^t(f)$ and $I_{sm,k}^t(f)$ are the co-tier interference and cross-tier interference at SAP k at each discrete instant t. An easy way to find the global optimum solution is the exhaustive search method. However, the search space is up to $|\mathcal{A}_k|^K$, which results in a prohibitive computational complexity. Another problem is that the global optimum solution depends on the private information of every SAP. Because a small cell is an undefined and user-plug-in system, this assumption is not practical. For this reason, an allocation scheme using a decentralized approach is needed to maximize global resources. $P_k^t(f)$ denotes the penalty term, which is defined as follows:

$$P_k^t(f) = \begin{cases} \log_2\left(1 - \left(\dfrac{I_{sm,k}^t(f)}{\sigma_k^2(f)}\right)^\alpha\right) & I_{sm,k}^t(f) < P_I^{\max}(f) \\ -\infty & I_{mf,k}^t(f) \geq P_I^{\max}(f) \end{cases}, \forall k. \tag{10}$$

Here, "local" means that each SUE in the cluster only has the information about its own action $\mathbf{a}_k^t$ but does not have any information about other SUEs' actions, which are $\left\{\mathbf{a}_i^t,\ i=1,\ldots,N_{SUE},\ i\neq k\right\}$.

Because the SUEs cannot affect the operation of the MBS, substitute the penalty function in (10) into the utility function to constrain the interference power such that it does not exceed the tolerable interference power. Thus, each SUE can selfishly maximize its own utility. In the proposed scheme, the power control and cross-tier interference reduction are realized by the penalty function. Here, a simple example is presented to illustrate the penalty term in the next paragraph.

Example: In the considered scenario, the number of devices is small (Hung & Krishnamurthy, 2011); thus, the interference power level is set to 10^{-15}. If the interference power level is equal to or becomes greater than the tolerable interference power, the penalty value changes to minus infinity, and the utility value changes to zero. In addition, the setting parameter α depends on the power levels of the SUEs that interfere with the MUEs and is set to 0.1 to start the interference constraint from approximately 10^{-15}.Fig. 2 shows that the penalty value would greatly decay and in turn affect the utility value in (8) when the transmit power is close to the constrained power level.

Power Allocation

The proposed resource allocation is distributed; thus, each SUE can instantaneously identify adaptive RB allocation and jointly consider the co- and cross-tier interference and noise. An optimal power al-

Figure 2. Penalty value when the transmit power is close to the power constraint

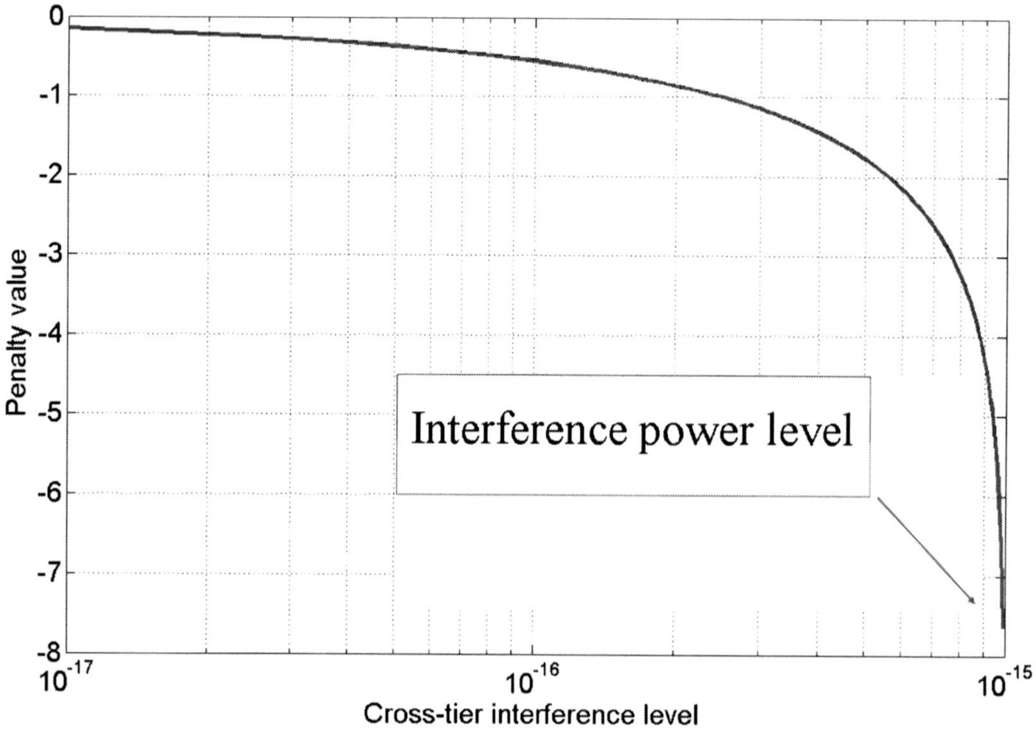

location scheme is carried out, which adopts the water filling algorithm (Tse & Viswanath, 2005) to allocate the power $p_k(f)$ to each RB after the spectrum allocation algorithm. Thus, the power allocation optimization problem is as follows:

$$\max_{\{p_1,\ldots p_{N_{acc}}\}} \sum_{n=1}^{N_{acc}} \log_2\left(1+\frac{p_n\|h_n\|^2}{\sigma_k^2(f)+I_{ss,k}(f)+I_{sm,k}(f)}\right), \text{s.t.} \sum_{n=0}^{N_{acc}} p_n \le P_{\text{total}}, \forall n. \tag{11}$$

The objective function in (11) is convex in the powers, and this optimization problem can be solved by the Lagrangian method. Note that the co- and cross-tier interference would be seen as noise. The power allocation can be expressed as

$$p_k^*(f)=\left(\frac{1}{\lambda}-\frac{\sigma_k^2(f)+I_{ss,k}(f)+I_{sm,k}(f)}{|h_k(f)|^2}\right)^+, \ \forall k, \tag{12}$$

where $(x)^+ \triangleq \max(x,0)$, and λ is a Lagrange multiplier.

Modified Regret-Matching Game

In this subsection, an algorithm that applies the original regret-matching procedure to our problem is designed. The regret is set to be the difference between the utility of the current action and the actions of others. Note that (13) is an expression for one iteration (Hung & Krishnamurthy, 2011).Then, every player updates the overall regret matrix and changes his or her action according to a probability, which is a function of the last overall regret matrix for all actions computed. The proposed modification algorithm and how to compute the instantaneous regret are described herein.

The instantaneous regret value is calculated as follows:

$$G_k^{(j,i)}(\mathbf{a}^t)=1_{(\mathbf{a}_k^t=S_j)}\times\left[U_k(S_i,\mathbf{a}_{-k}^t)-U_k(S_j,\mathbf{a}_{-k}^t)\right],\ k=1,2,\ldots,K, \tag{13}$$

where $1_{(\bullet)}$ is an indicator function. If $\mathbf{a}_K^t=S_j$, the element of column S_j is equal to one; otherwise, it is equal to zero. $G_k(\mathbf{a}^t)$ is the instantaneous regret matrix $|\mathcal{A}_k|\times|\mathcal{A}_k|$, and the elements of the matrix $G_k(\mathbf{a}^t)$ are $G_k^{(j,i)}(\mathbf{a}^t)$. S_j and S_i are the current and next action played by the SAPs, where $S_j,S_i\in\mathcal{A}_k$. The instantaneous regret matrix means that each SUE plays a strategy in which the probability of playing an action is proportional to the positive part of his regret for that action. Here, let μ is constant and non-negative real value, the probability distribution $\Pr(\mathbf{a}_k^t=S_i\,|\,\mathbf{a}_k^{t-1}=S_j,\theta_k^{t-1})$ at time instant t is defined as follows:

$$\Pr\left(\mathbf{a}_k^t = S_i \mid \mathbf{a}_k^{t-1} = S_j, \theta_k^{t-1}\right) = \begin{cases} \left(\frac{1}{\mu}\right)\left(\theta_k^{t-1}(i,j)\right)^+ & , S_i \neq S_j \\ 1 - \sum_{S_i \neq S_j} \left(\frac{1}{\mu}\right)\left(\theta_k^{t-1}(i,j)\right)^+ & , S_i = S_j \end{cases}, \forall k, \tag{14}$$

where θ_k^t is the overall regret value at time t, which is defined as follows:

$$\theta_k^t = \theta_k^{t-1} + \varepsilon^t \cdot \left(G_k(a^t) - \theta_k^{t-1}\right) \tag{15}$$

The instantaneous regret in (13)-(15) is exploited to build a modification algorithm, as summarized in Algorithm 1.

Correlated Equilibrium

Correlated equilibrium was originally applied in economics; it is the joint probability distribution over the possible action spaces. In particular, each player in this game receives a private signal that informs the player of how to determine his or her action, and different actions lead to different correlated equilibria. The notion of a correlated equilibrium originated from Nash equilibrium and was introduced by Aumann (1974) (Hart & Mas-Colell, 2000). Generally, it can be thought of as the distributions of action instructions that tell the players how to perform their action. Through these instructions, the players perform their actions independently and evenly, providing the best response to the received actions by others. The main difference between the Nash equilibrium and the correlated equilibrium (Lasaulce & Tembine 2011) is that signals in the former are independent, whereas they are not in the latter. The correlated equilibrium fits our scenario. Each SUE and MUE causes interference to others, and the interference observed by each SUE and MUE is not independent, which is how the players interact with each other. The definition of a correlated equilibrium was presented in Definition 1. Theorem 1 and Theorem 2 show that Algorithm 1 is able to converge to the set of correlated equilibria.

Algorithm 1. Modified regret-matching algorithm

1: Start (t= 0)
2: Identify the available RBs via cooperative sensing
3: Initialize $\mathbf{a}_k^0$ **and** $\theta_k^0 = \mathbf{G}_k(\mathbf{a}^0),\ \forall k \in K$
4: Repeat
5: $t := t + 1$
6: Calculate regret matrix
7: Observe: choose the action according to the probability in (14) to select a new action, attempt to minimize the regret value at every time instant.
8: After the new action S_j is chosen, each SUE adopts the water filling algorithm to allocate power $p_k(f)$ to each RB
9: Update regret matrices
10: $\theta_k^t = \theta_k^{t-1} + \varepsilon^t \cdot \left(\mathbf{G}_k(\mathbf{a}^t) - \theta_k^{t-1}\right)$
11: Until $\Pr\left(\mathbf{a}_k^t = S_i \mid \mathbf{a}_k^{t-1} = S_j, \theta_k^{t-1}\right) \geq \varepsilon,\ \forall k$
12: Output the joint action space a^t and allocation power $p_k(f)$

Theorem 1: Finite Games (Nash, 1950)

Assume an action forms a game, in which this game has a definite beginning and ending. If the set of action profiles $\mathcal{A} \triangleq \prod_{k \in K} \mathcal{A}_k$ is finite, at least one mixed Nash equilibrium exists in this game.

Note that every Nash equilibrium is a correlated equilibrium. In fact, the solution of a correlated equilibrium is more general than the Nash equilibrium. Hence, Theorem 1 means that at least one correlated equilibria exists in every finite game.

Theorem 2

For the proposed dynamic resource allocation game scheme (Algorithm 1 and Algorithm 3), a correlated equilibrium always exists. Moreover, our algorithms can converge to the set of correlated equilibria of the game with a probability of one.

Proof: In Hart and Schmeidler (1989), Theorem 1 proves that every finite game has a correlated equilibrium. In Hart and Mas-Colell (2000) andHart (2004), the authors provide a procedure to converge to the set of correlated equilibria. Therefore, the proposed game algorithm is obviously a finite game, and it has the same properties of the procedures in Hart and Mas-Colell (2000) and Hart (2004). Hence, this theorem can directly use the conclusions of Hart and Mas-Colell (2000), Hart (2004), and Hart and Schmeidler (1989).

A simple explanation of correlated equilibrium is that each player has chosen a prior action, and no player can increase their own local utility by changing their own action while the actions of the other players are not changed. Under this condition, the set of actions are called correlated equilibrium. In other words, a correlated equilibrium is a more general form of Nash equilibrium, which is a probability distribution over the set of action spaces.

θ_k^t in (15) may be expressed in detail as follows:

$$\theta_k^t = \theta_k^{t-1} + \varepsilon^t \cdot \left(\mathbf{G}_k(\mathbf{a}^t) - \theta_k^{t-1}\right) = (1-\varepsilon^t)\cdot\theta_k^{t-1} + \varepsilon^t \cdot \mathbf{G}_k(\mathbf{a}^t),\ k = 1,2,\ldots,K, \tag{16}$$

where $\varepsilon^t = 1/t$ is a decreasing step size used to accumulate the instantaneous regret value into the overall regret value. The entries of the overall regret matrix θ_k^t denote the averaged regret value of user k until time t, which is denoted as

$$\theta_k^t(j,i) = \frac{1}{t} \sum_{\tau < t, \mathbf{a}_k^t = S_j} \left(U_k\left(S_i, \mathbf{a}_{-k}^\tau\right) - U_k\left(S_j, \mathbf{a}_k^\tau\right)\right),\ k = 1,2,\ldots,K. \tag{17}$$

When the overall regret value has been computed, the action in next period will switch to action S_i from S_j with a probability that is proportional to the regret value as follows:

$$\begin{cases} \mathrm{P}(\mathbf{a}_k^t = S_i \mid \mathbf{a}_k^{t-1} = S_j, \theta_k^{t-1}) = \frac{1}{\mu}\left[\theta_k^{t-1}(j,i)\right]^+ & \text{, switch to } S_i \text{ from } S_j \\ \mathrm{P}(\mathbf{a}_k^t = S_j \mid \mathbf{a}_k^{t-1} = S_j, \theta_k^{t-1}) = 1 - \frac{1}{\mu}\sum_{i \neq j}\left[\theta_k^{t-1}(j,i)\right]^+ & \text{, stay in } S_j \end{cases}, \tag{18}$$

where $\mu > \sum_{i \neq j}\left[\theta_k^{t-1}(j,i)\right]^+$ is a constant to ensure that the probability is non-negative. Therefore, according to Theorem 1 and Theorem 2, Algorithm 1 can converge to the solution of the correlated equilibrium.

Interference Avoidance Strategy

In this subsection, an interference avoidance procedure is presented. Because each SUE is able to obtain the eigenvalue of the covariance matrix of the received signal with the MME sensing (Khoshkholgh *et al.*, 2010) method, the SUE may identify RBs that are already occupied by MUEs. Specifically, the core concept of MME-based spectrum sensing is to separate the information about the signal power of the MUEs from the noise power through an eigenvalue decomposition (EVD) function. The maximum eigenvalue contains the information about both the signal power of the MUEs and the noise power; in contrast, the minimum eigenvalue only contains the information about the noise power. Hence, an RB should be assigned to the SUE that has the smaller maximum eigenvalue. The details of the MME sensing method was presented in the papers of Khoshkholgh *et al.* (2010) and Lai *et al.* (2012). The proposed procedure for interference avoidance is summarized in Algorithm 2.

In the initialization phase, the RBs are surely occupied or unoccupied; therefore, these RBs are pre-assigned to the SUEs, where the detected signals have the highest path loss from an MUE to an SUE if the RBs are truly occupied. Further, N_{pre} is determined by N_{aRBx} and the desired reduction in interference at the cost of utility degradation. In summary, the main purpose of Algorithm 2 is to pre-assign some RBs to an SUE with lower chances of interference with the MUE. Fig. 3 illustrates how Algorithm 2 works. The greater eigenvalue $\lambda_{\max}^{k,i}$ means that the potential cross-tier interference is greater at RB i. Hence, RB i is assigned to SUE k if $\lambda_{\max}^{k,i}$ is smaller than another eigenvalue. The definitions of N_{pre}, N_{aRBc}, and N_{aRBx} are listed in Table 1 and shown in Fig. 1(b).

JOINT DISTRIBUTED DYNAMIC RESOURCE ALLOCATION AND USER ADMISSION CONTROL DESIGN

The operator now faces another challenge that is the limited resources to provide a certain QoS in densely deployed small cell environments. Hence, our regret-matching game method is extended to the joint resource allocation and user admission control problem, as described in this section.

Admission Control Strategy

The admission control problem is to find the largest subset of SUEs $\mathcal{S}$ that satisfies the QoS constraint:

Figure 3. Illustration of how Algorithm 2 works

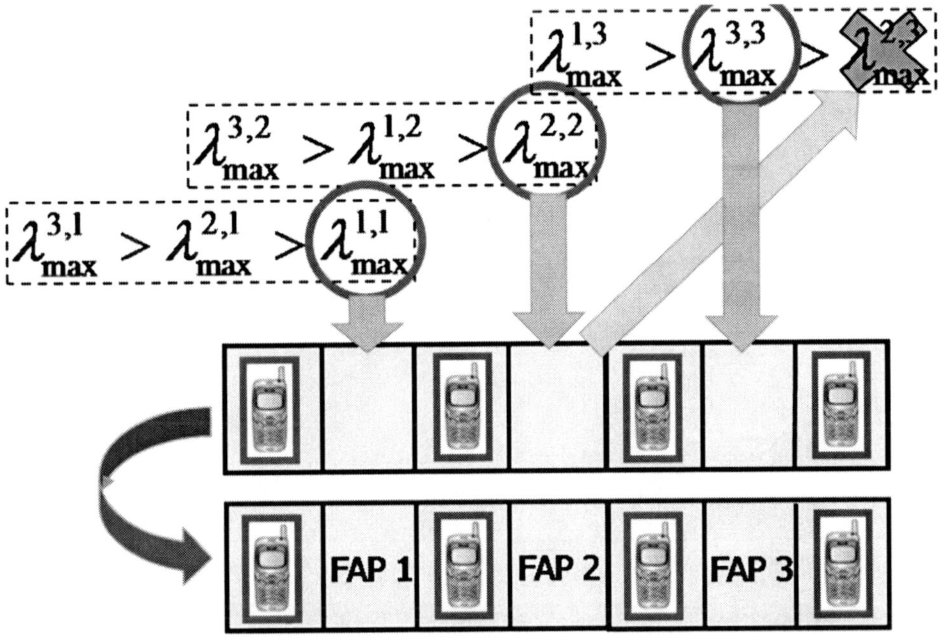

Algorithm 2. Interference avoidance

1: Start (t = 0)
2: On the basis of the sensing information, obtain N_{aRBx} and determine N_{pre} (this procedure could be performed after all N_{aRBc} RBs are fully accessed to produce less cross-tier interference)
3: Repeat
4: If $\lambda_{\max}^{k,i} = \max\left\{\lambda_{\max}^{k,i}, \ldots, \lambda_{\max}^{N_{SUE},i}\right\}$ of SUE *k* at RB *i* is the smallest among all the SUEs, RB*i* is then assigned to SUE *k*
5: If SUE *k* has acquired up to N_{pre} RBs, then exit the candidate space
6: Until all N_{aRBx} RBs have been allocated or each SUE already has RBs up to N_{pre}
7: Output the RB assigned as an initial point of Algorithm 1, then start Algorithm 1 to allocate the spectrum and power.

$$\mathcal{S} = \left\{k \mid R_k(\mathbf{a}_k^t) \geq R_{\min}, \forall k\right\}. \tag{19}$$

The maximization problem of the number of served SUEs is as follows:

$$\max_{\mathbf{a}_k^t, p_k(f)} |\mathcal{S}| \tag{20}$$

$$\text{s.t. } R_k(\mathbf{a}_k^t) \geq R_{\min}, \forall k, \tag{21}$$

where $R_{\min}$ is the required minimum rate. Our regret-based game algorithm is extended to the joint admission control problem in (20)-(21), as summarized in Algorithm 3. Here, suppose that each SUE can

Algorithm 3. Regret-matching algorithm with admission control

1: Start ($t=0$)
2: Given the set of SUEs $\mathcal{N}_{\text{SUE}}$, **let** $\mathcal{N}_{\text{QoS}} \leftarrow \mathcal{N}_{\text{SUE}}$
3: Identify the available RBs via cooperative sensing
4: Initialize $\mathbf{a}_k^0$ **and** $\theta_k^0 = \mathbf{G}_k(\mathbf{a}^0),\ \forall k \in K$
5: Repeat
6: Employ Algorithm 1 and Algorithm 2 to solve the resource allocation problem in (5) and obtain $\mathcal{A}_k$ and the allocation power $p_k(f)$.
7: Calculate the individual rate $R_k(A_k)$ in (1)
8: Send the individual rate to other SAPs
9: Let $R_j(\mathbf{A}_j)=\min_k R_k(\mathbf{A}_k)$
10: If $R_{\hat{j}}(\mathbf{A}_{\hat{j}}) \leq R_{\min}$, **drop the $\hat{j}$ th SUE and goto step 5**
11: Set $\mathcal{N}_{\text{QoS}} \leftarrow \mathcal{N}_{\text{QoS}} - \hat{j}$
12: Until $R_j(\mathbf{A}_j) \geq R_{\min}, \forall j \in \mathcal{N}_{\text{QoS}}$
13: Output $\mathcal{N}_{\text{QoS}}$ as the set of admissible SUEs and $\{p_k(f), \mathbf{a}\}$ as the associated transmit powers and joint action space.

send its rate to the other SUEs. Thus, Algorithm 1is exploited to solve the resource allocation problem individually. The small cells can independently and selfishly maximize their rates. If the rate of an SUE still does not satisfy the required minimum rate; the small cells can independently decide to drop the SUE.

SIMULATION RESULTS

In this section, simulation results for the throughput performance of the proposed algorithm with game-theoretic resource allocation are presented. The environmental parameters used in the simulation are listed in Table 2. The locations of the SAPs and SUEs are randomly distributed and under laid with a macrocell. Moreover, each SAP serves one SUE within the small cell coverage, which is shown in Fig. 4.Here,it is assumed that there are five SAPs and 10 MUEs that coexist and share {13, 15, 17, 19} RBs.

Here, the channel model is the indoor channel model in 3GPP TS 36.814v9.0.0. The path losses of an MUE to an MBS, an SUE to an MBS, an MUE to an SAP, and an SUE to an SAP in the same building are adopted according to 3GPP specification suggestions. Further, the penetration losses of the inner and outer aspects of a building and the number of walls and floors that the signal passes through in the simulation are also adopted from3GPP TS 36.814v9.0.0.

Figure 5 shows the benefit of the sensing-based mixed-access strategy, in which a higher spectrum efficiency is achieved. If the number of RBs is not high enough, some of the SUEs must use the same RB at a time instant. Fig. 6 shows the benefits of the sensing information obtained by the modified MME spectrum sensing. The interference to the MUEs from the SUEs can be reduced to one-fourth in the first few iterations and continues to be reduced in subsequent iterations, which can alleviate unavoidable and possibly harmful interference to the MUEs. In Figs. 7 and 8, cross-tier interference and co-tier interference are both reduced to approximately 3 dB in the beginning, and the reduction shown in Fig. 8is gradually enhanced as the number of iterations increases, confirming the effectiveness of cross-tier interference avoidance.

Table 2. Environmental parameters

Parameter	Value
Macrocell coverage	Radius 500 m
Small cell coverage	Radius 30 m
SUE transmit power	26 dBm
MUE transmit power	26 dBm
Path loss model	3GPP TR36.814 v9
Fading channel	Rayleigh
Average time of regret-matching algorithm	1000
Resource block bandwidth	180kHz
Thermal noise PSD	-174 dBm/Hz
Noise figure	9 dB
Number of antennas of the SAP	4
Number of antennas of the SUE and MUE	1

Figure 4. Distributions of the MBS, MUEs, SAPs, and SUEs

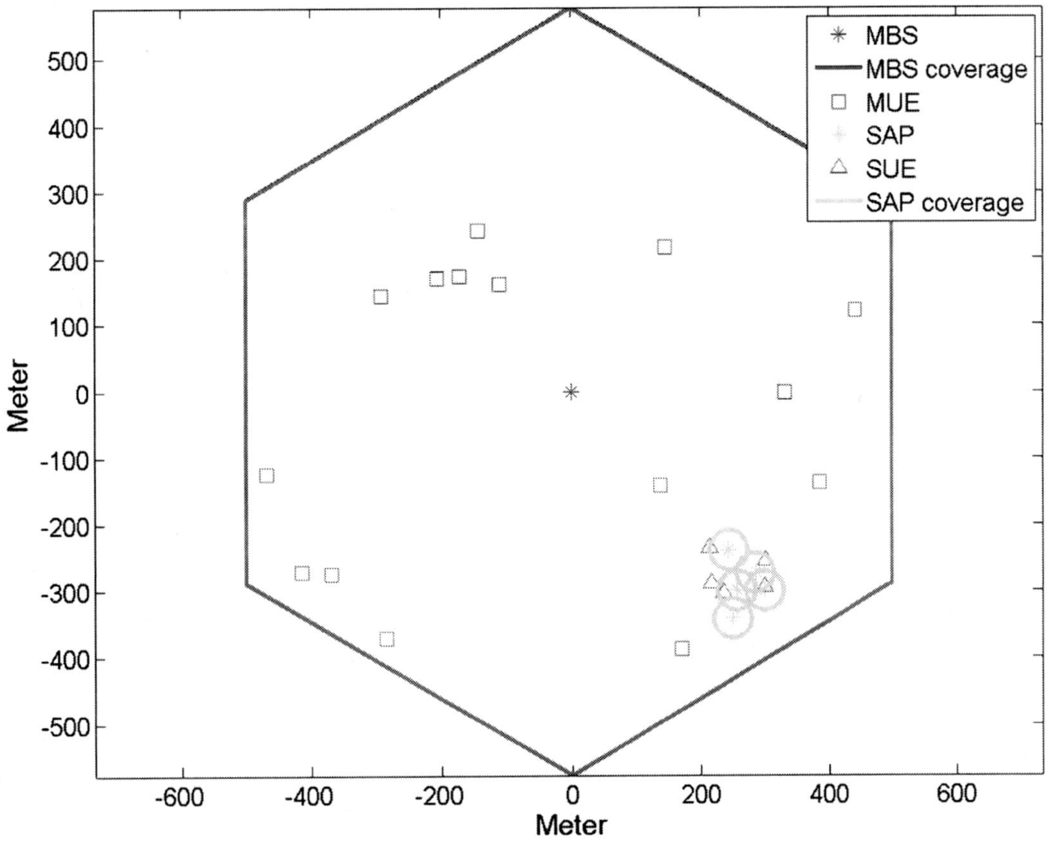

Figure 5. Comparison of the utility between regretmatching with overlay, underlay, and exhaustive search + waterfilling, where N_{SAP} and N_{MUE} are 5 and 10, respectively

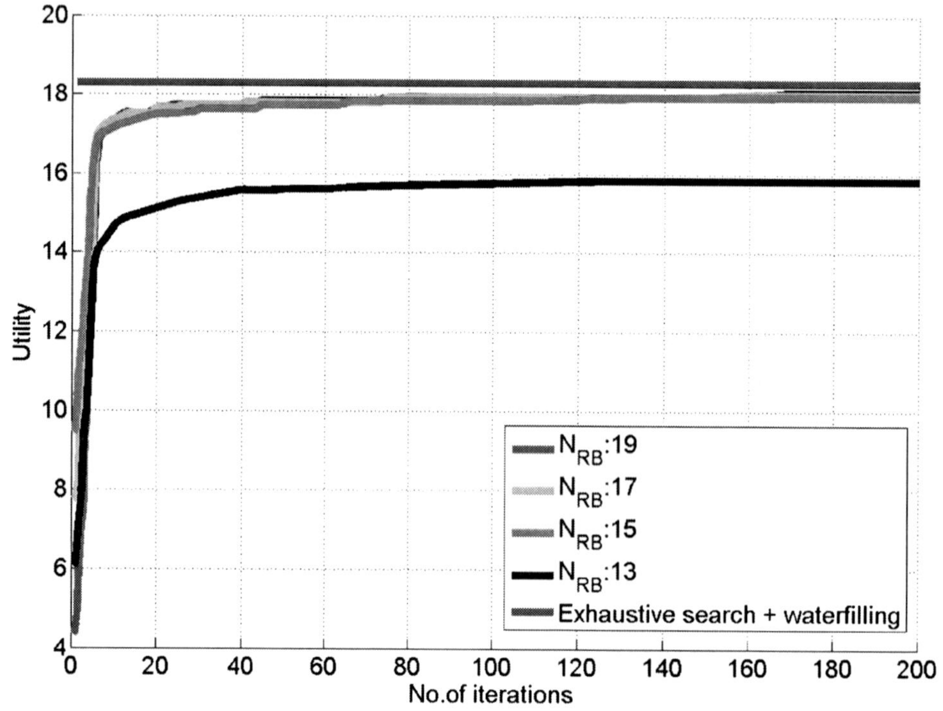

Figure 6. Comparison of the interference between allocation with and without sensing information

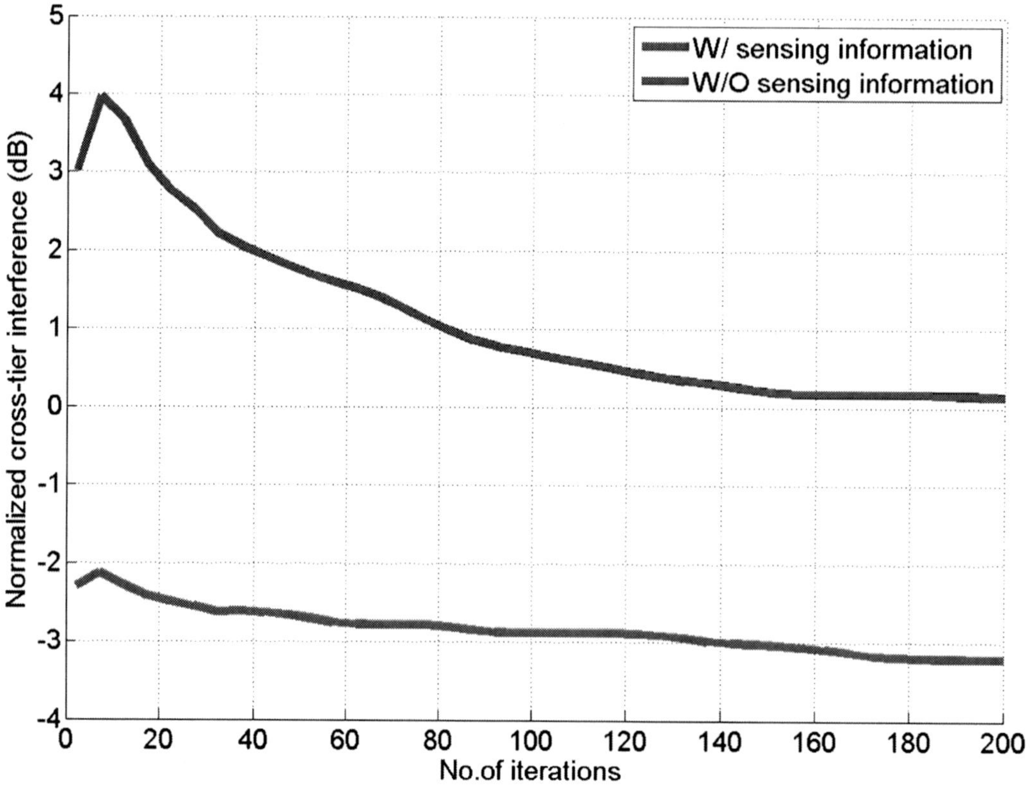

Figure 7. Comparison of the interference with and without penalty, where N_{SAP}, N_{MUE}, and N_{RB} are 2, 10, and 20, respectively; both are normalized by the minimum interference values obtained in the last iteration of the without penalty applied scheme

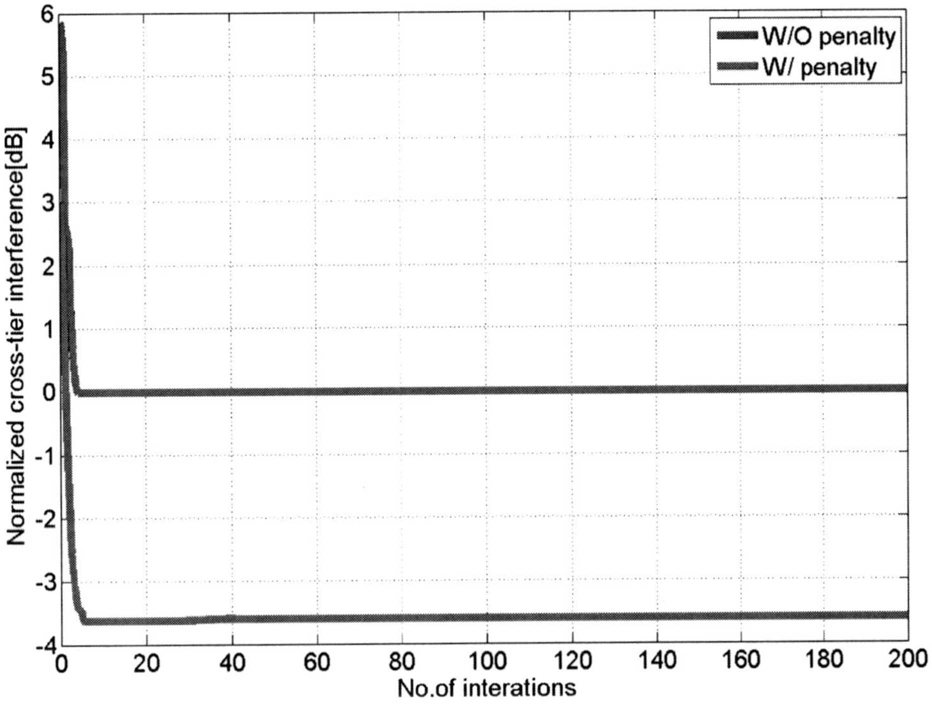

Figure 8. Comparison of the interference with and without interference avoidance, where N_{SAP}, N_{MUE}, N_{RB}, and N_{pre} are 2, 10, 20, and 2, respectively; both normalized by the minimum interference in the last iteration of the without interference avoidance applied scheme

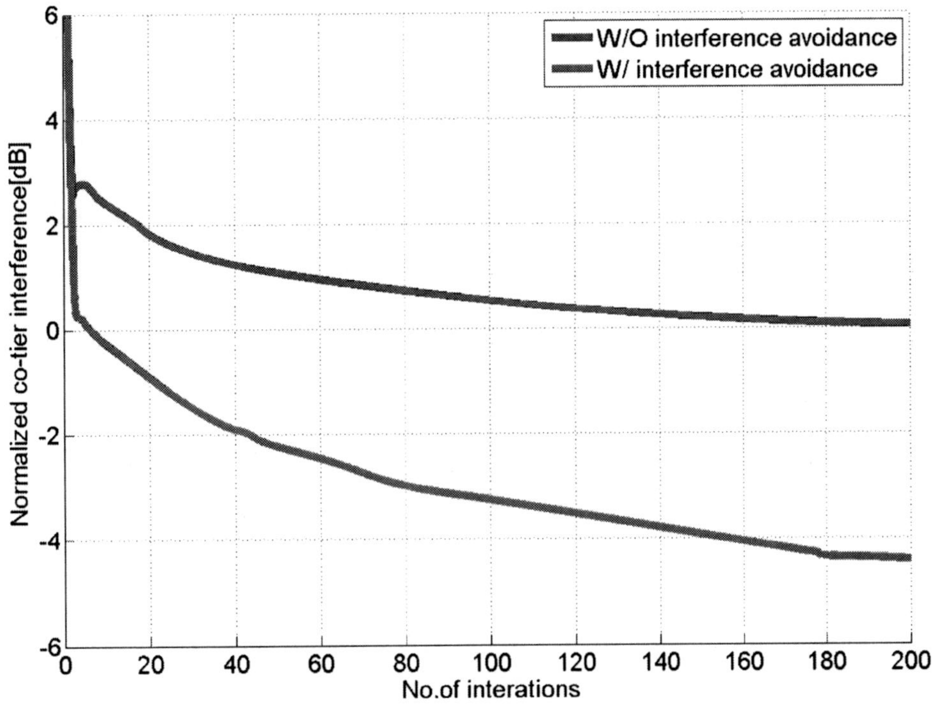

Figure 9 shows a comparison of the number of iterations needed for regret matching with and without interference avoidance. The performance can achieve 95% of the utility obtained by exhaustive search plus waterfilling (ESW). However, the regret-matching-only scheme needs approximately 4 to 10times the number of iterations for various values of N_{acc}. Further, for larger values of N_{acc}, more iterations are needed. The number of iterations needed for regret matching is not linear, and it depends on the variation in the utility value, the resource allocation parameters, and also heavily on the size of the strategy space (the combinations of accessed RBs).When interference avoidance is applied, the iteration process can save some time by avoiding access collisions on the same RB, which makes the convergence faster compared to ESW.

Figure 10 shows the performance (utility) when only interference avoidance is applied and when both penalty and interference avoidance are applied, with the optimal performance obtained by ESW. When the RBs are sufficiently allocated to the overall SUEs, the performance of the proposed scheme is close to the performance of ESW. However, with an increase in N_{acc} (especially if the number is five or six), there are obvious gaps in the performance for any two of the three schemes. This is because both co-tier interference and cross-tier interference considerably increase as N_{acc} increases, strengthening the effects from interference avoidance and penalty avoidance, thereby compromising utility performance.

Figure 9. Comparison of the number of required iterations between original regret matching and regret matching with interference avoidance, and the figure shows the number of iterations required to achieve 95% of exhaustive search plus waterfilling

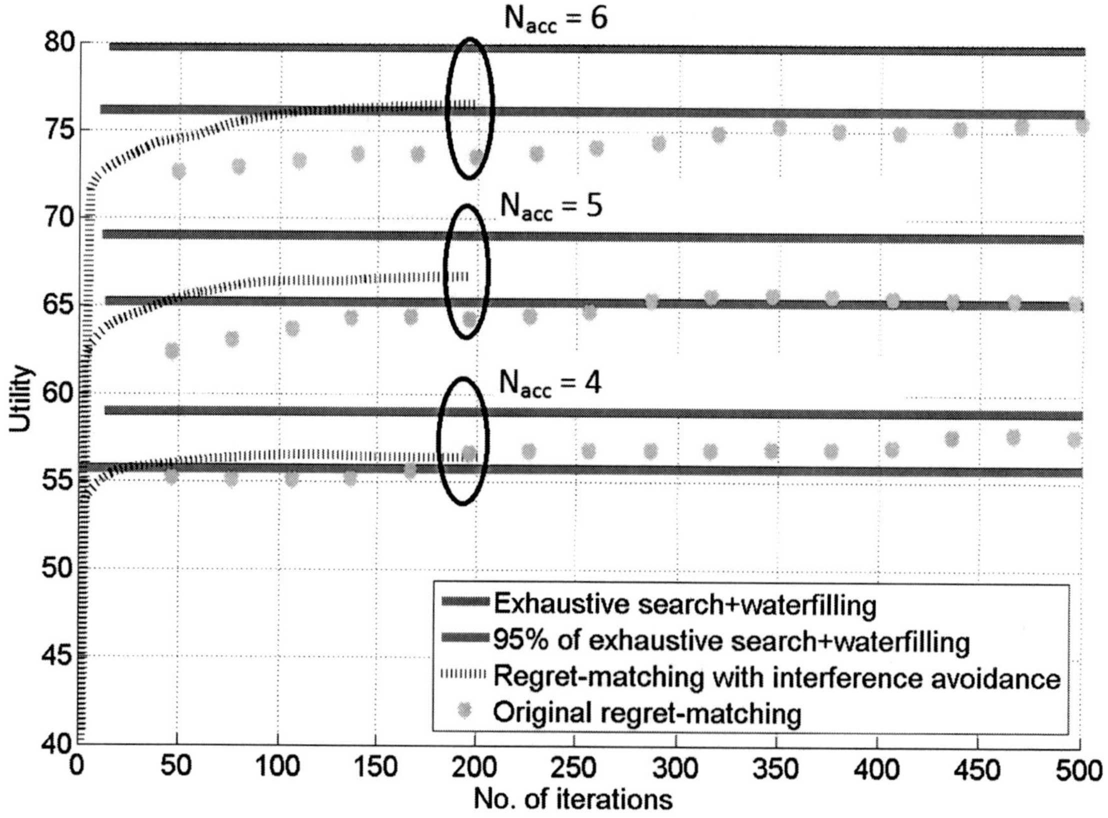

Figure 10. Comparison of the utility for various values of N_{acc}, where N_{SAP}, N_{MUE}, N_{RB}, and N_{pre} are 2, 10, 20, and 2, respectively (gradually insufficient RBs)

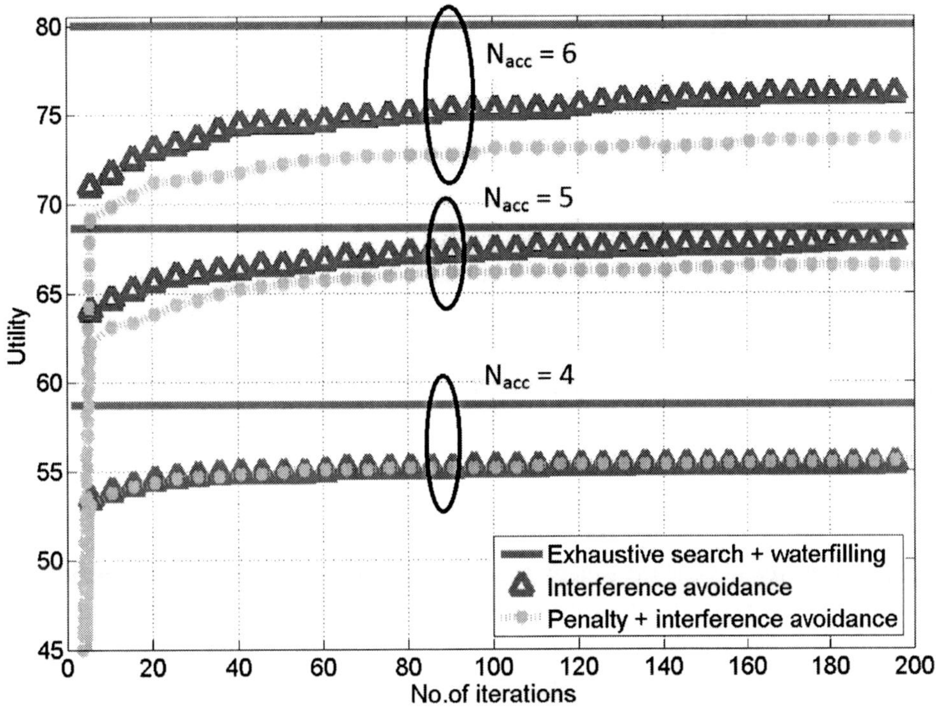

Figure 11. Comparison of the sum rates between Algorithm 1 and Algorithm 3, where N_{acc}, N_{SAP}, N_{MUE}, N_{RB}, and N_{pre} are 5, 2, 10, 20, and 2, respectively

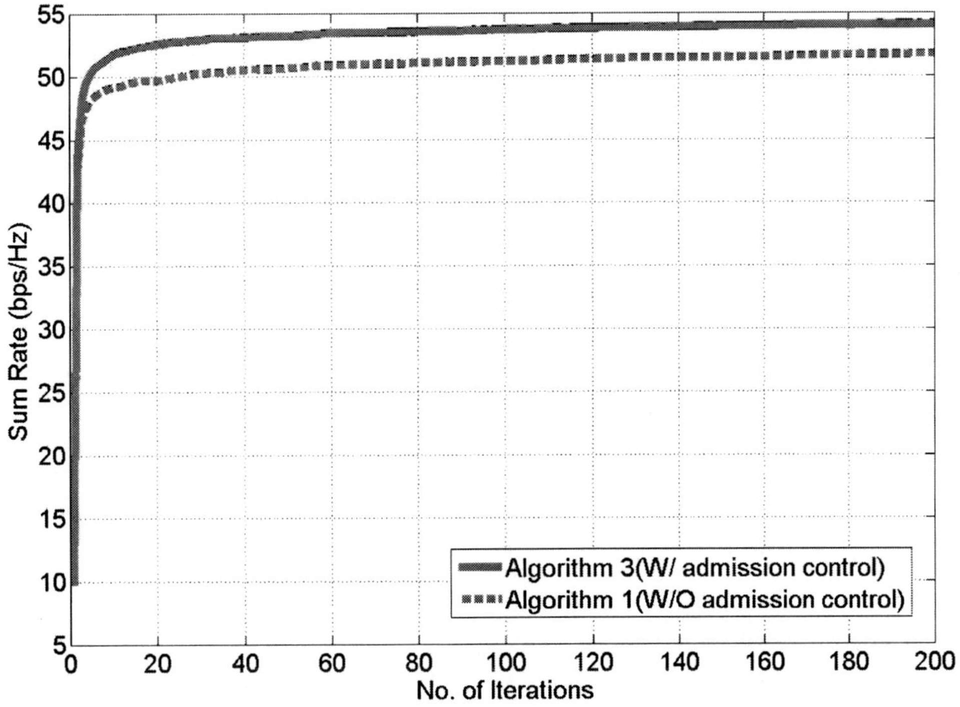

Figure 11 shows a comparison of the sum rate of Algorithm 1 and Algorithm 3 when an SUE is located in an environment in which an SAP cannot provide a guaranteed QoS to the SUE. Algorithm 3 would drop the SUE, but the SUE is not guaranteed to achieve the target QoS. Hence, co-tier interference is reduced by suitable admission control in the proposed algorithm.

CONCLUSION AND FUTURE TRENDS

The basic idea of introducing MME-based cooperative spectrum sensing information into the regret-matching procedure was discussed. The proposed resource allocation algorithm is a type of learning algorithm, and the basic idea of the regret-matching algorithm is to learn about the regret of the actions taken at every time instant and to minimize its regret value. A smaller regret value means that the utility is larger and also means that the co-tier interference or cross-tier interference becomes smaller; that is, the best distribution policy can be found to allocate RBs with the smallest interference. Furthermore, the properties of the eigenvalues are exploited in the MME-based sensing method to propose an interference avoidance strategy in Algorithm 2becausethe maximum eigenvalue contains information about both the signal power of the MUEs and the noise power.

In addition, the joint resource allocation and user admission control problem is considered in densely deployed small cell environments. Algorithm 3 extends the regret-matching game method to the joint resource allocation and user admission control problem. Further, Algorithm 3 can simultaneously provide a guaranteed QoS and maintain the largest number of served SUEs. Only some of the information exchanged between the SAPs is needed in Algorithm 3.The simulation results show that Algorithm 1 and Algorithm 2 can efficiently allocate resources and reduce co- and cross-tier interference. Moreover, Algorithm 3 can efficiently drop SUEs that do not meet the QoS for the admission control problem in densely deployed small cell environments.

The emerging communication trend is towards the highest spectral efficiency in a densely deployed heterogeneous network because the available spectrum is a limited resource. The proposed algorithm is suitable distributed method for opportunistically exploiting unlicensed bands to offload increased traffic demands. However, this research topic still has many issues. One of the important directions for future research is to develop algorithms with a lower computational time for the largest number of heterogeneous users. Furthermore, the scenario of multiple macrocell may be taken into account in future works.

REFERENCES

Al-Rubaye, S., Al-Dulaimi, A., & Cosmas, J. (2011). Cognitive femtocell. *IEEE Vehicular Technology Magazine*, *6*(1), 44–51. doi:10.1109/MVT.2010.939902

Bennis, M., Perlaza, S. M., & Debbah, M. (2012). Learning coarse correlated equilibrium in two-tier wireless networks. *In Proceedings of the IEEE International Conference on Communications* (pp. 1592-1596). Ottawa, Canada: IEEE.

3. GPP Technical Specification TS 36.814v9.0.0. (2012). *Evolved universal terrestrial radio access (E-UTRA) and evolved universal terrestrial radio access network (E-UTRAN); Further advancements for E-UTRA physical layer aspects*. Retrieved from http://www.3gpp.org/ftp/Specs/html-info/36814.htm

Hart, S. (2004). Adaptive heuristics: a little rationality goes a long way. *Walras–Bowley Lecture slides*. Retrieved from http://www.ma.huji.ac.il/~hart/papers/wb.pdf

Hart, S., & Mas-Colell, A. (2000). A simple adaptive procedure leading to correlated equilibrium. *Econometrica*, *68*(5), 1127–1150. doi:10.1111/1468-0262.00153

Hart, S., & Schmeidler, D. (1989). Existence of correlated equilibria. *Mathematics of Operations Research*, *14*(1), 18–25. doi:10.1287/moor.14.1.18

Huang, J. W., & Krishnamurthy, V. (2011). Cognitive base stations in LTE/3GPP small cells: A correlated equilibrium game-theoretic approach. *IEEE Transactions on Communications*, *59*(12), 3485–3493. doi:10.1109/TCOMM.2011.093011.100693

Kang, X., Liang, Y.-C., Garg, H. K., & Zhang, L. (2008). Sensing-based spectrum sharing in cognitive radio networks. In *Proceedings of the IEEE Global Telecommunications Conference*. New Orleans, LA: IEEE. doi:10.1109/GLOCOM.2008.ECP.844

Khoshkholgh, M. G., Navaie, K., & Yanikomeroglu, H. (2010). Access strategies for spectrum sharing in fading environment: Overlay, underlay, and mixed. *IEEE Transactions on Mobile Computing*, *9*(12), 1780–1793. doi:10.1109/TMC.2010.57

Lai, W. S., Lee, S. C., Chiang, M. E., & Lee, T. S. (2012).Low complexity eigenvalue-based cooperative sensing in cognitive femtocell. In *Proceedings of the 9th IEEE Vehicular Technology Society Asia Pacific Wireless Communications Symposium*, Kyoto, Japan: IEEE.

Lasaulce, S., & Tembine, H. (2011). *Game Theory and Learning for Wireless Networks: Fundamentals and Applications* (1st ed., pp. 153–180). Waltham, MA: Elsevier. doi:10.1016/B978-0-12-384698-3.00006-2

Maskery, M., Krishnamurthy, V., & Zhao, Q. (2009). Decentralized dynamic spectrum access for cognitive radios: Cooperative design of a non-cooperative game. *IEEE Transactions on Communications*, *57*(2), 459–469. doi:10.1109/TCOMM.2009.02.070158

Nash, J. F. (1950). Equilibrium points in *n*-person games. *Proceedings of the National Academy of Sciences of the United States of America*, *36*(1), 48–49. doi:10.1073/pnas.36.1.48 PMID:16588946

Nie, N., & Comaniciu, C. (2005). Adaptive channel allocation spectrum etiquette for cognitive radio networks. In *Proceedings of the1st IEEE International Symposium on New Frontiers in Dynamic Spectrum Access Networks* (pp. 269-278). Baltimore, MD: IEEE. doi:10.1007/s11036-006-0049-y

Tse, D., & Viswanath, P. (2005). *Fundamentals of Wireless Communications*. Cambridge, UK: Cambridge University Press. doi:10.1017/CBO9780511807213

Wang, B., & Liu, K. J. R. (2011). Advances in cognitive radio networks: A survey. *IEEE Journal of Selected Topics in Signal Processing*, *5*(1), 5–23. doi:10.1109/JSTSP.2010.2093210

Yaacoub, E. E., & Dawy, Z. (2012). *Resource Allocation in Uplink OFDMA Wireless Systems*. Hoboken, NJ: Wiley & Sons. doi:10.1002/9781118189627

Zemlianov, A., & de Veciana, G. (2005). Cooperation and decision-making in a wireless multi-provider setting. In *Proceedings of the 24th Annual Joint Conference of the IEEE Computer and Communication Societies* (vol. 1, 386-397). Miami, FL: IEEE. doi:10.1109/INFCOM.2005.1497908

Zeng, Y., & Liang, Y.-C. (2009). Eigenvalue-based spectrum sensing algorithms for cognitive radio. *IEEE Transactions on Communications*, *57*(6), 1784–1793. doi:10.1109/TCOMM.2009.06.070402

ADDITIONAL READING

FemtoForum White Paper. (2008). Interference management in UMTS femtocells. Small Cell Forum. Retrieved from http://www.smallcellforum.org/

Garetto, M., Nordio, A., Chiasserini, C.-F., & Leonardi, E. (2011). Information-theoretic capacity of clustered random networks. *IEEE Transactions on Information Theory*, *57*(11), 7578–7596. doi:10.1109/TIT.2011.2159574

Luo, Z.-Q., & Zhang, S. (2008). Dynamic spectrum management: Complexity and duality. *IEEE Journal of Selected Topics in Signal Processing*, *2*(1), 57–73. doi:10.1109/JSTSP.2007.914876

Quan, Z., Cui, S., & Sayed, A. H. (2008). Optimal linear cooperation for spectrum sensing in cognitive radio networks. *IEEE Journal of Selected Topics in Signal Processing*, *2*(1), 28–40. doi:10.1109/JSTSP.2007.914882

Senthuran, S., Anpalagan, A., & Das, O. (2012). Throughput analysis of opportunistic access strategies in hybrid underlay—overlay cognitive radio networks. *IEEE Transactions on Wireless Communications*, *11*(6), 2024–2035. doi:10.1109/TWC.2012.032712.101209

Stevenson, C., Chouinard, G., Lei, Z. D., Hu, W. D., Shellhammer, S. J., & Caldwell, W. (2009). IEEE 802.22: The first cognitive radio wireless regional area network standard. *IEEE Communications Magazine*, *47*(1), 130–138. doi:10.1109/MCOM.2009.4752688

Third Generation Partnership Project work items (2010). 3GPP work items on self-organizing networks. Retrieved from http://www.3gpp.org/ftp/Information

Yucek, T., & Arslan, H. (2009). A survey of spectrum sensing algorithms for cognitive radio applications. *IEEE Communications Surveys and Tutorials*, *11*(1), 116–130. doi:10.1109/SURV.2009.090109

Zhang, J., & de la Roche, G. (2010). *Femtocells: Technologies and Deployment*. West Sussex, UK: John Wiley & Sons. doi:10.1002/9780470686812

KEY TERMS AND DEFINITIONS

Heterogeneous Network: A network connecting devices with different operating systems and/or protocols. In telecommunications, a heterogeneous network is a very important technology that may have different connotations in wireless telecommunications. It may refer to the paradigm of seamless and ubiquitous interoperability between various multi coverage protocols (sometimes also called HetNet in LTE).

LTE: An abbreviation for Long-Term Evolution, which is a standard for the wireless communication of high-speed data for mobile phones and data terminals. The standard is developed by the3rd Generation Partnership Project (3GPP).

Quality of Service: A set of characteristics of a networking service that can be interpreted as representative of its communication quality.

Regret Matching: The players switch to a different action during the next time instant with a probability that is proportional to the regret for that action.

Regret: If each player plays a strategy, the player will choose the action of increased payoff (i.e., no-regret action) when the players switch to the next action.

Resource Block: In LTE, the grouping of subcarrier frequencies. Resource blocks are a basic allocation unit in the frequency dimension inside an LTE frame.

Small Cell: Low-powered radio access nodes that operate in licensed and unlicensed spectra that have a range of a few tens of meters. They are "small" compared to a mobile macrocell. In addition, the small cell can offload traffic to increase traffic demand in cellular communications.

Chapter 10
Self-Organizing Spectrum Access with Geo-Location Database

Xu Chen
University of Goettingen, Germany

Jianwei Huang
The Chinese University of Hong Kong, China

ABSTRACT

Dynamic spectrum access is envisioned as a promising paradigm for addressing the spectrum under-utilization problem. According to the recent ruling of Federal Communications Commission (FCC) for white-space spectrum access, white-space devices are required to query a geo-location database to determine the spectrum availability. This chapter adopts a game theoretic approach for the self-organizing white-space spectrum access network design. This chapter first models the distributed channel selection problem among the devices as a distributed spectrum access game, and shows that the game is a potential game. This chapter then designs a self-organizing spectrum access algorithm which can achieve a Nash equilibrium of the game without any information exchange among the devices. Numerical results demonstrate that the proposed algorithm is efficient and can adapt to the dynamical network context changing.

INTRODUCTION

Wireless spectrums are often licensed to service providers based on long-term agreements. A service provider will serve its primary users using its licensed spectrum. Due to the stochastic nature of primary users' traffic, the licensed spectrum may not be fully utilized at all locations and all times. Field measurements by Shared Spectrum Cooperation in Chicago and New York City showed that the overall average utilization of a wide range of different types of spectrum bands is below than 30% (McHenry et al., 2005). Dynamic spectrum access is hence proposed as a promising technique to alleviate the problem of spectrum underutilization (Akyildiz et al., 2006). Dynamic spectrum access enables unlicensed wireless users (secondary users) to opportunistically share the vacant licensed channels owned by legacy

DOI: 10.4018/978-1-4666-8642-7.ch010

spectrum holders (primary users), and thus can significantly improve the spectrum efficiency (Wu et al., 2014) (Duan et al., 2014) (Huang, 2013) (Yan et al.,2013) (Li et al., 2013) (Chen and Huang, 2013b) (Gao et al., 2013) (Huang et al., 2006).

The Federal Communications Commission (FCC) is now actively formulating policy and regulations for dynamic spectrum access. The most recent FCC ruling requires that secondary TV spectrum users (i.e., white-space devices) must rely on a geo-location database to determine the spectrum availability (FCC,2010). In such a database-assisted architecture, the incumbents (primary users of TV spectrum) provide the database with the up-to-date information including TV tower transmission parameters and TV receiver protection requirements. As illustrated in Figure 1, based on this information the database will be able to tell a white-space device the vacant TV channels and the allowable transmission power level at a particular location.

Although the database-assisted approach obviates the need of spectrum sensing (i.e., detection of spectrum opportunities by individual secondary users), the task of developing a comprehensive and reliable white-space spectrum access system remains challenging (Murty et al., 2011). A key challenge is how to choose proper vacant channel for each device in a distributed manner in order to avoid severe interference with other devices. In this chapter, we adopt the game theoretic approach to address the challenge. Game theory is a useful framework for designing distributed mechanisms, such that the white-space devices in the system can self-organize into a mutually acceptable channel allocation. The self-organizing feature can add autonomics into white-space networking and help to ease the heavy burden of complex centralized system management (Brunneret al., 2011).

Specifically, in this chapter we model the distributed channel selection problem among the devices as a distributed spectrum access game. We then propose a self-organizing spectrum access algorithm that can achieve the Nash equilibrium of the game. The main results and contributions of this chapter are as follows:

1. **General Game Formulation**: We formulate the distributed channel selection problem among the white-space devices as a distributed spectrum access game based on the general physical interference model.

Figure 1. Distributed spectrum access with geo-location database

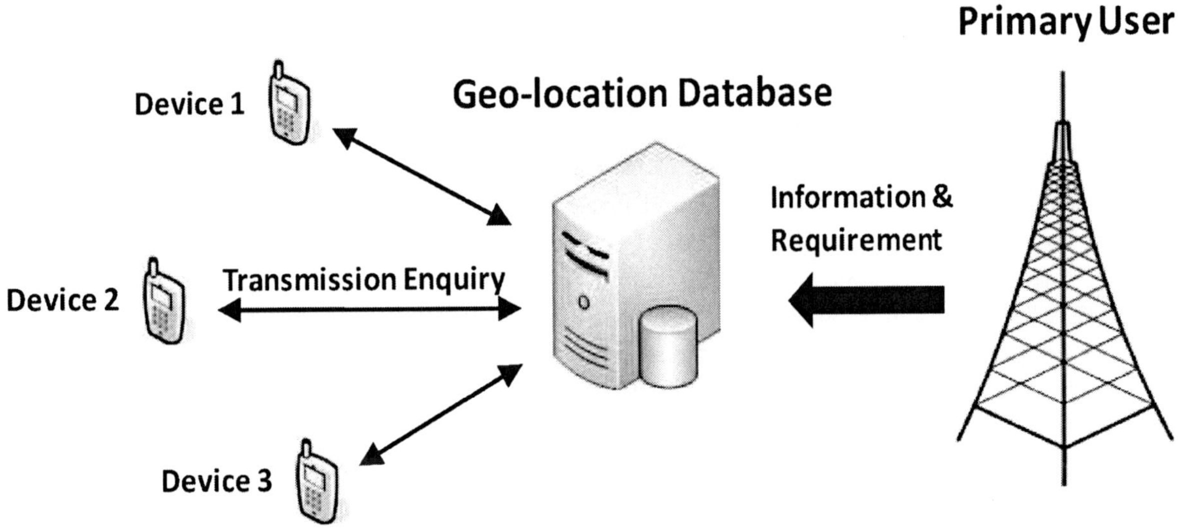

2. **Existence of Nash Equilibrium**: We show that the distributed spectrum access game is a potential game, which possesses a Nash equilibrium. The existence of Nash equilibrium implies that the game theoretic approach has the nice self-stability property, such that the white-space devices can self-organize into a mutually acceptable channel allocation.
3. **Self-Organizing Spectrum Access Algorithm**: We devise a self-organizing spectrum access algorithm that can achieve a Nash equilibrium of the distributed spectrum access game without any information exchange among the devices. Numerical results demonstrate that the proposed algorithm is very efficient with a less than 10% performance loss, compared with the centralized system throughput maximization solution.

RELATED WORK

For the white-space networking system design, many existing works focus on the experimental test-bed implementation. (Bahl et al., 2009) designed a single white-space AP system. (Murty et al., 2011) addressed the client bootstrapping and mobility handling issues in white-space AP networks. (Feng et al., 2011) considered the OFDM-based AP white-space network system design. (Deb et al., 2009) presented a centralized white-space spectrum allocation algorithm. (Chen and Huang, 2012a) (Chen and Huang, 2013a) investigated both cooperative and non-cooperative AP white-space channel allocation problems. Along a different line, in this chapter we focus on the infrastructure-free (i.e., without AP) white-space spectrum access mechanism design.

The economic issue of white-space spectrum sharing is also gaining increasing attention from the community. (Luo et al., 2012) (Luo et al., 2014a) proposed that the geo-location database acts as a spectrum broker reserving the white-space spectrum from spectrum licensees. (Feng et al., 2014) developed a hybrid pricing mechanism such that the database operator employs both the registration scheme and the service plan scheme to serve the white-space users. (Luo et al., 2014b) advocated novel business of establishing an information market for TV white space networks, where the spectrum database operator sells the information regarding TV white space to white-space users.

Game theory has been used to design non-white-space self-organizing networks.(Han et al., 2012) proposed a game based self-organizing scheme for femto cell networks. (Rogers et al., 2005) designed an energy-aware self-organizing routing algorithm that enables individual sensors to follow locally selfish strategies for wireless sensor networks. (Abdelkader et al., 2011) developed a Voronoi based strategic game such that the wireless sensor nodes can self-organize to reach uniform deployment. (Chen and Huang, 2012b) (Chen and Huang, 2014b) proposed a spatial spectrum access game framework for self-organizing spectrum access based on the protocol interference model. In this chapter, we study the self-organizing white-space spectrum access based on the physical interference model.

Besides game theory, biologically-inspired principles have also been applied for self-organizing network design. (Gunes et al., 2002) proposed an ant-colony based routing algorithm for mobile ad hoc networks. (Balasubramaniam et al., 2007) applied the chemotaxis, reaction diffusion, and hormone signaling model to design self-organizing mechanism for de-centralized routing. (Kruger and Dressler, 2005) developed autonomous network systems by mimicking structures and functionalities of organisms in molecular biology. (Chen and Huang, 2013c) developed a stable spectrum sharing mechanism based on the evolutionary rules observed from human and animal interactions. (Chen and Huang, 2014a) leveraged the social phenomenon of imitation to devise an imitative spectrum access mechanism for cognitive radio networks. (Lee and Suzuki, 2009) applied the principle of immune system to design self-organizing and evolvable network systems. In this chapter, we leverage the self-stability property of potential game for devising the self-organizing white-space spectrum access mechanism.

DISTRIBUTED SPECTRUM ACCESS GAME

In this part, we first introduce the system model of the distributed white-space spectrum access, and then formulate the distributed spectrum access problem among white-space devices as a game.

System Model

We consider a set of white-space devices $\mathcal{N} = \{1, 2, ..., N\}$,where N is the total number of devices. Let $\mathcal{M} = \{1, 2, ..., M\}$ denote the set of white-space channels, and W denote the bandwidth of each channel (e.g., W=6 MHz in U.S. and W=8 MHz in EU). To protect the incumbent primary white-space users, each device $n \in \mathcal{N}$ will first send a spectrum access request message containing its geo-location information to the Geo-location database. The database then feeds back the allowable transmission power level P_n and the set of feasible channels $\mathcal{M}_n \subseteq \mathcal{M}$ to device n.

Each device n then chooses a feasible channel $a_n \in \mathcal{M}_n$ for data transmission. We denote the channel selection profile of all the devices as

$$\mathbf{a} = (a_1, ..., a_N) \in \prod_{n=1}^{N} \mathcal{M}_n.$$

According to the physical interference model (Gupta and Kumar, 2000), we can compute the Signal-to-Interference-plus-Noise Ratio (SINR) of device n as

$$\eta_n(\mathbf{a}) = \frac{P_n d_n^{-\alpha}}{\omega_{a_n}^n + \sum_{i \in \mathcal{N} \setminus \{n\}: a_i = a_n} P_i d_{in}^{-\alpha}}. \tag{1}$$

Here α is the path loss factor, d_n is the distance from device n to its destination, and $d_n^{-\alpha}$ denotes the channel gain of device n due to free-space attenuation (Gupta and Kumar, 2000). Furthermore, $\omega_{a_n}^n$ denotes the noisy power including the interference from primary users on the channel a_n, d_{in} denotes the distance between devices i and n, and $\sum_{i \in \mathcal{N} \setminus \{n\}: a_i = a_n} P_i d_{in}^{-\alpha}$ denotes the accumulated interference from other devices to device n. Accordingly, the throughput of device n under the channel selection profile $\mathbf{a}$ can be computed as

$$U_n(\mathbf{a}) \;=\; W \log_2(1 + \eta_n(\mathbf{a})). \tag{2}$$

Game Formulation

We next consider the distributed spectrum access problem among the white-space devices. Let $a_{-n} = \{a_1,\ldots,a_{n-1},a_{n+1},\ldots,a_N\}$ be the set of channels chosen by all other devices except device n. Given other devices' channel selections a_{-n}, device n wants to choose a proper channel $a_n \in \mathcal{M}_n$ to maximize its throughput, i.e.,

$$\max_{a_n \in \mathcal{M}_n} U_n(a_n, a_{-n}), \forall n \in \mathcal{N}.$$

The distributed nature of the channel selection problem naturally leads to a formulation based on game theory, such that each device can self-organize into a mutually acceptable channel selection (**Nash equilibrium**) $\mathbf{a}^* = (a_1^*,\ldots,a_N^*)$ with

$$a_n^* \in \arg\max_{a_n \in \mathcal{M}_n} U_n(a_n, a_{-n}), \forall n \in \mathcal{N}.$$

We now formulate the distributed spectrum access problem among the white-space devices as a strategic game $\Gamma = (\mathcal{N}, \{\mathcal{M}_n\}_{n\in\mathcal{N}}, \{U_n\}_{n\in\mathcal{N}})$ where the set of devices $\mathcal{N}$ is the set of players, these of feasible channels $\mathcal{M}_n$ is the set of strategies for each device, and the throughput function U_n is the pay-off function of device n. We call the game Γ as the distributed spectrum access game in the sequel.

Game Property

We then study the existence of Nash equilibrium of the distributed spectrum access game. Here we resort to a useful tool of potential game (Monderer and Shapley, 1996).

Definition 1: A game is called a potential game if it admits a potential function $\Phi(\mathbf{a})$ such that for every $n \in \mathcal{N}$, $a_{-n} \in \prod_{i \neq n} \mathcal{M}_i$, and any $a_n, b_n \in \mathcal{M}_n$,

$$\mathrm{sgn}(U_n(b_n, a_{-n}) - U_n(a_n, a_{-n})) = \mathrm{sgn}(\Phi(b_n, a_{-n}) - \Phi(a_n, a_{-n})), \tag{3}$$

where sgn(•) is the sign function defined as

$$\mathrm{sgn}(z) = \begin{cases} 1 & \text{if } z > 0, \\ 0 & \text{if } z = 0, \\ -1 & \text{if } z < 0. \end{cases}$$

An appealing property of the potential game is that it always admits Nash equilibrium, and any strategy profile that maximizes the potential function $\Phi(\mathbf{a})$ is a Nash equilibrium (Monderer and Shapley, 1996).

To show that the distributed spectrum access game Γ is a potential game, we introduce a closely related game $\Lambda = (\mathcal{N}, \{\mathcal{M}_n\}_{n\in\mathcal{N}}, \{\gamma_n\}_{n\in\mathcal{N}})$ with the new payoff functions being

$$\gamma_n(\mathbf{a}) = -\omega_{a_n}^n - \sum_{i\in\mathcal{N}\setminus\{n\}:a_i=a_n} P_i d_{in}^{-\alpha}. \tag{4}$$

Here the physical meaning of $-\gamma_n(\mathbf{a})$ is the received interference level of device n. Thus, in the modified distributed spectrum access game Λ, each device n tries to minimize its received interference $-\gamma_n(\mathbf{a})$ in a distributed manner. For the modified game Λ, we have the following result.

Lemma 1: *If the modified distributed spectrum access game* Λ is a potential game, then the original game Γ is also a potential game with the same potential function.

Proof: According to (1), (2), and (4),we first have that

$$U_n(\mathbf{a}) = W\log_2\left(1 + \frac{P_n d_n^{-\alpha}}{-\gamma_n(\mathbf{a})}\right).$$

Since $f(z)=W\log_2(1+z)$ is a monotonically strictly increasing function, we have that

$$\operatorname{sgn}\left(U_n(b_n, a_{-n}) - U_n(a_n, a_{-n})\right) = \operatorname{sgn}\left(\gamma_n(b_n, a_{-n}) - \gamma_n(a_n, a_{-n})\right).$$

If the modified game Λ is a potential game with a potential function Φ, we must also have that

$$\operatorname{sgn}\left(\Phi(b_n, a_{-n}) - \Phi(a_n, a_{-n})\right) = \operatorname{sgn}\left(U_n(b_n, a_{-n}) - U_n(a_n, a_{-n})\right),$$

which completes the proof. □

For the modified game Λ, we can show that it is a potential game with the following potential function

$$\Phi(\mathbf{a}) = -\sum_i \sum_{j\neq i} P_i P_j d_{ij}^{-\alpha} I_{\{a_i=a_j\}} - 2\sum_{n=1}^{N} P_n \omega_{a_n}^n, \tag{5}$$

where $I_{\{\bullet\}}$ is an indicator function such that $I_{\{\bullet\}}=1$ if the event $\{\bullet\}$ is true and $I_{\{\bullet\}}=0$ otherwise.

Lemma 2: *The modified distributed spectrum access game* Λ is a potential game with the potential function $\Phi(\mathbf{a})$ in (5) satisfying that

$$\Phi(b_n, a_{-n}) - \Phi(a_n, a_{-n}) = 2P_n\left(\gamma_n(b_n, a_{-n}) - \gamma_n(a_n, a_{-n})\right). \tag{6}$$

Proof: We have that

$$\Phi(b_n, a_{-n}) - \Phi(a_n, a_{-n}) = -\sum_{i \neq n} P_n P_i d_{ni}^{-\alpha} I_{\{b_n = a_i\}}$$
$$-\sum_{i \neq n} P_i P_n d_{in}^{-\alpha} I_{\{a_i = b_n\}} + \sum_{i \neq n} P_n P_i d_{ni}^{-\alpha} I_{\{a_n = a_i\}} + \sum_{i \neq n} P_i P_n d_{in}^{-\alpha} I_{\{a_i = a_n\}} - 2P_n \omega_{b_n}^n + 2P_n \omega_{a_n}^n$$
$$= -2P_n \left(\sum_{i \neq n: a_i \neq b_n} P_i d_{in}^{-\alpha} + \omega_{b_n}^n \right) + 2P_n \left(\sum_{i \neq n: a_i \neq a_n} P_i d_{in}^{-\alpha} + \omega_{a_n}^n \right) = 2P_n \left(\gamma_n(b_n, a_{-n}) - \gamma_n(a_n, a_{-n}) \right),$$

which completes the proof. □

According to Lemmas 1 and 2,we know that the original distributed spectrum access game Γ is also a potential game.

Theorem 1: *The distributed spectrum access game* Γ is a potential game with the potential function $\Phi(\mathbf{a})$ in (5), and hence has a Nash equilibrium.

Theorem 1 implies that the distributed spectrum access game has the nice self-stability property such that the white-space devices can self-organize into a mutually acceptable channel allocation. We next design a self-organizing spectrum access algorithm that can achieve a Nash equilibrium of the distributed spectrum access game without any information exchange among devices.

SELF-ORGANIZING SPECTRUM ACCESS ALGORITHM

We next consider the self-organizing spectrum access algorithm design in this part.

Algorithm Design

According to the property of potential game, any channel selection profile that maximizes the potential function $\Phi(\mathbf{a})$ is a Nash equilibrium (Monderer and Shapley, 1996). We hence design a self-organizing spectrum access algorithm that achieves the Nash equilibrium of the distributed spectrum access game by maximizing the potential function $\Phi(\mathbf{a})$.

To proceed, we first consider the problem that the white-space devices collectively determine the optimal channel selection profile such that the potential function is maximized, i.e.,

$$\max_{\mathbf{a} \in \Theta \triangleq \prod_{n=1}^{N} \mathcal{M}_n} \Phi(\mathbf{a}). \tag{7}$$

The problem (7) is a combinatorial optimization problem of finding the optimal channel selection profile over the discrete solution space Θ. In general, such a problem is very challenging to solve exactly especially when the system size is large (i.e., the solution space Θ is large).

We then consider to approach the potential maximization solution approximately. To proceed, we first write the problem (7) into the following equivalent problem:

$$\max_{(q_{\mathbf{a}} \geq 0: \mathbf{a} \in \Theta)} \sum_{\mathbf{a} \in \Theta} q_{\mathbf{a}} \Phi(\mathbf{a}) \\ \text{s.t.} \sum_{\mathbf{a} \in \Theta} q_{\mathbf{a}} = 1, \tag{8}$$

where $q_{\mathbf{a}}$ is the probability that the channel selection profile **a** is adopted. Obviously, the optimal solution to problem (8) is to choose the optimal channel selection profiles with probability one. It is known from (Chen et al., 2010) that problem(8) can be approximated by the following convex optimization problem:

$$\max_{(q_{\mathbf{a}} \geq 0: \mathbf{a} \in \Theta)} \sum_{\mathbf{a} \in \Theta} q_{\mathbf{a}} \Phi(\mathbf{a}) - \frac{1}{\theta} \sum_{\mathbf{a} \in \Theta} q_{\mathbf{a}} \ln q_{\mathbf{a}} \\ \text{s.t.} \sum_{\mathbf{a} \in \Theta} q_{\mathbf{a}} = 1, \tag{9}$$

where θ is the parameter that controls the approximation ratio. We see that when $\theta \rightarrow \infty$, the problem (9) becomes exactly the same as problem (8). That is, when $\theta \rightarrow \infty$, the optimal solutions that maximize the potential function $\Phi(\mathbf{a})$ will be selected with probabilityone. A nice property of such an approximation in (9) isthat we can obtain the close-form solution, which enables the distributedalgorithm design later. More specifically, by the KKT condition (Boyd andVandenberghe, 2004) we can derive the optimal solution to problem (9) as

$$q_{\mathbf{a}}^{*} = \frac{\exp(\theta \Phi(\mathbf{a}))}{\sum_{\hat{\mathbf{a}} \in \Theta} \exp(\theta \Phi(\hat{\mathbf{a}}))}. \tag{10}$$

Based on (10), we then design a self-organizing algorithm such that the asynchronous channel selection updates of the devices form a Markov chain (with the system state as the channel selection profile **a** of all devices). As long as the Markov chain converges to the stationary distribution as given in (10), we can approach the Nash equilibrium channel selection profile that maximizes the potential function by setting a large enough parameter θ.

According to (6), we have that

$$\Phi(b_n, a_{-n}) - 2P_n \gamma_n(b_n, a_{-n}) = \Phi(a_n, a_{-n}) - 2P_n \gamma_n(a_n, a_{-n}). \tag{11}$$

Recall that $-\gamma_n(\mathbf{a})$ is the received interference level of device n. Then (11) implies that the increase (or decrease) of potential function $\Phi(\mathbf{a})$ equals to a weighted decrease (or increase) of each device's received interference. Since the received interference $-\gamma_n(\mathbf{a})$ can be measured locally, we propose a spectrum access algorithm without any information exchange in Algorithm 1. The proposed algorithm works in a self-organizing manner such that each device n measures its received interference $-\gamma_n(\mathbf{a})$ locally and updates its channel selection according to a timer value that follows the exponential distribution with a rate of τ_n. Since the probability density function of exponential distribution is continuous, the probability that more than one devices generate the same timer value and update their channels simultaneously equals zero.

Algorithm 1: Self-Organizing Spectrum Access

Initialization:
Set the parameter θ and the rate τ_n for channel selection update.
Choose **a channel** $a_n \in \mathcal{M}_n$ **randomly for each device** $n \in \mathcal{N}$.
End Initialization
Loop **for each device** $n \in \mathcal{N}$ **in parallel:**
Generate a timer value following the exponential distribution with the mean equal to $\frac{1}{\tau_n}$.
Count down until the timer expires.
If the timer expires
Measure **the total interference** $\gamma_n(a_n, a_{-n})$ **on the chosen channel** a_n.
Choose **a new channel** $b_n \in \mathcal{M}_n$ **randomly.**
Measure **the total interference** $\gamma_n(b_n, a_{-n})$ **on the new channel** b_n.
Stay in **the new channel** b_n **with probability** $\frac{\exp\left(-2\theta P_n \gamma_n(a_n, a_{-n})\right)}{\exp\left(-2\theta P_n \gamma_n(a_n, a_{-n})\right) + \exp\left(-2\theta P_n \gamma_n(b_n, a_{-n})\right)}$,
OR switch back **to the original channel** a_n **with probability**

$$\frac{\exp\left(-2\theta P_n \gamma_n(b_n, a_{-n})\right)}{\exp\left(-2\theta P_n \gamma_n(a_n, a_{-n})\right) + \exp\left(-2\theta P_n \gamma_n(b_n, a_{-n})\right)}.$$

End If
End Loop

Convergence Analysis

Since one device will activate for the channel selection update at a time, the direct transitions between two system states **a** and **b** are feasible if these two system states differ by one and only one device's channel selection. As an example, Figure 2 shows the system state transition diagram of the spectrum access Markov chain by two devices. We also denote the set of system states that can be transited directly from the state **a** as $\Delta_{\mathbf{a}} = \{\mathbf{b} \in \Theta : |\{\mathbf{b} \cup \mathbf{a}\} \setminus \{\mathbf{b} \cap \mathbf{a}\}| = 2\}$, where |•| denotes the size of a set.

Since a device n will randomly choose a new channel $b_n \in \mathcal{M}_n$ and adhere to this channel with a probability

$$\frac{\exp\left(-2\theta P_n \gamma_n(a_n, a_{-n})\right)}{\exp\left(-2\theta P_n \gamma_n(a_n, a_{-n})\right) + \exp\left(-2\theta P_n \gamma_n(b_n, a_{-n})\right)},$$

then the probability of transition from state (a_n, a_{-n}) to (b_n, a_{-n}) is

$$\frac{1}{|\mathcal{M}_n|} \frac{\exp\left(-2\theta P_n \gamma_n(a_n, a_{-n})\right)}{\exp\left(-2\theta P_n \gamma_n(a_n, a_{-n})\right) + \exp\left(-2\theta P_n \gamma_n(b_n, a_{-n})\right)}.$$

Figure 2. System state transition diagram of the spectrum access Markov chain by two devices

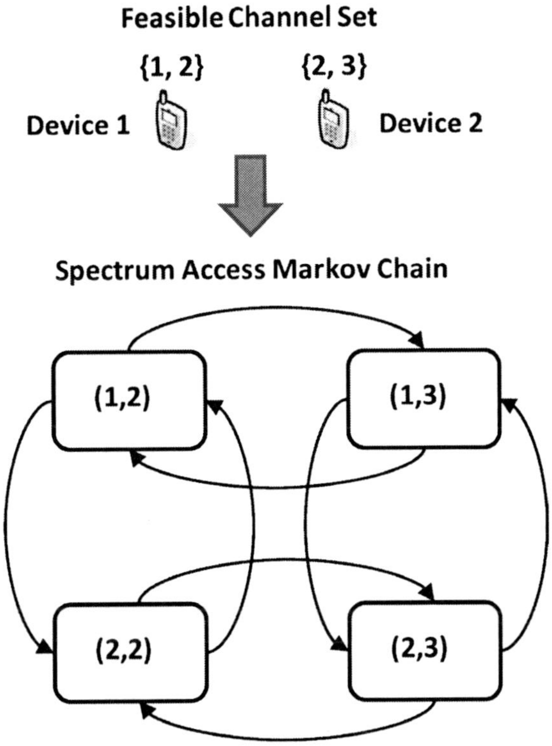

Since each device n activates its channel selection update according to the countdown timer mechanism with a rate of τ_n, hence if $\mathbf{b}\in\Delta_{\mathbf{a}}$, the transition rate from state $\mathbf{a}$ to state $\mathbf{b}$ is given as

$$q_{\mathbf{a},\mathbf{b}} = \frac{\tau_n}{|\mathcal{M}_n|}\frac{\exp\left(-2\theta P_n\gamma_n(a_n,a_{-n})\right)}{\exp\left(-2\theta P_n\gamma_n(a_n,a_{-n})\right)+\exp\left(-2\theta P_n\gamma_n(b_n,a_{-n})\right)}. \tag{12}$$

Otherwise, we have $q_{\mathbf{a},\mathbf{b}}=0$.We show in Theorem 2 that the spectrum access Markov chain is time reversible. Time reversibility means that when tracing the Markov chain backwards, the stochastic behavior of the reverse Markov chain remains the same. A nice property of a time reversible Markov chain is that it always admits a unique stationary distribution, which is independent of the initial system state. This implies that given any initial channel selections the self-organizing spectrum access algorithm can drive the system converging to the stationary distribution given in (10).

Theorem 2: *The self-organizing spectrum access algorithm induces a time-reversible Markov chain with the unique stationary distribution as given in (10).*

Proof: As mentioned, the system state of the spectrum access Markov chain is defined as the channel selection profile $\mathbf{a}\in\Theta$ of all devices. Since it is possible to get from any state to any other state within finite steps of transition, the spectrum access Markov chain is hence irreducible and has a stationary distribution.

We then show that the Markov chain is time reversible by showing that the distribution in (10) satisfies the following detailed balance equations:

$$q_{\mathbf{a}}^{*} q_{\mathbf{a,b}} = q_{\mathbf{b}}^{*} q_{\mathbf{b,a}}, \forall \mathbf{a}, \mathbf{b} \in \Theta.$$

To see this, we consider the following two cases:

1) If $\mathbf{b} \notin \Delta_{\mathbf{a}}$, we have $q_{\mathrm{a,b}}=q_{\mathrm{b,a}}=0$ and the equation (\ref{eq:phh7}) holds.
If $\mathbf{b} \in \Delta_{\mathrm{a}}$, according to(10) and (12), we have

$$q_{\mathbf{a}}^{*} q_{\mathbf{a,b}} = \frac{\tau_n}{|\mathcal{M}_n|} \frac{\exp(\theta\Phi(\mathbf{a}))}{\sum_{\hat{\mathbf{a}} \in \Theta} \exp(\theta\Phi(\hat{\mathbf{a}}))} \times \frac{\exp\left(-2\theta P_n \gamma_n(\mathbf{a})\right)}{\exp\left(-2\theta P_n \gamma_n(\mathbf{a})\right) + \exp\left(-2\theta P_n \gamma_n(\mathbf{b})\right)}$$

$$= \frac{\tau_n}{|\mathcal{M}_n|} \frac{\exp\left(\theta\left(\Phi(\mathbf{a}) - 2P_n \gamma_n(a_n, a_{-n})\right)\right)}{\sum_{\hat{\mathbf{a}} \in \Theta} \exp(\theta\Phi(\hat{\mathbf{a}}))} \times \frac{1}{\exp\left(-2\theta P_n \gamma_n(\mathbf{a})\right) + \exp\left(-2\theta P_n \gamma_n(\mathbf{b})\right)},$$

and similarly,

$$q_{\mathbf{b}}^{*} q_{\mathbf{b,a}} = \frac{\tau_n}{|\mathcal{M}_n|} \frac{\exp\left(\theta\left(\Phi(\mathbf{b}) - 2P_n \gamma_n(\mathbf{b})\right)\right)}{\sum_{\hat{\mathbf{a}} \in \Theta} \exp(\theta\Phi(\hat{\mathbf{a}}))} \times \frac{1}{\exp\left(-2\theta P_n \gamma_n(\mathbf{a})\right) + \exp\left(-2\theta P_n \gamma_n(\mathbf{b})\right)}.$$

Thus, according to (11), we must have $q_{\mathbf{a}}^{*} q_{\mathbf{a,b}} = q_{\mathbf{a}}^{*} q_{\mathbf{a,b}}$.

The spectrum access Markov chain is hence time-reversible and has the unique stationary distribution as given in (10).□

According to Theorem 2, we can approach the Nash equilibrium $\mathbf{a}^*$ that maximizes the potential function $\Phi(\mathbf{a})$ of the distributed spectrum access game by setting $\theta \rightarrow \infty$. Let $\bar{\Phi} = \sum_{a \in \Theta} q_a^{*} \Phi(\mathbf{a})$ bathe expected potential by Algorithm 1 and $\Phi^{*} = \max_{a \in \Theta} \Phi(\mathbf{a})$ be the maximum potential value. Numerical results demonstrate that when a large enough θ is adopted, the performance gap between $\bar{\Phi}$ and Φ^* is very small.

NUMERICAL RESULTS

In this section, we evaluate the proposed self-organizing spectrum access algorithm by numerical studies. We first consider a white-space system consisting of M=4 channels and N=8 devices, which are scattered across a square area of a length of 500m (see Figure 3). The bandwidth of each channel is 6

Figure 3. A square area of a length of 500m with 8 scattered white-space devices

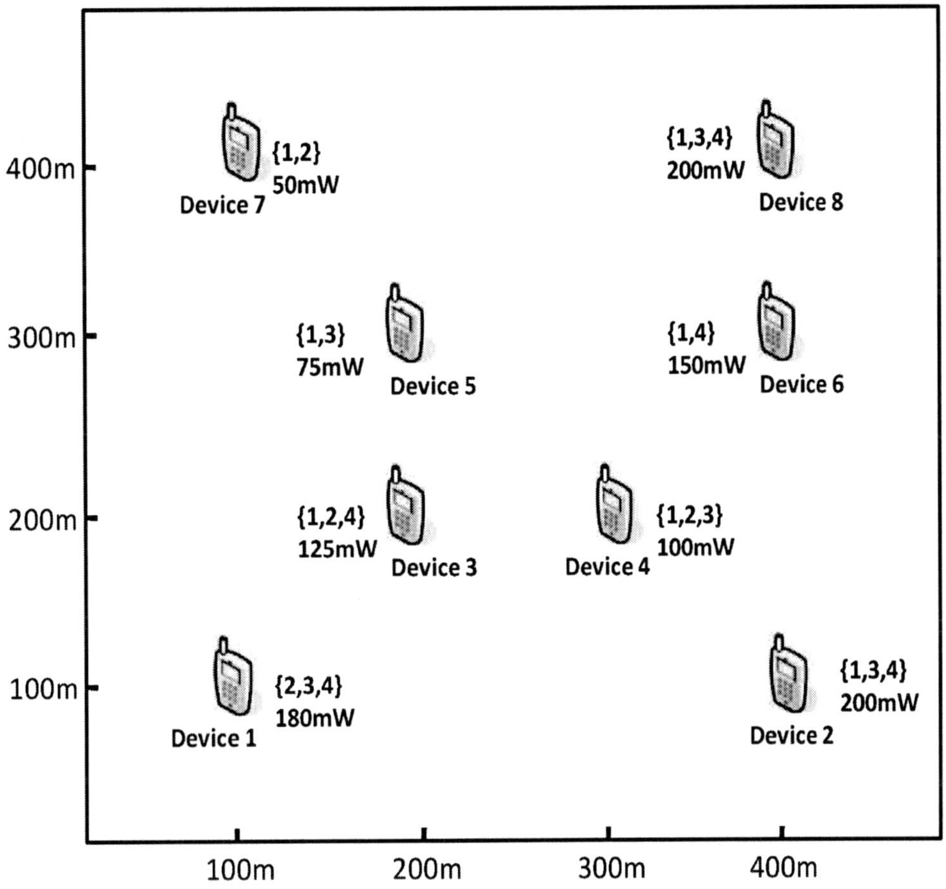

MHz, the noise power is $\omega_m^n = -100\,\text{dBm}$, and the path loss factor α=4. Each device *n* operates with a specific transmission power P_n and has a different setoff available channels by consulting the geo-location database (please refer to Figure 3 for the details of these parameters).

We implement the self-organizing spectrum access algorithm with different parameters θ in Figure 4.We see that the convergent potential function value Φ of the distributed spectrum access game increases as the parameter θ increases. When the parameter θ is large enough (e.g., θ=1.0×10⁶), the algorithm can approach the optimal potential function value $\Phi^* = \max_{\mathbf{a}\in\Theta} \Phi(\mathbf{a})$.

Figure 5 shows the dynamics of devices' time average throughputs when the parameter $\theta=1.0\times10^6$. It demonstrates the convergence of the self-organizing spectrum access algorithm. To verify that the algorithm can approach the Nash equilibrium of the distributed spectrum access game, we show the potential function value Φ in Figure 6. We see that the self-organizing spectrum access algorithm can drive the potential value Φ increasing and approach the maximum potential value Φ^*.According to the property of potential game, the algorithm hence can approach the Nash equilibrium of the distributed spectrum access game.

To evaluate the self-stability property of the proposed self-organizing spectrum access algorithm, we implement another experiment with perturbations. We let device 5 leave the system (i.e., cease the

Figure 4. Potential value Φ at equilibrium with different parameters θ

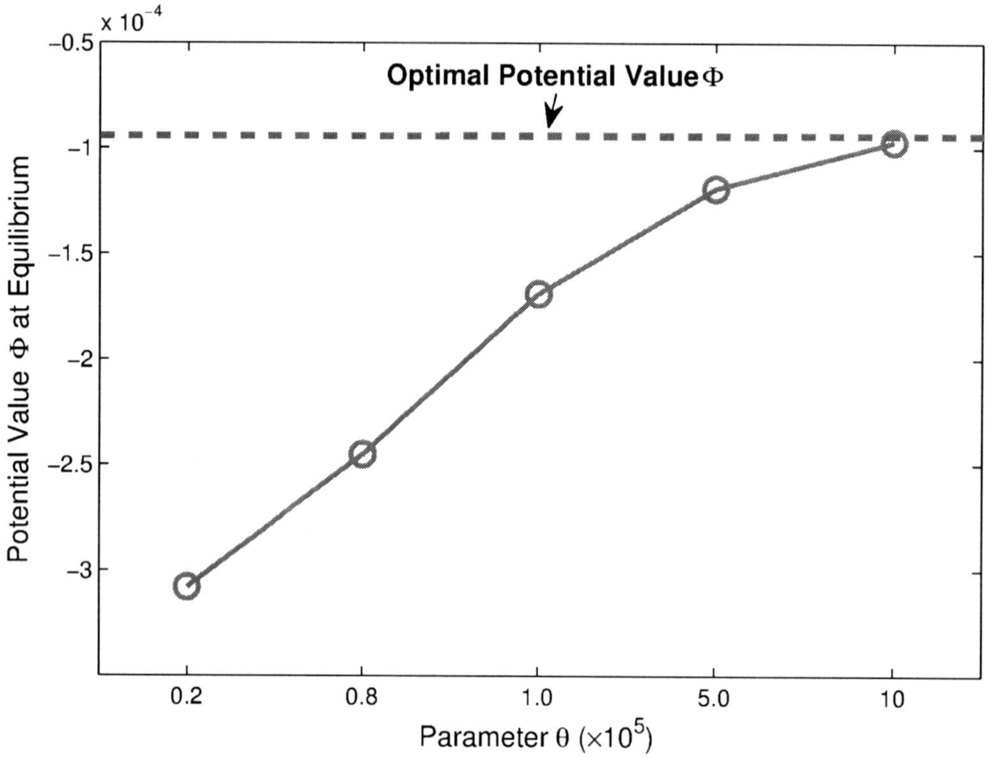

Figure 5. Devices' average throughputs when $\theta=1.0\times10^6$

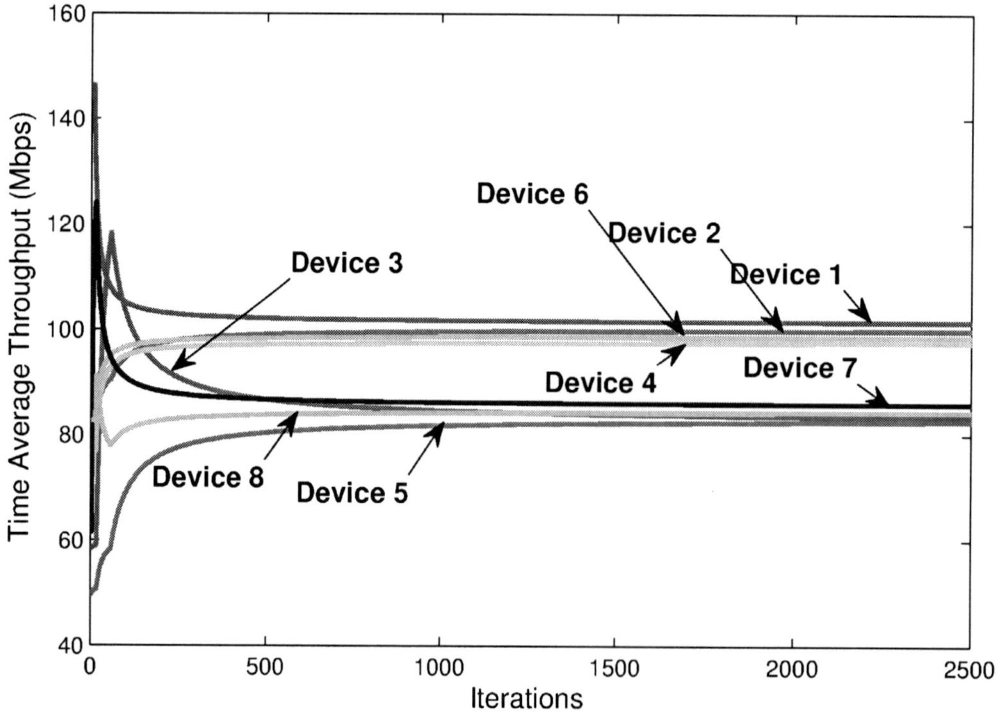

Figure 6. Dynamics of potential when $\theta=1.0\times10^6$

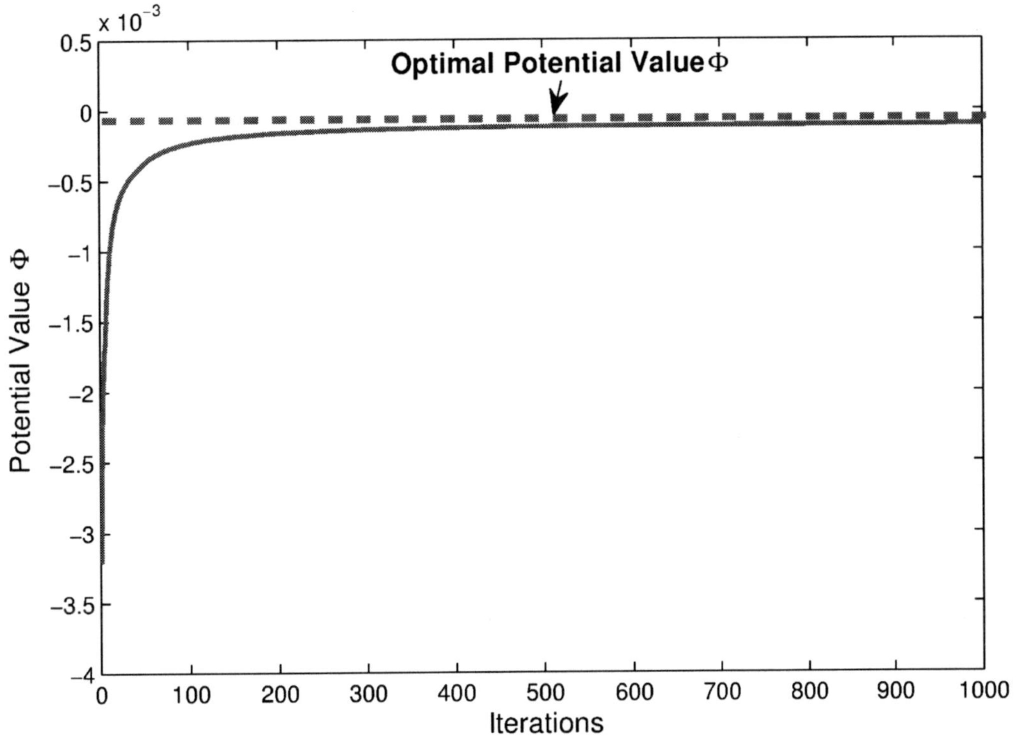

Figure 7. Dynamics of devices' average throughputs with perturbations

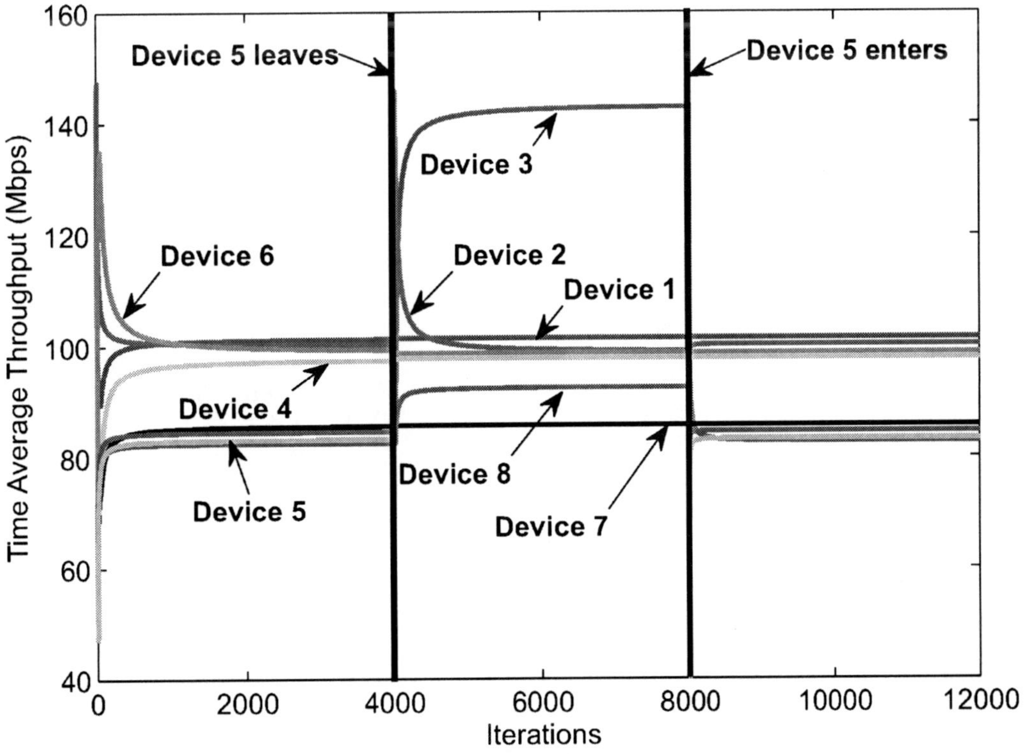

data transmission) at time t=4000 and enter the system again (i.e., recover the data transmission) at time t=8000. We show the dynamics of devices' time average throughputs and the potential function value in Figures7 and 8, respectively. We observe that, when device 5 leaves the system, the algorithm can adapt to such context change and the remaining devices can self-organize into a new stable Nash equilibrium. When device5 enters the system again, the algorithm can recover the previously convergent Nash equilibrium. This demonstrates that the self-organizing spectrum access algorithm can adapt to the dynamic network context changing and has the nice self-stability property.

To benchmark the performance of the self-organizing spectrum access algorithm, we also implement the system-wide throughput maximization solution by centralized optimization, i.e., $\max_{\mathbf{a}\in\Theta}\sum_{n=1}^{N}U_n(\mathbf{a})$ .Notice that the centralized optimization solution requires the complete network information, such as the geo-locations, the transmission power and the set of feasible channels of all devices. While the proposed self-organizing spectrum access algorithm only requires each device to measure its received interference locally. We implement experiments with N=20,…,50 devices being randomly scattered over the square area in Figure 3, respectively. The number of white-space channels M=10 and 5 channels out of these \$10\$channels will be randomly chosen as the set of vacant channels $\mathcal{M}_n$ for each device n. The transmission power P_n of each device n is randomly assigned from the set {50, 100, 150}mW. For each fixed number of devices N, we repeat the experiment \$100\$ times and show the average system throughput in Figure 9.We see that the performance loss of the self-organizing spectrum access algorithm is less than 10% in all cases, compared with the centralized optimization solution. This demonstrates the efficiency of the self-organizing spectrum access algorithm.

Figure 8. Dynamics of potential value with perturbations

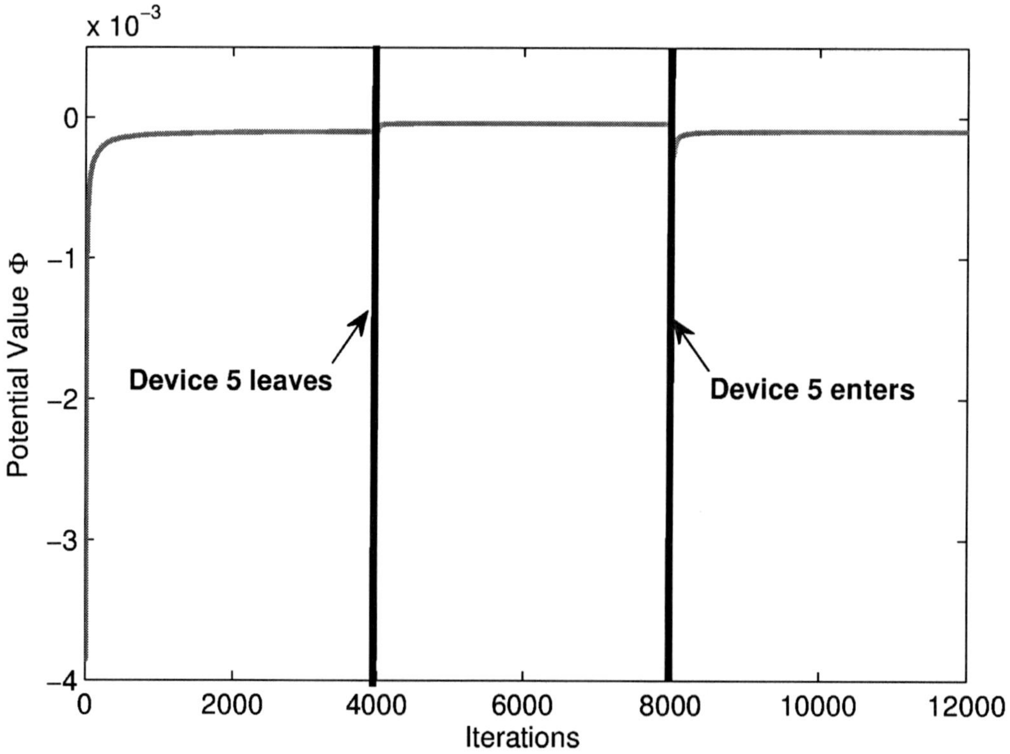

Figure 9. System throughput by the self-organizing spectrum access algorithm and centralized optimization solution

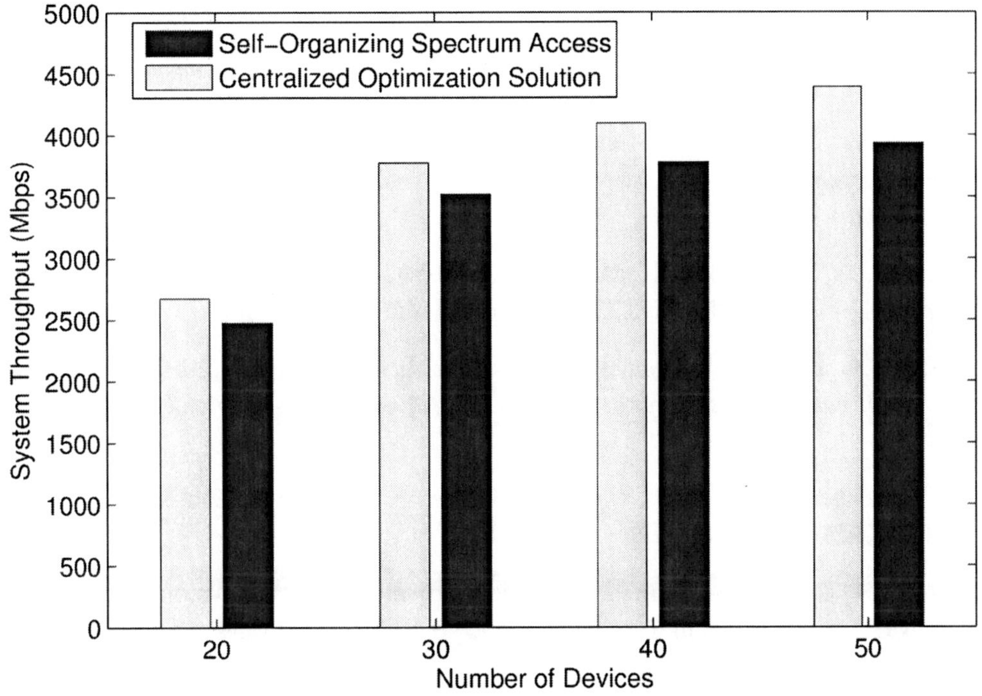

CONCLUSION

In this chapter, we consider self-organizing database-assisted white-space spectrum access network design. We address the distributed channel selection problem among the devices through the game theoretic approach. We propose a self-organizing spectrum access algorithm, which achieves a Nash equilibrium of the game. Numerical results demonstrate that the proposed algorithm is efficient and can adapt to the dynamical network context changing.

Regarding future research directions, one may consider investigating the scenario that some devices may run inelastic applications such as video streaming and have specific quality of service (QoS) requirements. The application cannot work properly when a device's QoS requirement (such as targeted data rate) is violated, but does not have much additional benefits when given more resources than needed. In this case, we can model the payoff function of each device as a threshold function with the threshold given by the QoS requirement. How to develop a self-organizing spectrum access algorithm to satisfy as many devices as possible will be very interesting and challenging.

ACKNOWLEDGMENT

This work is supported by the General Research Funds (Project Number CUHK 412713 and CUHK 14202814) established under the University Grant Committee of the Hong Kong Special Administrative Region, China. It is also partially supported by the funding from Alexander von Humboldt Foundation.

REFERENCES

Abdelkader, M., Hamdi, M., & Boudriga, N. (2011). Game theory and iteratedvoronoi diagrams for self-organized wireless sensor networks. In *International Wireless Communications and Mobile Computing Conference (IWCMC)* (pp. 1250–1255). IEEE.

Akyildiz, I., Lee, W., Vuran, M., & Mohanty, S. (2006). Next generation/dynamic spectrum access/cognitive radio wireless networks: A survey. *Computer Networks*, *50*(13), 2127–2159. doi:10.1016/j.comnet.2006.05.001

Bahl, P., Chandra, R., Moscibroda, T., Murty, R., & Welsh, M. (2009). *White space networking with wi-fi like connectivity* (pp. 27–38). SIGCOMM. doi:10.1145/1592568.1592573

Balasubramaniam, S., Botvich, D., Donnelly, W., Foghlu, M., & Strassner, J. (2007). *Bio-inspired framework for autonomic communication systems* (pp. 3–19). Advances in Biologically Inspired Information Systems.

Boyd, S., & Vandenberghe, L. (2004). *Convex optimization*. Cambridge: Cambridge university press. doi:10.1017/CBO9780511804441

Brunner, M., Eggert, L., Felis, S., Schmid, S., & Zimmermann, K. (2011). Autonomicwireless network management. IFIP Lecture Notes in Computer Science (LNCS), 3854(3854):57–70.

Chen, M., Liew, S., Shao, Z., & Kai, C. (2010). Markov approximation for combinatorial network optimization. In *Proceedings of IEEE INFOCOM* (pp. 1–9). IEEE.

Chen, X., & Huang, J. (2012a). Game theoretic analysis of distributed spectrum sharing with database. *In the 32nd International Conference on Distributed Computing Systems (ICDCS).*

Chen, X., & Huang, J. (2012b). Spatial spectrum access game: Nash equilibria and distributed learning. *In ACM International Symposium on Mobile Ad Hoc Networking and Computing (MobiHoc).* doi:10.1145/2248371.2248401

Chen, X., & Huang, J. (2013a). Database-assisted distributed spectrum sharing. *IEEE Journal on Selected Areas in Communications*, *31*(11), 2349–2361. doi:10.1109/JSAC.2013.131110

Chen, X., & Huang, J. (2013b). Distributed spectrum access with spatial reuse. *IEEE Journal on Selected Areas in Communications*, *31*(3), 593–603. doi:10.1109/JSAC.2013.130323

Chen, X., & Huang, J. (2013c). Evolutionarily stable spectrum access. *IEEE Transactions on Mobile Computing*, *12*(7), 1281–1293. doi:10.1109/TMC.2012.94

Chen, X., & Huang, J. (2014a). Imitation-based social spectrum sharing. *IEEE Transactions on Mobile Computing*.

Chen, X., & Huang, J. (2014b). Spatial spectrum access game. *IEEE Transactions on Mobile Computing*.

Deb, S., Srinivasan, V., & Maheshwari, R. (2009). *Dynamic spectrum accessing DTV whitespaces: design rules, architecture and algorithms* (pp. 1–12). MOBICOM.

Duan, L., Gao, L., & Huang, J. (2014). Cooperative spectrum sharing: A contract-based approach. *IEEE Transactions on Mobile Computing*, *13*(1), 174–187. doi:10.1109/TMC.2012.231

FCC. (2010, September 23). *Second memorandum opinion and order*. FCC.

Feng, X., Zhang, J., & Zhang, Q. (2011). Database-assisted multi-AP network on TV white spaces: Architecture, spectrum allocation and AP discovery. In *IEEE Symposia on New Frontiers in Dynamic Spectrum Access Netoworks*. DySpan. doi:10.1109/DYSPAN.2011.5936215

Feng, X., Zhang, Q., & Zhang, J. (2014). A hybrid pricing framework for TV white space database. *IEEE Transactions on Wireless Communications*, *13*(5).

Gao, L., Huang, J., Chen, Y.-J., & Shou, B. (2013). An integrated contract and auction design for secondary spectrum trading. *IEEE Journal on Selected Areas in Communications*, *31*(3), 581–592. doi:10.1109/JSAC.2013.130322

Gunes, M., Sorges, U., & Bouazizi, I. (2002). Ara-the ant-colony based routing algorithm for MANENTs. In *International Conference on Parallel Processing Workshops*, pages 79–85. IEEE. doi:10.1109/ICPPW.2002.1039715

Gupta, P., & Kumar, P. R. (2000). The capacity of wireless networks. *IEEE Transactions on Information Theory*, *46*(2), 388–404. doi:10.1109/18.825799

Han, K., Choi, S., Kang, D. H., & Choi, S. (2012). Game based self-organizing scheme for femto-cell networks. *Lecture Notes of the Institute for Computer Sciences. Social Informatics and Telecommunications Engineering*, *75*, 57–75.

Huang, J. (2013). Market mechanisms for cooperative spectrum trading with incomplete network information. *IEEE Communications Magazine*, *51*(10), 201–207. doi:10.1109/MCOM.2013.6619584

Huang, J., Berry, R. A., & Honig, M. L. (2006). Auction-based spectrum sharing. *ACM/Springer. Mobile Networks and Applications*, *11*(3), 405–418. doi:10.1007/s11036-006-5192-y

Kruger, B., & Dressler, F. (2005). Molecular processes as a basis for autonomous networking. *IPSI Transactions on Advances Research: Issues in Computer Science and Engineering*, *1*(1), 43–50.

Lee, C., & Suzuki, J. (2009). An immunologically-inspired autonomic framework for self-organizing and evolvable network applications. [TAAS]. *ACM Transactions on Autonomous and Adaptive Systems*, *4*(4), 22. doi:10.1145/1636665.1636668

Li, S., Huang, J., & Li, S.-Y. R. (2013). Dynamic profit maximization of cognitive mobile virtual network operator. *IEEE Transactions on Mobile Computing*.

Luo, Y., Gao, L., & Huang, J. (2012). Spectrum broker by geo-locationdatabase. In *IEEE Global Communications Conference (GLOBECOM)*, pages 5427–5432. IEEE. doi:10.1109/GLOCOM.2012.6503984

Luo, Y., Gao, L., & Huang, J. (2014a). Price and inventory competitionin oligopoly TV white space markets. *IEEE Journal on Selected Areas in Communications*.

Luo, Y., Gao, L., & Huang, J. (2014b). Trade information, not spectrum: A novel TV white space information market model. In *International Symposium on Modeling and Optimization in Mobile, Ad Hoc, and Wireless Networks (WiOpt)*, pages 405–412. IEEE. doi:10.1109/WIOPT.2014.6850326

McHenry, M. A., McCloskey, D., Roberson, D., & MacDonald, J. T. (2005). *Spectrum occupancy measurements. Technical report*. Shared Spectrum Company.

Monderer, D., & Shapley, L. S. (1996). Potential games. *Games and Economic Behavior*, *14*(1), 124–143. doi:10.1006/game.1996.0044

Murty, R., Chandra, R., Moscibroda, T., & Bahl, P. (2011). Senseless: A database-driven white spaces network. In IEEE Symposia on New Frontiers in Dynamic Spectrum Access Netoworks (DySpan). doi:10.1109/DYSPAN.2011.5936197

Rogers, A., David, E., & Jennings, N. (2005). Self-organized routing for wireless microsensor networks. *IEEE Transactions on Systems, Man, and Cybernetics. Part A, Systems and Humans*, *35*(3), 349–359. doi:10.1109/TSMCA.2005.846382

Wu, Y., Zhu, Q., Huang, J., & Tsang, D. H.-K. (2014). *Revenue sharing based resource allocation for dynamic spectrum access networks*. IEEE Journalon Selected Areas in Communications.

Yan, Y., Huang, J., & Wang, J. (2013). Dynamic bargaining for relay-based cooperative spectrum sharing. *IEEE Journal on Selected Areas in Communications*, *31*(8), 1480–1493. doi:10.1109/JSAC.2013.130812

KEY TERMS AND DEFINITIONS

Dynamic Spectrum Access: Is a new spectrum sharing paradigm that allows secondary users to access the abundant spectrum holes or white-spaces in the licensed spectrum bands. Dynamic spectrum access is a promising technology to alleviate the spectrum scarcity problem and increase spectrum utilization.

Geo-Location Database: Is an information infrastructure with specific operating requirements to determine not only if a TV channel is available for secondary usage at a particular geographical location, but also to determine a maximum allowed safe transmit power level (per channel) at the operating location to avoid causing interference to licensed incumbents.

Self-Organizing Spectrum Access: Is an automation technology designed to make the planning, configuration, management, optimization and healing of dynamic spectrum access networks simpler and faster.

Chapter 11
Auction-Based Resource Management in Multi-Cell OFDMA Networks

Seyed Mohamad Alavi
Illinois Institute of Technology, USA

Chi Zhou
Illinois Institute of Technology, USA

ABSTRACT

This chapter addresses the complex scenario of multi cell OFDMA network resource allocation and interference control by auction theory. An auction framework that meets the properties of efficiency and incentive compatibility is proposed. We consider a multi-cell OFDMA network with Fractional Frequency Reuse (FFR) implementation. An auction is presented to allocate subcarriers. According to the proposed auction framework, users will avoid bidding for the subcarriers where they have a relatively low chance of winning. Optimal bidding strategy based on Bayesian Nash Equilibrium (BNE) is obtained in which users are maximizing their net profit. A focal distance which classifies the users into cell-center and cell-edge users is characterized. The proposed approach maximizes the total auctioneer revenue, while balancing a tradeoff between system performance in terms of total system throughput and quality of service provisioning for cell-edge users.

INTRODUCTION

Recently, there have been substantial advances in research which use game theory for analysis of communication networks. This is basically due to the need for developing autonomous and distributed mobile networks where the users can make independent and rational strategic decisions. Moreover, there is a need for low complexity distributed methods that can efficiently represent competitive behavior among network entities. To this end, game theory provides a mathematical tool for analysis of the network. The Orthogonal Frequency Division Multiple Access (OFDMA) is one promising candidate for next generation wireless communication systems. The combination of OFDMA with adaptive modulation and

DOI: 10.4018/978-1-4666-8642-7.ch011

coding (AMC) and dynamic power allocation is of great prominence in the design of future broadband radio systems. Obviously resource allocation in which resources include subcarriers, power, and time slots and scheduling schemes play important roles in the design of OFDMA system.

In resource allocation there is always a trade-off between efficiency and fairness. Two end-point problems considering these factors are sum-rate maximization and max-min problem. To guarantee a trade-off between sum rate maximization and max-min fairness, proportional fairness tries to maximize the sum of logarithmic utility function. Even though proportional fairness can provide a balance between efficiency and fairness, but the provided fairness is not the result of users' competition. A proposed solution is that users are allowed to compete for available resources. In this way, each user is responsible for his/her own action. This kind of fairness is called competitive fairness, in which users are receiving a fraction of resources through their competition. Auction game theory is a decentralized market mechanism to address the competitive behavior of nodes for available resources. It mimics the competitive behavior of users in which users are receiving resources through their competition process. Auctions as a subset of game theory have recently been introduced into several areas of wireless communications to handle the problem of resource competition among selfish users.

In a multi-cell OFDMA system, a user not only competes for resources with the users in its own cell, but also generates interference for adjacent cells. Advanced and intelligent radio resource management (RRM) schemes which consider the dynamic co-channel interference have to be designed. Moreover, RRM must take advantage of the opportunities and flexibilities of OFDMA system, and address the critical notion of fairness. Taking these challenges into consideration, introduces additional complexity into the problem of resource allocation. Meanwhile, the conventional centralized approaches do not scale well to the multi-cell networks where distributed algorithms are preferred. The next generation cellular network is increasingly required to provide guaranteed quality of service regardless of the user location in the cell. So, there is a need for algorithms which would provide a tradeoff between the rate improvement and cell coverage for cell-edge users. Fractional Frequency Reuse (FFR) is an attractive approach toward the frequency planning for multi-cell OFDMA network due to low complexity of implementation and significant gains in performance who especially for the cell-edge users usually encounter service starvation. The basic idea of FFR scheme is to apply a small frequency reuse factor for cell center users who have better SINR and large frequency reuse factor for cell-edge users who have worse SINRs. In this way, cell-edge users of adjacent cells do not interfere with each other. Moreover, interference from inner-cell users is reduced. The result is that FFR can achieve good tradeoff between the average network throughput and the performance of the cell-edge users.

In this chapter, we propose a comprehensive auction framework for multi cell OFDMA Network which takes into account the interference from adjacent cells. We obtain optimal bidding strategy based on Bayesian Nash Equilibrium (BNE). In a Fractional Frequency Reuse (FFR) implementation of frequency planning, we will find a focal distance (R_f) which classifies the users into cell-center and cell-edge users. The proposed approach aims to maximize the total auctioneer revenue, while providing a tradeoff between system performance in terms of total system throughput and quality of service provisioning for cell-edge users.

Background

Resource allocation and scheduling schemes play important roles in the design of an OFDMA system. The resource allocation problem can be modeled as an optimization problem subject to various QoS constraints. The optimization problem of OFDMA resource allocation, in terms of subcarrier, time slots, and power assignment subject to various QoS constraints, is nonlinear in nature, and finding the solution tend to be very computationally intensive. The complexity of OFDMA resource allocation problems can be reduced by introducing time sharing property, and it can be proved that for large number of subcarriers optimal solutions can be obtained via dual decomposition techniques. Unfortunately, the complexities of the dual decomposition technique as well as proposed heuristic algorithms are still too much for practical cellular systems being considered for deployment. As a result, less computational demanding solutions than that pursued in previous works must be devised.

Recently, there have been substantial research works which apply game theory for analysis of communication networks. Also there is a need for an efficient approach with low complexity. While an efficient resource allocation aims to satisfy the users' QoS requirements, the crucial notion of fairness arises and must be dealt with at the same time.

Auctions have been introduced into several areas of wireless communications to handle the problem of resource competition among selfish users. To implement an auction approach, some challenging problems must be tackled. In combinatorial auction, the challenges come from the computational complexity along with the problem of incentive compatibility where the users reveal their true valuation for the demands. Moreover, there is a need for a clear and expressive bidding language.

Aforementioned problems have prompted rigorous effort which are applying the concepts of game and auction theory. (Han *et al.,* 2007) proposes a virtual referee for a noncooperative resource allocation game in multicell OFDMA networks. A referee monitors the network performance. In cases where the lack of cooperation leads to performance deterioration, a referee mediates the game rules, so that the performance can be enhanced. In (Stiakogiannakis and Kaklamani, 2011) an auction approach is proposed for allocation of subcarriers in single-cell and multicell OFDMA networks. In multicell scenario, the users are categorized as interior and edge cell users. To simplify the problem, it is assumed that there is no interference from inner users to the users of the neighboring cells. (Moretti *et al.*, 2011) have studied the problem of resource allocation in the downlink of a multicell OFDMA system, with a frequency reuse distance equal to one. The paper presents a layered architecture that integrates a packet scheduler with an adaptive resource allocator to reduce the interference and provide a fair distribution of resources to the users. Moreover, a load control will make sure that the procedure always converges to a stable allocation. Scutari *et al.* present decentralized non-cooperative approach for resource allocation in Gaussian frequency-selective interference channels (Scutari *et al.*, 2008). The Nash equilibrium is achieved through distributed algorithm called asynchronous iterative water-filling algorithm in a static noncooperative game of complete information. (Hui *et al.* 2006) propose a noncooperative transmit power control and cooperative subcarrier allocation in a multicell OFDMA system which aims to maximize the energy efficiency. Although OFDMA systems take advantage of inter symbol interference, but the co-channel interference (CCI) or so-called inter-cell interference (ICI) remains a big issue that affect the coverage and spectral efficiency. Fractional Frequency Reuse is an attractive approach toward the frequency planning for multicell OFDMA network due to low complexity of implementation and significant gains in performance especially for the cell-edge users who usually encounter service starvation. Interference avoidance techniques have been proposed for OFDMA systems to reduce the interference. Traditional cellular systems employ orthogonal subcarrier allocation within a cell and choose a frequency reuse distance larger than one. This approach provides interference robustness while scarifying the spectral efficiency. Moreover, Base Station (BS) cooperation or precise frequency planning is required. Dynamic resource allocation has been proposed to employ the full reuse of spectrum in order to improve the spectral efficiency.

It is known that effective reuse of resources in a cellular system can highly enhance the system capacity. With a smaller frequency reuse factor, more available bandwidth can be obtained by each cell. So, in this sense the reuse factor of 1 is desirable. However, with the reuse factor of 1, most users are affected with heavy ICI, especially the cell edge users which in turn causes low cell coverage and inferior system capacity. The conventional solution to this problem is by increasing the cluster-order, which can alleviate the ICI efficiently, but at the cost of available bandwidth shrink for each cell.

The Soft Frequency Reuse (SFR) scheme, which has been adopted in the 3GPP-LTE system, addresses this issue by increasing reuse factor and transmission power for cell edge users, so that the ICI from neighboring cells to those users can be decreased, and thereby to enhance their performance. The basic idea of the SFR scheme is applying reuse factor of 1 to cell center users and reuse factor of 3 to cell edge users.

Methods for an optimal design of FFR system with the objective to maximize the total system throughput using graph theory, Markov process and convex optimization are proposed in (Chang *et al.*, 2009) (Elayoubi *et al.*, 2008) (Assad, 2008), respectively. (Huang *et al.*, 2014) designed truthful auction mechanisms that maximize either the overall social efficiency of new users (a.k.a buyers) or the revenue of the spectrum owner (a.k.a seller). Given that the optimal conflict-free spectrum allocation problem is NP-hard, they proposed a series of near-optimal auction mechanisms. The proposed auction mechanisms are not only truthful but also provide theoretically provable performance guarantee. (G. Sun *et al.*, 2014) introduced the coalition double auction for efficient spectrum allocation in cognitive radio network. They proposed a fully-economic spatial spectrum allocation mechanism by incorporating the coalition formation approach with auction theory. With the proposed scheme, the primary operator acts as an auctioneer, who performs multiple virtual auctions to form a stable partition of secondary users and conducts a final auction to decide the winning users. (Al-Tous and Barhumi, 2014) have investigated joint resource allocation for multiple users multiple relays. Amplify and forward OFDMA cooperative communication system in the presence of a direct link between the source and destination nodes. The subcarriers are assigned based on the auction framework. They proposed subcarrier assignment algorithms based on sequential auction framework. The base-station receives bids from all the users for the subcarrier, then the subcarrier is assigned to the user with the highest bid. Sequentially, the procedure progresses until all subcarriers are allocated. The bidding strategies of these algorithms are designed either to maximize the sum data rate or maximize the fairness index.

Heuristic-based approaches are considered for frequency partitioning, sub-band allocation and user scheduling in (Ali and Leung, 2009) (Fang and Zhang, 2008) (Rahman and Yanikomeroglu, 2010). (Son *et al.*, 2009) investigates partial frequency reuse (PFR) and load-balancing schemes to improve cell edge performances and support a more balanced data rate among all users. A utility-maximization problem is formulated to provide proportional fairness throughout the network. In (Stolyar and Viswanathan, 2009) two algorithms, namely Multi sector gradient and sector autonomous are used to create efficient, soft fractional frequency reuse (FFR) patterns for enhancing performance of OFDMA based cellular systems for forward link best effort traffic. Xie and Walke investigate the capacity and PDFs of the SINR in the uplink and downlink of cellular networks using a modified SFR strategy which allows for more than two user classifications (Xie and Walke, 2010).

SYSTEM MODEL

Consider a multi-cell OFDMA system with M base stations that cover a geographic area. In each cell, there are a BS and K users competing for the available resources. Here resources mean the subcarriers to be allocated through auction process. The whole bandwidth is divided into N OFDM subcarriers to be allocated to the users. Each user k is assigned a subset of subcarriers N_k to send data over the wireless media. Note that subcarriers are not shared by different users. It is supposed that the subcarrier separation is narrower than the coherent bandwidth. Moreover, in each cell the base station knows the channel state information (CSI) and the amount of interference relative to all users within the cell which is then made known to the users via a dedicated feedback channel. On the basis of this information, users send their valuation to the base station as the offering bid.

The reuse factor determines the available frequency to be allocated to the users. Higher reuse factor allows cells to operate in orthogonal carriers to reduce the amount of inter-cell interference. On the other hand the substantial demand for bandwidth is growing as the new applications with higher data rate are emerging. As a result universal frequency reuse is usually the deployment choice for frequency planning.

The 4th generation cellular network is increasingly required to provide guaranteed quality of service regardless of the user location in the cell. So, there is a need for algorithms which would provide a tradeoff between the rate improvement and cell coverage for cell-edge users. The basic idea of FFR scheme is to apply a small frequency reuse factor for cell center users who have better SINR and large frequency reuse factor for cell-edge users who have worse SINRs. In this way, cell-edge users of adjacent cells do not interfere with each other. Moreover, interference from inner-cell users is reduced. The result is that FFR can achieve good tradeoff between the average network throughput and the performance of the cell-edge users.

Strict FFR deployment has been shown in Figure 1. The reuse factor of cell-edge users is chosen to be 3. Inner-cell users are allocated a common frequency band while cell-edge users' frequency band is partitioned across the cells according to the reuse factor of 3. Note that the interior users do not share the frequency bandwidth with edge users. The result of this allocation would be interference reduction for both interior users and cell-edge users.

Let $h_{k,i,j}$ be the radio channel gain between BS j and user k on subcarrier i. In a cellular OFDMA system, the computation of the SINR achieved at subcarrier i in the receiver of user k served by BS j, is obtained as follows

$$\mathrm{SINR}_{k,i,j} = \frac{p_{k,i,j}\left|h_{k,j}\left(i\right)\right|^2}{I_j\left(i\right) + BN_0} \tag{1}$$

where $I(j)_i$ is the total interference in subcarrier i, cell j which is

$$I_j\left(i\right) = \sum_{l=1,l\neq j}^{M} p_{U(l,i)}\left(i\right)\left|h_{U(l,i),j}\left(i\right)\right|^2 \tag{2}$$

Figure 1. Multicell OFDMA network with FFR frequency planning

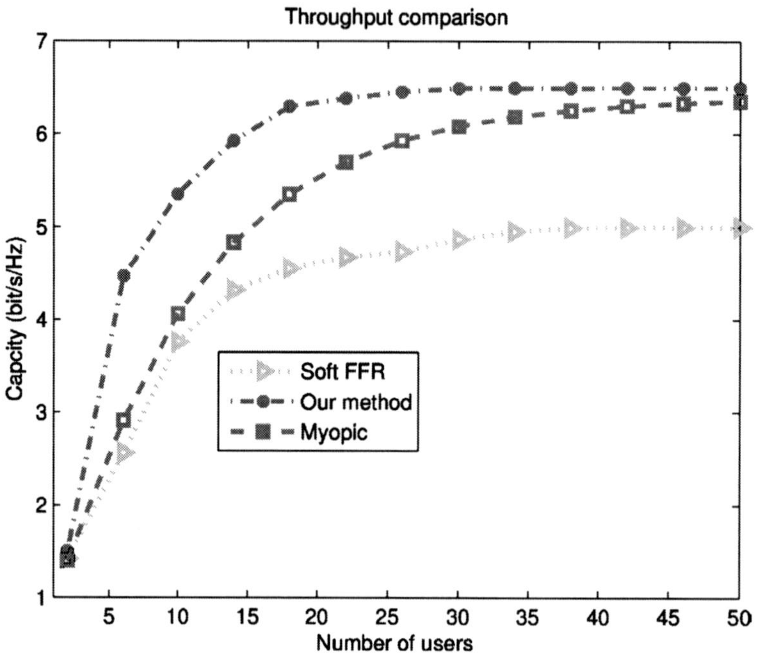

$p_{k,i,j}$ is the transmit power of BS j to user k on subcarrier i, $U(l,i)$ are the BS transmitting on subcarrier i, and $h_{U(l,i),j}(i)$ is the channel gain of BS in $U(l,i)$ on subcarrier i. Note that depending on the location of the users, the interfering cells, $U(l,i)$ are different. For a user in the cell center the interference will be from all the cells in the network, since all the cells use the same frequency band for the cell center. If the user is in the cell edge area, the interfering cell set would be smaller. For example considering cell 'a' in Figure 1, $U(l,i) = \{j,l,n,p,r,h\}$ for the cell edge user, while for the cell center user, all the cells contribute for the interference. The bit rate of user k on allocated subcarrier i of cell j would be

$$r_{k,i,j} = \log_2(1 + \mathrm{SINR}_{k,i,j}) \tag{3}$$

In the following, the auction framework is described which is based on the second price auction also known as Vickrey-Clarke-Grove (VCG) auction. Moreover, an approach based on Bayesian Nash Equilibrium is proposed to identify the bidding strategy and the properties of the auction framework. Particularly, the conditions according to which a user will decide whether to participate in the auction or not are drawn. The decision is made according to users' chance of winning a specific subcarrier. Before formulating the problem, the following difficulties should be considered:

Users' private evaluation of the subcarriers is based on the changing condition of the subcarrier market, as users must estimate the valuation of other users for a specific subcarrier and adjust their bids accordingly.

The set of strategies of users is typically large, as users can virtually submit any positive value as their bid and the parameters of the auction are based on the uncertain behavior of the other users.

Users submit their bids to the BS which acts as an auctioneer, then the highest bidders win the auction and get that specific subcarrier. As in the second price auction the winner pays the amount equal to the second highest bid. For fixed bidding strategy of other users, the user encounters a tradeoff between profit and chance of winning. As the bidder raises her bid, it will increase the probability of wining, while by higher bidding, she will reduce the profit from winning. In out auction, in addition to the payment for winning the subcarrier, there are two other fees, namely entry fee and monitoring fee. The entry fee is the amount a bidder must pay in order to submit a bid. Entry fees usually considered as the cost of preparing the strategic bid. Note that the entry fees are charged for all bidders who take part in the auction without considering the outcome of the auction. A user may decide not to participate in the auction in which there is no incurred entry fee. The monitoring fee is the cost for sensing the channel and doing the measurements. Consider a scenario in which BS wishes to sell N subcarrier to K potential users. The following rules define the auction mechanism:

- BS auctions off the subcarriers among potential buyers i.e., users
- Each user k knows its current valuation for each subcarrier.
- Each user according to its valuation and incurred fees decides to participate in the auction while bidding.
- The auction mechanism decides about the allocations and the payments.
- Users receive the allocation decision and make payment accordingly.

At the beginning of each scheduling interval, user k decides whether to participate in the action or not. If he chooses to take part in the auction, he submits a sealed bid $b_{k,i}$ to the BS which acts as an auctioneer. Let b_{-k} be the type (i.e., strategy) profile of the user k opponent:

$$b_{-k} = \{b_k \mid k \in \kappa \setminus k\} \tag{4}$$

User k sees the type profile of her opponent as an observation of a vector of random variables, and bid accordingly. In cases where the user k decides to stay out of the auction, we set $b_{k,i}$=0.

The cost of user k is consisting of several parts. The entry fee c, which is only occurs when the user chooses to participate in the auction, the monitoring fee, m which is the cost for sensing the channel and do the measurements, and finally the actual payment $\beta_{k,i}$ which has to be paid if the user k wins the subcarrier i. Let $\alpha_{k,i}$ be the indicator function which shows the allocation strategy showing that whether the user k wins the subcarrier i or not

$$\alpha_k = \{\alpha_{k,1}, \ldots, \alpha_{k,N} \mid \alpha_{k,i} \in \{0,1\}\} \tag{5}$$

Clearly, $\alpha_{k,i}$ is a random variable. According to the subcarrier exclusiveness property, a subcarrier cannot be allocated to more than one user, so

$$\sum_k \alpha_{k,i} \leq 1 \tag{6}$$

In the cases where user k decides not to participate in the auction, there is no allocation

$$\alpha_{k,i}(b_{k,i} = 0) = 0 \tag{7}$$

Note that, user k may decide not to take part in bidding when participating in the auction game is not worthy and he desires to save entry fee. This situation is more likely when the valuation of the user for a specific subcarrier is not convincing due to the incurred cost. It may be the result of low data rate or the large number of users competing for a specific band. Following this approach, users can save energy expenditure. Moreover, the proposed approach can achieve better performance in terms of fairness. Suppose two users with different channel conditions. The user with the good channel quality will bids a lot and will pay more, as a result the user experiencing a low quality channel will have the opportunity of winning.

According to the second price auction the payment which the winner must pay is the second highest bid for that subcarrier, specifically

$$\beta(b_{k,i}) = \max\{b_{-k}\} \tag{8}$$

Utility function of user k on subcarrier i is obtained as follows:

$$U_{k,i} = \alpha_{k,i}.\log\left(1 + \frac{h_{k,i}p_{k,i}}{N_0 + \sum_{j \neq k} h_j p_j}\right) - \beta(b_{k,i}) - \overline{c} - m \tag{9}$$

Note that in the above formulation, $\overline{c}$ is the average of entry fee c. So the total user utility would be the summation of utility from its allocated subcarriers

$$U_k = \sum_i U_{k,i} \tag{10}$$

INCOMPLETE INFORMATION GAME

In this section, the auction framework for the allocation of subcarriers as a Bayesian game is modeled. In game theory, the game of complete information denotes a situation where each user has complete information of the other users bidding strategy. However, in the proposed auction framework, this is not the case. Particularly, users have incomplete information meaning that each user knows only its bidding strategy, but does not know the bidding strategy of the other users. In such scenario the objective of the users is to maximize the expected utility. Note that in the following the auction analysis for cell center and cell edge users would be identical, so the notation $\hat{K}$ is used to denote the number of users in the area of interest. So $\hat{K}$ denotes whether K^C the number of cell center users, or K^E the number of cell edge users.

The $\hat{K}$-user auction with incomplete information can be shown as:

$$\Gamma = \left\langle \hat{K}, T, S, F, U \right\rangle \tag{11}$$

in which

$\hat{\mathsf{K}} = \left\{1, 2, ..., \hat{K}\right\}$ is the player set,

$T = v_1 \times v_2 \times ... \times v_{\hat{K}}$ is the type set. In which v_k is the valuation of user k and $\times$ stands for Cartesian product,

$S = \left\{b_1, b_2, ..., b_{\hat{K}}\right\}$ is the action or strategy set,

$F = F_1 \times F_2 \times ... F_{\hat{K}}$ is the probability set, and

$U = \{u_1 \times u_2 \times ... \times u_k\}$ is the payoff function set.

It is supposed that user k has private valuation for subcarrier i as $v_{k,i}$. In the Bayesian game, $v_{k,i}$ is called the type of bidder k. Users know the distribution of their type and also the distribution of their opponent types. Note that although the users are aware of the distribution of the other users, but they are uncertain about the true valuation of the opponents. The private information of a user is her own value for the subcarrier, and it does not depend on what the other users know. Moreover, the value $v_{k,i}$ is the maximum amount that user k is willing to pay for subcarrier i. Users type are independent and identically distributed (i.i.d) with cumulative distribution function (CDF) $F(.)$ and probability density function $f(.)$.

The probability that user k with bidding strategy $b_{k,i}$ wins the auction can be calculated through the order statistics. Let $b_{1,i}, b_{2,i}, ..., b_{\hat{K},i}$ be the independent drawn from a distribution $F(.)$ and corresponding density $f(.)$. Let $Y_i^{(1)}, Y_i^{(2)}, ..., Y_i^{(\hat{K})}$ be the rearrangement of these variable so $Y_i^{(1)} \geq Y_i^{(2)} \geq Y_i^{(\hat{K})}$. The random variable $Y_i^{(k)}$ is called order statistics. Let $G^{(k)}$ be the distribution of $Y_i^{(k)}$ and $g^{(k)}$ the corresponding probability density function. We are interested in the first order statistic and second order statistic given the highest order is known i.e., in second price auction, the winner is the user with highest bid and the payment of the winner is the second highest bid. The derivation of highest order statistics is straightforward. The probability of $Y_i^{(1)} \leq y$ is the same probability of $b_{k,i}^k \leq y \ \forall k$. Taking into account the independency of random variables

$$G^1(y) = \Pr(Y^{(1)} \leq y) = \Pr(Y_i^{(1)} \leq y)\Pr(Y_i^{(2)} \leq y)...Pr(Y_i^{(\hat{K})} \leq y) = (F(y))^{\hat{K}} \tag{12}$$

The corresponding probability density function is

$$g^1(y) = \hat{K}(F(y))^{\hat{K}-1} f(y) \tag{13}$$

So, in the auction scenario for the fixed the bid of the winning bidder, the probability that user k with bidding strategy $b_{k,i}$ is the winner would be

$$\Pr(\text{user } k \text{ wins}) = (F(b_{k,i}))^{\hat{K}-1} \tag{14}$$

Particularly, Equation 14 may interpreted as

$$\begin{aligned}&\Pr(user\ k\ wins) = \Pr(b_{k,i} > \max\{b_{m,i,m\neq k}\})\\ &= \underbrace{\Pr(b_{1,i} > b_{1,i}) \times \Pr(b_{2,i} > v_2) \times \ldots \times \Pr(b_{m,i} > v_j)}_{\hat{K}-1 \text{ terms}} = (F(b_{k,i})^{\hat{K}-1})\end{aligned} \tag{15}$$

So, $\alpha_{k,i}$ which is the subcarrier allocation index can be written as a random variable with the following distribution

$$\alpha_{k,i} = \begin{cases} 1 & \text{with probability of } (F(b_{k,i}))^{\hat{K}-1} \\ 0 & \text{with probability of } 1-(F(b_{k,i}))^{\hat{K}-1} \end{cases} \tag{16}$$

The expectation of $\alpha_{k,i}$ would be

$$E\left[\alpha_{k,i}\right] = 1 \times \Pr\left(1\right) + 0 \times \Pr\left(0\right) = 1 \times F^{\hat{K}-1}\left(b_{k,i}\right) = F^{\hat{K}-1}\left(b_{k,i}\right) \tag{17}$$

The expected payment for a winning user with strategy profile $b_{k,i}$ would be

$$\begin{aligned}\beta\left(b_{k,i}\right) &= \Pr\left(b_{k,i} \text{ wins}\right) \times \mathrm{E}\left[\text{second highest bid} \middle| b_{k,i} \text{ is the highest bid}\right]\\ &= \Pr\left(Y^{(2)} \le b_{k,i}\right) \times \mathrm{E}\left[Y^{(2)} \middle| Y^{(2)} \le b_{k,i}\right] = G^{(1)}\left(b_{k,i}\right) \times \mathrm{E}\left[Y^{(2)} \middle| Y^{(2)} \le b_{k,i}\right]\end{aligned} \tag{18}$$

Moreover the joint density function of the first and second order statistics can be written as

$$g^{1,2}\left(y_1, y_2\right) = \hat{K}\left(\hat{K}-1\right) f\left(y_1\right) f\left(y_2\right) \left(F\left(y_2\right)\right)^{\hat{K}-2} \tag{19}$$

So the conditional density of second highest bid, given the highest bid would be

$$g^{(2)}\left(q \middle| Y^{(1)} = b_{k,i}\right) = \frac{g^{1,2}\left(b_{k,i}, q\right)}{g^{1}\left(b_{k,i}\right)} = \frac{\hat{K}\left(\hat{K}-1\right) f\left(b_{k,i}\right) f\left(q\right) \left(F\left(q\right)\right)^{\hat{K}-2}}{\hat{K} f\left(b_{k,i}\right) \left(F\left(b_{k,i}\right)\right)^{\hat{K}-1}} = \frac{\left(\hat{K}-1\right) f\left(q\right) F\left(q\right)^{\hat{K}-2}}{\left(F\left(b_{k,i}\right)\right)^{\hat{K}-1}} \tag{20}$$

Substituting Equation 20 in Equation 18 results in

$$\beta\left(b_{k,i}\right) = F^{\hat{K}-1}\left(b_{k,i}\right)\int_0^{b_{k,i}} q \frac{\left(\hat{K}-1\right) f\left(q\right)\left(F\left(q\right)\right)^{\hat{K}-2}}{\left(F\left(b_{k,i}\right)\right)^{\hat{K}-1}} dq \tag{21}$$

Optimum User Strategy

In a scenario where every rational and selfish user wants to maximize his own payoff, what would be the expected outcome of the game? In a Bayesian game, each user must decide its strategy given its information. Bayesian equilibrium represents a solution for Bayesian games which no user can get more payoff by deviating from that point. It is essential to characterize the equilibrium point in such scenario.

Bayesian equilibrium: The strategy profile

$$b^*\left(.\right) = \left\{b_k^*\left(.\right)\right\}_{k\in\hat{K}}$$

is a Bayesian equilibrium if for all $k \in \hat{K}$ and for all $b_k(.)\in B_k$ and $b_{-k}(.)\in B_{-k}$

$$E\left[U_k\left(b_k^*, b_{-k}^*\right)\right] \geq E\left[U_k\left(b_k, b_{-k}^*\right)\right]$$

Theorem 1: For the auction framework defined in Equation 11, there exists a Bayesian equilibrium.
Proof: The strategy space $b_{k,i}$ is convex, compact and non-empty for each user *k* and the payoff of each user is concave with respect to that user bidding strategy:

$$\frac{\partial^2 U_k\left(b_i; b_{-i}\right)}{\partial b_i} < 0.$$

So, there is an equilibrium point for this $\hat{K}$-person Bayesian game.

Since users are selfish and rational, they are looking for maximizing their total payoff. It can be easily proved that in second price auction the dominating strategy for each user *k* is to bid her own valuation. Particularly, the following proposition is in place:

Proposition 1: In a second price sealed bid auction, it is a weakly dominant strategy for every bidder to bid her own valuation. For the proposed auction, the strategy of user *k* for the subcarrier *i* is

$$b_{k,i} = \log\left(1 + \frac{h_{k,i} p_{k,i}}{N_0 + \sum_{j \neq k} h_j p_j}\right) \tag{22}$$

Proposition 1 can be proved using second price auction properties.

As discussed, user k may decide not to participate in the auction when it is not beneficiary and he desires to save entry fee. Following lemma introduces the condition that a user k decides to stay out of the auction.

Lemma 1: Consider a user k with entry fee c and monitoring fee m, this user do not participate in the auction if

$$F^{K-1}(b_{k,i}) \log\left(1 + \mathrm{SINR}\right) - \beta\left(b_{k,i}\right) - \bar{c} - m < 0 \tag{23}$$

where $\bar{c}$ is the average entry fee.

So if the lower bound for the SINR of a subcarrier satisfied by the following condition, the user k stays in the auction.

$$\mathrm{SINR} > 10^{\frac{\beta\left(\mathrm{b}_{k,i}\right) + \bar{c} + m}{F^{K-1}(b_{k,i})}} - 1 \tag{24}$$

Note that in the proposed method equal power allocation among subcarriers is considered. It can be proved that the performance degradation due to rate loss as a result of suboptimal power allocation is bounded. The optimal power allocation is multi level water filling method which assigns the optimal power allocation over different subcarriers. The suboptimal power allocation which meets the power limitation has a small sub-optimality gap from the optimal rate which can be obtained by multi level water filling. Note that in practice due to multi user diversity the sub-optimality gap would be smaller. Moreover, equal power allocation reduces the complexity and it help the design of low-rate feedback which needs limited exchange of information, particularly in the implementation of multicell systems.

Focal Distance

In OFDMA networks due to the orthogonality of subcarriers in one cell, there is no intra-cell interference. However the primary source of interference is inter-cell interference, which is limiting factor for users near the edge of the cells.

In a strict FFR scenario there are several parameters which affect the overall system performance i.e., the frequency band for inner and cell-edge users, the frequency reuse factor, and the SINR threshold which corresponds to a distance threshold to identify the edge-cell users from inner-cell users.

In the following a method to find the focal distance is proposed which maximizes the total auctioneer revenue. Note that in order to simplify the analysis in the following small scale fading is ignored and only the path loss model is considered.

The path loss may be modeled as

$$PL\left(d\right)=\frac{C}{d^{\alpha}} \tag{25}$$

where d the distance of the user from BS, C is the path loss at the reference distance $d=1m$, and α is the path loss exponent depends on the environment.

Moreover, it is assumed that users are uniformly distributed in the entire cell area of radius R. The users in each cell are classified into two sections according to their location in the cell, so

$$\Omega^{C}=\left\{k\in K\left|d_{k}\leq R_{f}\right.\right\}\text{ and }\Omega^{E}=\left\{k\in K\left|R\geq d_{k}\geq R_{f}\right.\right\}$$

are cell-center and cell-edge users set, respectively. PDF of user location from the BS is

$$s\left(d\right)=\frac{2d}{R_{f}^{2}-R_{ref}^{2}}\qquad R_{ref}\leq d\leq R_{f} \tag{26}$$

where R_{ref} is the minimum distance of a user from its BS, and R_f is the focal distance.

The objective of the auctioneer is to maximize its total revenue, which is the summation of the payments of the cell-center and cell-edge users. In the following, the total auctioneer revenue is derived.

$$\begin{aligned}&\int_{R_{ref}}^{R_f}\sum_{k=1}^{K}k\left(k-1\right)F\left(y\left(x\right)\right)^{k-2}\binom{K}{k}P_C^kP_E^{K-k}\frac{2x}{R_f^2-R_{ref}^2}y\left(x\right)\left(1-F\left(y\left(x\right)\right)\right)dx\\&+\int_{R_f}^{R}\sum_{k=1}^{K}k\left(k-1\right)F\left(y\left(x\right)\right)^{k-2}\binom{K}{k}P_E^kP_C^{K-k}\frac{2x}{R^2-R_f^2}y\left(x\right)\left(1-F\left(y\left(x\right)\right)\right)dx\end{aligned} \tag{27}$$

where

$$P_{E}=\frac{R^{2}-R_{f}^{2}}{R^{2}-R_{ref}^{2}}\text{ and }P_{C}=\frac{R_{f}^{2}-R_{ref}^{2}}{R^{2}-R_{ref}^{2}}$$

are the probability that a user is in the cell-edge and cell-center respectively. After some mathematical manipulation the total BS revenue can be written as

$$N_C K(K-1) P_C^2 \int_{R_{ref}}^{R_f} \frac{2x}{R_f^2 - R_{ref}^2} y(x)\left(1 - F(y(x))\right)\left(P_C F(y(x)) + P_E\right)^K dx$$
$$+ N_E K(K-1) P_E^2 \int_{R_f}^{R} \frac{2x}{R^2 - R_f^2} y(x)\left(1 - F(y(x))\right)\left(P_E F(y(x)) + P_C\right)^K dx \quad (28)$$

where N_C and N_E are the number of center-cell and cell-edge subcarriers, respectively. Following theorem summarizes the relations among different network parameters.

Theorem 2: For a multi-cell OFDMA network with FFR implementation, and cell radius R, and focal distance R_f, where users' bid are i.i.d variables from density $F(.)$, the following relation is true when the number of cell users and edge users are the same:

$$\frac{N_C}{N_E} = \left(\frac{\alpha\left(R^2 - R_f^2\right) + R_f^2 - R_{ref}^2}{\mu\left(R_f^2 - R_{ref}^2\right) + R^2 - R_f^2}\right)^K \times \frac{R^2 - R_f^2}{R_f^2 - R_{ref}^2} \quad (29)$$

where N_C and N_E are the number of center-cell and cell-edge subcarriers, respectively, and α and μ are constants.

Proof: Taking derivative of auctioneer total revenue with respect to R_f and using the following relation for optimum R_f, the result is straightforward.

$$\left.\frac{d\Omega_{BS}}{dR_f}\right|_{R_f = R_f^*} = 0 \quad (30)$$

where Ω_{BS} is the total auctioneer (BS) revenue. Equation 29 is the direct implication of the above condition.

In the simulation results, using Monte Carlo simulation the focal distance which maximizes the total auctioneer revenue is characterized.

PERFORMANCE EVALUATION

In this section, the performance of the proposed resource allocation scheme in a multicell OFDMA system using Matlab simulation is examined and compared with other approaches. An OFDMA system consisting of 19 cells with radius of R=500m is considered where there is a base station in each cell located in the middle of the cell and have omnidirectional antennas. Users are randomly located in each cell, and the path loss exponent is α=3.5. The noise power spectral density is -174dBm/Hz. The number of subcarriers in the system is 48. Note that for soft FFR, we applied reuse factor of 1 to cell center users and reuse factor of 3 to cell edge users.

Figure 2 compares the spectral efficiency as the number of users varies. It can be seen that the spectral efficiency of soft FFR is the worse than that of myopic and the proposed method. Note that in myopic approach, the game is played repeatedly with each player selects its best strategy until the solution converges. The reason is due to reduced interference which results in higher SINR offered by the proposed scheme. This situation is clear for the performance of the soft FFR. Note that in soft FFR cell center users may be allocated the subcarriers which interfere with the cell edge users of the adjacent cells. In the case of myopic method in which all users bid for the resources, users would be penalized by entering the auction process and losing the auction which leads to lower spectral efficiency.

It is important to evaluate the effects of the algorithm on fairness among users. A commonly used measure of fairness is Jain fair index.

$$\text{Jain Index} = \frac{(\sum_{i=1}^{n} x_i)^2}{n\sum_{i=1}^{n} x_i^2} \tag{31}$$

Figure 2. Spectral efficiency vs. number of users

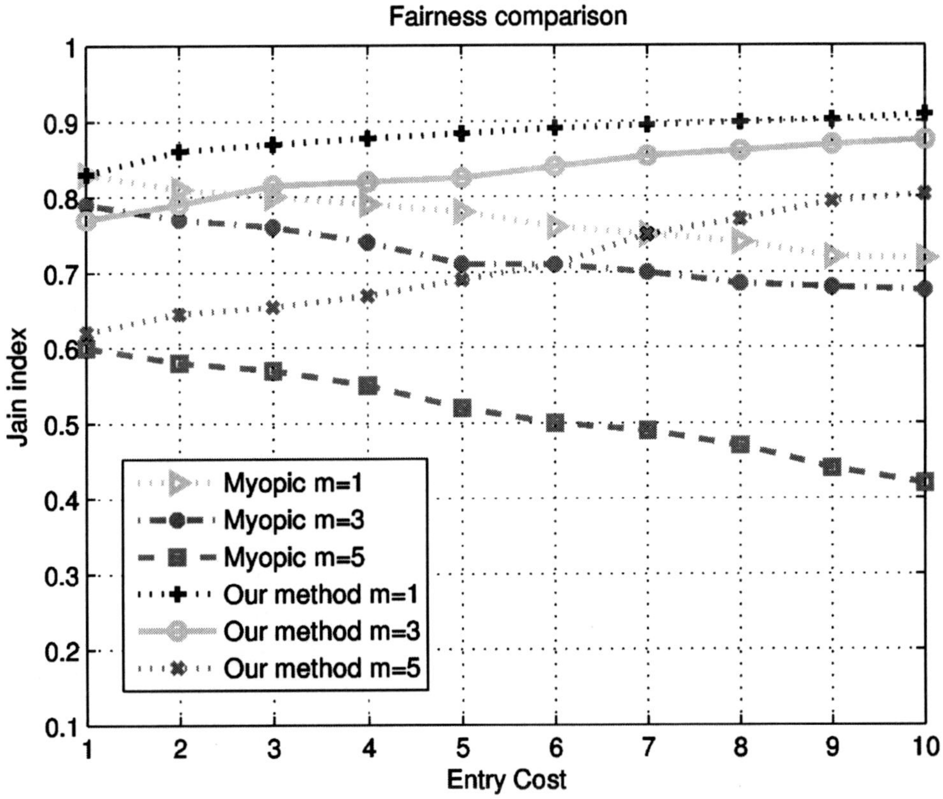

The Jain fairness index varies from 0 and 1. A rate allocation is perfectly fair if JainIndex=1. Higher Jain index indicates higher fairness among the users.

Figure 3 compares the fairness between myopic approach and the proposed method. The plots are drawn for different monitoring cost of 1, 3 and 5 when the entry cost is changing. Note that in the proposed method users may decide to stay out of the auction if the probability of winning the auction is small. This is not the case for myopic method where users always take part in auction process. It is seen that the proposed scheme results in more fair allocation of resources compared with the myopic bidding scheme. As the entry cost increases, the fairness of the myopic scheme also decreases. This is because users which experience low quality subcarriers are repeatedly penalized by losing the game and paying entry fees.

Figure 4 compares the average throughput of the cell edge users with soft FFR and the case where the system employs a reuse factor of 3. It is clear that the proposed method provides a better performance in terms of throughput for the cell edge users. The reuse factor 3 has the worst performance as it does not use the bandwidth efficiently by dividing the whole bandwidth into separated bands and then allocated each band to a cell. The soft FFR provides better throughput per bandwidth, by allowing users in center cell to use the bandwidth of adjacent cell edge users. However, its performance would be degraded by the additional interference from adjacent cells. The proposed method combines the efficient use of bandwidth while try to keep interference limited by a threshold.

Figure 5 shows the total cell throughput while the ratio of interior cell to the cell radius is changing. Monte Carlo simulation is used to keep track of the optimal ratio of the cell center area and cell adage area. The results is that the ration of 0.64 provides the best performance for out auction based scenario.

Figure 3. Fairness comparison vs. entry cost

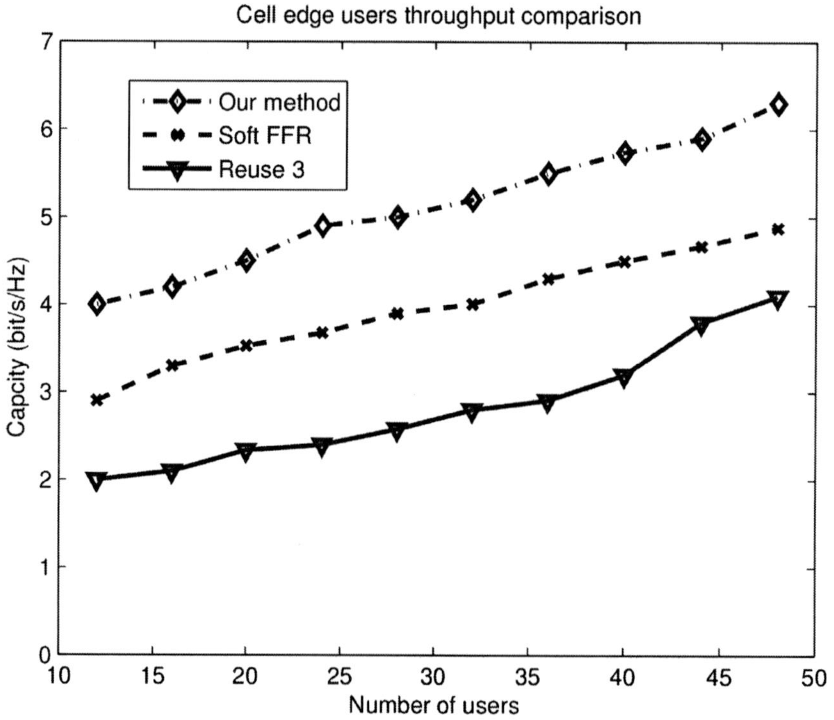

Figure 4. Cell edge users' throughput comparison

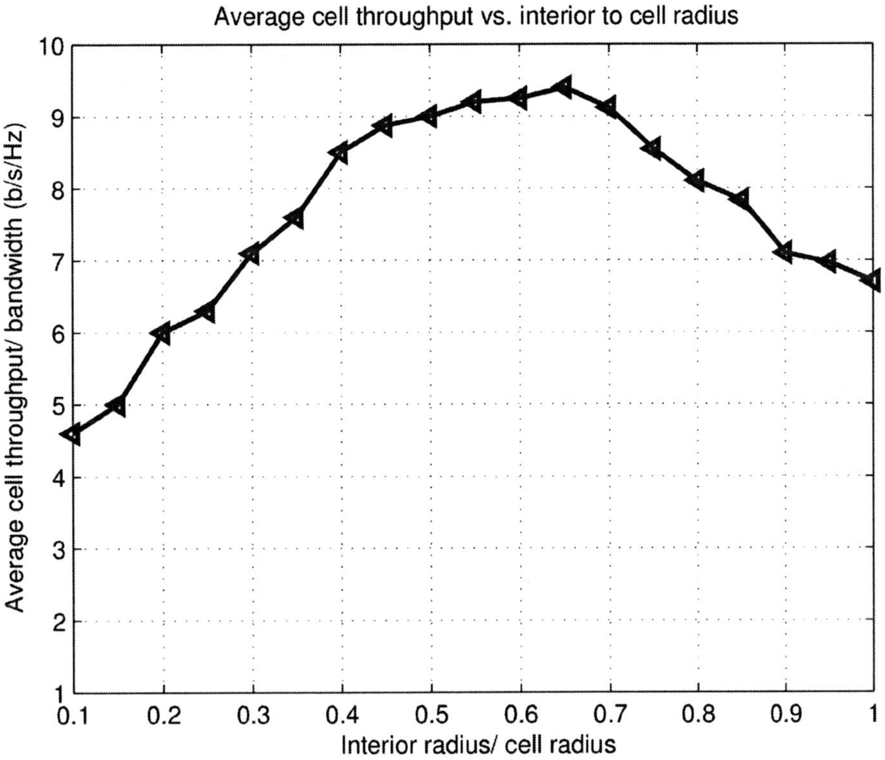

Figure 5. Average cell throughput vs. interior to cell radius

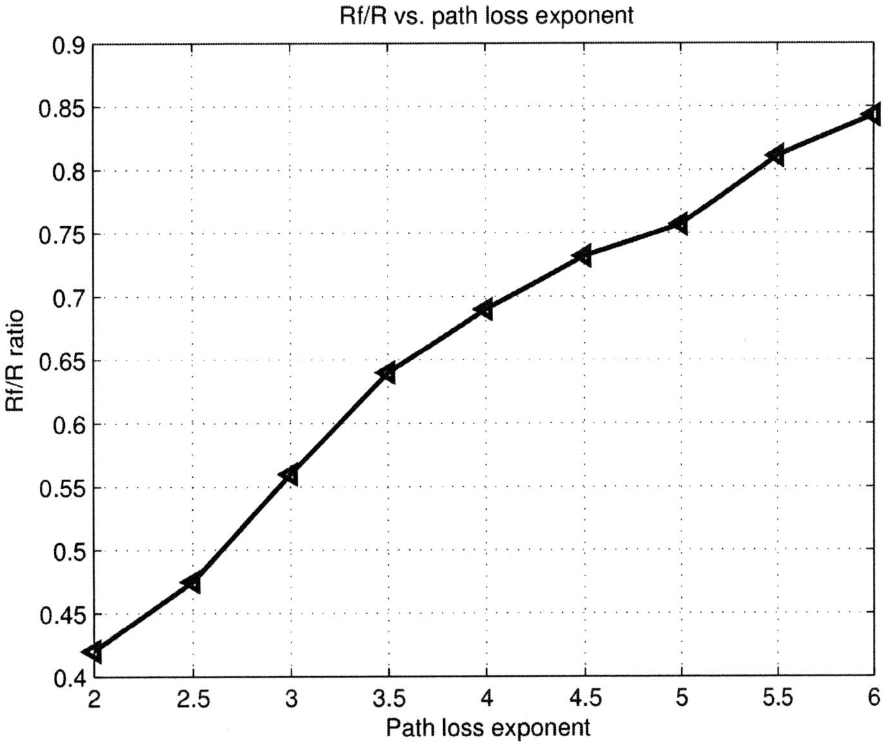

Figure 6 shows how the ratio of center-cell radius to cell radius is changing for different path loss exponent. The focal distance becomes larger when the path loss exponent is increasing. The higher path loss results in lower interference from neighboring cells. The result is that the optimum radius of inner-cell becomes larger leading to having more users in cell center radius which can use the whole center-cell frequency band.

The effect of frequency reuse is shown in Figure 7. Two frequency reuse 3, and 7 are considered for cell-edge users. For frequency reuse of 3, the whole bandwidth is divided into 4 bands, 1 for cell-center users and 3 bands for cell-edge users. For frequency reuse of 7, one frequency band is allocated to center-cell while 7 frequency band for cell-edge users. Note that the cell-edge users do not share the frequencies. The ratio of inner cell radius to cell radius is decreasing when the number of users is increasing. Since frequency reuse of 7 generates less interference than that of system with frequency reuse of 3, the optimal focal distance is larger. In this way, the system can benefit from less receiving interference from neighboring cells to increase the focal distance and take advantage of frequency re-allocation which leads to higher throughput.

Figure 8 demonstrates how the optimal focal distance changes for different number of cell users. The figure shows that for larger number of users, the system throughput is higher which a result of multi-user diversity is. The more important understanding is that for larger number of users in the system, the optimal focal distance has smaller value.

Figure 6. Focal distance vs. path loss exponent

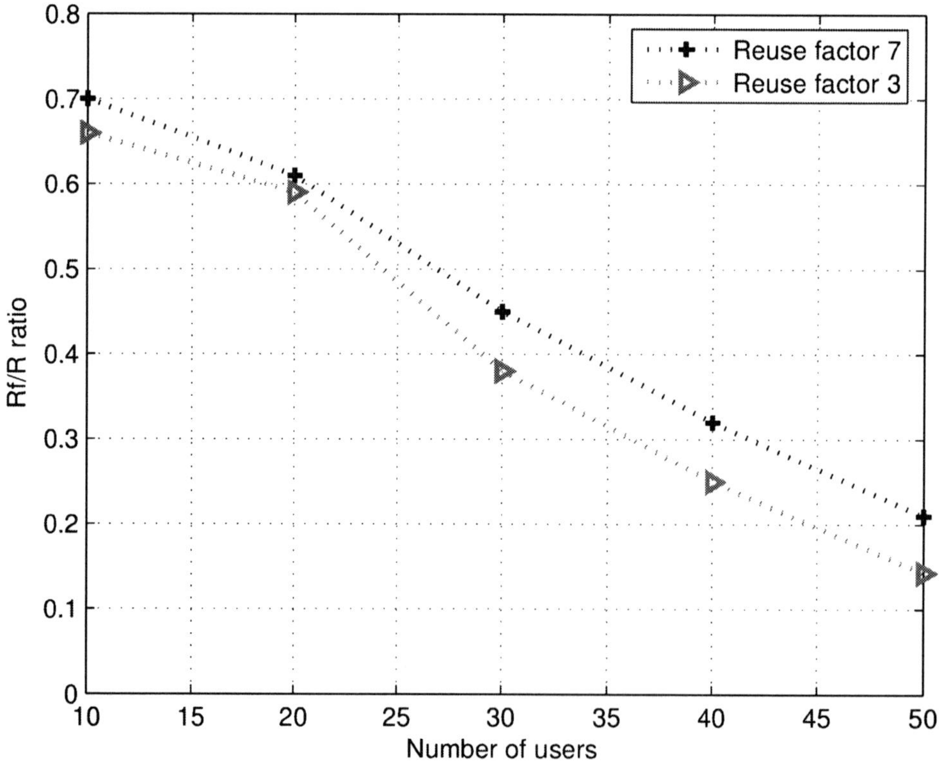

Figure 7. Focal distance for different frequency reuse

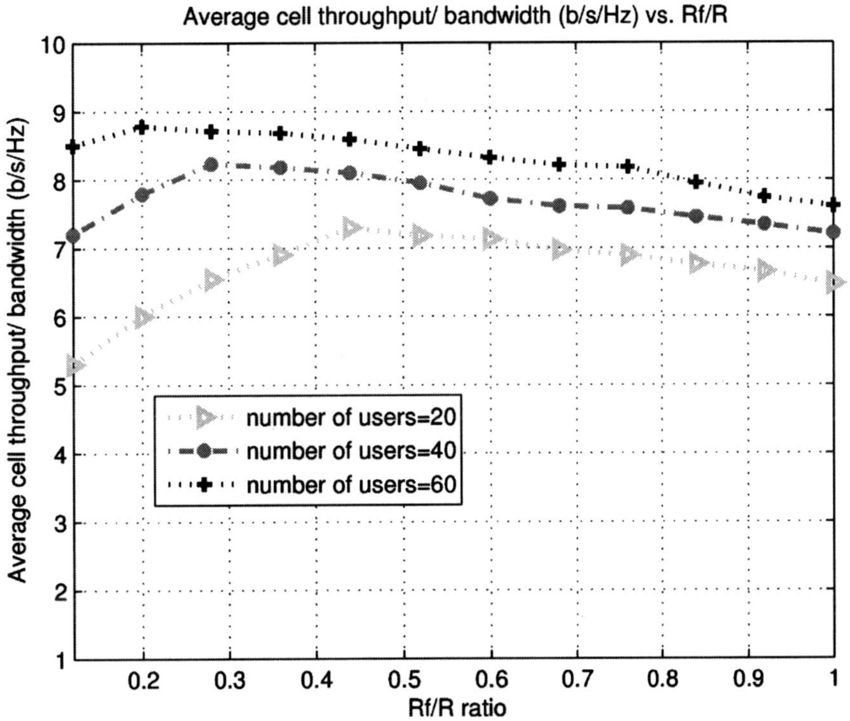

Figure 8. Focal distance changes with the number of users

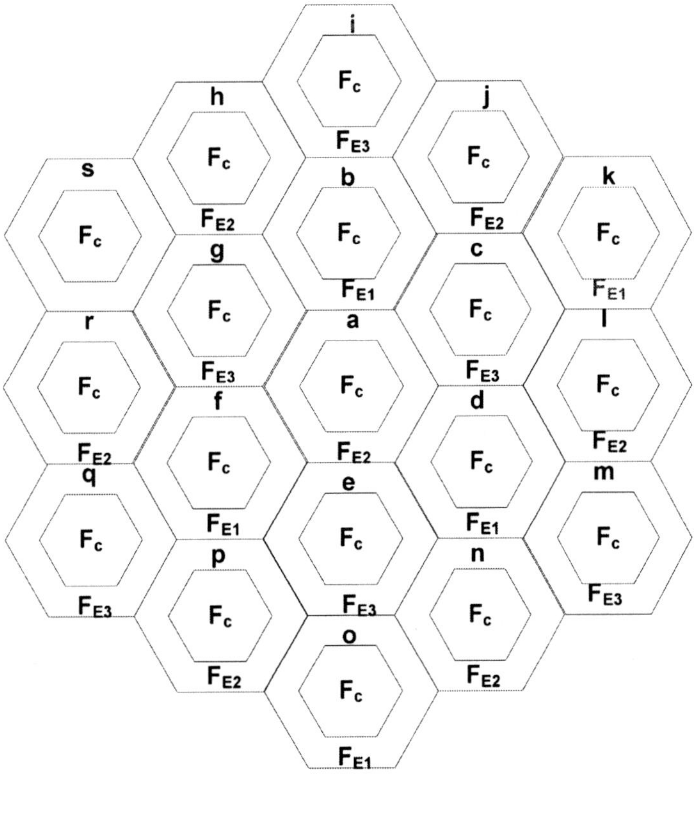

FUTURE RESEARCH DIRECTIONS

- **Admission Control for Channel-Aware Scheduling and MAC**: The resource allocation scheme at the link layer affect the available resources for admitting new call requests. Call admission control is responsible for admitting the users in the network while trying to keep the users requirements in a satisfactory level. Commercially it is desired to allow as many user as possible to access the resources simultaneously. Designing an admission control strategy that considers the efficiency, fairness, while trying to maximize the resource utilization is of great prominence. A theoretical study as well as commercial implementation of such admission control algorithm seems a potential extension of this work.
- **Characteristics of new-emerged Applications**: Resource utilization can be improved if traffic statistics are appropriately considered in the resource allocation scheme. For an example, consider commercial video applications, such as video on demand and internet protocol television (IPTV). Traffic characteristics of video, such as, importance of encoded video layers, burstiness of video content, and decoding dependency constraints of multimedia can be considered in the resource allocation scheme which leads to further improvement of resource utilization. Taking into account these parameters introduces new challenges on resource allocation problem. Nevertheless, the significant increase in video demand on wireless broadband networks asks for developing efficient resource allocation schemes, which provide the highest video quality given the limited available resources while delivering consistent and smooth service and maintaining fairness among users.
- **Multi-hop Wireless Networks**: The presented algorithms in this work are all concerned with single-hop transmission. In multi-hop networks, the co-existence of multiple links for transmission causes more complexity, because the active links in each resource allocation intervals should be determined in addition to the rate and sub-carrier allocation to the transmission on each link. This information is passed to the routing layer, which needs to determine routing paths to forward traffic from the source to the destination. In that case, the routing algorithm and the metrics used for routing decisions should capture the effects of channel assignment and transmission parameter. So the routed traffic are subject to end-to-end QoS guarantees, such as delay and throughput. Scheduling and routing resource management decisions should not be isolated. Routing can be viewed as the process of establishing efficient connectivity between nodes over multi-hop links. In communication systems with large number of nodes, there will be diverse link qualities associated with the various links along any route. Therefore, different routing schemes are expected to produce different performance results and overhead. Performing routing and scheduling jointly is known to produce superior performance result. Cross-layer protocol design that embraces the physical, MAC and routing layers is another broad research area that is open for future investigation.
- **Stochastic Learning Process**: The efficacy of the proposed auction mechanism could be improved if users be able to update their knowledge of the network based on the results of the past auctions. Particularly, a learning process, based on the previous outcomes helps users to bid efficiently so that the average profit over time increases. Each user makes decisions with the objective of increasing its long-term profit. Since each user's decision depends on the channel state information and also the on the actions of other users, a learning scheme can be used for updating the users' belief. The learning model acts in a way to enhance the users' knowledge as more data

is observed from the network. Bayesian nonparametric learning model is a candidate to model the distribution of winning bids for a given past auction outcomes. Then, each user then chooses its strategy to take part or staying in the auction according to this distribution.

CONCLUSION

A second price auction framework have been considered to model the competitive behavior of users in a multicell OFDMA network with FFR frequency planning. Users compete for the available resources to maximize their utility and minimize the total occurred cost. The utility is the throughput considering the interference from co-channel cells and the costs are the entry fee, monitoring fee and the fee upon winning the auction. The proposed algorithm provides a tradeoff between the rate improvement and cell coverage for cell-edge users. Users are sending their valuation for subcarriers to the base station, which acts as an auctioneer and makes the decisions on the allocation. Users will be informed the allocation decisions which are made by the BS. Note that users may avoid bidding for the subcarriers where they have a relatively low chance of winning the auction. In this way, they don't get penalized by paying the fees and not winning the auction. Optimal bidding strategy based on Bayesian Nash Equilibrium (BNE) is obtained. Also the users have been classified into cell-center and cell-edge users. The proposed approach aims to maximize the total auctioneer revenue, while providing a tradeoff between system performance in terms of total system throughput and quality of service provisioning for cell-edge users. Simulation results demonstrate the properties of the proposed framework which outperform the interference minimizing algorithm and provide a reasonable tradeoff between efficiency and fairness.

REFERENCES

Al-Tous, H., & Barhumi, I. (2014). (in press). Resource Allocation for Multiple-Users AF-OFDMA Systems Using the Auction Framework. *DOI.*

Ali, S. H., & Leung, V. C. M. (2009, August). Dynamic frequency allocation in fractional frequency reused OFDMA networks. *IEEE Transactions on Wireless Communications*, *8*(8), 4286–4295. doi:10.1109/TWC.2009.081146

Assad, M. (Sep. 2008). Optimal fractional frequency reuse (FFR) in multicellular OFDMA system. In *Proc. IEEE Vehicular Technology Conf.*, 1-5. doi:10.1109/VETECF.2008.381

Chang, R. Y., Tao, Z., Zhang, J., & Kuo, C. (Jun. 2009). A graph approach to dynamic fractional frequency reuse (FFR) in multi-cell OFDMA networks. In *Proc. IEEE Intl. Conf. on Communications*, Dresden, Germany, 1-6. doi:10.1109/ICC.2009.5198612

Elayoubi, S. E., Ben Haddada, O., & Fourestie, B. (2008, May). Performance evaluation of frequency planning schemes in OFDM-based networks. *IEEE Transactions on Wireless Communications*, *7*(5), 1623–1633. doi:10.1109/TWC.2008.060458.

Fang, L., & Zhang, X. (Oct. 2008). Optimal fractional frequency reuse in OFDMA based wireless networks. In *Proc. Intl. Conf. on Wireless Communications, Networking and Mobile Computing*, 1-4. doi:10.1109/WiCom.2008.166

Han, Z., Ji, Z., & Liu, K. J. R. (2007, August). Non-Cooperative Resource Competition Game by Virtual Referee in Multi-Cell OFDMA Networks. *IEEE Journal on Selected Areas in Communications*, *25*(6), 1079–1090. doi:10.1109/JSAC.2007.070803

Huang, H., Sun, Y., Li, X., Chen, S., Xiao, M., & Huang, L. (2014). (in press). Truthful Auction Mechanisms with Performance Guarantee in Secondary Spectrum. *IEEE Transactions on Mobile Computing*. doi:10.1109/TMC.2014.2343976

Hui, D. S. W., Lau, V. K. N., & Lam, W. H. (June 2006). Cross layer designs for OFDMA wireless system with heterogeneous delay requirements, In *Proc. of the IEEE International Conference on Communications*, *12*, 5325-5330.

Moretti, M., Todini, A., Baiocchi, A., & Dainelli, G. (2011, May). A Layered Architecture for Fair Resource Allocation in Multicellular Multicarrier Systems. *IEEE Transactions on Vehicular Technology*, *60*(4), 1788–1798. doi:10.1109/TVT.2011.2119501

Rahman, M., & Yanikomeroglu, H. (2010, April). A downlink dynamic interference avoidance scheme with inter-cell coordination. *IEEE Transactions on Wireless Communications*, *9*(4), 1414–1425. doi:10.1109/TWC.2010.04.090256

Scutari, G., Palomar, D. P., & Barbarossa, S. (2008, July). Asynchronous iterative water-filling for Gaussian frequency-selective interference channels. *IEEE Transactions on Information Theory*, *54*(7), 2868–2878. doi:10.1109/TIT.2008.924723

Son, K., Chong, S., & de Veciana, G. (2009, July). Dynamic association for load balancing and interference avoidance in multi-cell networks. *IEEE Transactions on Wireless Communications*, *8*(7), 3566–3576. doi:10.1109/TWC.2009.071140

Stiakogiannakis, & Kaklamani. (Jul. 2011). A Combinatorial Auction Based Subcarrier Allocation Algorithm for Multiuser OFDMA. *IEEE VTC Spring*, 1-5.

Stolyar & Viswanathan. (2009). Self-organizing dynamic fractional frequency reuse for best-effort traffic through distributed inter-cell coordination. In *Proceedings of IEEE INFOCOM* (pp. 1287-1295). IEEE.

Sun, G., Feng, X., Tian, X., Gan, X., Xu, Y., Wang, X., & Guizani, M. (2014, June). Coalitional Double Auction for Spatial Spectrum Allocation in Cognitive Radio Networks. *IEEE Transactions on Wireless Communications*, *13*(6), 3196–3206. doi:10.1109/TWC.2013.042114.130952

KEY TERMS AND DEFINITIONS

Auction Theory: A branch of game theory that deals with assigning the resources to bidders based on defined bidding framework.

Equilibrium: An outcome of the auction game where no player has an incentive to deviate from his or her chosen strategy.

Multi-Cell Network: Compared to single cell scenario the multi-cell models the practical modeling of cellular network considering the interference from neighbor cells.

OFDMA: Orthogonal Frequency-Division Multiple Access is a multi-user OFDM scheme. Multiple access is achieved by assigning subsets of subcarriers to individual users.

Quality of Service: A defined measure of performance in a communication network.

Resource Management: The process of assigning the resources in OFDMA network including frequency, power and time.

Subcarrier: Each of the resource units in an OFDMA network in frequency domain.

Section 3

Advanced Game Theory for Promising Wireless Techniques and Services

Chapter 12
Applications of Game Theory for Physical Layer Security

Kaizhi Huang
National Digital Switching System Engineering and Technological Research Center, China

Zhou Zhong
National Digital Switching System Engineering and Technological Research Center, China

Mingliang Li
National Digital Switching System Engineering and Technological Research Center, China

Hua Zhao
National Digital Switching System Engineering and Technological Research Center, China

ABSTRACT

This chapter provides a comprehensive review of the domain of game theory based physical layer security in wireless communications. By exploiting the wireless channel characteristic and secure cooperation of nodes, physical layer security is to enable the exchange of confidential messages over a wireless medium in the presence of unauthorized eavesdroppers, without relying on higher-layer encryption. However, the selfness of nodes seriously affects the secure cooperation; game theory can model the influence of the selfness on physical layer security. This chapter firstly describes the physical layer security issues in the wireless networks and the role of game theory in the research on physical layer security. And then the typical applications of game theory in physical layer security are subsequently covered, including zero-sum game, Stackelberg game, auction theory, coalition game. Finally, the chapter concludes with observations on potential research directions in this area.

INTRODUCTION

The broadcast and superposition nature are the two fundamental characteristics of the wireless medium, which extend the communication range and improve the performance of wireless system respectively, but make the wireless signal easier to be eavesdropped and jammed (Kashyap, Basar, & Srikant, 2004). Therefore the eavesdropping and jamming become the hidden dangers of the wireless communication. At present the high-layer enciphering techniques is also the main methods to solve these problems (Massey, 1988). However, in dynamic wireless networks this raises some issues such as key distribution for symmetric cryptosystems, and high computational complexity of asymmetric cryptosystems (Schneier, 1998).

DOI: 10.4018/978-1-4666-8642-7.ch012

In the wireless communication, different communication links have different channel characteristics, which is different from the wire communications. Inspired by it, in 1975, Wyner built the wiretap channel model considering the channel imperfections and theoretically proved the feasibility of physical layer security (Wyner, 1975). At present, the physical-layer security methods can be divided into internal security methods and external security methods. Internal security methods guarantee secure communication by exploiting the difference of the wireless channel and background noise observed by the different communication nodes (Leung-Yan-Cheong & Hellman, 1978). External security methods guarantee secure communication by using target-specified artificial jamming to selectively jam the eavesdropper (Negi & Goel, 2005).

But the traditional physical-layer security methods also have some limitations, for example, in a multi-node wireless network, there is usually one optimization target with the single policymaker responsible for the communication strategy design of all the nodes. However, the nodes operate under its own steam and the communication state of the nodes also dynamically change in the real network. According to the other nodes' communication strategies, each node can choose its strategy for maximizing its interest. So every node has the different but related optimization target. Obviously, the traditional physical layer security methods can not roundly model the real communication scenario and measure the dynamic secrecy performance of the wireless network.

Game theory (Osborne & Rubinstein, 1994) is a formal framework with a set of mathematical tools to study the complex interactions among interdependent rational players, which can model the confrontation behavior and selfishness of the nodes in the scenarios with secrecy requirement. Further, the secrecy communication strategy of every node can be selected based on the prediction of the game result. This chapter uses game theory to analyze the dynamic secrecy performance of the wireless network.

Background

As shown in Figure 1, there are N legitimate nodes (legitimate transmitter, friendly relay, and friendly jammer, legitimate receivers), and L malicious nodes (untrusted relay, malicious jammer and eavesdropper) in the network. We denote jammer and relay by helper. A non-cooperative game in strategic (or normal) form can be denoted by $\mathbf{G}=\{\mathbf{N},\mathbf{S},\mathbf{U}\}$, where $\lambda(0)$ denotes a finite set of players including the set A of legitimate nodes, and the set E of malicious nodes. $\mathbf{S}=\{\mathbf{S}^A,\mathbf{S}^E\}$ denotes the set of available strategies for all the players, where $\mathbf{S}^A$ is the set of strategies for legitimate nodes including sending useful signal, forwarding signal, sending friendly jamming, sending artificial noise and so on, $\mathbf{S}^E$ is the set of strategies for malicious nodes including eavesdropping and jamming. $\mathbf{U}=\{u_i(s)\}_{i\in\mathbf{N}}$ is the set of utility(payoff) function for all the players. For any player i, $\mathbf{s}=\{s_i,\mathbf{s}_{-i}\}$is referred to a strategy profile, where s_i is the strategy of player i, $\mathbf{s}_{-i}$ denotes the vector of strategies of all players except player i.

Although the utility function for the different players varies with the secrecy communication scenario, in general, the sum of payoff for legitimate nodes denoted by $\sum_{i\in\mathbf{A}} u_i$ reflects their tendency to improve the secrecy performance (e.g. secrecy rate outage probability, secrecy freedom degree), the sum $\sum_{i\in\mathbf{E}} u_i$ of payoff for malicious nodes reflects their tendency to break the secrecy performance.

According to $\mathbf{s}_{-i}$, the player i chooses the strategy maximizing its payoff, the optimal strategy for player i is $s_i^* = \max_{s_i\in\mathbf{S}^A} u_i(s_i,\mathbf{s}_{-i}), i\in\mathbf{A}$, $s_i^* = \max_{s_i\in\mathbf{S}^E} u_i(s_i,\mathbf{s}_{-i}), i\in\mathbf{E}$. The strategy profile $\mathbf{s}^*$ is Nash equilib-

Figure 1. Secure wireless network model

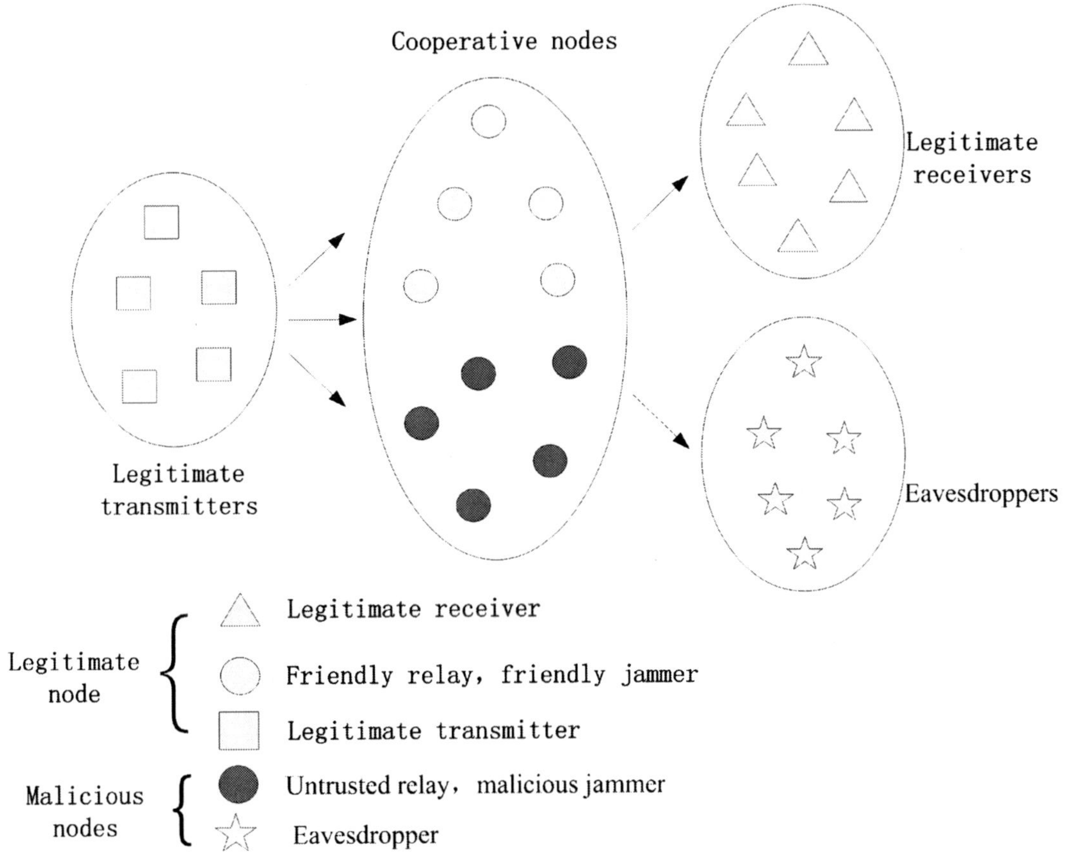

rium if $u_i(s_i^*, s_{-i}^*) \geq u_i(s_i, s_{-i}^*)$, $\forall i \in \mathbf{N}$. The secrecy performance of the network is analyzable only if the network achieves the equilibrium state. Therefore, the core of using game theory to solve the physical layer security issues can be summarized as whether the equilibrium state exists in the network, whether the equilibrium is unique, and how the network achieve the existing equilibrium state.

The aforementioned generalized secure game model can not solve all the practical issues. We should build the specific game model for the practical issue. As shown in Table 1, according to whether the players can conclude a binding bargain, the game theory can be divided into the cooperative games and the non-cooperative games. The applications of the zero-sum game, the Stackelberg game (Ba & Srikant, 2002) and the auction (Krishna, 2009) are the hot areas of the research on the physical layer security based on the non-cooperative games. The zero-sum game can model the interaction of the hostile players, which can be used to analyze the best defense strategy coping with the jammer and eavesdropper. The Stackelberg game is a seller-buyer game, which can be used to analyze the centralized secure resource allocation strategy of the friendly helper in the case of only one legitimate transmitter. An auction is a distributed market mechanism for allocating resources, which can be used to decentralized secure resource allocation strategy of the friendly helper in the present of multiple legitimate transmitters. The application of the coalition game (Ray, 2007) is the hot area of the research on the physical layer security based on the cooperative games, which can be used to analyze the cooperation among the legitimate nodes and the cooperation among the malicious nodes how to affect the secrecy performance of the network.

Table 1. The application of game theory for secure wireless network

Game Models		Players	Main Results
Non-cooperative games	Zero-sum game	Legitimate node, malicious node	Best defense strategy of the legitimate nodes
	Stackelberg game	Legitimate nodes	Centralized secure resource allocation strategy
	Auction	Legitimate nodes	Distributed secure resource allocation strategy
Cooperative games	Coalition game	Malicious nodes	Coalition formation algorithm for the malicious nodes
		Legitimate nodes	Coalition formation algorithm for the legitimate nodes

Instead of proceeding in a strictly chronological order, the chapter aim to introduce the typical research of the field along with the most pertinent references. The structure of the chapter is present as follows. Firstly, non-cooperation games is used to analyze the influence of nodes' selfishness and confrontation behavior on secrecy performance of the network. Secondly, coalition game is used to analyze the system security when the communication nodes cooperate with each other. Finally, the future direction is concluded.

NON-COOPERATIVE GAMES BASED PHYSICAL LAYER SECURITY

This section mainly refers to using non-cooperation games to analyze the influence of nodes' selfishness and confrontation behavior on secrecy performance of the network. At present, non-cooperation games based physical layer security mainly focuses on using the zero-sum game to analyze the optimal communication strategy for legitimate nodes when eavesdropping and jamming exist, using the Stackelberg game and auction to obtain the optimal secure resource allocation strategy of legitimate nodes.

Zero-Sum Game Based Physical Layer Security

In a secure communication network consisting of an active eavesdropper, the legitimate and malicious nodes are natural competitors with opposite goals, constituting a zero-sum game. Mukherjee and Swindlehurst (2013) firstly models the legitimate transmitter and the eavesdropper as a zero-sum game with the secrecy rate as the payoff function. Then, the optimal strategy of the legitimate transmitter is obtained based on the Nash equilibrium.

As shown in Figure 2, consider a three-node complex Gaussian system model consisting of one legitimate transmitter, one legitimate receiver and one eavesdropper. Assume that they are equipped with N_A, N_B, and N_E antennas respectively. The half-duplex eavesdropper acts either as a passive eavesdropper or as a jammer, while the legitimate transmitter can transmit the signal with full power or transmit the signal and artificial noise simultaneously.

When the eavesdropper only eavesdrops, the received signals at the legitimate receiver and the eavesdropper can be expressed as

$$\mathbf{y}_B = \sqrt{\beta_{AB}}\mathbf{h}_{AB}\mathbf{x}_A + \mathbf{n}_B \tag{1}$$

Figure 2. Complex gaussian system model A: legitimate transmitter, B: legitimate receiver, E: eavesdropper

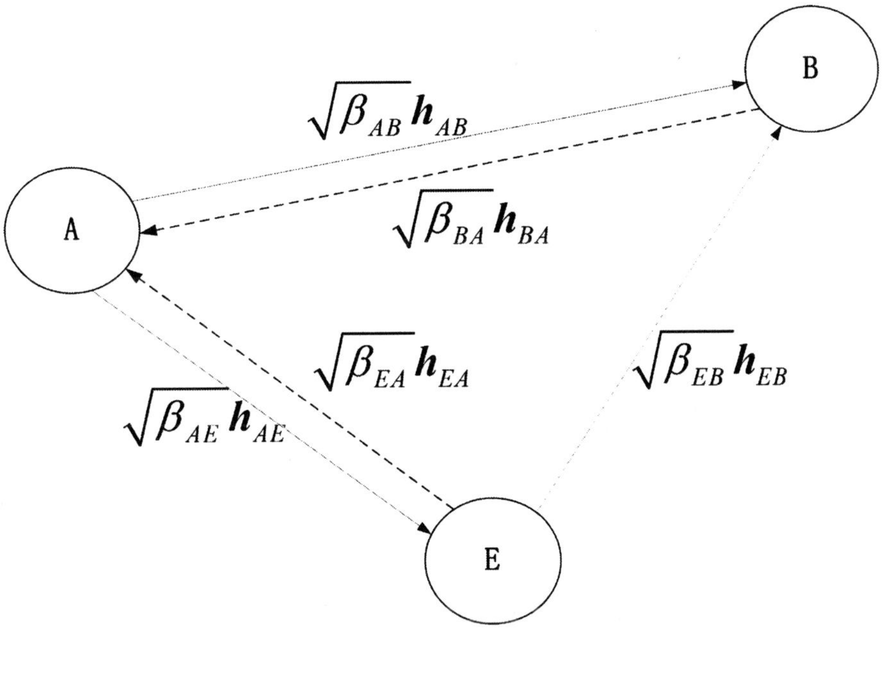

$$\mathbf{y}_E = \sqrt{\beta_{AE}}\mathbf{h}_{AE}\mathbf{x}_A + \mathbf{n}_E \quad (2)$$

When the eavesdropper acts as an active jammer, the received signal at the legitimate receiver is

$$\mathbf{y}_B = \sqrt{\beta_{AB}}\mathbf{h}_{AB}\mathbf{x}_A + \sqrt{\beta_{EB}}\mathbf{h}_{EB}\mathbf{x}_E + \mathbf{n}_B \quad (3)$$

where $\sqrt{\beta_{kl}}\mathbf{h}_{kl}$, $k=A,B,E$, $l=A,B,E$, $k\neq l$ models the CSI from node k to node l. $\sqrt{\beta_{kl}}$ denotes the large scale path loss coefficient. $\mathbf{h}_{kl}$ is the small-scale fading matrix, whose elements are independent and identically distributed zero-mean complex Gaussian random variables with unit variance. $\mathbf{n}_B$ and $\mathbf{n}_E$ represent the zero-mean complex Gaussian noise at the legitimate receiver and the eavesdropper respectively with variance σ_B^2 and σ_E^2. $\mathbf{x}_A$ and $\mathbf{x}_E$ are the transmission signals from the legitimate transmitter and the eavesdropper. When the legitimate transmitter sends both signal and artificial noise to increase the system security, the precoding signal and noise $\mathbf{x}_A$ is given by

$$\mathbf{x}_A = \mathbf{Tz} + \mathbf{T}'\mathbf{z}' \quad (4)$$

where $\mathbf{T}$ is $N_A \times d$ precoding matrix of the signal $\mathbf{z}$, $\mathbf{T}$ is $N_A \times (N_A-d)$ precoding matrix of the noise $\mathbf{z}'$. In order to make the artificial noise only jam the eavesdropper, the precoding matrix $\mathbf{T}$ and $\mathbf{T}'$ are chosen as disjoint sets of the right singular vectors of $\mathbf{h}_{A,B}$ (Goel & Negi, 2008). P_a is the transmission power of the legitimate transmitter. ρP_a is the transmission power of the signal and $(1-\rho)P_a$ is the transmission power of the noise. Then, the secrecy rate under eavesdropping and jamming are respectively

$$R_{iE} = \left[\log_2\left|\mathbf{I} + \rho P_a \beta_{AB} \mathbf{h}_{AB} \mathbf{T}\mathbf{T}^H \mathbf{h}_{AB}^H / (d * \sigma_B^2)\right| - \log_2\left|\mathbf{I} + \rho P_a \beta_{AE} \mathbf{h}_{AE} \mathbf{T}\mathbf{T}^H \mathbf{h}_{AE}^H \mathbf{Q}_E^{-1} / d\right|\right]^+ \tag{5}$$

$$R_{iJ} = \log_2\left|\mathbf{I} + \rho P_a \beta_{AB} \mathbf{h}_{AB} \mathbf{T}\mathbf{T}^H \mathbf{h}_{AB}^H \mathbf{Q}_B^{-1} / (d * \sigma_B^2)\right| \tag{6}$$

where $\mathbf{Q}_E = (1-\rho) P_a \beta_{AE} \mathbf{h}_{AE} \mathbf{T}'\mathbf{T}'^H \mathbf{h}_{AE}^H / (N_A - d)$. $\mathbf{Q}_B = \rho P_e \beta_{EB} \mathbf{h}_{EB} \mathbf{h}_{EB}^H + \sigma_B^2 \mathbf{I}$, P_e is the jamming power of the eavesdropper. $(\cdot)^+ \triangleq \max(\cdot, 0)$. $i \in (F,A)$ denotes two strategies of the legitimate transmitter.

According to the above system model, the strategies of the legitimate transmitter are transmitting the signal with full power or transmitting the signal and the artificial noise simultaneously. The strategies of the eavesdropper are passive eavesdropping or jamming. Based on the property that the sum of the players' payoffs is zero, we can obtain the payoff of the legitimate transmitter is the secrecy rate and the payoff of the eavesdropper is its negation. Then the payoff matrix is shown as Table 2.

Pure Strategy Nash Equilibrium

The pure strategy Nash equilibrium refers that the players select their optimal strategies with a probability of 1 simultaneously. According to the general properties of the wiretap channel, when the eavesdropper eavesdrops, the secrecy performance of transmitting the signal and the artificial noise simultaneously is better than that of transmitting the signal with full power. However, the conclusion is opposite when the eavesdropper jams, i.e. $R_{AE} \geq R_{FE}$ and $R_{FJ} \geq R_{AJ}$.

If the secrecy rates in the payoff matrix satisfy the condition of pure strategy Nash equilibrium $R_{AE} \leq R_{AJ}$, we will get $R_{FJ} \geq R_{AJ} \geq R_{AE} \geq R_{FE}$. In this case, R_{AE} represents an Nash equilibrium since neither the legitimate transmitter nor the eavesdropper can improve their respective payoffs by unilaterally switching strategies, i.e., the secrecy rate will decrease if the legitimate transmitter chooses to transmit the information signal with full power, and the secrecy rate will increase if the eavesdropper decides to jam. Similarly, when $R_{FJ} \leq R_{FE}$, we will obtain $R_{AE} \geq R_{FE} \geq R_{FJ} \geq R_{AJ}$, and R_{FE} will be the mutual best response for both players. Based on the above analysis, the payoff of the pure strategy Nash equilibrium is given by

$$R(x^*, y^*) = \begin{cases} R_{AE}, R_{AE} \leq R_{AJ} \\ R_{FJ}, R_{FJ} \leq R_{FE} \end{cases} \tag{7}$$

Table 2. Payoff matrix $\mathbf{R}$

		Eavesdropper	
		Eavesdropping	**Jamming**
Legitimate Transmitter	full power	$(R_{FE}, -R_{FE})$	$(R_{FJ}, -R_{FJ})$
	artificial noise	$(R_{AE}, -R_{AE})$	$(R_{AJ}, -R_{AJ})$

Mixed Strategy Nash Equilibrium

When the secrecy rates do not satisfy the conditions of the pure strategy equilibrium $R_{AE} \leq R_{AJ}$ and $R_{FJ} \leq R_{FE}$, according to the minimax theorem that any finite zero-sum game has a saddle-point in randomized strategies (Myerson, 1997), the legitimate transmitter and the eavesdropper should adopt the mixed strategies. The mixed strategy equilibrium refers that the players maximize their individual payoffs with a certain probability of their selected strategies. Let $\mathbf{p}=(p,1-p)$ and $\mathbf{q}=(q,1-q)$ represent the probabilities with which the legitimate transmitter and the eavesdropper randomize over their strategy sets, respectively. Then, the legitimate transmitter obtains the optimal strategy by solving

$$\max_{\mathbf{p}} \min_{\mathbf{q}} \mathbf{p}^T \mathbf{R} \mathbf{q} \tag{8}$$

Then, the optimal mixed strategies and unique Nash equilibrium value of the game can be easily derived as

$$\begin{aligned} (p^*, 1-p^*) &= \frac{(R_{AJ} - R_{AE}, R_{FE} - R_{FJ})}{D} \\ (q^*, 1-q^*) &= \frac{(R_{AJ} - R_{FJ}, R_{FE} - R_{AE})}{D} \\ R(p^*, q^*) &= \frac{(R_{FE} R_{AJ} - R_{FJ} R_{AE})}{D} \end{aligned} \tag{9}$$

where $D = R_{FE} + R_{AJ} - R_{FJ} - R_{AE}$. p^* and q^* are the optimal probability at the legitimate transmitter and the eavesdropper. $R(p^*,q^*)$ is the secrecy rate of the mixed strategy Nash equilibrium (Osborne & Rubinstein, 1994).

For the strategic game, we set $P_e=4P_a$, $N_A=N_E=8$, $N_B=6$, $d=4$. Before the cross-over in Figure 3, a pure strategy Nash equilibrium in R_{AE} is obtained, however after the cross-over, the mixed strategies are optimal for both players. In this case, randomizing strategies clearly leads to better payoffs for the players as the jamming power increases, compared to adopting a pure strategy. The optimal mixing probabilities are shown in Figure 4 with a clear division between pure and mixed strategy Nash equilibrium regions. When the transmission power is low, the players choose their optimal strategy with a probability of 1, indicating the existence of pure strategy Nash equilibrium. When the transmission power increases, the players choose the optimal strategy probability between 0 and 1.

Stackelberg Game Based Physical Layer Security

Stackelberg game is used to model the interaction between the buyer and the seller. The sellers maximize their own profits by set proper prices, while the buyers choose a proper number of the goods based on the prices. It can effectively solve the problem of selfishness and the competition problem with master-slave relationship. Wang et al. (2013) analyzes the secure power allocation issue of a friendly jammer based on the Stackelberg game. In the game, the friendly jammer plays as the seller and the legitimate transmitters are the buyers. Then, the Stackelberg equilibrium is analyzed and a centralized power allocation algorithm is proposed.

Figure 3. The secrecy rate versus the transmission power of the legitimate transmitter

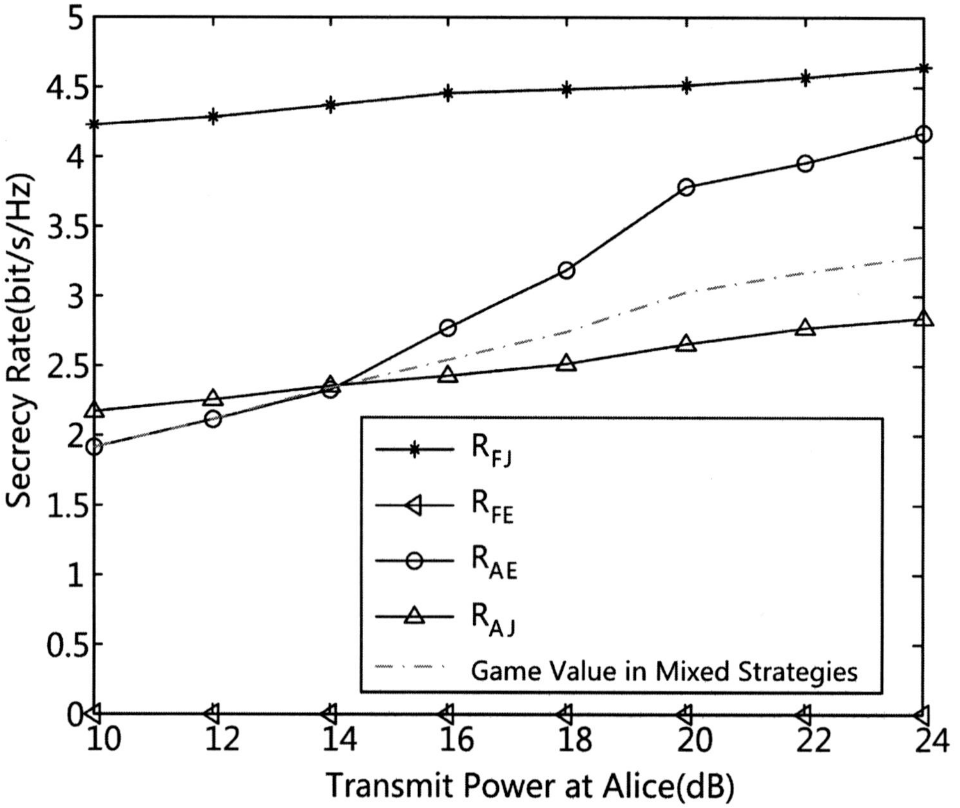

We consider a multiuser OFDMA communication network where there are N legitimate transmitters S_i and legitimate receivers D_i, an eavesdropper E and a friendly jammer J, as shown in Figure 5. The friendly jammer jams the orthogonal channels with artificial interference to help legitimate nodes. However, the jamming service is not free and the friendly jammer charges for some cost. The secrecy rates from S_i to D_i and E can be written as

$$I_{sd}^{i} = \log_2(1 + p_s \gamma_{1,i}) \tag{10}$$

$$I_{se}^{i} = \log_2(1 + \frac{p_s \gamma_{2,i}}{1 + p_i \gamma_{j,i}}) \tag{11}$$

where $\gamma_{1,i} = \left|h_{SD}^{i}\right|^2 / \sigma^2$ is the signal-to-noise ratio (SNR) of the S_i at the destination D_i. $\gamma_{2,i} = \left|h_{SE}^{i}\right|^2 / \sigma^2$ is SNR of the transmitter S_i at the destination E. $\gamma_{j,i} = \left|h_{JE}\right|^2 / \sigma^2$ is the SNR of the jammer$J$ at the des-

Figure 4. The optimal mixing probabilities versus the transmission power of the legitimate transmitter

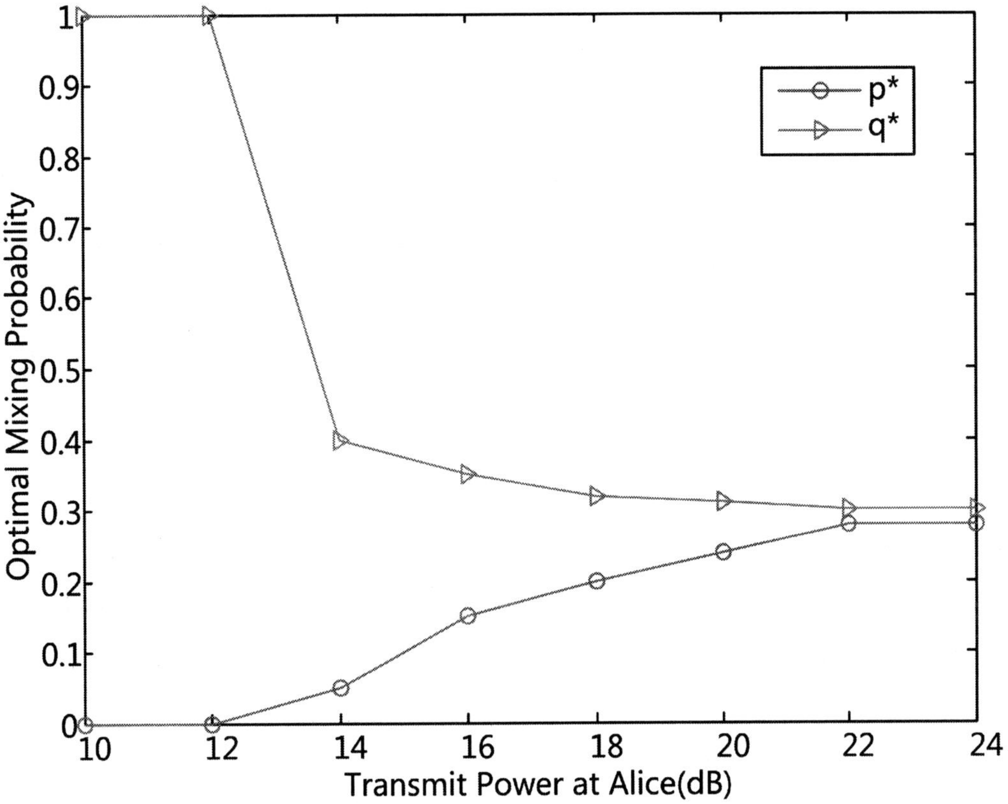

tination E. h_{SD}^{i}, h_{JE} and h_{SE}^{i} are the fading channel coefficients. σ^2 is the noise variance. p_s represents the transmission power of the legitimate transmitter, p_i is jamming power from the friendly jammer with a total power of p_J. In consequence, the secrecy rate of the legitimate transmitter S_i and the system can be expressed as

$$R_i^S = \left[I_{sd}^i - I_{se}^i\right]^+ = \left[\log_2(1 + p_s\gamma_{1,i}) - \log_2(1 + \frac{p_s\gamma_{2,i}}{1 + p_i\gamma_{j,i}})\right]^+ \tag{12}$$

$$R_S = \sum_{i\in N} R_i^S \tag{13}$$

The legitimate transmitter S_i aims to acquire the most benefits at the least possible cost. As a result, the Stackelberg secrecy game of the buyers' side can be formulated as

Figure 5. The system model

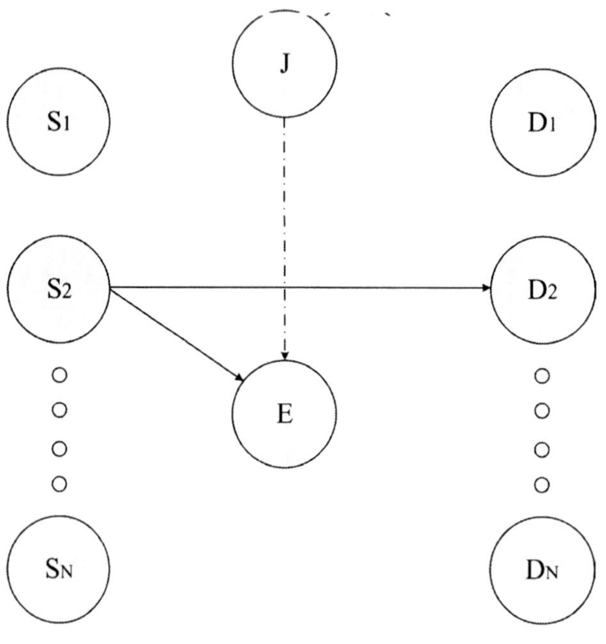

$$\max_{\{p_i\}} U_i = R_i^S - \lambda_i p_i \quad s.t.\ 0 \le \{p_i\}_{i\in N} \le p_J \tag{14}$$

where $\lambda_i p_i$ is the cost paid to the friendly jammer and λ_i is the price of unit power which the friendly jammer sets for the legitimate transmitter S_i. The friendly jammer charges for the jamming service and tries to gain as much profit as possible. So we can define the Stackelberg secrecy game of the friendly jammer's side as

$$U_J = \sum_{i\in N} \lambda_i p_i^{c_i} \tag{15}$$

where $c_i \ge 1$ is a constant to balance the payment and the transmission power p_i.

Then, we give the definition of the Stackelberg Equilibrium in a mathematical way. If λ_i^{SE} and p_i^{SE} are the Stackelberg Equilibrium of the proposed game, the following conditions should be satisfied:when λ_i^{SE} is fixed,

$$U_i(\{p_i^{SE}\}) = \sup_{0\le p_i \le p_J} U_i(\{p_i\}),\ \forall i \in N \tag{16}$$

when p_i^{SE} is fixed,

$$U_J(\{\lambda_i^{SE}\}) = \sup_{\lambda_i} U_J(\{\lambda_i\}),\ \forall i \in N \tag{17}$$

When λ_i^{SE} is fixed, from the first order derivative and second order derivative of U_i with respect to p_i, we obtain that the payoff function of the seller is a convex function. Let the first order derivative be 0 and the optimal power is

$$p_i^* = -u_i + \sqrt{w + v_i / \lambda_i} \tag{18}$$

where $u_i = \dfrac{1 + p_s\gamma_{2,i}}{2\gamma_{j,i}}$, $w = p_s^2 / 4$, $v_i = \dfrac{p_s\gamma_{2,i}}{\ln 2\gamma_{j,i}}$.

The friendly jammer's goal is to maximize his profit through setting optimal unit price for each legitimate transmitter. By differentiating the friendly jammer's utility function and setting it to zero, we have

$$\frac{\partial U_J}{\partial \lambda_i} = (p_i^*)^{c_i} + \lambda_i c_i (p_i^*)^{c_i - 1} \frac{\partial p_i^*}{\partial \lambda_i} = 0 \tag{19}$$

From the closed form expression of p_i^*, we can get the solution of the optimal price of unit power

$$\lambda_i^* = \lambda_i^*(u_i, w, v_i, c_i) \tag{20}$$

Due to the existence of the Stackelberg equilibrium, there are optimal prices of unit power and optimal power (λ_i^*, p_i^*), satisfying Equations 18 and 19. In order to reduce the computation complexity, this section proposes a centralized algorithm to achieve the optimal value.

From Equation 19, we will get

$$\lambda_i = -\frac{p_i^*}{c_i \partial p_i^* / \partial \lambda_i} \tag{21}$$

Equations 18 and 21 are the updating function of the centralized algorithm. We can write the updating function in a vector form as

$$\begin{aligned} \boldsymbol{\lambda}(t+1) &= -\frac{\mathbf{p}(t)}{\mathbf{c}\partial \mathbf{p}(t) / \partial \boldsymbol{\lambda}(t)} \\ \mathbf{p}(t+1) &= -\mathbf{u} + \sqrt{\mathbf{w} + \mathbf{v} / \boldsymbol{\lambda}(t+1)} \end{aligned} \tag{22}$$

where $\lambda = [\lambda_1, \ldots, \lambda_N]^T$ and $\mathbf{p} = [p_1, \ldots, p_N]^T$ denote the vectors of the optimal prices of unit power and the optimal powers. So the proposed Stackelberg game based algorithm can be described in Table 3.

Table 3. A centralized power allocation algorithm based on Stackelberg game

A CENTRALIZED POWER ALLOCATION ALGORITHM
Step1: Initialization: For the friendly jammer, initialize the price vector $\boldsymbol{\lambda}(0) = [\lambda_1^0, ..., \lambda_N^0]^T$ and broadcast it to all the legitimate transmitters; For each legitimate transmitter, initialize its optimal power $\mathbf{p}(0) = [p_1^0, ..., p_N^0]^T$ .The iteration index is t=1. **Step2:** transmitters' update: In each iteration, we obtain the optimal power for each legitimate transmitter $\mathbf{p}(t+1) = [p_1^*(t+1), ..., p_N^*(t+1)]$. **Step3:** Friendly jammer's update: Use **p**(t+1) obtained in Step2 to calculate the optimal price λ_i^* according to $\boldsymbol{\lambda}(t+1) = [\lambda_1^*(t+1), ..., \lambda_N^*(t+1)]^T$. Then, the iteration index changes from t to t+1. **Step4:** Repeat **Step2** and **Step3** until both **p** and $\boldsymbol{\lambda}$ do not change any longer. And the final Stackelberg Equilibrium is reached.

As shown in Figure 6, it is obvious that the optimal jamming power is monotonically decreasing with respect to the price of unit power. This conclusion also verifies the analysis above: the higher the price is, the less power the legitimate transmitter buys. If the perfect secrecy rate cannot be obtained, the transmitter-receiver pairs will buy nothing. Figure 7 shows that how the optimal power of each pair

Figure 6. The optimal power one transimitter buys versus the price of unit power (Pair represents an orthogonal link between the legitimate transmitter and receiver)

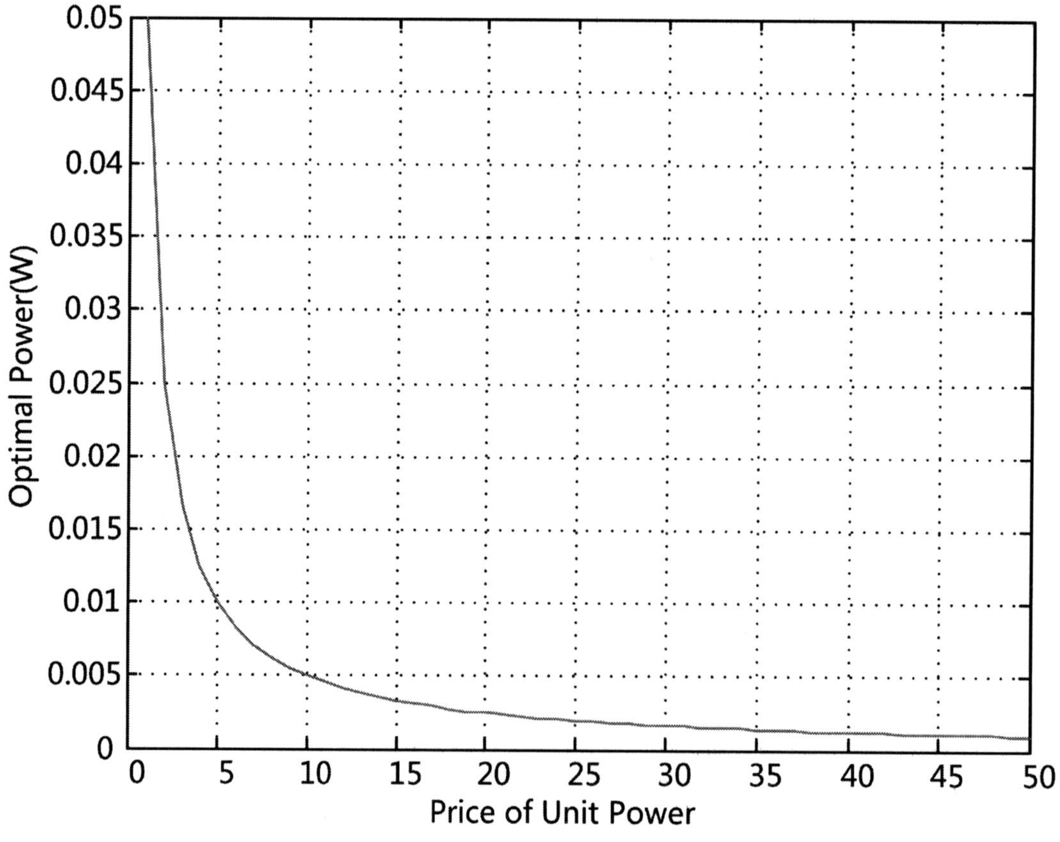

Figure 7. The optimal power change of each pair versus iteration number

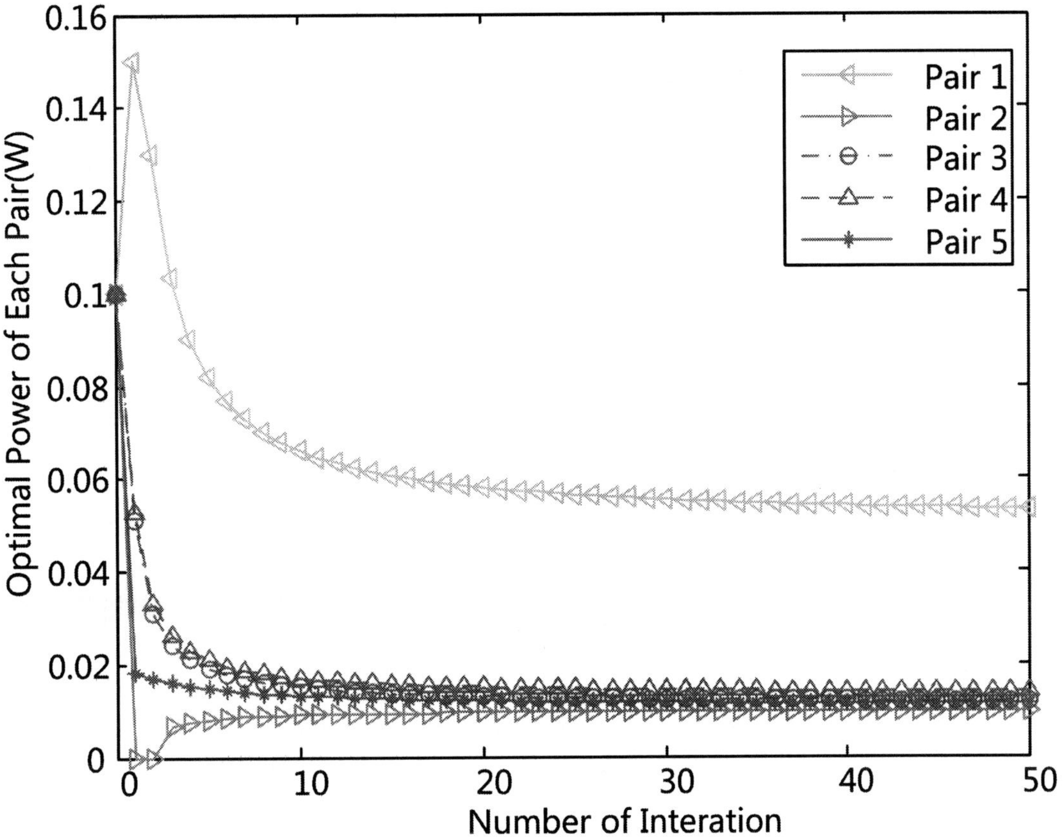

changes with the iteration. In the simulation experiments, the iteration process will not stop until the sum of the absolute value of optimal power change in the adjoining iterations is less than 10^{-3}. And we can see that the iteration stops within about 40 times when the number of pairs is five. Therefore, the proposed algorithm occupies a desirable convergence performance and can converge within a finite number of iterations

Auction Based Physical Layer Security

In the above section, we use Stackelberg game to realize centralized jammer power allocation, where legitimate transmitters share their channel state information and the resource allocation is dominated by a single policymaker. However, when legitimate transmitters are located in the different geographical locations and unable to exchange information with each other, centralized secrecy resource allocation will be no longer feasible.

Auction theory is a distributed market mechanism to solve resource allocation problems, which can model and solve resource allocation problems between buyers and sellers with incomplete information. Typical auctions include English auction, Dutch auction, first-price auction, and second-price auction etc.

Aiming at the distributed communication scenario of the above Section, Zhang, et al. (2011) proposes a distributed secrecy resource allocation solution, where legitimate transmitters are selfish and unable to exchange information with each other. Firstly, an auction model is established, and then two distributed auction-based power allocation mechanisms are proposed. Finally, simulation shows the convergence of the two mechanisms.

From Equation 12, we can easily find the jamming power allocation of jammer directly affects the system secrecy performance. To allocate the jamming power in a distributed and efficient way, we model the power allocation problem as an auction game in which the auctioneer is the friendly jammer and the bidders are the transmitters. Legitimate transmitters submit bids to compete for the jamming power from the friendly jammer, and the payment is determined by the amount of the jamming power and its unit price. Therefore, the utility function of the legitimate transmitter S_i can be defined as

$$U_i\left(p_i,\lambda\right) = g\left(p_i\right) - P\left(p_i,\lambda\right) \tag{23}$$

Where $g(p_i)$ is the secrecy performance gain stem from the jammer J allocating jamming power p_i to S_i. $P(p_i,\lambda)$ is the payment paid for the friendly jammer, λ is the unit price of jammer power defined by the friendly jamming during the auction. Therefore, the payment function of S_i can be written as

$$P\left(p_i,\lambda\right) = \lambda p_i \tag{24}$$

The secrecy rate of S_i without jammer power can be obtained by setting p_i=0, i.e.,

$$\tilde{R}_i^S = \left[\log_2(1 + p_s\gamma_{1,i}) - \log_2(1 + p_s\gamma_{e,i})\right]^+ \tag{25}$$

Combined with Equation 12, $g(p_i)$ can be defined as

$$g\left(p_i\right) = R_i^s - \tilde{R}_i^S = \log_2(1 + p_s\gamma_{e,i}) - \log_2(1 + \frac{p_s\gamma_{e,i}}{1 + p_i\gamma_{j,i}}) \tag{26}$$

The friendly jammer charges legitimate transmitters for the jamming service at a price λ for every unit of jamming power. Provided the maximum power is bounded by $p_{\max}$, we have the utility function of friendly jammer J as

$$\begin{aligned} &U_J\left(\left\{p_i\right\},\lambda\right) = \lambda\sum_i p_i \\ &s.t.\ \ 0 \le \sum_i p_i \le p_{\max} \end{aligned} \tag{27}$$

Zhang (2011) proposed two auction-based distributed power allocation schemes, i.e. Traditional Ascending Clock Auction (ACA-T) and Alternative Ascending Clock Auction (ACA-A) based power allocation. In these auctions, the auctioneer firstly announces an initial price, then the bidders feedback the auctioneer their demands at the price, and the auctioneer raises the price until the total demands meet the power constrain. The detailed processes of two auction schemes are as followed.

The ACA-T scheme is proposed based on traditional ascending clock auction (Krishna, 2009), where each legitimate transmitter is allowed to bid any power demand between 0 and $p_{\max}$ at every round. The auction procedure is shown in Table 4, before the auction, the friendly jammer sets up the price step $\delta>0$, as well as the initial asking price λ^0. In iteration t, each legitimate transmitter submits its optimal bid p_i by computing

$$p_i\left(t\right) = \arg\max_{p_i} U_i\left(p_i, \lambda\left(t\right)\right) \tag{28}$$

The friendly jammer sums up all the bids from the legitimate transmitters $p_{total}\left(t\right) = \sum_i p_i\left(t\right)$ and compares $p_{total}(t)$ with $p_{\max}$. If $p_{total}(t) \le p_{\max}$, the friendly jammer concludes the auction and allocates the rest of power to every transmitter proportionally. Then the final allocated jamming power of legitimate transmitter S_i can be given as

$$p_i^* = p_i\left(t\right) + \frac{p_i\left(t-1\right) - p_i\left(t\right)}{\sum_i p_i\left(t-1\right) - \sum_i p_i\left(t\right)}\left(p_{\max} - \sum_i p_i\left(t\right)\right) \tag{29}$$

where $\sum_i p_i^* = p_{\max}$.

Table 4. ACA-T

ALOGORITHM 1: ACA-T
Step1: Given the available jamming power $p_{\max}$, price step $\delta>0$, and iteration index t=0, the friendly jammer initializes the asking price with the reserve price $\lambda(0)$.
Step2: Legitimate transmitter S_i computes $p_i\left(0\right) = \arg\max_{p_i} U_i\left(p_i, \lambda\left(0\right)\right)$ and submits its optimal bid $p_i(0)$.
Step3: The friendly jammer sums up all the bids from the transmitters, $p_{total}\left(0\right) = \sum_i p_i\left(0\right)$ and compares $p_{total}(0)$ with $p_{\max}$: • If $p_{total}(0) \le p_{\max}$, the friendly jammer concludes the auction and chooses not to participate in the trade. • Else, set $\lambda(t+1) = \lambda(t)+\delta$, $t=t+1$ and repeat: o The friendly jammer announces $\lambda(t)$ to all the legitimate transmitters. o Legitimate transmitter S_i computes $p_i\left(t\right) = \arg\max_{p_i} U_i\left(p_i, \lambda\left(t\right)\right)$ and submits its optimal bid $p_i(t)$. o The friendly jammer sums up all the bids from the legitimate transmitters $p_{total}\left(t\right) = \sum_i p_i\left(t\right)$ and compares $p_{total}(t)$ with $p_{\max}$: ▪ If $p_{total}(t) > p_{\max}$, set $\lambda(t+1) = \lambda(t)+\delta$, and continue the auction. ▪ Else, conclude the auction, set T=t, and allocate $p_i^* = p_i\left(t\right) + \frac{p_i\left(t-1\right) - p_i\left(t\right)}{\sum_i p_i\left(t-1\right) - \sum_i p_i\left(t\right)}\left(p_{\max} - \sum_i p_i\left(t\right)\right)$ to the transmitter S_i.
Step4: Finally, the utility function of legitimate transmitter S_i is $U_i^*\left(p_i^*, \lambda\left(T\right)\right) = g\left(p_i^*\right) - \lambda\left(T\right) p_i^*$.

The selfish transmitter may make a fake bid in order to get more profits, which will decrease the system secrecy performance. ACA-T can achieve efficient power allocation, but is not cheat-proof. To solve this problem, the ACA-A scheme introduces the correlation of the auction rounds into the payoff function (Ausubel, 2004). The procedures of ACA-A are shown in Table 5. Different from payment function of ACA-T in Equation 24, when the auction is finished in iteration T, the payment function of legitimate transmitter S_i is

$$P\left(p_i^*, \lambda(T)\right) = \lambda^0 L_i^0 + \sum_{t=1}^{T} \lambda(t)\left(L_i^t - L_i^{t-1}\right) \quad (30)$$

where,

$$L_i^t = \begin{cases} \max\left(0, p_{\max} - \sum_{j\neq i} p_i(t)\right), t < T \\ p_i(t) + \dfrac{p_i(t-1) - p_i(t)}{\sum_i p_i(t-1) - \sum_i p_i(t)}\left(p_{\max} - \sum_i p_i(t-1)\right), t = T \end{cases}.$$

Table 5. ACA-A

ALOGORITHM 2: ACA-A
Step1: Given the available jamming power $p_{\max}$, price step δ>0, and iteration index t=0, the friendly jammer initializes the asking price with the reserve price λ(0).
Step2: Legitimate transmitter S_i computes $p_i(0) = \arg\max_{p_i} U_i\left(p_i, \lambda(0)\right)$ and submits its optimal bid $p_i(0)$.
Step3: The friendly jammer sums up all the bids from the transmitters, $p_{total}(0) = \sum_i p_i(0)$ and compares $p_{total}(0)$ with $p_{\max}$:
• If $p_{total}(0) \le p_{\max}$, the friendly jammer concludes the auction and chooses not to participate in the trade.
• Else, set $\lambda(t+1) = \lambda(t)+\delta$, $t=t+1$ and repeat:
o The friendly jammer announces $\lambda(t)$ to all the legitimate transmitters.
o Legitimate transmitter S_i computes $p_i(t) = \arg\max_{p_i} U_i\left(p_i, \lambda(t)\right)$ and submits its optimal bid $p_i(t)$.
o The friendly jammer sums up all the bids from the legitimate transmitters $p_{total}(t) = \sum_i p_i(t)$ and compares $p_{total}(t)$ with $p_{\max}$:
▪ If $p_{total}(t) > p_{\max}$, first computer $L_i^t = \max\left(0, p_{\max} - \sum_{j\neq i} p_i(t)\right)$, then set $\lambda(t+1) = \lambda(t)+\delta$, $t=t+1$, and continue the auction.
▪ Else, conclude the auction, set T=t, computer $L_i^T = p_i(t) + \dfrac{p_i(t-1) - p_i(t)}{\sum_i p_i(t-1) - \sum_i p_i(t)}\left(p_{\max} - \sum_i p_i(t-1)\right)$, and allocate $p_i^* = L_i^T$ to the legitimate transmitter S_i.
Step4: Finally, the utility function of legitimate transmitter S_i is $U_i^*\left(p_i^*, \lambda^T\right) = g\left(p_i^*\right) - P_i^*$, where P_i^* is the payment of S_i and can be expressed as $P_i^* = \lambda^0 L_i^0 + \sum_{t=1}^{T} \lambda^t \left(L_i^t - L_i^{t-1}\right)$

To evaluate the performance of the proposed schemes, assuming there is a cheat transmitter in the system, who report a false power demand $\tilde{p}_i(t)$ by scaling the true optimal demand $p_i(t)$ with a cheat factor k, i.e., $\tilde{p}_i(t) = \min\left(p_{\max}, \max\left(0, kp_i(t)\right)\right)$, where k represents the probability of transmitter cheats. Thus, the cheat-proof property and secrecy rate using ACA-T and ACA-A are shown as Figure 8 and Figure 9, respectively.

In Figure 8, the final utilities of the cheat transmitter as a function of the cheat factor k in the ATA-T and ATA-A schemes are shown, respectively. We can see that with ACA-T, the cheat transmitter achieves the maximal utility when k is around 0.7. Thus, the ACA-T scheme is not cheat-proof. With ACA-A, we can see that the cheat transmitter achieves the maximal utility when k=1, which means that no source has the incentive to cheat since any cheating will lead to a loss in its utility. Thus, the ACA-A scheme is cheat-proof. In Figure 9, the system secrecy rates with the ACA-T and ACA-A schemes as well as the no-jamming case are shown, respectively. We can see that the two proposed schemes can always achieve efficient jamming power allocation and obtain the optimal system secrecy rate.

In this section, in order to solve the problem of active eavesdropping, a zero-sum game is firstly established between the legitimate transmitter and the active eavesdropper considering the secrecy rate as the utility function. Through analyzing the system secrecy rates corresponding to different schemes, we study the condition of Nash equilibrium (Mukherjee & Swindlehurst, 2013). Then Stackelberg game

Figure 8. Cheat-proof property of ACA-T and ACA-A schemes

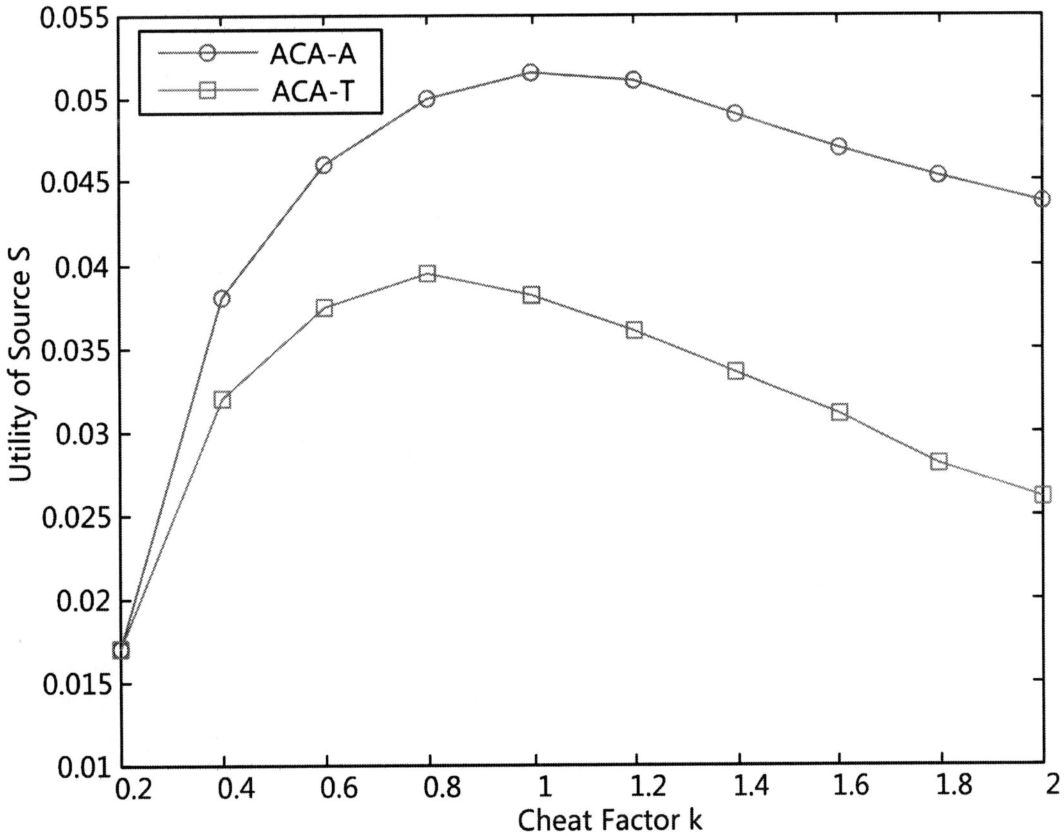

Figure 9. System secrecy rates of the ACA-T and ACA-A schemes as well as no-jamming case

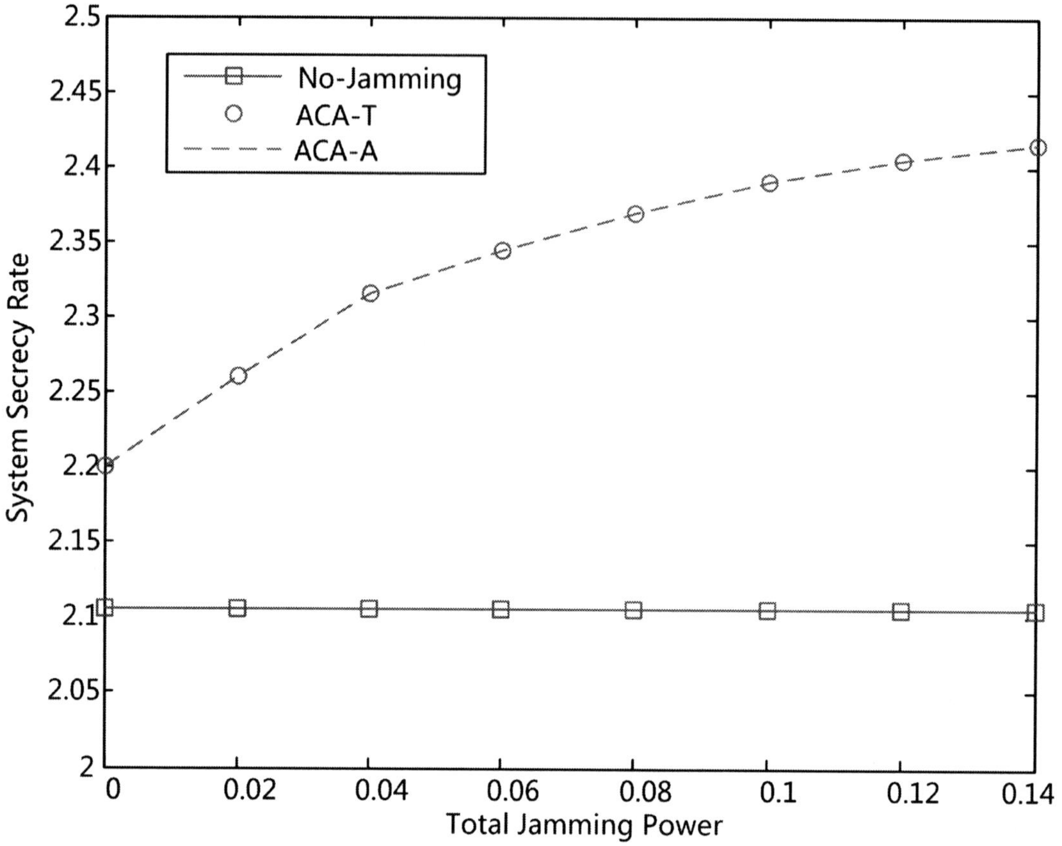

is used to model and solve the jammer power allocation problem among the cooperative jammer and legitimate transmitters. Through analyzing the Stackelberg equilibrium, we introduce a kind of centralized jamming power allocation algorithm (Wang, A., Cai, Yang, & Hou, 2013). At last, considering the distributed communication scenario, we introduce two auction based power allocation schemes: ACA-T and ACA-A (Zhang, Song, Han, & Jiao, 2011).

COALITION GAME BASED PHYSICAL LAYER SECURITY

In a secure communication network consisting of multiple nodes, the legitimate nodes can transmit artificial interference, in order to help each other improving the secrecy rate. The malicious nodes can cooperate with each other to decrease the secrecy rate, through Maximal Ratio Combine(MRC) reception. But, the cooperation of communication nodes is restricted by some factors, such as the relative location and channel condition of the nodes. Coalition game studies the coalition formation of the players, and the distribution of the coalition utility. In this section, we use coalition game to analyze the system security when the communication nodes cooperate with each other. The existing study of the physical security based on coalition game can be divided into three types: the coalition formation game of malicious nodes, friendly relays and legitimate transmitters.

Coalitional Game of the Malicious Nodes

The untrusted relay can jam both legitimate nodes and eavesdroppers. In this part, we study the cooperation of eavesdroppers and untrusted relays from the perspective of eavesdropper, in order to find out the impact to system security of this kind of coalition. Cao and Wei (2012) formulate this problem as a coalition game with non-transferable utility, and use job-hopping game to analyze how the eavesdropper join or leave a coalition. The eavesdropper's objective is to optimize its utility.

As shown in Figure 10, there are N users in the network as the legitimate transmitters, expressed as $\mathbf{N}=\{1,\ldots,n,\ldots,N\}$, sending information to the legitimate receiver D. There exists E eavesdroppers expressed as $\boldsymbol{\varepsilon}=\{1,\ldots,i,\ldots,E\}$. The channel noise is independent and identically distributed according to the Gaussian white noise. To help eavesdroppers, J relays play the role of jammer.

The secrecy rate of legitimate transmitter n is the difference of the channel capacity at receiver D and the channel capacity at eavesdropper i:

$$R_{i,n} = \left(I_{n,D} - I_{n,i}\right)^{+} \tag{31}$$

Where

$$I_{n,D} = \log_2\left(1 + \frac{\left|h_{n,D}\right|^2 \bullet P_T}{\sigma^2}\right),\ I_{n,i} = \log_2\left(1 + \frac{\left|h_{n,i}\right|^2 \bullet P_T}{\sigma^2}\right),$$

Figure 10. An illustrative example of system with eavesdropping coalitions

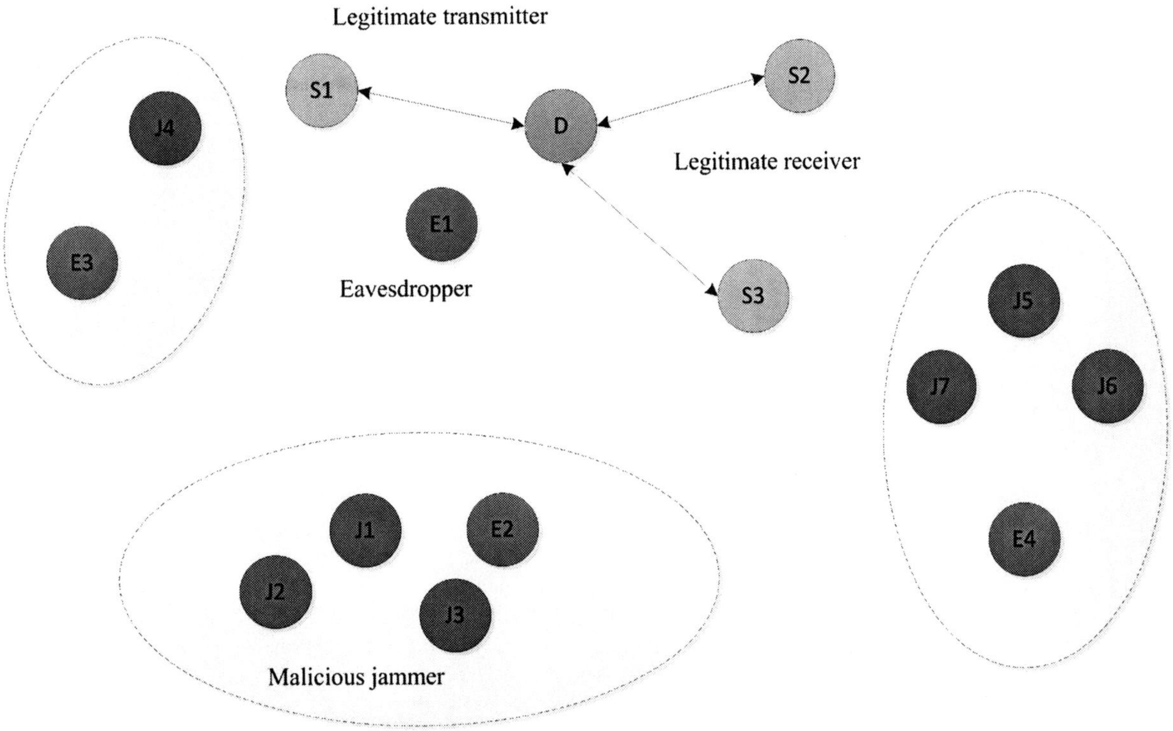

$h_{n,D}$ is the channel gain between legitimate transmitter n and legitimate receiver D, and $h_{n,i}$ is the channel gain between legitimate transmitter n and eavesdropper i. The legitimate transmitter's transmitting power is P_T, $(\cdot)^+ \triangleq \max(\cdot, 0)$.

From (31), it is obvious that the eavesdropper's performance is strictly restrained by the channel between the eavesdropper and the legitimate transmitter. To further decrease the secrecy rate, the eavesdropper i will ask jammers for cooperation to form a malicious coalition S. There is only one eavesdropper and some jammers in a malicious coalition. But while the jammers interfere the legitimate receiver, they cause interference to the eavesdropper as well. So with the existence of the malicious coalition, the legitimate transmitter n's secrecy rate is the channel capacity of legitimate transmitter minus the channel capacity of eavesdropper:

$$R_{i,n}^S = \left(I_{n,D}^S - I_{n,i}^S\right)^+ \quad (32)$$

Where

$$I_{n,D}^S = \log_2\left(1 + \frac{\left|h_{n,D}\right|^2 \bullet P_T}{\left|W_{n,S}^H h_{j,D}\right|^2 + \sigma^2}\right), I_{n,i}^S = \log_2\left(1 + \frac{\left|h_{n,i}\right|^2 \bullet P_T}{\left|W_{n,S}^H h_{j,i}\right|^2 + \sigma^2}\right),$$

$h_{j,D}$ is the channel gain between jammer j and legitimate receiver D, and $h_{j,i}$ is the channel gain between jammer j and eavesdropper i, $W_{n,S}^H$ is the conjugate transpose of |S|-1 numbers of jammers' antenna weights, $\left|W_{n,S}^H h_{j,D}\right|^2$ is the interference to legitimate transmitters caused by the jammers. When eavesdropper eavesdrops, transmitting power of the interference signal is expressed as: $P^J = p_0 - P_T$, in which P_0 is the total power of legitimate transmitters and the jammers (Dong, Han, Petropulu, & Poor, 2010).

When the total power is limited, in order to completely eliminate the interference signal at eavesdroppers, let $W_{n,S}^H h_{j,i} = 0$, construct the optimal weight vector $W_{n,S}^{opt}$ which satisfies:

$$W_{n,S}^{opt} = \arg\max_{W_{n,S}} \left|W_{n,S}^H h_{j,D}\right|^2, s.t. \begin{cases} W_{n,S}^H h_{j,i} = 0 \\ \left\|W_{n,S}\right\|^2 \le P^J \end{cases} \quad (33)$$

When there is only one eavesdropper, the requirement $W_{n,S}^{\dagger} h_{j,i} = 0$ can't be reached. The channel capacity of legitimate transmitter n and eavesdropper i can be rewrite as: $I_{n,D}^S = \log_2\left(1 + \frac{\left|h_{n,D}\right|^2 \bullet P_T}{P_J \left|h_{j,D}\right|^2 + \sigma^2}\right)$ and $I_{n,i}^S = \log_2\left(1 + \frac{\left|h_{n,i}\right|^2 \bullet P_T}{P_J \left|h_{j,i}\right|^2 + \sigma^2}\right)$. Bring these two equations into Equation 32, we will get the secrecy rate with one jammer.

Coalition Game

When the eavesdropper i forms a coalition S with the jammers, the coalition S can eavesdrop and jam the legitimate transmitter, the loss of the secrecy rate of the legitimate transmitter can be defined as $-\sum_{n=1}^{N} R_{i,n}^{S}$, which is the sum of the loss of the secrecy rate of the N legitimate transmitters. Because of the selfishness of the nodes, the eavesdropper has to provide some payment to the jammers for the interference power with unit price λ. Therefore, the utility function of eavesdropper i can be defined as:

$$v_i(S) = -\sum_{n=1}^{N} R_{i,n}^{S} - \sum_{n=1}^{N}\sum_{j \in S} P_{j,n}(S) \tag{34}$$

Where the number of jammers within the coalition is |S|-1, the payoff of the jammers is $P_{j,n}(S) = \begin{cases} \lambda \cdot \left|W_{n,S}^{opt}(j)\right|^2 & if\left|S\right| > 2 \\ \lambda \cdot P^J & if\left|S\right| = 2 \\ 0 & otherwise \end{cases}$, $\left|W_{n,S}^{opt}(j)\right|^2$ is the interference power which jammer j distributes to the legitimate transmitter n, and P^J is the interference power when |S|=2. $\sum_{n=1}^{N}\sum_{j \in S} P_{j,n}(S)$ is the sum of the payment which eavesdropper provides to all jammers. Obviously, the utility function of the jammer j is:

$$v_j(S) = \sum_{n=1}^{N} P_{j,n}(S) \tag{35}$$

The non-transferable utility coalition game is defined as (**N**,**U**). Where **N** is the collection of game players, and **U** is a collection of coalition's utility vectors:

$$\mathbf{U}(S) = \left\{ \phi(S) \middle| \begin{matrix} \phi(S) = -\infty & if \;\; S = \{i, i', \mathbf{J}\} \\ \phi_i(S) = v_i(S), \phi_j(S) = v_j(S) & \forall i,j \;\; otherwise \end{matrix} \right\} \tag{36}$$

Where ϕ(*S*) is the |*S*|×1 utility vector for all players in coalition S. Each element of ϕ(*S*) is the corresponding player' utility. Due to the limitation that each coalition has only one eavesdropper, when a coalition contains two eavesdroppers *i* and i', the utility of the coalition has to be set as -∞.

Job-Hopping Game

The hedonic game (Saad, Han, Debbah, Hjøungnes, & Basar, 2009) provides a method of researching the nodes' selfishness in the coalition formation process, and it is suitable to analyze the problems in this section. In the hedonic game, game players establish the preference relation:

$$S_\alpha \rhd_k S_\beta \Leftrightarrow f_k(S_\alpha) \rhd f_k(S_\beta) \tag{37}$$

Equation 37 shows that player k is more inclined to join, where S_α,S_β∈**N**are the two coalitions which include player k, and f_k is the preference relation function of player k:

$$f_k(S) = \begin{cases} \phi_k(S) & \begin{array}{l} if \phi_i(S) \geq \phi_i(S \setminus \{k\}), i \in S \setminus \{k\} \\ \& i \in \boldsymbol{\mu}, S \notin h(k) \end{array} \\ -\infty & otherwise \end{cases} \tag{38}$$

Where $\phi_k(S)$ is from Equation 36, and $h(k)$ is denoted as a coalition which player k has ever joined. $f_k(S)$ means a kind of participate or departure mechanism, which is related to the contribution of player k to the coalition utility. with this preference relation, the eavesdropper is the leader of the coalition, the jammers play as follower roles. The relationship between the eavesdropper and jammers is similar to the relationship between an employer and employees. Jammers can carry out "job-hopping" (Cao et al., 2012) based on their payoffs, and select the eavesdropper who pays the highest "salary".

Based on the above analysis, the job-hopping rules are defined as follows: the collection **N** is divided into several coalition collections: $\Pi=\{S_1,\ldots,S_r\}$, and the game player k belong to the coalition $S_\Pi(k)\in\Pi$. If there is a coalition $S_g\in\Pi$, and $S_g \cup \{k\} \rhd_k S_\Pi(k)$, then the game player k chooses to leave the current coalition $S_\Pi(k)$, and join another coalition $S_g\in\Pi\cup\{\phi\}$. The procedure of the algorithm is shown in Table 6.

For simulations, a square network of $0.5km\times0.5km$ is set up with the legitimate receiver located at the origin of coordinate and N legitimate transmitters are randomly distributed in the network. As the adversaries are generally not easy to approach to the sources, we consider a bigger scenario of $2km\times2km$ surrounding the destination where E eavesdroppers and J jammers are randomly deployed. To highlight the effects of distances, channels between any two nodes in simulations are modeled by a line-of-sight channel model including the path loss effect and a random phase. The simulation parameters used are as follows: The power constraint $P_0=0.01W$ and we take $P_T=0.2P_0$. For channels, the propagation loss $\kappa=3.5$, and the bandwidth $B=1$. Finally, the price of unit power is taken as $\theta=1$ and noise lever in the network is 100 dBm.

Figure 11 shows the final partition (i.e. the cooperative eavesdropping structure) through the job-hopping algorithm for a randomly deployed network with a single legitimate transmitter, and E=5, J=15. All untrusted relays are subject to the eavesdropper "employment" and form five coalitions. Figure 12 shows the case where N=5, E=5, and depicts each eavesdropper's eavesdropping capacity curve in one round. The eavesdropping capacity is the eavesdropper's channel capacity. And we do a comparison of

Table 6. Distributed Hedonic job-hopping coalition formation algorithm

Coalition Formation Algorithm
Step 1: Initialization:all malicious nodes are divided into $\Pi_t=\{S_1,\ldots,S_r\}$ at the time t. Π_t=N, when t=0.
Step 2: Coalition formation stage 1: under the current division Π_t, each game player randomly carries out "job-hopping" based on their preference relation;
Step 3: Coalition formation stage 2: if any player carries out the "job-hopping", update the coalition division Π_t
Step 4: Coalition formation stage 3: set the time to $t=t+1$, repeat steps 2 and 3, until it reaches a stable coalition division state.

Figure 11. The final structure of collaboration eavesdropping by job-hopping coalition formation game

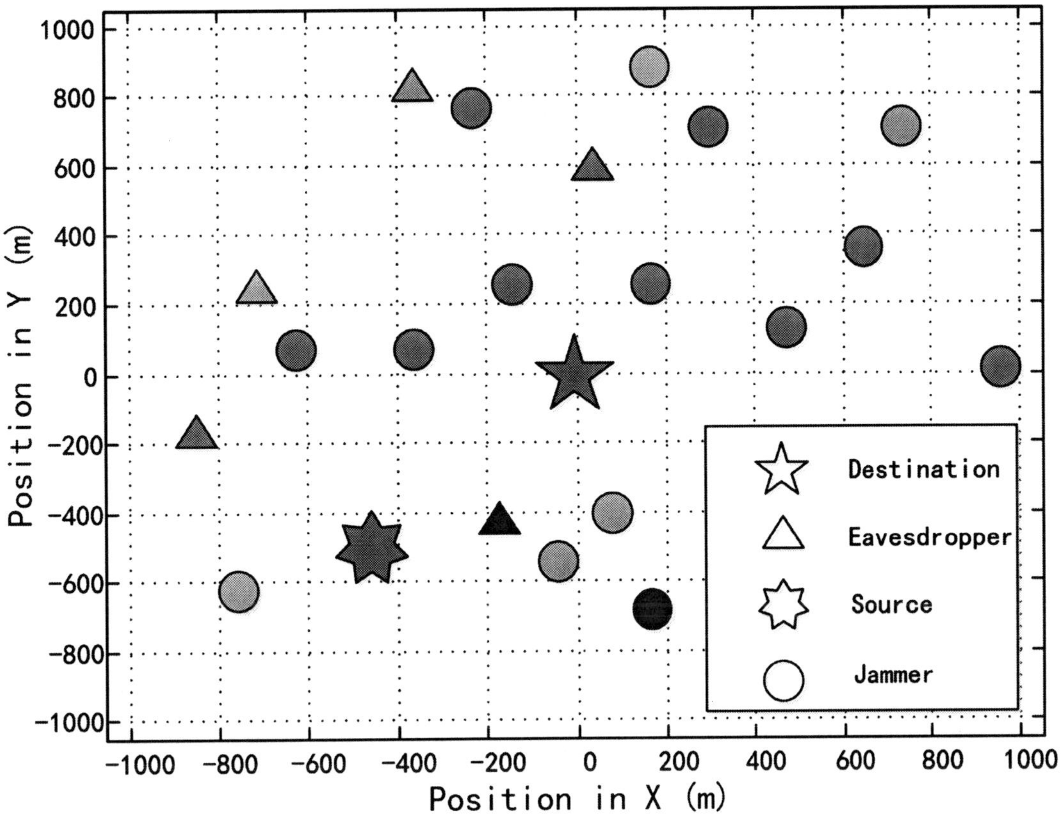

eavesdropping performance between using eavesdropping job-hopping coalition formation algorithm and not. The results show that the coalition formation in this section to significantly boost the performance compared to the non-cooperative algorithm.

Coalitional Game of the Friendly Helpers

When the friendly cooperators exist in the system, the secrecy performance of the system can be improved through relaying signals and sending artificial interference. Zhang (2011) uses coalition formation game to analyze the influence of the friendly cooperation behavior on the system security.

As shown in Figure 13, the network has a legitimate transmitter S, a legitimate receiver D, a eavesdropper E and the number of the friendly cooperators is N, These cooperators can act as relay to forward confidential signal or as jammer to jam the eavesdropper. The transmission process of confidential information from legitimate transmitter to legitimate receiver can be divided in two phases:

Phase 1: legitimate transmitter sends confidential information to all the friendly cooperative nodes;

Phase 2: nodes are selected from the friendly cooperators who can correctly decode the received signal. One of these selected nodes acts as relay to forward the confidential signal, while the remaining selected nodes act as jammers to jam the eavesdropper.

Figure 12. Average eavesdropping capacity

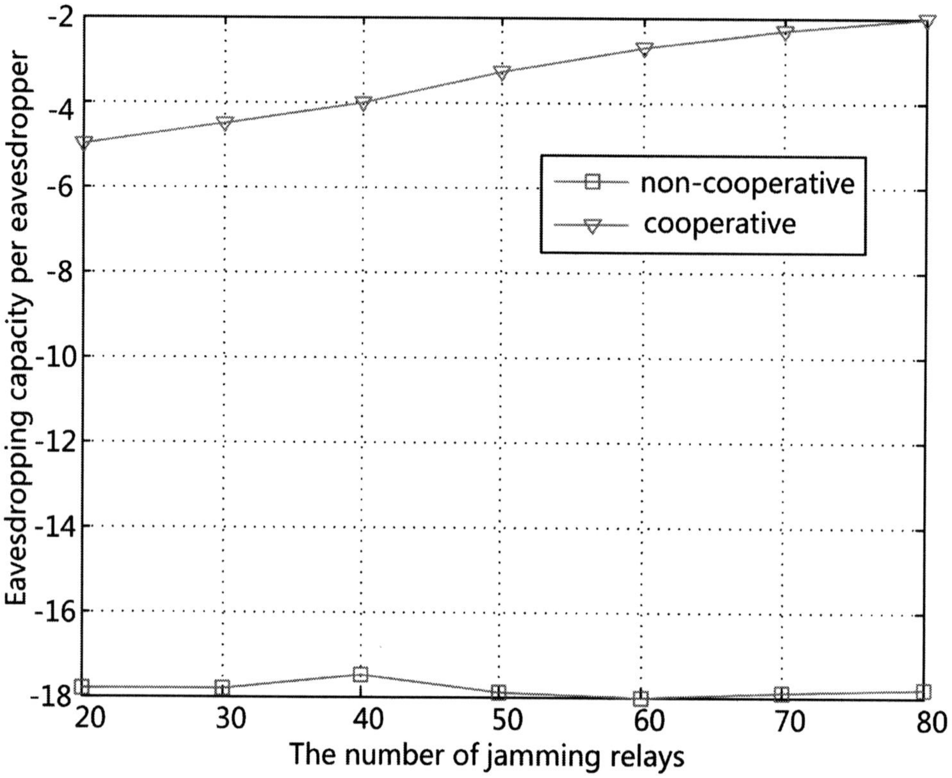

Figure 13. System model for relay and jammer assisted secure cooperative network.

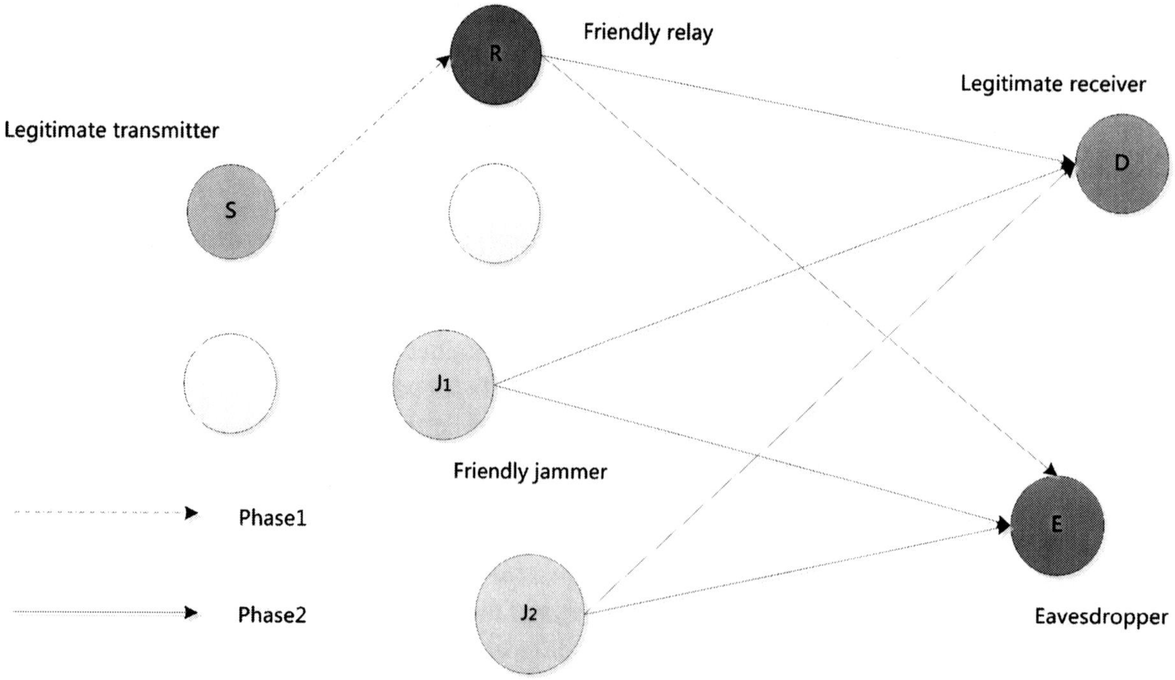

Assuming that the friendly cooperator Z is selected as the relay, the cooperator $J_i, i\in\{1,2,\ldots,K\}$ is selected as the friendly jammer, transmitting power is .. and p_i^J respectively. The channel gain of relay Z to legitimate receiver D and eavesdropper E are $h_{Z,D}$ and $h_{Z,E}$ respectively, the channel gains of friendly jammer J_i to legitimate receiver and eavesdropper are $h_{J,D}$ and $h_{J,E}$ respectively.

When only relay Z is engaged in communication, the secrecy rate of legitimate transmitter can be expressed as:

$$R_s\left(Z\right)=\left(I_{Z,D}-I_{Z,E}\right)^+=\left[\log\left(1+\frac{p_Z h_{Z,D}}{\sigma^2}\right)-\log\left(1+\frac{p_Z h_{Z,E}}{\sigma^2}\right)\right]^+ \tag{39}$$

With the participation of friendly jammer $J_i, i\in\{1,2,\ldots,K\}$ and relay Z, the channel capacities of relay to legitimate receiver and eavesdropper are $I_{Z,D}(J_i)$ and $I_{Z,E}(J_i)$ respectively, so the secrecy rate of legitimate transmitter is:

$$R_s\left(Z,J_i\right)=\left(I_{Z,D}(J_i)-I_{Z,E}(J_i)\right)^+=\left[\log\left(1+\frac{p_z h_{Z,D}}{\sigma^2+\sum_i p_i^J h_{J_i,D}}\right)-\log\left(1+\frac{p_z h_{Z,E}}{\sigma^2+\sum_i p_i^J h_{J_i,E}}\right)\right]^+ \tag{40}$$

Equation 40 shows that the system could improve its secrecy rate by reasonably choosing friendly cooperators and assigning the cooperators' roles. Each friendly cooperator has two identities to select: when the friendly cooperators can not successfully decode the broadcast signal, the node can only act as jammers; When the friendly cooperators can successfully decode the broadcast signal, from Equation 39 and Equation 40, it is obvious that the friendly cooperator who has the best channel quality should act as a relay while the other nodes should act as jammers.

Zhang (2011) formulates the cooperators' cooperation for data transmission described above as a coalitional game with transferable utility (TU). Some cooperators form coalition S to provide security communication service, the legitimate transmitter pay for this service. The cooperators choose to be relays or jammers according to the payoff. The payoff is determined by the unit price and the secrecy rate offered by coalition S. Assume that there exists only one relay in a coalition. The utility function of coalition S is:

$$u\left(S\right)=\left[\log\left(1+\frac{\varepsilon_z p_z h_{Z,D}}{\sigma^2+\sum_i p_i^J h_{J_i,D}}\right)-\log\left(1+\frac{\varepsilon_z p_z h_{Z,E}}{\sigma^2+\sum_i p_i^J h_{J_i,E}}\right)\right]^+ \tag{41}$$

Where ε_z is a switching factor that $\varepsilon_z=1$ when there exists a conventional relay in S and $\varepsilon_z=0$ when there is no conventional relay in S. The coalition members' utility increases when the coalition's utility increases. Assume that the unit price of relays' service is α, and the unit price of jammers' service is β. The utility function of relay and jammer is:

$$u_Z(S) = \alpha R_s(Z) = \alpha\left(I_{Z,D} - I_{Z,E}\right)^+ \tag{42}$$

$$u_{J_i}(S) = \beta\left[R_s\left(Z,\{J_i\}_{i\in S\setminus Z}\right) - R_s\left(Z,\{J_k\}_{k\in S\setminus\{i,Z\}}\right)\right] \tag{43}$$

Preference Relation

Assume that collection $P=\{S_1,\ldots,S_P\}$ and $L=\{S_1^*,\ldots,S_l^*\}$ are the different division of cooperator collection **N**. The preference relation $P \rhd L$, means division P is better than division L. Due the fact that the legitimate transmitter needs only one coalition whose utility is highest, the preference relation can be defined as:

$$P \rhd L \Leftrightarrow \left\{\max\left\{u(S_1),\ldots,u(S_P)\right\}\right\} \geq \left\{\max\left\{u(S_1^*),\ldots,u(S_P^*)\right\}\right\} \tag{44}$$

Merge-and-Split Rules

Merge rule: when $\{\cup_{j=1}^{l} S_j\} \rhd \{S_1,\ldots,S_l\}$, then $\{S_1,\ldots,S_l\} \rightarrow \{\cup_{j=1}^{l} S_j\}$;

Split rule: when $\{S_1,\ldots,S_l\} \rhd \{\cup_{j=1}^{l} S_j\}$, then $\{\cup_{j=1}^{l} S_j\} \rightarrow \{S_1,\ldots,S_l\}$.

The procedure of the algorithm is shown in Table 7.

Table 7. The secure coalition formation algorithm

Coalition Formation Algorithm
Step 1: Initialization:The set of friendly cooperators is partitioned by $S_t=\{S_1,S_2,\ldots,S_N\}$ at the beginning with non-cooperation. **Step 2:** Coalition Formation Algorithm: Phase 1 - Identity Choice: The transmitter transmits its data to the cooperator. Each cooperator chooses its optimal identity (to be a relay or a friendly jammer) based on its channel conditions to the destination and the eavesdropper. For an friendly cooperator S_i, if $h_{T_i,D} > h_{T_i,E}$, then it will choose to be a relay. Else, it will choose to be a friendly jammer. The friendly cooperator who cannot decode the source messages correctly has no choice but to be a friendly jammer. Phase 2 - Adaptive Coalition Formation: Each cooperator with an identity can be regarded as a disjoint independent coalition, then coalition formation using merge-and-split rules occurs: - Clock index t=0. - **Repeat** (a) P_t=*Merge*(S_t): coalitions in S_t decide to merge based on the merge rules. (b) L_t=*Split*(P_t): coalitions in P_t decide to split based on the split rules. (c) Update clock index $t=t+1$. (d) Value assignment $S_t=L_t$. - **Each** coalition reports the secrecy rate of data transmission that it can achieve. Phase 3 - Secure Transmission: The source chooses the coalition who can provide the highest secrecy rate for secure data transmission and pays for it.

For simplicity and without loss of generality, we consider a cooperative network with a couple of intermediate nodes in a 0.5×0.5 square area, of which center is (0,0). Legitimate transmitter S, legitimate receiver D, and eavesdropper E are located at (−2,0), (2,0), and (2,−1), respectively.

In Figure 14, we show the network scenarios before and after coalition formation based on the proposed merge-and-split algorithm. We can see that cooperators automatically choose their roles and form coalitions based on the algorithm. In Figure 15, we compare the secrecy performance of the proposed algorithm, centralized relay and jammer selection algorithm and optimal selection with jamming (OSJ) scheme (Krikidis, Thompson, & McLaughlin, 2009), and the non-cooperative with friendly jammers case. The results show that the proposed algorithm can secure the communication better.

Coalitional Game of the Malicious Transmitters

When there are no cooperators in the system, multiple legitimate transmitters can secure the communication through cooperation among them. Liang et al. (2011) analyze the legitimate transmitters' cooperation strategies and the impact of the topology's change and the cost in the information exchange on the legitimate transmitters' cooperation.

Figure 14. Network scenarios before and after coalition formation

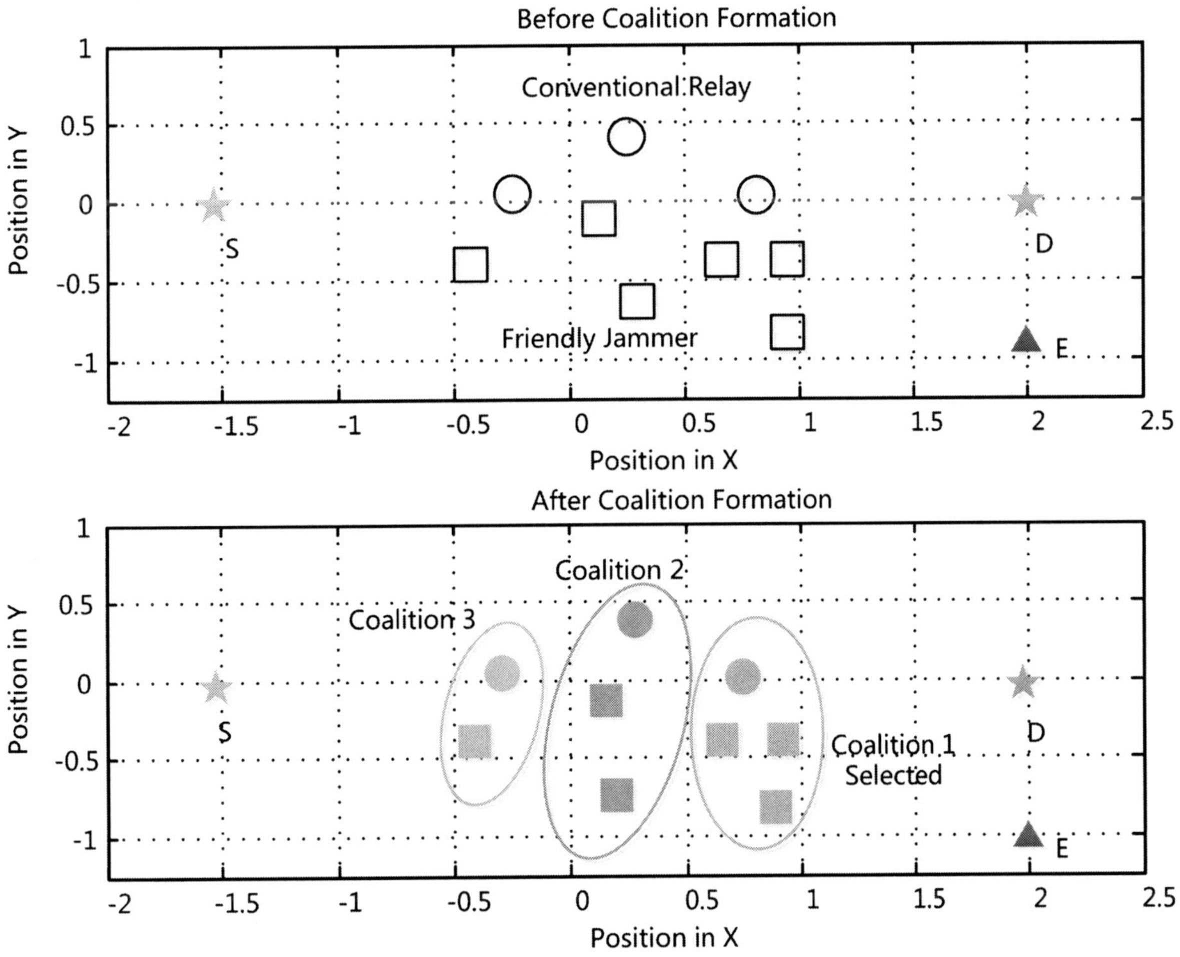

Figure 15. Secrecy rate of different cases

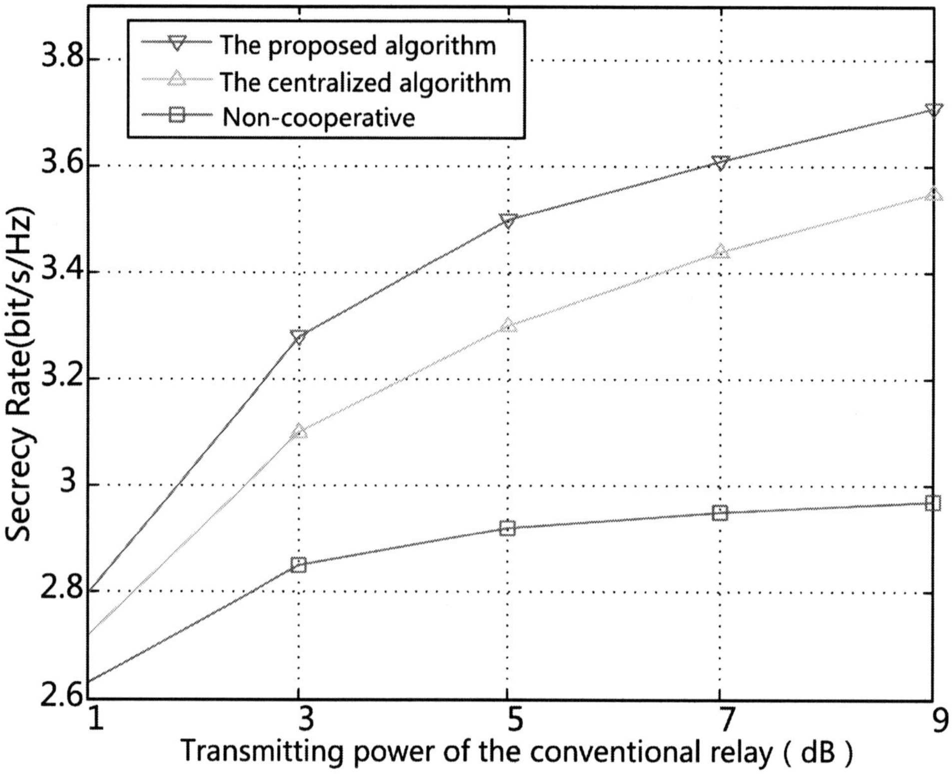

As shown in Figure 16, there exists N legitimate transmitters **N**={1,2,…,*N*}, and M legitimate receivers and E eavesdroppers, Time Division Multiple Access(TDMA) is adopted to prevent the inter-interference, meanwhile the legitimate transmitters utilize cooperative beamforming techniques to improve their secrecy rates (Liang, Poor, & Shamai, 2008). The transmission can be divided into two phases (Dong et al., 2008):

1) Broadcast phase: in each time slot, the node who owns the slot broadcasts its confidential information to other nodes;
2) Beamforming phase: the legitimate transmitters form coalition, the nodes who do not own the slot offer the beamforming service.

At the broadcast phase, the transmitter who seeks for cooperation will leak its information. This security loss quantified by the capacity at the eavesdroppers resulting from the information exchange can be expressed as:

$$I_{i,E} = \max_{1 \le e \le E} \log_2 \left(1 + \frac{P_i^{broadcast} \left| h_{i,e} \right|^2}{\sigma^2} \right) \tag{45}$$

Figure 16. System Model for Physical Layer Security Coalitional Game

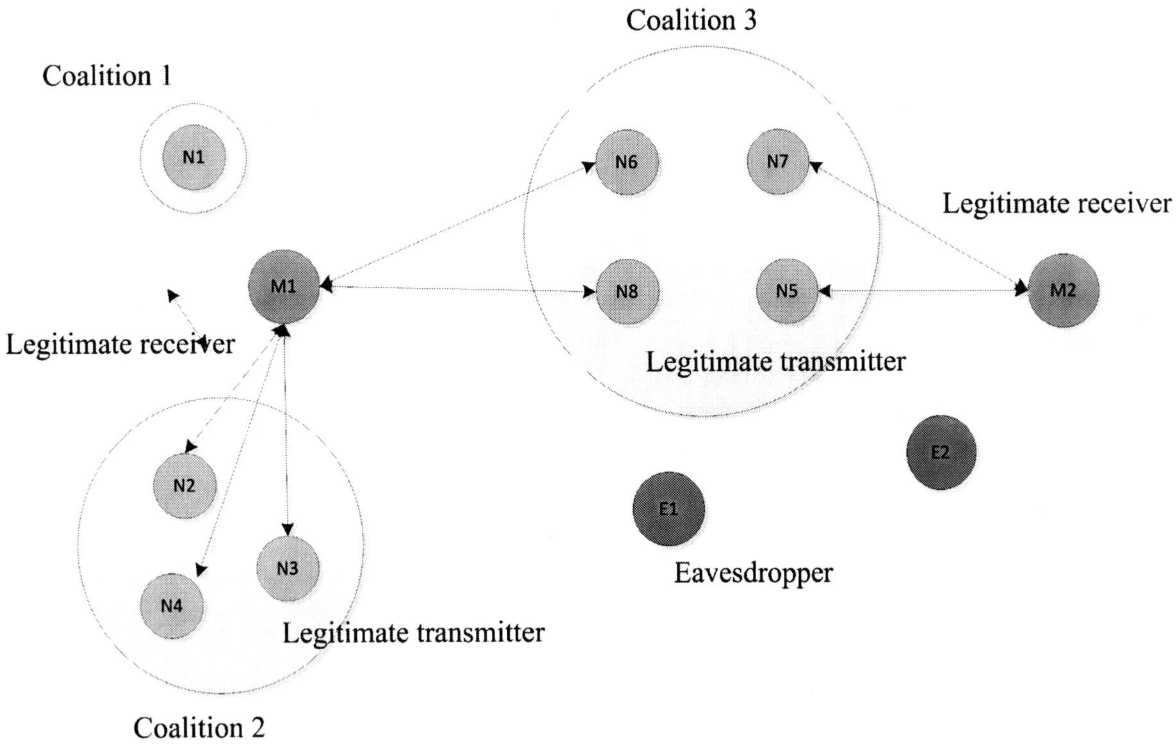

Where $h_{i,e}$ is the channel gain between legitimate transmitter i and eavesdropper e. σ^2 is the channel noise variance. $P_i^{broadcast} = P^{total} - P_i^s$ is the power consumed by legitimate transmitter i during its broadcast phase, and P^{total} is the total power of legitimate transmitter i, and P_i^s is residual power used in beamforming phase. Obviously, the larger $P_i^{broadcast}$ is, the larger the communication distance of broadcast phase is and the higher the probability of forming coalition is, but the larger $I_{i,e}$ is. During this procedure, $P_i^{broadcast}(S)$ is defined as the consumed power of the legitimate transmitter i to inform the farthest node in coalition S. From Equation 45, the legitimate transmitter i's cost function is defined as:

$$I_{i,E}(S) = \max_{1 \le e \le E} \log_2 \left(1 + \frac{P_i^{broadcast}(S) \left| h_{i,e} \right|^2}{\sigma^2} \right) \tag{46}$$

In the beamforming phase, in order to restrain the eavesdropper, coalition S construct $|S| \times 1$ beamforming vector, which satisfies the condition:

$$w_S^H \mathbf{h}_{S,E} = 0 \tag{47}$$

Where $\mathbf{h}_{S,E}$ is a $|S|\times E$ channel matrix, which contains the channel gain from the legitimate transmitter to the eavesdroppers. By nulling the signals at the eavesdropper through cooperation within coalition S, the legitimate transmitter i's secrecy rate is:

$$R_{i,m}\left(s\right) = \log_2\left(1 + \frac{P_i^s\left|w_S^H h_{S,m}\right|^2}{\sigma^2}\right) \tag{48}$$

Where $h_{S,m}$ is a $|S|\times 1$ vector which contains the channel from legitimate transmitters in S to the legitimate receiver m.

The cooperation of transmitters can improve the secrecy performance, and this improvement is related to the coalition's scale and formation. In this section, the cooperation is defined as (**N**,**U**) coalitional game with a non-transferable utility. The utility function is:

$$\mathbf{U}\left(S\right) = \left\{\phi\left(S\right)\middle| \begin{array}{c}\phi_i\left(S\right) = \left(R_{i,m}\left(S\right) - I_{i,E}\left(S\right)\right)^+,\ P_i^s > 0 \\ \phi_i\left(S\right) = -\infty,\ otherwise\end{array}\right\}_{\forall i \in S} \tag{49}$$

Where $\phi(S)$ is a $|S|\times 1$ coalition payoff vector, and $\phi_i(S)$ is the legitimate transmitter i's payoff function. Equation 49 shows that only when the coalition members help legitimate transmitter i relay its information, i.e. $P_i^s > 0$, the legitimate transmitter i can get payoff.

We assume that $P=\{S_1,\ldots,S_p\}$ and $L = \left\{S_1^*,\ldots,S_l^*\right\}$ are different division of transmitters' collection **N**. $\forall i \in S_k, S_k \in P, j \in S_k^*, S_k^* \in L$, $\phi_i(P)=\phi_i(S_k), \phi_j\left(L\right) = \phi_j\left(S_k^*\right)$. The preference relation $P \rhd L$ means division P is better than division L. Because the coalition's utility is non-transferable, the preference relation can be defined as:

$$P \rhd L \Leftrightarrow \left\{\phi_i\left(P\right) \geq \phi_i\left(L\right), \forall i \in P, L\right\} \tag{50}$$

Based on the Merge-and-Split Rules defined in the above Section, the procedure of the algorithm is shown in Table 8.

Due to the large-scale fading of wireless channel, the cooperation cost $I_{i,E}(S)$ will increase as the coalition scale increases. Therefore, the coalition formation will change with coalition member's mobility.

For simulations, a square network of 2.5km×2.5km is set up with 15 legitimate transmitters (N=15), 2 eavesdroppers (E=2), and 2 legitimate receivers (M=2) randomly deployed within this area. Figure 17 depicts how the legitimate transmitters form 6 coalitions in a self-organized way. The coalition member should be more than 2, because of the number of the eavesdropper is 2. Transmitter 4 and 15, having no suitable partners for forming a coalition of size larger than 2, do not cooperate. The result of coalition formation is namely the result of merge-and-split.

Table 8. The secure coalition formation algorithm

Coalition Formation Algorithm
Step 1: Initialization:The network is partitioned by $P=\{S_1,S_2,\ldots,S_p\}$ (At the beginning of all time $P=N=\{1,\ldots,N\}$ with non-cooperative users).
Step 2: Neighbor Discovery:
a) Each coalition surveys its neighborhood for candidate partners.
b) For every coalition S_i, the candidate partners lie in the area represented by the intersection of $\|S_i\|$ circles with centers $j\in S_i$ and radii determined by the distance where the power for information exchange does not exceed $P_i^{broadcast}\left(S_i\right)$ for any user.
Step 3: Adaptive Coalition Formation:
In this phase, coalition formation using merge-and-split occurs.
Repeat
a) $F=Merge(P)$: coalitions in P decide to merge based on the Merge-and-Split Rules and Equation 50.
b) $P=Split(F)$: coalitions in F decide to split based on the Merge-and-Split Rules and Equation 50.
Until merge-and-split terminates.
Step 4: Secure Transmission:
Each coalition's users exchange their information and transmit their data within their allotted slots.

Figure 17. A snapshot of a coalitional structure resulting from our proposed coalition formation algorithm for a network with N =15 users, M =2 destinations and, K =2 eavesdroppers.

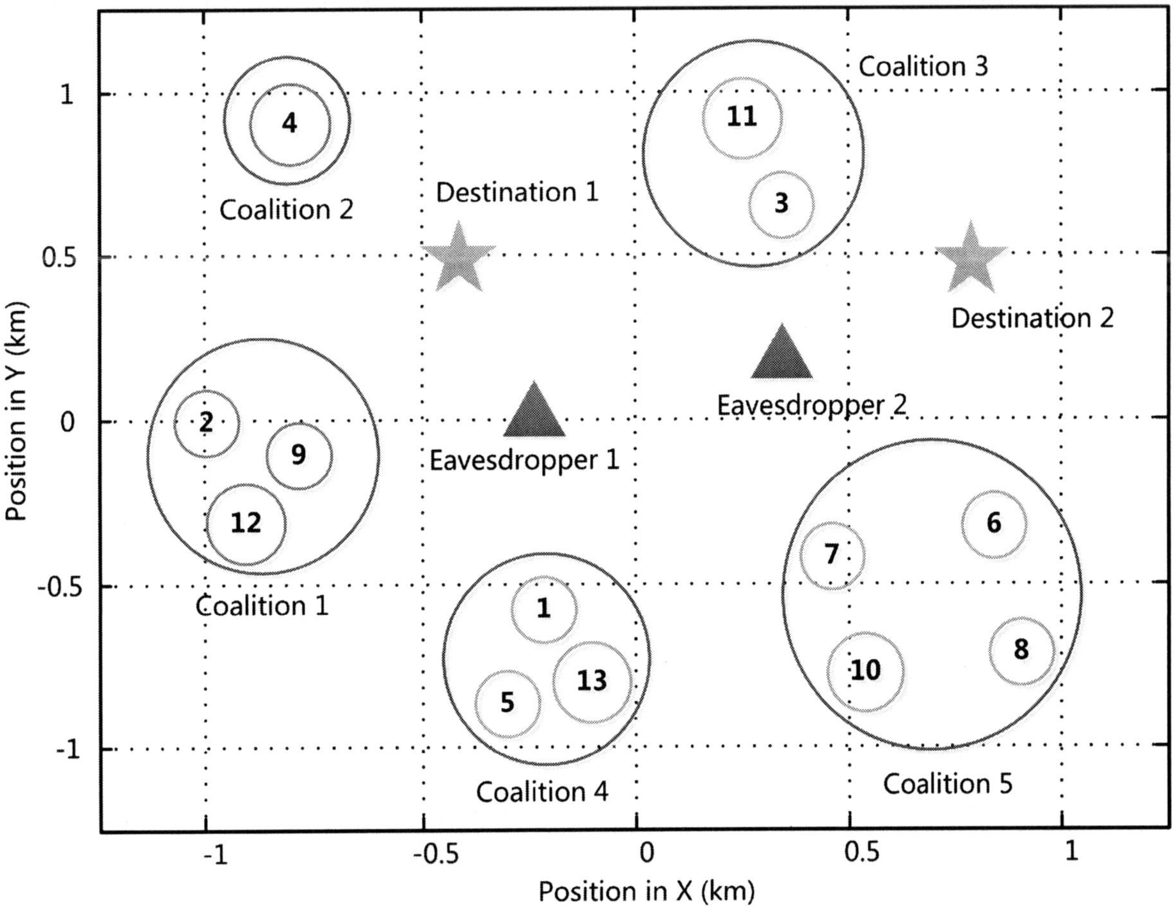

In Figure 18 we show how the algorithm handles mobility through appropriate coalition formation decisions. The transmitter 12 is moving horizontally for 1.1 km in the direction of the negative x-axis. Transmitter 12 starts getting closer to its receiver (legitimate receiver 2), and hence, it improves its utility. In the meantime, the utilities of transmitter 12's partners (transmitters 3 and11) drop due to the increasing cost. As long as the distance covered by transmitter 12 is less than 0.2 km, the coalition of transmitters 1 and 6 can still bring mutual benefits to all three users. After that, splitting occurs by a mutual agreement and all three transmitters transmit independently. When transmitter 12 moves about 0.8 km, it begins to distance itself from its receiver and its utility begins to decrease. When the distance covered by transmitter 12 reaches about 1 km, it will be beneficial to transmitters 12, 4, and 15 to form a 3-user coalition through the merge rule.

In Figure 19 we show the performance, in terms of average utility (secrecy rate) per transmitter, as a function of the network size N. For cooperation with coalitions, the average individual utility increases with the number of transmitters. This is interpreted by the fact that as the number of transmitters N increases, the probability of finding candidate partners to form coalitions increases for every transmitter. In contrast, for the non-cooperative approach an almost constant performance is noted.

In this section, job-hopping game is used to analyze how eavesdropper form a coalition with jammers firstly, then we the impact of malicious cooperation to system secrecy performance is analyzed. After that, a coalition game model with transferrable-utility is introduced to study the users' identity

Figure 18. Self-adaptation of the network's topology to mobility as User12 in Figure 17 moves horizontally on the negative x-axis

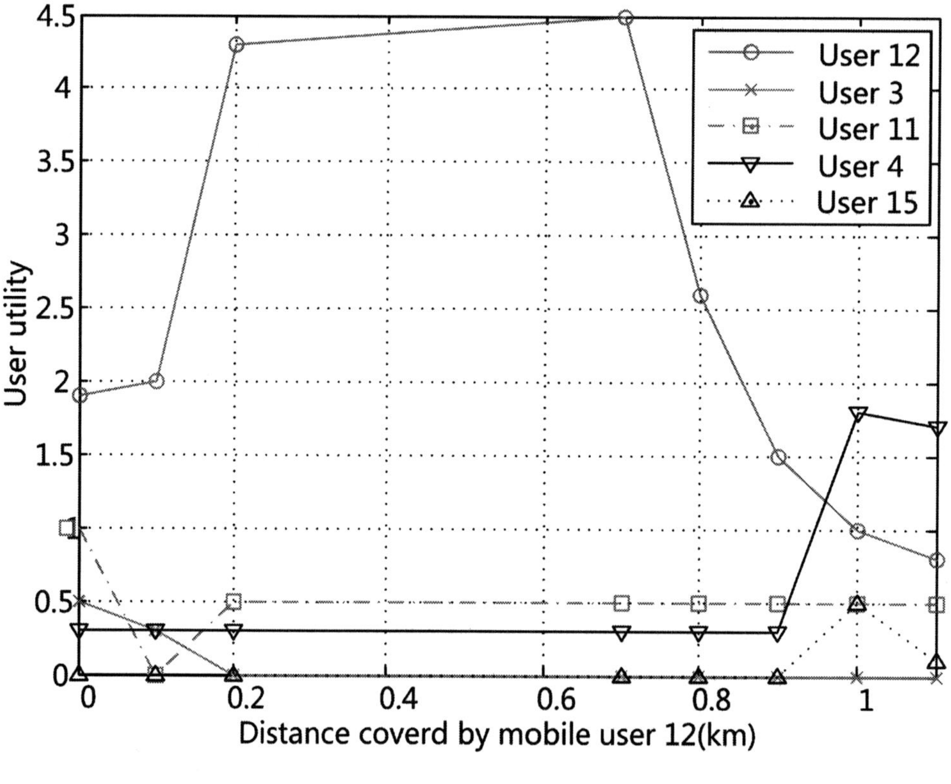

Figure 19. Performance in terms of the average individual user utility (secrecy rate) as a function of the network size N for M =2 destinations and K =2 eavesdroppers.

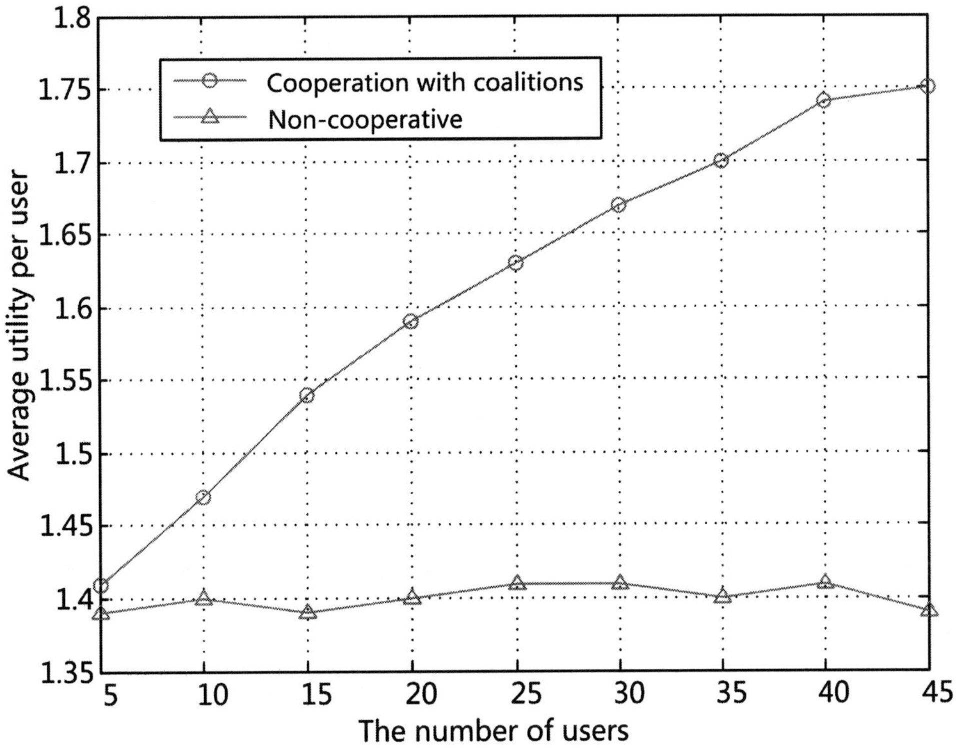

choice and the impact of cooperation to the secure communication. Finally, a coalition model with non-transferrable utility is introduced to study the problem of multiple legitimate transmitters' cooperation. The transmitters work at TDMA model, and use cooperative beamforming to restrain the eavesdroppers. Meantime, the cost and information leak during the information exchange is considered.

CONCLUSION AND FUTURE DIRECTIONS

As a mathematical tool, game theory can model the selfishness of nodes and its influence on the physical layer security performance of wireless networks. The chapter introduces the typical applications of game theory in physical layer security. Firstly, Zero-sum game is used to analysis the legitimate node's optimal secure communication strategy, when the malicious nodes can jam and eavesdrop. Secondly, the Stackelberg game is used to design a centralized secure interference power allocation algorithm among multiple legitimate transmitters and a friendly jammer. Thirdly, when the legitimate transmitters can not cooperate with each other, auction theory is used to design a distributed secure interference power allocation algorithm among multiple legitimate transmitters and a friendly jammer. Finally, the coalition game is used to analysis the cooperation among the legitimate nodes, the cooperation among the malicious nodes and their influences on the physical layer security performance of wireless networks.

With the development of wireless communications, some new issues emerge in the field of physical layer security. The application of large scale MIMO, overlay cognitive radio networks (Eun-Kyu, Gerla, & Oh, 2012), heterogeneous network (Budapest et al., 2013) bring new demand on the secure communication; the problem of potential eavesdroppers (Geraci et al., 2014) poses new challenges to the design of physical layer security algorithm; cross layer optimization (Liang, Poor, & Ying, 2011) and algorithm design under the non-ideal conditions of the physical layer security have important application value. How to choose the right game theory tools to analysis these problems is a potential research direction in the physical layer security field.

REFERENCES

Ausubel, L. M. (2004). An efficient ascending-bid auction for multiple objects. *The American Economic Review*, *94*(5), 1452–1475. doi:10.1257/0002828043052330

Ba, T., & Srikant, R. (2002). A Stackelberg network game with a large number of followers. *Journal of Optimization Theory and Applications*, *115*(3), 479–490. doi:10.1023/A:1021294828483

Budapest, Z. H. M., Xue, R., Yu, H., Luo, H., & Chen, W. (2013, June). Secrecy capacity optimization in coordinated multi-point processing. In *Proceedings of International Conference on Communications (ICC '13)*. Budapest, Hungary: IEEE.

Cao, Y., & Wei, J. (2012, October). Distributed coalition formation for selfish relays and eavesdroppers in wireless networks: A job-hopping game. In *Proceedings of Wireless Communications & Signal Processing (WCSP '12)*. Huangshan, China: IEEE. doi:10.1109/WCSP.2012.6542873

Dong, L., Han, Z., Petropulu, A. P., & Poor, H. V. (2008, September). Secure wireless communications via cooperation. In *Proceedings of* 46th *Annual Allerton Conference on Communication, Control, and Computing*. Urbana-Champaign, IL: IEEE. doi:10.1109/ALLERTON.2008.4797687

Dong, L., Han, Z., Petropulu, A. P., & Poor, H. V. (2010). Improving wireless physical layer security via cooperating Relays. *IEEE Transactions on Signal Processing*, *58*(3), 1875–1888. doi:10.1109/TSP.2009.2038412

Eun-Kyu, L., Gerla, M., & Oh, S. Y. (2012). Physical layer security in wireless smart grid. *IEEE Communications Magazine*, *50*(8), 46–52. doi:10.1109/MCOM.2012.6257526

Geraci, G., Singh, S., & Andrews, J. G. (2014). Secrecy Rates in Broadcast Channels with Confidential Messages and External Eavesdroppers. *IEEE Transactions on Wireless Communications*, *13*(5), 2931–2943. doi:10.1109/TWC.2014.041014.131101

Goel, S., & Negi, R. (2008). Guaranteeing secrecy using artificial noise. *IEEE Transactions on Wireless Communications*, *7*(6), 2180–2189. doi:10.1109/TWC.2008.060848

Kashyap, A., Basar, T., & Srikant, R. (2004). Correlated jamming on MIMO Gaussian fading channels. *IEEE Transactions on Information Theory*, *50*(9), 2119–2123. doi:10.1109/TIT.2004.833358

Krikidis, I., Thompson, J. S., & McLaughlin, S. (2009). Relay selection for secure cooperative networks with jamming. *IEEE Transactions on Wireless Communications*, *8*(10), 5003–5011. doi:10.1109/TWC.2009.090323

Krishna, V. (2009). *Auction Theory*. New York, NY: Academic Press.

Leung-Yan-Cheong, S. K., & Hellman, M. E. (1978). The Gaussian wiretap channel. *IEEE Transactions on Information Theory*, *24*(4), 451–456. doi:10.1109/TIT.1978.1055917

Li, J., Petropulu, A. P., & Weber, S. (2011). On cooperative relaying schemes for wireless physical layer security. *IEEE Transactions on Signal Processing*, *59*(10), 4985–4997. doi:10.1109/TSP.2011.2159598

Liang, Y., Poor, H. V., & Shamai, S. (2008). Secure communication over fading channels. *IEEE Transactions on Information Theory*, *54*(6), 2470–2492. doi:10.1109/TIT.2008.921678

Liang, Y., Poor, H. V., & Ying, L. (2011). Secure Communications Over Wireless Broadcast Networks: Stability and Utility Maximization. *IEEE Transactions on Information Forensics and Security*, *6*(3), 682–692. doi:10.1109/TIFS.2011.2158311

Liang, Y., Poor, H. V., & Ying, L. (2011). Secure communications over wireless broadcast networks: Stability and utility maximization. *IEEE Transactions on Information Forensics and Security*, *6*(3), 682–692. doi:10.1109/TIFS.2011.2158311

Massey, J. L. (1988). An introduction to contemporary cryptology. *Proceedings of the IEEE*, *76*(5), 533–549. doi:10.1109/5.4440

Mukherjee, A., & Swindlehurst, A. L. (2013). Jamming Games in the MIMO Wiretap Channel with an Active Eavesdropper. *IEEE Transactions on Signal Processing*, *61*(1), 82–91. doi:10.1109/TSP.2012.2222386

Myerson, R. (1997). *Game Theory: Analysis of Conflict*. Cambridge, MA: Harvard University Press.

Negi, R., & Goel, S. (2005, September). Secret communication using artificial noise. In *Proceedings of Vehicular Technology Conference (VTC '05)*. Tokyo, Japan: IEEE.

Osborne, M. J., & Rubinstein, A. (1994). *A course in game theory*. Cambridge, MA: MIT press.

Ray, D. A. (2007). *Game-Theoretic Perspective on Coalition Formation*. New York, NY: Oxford University Press. doi:10.1093/acprof:oso/9780199207954.001.0001

Saad, W., Han, Z., Debbah, M., Hjørungnes, A., & Basar, T. (2009). Coalition game theory for communication networks: A tutorial. *IEEE Signal Processing Magazine*, *26*(5), 77–97. doi:10.1109/MSP.2009.000000

Schneier, B. (1998). Cryptographic design vulnerabilities. *IEEE Computer*, *31*(9), 26–33. doi:10.1109/2.708447

Wang, A., Cai, Y., Yang, W., & Hou, Z. (2013). A Stackelberg security game with cooperative jamming over a multiuser OFDMA network. In *Proceedings of Wireless Communications and Networking Conference (WCNC '13)*. Shanghai, China: IEEE. doi:10.1109/WCNC.2013.6555246

Wyner, A. D. (1975). The wire-tap channel. *The Bell System Technical Journal, 54*(8), 1355–1387. doi:10.1002/j.1538-7305.1975.tb02040.x

Zhang, R., Song, L., Han, Z., & Jiao, B. (2011, April). Improve physical layer security in cooperative wireless network using distributed auction games. In *Proceedings of Conference on Computer Communications Workshops (INFOCOM WKSHPS '11).* Shanghai, China: IEEE.

Zhang, R., Song, L., Han, Z., & Jiao, B. (2011, June). Distributed coalition formation of relay and friendly jammers for secure cooperative networks. In *Proceedings of International Conference on Communications (ICC '11).* Milan, Italy: IEEE. doi:10.1109/icc.2011.5962513

KEY TERMS AND DEFINITIONS

Artificial Noise: The gaussian noise transmitted in the null space of the legitimate user.

Beamforming: A signal processing technique used in sensor arrays for directional signal transmission or reception.

Nash Equilibrium: A solution concept of a non-cooperative game involving two or more players, in which each player is assumed to know the equilibrium strategies of the other players, and no player has anything to gain by changing only their own strategy.

OFDMA: A multi-user version of the popular orthogonal frequency-division multiplexing (OFDM) digital modulation scheme. Multiple access is achieved in OFDMA by assigning subsets of subcarriers to individual users.

Perfect Security: An encryption algorithm is perfectly secure if a ciphertext produced using it provides no information about the plaintext without knowledge of the key.

Secrecy Rate: The communication rate for perfect security.

Utility: A measure of how good a situation or solution is for a user.

Chapter 13
A Game Theoretic Framework for Green HetNets Using D2D Traffic Offload and Renewable Energy Powered Base Stations

Elias Yaacoub
Strategic Decisions Group (SDG) and Arab Open University (AOU), Lebanon

Hakim Ghazzai
Qatar Mobility Innovations Center (QMIC), Qatar

Mohamed-Slim Alouini
King Abdullah University of Science and Technology (KAUST), Saudi Arabia

ABSTRACT

This chapter investigates the interplay between cooperative device-to-device (D2D) communications and green communications in LTE heterogeneous networks (HetNets). Two game theoretic concepts are studied and analyzed in order to perform dynamic HetNet base station (BS) on/off switching. The first approach is a coalition-based method whereas the second is based on the Nash bargaining solution. Afterwards, a method for coupling the BS on/off switching approach with D2D collaborative communications is presented and shown to lead to increased energy efficiency. The savings are additionally increased when a portion of the small cell BSs in a HetNet are powered by renewable energy sources. Different utility functions, modeling the game theoretic framework governing the energy consumption balance between the cellular network and the mobile terminals (MTs), are proposed and compared, and their impact on MT quality of service (QoS) is analyzed.

INTRODUCTION

Energy efficiency is representing an increasing concern for cellular network operators. Although the main purpose is to minimize their electricity costs and maintain profitability, reducing negative effects on the environment is also an important objective (Hasan, 2011).

A large portion of the energy dissipated in a cellular system is actually consumed at the base stations (BSs). Hence, putting certain BSs in sleep mode, or switching them off in light traffic conditions,

DOI: 10.4018/978-1-4666-8642-7.ch013

is an efficient technique to save energy in wireless networks, e.g., see (Bousia, 2012; Han, 2013). In (Niu, 2010), the cell size is adjusted dynamically depending on the traffic load using a technique called "cell zooming" for the purpose of reducing energy consumption. The power ratio, corresponding to the ratio between the dynamic and the fixed power part of a BS power consumption model, is introduced in (Xiang, 2011). This ratio is used to propose a solution based on traffic load balancing.

Several enhancements incorporated in next generation cellular systems, e.g., LTE-Advanced (LTE-A), consist of reducing effective cell sizes by using combinations of microcells (Lan, 2012), distributed antenna systems (Zhao, 2012), relays (Salem, 2010), and indoor femtocells (Hoydis, 2010). In this chapter, we use the term "small cells" to refer to a combination of these cells. Together with macrocells, they form a heterogenous network (HetNet), with HetNets expected to constitute a paradigm shift in state-of the-art cellular networks (Andrews, 2013). HetNets are one of the possible solutions to address the load increase on state-of-the-art cellular systems, such as LTE-A, due to the exponential increase in multimedia and data traffic. The operation of such HetNets is optimized through the use of advanced interference coordination/mitigation techniques, heterogeneous fractional frequency reuse patterns, and cooperative multipoint transmission/reception techniques. However, network densification brings with it the challenge of increased energy consumption in cellular BSs, which makes BS on/off switching techniques of high importance in HetNets (Bousia, 2012; Han, 2013). Furthermore, renewable energy sources, such as solar panels or wind turbines, could be used to power small cell BSs (SCBSs) whenever possible. This would be difficult to apply with macrocell BSs (MBSs) due to their larger energy consumption. Hence, renewable energy SCBSs (RE-SCBSs) could be used in conjunction with BS switch off techniques.

In addition, device-to-device (D2D) communications have been receiving significant research attention recently, due to their planned incorporation in future releases of LTE-Advanced (LTE-A) in beyond 4G and 5G cellular systems. D2D communications in LTE-A would allow a device to use the cellular spectrum in order to be connected directly to another device. This short range (SR) D2D connectivity allows offloading some traffic from the cellular network. Several previous work focused on using D2D as an underlay network to the traditional LTE/LTE-A cellular communication network (Ma, 2013; Wang & Chu, 2012; Wang, 2013). Other work considers maximizing the spatial reuse while using D2D communications (Lee, 2013). In (Han, 2012), D2D connections are allowed to use the same wireless resources as long range (LR) cellular links in order to maximize the spatial reuse gain. A resource allocation scheme to achieve this objective is proposed. Most of the previously cited references consider D2D in unicast mode. Multicasting in D2D underlay networks is investigated in (Wang & Wang, 2012; Chen, 2013), where the challenge is to group mobile terminals (MTs) into clusters receiving the multicast data through D2D communications. However, efficient cooperative D2D clustering schemes under LTE-A are still an active research topic.

In this chapter, a game theoretic framework based on utility maximization is proposed in order to ensure that green LTE/LTE-A HetNets operate with the least required number of BSs while maintaining a certain degree of quality of service (QoS). The main novelty of this work consists in using the combination of RE-SCBSs and D2D offload in order to achieve green LTE-A HetNets. This is done through a utility maximizing algorithm, along with the selection of utility functions that balance between the energy consumption in the BSs and the MTs to achieve mutual benefit and energy savings.

Starting from the scenario of reaching energy efficiency in green HetNets without D2D, two game theoretic concepts are studied and analyzed in order to perform dynamic BS on/off switching. The first approach is a coalition-based method that assumes a group of BSs in a network form a coalition. The second approach is based on the Nash bargaining solution and considers that BSs play a bargaining game where each BS attempts to maximize its own utility. Then, we extend the approach to a scenario with D2D communications. The objective is to use D2D traffic offload in order to reduce the load on cellular BSs, which allows putting more of them in sleep mode. An efficient approach for grouping MTs into coalitional groups, or cooperative clusters, is described. In each cluster, MTs cooperate via D2D communications to share content of common interest. When coupled with the on/off BS switching approach, the D2D clustering method leads to increased energy efficiency in the network. Furthermore, we investigate the scenario where small cell BSs in a HetNet are powered by renewable energy sources. In fact, due to their small size and limited coverage, certain small cell BSs can be powered by renewable energy sources, e.g., solar panels or wind turbines. In addition, the power consumption of mains powered small cell BSs would be less than that of macrocell BSs. This will favor having more small cells switched on at reduced cost since BSs powered by renewable energy powered do not incur power consumption costs.

After defining game theoretic frameworks for green HetNets and for D2D clustering, different utility functions, modeling the game theoretic framework governing the energy consumption balance between the cellular network and the MTs, are proposed and investigated. This is one of main novel aspects of the chapter, since previous work in the literature investigates separately green HetNets and D2D communications. Since most of the energy consumption in the network occurs at the BSs, and since D2D collaboration for traffic offload could entail and increase in the energy consumption of some MTs, the utility functions should reflect a sort of fairness towards MTs, so that their batteries are not depleted in order to increase the energy savings of cellular BSs. The results show that the joint use of D2D traffic offload and renewable energy small cell BSs leads to significant energy savings in the network while respecting the QoS requirements of the MTs.

SYSTEM MODEL

HetNet deployments are considered in this chapter, as shown in Figure 1. The game theoretic techniques are first proposed to a scenario without D2D communications, then extended to the case where D2D can be used to offload traffic and reduce the number of active BSs.

The game theoretic techniques discussed in this chapter will be used in conjunction with several presented utility functions in order to switch-off certain BSs and achieve energy efficiency in the network. The proposed methods can be applied to any combination of macrocells and small cells in the network, independently of the geometric layout of the BSs. Nevertheless, to simplify the simulations, we consider a geographical area of interest with uniform user distribution. The area is covered by a heterogeneous LTE network, consisting of macrocell BSs with a cell radius R_M, and small cell BSs with a smaller cell radius $R_S < R_M$, both placed on a square grid.

In the downlink (DL) direction of LTE, orthogonal frequency division multiple access (OFDMA) is used, whereas single carrier frequency division multiple access (SCFDMA) is used in the uplink (UL) direction (Myung, 2008). The LTE spectrum is subdivided into resource blocks (RB) where each RB consists of 12 adjacent subcarriers. The assignment of a single RB takes place every 1 ms, which is the duration of one transmission time interval (TTI), or the duration of two 0.5 ms slots in LTE (3GPP TS 36.213, 2014).

The data rates depend on the channel gain of each user on each subcarrier. The channel model adopted in this paper includes pathloss, lognormal shadowing, and Rayleigh fading. Intercell interference

Figure 1. HetNet deployment example

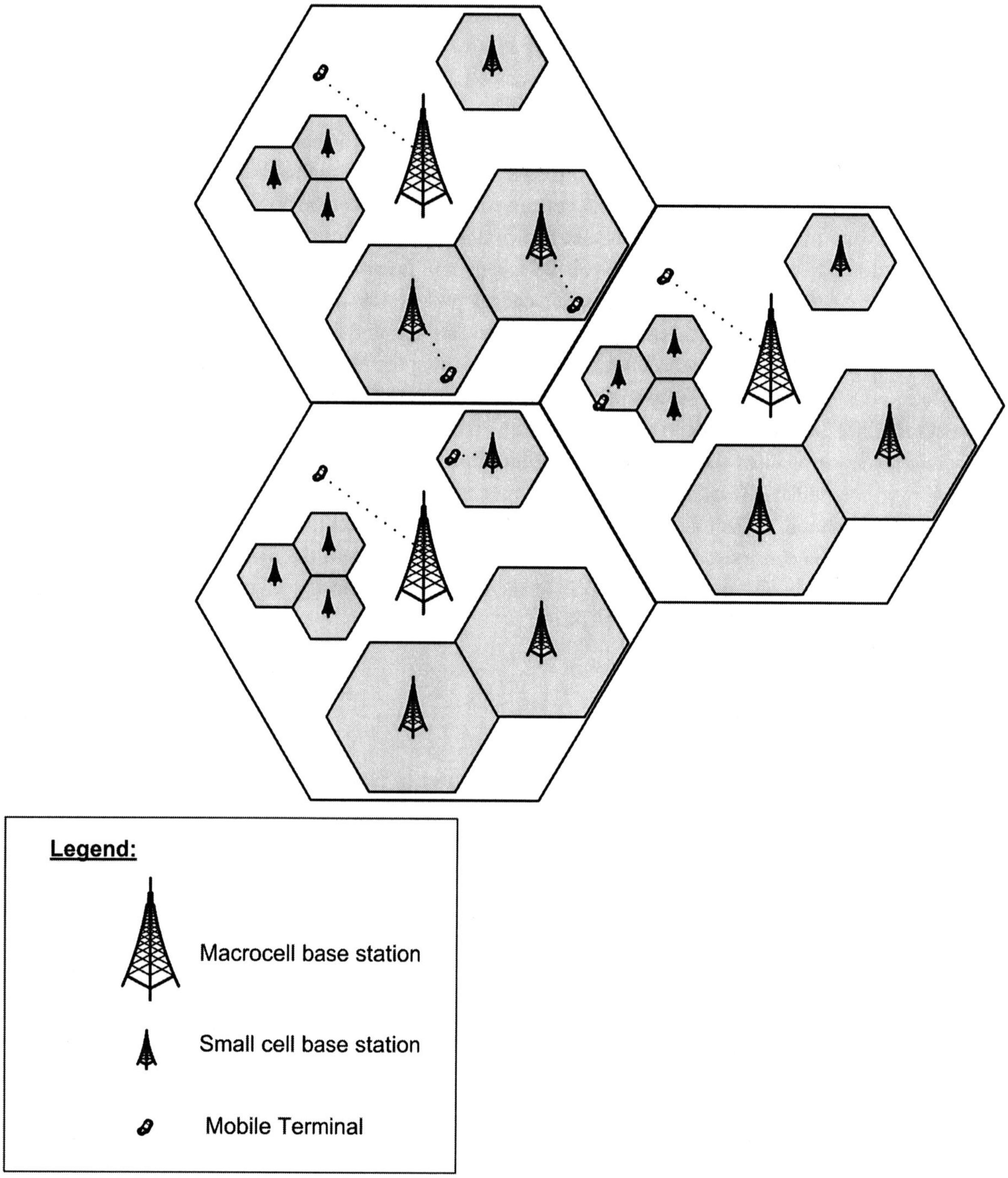

is also taken into account in the calculation of the signal to interference plus noise ratio (SINR), and thus affects the data rates achieved and consequently the resource allocation process. The details of the channel model, uplink and downlink SINR, interference, and data rate calculations are presented next.

Channel Model

This section presents the channel model adopted in this chapter. The channel gain of user k_l in cell l over subcarrier i on the link with BS j can be expressed as follows (Goldsmith, 2005):

$$H_{k_l,i,j,\mathrm{dB}} = (-\kappa - \upsilon \log_{10} d_{k_l,j}) - \xi_{k_l,j} + 10\log_{10} F_{k_l,i,j}, \tag{1}$$

where the first factor in (1) captures propagation loss, with κ the pathloss constant, $d_{k_l,j}$ the distance in km from user k_l to BS j, and v the path loss exponent. The second factor in (1), $\xi_{k_l,j}$, captures log-normal shadowing with zero-mean and a standard deviation σ_ξ, whereas the last factor in (1), $F_{k_l,i,j}$, corresponds to Rayleigh fading with a Rayleigh parameter a (usually selected such that $E[a^2]=1$). It should be noted that the above analysis applies to both the UL and DL, depending on whether i is an UL or DL subcarrier, respectively. In the rest of the chapter, whenever there is a need to distinguish between the uplink and downlink directions, the notation $H^{(\mathrm{UL})}_{k_l,i,j}$ and $H^{(\mathrm{DL})}_{k_l,i,j}$ will be used, respectively.

Data Rates in the Downlink

Letting $I^{(\mathrm{DL})}_{\mathrm{sub},k_l}$ be the set of DL subcarriers allocated to user k_l in cell l, $I^{(\mathrm{DL})}_{\mathrm{RB},k_l}$ the set of RBs allocated to user k_l in the DL, $N^{(\mathrm{DL})}_{\mathrm{RB}}$ the total number of DL RBs, $P^{\mathrm{tx}}_{i,l}$ the total power transmitted by the BS l over subcarrier i, .. the maximum transmission power of BS l, N_{BS} the total number of deployed BSs, K_l the total number of subscribers in th ecell covered by BS l and $R^{(\mathrm{DL})}_{k_l}$ the achievable DL rate of k_l. Then, the OFDMA throughput of user k_l is given by:

$$R^{(\mathrm{DL})}_{k_l}\left(P^{(DL)}_{l,\max}, I^{(\mathrm{DL})}_{\mathrm{sub},k_l}\right) = \sum_{i\in I^{(DL)}_{sub,k_l}} B^{(\mathrm{DL})}_{\mathrm{sub}} \cdot \log_2(1+\Gamma^{(\mathrm{DL})}_{k_l,i,l}), \tag{2}$$

where $\Gamma^{(\mathrm{DL})}_{k_l,i,l}$ is the SINR of user k_l over subcarrier i in cell l, and $B^{(\mathrm{DL})}_{\mathrm{sub}}$ is the subcarrier bandwidth. It is expressed as:

$$B^{(\mathrm{DL})}_{\mathrm{sub}} = \frac{B^{(\mathrm{DL})}}{N^{(\mathrm{DL})}_{\mathrm{sub}}}, \tag{3}$$

with $B^{(\mathrm{DL})}$ the total usable DL bandwidth, and $N^{(\mathrm{DL})}_{\mathrm{sub}}$ the total number of DL subcarriers.In this chapter, we consider equal power transmission over the subcarriers, i.e., for all i, we have:

$$P_{i,l}^{\mathrm{tx}} = \frac{P_{l,\max}^{(DL)}}{N_{\mathrm{sub}}^{(\mathrm{DL})}}. \tag{4}$$

The DL-SINR of user k_l over subcarrier i in cell l, $\Gamma_{k_l,i,l}^{(\mathrm{DL})}$, is given by:

$$\Gamma_{k_l,i,l}^{(\mathrm{DL})} = \frac{P_{i,l}^{\mathrm{tx}} H_{k_l,i,l}^{(\mathrm{DL})}}{I_{i,k_l}^{(\mathrm{DL})} + \sigma_{i,k_l}^2}, \tag{5}$$

where $H_{k_l,i,l}^{(\mathrm{DL})}$ is the channel gain of user k_l over subcarrier i in cell l, σ_{i,k_l}^2 is the noise power over subcarrier i in the receiver of user k_l, and $I_{i,k_l}^{(\mathrm{DL})}$ is the interference on subcarrier i measured at the receiver of user k_l. The expression of the interference is given by:

$$I_{i,k_l}^{(\mathrm{DL})} = \sum_{j=1, j\neq l}^{N_{BS}} \left(\sum_{k_j=1}^{K_j} \alpha_{k_j,i,j}^{(\mathrm{DL})} \right) \cdot P_{i,j}^{\mathrm{tx}} H_{k_l,i,j}^{(\mathrm{DL})}, \tag{6}$$

where $\alpha_{k_j,i,j}^{(\mathrm{DL})} = 1$ if DL subcarrier i is allocated to user k_j in cell j, i.e., $i \in I_{\mathrm{sub},k_j}^{(\mathrm{DL})}$. Otherwise, $\alpha_{k_j,i,j}^{(\mathrm{DL})} = 0$. In each cell, an LTE RB, and hence the subcarriers constituting that RB, can be allocated to a single user at a given Time Transmission Interval (TTI). Hence, in each cell j, we have:

$$\sum_{k_j=1}^{K_j} \alpha_{k_j,i,j}^{(\mathrm{DL})} \leq 1. \tag{7}$$

Data Rates in the Uplink

Let $I_{\mathrm{sub},k_l}^{(\mathrm{UL})}$ be the set of UL subcarriers allocated to user k_l, $I_{\mathrm{RB},k_l}^{(\mathrm{UL})}$ the set of UL RBs allocated to user k_l, $N_{\mathrm{RB}}^{(\mathrm{UL})}$ the total number of RBs in the UL, $P_{k_l}^{(UL)}$ the total transmit power of user k_l, and $R_{k_l}^{(\mathrm{UL})}$ its achievable rate in the UL. Then, the SCFDMA throughput of user k_l is given by:

$$R_{k_l}^{(\mathrm{UL})}\left(P_{k_l}^{(UL)}, I_{\mathrm{sub},k_l}^{(\mathrm{UL})}\right) = \frac{B^{(\mathrm{UL})} \mid I_{\mathrm{sub},k_l}^{(\mathrm{UL})} \mid}{N_{\mathrm{sub}}^{(\mathrm{UL})}} \cdot \log_2\left(1 + \Gamma_{k_l}^{(\mathrm{UL})}\left(P_{k_l}^{(UL)}, I_{\mathrm{sub},k_l}^{(\mathrm{UL})}\right)\right), \tag{8}$$

where $B^{(\mathrm{UL})}$ is the total UL bandwidth, $\mid I_{\mathrm{sub},k_l}^{(\mathrm{UL})} \mid$ is the cardinality of $I_{\mathrm{sub},k_l}^{(\mathrm{UL})}$ and $N_{\mathrm{sub}}^{(\mathrm{UL})}$ is the number of UL subcarriers. Finally, $\Gamma_{k_l}^{(\mathrm{UL})}\left(P_{k_l}^{(UL)}, I_{\mathrm{sub},k_l}^{(\mathrm{UL})}\right)$ is the SINR of user k_l after Minimum Mean Squared Error (MMSE) frequency domain equalization at the receiver (Myung, 2008):

$$\Gamma_{k_l}^{(\mathrm{UL})}\left(P_{k_l}^{(UL)}, I_{\mathrm{sub},k_l}^{(\mathrm{UL})}\right) = \left(\frac{1}{\frac{1}{\mid I_{\mathrm{sub},k_l}^{(\mathrm{UL})} \mid} \sum_{i \in I_{\mathrm{sub},k_l}^{(\mathrm{UL})}} \frac{\Gamma_{k_l,i,l}^{(\mathrm{UL})}}{\Gamma_{k_l,i,l}^{(\mathrm{UL})}+1}} - 1\right)^{-1}. \quad (9)$$

In (9), $\Gamma_{k_l,i,l}^{(\mathrm{UL})}$ is the UL SINR of user k_l over subcarrier i served by BS l. It is given by:

$$\Gamma_{k_l,i,l}^{(\mathrm{UL})} = \frac{P_{k_l,i,l}^{(\mathrm{UL})} H_{k_l,i,l}^{(\mathrm{UL})}}{I_{i,l}^{(\mathrm{UL})} + \sigma_{i,l}^2}, \quad (10)$$

where $H_{k_l,i,l}^{(\mathrm{UL})}$ is the channel gain between user k_l and BS l over subcarrier i, $\sigma_{i,l}^2$ is the noise power over subcarrier i at BS l, $P_{k_l,i,l}^{(\mathrm{UL})}$ is the power transmitted by user k_l over subcarrier i in cell l, and $I_{i,l}^{(\mathrm{UL})}$ is the UL interference on subcarrier i, measured at BS l. The expression of the interference is given by:

$$I_{i,l}^{(\mathrm{UL})} = \sum_{j=1, j\neq l}^{N_{BS}} \sum_{k_j=1}^{K_j} \alpha_{k_j,i,j}^{(\mathrm{UL})} P_{k_j,i,j}^{(\mathrm{UL})} H_{k_j,i,l}^{(\mathrm{UL})}, \quad (11)$$

where $\alpha_{k_j,i,j}^{(\mathrm{UL})} = 1$ if subcarrier i is allocated to user k_j served by BS j, i.e., $i \in I_{\mathrm{sub},k_j}^{(\mathrm{UL})}$. Otherwise, $\alpha_{k_j,i,j}^{(\mathrm{UL})} = 0$.

The LTE standard imposes the constraint that the RBs allocated to a single user should be consecutive with equal power allocation over the RBs (3GPP TS 36.211, 2014; 3GPP TS 36.213, 2014; Myung, 2008). Hence, we set:

$$P_{k_l,i,l}^{(\mathrm{UL})} = \frac{P_{k_l,\max}^{(UL)}}{\mid I_{\mathrm{sub},k_l}^{(\mathrm{UL})} \mid}. \quad (12)$$

User Admission and QoS Requirements

When a user k joins the network, it is associated with the best serving cell l^*, i.e the cell having the available UL and DL RBs that maximize the user's performance. Assuming one UL RB and one DL RB are allocated for each user, this corresponds to the RBs for which the UL and DL subcarriers, $i^{*(UL)}$ and $i^{*(DL)}$, satisfy (13) and (14), respectively:

$$(i^{*(\mathrm{UL})}, l^*) = \arg\max_{(i,l)} \left(1 - \sum_{k_l=1; k_l \neq k}^{K_l} \alpha_{k_l,i,l}^{(\mathrm{UL})}\right) R_{k,i,l}^{(\mathrm{UL})}, \quad (13)$$

$$i^{*(\mathrm{DL})} = \arg\max_{i}\left(1 - \sum_{k_l=1;k_l\neq k}^{K_l} \alpha_{k_l,i,l^*}^{(\mathrm{DL})}\right) R_{k,i,l^*}^{(\mathrm{DL})}, \tag{14}$$

where $R_{k,i,l}^{(\mathrm{UL})}$ and $R_{k,i,l}^{(\mathrm{DL})}$ represent the UL and DL achievabledata rates, respectively, of user k over subcarrier i in cell l. The first term in the multiplications of (1) and (2) indicates that the search is on the RBs that are not yet allocated to other users, where $\alpha_{k_l,i,l}^{(\mathrm{UL})} = 1$ if UL subcarrier i is allocated to user k_l in cell l. Otherwise, $\alpha_{k_l,i,l}^{(\mathrm{UL})} = 0$. The same rules apply for DL subcarriers.

A user is considered to be successfully served if the following conditions are satisfied:

$$\begin{Bmatrix} R_{k_l}^{(\mathrm{UL})} \geq R_{\mathrm{Target},\ k_l}^{(\mathrm{UL})} \\ R_{k_l}^{(\mathrm{DL})} \geq R_{\mathrm{Target},\ k_l}^{(\mathrm{DL})} \end{Bmatrix}, \tag{15}$$

where $R_{k_l}^{(\mathrm{UL})}$ and $R_{k_l}^{(\mathrm{DL})}$ are the respective UL and DL data rates of user k_l in cell l, aggregated over all its allocated UL and DL subcarriers, respectively. $R_{\mathrm{Target},\ k_l}^{(\mathrm{UL})}$ and $R_{\mathrm{Target},\ k_l}^{(\mathrm{DL})}$ are the UL and DL target data rates, respectively, representing the QoS constraints. They can vary depending on the service used by the user. Hence, a user is considered to be in outage if at least one of the conditions in (15) is not met.

PROPOSED GAME THEORETIC TECHNIQUES

This section describes two game theoretic methods proposed in this chapter and implemented first in HetNets without using D2D for energy efficiency. The first approach is a cooperative coalition-based method where a group of BSs in a network are considered to form a coalition. The second approach is based on BS competition where each BS aims to maximize its own utility.

BS Coalition Approach

BSs in a certain geographical area are assumed to cooperate together by forming a coalition. Hence, the objective would be to maximize the benefits of the coalition as a whole, not of individual BSs. Considering there are N_{BS} BSs in the coalition, each having its own payoff function or utility, such that the utility of BS l is denoted by U_l, then the objective is to maximize the total utility of the coalition as follows:

$$\max_{\alpha_{k_l,i,l}^{(DL)},\alpha_{k_l,i,l}^{(UL)},P_l^{(DL)},P_{k_l}^{(UL)}} \left(\sum_{l=1}^{N_{BS}} U_l\right) \tag{16}$$

Subject to:

$$P_{k_l}^{(UL)} \leq P_{k_l,\max}^{(UL)}; \forall k_l = 1,\cdots,K_l; \forall l = 1,\cdots,N_{BS} \tag{17}$$

$$P_l^{(DL)} \leq P_{l,\max}^{(DL)}; \forall l = 1, \cdots, N_{BS} \tag{18}$$

$$\sum_{k_l=1}^{K_l} \alpha_{k_l,i,l}^{(UL)} \leq 1; \forall i = 1, \cdots, N_{sub}^{(UL)}; \forall l = 1, \cdots, N_{BS} \tag{19}$$

$$\sum_{k_l=1}^{K_l} \alpha_{k_l,i,l}^{(DL)} \leq 1; \forall i = 1, \cdots, N_{sub}^{(DL)}; \forall l = 1, \cdots, N_{BS} \tag{20}$$

$$\sum_{l=1}^{N_{BS}} \frac{N_{out,l}}{N_{served,l} + N_{out,l}} \leq P_{out,th} \tag{21}$$

The constraints in (17) and (18) indicate that the transmit power cannot exceed the maximum power for the UL and DL, respectively. The constraints in (19) and (20) correspond to the exclusivity of subcarrier allocations in each cell for the UL and DL, respectively, since in each cell, a subcarrier can be allocated at most to a unique user at a given scheduling instant. Finally, the constraint in (21) is related to enforcing QoS, where $N_{out,l}$ corresponds to the number of users in outage in cell l, i.e., the users associated with cell l as their best serving cell according to (13) and (14), but that were not able to satisfy their QoS requirements in (15). $N_{served,l}$ indicates the number of users served successfully in cell l. Hence, this constraint indicates that the total outage rate in the network should not exceed a tolerated outage threshold $P_{out,th}$.

To perform this sum-utility maximization, Algorithm 1 is implemented. In this algorithm, we introduce two indicator variables: I_j^{ON} j indicates if a BS j is on or off, by setting its value to 1 or 0, respectively, whereas I_j^{attempt} is a tracking parameter that indicates whether an attempt has been made to switch BS j off in the current iteration or not. It is set to 1 if the attempt was made and to 0 otherwise. The loop at Lines 1-4 is an initialization phase. In the Loop at Lines 5-23, the algorithm finds the BS having the weakest individual utility, then checks if the reassignment of its users to other BSs and putting it in sleep mode leads to an enhancement for the coalition's utility. If an enhancement is reached, the BS is switched off. Otherwise it is kept on. Then, the algorithm moves to the next BS, and so on. The iterations are repeated until no improvement can be made in the sum-utility, even if an attempt is made on all the BSs that remained "on" (which in this case will lead to $\prod_{j=1;I_j^{ON}=1}^{N_{BS}} I_j^{\mathrm{attempt}} = 1$ and allows exiting the loop at Line 5).

In this approach, the BS acts in the benefit of the coalition by allowing its utility to be set to zero if this leads to an increase in the utility of the coalition. This can be implemented in the case of centralized control in the coalition, or in the case of a single operator that would hardwire the intelligence of Algorithm 1 in all its BSs.

Algorithm 1: Uitlity Maximization Algorithm

1: **for all** BS *j* **do**
2: $I_j^{\text{ON}} = 1$
3: $I_j^{\text{attempt}} = 1$
4: **end for**
5: **while** $\prod_{j=1;I_j^{ON}=1}^{N_{BS}} I_j^{\text{attempt}} = 0$ do
6: Find: $j^* = \arg\min_{I_j^{ON}=1, I_j^{\text{attempt}}=0} U_j$
7: **for all** k_{j*} served by BS *j** **do**
8: Implement (13) and (14) after excluding *j** from the BS search in (13); i.e. the search is done over BSs $l \neq j^*$
9: **end for**
10: **for all** $j \neq j^*$ such that $I_j^{\text{ON}} = 1$ **do**
11: Compute $U_j^{(\text{new})}$ obtained after moving the users k_{j*} as described above
12: Set $U_{j^*}^{(\text{new})} = 0$
13: **end for**
14: **if** $\sum_{j=1}^{N_{BS}} U_j^{(\text{new})} > \sum_{j=1}^{N_{BS}} U_j$ then
15: **for all** *j* such that $I_j^{\text{ON}} = 1$ **do**
16: Set $U_j = U_j^{(\text{new})}$ and $I_j^{\text{attempt}} = 0$
17: **end for**
18: Set $I_{j^*}^{\text{ON}} = 0$
19: **else**
20: No changes are made (keep the old utility values)
21: Set $I_{j^*}^{\text{attempt}} = 1$
22: **end if**
23: **end while**

BS Competition Approach

In this section, we consider that BSs are competing selfishly to maximize their individual utilities. However, we consider that they negotiate the allocation of resources and users by communicating with each other. This corresponds in game theory to a bargaining game, where each player attempts to maximize its payoff (or utility) by bargaining with other players to share the resources in a way they cannot jointly improve on. The solution in game theory to this bargaining game consists of maximizing the Nash product, and it is known as the Nash bargaining solution (NBS) (Carmichael, 2005). Hence, this translates into:

$$\max_{\alpha_{k_l,i,l}^{(DL)}, \alpha_{k_l,i,l}^{(UL)}, P_l^{(DL)}, P_{k_l}^{(UL)}} \left(\prod_{l=1}^{N_{BS}} U_l \right), \tag{22}$$

subject to the constraints (17)-(21). Since the logarithm is a continuous strictly increasing function, (22) is equivalent to:

$$\max\left(\prod_{l=1}^{N_{BS}} U_l\right) \Leftrightarrow \max \ln\left(\prod_{l=1}^{N_{BS}} U_l\right) = \max \sum_{l=1}^{N_{BS}} \ln\left(U_l\right). \tag{23}$$

Interestingly, the algorithmic implementation of (22) can be handled by Algorithm 1, by using, in that algorithm, $U_l^{(\mathrm{NBS})} = \ln\left(U_l\right)$ as the BS utility instead of U_l. Thus, to model a bargaining game in a practical network, bargaining "negotiations" do not need to take place between BSs, and the implementation of Algorithm 1 with $U_l^{(\mathrm{NBS})} = \ln\left(U_l\right)$ is sufficient to lead to the NBS, i.e., to the equilibrium solution of the bargaining problem. Hence, with this approach, the BS knows it is achieving its best possible utility, given the utilities of the other BSs and the conditions of the network. Thus, the NBS solution can be hardwired in the BSs, even if multiple operators are involved (in this case, pricing and billing issues due to potentially moving subscribers of one operator to the BS of another by Algorithm 1 should be taken into account, e.g., by being included in a suitable utility).

UTILITY CALCULATIONS

This section presents utility metrics that can be used with the proposed game theoretic methods of the previous section. The two utilities presented focus either on traffic load and QoS, or on the QoS versus the consumed power in the network.

Utility Based on Load and QoS Performance

In this section, we define a utility that depends on the traffic load and QoS performance of each BS. It is selected as follows:

$$U_l = N_{\mathrm{served},l} \cdot \exp\left(P_{\mathrm{out,th}} - \frac{N_{\mathrm{out},l}}{N_{\mathrm{served},l} + N_{\mathrm{out},l}}\right). \tag{24}$$

The utility in (24) increases with the number of served users and decreases with the number of users in outage. When the outage rate exceeds the tolerated threshold $P_{\mathrm{out,th}}$, the exponential term in (24) becomes negative and the utility decreases quickly towards zero. If no users are served by a certain BS, then $U_l = 0$ and the BS will be switched off by Algorithm 1.

In the NBS case, we have: $U_l^{(\mathrm{NBS})} = \ln\left(U_l\right)$. However, to avoid having ln(0) in computer implementations, the utility needs to be redefined for boundary conditions. Thus, when a BS is "on", we set the utility to:

$$U_l^{(\mathrm{NBS})} = \left\{ \begin{array}{l} \ln\left(N_{\mathrm{served},l}\right) + \left(P_{\mathrm{out,th}} - \dfrac{N_{\mathrm{out},l}}{N_{\mathrm{served},l} + N_{\mathrm{out},l}}\right); N_{\mathrm{served},l} > 0 \\ -1 + \left(P_{\mathrm{out,th}} - \min\left(N_{\mathrm{out},l}, 1\right)\right); N_{\mathrm{served},l} = 0 \end{array} \right\}. \quad (25)$$

When a BS is switched-off, we set U_l=0 and $U_l^{(\mathrm{NBS})} = 0$. In the NBS case with $N_{\mathrm{serverd},l}$=0, the utility in (25) will have a negative value. This will favor the switching-off of the corresponding BS by Algorithm 1, since this will lead to an increase in its utility (which will become zero).

Utility Based on Load and Power Consumption

The utility defined in (24) does not explicitly depend on the consumed power at the BS. In this section, we define a utility that reflects the number of served users versus the power consumption in the network. The selected utility is given by:

$$U_l = \frac{N_{\mathrm{served},l}}{P_{C,l}}, \quad (26)$$

where $P_{C,l}$ is the total power consumed by BS l (not to be confused with the transmit power at the antenna, which is included as a fraction of this power term). In general, utilities aiming at maximizing the sum rate while being concerned with energy efficiency can be defined in terms of bps/Hz/Watt. However, in this chapter, the interest is in maximizing the number of served users satisfying the constraints in (15), while minimizing the power consumption in the network.

In the NBS case, we have: $U_l^{(\mathrm{NBS})} = \ln\left(U_l\right)$. Thus, when a BS is "on", we obtain:

$$U_l^{(\mathrm{NBS})} = \left\{ \begin{array}{l} \ln\left(N_{\mathrm{served},l}\right) - \ln\left(P_{C,l}\right); N_{\mathrm{served},l} > 0 \\ -1 - \ln\left(P_{C,l}\right); N_{\mathrm{served},l} = 0 \end{array} \right\}. \quad (27)$$

When a BS is switched-off, we have U_l=0 and $U_l^{(\mathrm{NBS})} = 0$.

EXTENSION TO THE CASE WITH D2D COMMUNICATIONS

In this part of the chapter, we extend the proposed method to the case where D2D communications are implemented for the purpose of energy efficiency. In this case, the concern is not only for saving energy in the cellular networks, but also in reducing the energy consumed from the MTs' batteries. The presented method is presented in a general framework first then customized to a multicasting scenario as a special case. In this scenario, content of common interest is to be delivered from the BS to requesting MTs distributed throughout the cell area of the BS. MTs communicate with the BS using LTE, or with neighboring MTs using D2D communications. MTs form cooperating clusters for the purpose of energy minimization during cooperative content distribution. Within each cooperating cluster, the content is delivered on the LR to a single MT, the cluster head (CH), which in turn sends the data to the other MTs in that cluster using SR collaboration. Figure 2 shows the scenario considered.

At a given fading realization, each MT receives the data content from a single source, which could be either the BS or another MT. We assume that MTs form coalitions where the energy consumption in

Figure 2. D2D collaboration used to enhance BS switch off

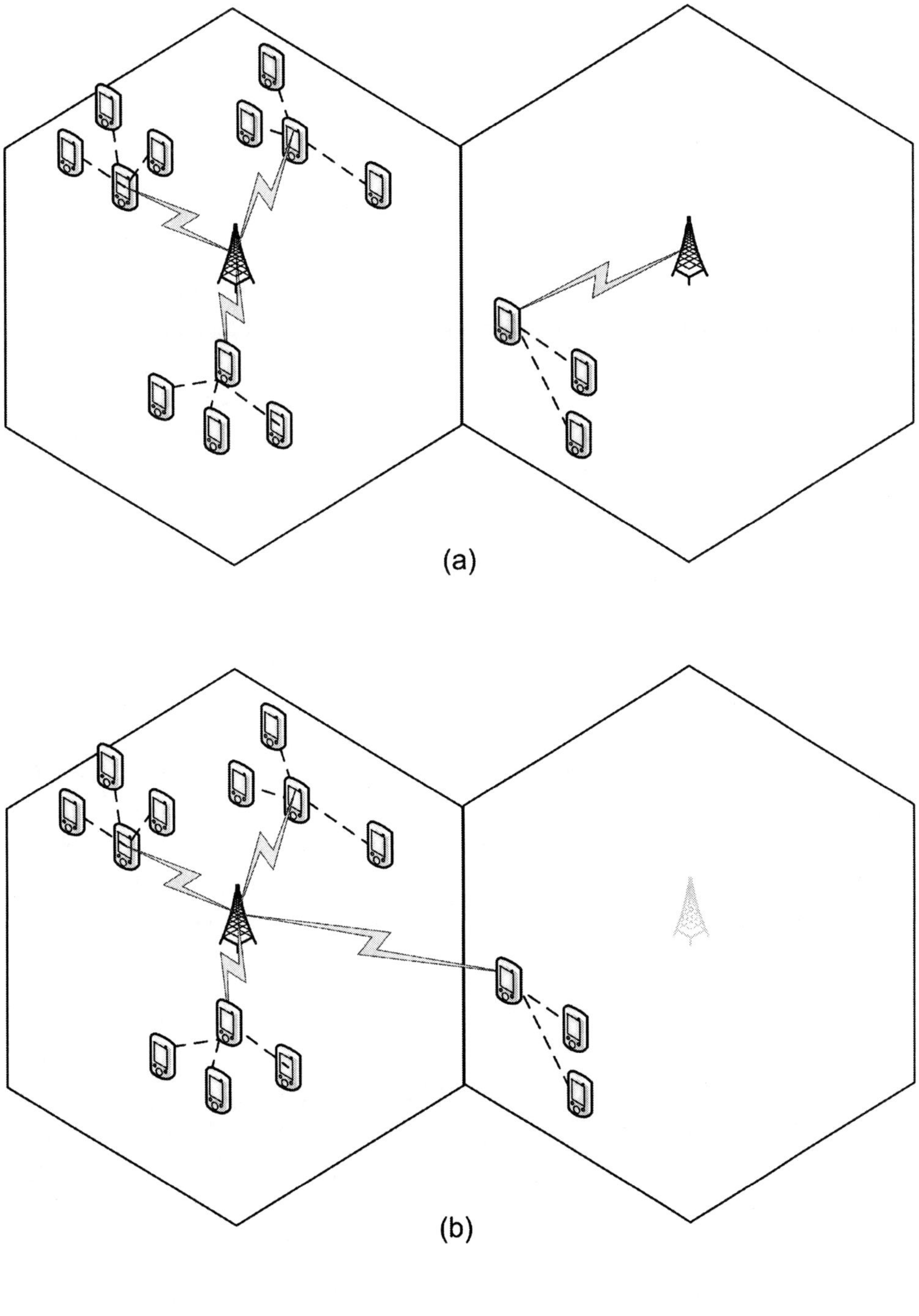

the coalition is lower than the sum of the individual energy consumptions of the coalition members. In a coalition cluster C_k, n_k is the cluster head communicating on the LR with the BS on behalf of all the cluster members. After forming the clusters as shown in Figure 2(a), the BS switch off techniques can be implemented by considering only the CHs that are still connected to the BSs on the LR. This allows enhancing the switch off process and increasing energy savings since a reduced number of MTs (CHs only) have to be offloaded from a given BS in order to put it in sleep mode, as it is shown in Figure 2(b).

Rate and Interference Calculations

In this section, we adapt the notation to the scenario where D2D communications are involved in the green operation of the network. We use the term 'node' to refer to either an MT or a BS. The N_{BS} BSs are numbered from node n_1 to node $n_{N_{BS}}$. Having K MTs in the system ($K = \sum_{j=1}^{N_{BS}} K_j$), they are numbered from node $n_{N_{BS}+1}$ to node $n_{N_{BS}+K}$. We denote by C_k the coalition of nodes forming a single cooperative cluster with n_k as CH communicating on the LR with a BS on behalf of all the cluster members. We generalize the notation of Section "System Model" by using the notation X_{kij} to denote the value of a quantity X between nodes n_k and n_j over subcarrier i (X could be, for example, the data rate R, the channel gain H, the transmit power P, etc.) .

With this generalized notation, the interference expression can be made more general in order to take D2D communications into account. In fact, the transmission from node n_k to node n_l over subcarrier i is subjected to the following interference, measured at n_l:

$$I_{k,i,l} = \sum_{\substack{m=1,\\ m\neq k}}^{N_{BS}+K} \sum_{\substack{j=1,\\ j\neq l}}^{N_{BS}+K} \alpha_{m,i,j} P_{m,i,j} H_{m,i,l}. \tag{28}$$

However, in the scenario of this chapter, a link is broken with the BS in order to use D2D communications instead. Thus, interference between SR D2D communications and the LR communications with the BS in a single cell can be avoided by allocating the freed subcarriers from the LR MT-BS link to be used for the SR D2D communications instead. Furthermore, selecting UL subcarriers for D2D allows reducing the interference from D2D transmissions to cell edge users in neighboring cells. Instead, interference from D2D will be received at the BSs of other cells, which are relatively at a larger distance. Coupled with reduced transmission power on the SR due to the short distances, this interference will be significantly reduced.

Energy Calculations

In this section, we discuss the energy consumption of MTs in order to take it into account in the green operation of the network, in addition to the energy consumption of BSs. The time needed to transmit a content of size S_T bits on a link between nodes n_k and n_j having an achievable rate R_{kj} bps is given by S_T/P_{kj}. Denoting the power drained from the battery of node n_j to receive the data from node n_k by $P_{Rx,kj}$, then the energy consumed by n_j to receive the data from n_k is given by $S_T \bullet P_{Rx,kj}/R_{kj}$. Similarly, denoting by $P_{Tx,k}$

the power drained by the battery of n_k to transmit the data via multicasting, then the energy consumed by n_k to transmit the content to n_j is given by $S_T \bullet P_{Tx,k}/R_{kj}$. It should be noted that $P_{Tx,k}$ can be expressed as:

$$P_{\mathrm{Tx},k} = P_{\mathrm{Tx}_{\mathrm{ref}},k} + P_{t,k}, \tag{29}$$

where $P_{\mathrm{Tx}_{\mathrm{ref}},k}$ corresponds to the power consumed by the circuitry of node n_k during transmission on the communication interface, and $P_{t,k}$ corresponds to the power transmitted over the air interface by n_k.

We present the formulation first in a general case where different data are transmitted to the different users and where both UL and DL directions are taken into account. Denoting by E_{C_k} the energy consumed by the MTs that are members of cluster C_k with node n_k as cluster head, and assuming, without loss of generality, that we are considering cell l with n_l as BS in this section, then the energy consumed in C_k is given by: $E_{C_k} = E_{C_k}^{(UL)} + E_{C_k}^{(DL)}$.

The cluster energy consumption for UL transmission is given by:

$$E_{C_k}^{(UL)} = \sum_{j \neq k; n_j \in C_k} S_{jk} \cdot \left(\frac{P_{\mathrm{Tx},jk}}{R_{jk}} + \frac{P_{\mathrm{Rx},jk}}{R_{jk}} + \frac{P_{\mathrm{Tx},kl}}{R_{kl}} \right), \tag{30}$$

where the first term corresponds to the energy consumed by the nodes to transmit their data to node n_k on the SR via D2D communications, the second term corresponds to the energy consumed by node n_k for reception on the SR, and the last term corresponds to the energy consumed by node n_k for transmission to the BS on the LR cellular link. S_{jk} denotes the number of bits transmitted from n_j to BS l through CH n_k. The summation is used to indicate that n_k aggregates the data of the MTs within its coalition cluster before transmitting it to the BS.

Similarly, the cluster energy consumption for DL transmission is expressed as:

$$E_{C_k}^{(DL)} = \sum_{j \neq k; n_j \in C_k} S_{kj} \cdot \left(\frac{P_{\mathrm{Rx},lk}}{R_{lk}} + \frac{P_{\mathrm{Tx},kj}}{R_{kj}} + \frac{P_{\mathrm{Rx},kj}}{R_{kj}} \right). \tag{31}$$

In (31), S_{kj} denotes the number of bits transmitted to n_j through CH n_k. The first term corresponds to the energy consumed by node n_k to receive the data destined to the various nodes in C_k from BS l on the LR link, the second term corresponds to the energy consumed by node n_k for transmission on the SR via D2D communications, and the last term corresponds to the energy consumed by the nodes in C_k to receive their data from node n_k on the SR via D2D communications.

In the results of this chapter, we consider the DL scenario (31). Without loss of generality, we assume that all MTs are interested in downloading files of similar size $S_{kj}=S_T$ (set to S_T=1Mbits in the simulations). In (Yaacoub, 2013), the scenario where users are interested in the same content was considered. In this case, the equation (31) can be customized to a scenario where multicasting is used in the DL. Hence, the energy consumed in C_k with multicasting is given by:

$$E_{C_k} = \frac{S_T \cdot P_{\text{Rx},lk}}{R_{lk}} + \frac{S_T \cdot P_{\text{Tx},k}}{\min_{i \neq k; n_i \in C_k} R_{ki}} + \sum_{j \neq k; n_j \in C_k} \frac{S_T \cdot P_{\text{Rx},kj}}{\min_{i \neq k; n_i \in C_k} R_{ki}}, \tag{32}$$

where the first term corresponds to the energy consumed by node n_k to receive the data from the BS on the LR cellular link, the second term corresponds to the energy consumed by node n_k to transmit the data to the other nodes in its cluster on the SR via D2D communications, and the last term corresponds to the energy consumed by the nodes to receive their data from node n_k on the SR. To avoid multiple transmissions among nodes of the same cluster, MT n_k transmits using multicasting. Therefore, the second term does not involve any summation over the MTs, conversely to the third term. In addition, with SR multicasting, $R_{kj} = \min_{i \neq k; n_i \in C_i} R_{ki}$, since transmission should take place at the minimum achievable rate in the cluster in order to guarantee that all MTs in the cluster receive the desired information. When all MTs have similar characteristics in terms of power consumption, then we have: $P_{Rx,lk}=P_{Rx,LR} \forall k,j$ for the power drained from the batteries of the MTs during reception on the LR from the BS, $P_{Rx,kj}=P_{Rx,SR} \forall k,j$ for the power drained from the batteries of the MTs during reception on the SR from the CHs, and $P_{Tx,k}=P_{Tx,SR} \forall k$ for the power drained from the batteries of the CHs during multicasting on the SR to the other MTs. In this case, (32) can be simplified as follows:

$$E_{C_k} = S_T \cdot \left(\frac{P_{\text{Rx,LR}}}{R_{lk}} + \frac{P_{\text{Tx,SR}} + (|C_k| - 1) P_{\text{Rx,SR}}}{\min_{i \neq k; n_i \in C_k} R_{ki}} \right), \tag{33}$$

where |•| denotes set cardinality.

It should be noted that the work in (Yaacoub, 2013) did not consider neither HetNets nor RE-SCBSs as in this chapter. In addition, it did not compare various utility functions derived based on a game theoretical frawmework. Although the results presented in this chapter are focused on the DL transmission, from the BS to the MTs, the proposed approach can be applied to the joint UL/DL scenario, by simply replacing, in the remaining sections, the energy expression of (31) by $E_{C_k} = E_{C_k}^{(UL)} + E_{C_k}^{(DL)}$, that is, the sum of energies in (30) and (31). It should be noted that, because different data are being transmitted on each SR link via unicasting in (30) and (31), the actual rates R_{kj} are used instead of the minimum SR rate in the cluster as in the multicasting case of (32) and (33). With the scenarios of (30) and (31), a CH needs to aggregate the data of all the MTs in its cluster before transmitting the data to/from the BS. Although this leads to additional energy consumption, it was still shown to lead to significant gains in (Yaacoub, 2012), where UL clustering was considered in wireless sensor networks, but within a single macrocell (i.e., without considering green networking in a multiple cell scenario).

Algorithms to Implement the Proposed Approach

This section describes the proposed method based on joint D2D offload and green networking through BS switch off. First, D2D clustering is performed according to Algorithm 2. Then, Algorithm 1 is implemented after D2D clustering is performed with Algorithm 2. Algorithm 3 summarizes the procedure, which is based on Algorithms 1 and 2.

Algorithm 2 describes the D2D cluster formation approach. After finding the best candidate clusters at Lines 3 and 4, they are merged together if the merger leads to a cluster with better energy efficiency (Lines 5-8). Otherwise no changes are made. The process is repeated (Loop at Lines 2-10) until no additional improvement is reached. The step at Line 1 of Algorithm 3 leads to offloading traffic from the cellular network via D2D communications, while the step at Line 2 leads to switching redundant BSs off using Algorithm 1, which is facilitated by having to consider only the links between the CHs and the BSs instead of the links between all MTs and the BSs.

Algorithm 2: D2D Cluster Formation Algorithm

1: At the start of the proposed method, all the nodes n_k are directly connected to the BS via LR LTE links; i.e., each cluster consists of a single node such that $C_k = \{n_k\}$ and $|C_k| = 1$. This is equivalent to the scenario without D2D collaboration. All clusters are in the search space $S = \{k;\ C_k \neq \varnothing\}$.
2: **while** $S \neq \varnothing$ **do**
3: Find the cluster having the highest energy consumption per node: $k = \arg\max_{m \in S} E_{C_m} / |C_m|$.
4: Find the cluster C_j that leads to the lowest energy consumption when merged with cluster C_k: $j = \arg\min_{m \neq k} E_{C_m \cup C_k}$.
5: **if** $E_{C_j \cup C_k} \leq E_{C_j} + E_{C_k}$ then
6: Set $C'_j = C_j \cup C_k$
7: Set $C_j = C'_j$ and $C_k = \varnothing$
8: **end if**
9: Remove k from the search space: $S = S \setminus \{k\}$.
10: **end while**

Algorithm 3: Joint Green/D2D Algorithm

1: Form D2D clusters by implementing Algorithm 2.
2: Implement the green communications method of Algorithm 1.

Utility Calculation for Green Networking with D2D

As the purpose in green networks is energy efficiency, the simplest way to set the utility value is to consider minimizing the energy consumption in the network. However, considering only BS energy is unfair towards MTs when D2D offload is used. In fact, CHs consume additional energy to relay data to other MTs instead of just receiving their own. Thus, MT energy consumption should be taken into account in addition to BS energy consumption. But it should be noted that MT battery and transmit power are much smaller than those of BSs. Therefore, taking the sum of consumed energies at the BSs and MTs will not make a significant difference for the MTs, since the sum will be dominated by the BS energy consumption.

Hence, there is a need for a balanced utility function that takes into account the role of all components in the network: the cellular BSs handling LR communications and the collaborating MTs communicating via D2D. Consequently, the proposed utility functions used in this chapter are defined according to the form:

$$U_l = f\left(\frac{N_{\text{served},l}}{E_{C,l}}, \frac{N_{\text{served},C_{k_l}}}{E_{C_{k_l}}}\right), \tag{34}$$

where U_l is the utility of cell l, $E_{C,l}$ is the energy consumed by BS n_l during the transmission of the data in cell l, $E_{C_{k_l}}$ is the energy consumed by cluster C_{k_l} in cell l, $N_{served,l}$ is the number of MTs served directly by BS n_l on the LR (including CHs), and $N_{\text{served},C_{k_l}}$ is the number of MTs served by D2D communications on the SR in cluster C_{k_l}. The function $f(x,y)$ is selected such that it is increasing with x and y. Thus, the first term in (34), $N_{\text{served},l} / E_{C,l}$, corresponds to the network of BSs, and the second term, $N_{\text{served},C_{k_l}} / E_{C_{k_l}}$, corresponds to the D2D clusters. By considering the ratio of the number of served users to the energy consumed, the components for the network and for the MTs in the utility become within the same order of magnitude, conversely to taking the sum of energies for example. Furthermore, for RE-SCBSs, the energy is obtained at no cost. For such BSs, we set $E_{C,l}$=1 (to avoid division by zero), which makes it much less than the energy consumed by mains powered BSs (MP-BSs). This will favor the offload of MTs to RE-SCBSs and switch off MP-BSs, since this would lead to a significant increase in the first term of (34). In addition, since the D2D energy consumption on the SR consumes significantly less energy than the BS energy, an increase in D2D offload will lead to a significant increase of the second term of (34). Hence, the combined use of RE-SCBSs and D2D offload will increase (34), thus favoring the switch off of significant numbers of MP-BSs. In this chapter, we consider the transmission of a file of size S_T bits to all MTs. Denoting by $T_{max,l}$ the time needed to transmit the content by BS n_l to all the MTs served by the BS (which depends on their achievable data rates), then we have:

$$E_{C,l} = P_{C,l} \cdot T_{\max,l}, \tag{35}$$

with $P_{C,l}$ denoting the total power consumed to operate BS n_l (including circuitry power, transmit power, processing power, etc.).

Based on (34), we define three cell utilities in this section. The network utility to be maximized is taken as the sum of individual cell utilities.

Utility 1

The first cell utility is defined as follows:

$$U_{l,1} = \frac{N_{\text{served},l}}{E_{C,l}} + \frac{N_{\text{served},C_{k_l}}}{E_{C_{k_l}}}. \tag{36}$$

The corresponding network utility is thus defined as:

$$U_1 = \sum_{l=1}^{N_{BS}} U_{l,1} = \sum_{l=1}^{N_{BS}} \left(\frac{N_{\text{served},l}}{E_{C,l}} + \frac{N_{\text{served},C_{k_l}}}{E_{C_{k_l}}} \right). \tag{37}$$

This utility corresponds to a cooperative coalitional utility between all BSs and MT clusters. The BSs and MT clusters are treated similarly in this utility, and the objective is to minimize the energy consumption of both BSs and MTs jointly by considering the overall sum of utilities. This utility extends (26) by assuming a coalition between all network entities, whether BSs or MTs, for the sake of energy minimization while satisfying QoS. Hence, (37) also extends (16) to green networking in a D2D scenario.

Utility 2

The second cell utility is defined as follows:

$$U_{l,2} = \frac{N_{\text{served},l}}{E_{C,l}} \cdot \frac{N_{\text{served},C_{k_l}}}{E_{C_{k_l}}}. \tag{38}$$

The corresponding network utility is thus defined as:

$$U_2 = \sum_{l=1}^{N_{BS}} U_{l,2} = \sum_{l=1}^{N_{BS}} \left(\frac{N_{\text{served},l}}{E_{C,l}} \cdot \frac{N_{\text{served},C_{k_l}}}{E_{C_{k_l}}} \right). \tag{39}$$

Utility 2 represents a hybrid scenario involving both coalition and competition: In each cell, a Nash Bargaining Game is played between the BS and MTs served by CHs in its coverage area, since the utility of the cell is the product of the BS utility and the utility of the MT cluster(s) in that cell, as it is expressed in (38). Then, the network utility in (39) considers the sum of cell utilities, i.e., the cells in the network are assumed to cooperate by forming a coalition. Consequently, with this network utility, the purpose is to maximize the sum of cell utilities; i.e., the Nash Bargaining Game is played separately in each cell, but not across cells. This latter situation corresponds to Utility 3, defined next.

Utility 3

The third cell utility is defined as follows:

$$U_{l,3} = \ln\left(\frac{N_{\text{served},l}}{E_{C,l}}\right) + \ln\left(\frac{N_{\text{served},C_{k_l}}}{E_{C_{k_l}}}\right). \tag{40}$$

The corresponding network utility is defined as:

$$U_3 = \sum_{l=1}^{N_{BS}} U_{l,3} = \sum_{l=1}^{N_{BS}} \left(\ln\left(\frac{N_{\text{served},l}}{E_{C,l}}\right) + \ln\left(\frac{N_{\text{served},C_{k_l}}}{E_{C_{k_l}}}\right)\right). \tag{41}$$

Maximizing U_3 is equivalent to:

$$\max U_3 \Leftrightarrow \max \prod_{l=1}^{N_{BS}} \left(\frac{N_{\text{served},l}}{E_{C,l}} \cdot \frac{N_{\text{served},C_{k_l}}}{E_{C_{k_l}}} \right). \tag{42}$$

The utility in (40) extends (26) by assuming a competition between all network entities, whether BSs or MTs, for the sake of energy minimization while satisfying QoS. Hence, (41) and (42) also extend (22) and (23) to green networking in a D2D scenario. Using this utility is equivalent to playing a Nash Bargaining Game between all the BSs and MTs. In this case, each BS acts selfishly to minimize its energy consumption, whereas each cluster of MTs forms a coalition aiming at reducing its energy consumption. This utility represents the NBS to this bargaining problem in a game theoretical framework.

SIMULATION RESULTS AND DISCUSSION

This section presents the Matlab simulation results. In the simulations, a uniform user distribution is considered over a coverage area of size 5 × 5 km^2. BSs are placed on a rectangular grid uniformly in the area. The cell radii are set to $R_M = 0.5$ km and $R_S = 0.1$ km for macrocells and small cells, respectively. The transmit power is set to $P_{l,\max}^{(DL)} = 10$ W for macrocell BSs, $P_{l,\max}^{(DL)} = 1$ W for small cell BSs, and $P_{k_l,\max}^{(UL)} = 0.125$ W for mobile devices. The power consumption is set to $P_{C,l} = 500$ W for macrocell BSs and $P_{C,l} = 100$ W for small cell BSs. We consider an LTE bandwidth of 10 MHz for each of the UL and DL directions, subdivided into 50 RBs. LTE parameters are obtained from (3GPP TS 36.211, 2014; 3GPP TS 36.213, 2014), and channel parameters are obtained from (3GPP TR 25.814, 2006). The complete list of parameters is shown in Table 1.

Green Networking Results without D2D

This section describes the proposed method based on green networking through BS switch off without resorting to D2D communications. Different services are analyzed depending on their UL and DL target data rates. They are presented in Table 2. Service 1 could correspond to a symmetric voice service,

Table 1. Simulation parameters

Parameter	Value	Parameter	Value
κ	-128.1 dB	v	3.76
σ_ξ (dB)	8 dB	Rayleigh parameter a	$E[a^2]=1$
Macrocell Radius R_M	0.5 km	Small cell Radius R_S	0.1 km
B	10 MHz	B_{sub}	15 kHz
N_{RB}	50	$P_{Rx,LR}$	1.8 Joules/s
$P_{Tx,SR}$	1.425 Joules/s	$P_{Rx,SR}$	0.925 Joules/s
S_T	1 Mbits	Area size	5×5 km^2
Max. macro BS transmit power	10 W	Max. small BS transmit power	1 W
Power Consumption of MBSs	500 W	Power Consumption of MP-SCBSs	100 W
Max. MT transmit power	0.125 W	Percentage of RE-SCBSs	50% (η=0.5)

Table 2. Studied scenarios

Scenario	$R^{(UL)}_{Target,\, k_l}$ (kbps)	$R^{(DL)}_{Target,\, k_l}$ (kbps)
Service 1	64	64
Service 2	56	256
Service 3	384	384
Service 4	384	1000

Services 2 and 4 are asymmetric services with different rates (e.g., comparable to fixed ADSL services), and Scenario 3 can represent a symmetric service with rates sufficient for video conferencing. It should be noted that significantly higher data rates can be reached compared to these services when the whole LTE bandwidth of 20 MHz (100 RBs) is allocated to a single user in the absence of interference. But these services are more realistic in the case of one RB allocated per user in a loaded network with high interference levels.The outage threshold is set to $P_{out,th}$=0.05.

The simulation results are shown in Figures 3 to 10. Figures 3 and 4 show the number of macrocell BSs that are active for the utilities (24) and (26), respectively, whereas Figures 5 and 6 show the number of SCBSs that are active for the utilities (24) and (26), respectively. In addition, Figures 7 and 8 show the network power consumption while using the utilities (24) and (26), respectively.

It can be seen that the coalitional approach outperforms the NBS approach since it leads to a lower number of active BSs (as shown in Figures 3 to 6), which leads to a lower power consumption in the network, as shown in Figures 7 and 8, for both utilities (24) and (26).

Figures 9 and 10 show the number of users in outage for the utilities (24) and (26), respectively. These figures show that both game theoretic techniques (coalition and NBS), when used with either of the utilities (24) or (26), satisfy the QoS requirements by leading to an outage rate below $P_{out,th}$. However,

Figure 3. Number of macrocell BSs that are "on" using the utility (24)

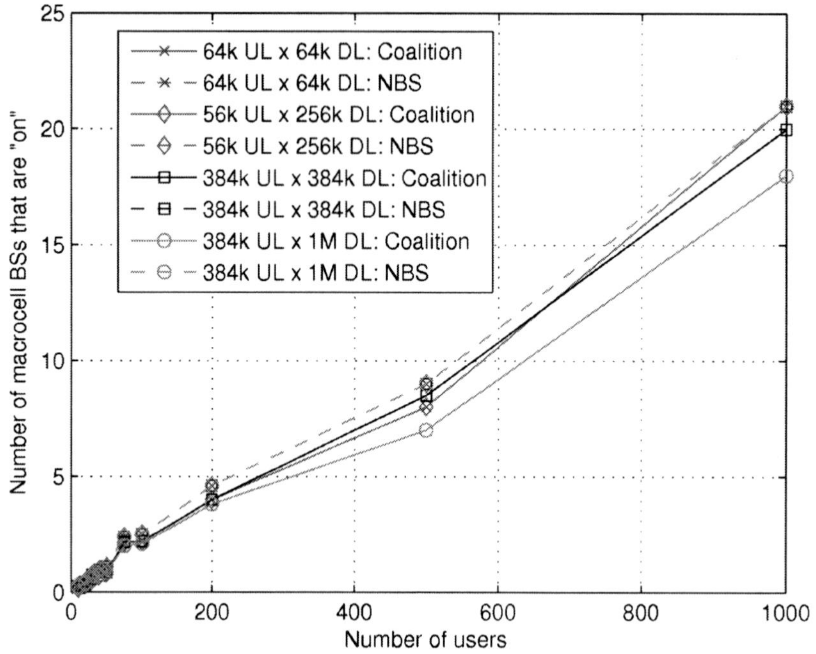

Figure 4. Number of macrocell BSs that are "on" using the utility (26)

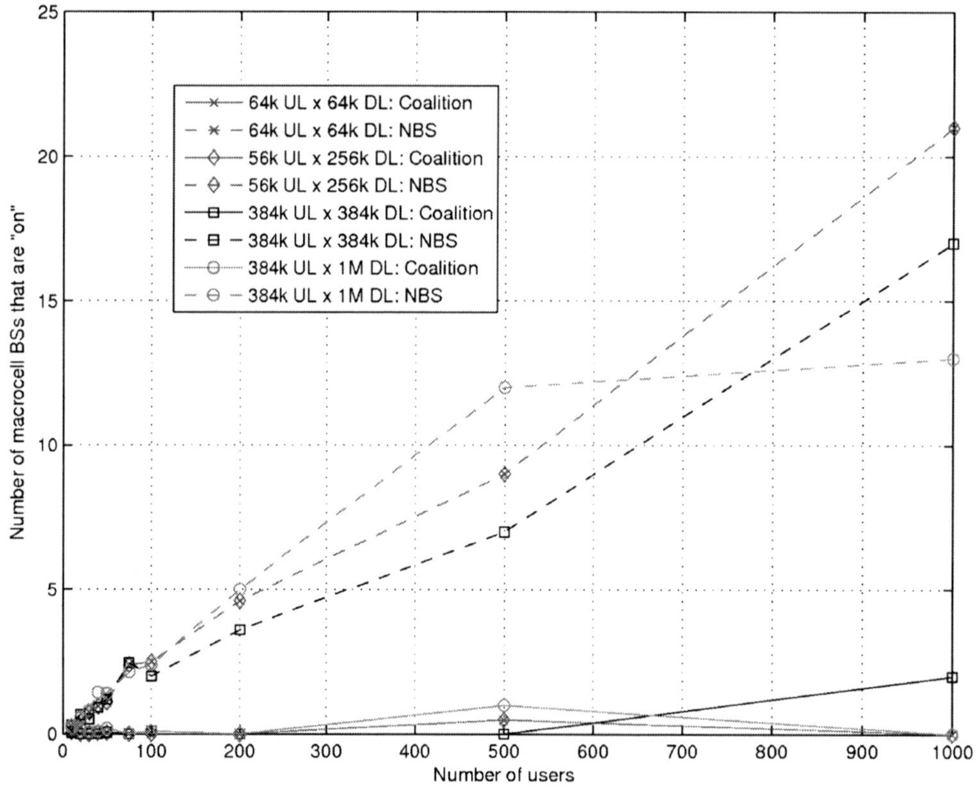

Figure 5. Number of small cell BSs that are "on" using the utility (24)

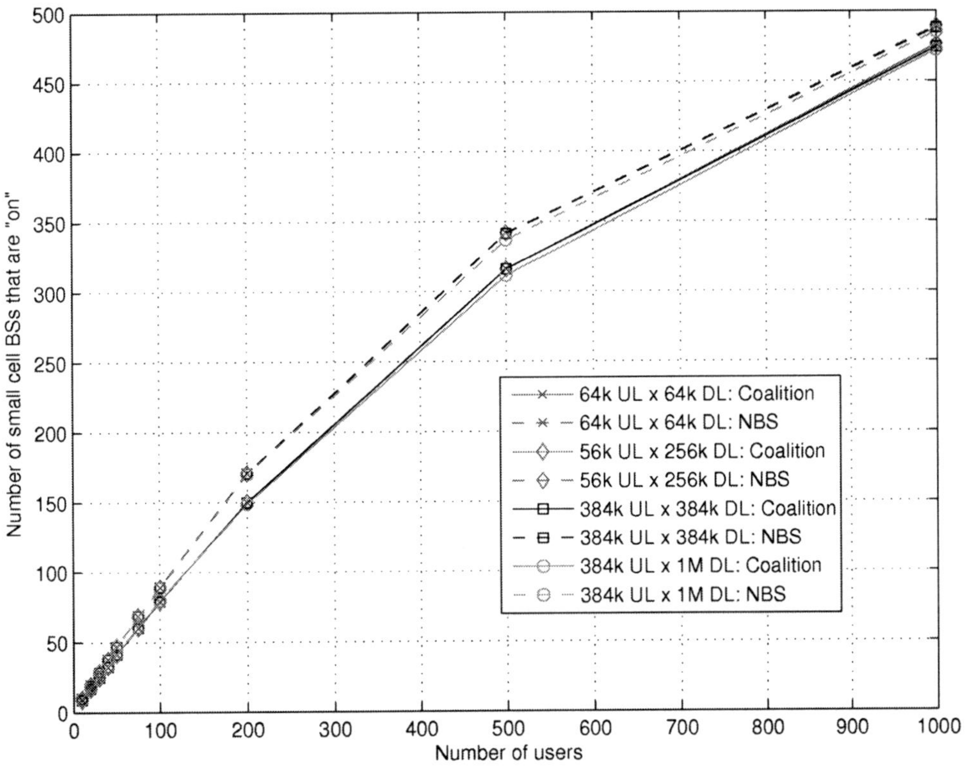

it should be noted that the utility in (26) leads to a slightly worse outage performance, since it is not explicitly dependent on $N_{out,l}$, conversely to the utility in (24), which is highly dependent on the increase of the number of users in outage.

On the other hand, the utility in (26) is more sensitive to the power consumption. This is clearly seen in Figures 3 to 8, where, although the number of active small cell BSs is comparable between the two utilities (see Figures 5 and 6), the number of active macrocell BSs is obviously less with the utility in (26), as it is shown by comparing Figures 3 and 4. This is particularly true with the coalitional game theoretic approach, where power consuming macrocell BSs are replaced by lower power small cell BSs, to an extent that the number of active macrocell BSs is zero most of the time. This is due to the "altruistic" behavior of BSs in the coalitional model, where the players act in the interest of the coalition as a whole.

In the NBS case, where each player "bargains" to increase its own utility, the overall performance becomes worse than the coalition scenario. This translates into a higher number of active macro and small cell BSs, and a higher power consumption, although large gains are achieved compared to the "traditional" scenario where no BSs are switched-off: in such a scenario, the network power consumption in the simulated model is around 75 kW, even when the number of users is reduced.

In the following section, we analyze the performance in the scenario where D2D communications are used in the network to enhance the energy efficiency for both BSs and MTs.

Figure 6. Number of small cell BSs that are "on" using the utility (26)

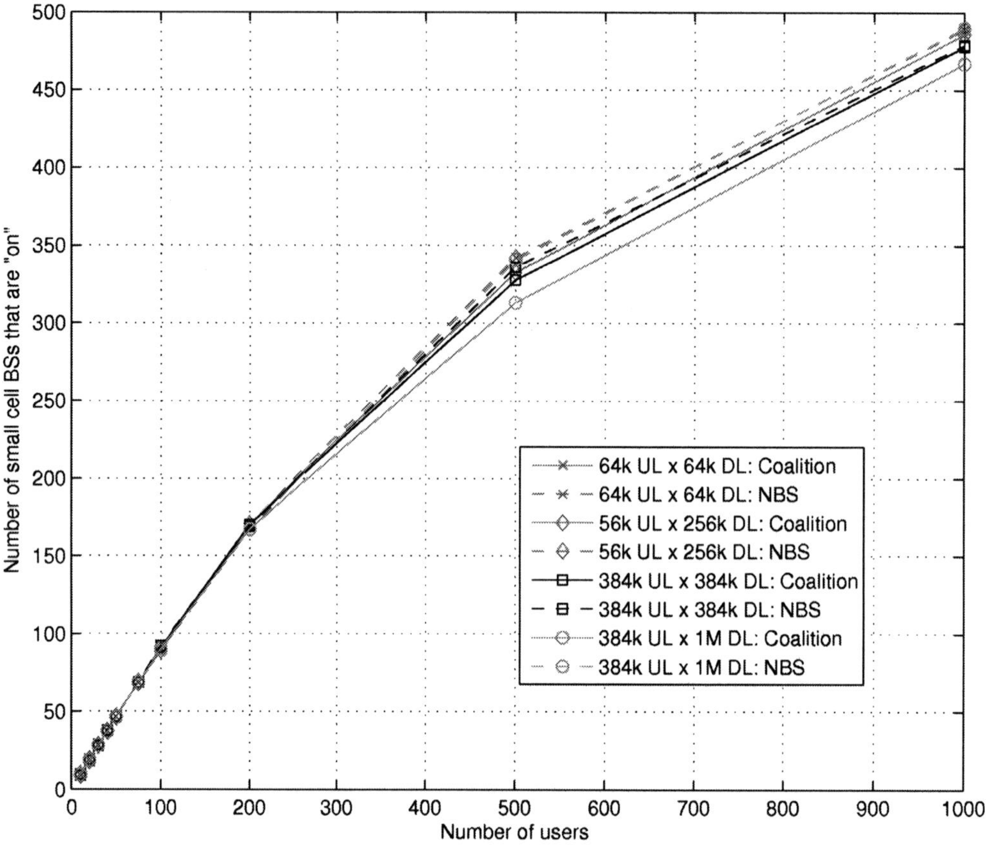

Green Networking Results with D2D

This section describes the proposed method based on joint D2D offload and green networking through BS switch off. First, D2D clustering is performed according to Algorithm 2 in order to reduce the number of LR links connected to BSs. Then, Algorithm 1 is implemented, considering that only CHs need to be handed over to neighboring BSs in order to put their serving BS into sleep mode.

Figure 11 shows the BS energy consumption results in the simulated area. The scenarios without green communications, where all BSs are active, correspond to a power consumption of 75 kW when all BSs are MP-BSs, and of 43.75 kW when RE-SCBSs are used with $\eta = 0.5$, regardless of the number of users. They are not plotted for clarity purposes. Figures 12 and 13 show the number of active MBSs and SCBSs, respectively. The results of Figure 11 show that implementing Algorithm 1 ("Green Hetnet") leads to significant energy savings. Furthermore, the use of RE-SCBSs (denoted "Green Hetnet + renewable energy small cells (RESC)" in the figures) leads to large gains compared to the scenario where only MP-BSs are used, especially when the number of users increases, i.e., when the load in the network becomes high. Moreover, D2D traffic offloading, i.e., the implementation of Algorithm 3, consisting of the sequential implementation of Algorithms 2 and 1 (case "Green Hetnet + D2D" without RE-SCBSs and "Green Hetnet + RESC + D2D" when RE-SCBSs are available), leads to additional energy savings. The largest savings are obtained when D2D traffic offloading is combined with green BS switch-off

Figure 7. Network power consumption with the utility (24)

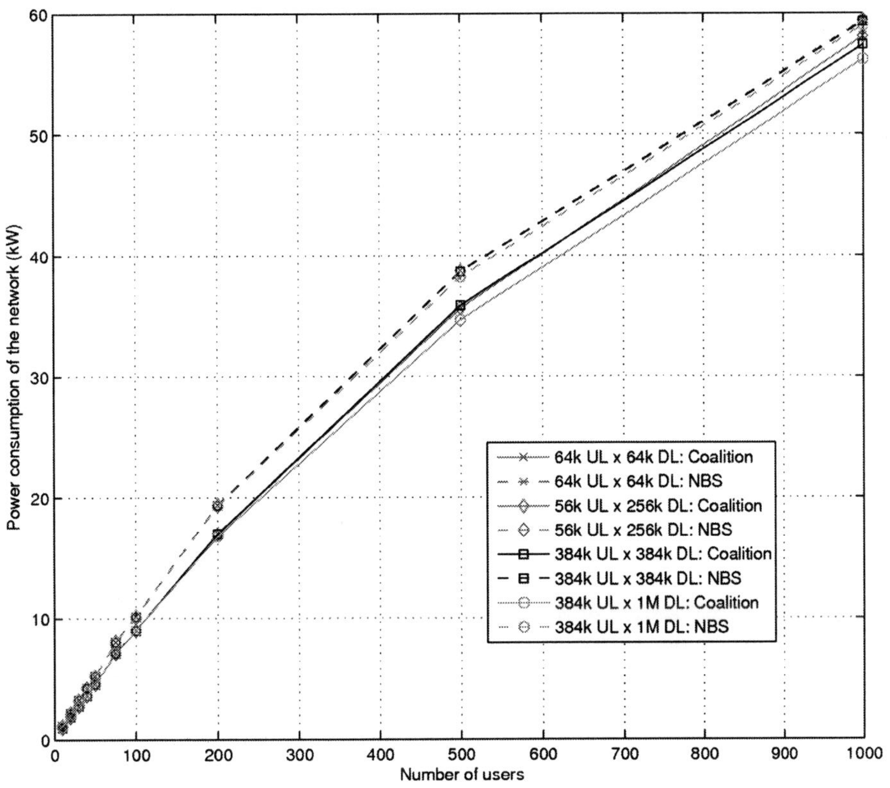

Figure 8. Network power consumption with the utility (26)

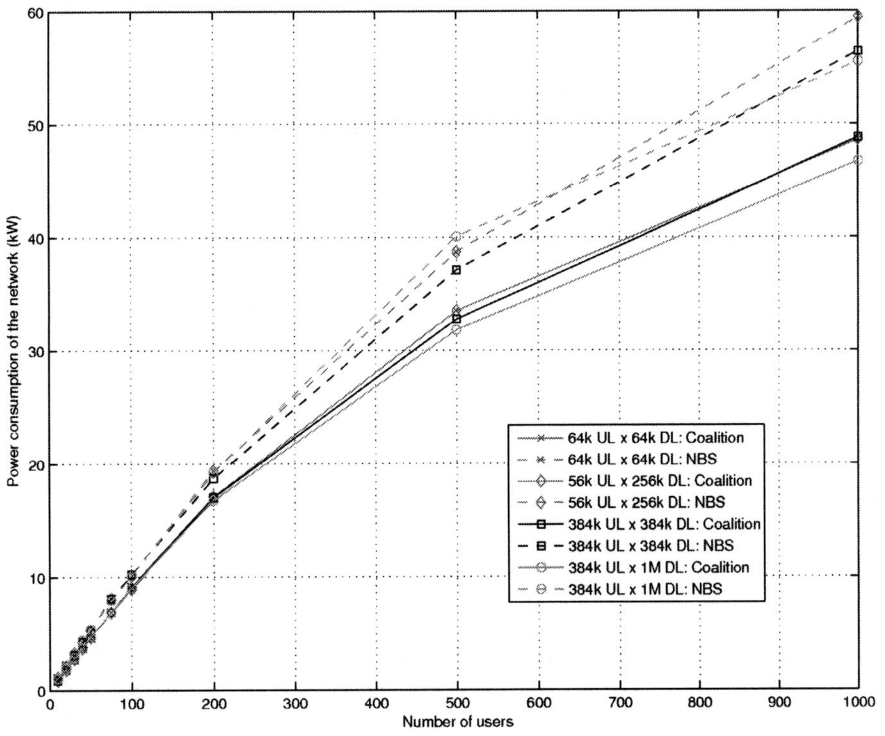

Figure 9. Percentage of users in outage with the utility (24)

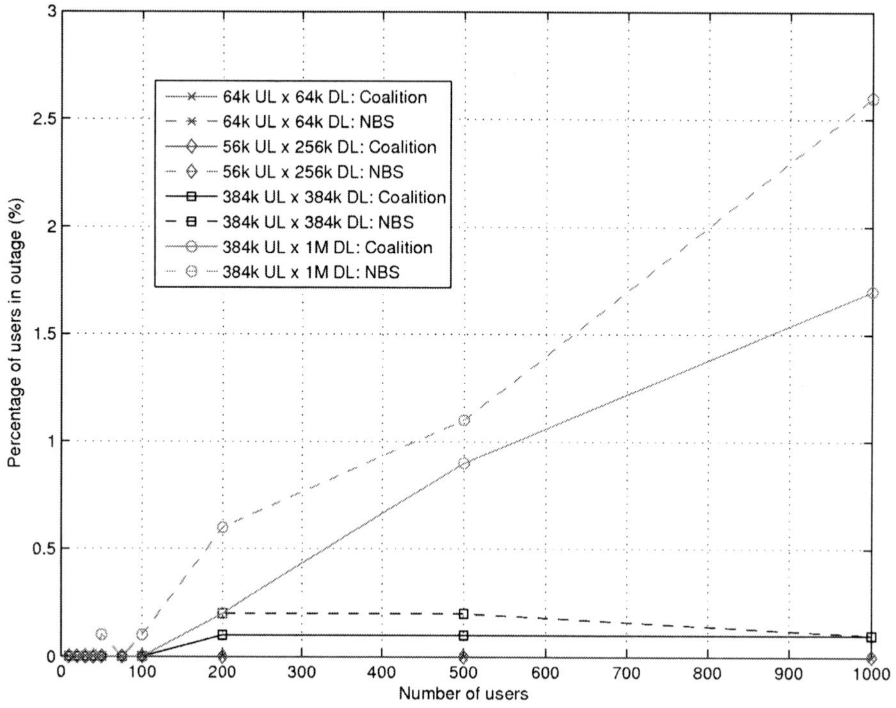

Figure 10. Percentage of users in outage with the utility (26)

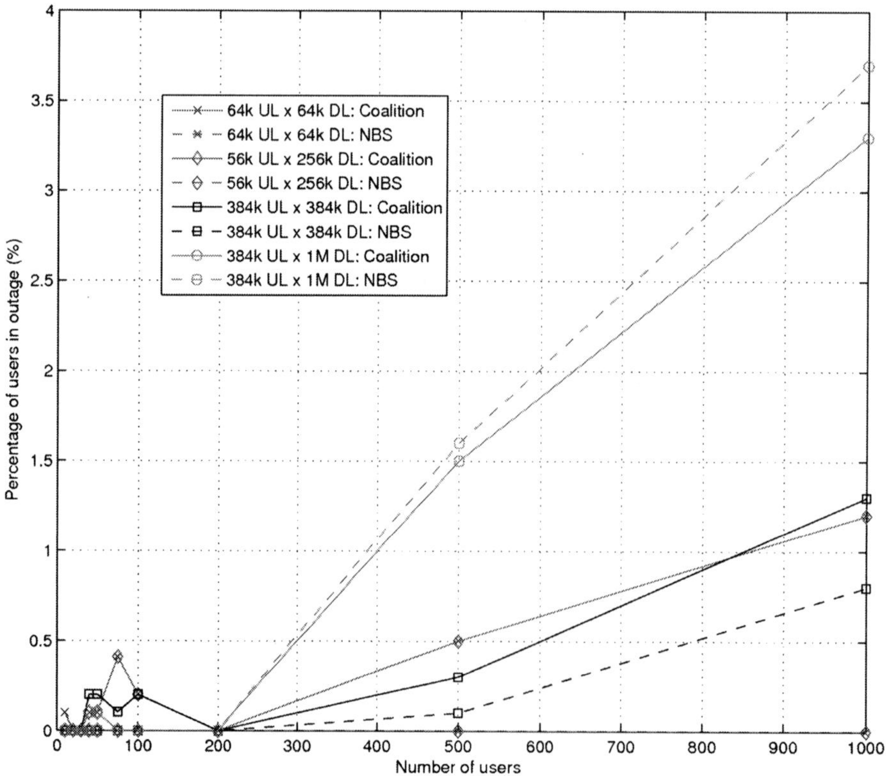

Figure 11. BS power consumption

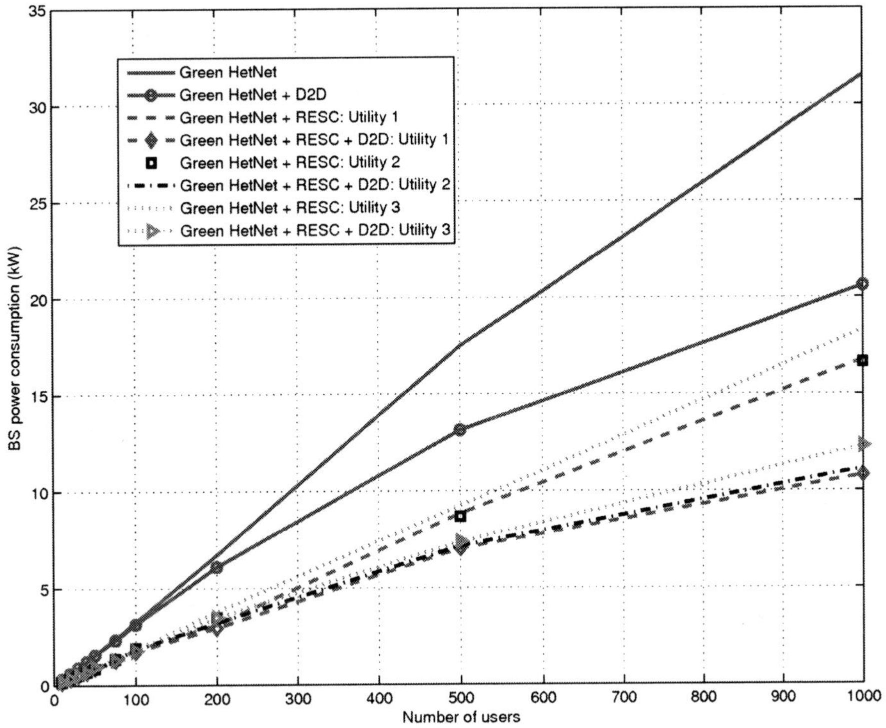

Figure 12. Number of macrocell BSs (MBSs) that are "on"

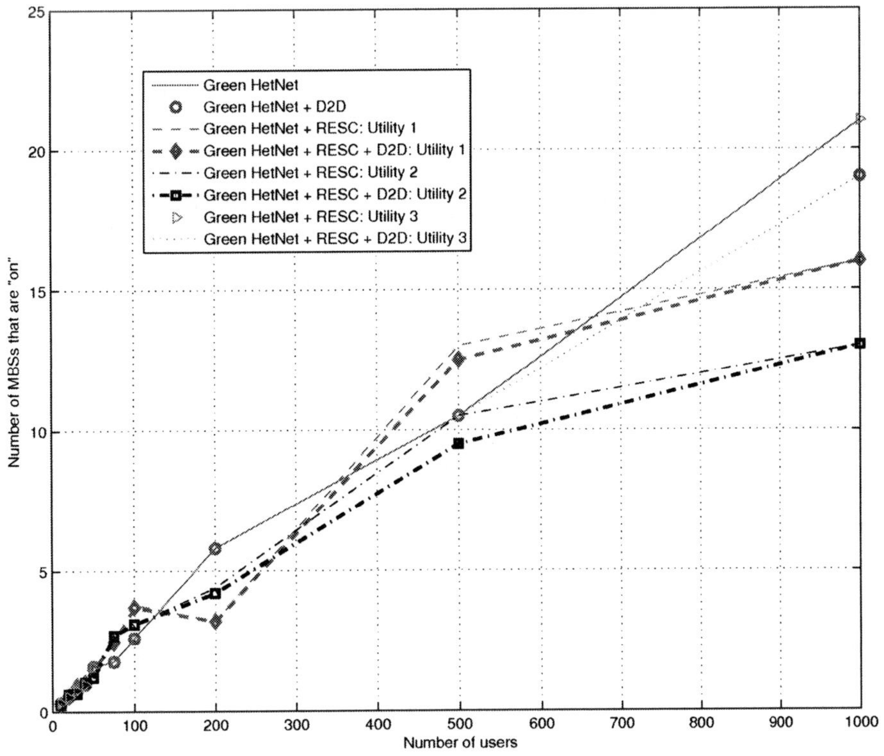

Figure 13. Number of small cell BSs that are "on"

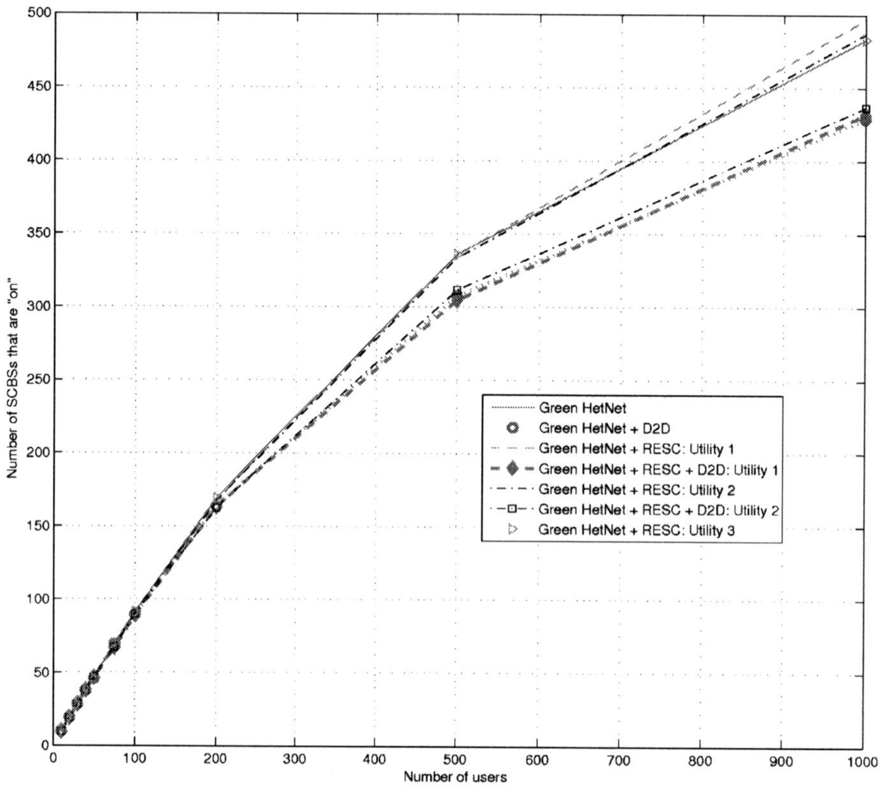

techniques in the presence of RE-SCBSs. Indeed, with 1000 users, the power consumption with the scenario "Green Hetnet + RESC + D2D" is around 10 kW, with savings of more than 85% compared to the traditional scenario where all BSs are on without D2D offload (having a consumption of 75 kW).

The MT power consumption is shown in Figure 14. It can be seen that the use of D2D communications according to Algorithm 2 leads to energy savings. Figure 15 shows the minimum data rates in the network, i.e., the data rates of the MT that is the last to finish receiving the content. It can be seen that when D2D clustering is implemented, the data rates increase in the network. In fact, CHs that are selected according to Algorithm 2 have relatively high LR data rates, in order to minimize (31). Due to the short distances during D2D communications, higher data rates can be achieved compared to the case where the MTs receive the content directly from the BS. Hence, the energy savings obtained by implementing the proposed approach do not cause any QoS degradation, but rather an increase in the data rate of the MTs.

Comparing the performance of the three utilities in Figures 11 to 15, it can be seen that Utility 1 leads to the lowest BS energy consumption (Figure 11), whereas the three utilities have comparable performance in terms of MT energy consumption (Figure 14). Hence, performing a single coalition between BSs and MTs leads to the best results. This validates the conclusion reached in the previous section: cooperation leads to gains for all involved players within the proposed game theoretical framework. It should be noted that from Figures 12 and 13, it can be seen that the number of active macro and small cell BSs with Utility 1 is not always less than the scenarios with Utilities 2 and 3. However, the implementation of Algorithms 1 and 3 leads to a judicious selection of active BSs by letting more RE-SCBSs be active

Figure 14. MT power consumption in the network

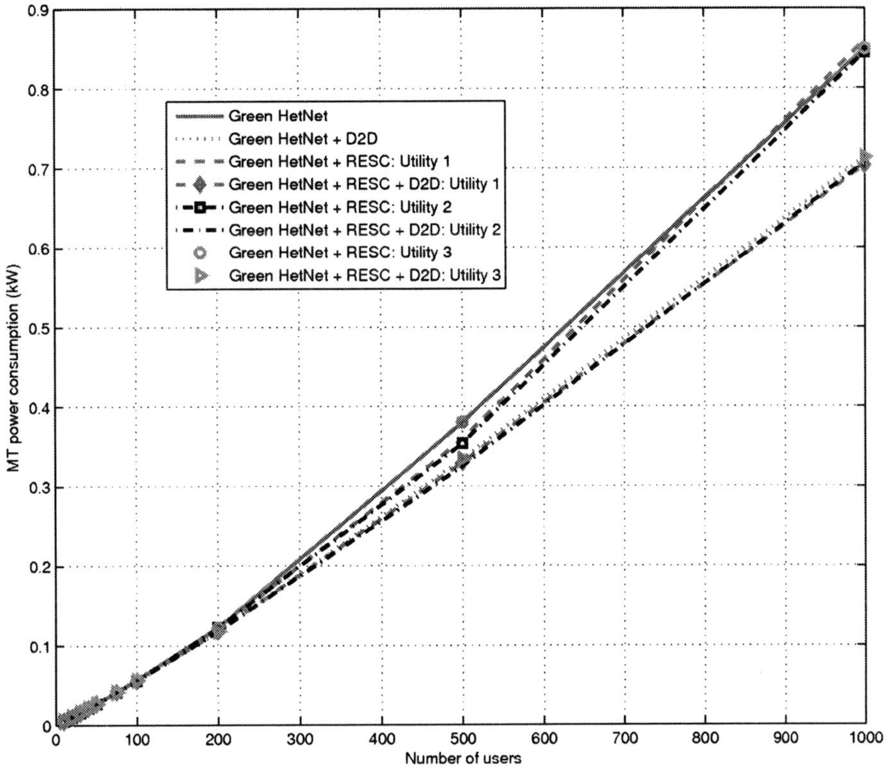

Figure 15. Minimum data rate achieved by MTs in the network

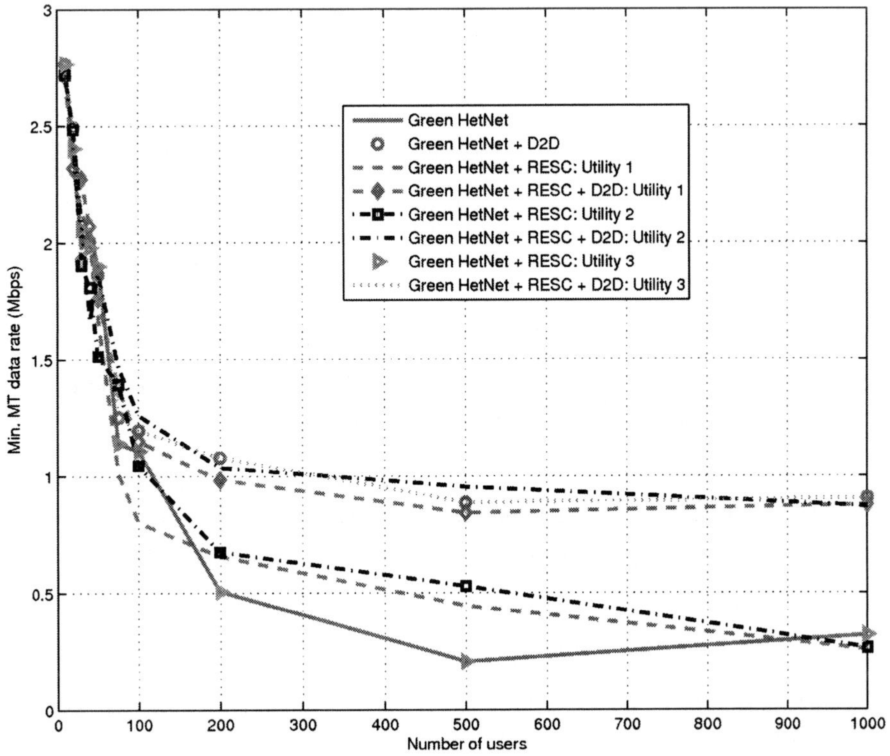

and reducing the number of MP-BSs. This is due to the "altruistic" behavior of Utility 1, which allows RE-SCBSs to be active for the benefit of the network, which optimizes performance. The "selfishness" in the NBS solution to the competition between cells in Utilities 2 and 3 reduces the gains, since RE-SCBSs in this case attempt to maximize their own cell utility, regardless of the fact that their energy costs are much less than for MP-BSs.

CONCLUSION

A game theoretical framework was proposed for ensuring green energy efficient communications in heterogeneous LTE cellular networks. Two game theoretic concepts were studied within this framework: a coalition-based approach and a competition approach based on the Nash bargaining solution. Within each approach, the tradeoffs between QoS and energy consumption were investigated, depending on the utility selected. In fact, the game theoretic techniques were implemented with utilities focusing either on traffic load and QoS, or on the QoS versus the consumed power in the network. The game theoretical framework was then extended to a scenario with device-to-device communications, where energy aware coordination between mobile devices and the cellular network infrastructure was investigated. The purpose is to reach a green cellular network through the joint use of device-to-device communications and small cell base stations powered by renewable energy. The benefits of traffic offloading and energy savings were captured by balanced utility functions that take into account the role of all components in the network: the cellular BSs handling long range communications and the collaborating devices communicating via device-to-device communications. The results showed significant energy savings in the network, in addition to savings for the mobile devices, without compromising their QoS, especially when cooperative utilities involving coalitions between the various network entities (base stations and mobile terminals) were adopted.

ACKNOWLEDGMENT

This work was made possible by NPRP grant # 6-001-2-001 from the Qatar National Research Fund (A member of The Qatar Foundation). The statements made herein are solely the responsibility of the authors.

REFERENCES

Andrews, J. G. (2013). Seven Ways that HetNets Are a Cellular Paradigm Shift. *IEEE Communications Magazine*, *51*(3), 136–144. doi:10.1109/MCOM.2013.6476878

Bousia, A., Antonopoulos, A., Alonso, L., & Verikoukis, C. (2012). Green Distance-Aware Base Station Sleeping Algorithm in LTE-Advanced. In *Proceedings of the IEEE International Conference on Communications (ICC 2012)* (pp. 1347 – 1351). Ottawa, Canada: IEEE. doi:10.1109/ICC.2012.6364240

Carmichael, F. (2005). *A Guide to Game Theory*. Harlow, UK: Prentice Hall.

Chen, Y., Xu, X., & Lei, Q. (2013). Joint Subcarriers and Power Allocation with Imperfect Spectrum Sensing for Cognitive D2D Wireless Multicast. *Transactions on Internet and Information Systems (Seoul)*, *7*(7), 1533–1546. doi:10.3837/tiis.2013.07.001

Goldsmith, A. (2005). *Wireless Communications*. New York, NY: Cambridge University Press. doi:10.1017/CBO9780511841224

Han, M.-H., Kim, B.-G., & Lee, J.-W. (2012). Subchannel and Transmission Mode Scheduling for D2D Communication in OFDMA Networks. In *Proceedings of the IEEE Vehicular Technology Conference (VTC Fall 2012)*. Quebec City, Canada: IEEE. doi:10.1109/VTCFall.2012.6399205

Han, T., & Ansari, N. (2013). On Greening Cellular Networks via Multicell Cooperation. *IEEE Wireless Communications*, *20*(1), 82–89. doi:10.1109/MWC.2013.6472203

Hasan, Z., Boostanimehr, H., & Bhargava, V. K. (2011). Green Cellular Networks: A Survey, Some Research Issues and Challenges. *IEEE Communications Surveys and Tutorials*, *13*(4), 524–540. doi:10.1109/SURV.2011.092311.00031

Hoydis, J., & Debbah, M. (2010). Green, Cost-Effective, Flexible, Small Cell Networks. *IEEE ComSoc MMTC E-Letter Special Issue on. Multimedia Over Femto Cells*, *5*(5), 23–26.

Lan, Y., & Harada, A. (2012). Interference Analysis and Performance Evaluation on the Coexistence of Macro and Micro/Pico Cells in LTE Networks. In *Proceedings of the IEEE Vehicular Technology Conference (VTC Spring 2012)*. Yokohama, Japan: IEEE. doi:10.1109/VETECS.2012.6239907

Lee, D. H., Choi, K. W., Jeon, W. S., & Jeong, D. G. (2013). Resource Allocation Scheme for Device-to-Device Communication for Maximizing Spatial Reuse. In *Proceedings of the IEEE Wireless Communications and Networking Conference (WCNC 2013)*. Shanghai, China: IEEE. doi:10.1109/WCNC.2013.6554548

Ma, C., Yue, J., Yu, H., Luo, H., Zhou, W., & Sun, X. (2013). An Interference Coordination Mechanism Based on Resource Allocation for Network Controlled Device-to-Device Communication. In *Proceedings of the IEEE/CIC International Conference on Communications in China - Workshops (CIC/ICCC 2013)*. Xi'an, China: IEEE.

Myung, H. G., & Goodman, D. J. (2008). *Single Carrier FDMA: A New Air Interface for Long Term Evolution*. Chichester, UK: John Wiley and Sons. doi:10.1002/9780470758717

Niu, Z., Wu, Y., Gong, J., & Yang, Z. (2010). Cell Zooming for Cost-Efficient Green Cellular Networks. *IEEE Communications Magazine*, *48*(11), 74–79. doi:10.1109/MCOM.2010.5621970

Third Generation Partnership Project (3GPP). (2006). *3GPP TR 25.814 3GPP TSG RAN Physical Layer Aspects for Evolved UTRA, v7.1.0, 2006*. Author.

Third Generation Partnership Project (3GPP). (2014a). *3GPP TS 36.211 3GPP TSG RAN Evolved Universal Terrestrial Radio Access (E-UTRA) Physical Channels and Modulation, version 12.3.0, Release 12, 2014*. Author.

Third Generation Partnership Project (3GPP). (2014b). *3GPP TS 36.213 3GPP TSG RAN Evolved Universal Terrestrial Radio Access (E-UTRA) Physical layer procedures, version 12.3.0, Release 12, 2014*. Author.

Salem, M., Adinoyi, A., Rahman, M., Yanikomeroglu, H., Falconer, D., Kim, Y.-D., & Cheong, Y.-C. et al. (2010). An Overview of Radio Resource Management in Relay-Enhanced OFDMA-Based Networks. *IEEE Communications Surveys and Tutorials*, *12*(3), 422–438. doi:10.1109/SURV.2010.032210.00071

Wang, D., Wang, X., & Zhao, Y. (2012). An Interference Coordination Scheme for Device-to-Device Multicast in Cellular Networks. In *Proceedings of the IEEE Vehicular Technology Conference (VTC Fall 2012)*. Quebec City, Canada: IEEE. doi:10.1109/VTCFall.2012.6398945

Wang, F., Song, L., Han, Z., Zhao, Q., & Wang, X. (2013). Joint Scheduling and Resource Allocation for Device-to-Device Underlay Communication. In *Proceedings of the IEEE Wireless Communications and Networking Conference (WCNC 2013)*. Shanghai, China: IEEE. doi:10.1109/WCNC.2013.6554552

Wang, H., & Chu, X. (2012). Distance-Constrained Resource-Sharing Criteria for Device-to-Device Communications Underlaying Cellular Networks. *Electronics Letters*, *48*(9), 528–530. doi:10.1049/el.2012.0451

Xiang, L., Pantisano, F., Verdone, R., Ge, X., & Chen, M. (2011). Adaptive Traffic Load-Balancing for Green Cellular Networks. In *Proceedings of the IEEE International Symposium on Personal Indoor and Mobile Radio Communications (PIMRC 2011)*. Toronto, Canada: IEEE. doi:10.1109/PIMRC.2011.6139995

Yaacoub, E., & Abu-Dayya, A. (2012). Multihop routing for energy efficiency in wireless sensor networks. In M. Matin (Ed.), *Wireless Sensor Networks - Technology and Protocols* (pp. 165–188). Rijeka, Croatia: InTech Academic Publishers. doi:10.5772/39221

Yaacoub, E., Ghazzai, H., Alouini, M.-S., & Abu-Dayya, A. (2013). Achieving Energy Efficiency in LTE with Joint D2D Communications and Green Networking Techniques. In *Proceedings of the IEEE International Wireless Communications and Mobile Computing Conference (IWCMC 2013)*. Cagliari, Italy: IEEE. doi:10.1109/IWCMC.2013.6583571

Zhao, X., & Yang, X. (2012). Downlink Ergodic Capacity Analysis for Wireless Networks with Cooperative Distributed Antenna Systems. In *Proceedings of the IEEE International Conference on Communications (ICC 2012)*. Ottawa, Canada: IEEE. doi:10.1109/ICC.2012.6364775

ADDITIONAL READING

Astély, D., Dahlman, E., Furuskär, A., Jading, Y., Lindström, M., & Parkvall, S. (2009). LTE: The Evolution of Mobile Broadband. *IEEE Communications Magazine*, *47*(4), 44–51. doi:10.1109/MCOM.2009.4907406

Bae, C., & Cho, D.-H. (2007). Fairness-Aware Adaptive Resource Allocation Scheme in Multi-Hop OFDMA Systems. *IEEE Communications Letters*, *11*(2), 134–136. doi:10.1109/LCOMM.2007.061381

Bhat, P., Nagata, S., Campoy, L., Berberana, I., Derham, T., Guangyi, L., & Jin, Y. et al. (2012). LTE-advanced: An Operator Perspective. *IEEE Communications Magazine*, *50*(2), 104–114. doi:10.1109/MCOM.2012.6146489

Choi, W., & Andrews, J. G. (2007). Downlink Performance and Capacity of Distributed Antenna Systems in a Multicell Environment. *IEEE Transactions on Wireless Communications*, *6*(1), 69–73. doi:10.1109/TWC.2007.05207

Choi, W., Andrews, J. G., & Yi, C. (2005). The Capacity of Multicellular Distributed Antenna Networks. *Proceedings of the International Conference on Wireless Networks, Communications and Mobile Computing*, pp. 1337-1342, Maui, Hawaii, USA.

Doumi, T., Dolan, M. F., Tatesh, S., Casati, A., Tsirtsis, G., Anchan, K., & Flore, D. (2013). LTE for Public Safety Networks. *IEEE Communications Magazine*, *51*(2), 106–112. doi:10.1109/MCOM.2013.6461193

Ekström, H., Furuskär, A., Karlsson, J., Meyer, M., Parkvall, S., Torsner, J., & Wahlqvist, M. (2006). Technical Solutions for the 3G Long-Term Evolution. *IEEE Communications Magazine*, *44*(3), 38–45. doi:10.1109/MCOM.2006.1607864

Jorguseski, L., Le, T. M. H., Fledderus, E. R., & Prasad, R. (2008, September). Downlink Resource Allocation for Evolved UTRAN and WiMAX Cellular Systems. *IEEE Personal, Indoor, and Mobile Radio Communications Conference (PIMRC 2008)*, pp. 1-5, Cannes, France.

Kaneko, M., & Popovski, P. (2007, June). Radio Resource Allocation Algorithm for Relay-Aided Cellular OFDMA System. *IEEE International Conference on Communications (ICC 2007)*, pp. 4831-4836, Glascow, Scotland. doi:10.1109/ICC.2007.797

Kim, M., & Lee, H. (2007, July). Radio Resource Management for a Two-Hop OFDMA Relay System in Downlink. *IEEE Symposium on Computers and Communications (ISCC 2007)*, pp. 25-31, Aveiro, Portugal. doi:10.1109/ISCC.2007.4381489

Lee, J., Park, S., Wang, H., & Hong, D. (2007, June). QoS-Guaranteed Transmission Scheme Selection for OFDMA Multi-Hop Cellular Networks. *IEEE International Conference on Communications (ICC 2007)*, pp. 4587-4591, Glascow, Scotland. doi:10.1109/ICC.2007.757

Li, G., & Liu, H. (2006). Resource Allocation for OFDMA Relay Networks with Fairness Constraints. *IEEE Journal on Selected Areas in Communications*, *24*(11), 2061–2069. doi:10.1109/JSAC.2006.881627

Lu, S., Cai, Y., Zhang, L., Li, J., Skov, P., Wang, C., & He, Z. (2009, April). Channel-Aware Frequency Domain Packet Scheduling for MBMS in LTE. *Proceedings of the IEEE Vehicular Technology Conference-Spring (VTC – Spring 2009)*, pp. 1-5, Barcelona, Spain. doi:10.1109/VETECS.2009.5073439

Lunttila, T., Lindholm, J., Pajukoski, K., Tiirola, E., & Toskala, A. (2007, February). EUTRAN Uplink Performance. *International Symposium on Wireless Pervasive Computing (ISWPC 2007)*, pp. 515-519, San Juan, Puerto Rico.

Nam, W., Chang, W., Chung, S.-Y., & Lee, Y. (2007, June). Transmit Optimization for Relay-Based Cellular OFDMA Systems. *IEEE International Conference on Communications (ICC 2007)*, pp. 75-81, Glascow, Scotland. doi:10.1109/ICC.2007.947

Oyman, O. (2007, November). Opportunistic Scheduling and Spectrum Reuse in Relay-Based Cellular OFDMA Networks. [Washington DC, USA.]. *IEEE GlobeCom, 2007*, 3699–3703.

Third Generation Partnership Project (3GPP). 3GPP TR 36.913 3GPP TSG RAN Requirements for Further Advancements for Evolved Universal Terrestrial Radio Access (E-UTRA) (LTE-Advanced), version 8.0.1, Release 8, 2008.

Yaacoub, E. (2012). Green Communications in LTE Networks with Environmentally Friendly Small Cell Base Stations. In *Proceedings of the IEEE Online Conference on Green Communications (GreenCom 2012)*, pp. 110-115. doi:10.1109/GreenCom.2012.6519625

Yaacoub, E., & Dawy, Z. (2009). A Game Theoretical Formulation for Proportional Fairness in LTE Uplink Scheduling. *IEEE Wireless Communications and Networking Conference (WCNC 2009)*, pp. 1-5, Budapest, Hungary. doi:10.1109/WCNC.2009.4917504

Yaacoub, E., & Dawy, Z. (2010, April). A Comparison of Uplink Scheduling in OFDMA and SCFD-MA. *IEEE International Conference on Telecommunications (ICT 2010)*, pp. 466-470, Doha, Qatar. doi:10.1109/ICTEL.2010.5478799

Yaacoub, E., & Dawy, Z. (2011). Achieving the Nash Bargaining Solution in OFDMA Uplink Using Distributed Scheduling with Limited Feedback. *International Journal of Electronics and Communication AEU (Elsevier)*, *65*(4), 320–330. doi:10.1016/j.aeue.2010.03.007

Yaacoub, E., & Dawy, Z. (2012). A Survey on Uplink Resource Allocation in OFDMA Wireless Networks. *IEEE Communications Surveys and Tutorials*, *14*(2), 322–337. doi:10.1109/SURV.2011.051111.00121

Yaacoub, E., & Dawy, Z. (2012). *Resource Allocation in Uplink OFDMA Wireless Systems: Optimal Solutions and Practical Implementations*. Hoboken, NJ: John Wiley and Sons / IEEE press. doi:10.1002/9781118189627

Yaacoub, E., & Kubbar, O. (2012, December). Energy-Efficient Device-to-Device Communications in LTE Public Safety Networks, *Globecom Workshops 2012 (International Workshop on Green Internet of Things)*, pp. 391-395, Anaheim, CA, USA.

KEY TERMS AND DEFINITIONS

Device-to-Device Communications: Technique whereby mobile terminals can communicate directly with each other whenever they are in proximity without relying on base stations for routing their traffic.

Green Communications: Approach to achieve energy efficiency in the network by reducing the redundant power consumption whenever possible.

Heterogeneous Networks: Networks consisting of a combination of macrocells and small cells.

LTE: Long Term Evolution. It is the state of the art cellular system being deployed for commercial cellular communications.

LTE-A: Long Term Evolution - Advanced.

OFDMA: Orthogonal Frequency Division Multiple Access. It is the accessing scheme used for LTE downlink. It consists of subdividing the bandwidth into a large number of subcarriers, each having a small bandwidth where the fading can generally be considered flat.

Quality of Service (QoS): Measure to assess the network performance.

Radio Resource Management: Algorithms and techniques to allocate the wireless resources and set certain parameters of the wireless system depending on the channel characteristics and interference levels, with the objective of optimizing the network performance.

SCFDMA: Single Carrier FDMA. It is the accessing scheme used for LTE uplink. It is a modified form of OFDMA that allows the subcarriers to be transmitted sequentially rather than in parallel. Its main advantage is a reduction in the signal peak to average power ratio (PAPR), which allows cheaper power amplifiers to be used in LTE handsets (UEs).

Chapter 14
Game Theoretic Infrastructure Sharing in Wireless Cellular Networks

Alexandra Bousia
Technical University of Catalonia (UPC), Spain

Elli Kartsakli
Technical University of Catalonia (UPC), Spain

Angelos Antonopoulos
Telecommunications Technological Center of Catalonia (CTTC), Spain

Luis Alonso
Technical University of Catalonia – BarcelonaTECH (UPC), Spain

Christos Verikoukis
Telecommunications Technological Center of Catalonia (CTTC), Spain

ABSTRACT

The emerging traffic demand has fueled the rapid densification of cellular networks. The increased number of Base Stations (BSs) leads to augmented energy consumption and expenditures for the Mobile Network Operators (MNOs), especially during low traffic, when many of the BSs remain underutilized. Hence, the MNOs are encouraged to provide "green" and cost effective solutions for their networks. In this chapter, an innovative algorithm for infrastructure sharing in two-operator environments is proposed, based on BSs switching off during low traffic periods. Motivated by the conflicting interests of the operators, the problem is formulated in a game theoretic framework that enables the MNOs to act individually to estimate the switching off probabilities that reduce their financial cost. The authors analytically and experimentally estimate the potential energy and cost savings that can be accomplished. The obtained results show a significant reduction in both energy consumption and expenditures, thus giving the operators the necessary incentives for infrastructure sharing.

DOI: 10.4018/978-1-4666-8642-7.ch014

INTRODUCTION

The increasing growth of mobile telecommunication services along with the emerging demand for multimedia applications, driven by the widespread use of laptops, tablets and smart-phones, have led to an impressive data traffic increase during the last few years. According to recent market predictions (Cisco, 2012), mobile video and web traffic will experience an increase of 70% and 20%, respectively, by the year 2016. To meet these demands, Mobile Network Operators (MNOs[1]) extend their infrastructure by installing more Base Stations (BSs). The densely deployed networks not only imply a rise in the Capital Expenditures (CapEx) of the MNOs, but also have a direct impact on the total energy consumption of the network, thus resulting in higher Operational Expenditures (OpEx) as well (Chen, Zhang, Zhao, & Chen, 2010). In particular, the use of information and communication technology currently accounts for 5.7% of the world's electricity consumption and 1.8% of global carbon emissions, something that translates into electricity bills in the order of $10 billions for the MNOs worldwide (Fettweis, & Zimmermann, 2008). Hence, network operators are strongly motivated to investigate energy efficient solutions that will bring down the energy consumption and the cost of cellular networks.

Since BSs are responsible for the major part of energy consumption in mobile networks (Hasan, Boostanimehr, & Bhargava, 2011), a lot of effort has been devoted on designing methods to reduce their power consumption. In (Correia et al., 2012), the authors focus on hardware enhancement by designing energy efficient power amplifiers and by employing renewable energy resources, whereas other works focus on accomplishing the optimal BS deployment strategies (Hansly, & Mathar, 2002; Richter, Fehske, & Fettweis, 2009; Suryaprakash, Fonseca dos Santos, Fehske, & Fettweis, 2012; Tseng, & Huang, 2012). Recently, in an effort to achieve more drastic energy saving gains, the research community has shifted towards the investigation of BS switching off schemes, since BSs still consume a significant amount of energy for operation, even when their traffic is low. Subsequently, under off-peak traffic conditions, energy can be saved and cost can be reduced by switching off the unnecessary BSs, while the remaining active BSs can extend their coverage range in order to serve the whole network area.

In this chapter, motivated by the aforementioned issues, the authors propose a roaming-based infrastructure sharing strategy, applicable during low traffic periods. A game theoretic framework that captures the conflicting interests of the MNOs, models the switching off decision process for two operators that are willing to share their resources. Besides the expected energy efficiency benefits, the proposed scheme allows the MNOs to significantly reduce their financial costs independently of the strategies of the coexisting MNOs, providing them the necessary incentives to participate in the game. The contribution of the work is threefold and is summarized as follows:

1. A game theoretic switching off algorithm is proposed that aims at minimizing the individual cost of two competing operators that are willing to share their infrastructure in a distributed way. The authors define realistic cost-oriented functions by explicitly taking into account actual roaming and operational costs of each operator. To this end, each MNO is able to estimate a switching off probability (i.e., Nash Equilibrium (NE)) that provides significant gains, independently of the other MNO actions.
2. To address the heterogeneous nature of voice and data traffic in current and future cellular networks, an analytical model for evaluating the network throughput and the energy efficiency for two types of traffic based on a two-dimensional Markov chain is presented.

3. The authors evaluate the effectiveness of the algorithm in terms of energy efficiency through analysis and simulation experiments and we study the actual costs and the profits for each operator, while they introduce a new metric, called cost efficiency, for further evaluating the performance of our proposal.

The remainder of the chapter is organized as follows. The background information concerning the state-of-the-art approaches for switching off algorithms are presented in the next section. Then, the system model and the network configuration are described. In continuation, the authors introduce the infrastructure sharing scheme for two MNOs, along with the game formulation of the switching off decision. The cost models are also presented in the same section. The analytical model for the network throughput and the energy efficiency metrics are given. The validation of the model along with simulation results are provided and, finally, the chapter is concluded along with the future directions and open issues.

BACKGROUND

In the context of energy efficient network planning for single operator wireless cellular systems, several works appear in the literature (Marsan, Chiaraviglio, Ciullo, & Meo, 2009; Niu, Zhou, Hua, Zhang, & Cao, 2002; Koudouridis, & Li, 2012; Bousia, Antonopoulos, Alonso, & Verikoukis, 2012; Bousia, Kartsakli, Alonso, & Verikoukis, 2012a; Bousia, Kartsakli, Alonso, & Verikoukis, 2012b). Beyond the traditional switching off schemes, the coexistence of multiple MNOs in the same geographical area, due to legal regulations that obligate them to install their antennas on the same buildings, has motivated a new business model, known as infrastructure sharing. The concept of infrastructure sharing is an upcoming promising topic that involves research and requires extensive study. The current technological solutions and regulatory constraints of infrastructure sharing are examined in (Meddour, Rasheed, & Gourhant, 2011). The authors present the three categories of infrastructure sharing: *i*) passive sharing (Frisanco, Tafertshofer, Laurin, & Ang, 2008; Leibner, 2004; Hasbani, El-Darwiche, Mourad, & Chanab, 2007; Khan, Kellerer, Kozu, & Yabusaki, 2011), limited to the joint use of sites, masts and building premises among MNOs, *ii)* active sharing (Nokia, 2001; Beckman, & Smith, 2005), where the MNOs share the active network components such as antennas, switches and backhaul equipment, and *iii)* roaming-based sharing (3GPP, 2002; 3GPP, 2001; Zhang, Zhang, Li, & Hanzo, 2012), where the MNOs share the cell coverage for a pre-negotiated period of time.

Among these approaches, roaming-based infrastructure sharing has attracted significant attention, since, currently, most of the MNOs deploy and own their network infrastructure. Oh *et al.* (Oh, Krishnamachari, Liu, & Niu, 2011) have studied the potential energy savings that can be achieved by opportunistically switching off part of the network during low traffic conditions in real-world scenarios. The authors investigate the challenges for operator cooperation to exploit the network capabilities to their limits. Marsan and Meo in (Marsan, & Meo, 2009) examine a network of two operators providing service in the same area. The authors propose four different cooperation strategies to switch off BSs according to different criteria: *i*) equal switching off time periods, *ii)* equal roaming costs, *iii)* equal energy gains, *iv)* maximum energy savings. In all cases, whenever a BS of an operator is switched off, its traffic is served by the collocated BS of the other MNO. In (Marsan, & Meo, 2011), the authors extend the algorithm that maximizes the energy savings (proposed in (Marsan, & Meo, 2009) for multi-operator environments with various traffic types and Quality of Service (QoS) requirements (i.e., throughput, lost calls). In another work (Bousia, Kartsakli, Antonopoulos, Alonso, & Verikoukis 2013), a game theoretic switching off strategy was introduced in networks with two operators, providing analytical expressions for the calculation of the system throughput and energy efficiency, assuming only one type of traffic (i.e., voice). Nonetheless, despite their novel insights in the infrastructure sharing concept, the existing literature leaves many open issues in the context of infrastructure sharing, thus encouraging further research efforts for the proposal of innovative solutions. The aforementioned works study only particular aspects of the problem (e.g., switching off time, roaming cost, energy savings) (Marsan, & Meo, 2009; Marsan, & Meo, 2011). In addition, the consideration of only voice traffic in some works is not realistic, since data traffic forms a significant part of the total traffic load in current cellular networks.

SYSTEM MODEL

The system model, depicted in Fig. 1a, considers clusters of two-operator cells. In particular, each cluster is formed by one central cell, surrounded by 6 peripheral cells, while each cell includes 2 BSs of different MNOs. Therefore, the term $BS_{i,j}$ is used to denote the BS of the *i*th operator in the *j*th macro cell, with $i \in [1,2]$ and $j \in [1,2,\ldots,6,cnt]$, where 1,…,6 refer to the peripheral cells and *cnt* refers to the central cell. As it will be explained in detail in the next section, part of the BS infrastructure in the 6 surrounding cells may be switched off during low traffic conditions, motivating the operators to reach roaming agreements and share the resources of the remaining active BSs in the same cell. On the other hand, the BSs of the central cell always remain active and can extend their transmission power and range in order to form an umbrella cell, in the extreme case where all the BSs of a peripheral cell are switched off.

The authors assume that the User Equipments (UEs[2]) of each operator are uniformly distributed within each cell. Each UE generates traffic according to a Poisson process and we consider two types of traffic, voice and data. They adopt a typical daily traffic profile for the voice and data traffic of each operator

Figure 1. (a) An example of 7-cell cluster network model with two operators, (b) Daily voice and data traffic load variations during a 24-h weekday

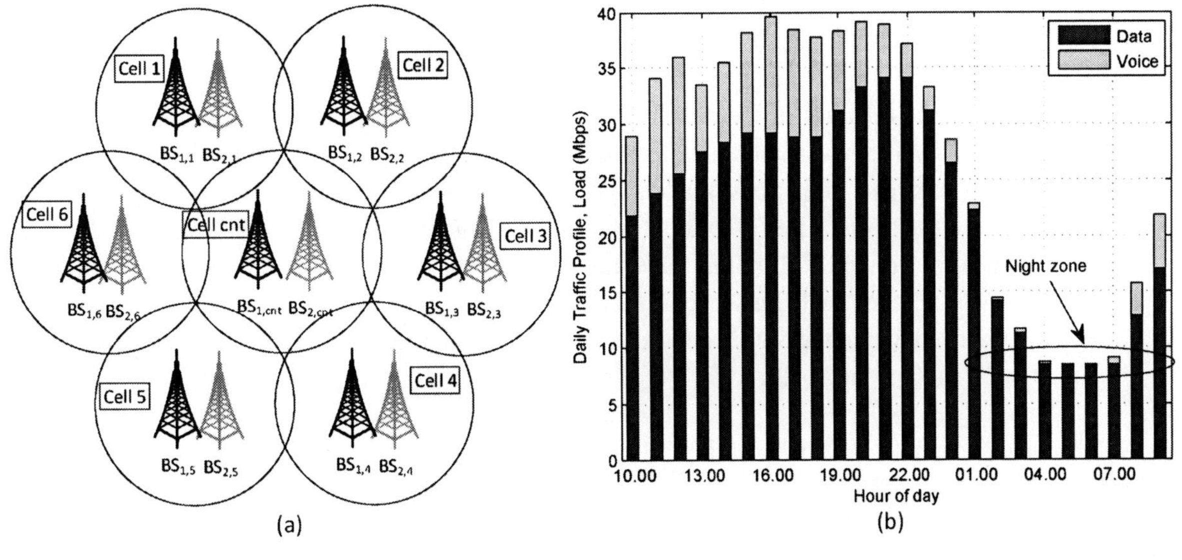

in a single cell based on real traffic information, illustrated in Fig. 1b. It is onserved that that the peak traffic hours occur in the morning and during early afternoon, while during night hours the traffic is low. In addition, the difference between high and low traffic load is significant. Based on (3GPP, 2010), voice traffic varies through time, having its peak reaching the 10% of the whole network capacity, and on the other hand, based on recent reports (Sandvine, 2011; UMTS Forum, 2011), the peak of data traffic, that is higher than voice traffic, corresponds to the 30% of the overall bandwidth[3]. In addition, for the sake of generality, they assume that the traffic volumes of different operators may be different, although they follow the same pattern.

INFRASTRUCTURE SHARING ALGORITHM WITH GAME THEORETIC SWITCHING OFF DECISION

In this section, the proposed algorithm is described that determines the set of BSs that should be switched off in a wireless cellular network under low traffic conditions. The strategy consists of an infrastructure sharing algorithm and a game theoretic switching off decision process.

Infrastructure Sharing Algorithm

Let us recall that the considered system model (Section 2) includes 2 MNOs that provide coverage to a cluster of cells formed by one central and 6 peripheral cells. For the low-traffic night zone, each operator switches off a subset of its BSs in the peripheral cells. According to the subset of switched off BSs in a cell, the proposed scheme determines how the traffic will be served by the remaining active infrastructure, taking into account the corresponding operation and roaming costs.

The proposed infrastructure sharing algorithm is applied in the network, after the execution of the independent switching off decisions. According to the outcome of the decision process, there are three possible outcomes in a given peripheral cell *j*:

1. If all the BSs remain active, no infrastructure sharing takes place in the cell. Hence, each BS consumes energy for operation and service of its own traffic.
2. If one of the BSs remains active, then it undertakes the service of the traffic of the switched off BS in the same cell. In particular, the traffic of the switched off BS is roamed to the active BS of the same cell. The operator of the deactivated BS should pay the corresponding roaming cost to the active one. On the other hand, the increased energy consumption (due to the higher traffic) of the active BS implies a higher cost that should also be considered.
3. If no BSs remain active, the BSs of the central cell ($BS_{i,cnt}$) increase their transmission power to cover the area of the peripheral cell. In this case, there is no collaboration between operators, since the traffic of each switched off BS is served by the central BS of the same operator. Hence, no roaming costs are involved, while the MNOs should take into account the extra cost for the increased power consumption in the central cell.

A flowchart of the infrastructure sharing scheme is given in Fig. 2.

Figure 2. Flowchart of the switching off algorithm

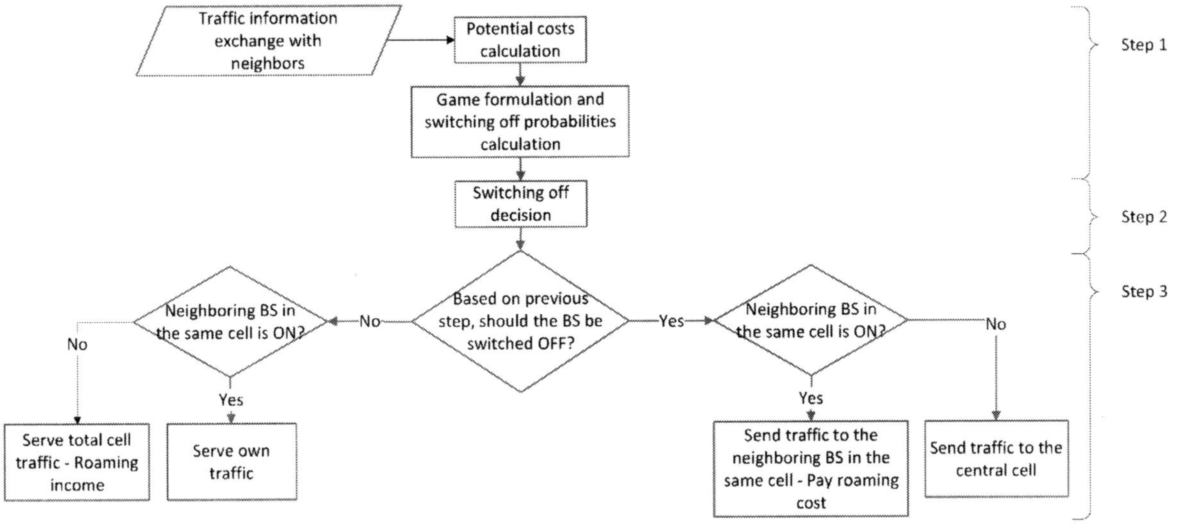

Game Theoretic Switching Off Strategy

Game Theory is the study of multi-person decision problems (Osborne, & Rubinstein, 1994) and consists of analytical tools and models designed to help us understand the actions of entities with conflicting objectives and preferences. Considering the basic concepts of Game Theory, a game theoretic approach is suitable for solving the problem of deciding which BSs should be switched off in a network with two operators. The common final goal for both operators is to have their traffic served, while minimizing their individual cost. In more details, keeping active many BSs implies significant cost and energy waste for the operators and, on the contrary, if a large number of BSs is switched off, the remaining active BSs may not be able to support the existing traffic. In order to analyze this trade-off, the authors model the algorithm as a static non-cooperative game with imperfect information (Fudenberg, & Tirole, 1994), where each MNOs selects the strategy that minimizes its own cost. The algorithm is applied in each of the peripheral cells by the two BSs, owned by each operator.

The proposed non-cooperative game Γ is represented by a tuple $\Gamma=(N,S,E[C_{i,j}])$, with $i \in N$ and $j \in M$, and it is characterized by the three following components:

- *Set of Players:* The player set corresponds to $N=\{1,2\}$, where 1 and 2 are the identification numbers of each MNO.
- *Set of Strategies:* $S_{i,j}$=(ON,OFF) is the set of two possible strategies of each $BS_{i,j}$, i.e., $BS_{i,j}$ can be active (state ON) or switched off (state OFF).
- *Set of Cost Functions:* The cost function represents the total cost of one action as a real number. The individual average cost $E[C_{i,j}]$ includes both the roaming and the energy consumption cost of each BS.

The straightforward representation of the game in its strategic form is given in Table 1, where ON and OFF are the two pure strategies and the contents of the table are the costs for each case that each player wishes to minimize.

The four states for the BSs in a cell with two operators, shown in Table 1, represent all the possible combinations of pure strategies of the game for the two operators and constitute the specific actions that can be followed. The four states depicted in Fig. 3 are explained in further detail in continuation:

Table 1. Cost matrix representation of the proposed game

Op.1 \ Op.2	ON	OFF
ON	$C_{BS}^{(1,j)}$ $C_{BS}^{(2,j)}$	$C_{new}^{(1,j)} - C_{roam}^{(2,j)}$ $C_{roam}^{(2,j)}$
OFF	$C_{roam}^{(1,j)}$ $C_{new}^{(2,j)} - C_{roam}^{(1,j)}$	$C_{incpwr}^{(1,cnt)}$ $C_{incpwr}^{(2,cnt)}$

Figure 3. Four different cases of the game theoretic scheme

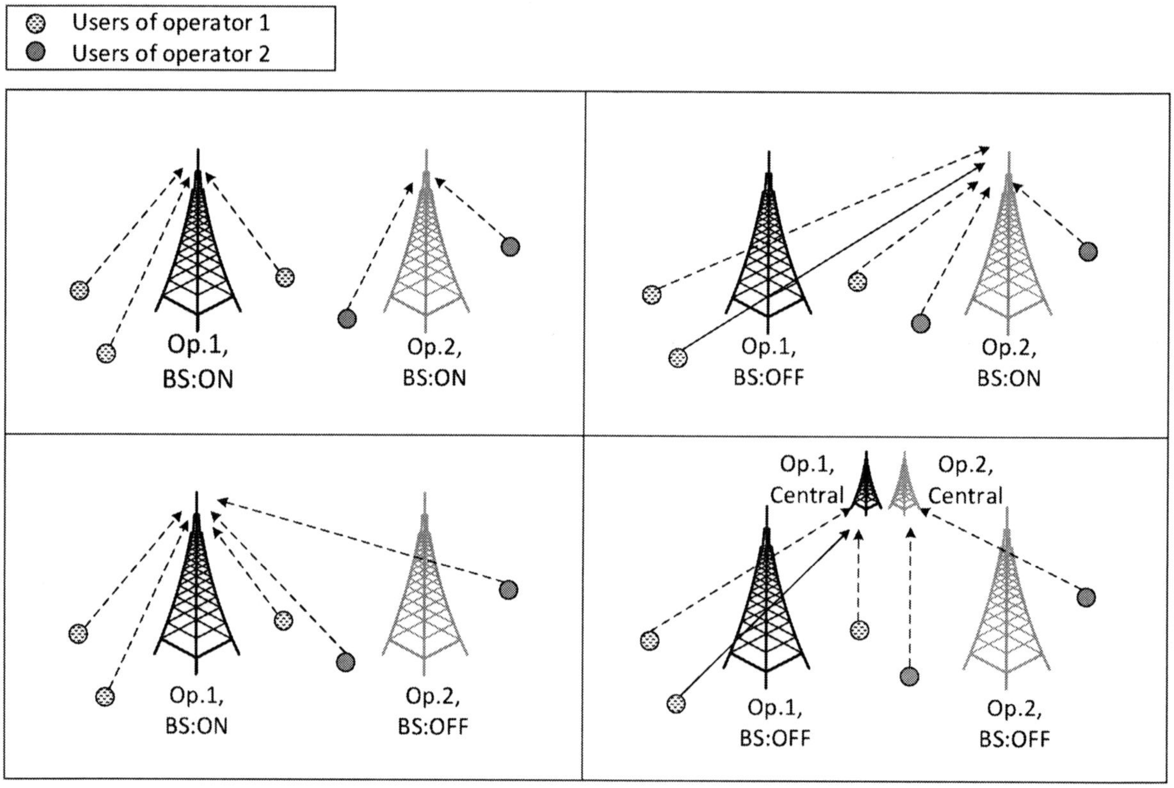

- $BS_{1,j}$ remains ON and $BS_{2,j}$ remains ON: Both BSs are active and consume energy for operation and service of their own traffic, paying cost equal to $C_{BS}^{(i,j)}$.
- $BS_{1,j}$ remains ON and $BS_{2,j}$ is switched OFF: In this case, $BS_{2,j}$ has no operational cost but it must roam its traffic to $BS_{1,j}$, providing the corresponding roaming cost, denoted by $C_{roam}^{(2,j)}$, to operator 1. On the other hand, $BS_{1,j}$ serves the total traffic of the cell, hence consuming additional energy for operation and transmission reflected by the increased cost equal to $C_{new}^{(1,j)}$, but also receives the roaming cost ($C_{roam}^{(2,j)}$) paid by operator 2.
- $BS_{1,j}$ is switched OFF and $BS_{2,j}$ remains ON: With respect to the previous case, this case is symmetrical and the roles of the two MNOs have been exchanged. Now, $BS_{1,j}$ must roam its traffic to $BS_{2,j}$, paying the roaming cost $C_{roam}^{(1,j)}$, whereas $BS_{2,j}$ pays the increased operational cost $C_{new}^{(2,j)}$ for serving the total traffic of the cell and receives the roaming cost from operator 1.
- $BS_{1,j}$ is switched OFF and $BS_{2,j}$ is switched OFF: Both BSs are switched off and their traffic is served by the corresponding central BSs of the same operator, i.e., $BS_{1,cnt}$ and $BS_{2,j}$, respectively. An increased cost, equal to $C_{incpwr}^{(i,cnt)}$, must be considered to account for the extra energy consumed by the central BS in order to increase its transmission power to cover a larger area and serve the traffic of the switched off BSs. Nevertheless, no roaming cost is involved.

The formulation of our problem in a strategic form reveals one NE in pure strategies, corresponding to the minimization of the number of active BS, i.e., the case where both operators switch off their BSs. However, the selection of a pure strategy for MNO *i* in all the peripheral cells would require major transmission power increase of the central BSs and would lead to QoS degradation, since due to the limited capacity of the BSs, the central cells would not be able to support all the traffic of the peripheral cells. Thus, the limitation of the pure strategies NE, along with the motivation of the MNOs to achieve an energy efficient but at the same time ubiquitous service network, have motivated us to study the problem in the mixed strategies domain, in order to provide feasible and applicable solutions for distributed systems. Therefore, we introduce our game theoretic policy as a system where the MNO *i* could randomize over the possible actions with a certain probability distribution.

Through the game theoretic formulation, the MNOs individually estimate the NE switching off probabilities according to the adopted cost-based functions. In the proposed algorithm, for each cell *j*, the $BS_{i,j}$ of the *i*th operator selects a switching off probability $s_{i,j}$ independently of the $BS_{i,j}$ of the other MNO, where $\{i, i' \mid i = \{1,2\}, i' = \{2,1\}\}$. Thus, the expected cost to be paid by each of the two BSs is given by[4]:

$$E[C_{i,j}] = \bar{s}_{i,j} \cdot \bar{s}_{i',j} \cdot C_{BS}^{(i,j)} + \bar{s}_{i,j} \cdot s_{i',j} \cdot \left(C_{new}^{(i,j)} - C_{roam}^{(i',j)}\right) + s_{i,j} \cdot \bar{s}_{i',j} \cdot C_{roam}^{(i,j)} + s_{i,j} \cdot s_{i',j} \cdot C_{incpwr}^{(i,cnt)}. \quad (1)$$

In the above equation, the first term represents the average cost when both BSs remain active, the second and third term is the average cost when one of the BSs is switched off, and finally, the fourth term is the average cost when both BSs are switched off.

Let us recall that the goal of each operator to estimate its individual switching off probability that minimizes its cost. To that end, the best response of $s_{i,j}$ to the strategy $s_{i',j}$ is given by calculating the roots of the partial derivative of the expected cost function with respect to $s_{i,j}$. Thus, it is concluded that:

$$s_{i,j} = \frac{C_{new}^{(i,j)} - C_{roam}^{(i',j)} - C_{BS}^{(i,j)}}{C_{new}^{(i,j)} - C_{incpwr}^{(i,cnt)} + C_{roam}^{(i,j)} - C_{roam}^{(i',j)} - C_{BS}^{(i,j)}}. \tag{2}$$

This probability refers to the outcome of the game that is used for the decision of the BSs to be switched off in *Step 2* of the proposed scheme (Fig. 2).

Cost Model

Infrastructure sharing promises significant reduction in the financial expenses from the operators' point of view. In order to evaluate the cost reduction, the economic model should be clarified. Referring to Table 1 (mentioned in Section 3.2), the costs of an operator can be categorized in two classes: (i) operational cost and (ii) roaming cost. The operational cost of a BS consists of the expenditures spent for being active and serving its traffic, while the roaming cost refers to the expenses of a MNO for having its traffic served by the other MNO.

Provided that c_1 is the electricity charge per energy unit, in [euros/kWh], and $E[E_{i,j}]$ is the expected BS energy consumption during the night zone, further explained in detail in Section 4.2, the total cost of $BS_{i,j}$ is given by (Tombaz, Vastberg, &. Zander, 2011):

$$C_{BS}^{(i,j)} = c_1 \cdot E\left[E_{i,j}\right]. \tag{3}$$

The costs $C_{new}^{(i,j)}$ and $C_{incpwr}^{(i,cnt)}$ are calculated in a similar way, since they refer to the cost for operation and transmission in different cases. The authors assume that the roaming cost should be a function of the traffic load that a BS has to serve, thus the roaming cost of an operator with high traffic exceeds the roaming cost of an operator with lower traffic. Thus, they assume that the roaming cost should be proportional to the energy consumed by a BS when being active and serving its traffic, $E[E_{i,j}]$. Furthermore, the roaming cost should be less than the operational cost in order to give MNOs an incentive to switch off their BSs. Otherwise, if the roaming cost was equal or higher than the operational cost, MNOs would not gain any profit by switching off their BSs. Based on these assumptions, the roaming cost is given by the following equation:

$$C_{roam}^{(i,j)} = \alpha \cdot C_{BS}^{(i,j)} = \alpha \cdot c_1 \cdot E\left[E_{i,j}\right], \tag{4}$$

where $\alpha \in [0,1]$ depicts the percentage of the OpEx that corresponds to the roaming cost.

ANALYTICAL MODEL

In this section, the authors develop an analytical model to calculate the throughput performance and the energy efficiency achieved with the application of the game theoretic infrastructure sharing scheme.

Two service classes are considered in our scenario: (i) voice calls and (ii) data sessions, as described in Section 2. Constant bit rate for both voice calls is assumed, denoted as R_V, and data sessions, defined as R_D. To formulate the network traffic load, the authors use a teletraffic model based on a multi-server M_1, M_2/G_1, $G_2/N/N_1$, N_2 queue. The two-dimensional Markov model, illustrated in Fig. 6, is used to analyze the performance of the proposed scheme and the important parameters are defined as follows:

- $\lambda_V^{(i,j)}$ Poisson arrival rate of voice calls in [call/s]
- $\lambda_D^{(i,j)}$ Poisson arrival rate of data sessions in [session/s]
- $1/\mu_V^{(i,j)}$ Exponential service time for voice calls in [s/call]
- $1/\mu_D^{(i,j)}$ Exponential service time for data sessions in [s/session]
- BW Overall network bandwidth in [Mbps]
- N_V Maximum number of voice calls served simultaneously, given as $N_V = BW/R_V$
- N_D Maximum number of data sessions served simultaneously, given as $N_D = BW/R_D$

Each state of the system is characterized by the number of active voice and data sessions. Being in the state (v,d) means that the system has v voice calls and d data sessions to serve with traffic generation rates $\lambda_V^{(i,j)}$ and $\lambda_D^{(i,j)}$ for voice and data traffic respectively. The state space of this Markov model, along with the bandwidth restrictions, is:

$$A = \{(\nu, d) \mid 0 \le \nu \le N_V, 0 \le d \le N_D, \nu \cdot R_V + d \cdot R_D \le BW\}. \tag{5}$$

By analyzing the state transition diagram (Fig. 4), the authors can obtain the system of linear equations for steady state probabilities, $p_{(v,d)}$. The balance equations must satisfy the normalization equation:

$$\sum_{\nu=0}^{N_V} \sum_{d=0}^{\lfloor \frac{BW-\nu \cdot R_V}{R_D} \rfloor} p_{(\nu,d)} = 1, \tag{6}$$

and the balance equations that represent the valid transitions, according to the predefined bandwidth restrictions, are given as follows:

$$\begin{aligned} &p_{(\nu,d)} \cdot \left(\lambda_V^{(i,j)} \cdot \phi_{\nu+1,d} + \lambda_D^{(i,j)} \cdot \phi_{\nu,d+1} \cdot \vartheta_{\nu,d+1} + \nu \cdot \mu_V^{(i,j)} \cdot \phi_{\nu-1,d} + d \cdot \mu_D^{(i,j)} \cdot \phi_{\nu,d-1}\right) = \\ &= \lambda_V^{(i,j)} \cdot p_{(\nu-1,d)} \cdot \phi_{\nu-1,d} + \lambda_D^{(i,j)} \cdot p_{(\nu,d-1)} \cdot \phi_{\nu,d-1} \cdot \vartheta_{\nu,d} + (\nu+1) \cdot \mu_V^{(i,j)} \cdot p_{(\nu+1,d)} \cdot \phi_{\nu+1,d} + \\ &+ (d+1) \cdot \mu_D^{(i,j)} \cdot p_{(\nu,d+1)} \cdot \phi_{\nu,d+1}, \end{aligned} \tag{7}$$

where $\phi_{v,d}$, $\vartheta_{v,d}$ denote the characteristic functions:

$$\phi_{\nu,d} = \begin{cases} 1, (\nu, d) \in A, \\ 0, otherwise, \end{cases} \tag{8}$$

and

Figure 4. Two-dimensional Markov model's state transition diagram for a network serving two classes of traffic

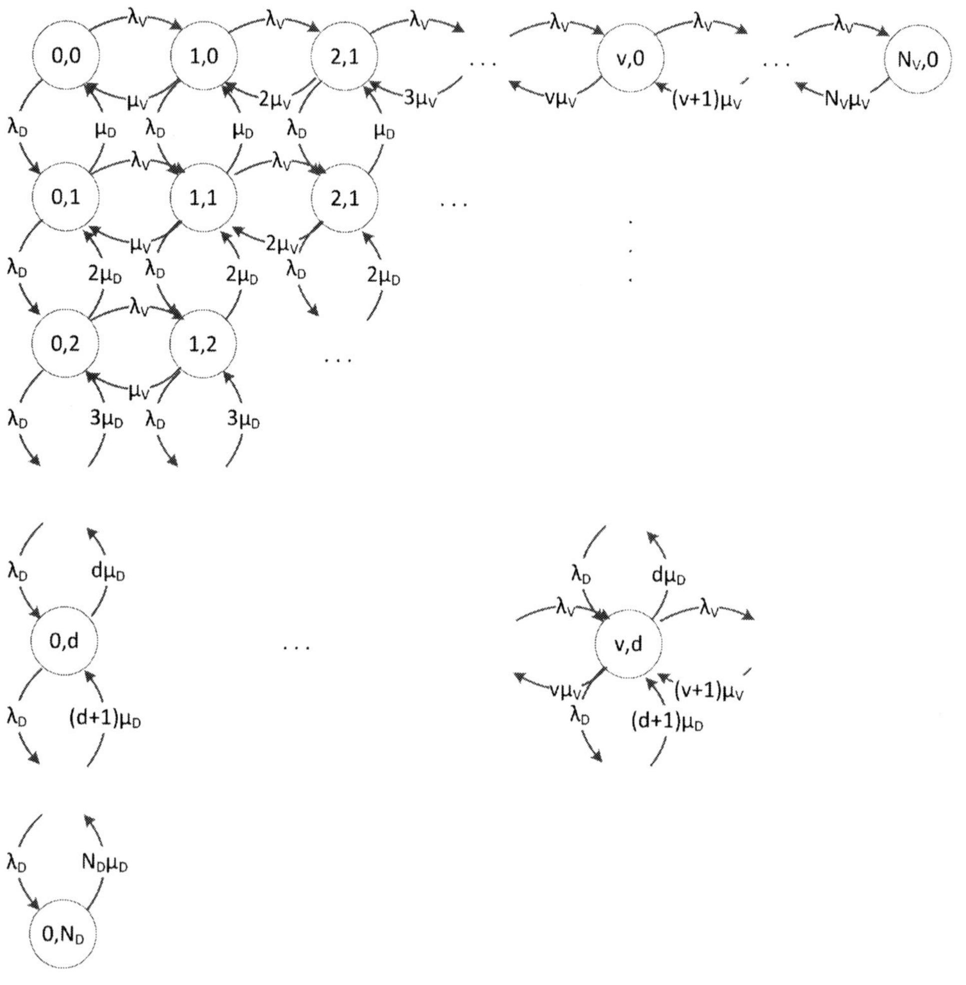

$$\vartheta_{\nu,d} = \begin{cases} 1, \nu \cdot R_V + d \cdot R_D \le BW, \\ 0, otherwise. \end{cases} \quad (9)$$

Throughput Analysis

Throughput for a Single Operator in a Cell

The expected throughput $E\left[T_{i,j}\left(\lambda_V^{(i,j)}, \lambda_D^{(i,j)}\right)\right]$ of a single BS depends on the traffic load of the network and is equal to the average number of served sessions in the system multiplied by the transmission rate of each session. The average throughput of a $BS_{i,j}$ with voice and data traffic equal to $\lambda_V^{(i,j)}$, $\lambda_D^{(i,j)}$, respectively, is calculated by the following equation:

$$E\left[T_{i,j}\left(\lambda_V^{(i,j)},\lambda_D^{(i,j)}\right)\right]=\sum_{\nu=0}^{N_V}\sum_{d=0}^{\lfloor\frac{BW-\nu\cdot R_V}{R_D}\rfloor}\left(\nu\cdot R_V+d\cdot R_D\right)\cdot p_{(\nu,d)}. \tag{10}$$

Throughput for the Network

In continuation, the authors calculate the thoughput for the network of 2 operators in a cluster of one central and 6 peripheral cells, for the proposed infrastructure sharing algorithm.

Firstly, they calculate the total throughput in one peripheral cell, *j*, in the four cases of the game theoretic scheme.

- Both BSs are ON:

$$E[T_{tot,j}^{(1)}]=\sum_{i=1}^{N}E\left[T_{i,j}\left(\lambda_V^{(i,j)},\lambda_D^{(i,j)}\right)\right], \tag{11}$$

- $BS_{1,j}$ is ON, and $BS_{2,j}$ is OFF:

$$E[T_{tot,j}^{(2)}]=E\left[T_{1,j}\left(\lambda_V^{(1,j)}+\lambda_V^{(2,j)},\lambda_D^{(1,j)}+\lambda_D^{(2,j)}\right)\right], \tag{12}$$

- $BS_{1,j}$ is OFF, and $BS_{2,j}$ is ON:

$$E[T_{tot,j}^{(3)}]=E\left[T_{2,j}\left(\lambda_V^{(1,j)}+\lambda_V^{(2,j)},\lambda_D^{(1,j)}+\lambda_D^{(2,j)}\right)\right], \tag{13}$$

- Both BSs are OFF:

$$E[T_{tot,j}^{(4)}]=0. \tag{14}$$

Therefore, the expected total throughput in one peripheral cell, *j*, is the summation of the throughput in each of the four cases multiplied by the probability of being in each case:

$$E[T_{tot,j}]=\overline{s}_{(1,j)}\cdot\overline{s}_{(2,j)}\cdot E[T_{tot,j}^{(1)}]+\overline{s}_{(1,j)}\cdot s_{(2,j)}\cdot E[T_{tot,j}^{(2)}]+s_{(1,j)}\cdot\overline{s}_{(2,j)}\cdot E[T_{tot,j}^{(3)}]+s_{(1,j)}\cdot s_{(2,j)}\cdot E[T_{tot,j}^{(4)}]. \tag{15}$$

Secondly, for the central cell, there are two cases concerning the throughput:

- The central cell serves only its traffic, meaning that at least one BS in each peripheral cell is active:

$$E[T_{tot,cnt}^{(1)}] = \sum_{i=1}^{N} E\left[T_{i,cnt}\left(\lambda_V^{(i,cnt)}, \lambda_D^{(i,cnt)}\right)\right], \tag{16}$$

- The central cell has to serve the traffic of $m \in \{1,2,\ldots,6\}$ of the M peripheral cells, meaning that in at least one peripheral cell both BSs are switched off:

$$E[T_{tot,cnt}^{(2)}] = \sum_{i=1}^{N} E\left[T_{i,cnt}\left((m+1)\cdot\lambda_V^{(i,cnt)}, (m+1)\cdot\lambda_D^{(i,cnt)}\right)\right]. \tag{17}$$

Therefore, the expected throughput of the central cell, *cnt*, is:

$$E[T_{tot,cnt}] = \prod_{j=1}^{M}\left(1 - s_{(1,j)}\cdot s_{(2,j)}\right)\cdot E[T_{tot,cnt}^{(1)}] + \sum_{m=1}^{M}\binom{6}{m}\cdot s_{(1,j)}\cdot s_{(2,j)}\cdot E[T_{tot,cnt}^{(2)}]. \tag{18}$$

Energy Efficiency Analysis

Energy Efficiency for a Single Operator in a Cell

The expected energy efficiency $E[\eta_{i,j}]$ is defined as the ratio of the average transmitted bits over the average energy consumption and is measured in *[bits/joule]*. The transmitted bits and the energy consumption during the night zone are directly dependent on the traffic rates of the voice and data packets and are denoted by $E\left[N_{bits,i,j}\left(\lambda_V^{(i,j)}, \lambda_D^{(i,j)}\right)\right]$ and $E\left[E_{i,j}\left(\lambda_V^{(i,j)}, \lambda_D^{(i,j)}\right)\right]$, respectively. Therefore, the energy efficiency of $BS_{i,j}$ is calculated by:

$$E[\eta_{i,j}] = \frac{E\left[N_{bits,i,j}\left(\lambda_V^{(i,j)}, \lambda_D^{(i,j)}\right)\right]}{E\left[E_{i,j}\left(\lambda_V^{(i,j)}, \lambda_D^{(i,j)}\right)\right]}. \tag{19}$$

The transmitted bits of $BS_{i,j}$ during t_{night} that corresponds to the total time that a BS remains active during the night zone are given by:

$$E\left[N_{bits,i,j}\left(\lambda_V^{(i,j)}, \lambda_D^{(i,j)}\right)\right] = t_{night}\cdot\left(\sum_{\nu=0}^{N_V}\sum_{d=0}^{\lfloor\frac{BW-\nu\cdot R_V}{R_D}\rfloor}\left(\nu\cdot R_V + d\cdot R_D\right)\cdot p_{(\nu,d)}\right). \tag{20}$$

For any given $BS_{i,j}$, the energy consumption, given in Eq. (21), is modeled as a linear function, consisting of a constant part of energy required to feed the antenna and for cooling, denoted as $E[E_{const}]$, and two variable parts, which depend on the traffic load, $E\left[E_{idle}\left(\lambda_V^{(i,j)},\lambda_D^{(i,j)}\right)\right]$, referring to the energy consumed when the BS is idle, and $E\left[E_{TX}\left(\lambda_V^{(i,j)},\lambda_D^{(i,j)}\right)\right]$, which corresponds to the energy for serving its traffic.

$$E\left[E_{i,j}\left(\lambda_V^{(i,j)},\lambda_D^{(i,j)}\right)\right] = E[E_{const}] + E\left[E_{idle}\left(\lambda_V^{(i,j)},\lambda_D^{(i,j)}\right)\right] + E\left[E_{TX}\left(\lambda_V^{(i,j)},\lambda_D^{(i,j)}\right)\right]. \tag{21}$$

In order to evaluate the energy consumption of a BS, the authors further analyze the terms that it consists of. To begin with, P_{const} refers to the constant power for BS operation that multiplied by the t_{night} gives the energy consumed over the night, defined as $E[E_{const}]$. During the active period of a BS, there exists a fraction of time that the BS remains idle and a portion of time that the BS is transmitting and serving its users. In the state (0.0), the BS is idle, since there are no ongoing sessions to be served. The power level in the idle state is represented by P_{idle} and the fraction of time that the BS remains idle is given by $t_{night} \bullet p_{(0,0)}$. In non-idle states when there is traffic to be served, the energy consumption for transmission must be calculated. The transmission power $P_{tx,V}$ refers to the power consumed for serving one voice call and $P_{tx,I}$ refers to the power consumed for serving one data session. The total time that a BS is transmitting is given by the summation of the periods that there are calls to be served. Thus, the time of the BS transmission is defined by the summation of the multiplication of the t_{night} by the $p_{(v,d)}$ that refers to the probability of the BS having v voice calls and d data sessions to serve. Summarizing, the expected energy consumption of a $BS_{i,j}$ is:

$$E\left[E_{i,j}\left(\lambda_V^{(i,j)},\lambda_D^{(i,j)}\right)\right] = t_{night} \cdot \left(P_{const} + P_{idle} \cdot p_{(0,0)} + \sum_{\nu=0}^{N_V} \sum_{d=0}^{\left\lfloor \frac{BW-\nu \cdot R_V}{R_D} \right\rfloor} \left(P_{tx,V} \cdot \nu + P_{tx,D} \cdot d\right) \cdot p_{(\nu,d)}\right). \tag{22}$$

Energy Efficiency for the Network

In continuation, the authros calculate the energy efficiency metric for the network of 2 operators in a cluster of one central and 6 peripheral cells.

Firstly, they calculate the total energy efficiency for serving the overall traffic in one peripheral cell, j, in the four cases of the game theoretic scheme.

- Both BSs are ON:

$$E[\eta_{tot,j}^{(1)}] = \frac{\sum_{i=1}^{N} E\left[N_{bits,i,j}\left(\lambda_V^{(i,j)},\lambda_D^{(i,j)}\right)\right]}{\sum_{i=1}^{N} E\left[E_{i,j}\left(\lambda_V^{(i,j)},\lambda_D^{(i,j)}\right)\right]}, \tag{23}$$

- $BS_{1,j}$ is ON, and $BS_{2,j}$ is OFF:

$$E[\eta_{tot,j}^{(2)}] = \frac{E\left[N_{bits,1,j}\left(\lambda_V^{(1,j)} + \lambda_V^{(2,j)}, \lambda_D^{(1,j)} + \lambda_D^{(2,j)}\right)\right]}{E\left[E_{1,j}\left(\lambda_V^{(1,j)} + \lambda_V^{(2,j)}, \lambda_D^{(1,j)} + \lambda_D^{(2,j)}\right)\right]}, \tag{24}$$

- $BS_{1,j}$ is OFF, and $BS_{2,j}$ is ON:

$$E[\eta_{tot,j}^{(3)}] = \frac{E\left[N_{bits,2,j}\left(\lambda_V^{(1,j)} + \lambda_V^{(2,j)}, \lambda_D^{(1,j)} + \lambda_D^{(2,j)}\right)\right]}{E\left[E_{2,j}\left(\lambda_V^{(1,j)} + \lambda_V^{(2,j)}, \lambda_D^{(1,j)} + \lambda_D^{(2,j)}\right)\right]}, \tag{25}$$

- Both BSs are OFF:

$$E[\eta_{tot,j}^{(4)}] = 0. \tag{26}$$

Therefore, the expected energy efficiency of one peripheral cell, j, is the summation of the energy efficiency in each of the four cases multiplied by the probability of being in each case:

$$E[\eta_{tot,j}] = \bar{s}_{(1,j)} \cdot \bar{s}_{(2,j)} \cdot E[\eta_{tot,j}^{(1)}] + \bar{s}_{(1,j)} \cdot s_{(2,j)} \cdot E[\eta_{tot,j}^{(2)}] + s_{(1,j)} \cdot \bar{s}_{(2,j)} \cdot E[\eta_{tot,j}^{(3)}]. + s_{(1,j)} \cdot s_{(2,j)} \cdot E[\eta_{tot,j}^{(4)}]. \tag{27}$$

Secondly, for the central cell, there are two cases concerning the energy efficiency:

- The central cell serves only its traffic, meaning that at least one BS in each peripheral cell is active:

$$E[\eta_{tot,cnt}^{(1)}] = \frac{\sum_{i=1}^{N} E\left[N_{bits,i,cnt}\left(\lambda_V^{(i,cnt)}, \lambda_D^{(i,cnt)}\right)\right]}{\sum_{i=1}^{N} E\left[E_{i,cnt}\left(\lambda_V^{(i,cnt)}, \lambda_D^{(i,cnt)}\right)\right]}, \tag{28}$$

- The central cell has to serve the traffic of $m \in \{1,2,\ldots 6\}$ of the M peripheral cells, meaning that in at least one peripheral cell both BSs are switched off:

$$E[\eta_{tot,cnt}^{(2)}] = \frac{\sum_{i=1}^{N} E\left[N_{bits,i,cnt}\left((m+1)\cdot \lambda_V^{(i,cnt)}, (m+1)\cdot \lambda_D^{(i,cnt)}\right)\right]}{\sum_{i=1}^{N} E\left[E_{i,cnt}\left((m+1)\cdot \lambda_V^{(i,cnt)}, (m+1)\cdot \lambda_D^{(i,cnt)}\right)\right]}. \tag{29}$$

Therefore, the expected energy efficiency of the central cell, *cnt*, is:

$$E[\eta_{tot,cnt}] = \prod_{j=1}^{M}\left(1 - s_{(1,j)} \cdot s_{(2,j)}\right) \cdot E[\eta^{(1)}_{tot,cnt}] + \sum_{m=1}^{M}\binom{6}{m} \cdot s_{(1,j)} \cdot s_{(2,j)} \cdot E[\eta^{(2)}_{tot,cnt}]. \quad (30)$$

PERFORMANCE EVALUATION

The authors have developed a custom-made C++ simulator for the network operation to validate the analytical expressions and assess the performance of the proposed infrastructure sharing scheme. In this section, the simulation setup is presented, along with the analytical and experimental results.

Simulation Scenario

To evaluate the performance of the proposed Game Theoretic algorithm (defined as GT in the rest of the paper), the authors consider a typical urban network, formulated into clusters of 7 cells. Each cell is covered by two BSs owned by different MNOs (Fig. 1), as described in Section 2. They assume that operator 1 has traffic equal to the values presented in Fig. 2. For operator 2, they assume that the traffic generation for voice and data traffic correspond to $\lambda_V^{(2,j)} / \lambda_V^{(1,j)} \leq 1$ and $\lambda_D^{(2,j)} / \lambda_D^{(1,j)} \leq 1$ respectively. Therefore, the total traffic intensities, consisting of the voice and data sessions, of the two operators, ρ_1, ρ_2, are connected by the expression $\rho=\rho_2/\rho_1\leq 1$. The parameters used in our analysis and simulations are summarized in Table 2.

Based on the analysis explained in Section 4, the parameters that play a crucial role in the behavior of the energy efficient scheme are summarized. The GT strategy depends directly on the traffic variations, since the calculated switching off probabilities, given by Eq. (2), are affected by the variations of the operational and the roaming costs that, in turn, are functions of the traffic. Therefore, it is important to observe the performance of GT algorithm as the network traffic changes. In addition, a critical parameter

Table 2. Simulation Parameters

Name	Value
Bandwidth, *BW* [Mbps]	115
Traffic load [Mbps]	Figure 2
Service rates, μ_V,μ_D [calls/s]	Mean: 1/50
Transmission rates, R_V, R_D [kbps]	64, 256
Voice transmission power, $P_{tx,V}$ [Watt]	[0.32, 0.37]
Data transmission power, $P_{tx,D}$ [Watt]	[1.29, 1.5]
Idle power, P_{idle} [Watt]	[0.34, 1.39]
Electricity charge, c_1 [/kWh]	0.1
Roaming cost variable, α	[0.1, 0.9]

in the analysis is the roaming cost value, as defined in Section 3.3. Different values of roaming cost are considered, in order to examine how the proposed algorithm is affected, namely High Roam (α=0.7), Medium Roam (α=0.5) and Low Roam (α=0.3) values. For low values of α, the roaming cost is very small compared to the operational cost, consisting of the expenditures of a BS spent for being active and for serving its traffic, whereas the opposite occurs, when α is bigger. The energy efficiency performance is evaluated and additional results about the trade-off between expenditure and profits are given. In addition to energy efficiency, the authors consider the cost efficiency measured in [*M*bits/euros], defined as the ratio of the transmitted bits over the operators' expenses for operation, transmission and roaming.

To assess the performance of the proposed scheme, a comparison to the state-of-the-art approaches is needed. Two state-of-the-art schemes (Marsan, & Meo, 2011) are considered: *(i)* a Roaming-to-One (R-to-1), where the MNO with the highest traffic serves the total traffic of the networks, while the MNO with the lowest traffic switches off its BSs during the entire night zone, and *(ii)* a Roaming-balanced (R-bal) scheme, where the MNOs switch off their BSs for different portions of time in order to balance their roaming costs. Moreover, a baseline approach (No Switch), where all BSs are active, is considered.

Performance Results

In this section, the authors provide the analytical and simulation results in terms of energy efficiency, network expenses, financial gains and cost efficiency.

In Fig. 5 , they illustrate the feasible switching off probabilities of a game with two operators with increasing total network traffic, while different roaming cost values are considered. The curves are examined in pairs and the switching off probabilities of the two MNOs represent how likely it is for the BSs to be switched off. The following three remarks are obtained from the figure: *i*) The switching off probabilities of the two operators decrease as the total network traffic increases, independently of the different roaming cost values. Even though the motivation of the MNOs to switch off their BSs is strong, the traffic increase leads to augmented roaming cost. As a result, the MNOs tend to switch off less BSs. *ii)* Comparing the switching off probabilities versus different values of roaming cost, it is observed that as the roaming cost value decreases, the switching off probabilities increase. In the High Roam case, the MNOs are not willing to switch off a high number of BSs due to the increased roaming cost, thus prefering to serve their own traffic. On the other hand, in the Low Roam case, where the roaming cost is rather small compared to the operational cost, the MNOs choose with higher probability to apply the switching off policy and to have their traffic served either by the central BSs, where no roaming is paid, or by the BS of the other MNO in the same cell. Depending on the roaming cost values, the switching off probabilities of the two MNOs have different behavior. For the Low Roam case, the switching off probability of operator 2 (with lower traffic) is higher than the one of operator 1, whereas it is almost equal for the Medium Roam case, and it is lower for the High Roam case. The selection of the roaming cost value is very critical. A small value of α leads to reduced roaming cost and, as a result, both operators have the tendency to switch off their active BSs. In addition, operator 2 has a higher switching off probability, given that its traffic load is lower with respect to operator 1. The reverse conclusion is observed when α is higher and the roaming cost approaches the operational cost, when operator 1 tends to keep less BSs active than operator 2.

Figure 6 demonstrates the total energy efficiency with regard to various traffic (Fig. 6a) and roaming cost values (Fig. 6b). In both figures, it is observed that the mathematical model, presented in Section 4, is validated, since there is a match between analytical and simulation results, thus demonstrating that

Figure 5. The switching off probabilities for increasing network traffic and variable roaming costs

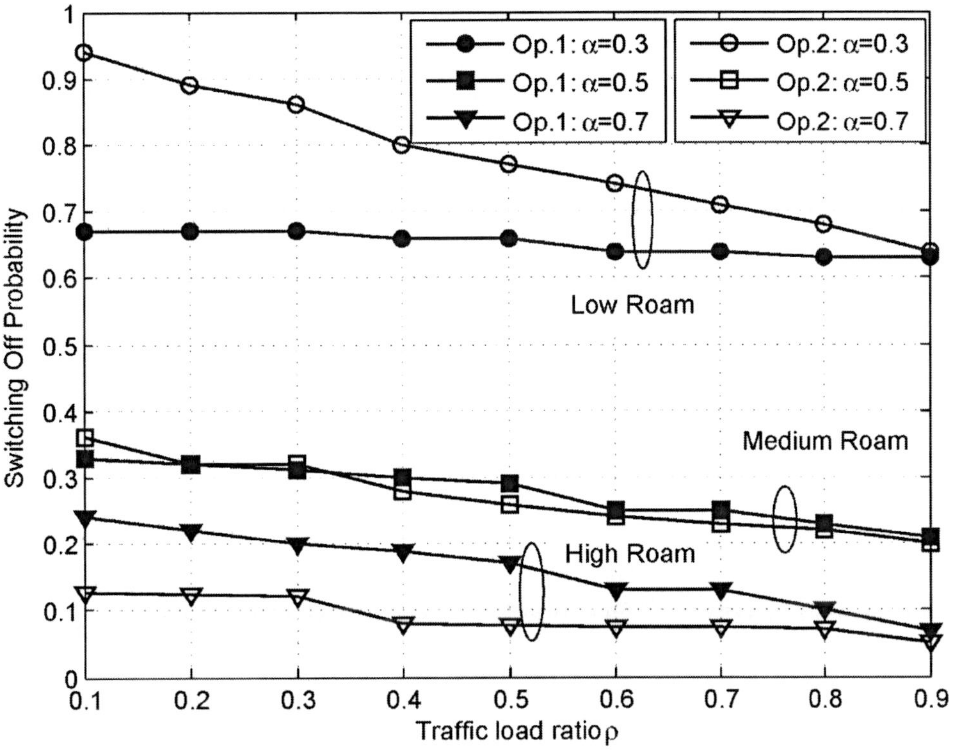

Figure 6. Energy efficiency validation for variable values of (a) Different traffic load values and Medium Roam (b) Different roaming cost values and ρ=0.1

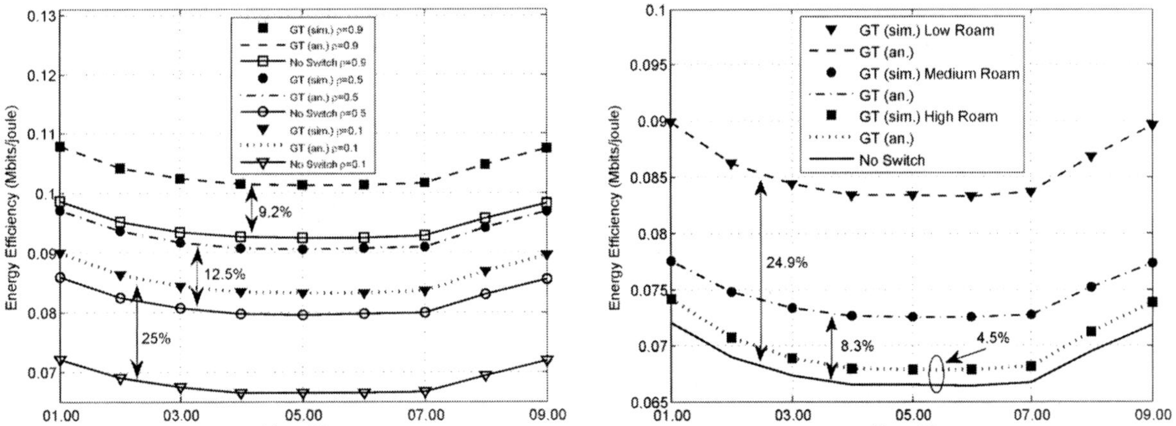

the theoretical approach provides an accurate estimation of energy efficiency. Moreover, from Fig. 6a, comparing the proposal to the baseline scheme where all of the BSs are active (No Switch), significant energy efficiency gain is spotted. A gain of almost 25% in energy efficiency is achieved when the total network traffic load is low and the ratio between the traffic load of the two operators is low (ρ=0.1).

However, when the network is overloaded, meaning that the traffic load ratio of operator 2 over operator 1 is high (ρ=0.9), the energy efficiency gain decreases and is about 9.2%, which is explained by the decrease of switching off probabilities shown in Fig. 5. The percentage of the energy efficiency gain is also dependent on the roaming cost values. In Fig. 6b, it is illustrated that whilst the roaming cost is high (High Roam), the operators do not have strong motivation to switch off their BSs, whereas when the roaming cost is low (Low Roam), more BSs are switched off and a 24.9% gain in energy efficiency is accomplished.

In Fig. 7, the game theoretic algorithm is compared to the state-of-the-art solutions, discussed in a previous section, in terms of energy efficiency under different conditions of traffic load and variable roaming cost values. Concerning the R-bal scheme, the MNOs select the number of switched off BSs in such a way so as to balance their roaming cost, independently of the exact value of the romaing cost. When applying the R-to-1 solution, the operator with the lowest traffic (i.e., operator 2 in this scenario) switches off its network for the entire night zone regardless of the roaming cost. Hence, given that the performance of the both state-of-the-art schemes is indifferent to the roaming cost, a single line is plotted for each algorithm in Fig. 7. The GT scheme outperforms the R-bal approach and compared to the R-to-1 algorithm, a better performance is achieved for the Low Roam case, where the application of GT scheme leads to a small number of active BSs. Thus, the consumed energy is reduced leading to significant energy efficiency gain up to 14.1%. In the Medium Roam case, the GT scheme accomplishes approximately the same energy efficiency gain as the R-to-1. On the other hand, with high roaming cost, the performance of the proposed approach is degraded but only a 5% loss in energy efficiency is observed.

Figure 7. Energy efficiency for variable values of traffic load and roaming cost: Comparison to the state-of-the-art solutions

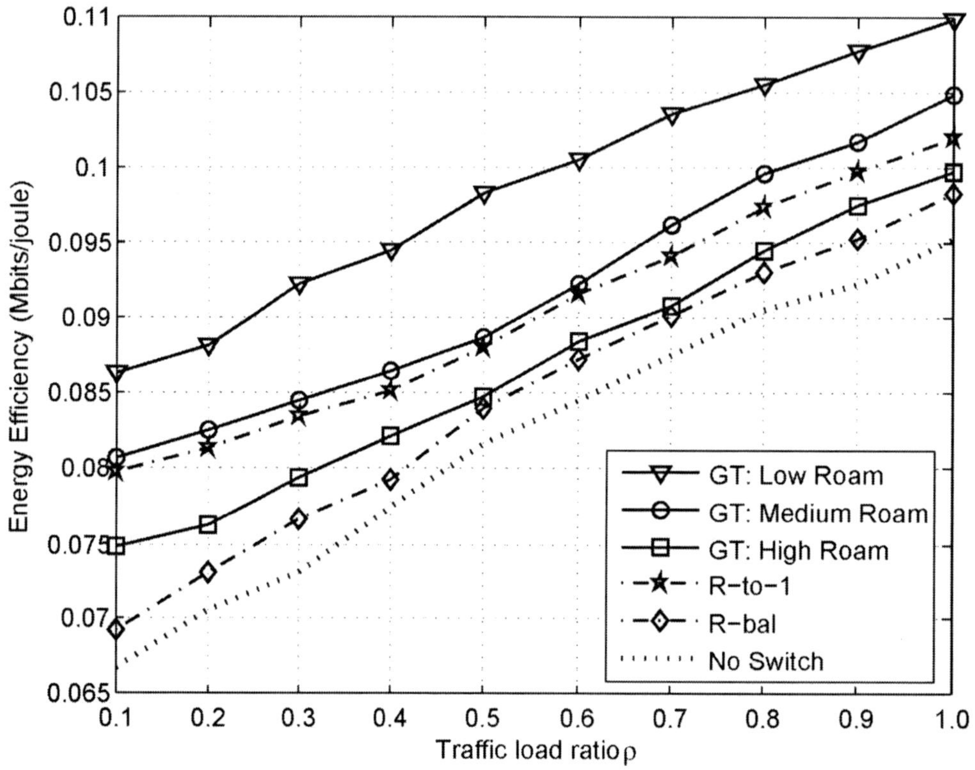

Aside from the total energy efficiency of both MNOs, an issue of major importance is the study of the energy efficiency that each MNO separately accomplishes, since it is this individual gain that will ultimately motivate them to apply a switching off policy. Therefore, a fair solution must be assured, to provide both MNOs incentives to share their resources. In other words, the fairness of the approach relies on attaining profits for both MNOs, despite the different traffic load values. The energy efficiency achieved by the two MNOs when applying our proposal versus the No Switch and the R-bal approaches is displayed in Fig. 8. By observing the three subfigures, a different behavior of the energy efficiency curves of the two MNOs is noticed. Firstly, the energy efficiency ofoperator 1 decreases when the traffic load grows. This can be explained by referring back to Fig. 5, where the switching off probabilities for each operator have been plotted. We can observe that the switching off probability of MNO 1 does not fluctuate with the traffic, leading to almost the same number of switched off BSs. On the other hand, more BSs of MNO 2 remain active as traffic increases, because the switching off probability decreases. Thus, with the same number of active BSs, MNO 1 serves less traffic and, as a consequence, its energy

Figure 8. Energy efficiency for the two operators for variable values of traffic and roaming cost

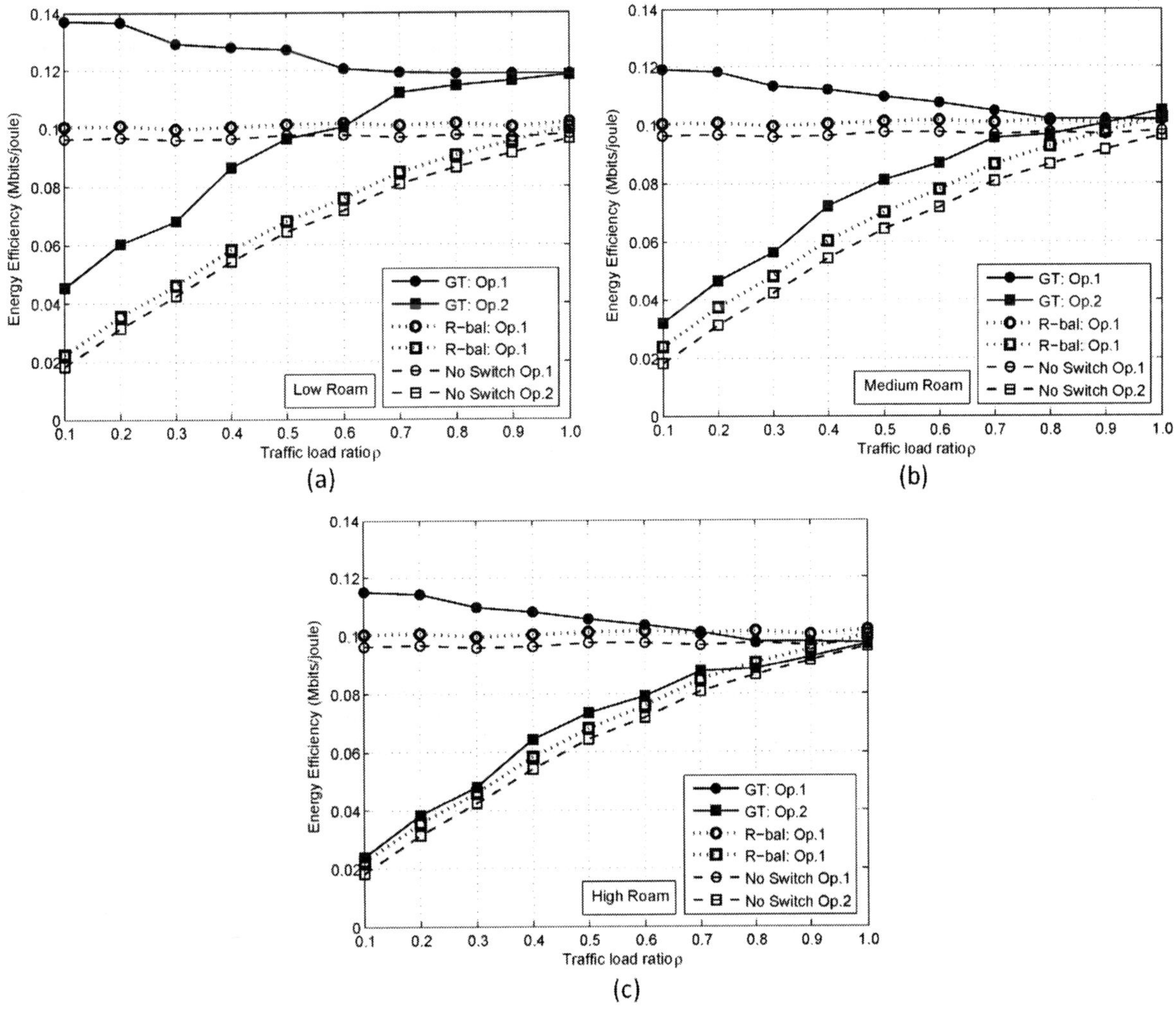

efficiency drops. The energy efficiency of MNO 2, on the contrary, increases with the traffic load growth. Comparing the three plots in Fig. 15, the authors can also observe that the energy efficiency of both MNOs increases with the decrease of roaming cost, a remark that was also concluded from the previous figures.

Finally, to facilitate the comparison between the GT algorithm and the state-of-the-art schemes, the individual energy efficiency gains of the two operators have been summarized in Table 3. The gains of each algorithm have been calculated with respect to the No Switch scheme. For the GT algorithm, it is shown that both MNOs have a positive energy efficiency gain for all values of the roaming cost. In particular, the gain for the Low Roam case is roughly 41% and 33% for MNO 1 and 2, respectively, 27% and 18.6%, respectively, for the Medium Roam case, and finally, 19% and 10.3% for the High Roam case. It can be observe that the respective energy efficiency gains of the R-bal approach are also positive but are clearly lower than the ones achieved by the GT algorithm. Finally, for the R-to-1 scheme the energy efficiency gains show a different behavior. In particular, MNO 2 has an infinite energy efficiency gain, whereas operator 1 suffers from energy efficiency loss (i.e., has a negative gain). This occurs because the R-to-1 scheme proposes that the operator with the lowest traffic (i.e., operator 2) completely switches off its network, thus maximizing its energy efficiency, while the BSs of MNO 1 remain always active and serve the total traffic (thus achieving a negative energy efficiency gain with respect to the No Switch case). This observation emphasizes the unfairness of the R-to-1 approach. The proposed approach is an appealing solution for the two operators since it yields significant profits to both of them. In contrast, in the case of the R-to-1 scheme, MNO 1 would be unwilling to apply the switching off policy. The fairness of the GT scheme gives the necessary incentives and conditions, under which two different MNOs can apply the deactivation decision, since both win in terms of energy savings. The gain of only one of the MNOs would not lead to the application of a switching off policy and energy savings would not be attained.

(a) Low Roaming (b) Medium Roaming (c) High Roaming

Apart from the energy efficiency, great attention should be paid to the cost of the network, consisting of the operational cost (cost when a BS is active and is serving the traffic) and the roaming expenditure, calculated by Eq. (1). The motivation of the MNOs for network planning is raised with the cost decrease, along with the energy savings. Figure 9 depicts the annual cost of the network under variable roaming conditions with increasing traffic load. The study of the two figures leads to two direct points: (i) The cost increases with the traffic growth due to the operational cost for serving the rising traffic, and (ii) the expenses decrease for lower roaming cost values. In Fig. 9a, the comparison of the GT scheme to the R-to-1 solution is illustrated and our algorithm clearly outperforms the R-to-1 scheme. Since the number of switched off BSs of the R-to-1 solution is fixed and as the roaming cost and the traffic of the network increase, MNO 1 faces a rising operational expenditure and MNO 2 is burdened by an increased roaming cost. Thus, the cost of the network augments. On the other hand, the game theoretic approach

Table 3. Energy Efficiency Gain with Respect to the No Switch Scheme

	Low Roam		Medium Roam		High Roam	
Algorithm	**Op.1**	**Op.2**	**Op.1**	**Op.2**	**Op.1**	**Op.2**
GT	41.0	33.0	27.1	18.6	19.2	10.3
R-bal	4.07	4.14	4.07	4.14	4.07	4.14
R-to-1	-32.6	∞	-32.5	∞	-32.3	∞

Figure 9. Cost under variable traffic and roaming conditions: Comparison to the state-of-the-art solutions
(a) GT compared to R-to-1 scheme (b) GT compared to R-bal scheme

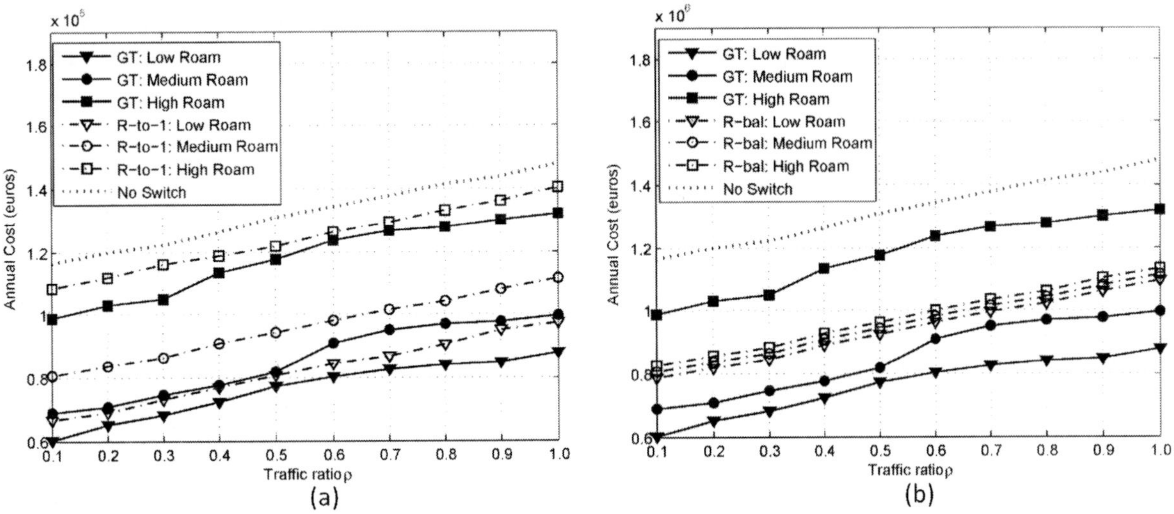

proposes to switch off a different number of BSs according to the corresponding values of traffic load and roaming cost, based on cost-oriented functions, leading to a dynamic switching off decision. The GT algorithm is compared to the R-bal solution in Fig. 9b. The GT approach outperforms the R-bal solution in the Medium Roam and Low Roam cases, but its behavior is degraded when the roaming cost is high. This performance degradation occurs because when the roaming cost is high, fewer BSs are switched off compared to the R-bal scheme and, thus, the cost of the network is higher.

Another very important metric, apart of the annual network cost, is the individual revenue of each MNO, plotted in Fig. 10. We consider the Medium Roam case, given that the Low Roam and High Roam

Figure 10. Money gained for each MNO under variable traffic conditions and medium roaming cost: Comparison to the state-of-the-art solutions
(a) GT compared to R-to-1 scheme (b) GT compared to R-bal scheme

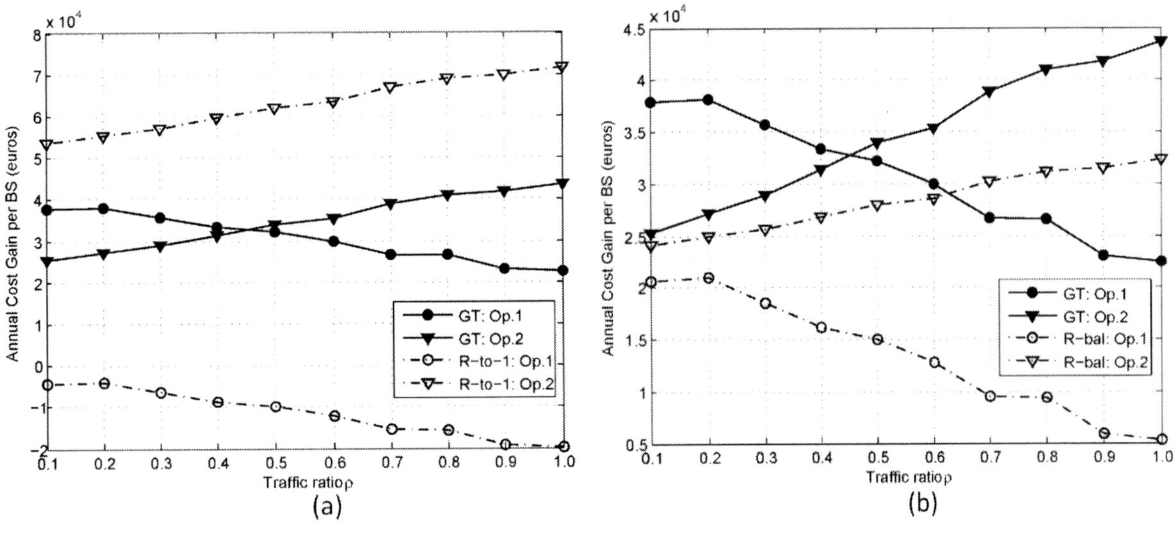

cases have similar behavior. In Fig. 10a and Fig. 10b our approach is compared to the R-to-1 and R-bal algorithms, respectively. By observing Fig. 10a, it is pointed out that with the GT approach, the two MNOs gain balanced incomes, whereas with the R-to-1 scheme, the revenue of operator 2 that switches off its whole network is significant, but the losses of operator 1 that remains active for the entire night zone are equally noteworthy. Thus, the fairness of our algorithm is demonstrated. The balanced savings in terms of both energy efficiency (Table 3) and income (Fig. 10a) encourage the operators to switch off their BSs through a game theoretic approach. From Fig. 10b, it is observed that in the R-bal solution, both MNOs receive profit by switching off their networks for different time periods, but the GT algorithm has a better performance by providing both operators with higher income. The proposed scheme outperforms the state-of-the-art approaches by better adapting the switching off decision to the varying conditions of the network and taking into account the interests of both MNOs.

In this chapter, the authirs also introduce an important metric, named cost efficiency. Figures 11 and 12 depict the cost efficiency of each MNO, in comparison to the R-to-1 and R-bal scheme, respectively. Two observations can be drawn by Fig. 11: (i) The GT algorithm always outperforms the No Switch scheme, while on the contrary, the R-to-1 scheme achieves better performance for operator 2, but degraded performance (even when compared to the No Switch approach) for operator 1. The proposal of switching off the whole network that belongs to the MNO with the lower traffic (R-to-1) results to great cost efficiency for operator 2 who does not consume any energy at all and is encumbered only with the compensation of the roaming cost to the other MNO. Contrariwise, operator 1 serves the whole traffic of the network and consumes significant energy leading to an increased cost that is not indemnified by the received roaming. Considering the proposed approach, the GT scheme achieves an equilibrium between the gain of the two MNOs in terms of cost efficiency. This balance is a consequence of the switching off policy since both MNOs are motivated to switch off some of their BSs. (ii) The cost efficiency of the two MNOs grows with the decrease of the roaming cost value, since a higher number of BSs is switched off and, therefore, less energy is needed to serve the existing traffic.

Likewise, in Fig. 12, GT is compared to the R-bal scheme in terms of cost efficiency. The GT algorithm outperforms the R-bal scheme concerning the MNO with the highest traffic and has better performance in some cases also for the operator with the lowest traffic. The difference of the performance of the two compared algorithms relies on the different number of active and switched off BSs. The R-bal strategy forces operator 2 to switch off its BSs for a longer period of time in order to balance the roaming expenditures. On the other hand, when applying our approach, both MNOs switch off some of their BSs for the entire night zone.

FUTURE RESEARCH DIRECTIONS

The energy consumption problem remains crucial and energy efficiency in cellular networks has to be significantly improved. Therefore, fundamental research for green wireless communications must be done. In the current analysis, the focused on achieving energy savings through network planning by switching off BSs during periods with low activity in a network with two operators. However, more aspects should be considered and studied towards this direction.

First of all, referring to the aforementioned works of the literature and the proposed switching off algorithms, energy efficient solutions were presented for deactivating the redundant BSs during low traffic periods. The coverage "gap" of the switched off BSs was covered by the remaining active BSs that

Figure 11. Cost efficiency for the two MNOs in terms of variable values of traffic load and roaming cost: Comparison to the R-to-1 scheme
(a) Low Roaming (b) Medium Roaming (c) High Roaming

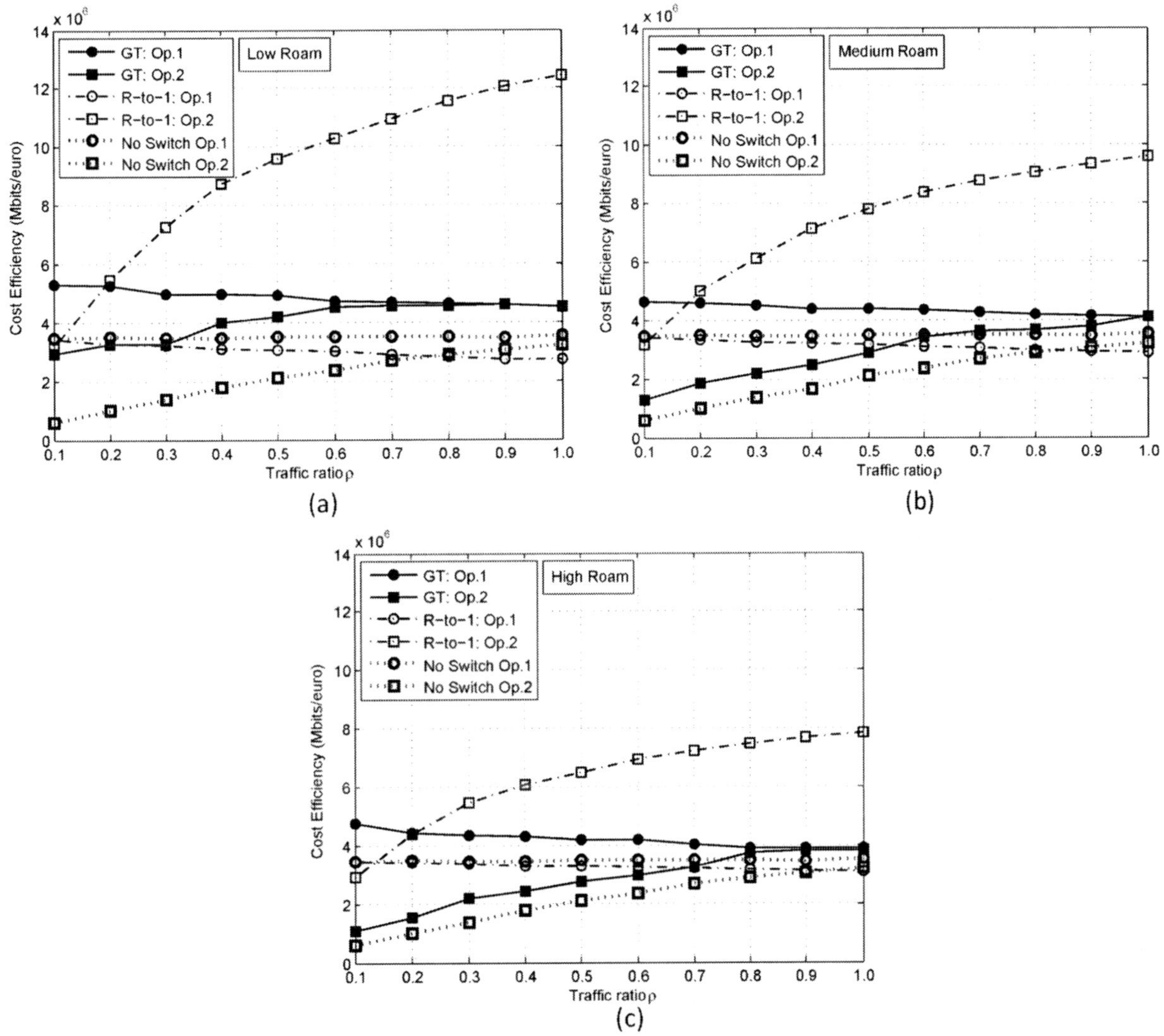

increased their transmission power. Still the details of the switching off schemes ought to be examined, because there might be cases where the deactivation of BSs leads to coverage holes. For example, in high traffic regions, a switching off policy may lead users to outage due to lack of network resources. The research works so far study scenarios with uniform traffic, in order to simplify the analysis. Therefore, more realistic scenario and networks configurations should be examined, because operators are willing to save energy, if and only if the service area is preserved and the QoS is not degraded. The question that remains is if it is feasible in practice for the active BSs to increase their transmission power in order to increase their coverage and what are the limitations of the coverage extension. To reduce the outage probability, relay nodes can be used to improve the network coverage and low consumption BSs can be used to maintain the QoS.

Figure 12. Cost efficiency for the two MNOs in terms of variable values of traffic load and roaming cost: Comparison to the R-bal scheme
(a) Low Roaming (b) Medium Roaming (c) High Roaming

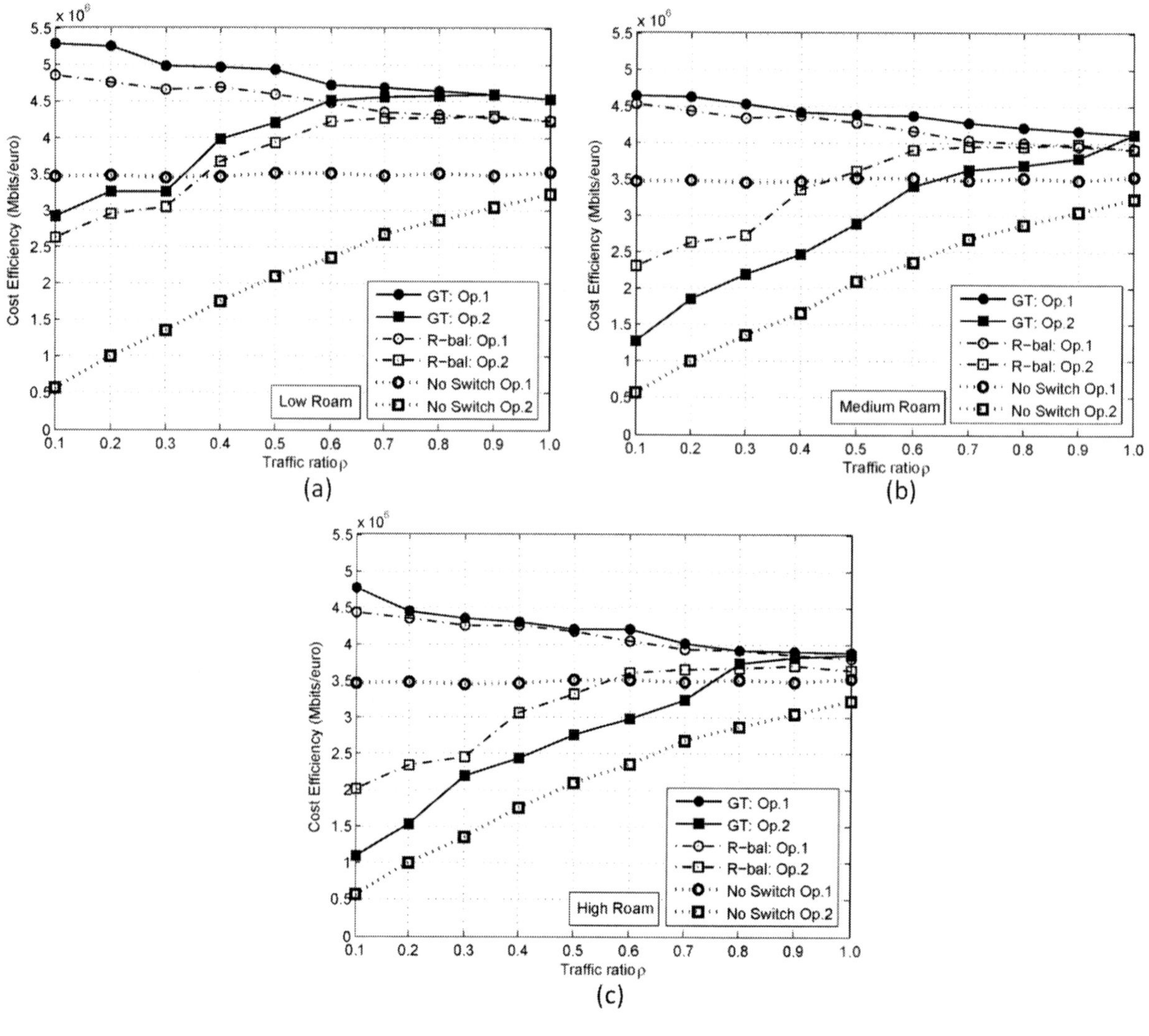

In addition, the innovative trend of heterogeneous networks can be exploited. The trend nowadays is to install low consumption nodes in order to provide service in areas where traffic load is concentrated. The existence of small cells will encourage the operators to switch off their BSs even during peak traffic hours and have their traffic served by small cells. Concerning the heterogeneous characteristic of networks, deployment strategies should be investigated since the coexistence of macro and small cells introduces additional complexity in the switching off techniques.

Another important issue that needs to be further studied consists of the traffic load variations during the days. As it was shown in this chapter, the majority of the energy efficient solutions are based on the traffic load information that unfortunately cannot be predicted. So far, we use average values of traffic load and information about real traffic data. The traffic-aware schemes provide promising results, but their behavior is degraded if the traffic load does not follow the predicted patterns. For these cases and to overcome this obstacle, either more advanced strategies must be proposed; or the energy efficient approaches must focus on real time traffic information, which implies more complicated scenarios.

Finally, significant energy can be saved if the operators cooperate and share their resources. The cooperation between operators through joint network management and roaming policies entails the proposal of advanced switching off schemes where even more BSs can be switched off. When designing cooperation algorithms, the agreements and the conditions under which the operators will be willing to cooperate formulate a problem that must be thoroughly examined.

Considering the above aspects, it is concluded to the fact that there are still many open problems in the concept of energy efficient network planning, and the proposal of innovative solutions is more than compulsory.

CONCLUSION

In this chapter, the authors studied the energy saving and cost reduction potentials that can be achieved by switching off redundant BSs in a network covered by two MNOs. They proposed an energy efficient game theoretic strategy, in order to decide which BSs are more suitable to be switched off. Individual energy efficient cost oriented functions were used and the tendency of MNOs to minimize their costs led to a fair switching off decision. One of the key advantages of the proposed approach is that it considers the competing preferences of the MNOs, thus providing impartiality among different operators and adaptability to the network traffic load conditions. Analytical and simulation results for a network with two classes of traffic were provided in terms of variable traffic conditions and roaming cost values. Given the results, it was concluded that our algorithm provides a balanced performance for the two operators in terms of energy savings and cost reduction compared to previous works in the literature. In the majority of scenarios, the game theoretic approach surpasses the works appearing in the literature. The reasonable conclusion of this work is that the coexistence of different MNOs leads to energy inefficiency and network planning approaches are more than compulsory to be applied.

ACKNOWLEDGMENT

This work has been funded by the Research Projects GREENET (PITN-GA-2010-264759), AGAUR (2014-SGR-1551) and GEOCOM (TEC2011-27723-C02-01).

REFERENCES

3GPP. (2001). *3GPP TS 23.236, 3rd Generation Partnership Project; Shared Networks: An operator alternative to reduce initial investments, increase coverage and reduce Time To Market for WCDMA by sharing infrastructure*. Author.

3GPP. (2002). *3GPP TS 23.236. 3rd Generation Partnership Project; Technical Specification Group Services and System Aspects; Intra-domain connection of Radio Access Network (RAN) nodes to multiple Core Network (CN) nodes. Release 5. Technical Report*. Author.

3GPP. (2010). *3GPP TR 32.826 V10.0.0. 3rd Generation Partnership Project; Technical Specification Group Services and System Aspects; Telecommunication management; Study on Energy Savings Management (ESM); Release 10, Technical Report*. Author.

Beckman, C., & Smith, G. (2005). Shared networks: Making wireless communication affordable. *IEEE Wireless Communications*, *12*(2), 78–85. doi:10.1109/MWC.2005.1421931

Bousia, A., Antonopoulos, A., Alonso, L., & Verikoukis, C. (2012). "Green" distance-aware base station sleeping algorithm in LTE-Advanced. *Proceedings from ICC'12: The IEEE International Conference on Communications*. Ottawa, ON, Canada.

Bousia, A., Kartsakli, E., Alonso, L., & Verikoukis, C. (2012a). Dynamic energy efficient distance-aware base station switch on/off scheme for LTE-Advanced. *Proceedings from GLOBECOM'12: The IEEE Global Communications Conference*. Anaheim, CA, USA.

Bousia, A., Kartsakli, E., Alonso, L., & Verikoukis, C. (2012b). Energy efficient base station maximization switch off scheme for LTE-Advanced. *Proceedings from CAMAD'12: The IEEE 17th International Workshop on Computer Aided Modeling and Design of Communication Links and Networks*. Barcelona, Spain.

Bousia, A., Katsakli, E., Antonopoulos, A., Alonso, L., & Verikoukis, C. (2013). Game Theoretic Approach for Switching Off Base Stations in Multi-Operator Environments, *Proceedings from ICC'13: The IEEE International Conference on Communications*. Budapest, Hungary. doi:10.1109/ICC.2013.6655262

Chen, T., & Zhang, H. Zhao, & Chen, X. (2010). Towards green wireless access networks. *Proceeding from CHINACOM'10: The 5th International ICST Conference on Communications and Networking in China*. Beijing, China.

Cisco (2012, February 14). *Cisco Visual Networking Index: Global Mobile Data, Traffic Forecast Update, 2011-2016*. Retrieved from: http://www.cisco.com/en/US/solutions/collateral/ns341/ns525/ns537/ns705/ns827/white_paper_c11-481360_ns827_Networking_Solutions_White_Paper.html

Correia, L., Zeller, D., Blume, O., Ferling, D., Jading, Y., Godor, I., & Van der Perre, L. et al. (2012). Challenges and enabling technologies for energy aware mobile radio networks. *IEEE Communications Magazine*, *48*(11), 66–72. doi:10.1109/MCOM.2010.5621969

Fettweis, G., & Zimmermann, E. (2008). ICT Energy Consumption - Trends and Challenges, *Proceedings from WPMC'08: The 11th International Symposium on Wireless Personal Multimedia Communications*. Lapland, Filand.

UMTS Forum. (2011). *Mobile traffic forecasts 2010-2020 report*. Author.

Frisanco, T., Tafertshofer, P., Laurin, P., & Ang, R. (2008). Infrastructure Sharing and Shared Operations for Mobile Network Operators-From a Deployment and Operations View, *Proceeding from NOMS'08: The IEEE Network Operations and Management Symposium*. Salvadro, Bahia.

Fudenberg, D., & Tirole, J. (1994). *Game Theory*. Cambridge, MA: MIT Press.

Hansly, S., & Mathar, R. (2002). On the Optimal Base-Station Density for CDMA Cellular Networks. *IEEE Transactions on Communications*, *50*(8), 1274–1281. doi:10.1109/TCOMM.2002.801523

Hasan, Z., Boostanimehr, H., & Bhargava, V. K. (2011). Green cellular networks: A survey, some research issues and challenges. *IEEE Communications Surveys and Tutorials*, *13*(4), 524–540. doi:10.1109/SURV.2011.092311.00031

Hasbani, G., El-Darwiche, B., Mourad, M., & Chanab, L. A. (2007). *Telecom Infrastructure sharing-Regulatory Enablers and economic benefits. Perspective report.* Booz Allen Hamilton.

Khan, A., Kellerer, W., Kozu, K., & Yabusaki, M. (2011). Network sharing in the next mobile network: TCO reduction, management flexibility, and operational independence. *IEEE Communications Magazine*, *49*(10), 134–142. doi:10.1109/MCOM.2011.6035827

Koudouridis, G. P., & Li, H. (2012). Distributed Power On-Off Optimisation for Heterogeneous Networks. *Proceedings from CAMAD'12: The IEEE 17th International Workshop on Computer Aided Modeling and Design of Communication Links and Networks*. Barcelona, Spain.

Leibner, T. (2004). *Network and Infrastructure Sharing in 2G networks*. Munich: Siemens.

Marsan, M. A., Chiaraviglio, L., Ciullo, D., & Meo, M. (2009). Optimal energy savings in cellular access networks, *Proceedings from ICC Workshops '09: The IEEE International Conference on Communications Workshops*. Dresden, Germany.

Marsan, M. A., & Meo, M. (2009). Energy efficient management of two cellular access networks. *Performance Evaluation Review*, *37*(4), 69–73. doi:10.1145/1773394.1773406

Marsan, M. A., & Meo, M. (2011). Energy efficient wireless Internet access with cooperative cellular networks. *Elsevier Journal Computer Networks*, *55*(2), 386–398. doi:10.1016/j.comnet.2010.10.017

Meddour, D. E., Rasheed, T., & Gourhant, Y. (2011). On the role of infrastructure sharing for mobile network operators in emerging markets. *Elsevier Computer Networks.*, *55*(7), 1567–1591. doi:10.1016/j.comnet.2011.01.023

Niu, Z., Zhou, S., Hua, Y., Zhang, Q., & Cao, D. (2002). Energy-aware network planning for wireless cellular system with inter-cell cooperation. *IEEE Transactions on Wireless Communications*, *11*(4), 1412–1423. doi:10.1109/TWC.2012.021412.110147

Nokia. (2001). *Nokia launches Multi-Operator Radio Access Network for controlled 3G network sharing*. Press release (23.05.2001).

Oh, E., Krishnamachari, B., Liu, X., & Niu, Z. (2011). Towards dynamic energy-efficient operation of cellular network infrastructure. *IEEE Communications Magazine*, *49*(6), 56–61. doi:10.1109/MCOM.2011.5783985

Osborne, M. J., & Rubinstein, A. (1994). *A Course in Game Theory*. Cambridge, MA: MIT Press.

Richter, F., Fehske, A. J., & Fettweis, G. P. (2009). Energy efficiency aspects of base station deployment strategies in cellular networks. *Proceedings from VTC Fall'09: The IEEE 70th Vehicular Technology Conference Fall*. Anchorage, AK. doi:10.1109/VETECF.2009.5379031

Sandvine. (2011). *Intelligent Broadband Networks Global Internet phenomena report*. Author.

Suryaprakash, V., Fonseca dos Santos, A., Fehske, A. J., & Fettweis, G. P. (2012). Energy Consumption Analysis of Wireless Networks using Stochastic Deployment Models. *Proceedings from GLOBECOM'12: The IEEE Global Communications Conference*. Anaheim, CA. doi:10.1109/GLOCOM.2012.6503603

Tombaz, S., Vastberg, A., & Zander, J. (2011). Energy- and cost-efficient ultra-high-capacity wireless access. *IEEE Wireless Communications.*, *18*(5), 18–24. doi:10.1109/MWC.2011.6056688

Tseng, Y., & Huang, C. (2012). Analysis of Femto Base Station Network Deployment. *IEEE Transactions on Vehicular Technology*, *50*(2), 748–757. doi:10.1109/TVT.2011.2176355

Zhang, J., Zhang, R., Li, G., & Hanzo, L. (2012). Remote Coalition Network Elements for Base Station Cooperation Aided Multicell Processing. *IEEE Transactions on Vehicular Technology*, *61*(3), 1406–1415. doi:10.1109/TVT.2012.2186600

ADDITIONAL READING

3GPP TR 25.912, v8.0.0 (2008). *Feasibility Study for Evolved Universal Terrestrial Radio Access (UTRA) and Universal Terrestrial Radio Access Network (UTRAN).*

3GPP TR 25.913, v8.0.0 (2008). *Requirements for Evolved UTRA (E-UTRA) and Evolved UTRAN (E-UTRAN).*

3GPP TR 25.913, v. 9.0.0 (2009). *Requirements for Evolved UTRA (EUTRA) and Evolved UTRAN (E-UTRAN).*

3GPP TR 36.814, v9.0.0 (2010). *Further Advancements for EUTRA Physical Layer Aspects.*

3GPP TR 36.913, v9.0.0 (2010). *Requirements for Further Advancements for Evolved Universal Terrestrial Radio Access (E-UTRA)* (LTE-Advanced).

Astély, D., Dahlman, E., Frenger, P., Ludwig, R., Meyer, M., Parkvall, S., & Wiberg, N. et al. (2006). A Future Radio-Access Framework. *IEEE Journal on Selected Areas in Communications*, *24*(3), 693–706. doi:10.1109/JSAC.2005.862420

Badic, B., O'Farell, T., Loskot, P., & He, J. (2009). Energy efficient radio access architectures for green radio: large versus small cell size deployment. Proceedings from VTC'09 Fall: *The IEEE 70th Vehicular Technology Conference Fall*. Anchorage, AK, USA. doi:10.1109/VETECF.2009.5379035

Bousia, A., Antonopoulos, A., Alonso, L., & Verikoukis, C. (2012). "Green" distance-aware base station sleeping algorithm in LTE-Advanced. Proceedings from ICC'12: *The IEEE International Conference on Communications*. Ottawa, ON, Canada.

Bousia, A., Kartsakli, E., Alonso, L., & Verikoukis, C. (2012). Dynamic energy efficient distance-aware base station switch on/off scheme for LTE-Advanced. Proceedings from GLOBECOM'12: *The IEEE Global Communications Conference*. Anaheim, CA, USA.

Bousia, A., Kartsakli, E., Alonso, L., & Verikoukis, C. (2012). Energy efficient base station maximization switch off scheme for LTE-Advanced. Proceedings from CAMAD'12: *The IEEE 17th International Workshop on Computer Aided Modeling and Design of Communication Links and Networks*. Barcelona, Spain.

Bousia, A., Katsakli, E., Antonopoulos, A., Alonso, L., & Verikoukis, C. (2013). Game Theoretic Approach for Switching Off Base Stations in Multi-Operator Environments, Proceedings from ICC'13: *The IEEE International Conference on Communications*. Budapest, Hungary. doi:10.1109/ICC.2013.6655262

Bousia, A., Katsakli, E., Antonopoulos, A., Alonso, L., & Verikoukis, C. (2013). Sharing the Small Cells for Energy Efficient Networking: How much does it cost? Proceedings from GLOBECOM'14: *The IEEE Global Communications Conference*. Austin, Texas, USA.

Bousia, A., Katsakli, E., Antonopoulos, A., Alonso, L., & Verikoukis, C. (2013). Energy Efficient Schemes for Base Station Management in 4G Broadband Systems, Chapter in Broadband Wireless Access Networks for 4G: Theory, Application and Experimentation, IGI Global publication, pp. 101-121, Chapter 6, ISBN13: 978146648883; EISBN13: 9781466648890

Dahlman, E., & Parkvall, S. (2011). *LTE/LTE-Advanced for Mobile Broadband*. Oxford, OX: Academic Press.

Dahlman, E., Parkvall, S., Skold, J., & Beming, P. (2008). *3G Evolution: HSPA and LTE for Mobile Broadband* (2nd ed.). Oxford, OX: Academic Press.

Kawai, H., Morimoto, A., Higuchi, K., & Sawahashi, M. (2010). Investigations on Inter-Node B Macro Diversity for Single-Carrier Based Radio Access in Evolved UTRA Uplink. *IEICE Transactions on Communications*, *E93-B*(1), 125–134. doi:10.1587/transcom.E93.B.125

Koudouridis, G. P., & Li, H. (2012). Distributed Power On-Off Optimisation for Heterogeneous Networks. Proceedings from CAMAD'12: *The IEEE 17th International Workshop on Computer Aided Modeling and Design of Communication Links and Networks*. Barcelona, Spain.

Larmo, A., Lindstrom, M., Meyer, M., Pelletier, G., Torsner, J., & Wiemann, H. (2009). The LTE Link Layer Design. *IEEE Communications Magazine*, *47*(4), 52–59. doi:10.1109/MCOM.2009.4907407

Marsan, M. A., & Meo, M. (2009). Energy efficient management of two cellular access networks. *Performance Evaluation Review*, *37*(4), 69–73. doi:10.1145/1773394.1773406

Marsan, M. A., & Meo, M. (2011). Energy efficient wireless Internet access with cooperative cellular networks. *Elsevier Journal Computer Networks*, *55*(2), 386–398. doi:10.1016/j.comnet.2010.10.017

Sesia, S., Toufik, I., & Baker, M. (2011). *LTE - The UMTS Long Term Evolution: From Theory to Practice* (2nd ed.). West Sussex: Wiley Publishing. doi:10.1002/9780470978504

Son, K., Oh, E., & Krishnamachari, B. (2011). Energy-aware hierarchical cell configuration: from deployment to operation. Proceedings from INFOCOM WKSHPS'11: *The IEEE Conference on Computer Communications Workshops*. Shanghai, China. doi:10.1109/INFCOMW.2011.5928825

Tombaz, S., Usman, M., & Zander, J. (2011). Energy efficiency improvements through heterogeneous networks in diverse traffic distribution scenarios. Proceedings from CHINACOM'11: *The 6th International ICST Conference on Communications and Networking in China*. Harbin, China. doi:10.1109/ChinaCom.2011.6158246

KEY TERMS AND DEFINITIONS

Energy Efficiency: Metric that is defined as the ratio of the transmitted bits over the average energy consumption, measured in [bits/joule].

Green Communications: Field of cellular networks that deals with the proposal of energy efficient solutions.

LTE-Advanced Standard: Mobile communication standard, formally submitted as a candidate 4G system to ITU-T in late 2009.

Switching Off Schemes: Algorithms for energy saving in cellular networks where operators switch off the BSs which become redundant due to low traffic variations.

ENDNOTES

1. Please note that the terms "MNO" and "operator" will be used interchangeably in the rest of this paper.
2. Please note that the terms "UE", "call", and "session" are used interchangeably in this paper.
3. The cell capacity is assumed to be equal to $BW=115Mbps$, that is a bandwidth value corresponding to an LTE-A network, but the traffic profile can be adapted to different bandwidth values.
4. $\overline{s}_{i,j}$ is the complementary probability equal to $\overline{s}_{i,j} = 1 - s_{i,j}$.

Chapter 15
Game Theoretic Approaches for Wireless Cooperative Content-Sharing

Leonardo Militano
Mediterranea University of Reggio Calabria, Italy

Antonio Iera
Mediterranea University of Reggio Calabria, Italy

Francesco Scarcello
University of Calabria, Italy

Antonella Molinaro
Mediterranea University of Reggio Calabria, Italy

Giuseppe Araniti
Mediterranea University of Reggio Calabria, Italy

ABSTRACT

This chapter provides an overview of game theoretic solutions in a wireless application of particular interest for multimedia and social interactions in modern telecommunication systems. In particular, coalitional games, bargaining solutions and fairness and stability issues will be investigated within wireless cooperative content-sharing. According to this paradigm, users download portions of data of common interest over long-range cellular links while exchanging the downloaded portions over short-range radio links. Expected benefits, which are natural incentives to cooperation, may be in terms of content price, energy consumption, and transfer delay reduction. Significant research activity has been conducted to design strategies that simultaneously exploit the multiple radio interfaces of modern wireless devices and maximize the gain. A valid solution for the cooperative content-sharing application should be based on fairness in the utility distribution among the involved nodes which can be eased by exploiting the game theoretic approaches presented in this chapter.

DOI: 10.4018/978-1-4666-8642-7.ch015

INTRODUCTION

Wireless cooperative networking, guaranteeing performance enhancements to handheld devices (e.g., smartphones, PDAs), is a well investigated research field (Iera, Militano, Romeo, & Scarcello, 2011). More specifically, cooperative content sharing is gaining momentum in the wireless communication arena thanks to its easy implementation in modern multi-interface mobile devices and the many applications for social and multimedia services. According to this paradigm, users download portions of data of common interest over costly long-range cellular links while exchanging the downloaded portions over short-range radio links. Significant research activity has been conducted to design strategies that simultaneously exploit the multiple radio interfaces of modern wireless devices and maximize the gains. As an example, the beneficial effects of integrating cellular and Wi-Fi networks are shown in Bhatia, Li, Luo, & Ramjee (2006) and Cavalcanti, Agrawal, Cordeiro, Xie, & Kumar (2005). The rewards of cooperation in terms of energy consumption and transfer delay are demonstrated in Militano, Fitzek, Iera, & Molinaro (2011) and Campolo, Iera, Militano, & Molinaro (2012) while the advantages of evolutionary theory and network coding techniques applied to cooperative scenarios are studied, e.g., in Militano, Fitzek, Iera, & Molinaro (2010a) and Militano, Fitzek, Iera, & Molinaro (2010b). Expected benefits, which are natural incentives to cooperation, may be in terms of content price, energy consumption, and transfer delay reduction. However, different file-sharing and cost-sharing policies may result in dissimilar performance levels experienced by the cooperating entities. Therefore, approaching the problem by relying on conventional tools based on analytical computation and linear programming to maximize the performance according to one key parameter only (e.g., energy or delay) could be unsatisfactory or too simplistic for some applications. The reason is that such tools distribute service costs disregarding any fairness among cooperating users. On the other hand, even a fair cost distribution could be unsatisfactory, in case a node (or a group of nodes) realizes that its reward increases by leaving the proposed cooperating group to form alternative cooperative groups making the solution unstable. A valid solution for the cooperative content-sharing application should indeed be based on fairness in the utility distribution among the involved nodes. The achievement of an agreement among cooperating users can be eased by exploiting game theory (GT). In fact, game theory is an analytical framework that attempts to analyze the behavior of rational entities with their own interests in reciprocal interactions. By starting from the economic field, during the last decades it has found successful applications to several other areas such as wireless networking. In Han, Niyato, Saad, Basar, & Hjørungnes (2011) a survey of GT-related studies is reported, which gives the reader an idea of the significance that game theory is gaining within the research community. Among others, fields of application of GT-based models are: Call admission control (Zhang & Fang, 2007), flow control (Tang & Andrew, 2008), radio resource management (Militano, Condoluci, Araniti, & Iera, 2013; Militano, Condoluci, Araniti, & Iera, 2012), content distribution (Goemans, Li, Mirrokni, & Thottan, 2006), malicious node detection in sensor networks (Yenumula, 2009), green networking (Militano, Molinaro, Iera, & Petkovics, 2013) and cooperation in wireless ad hoc networks (Srinivasan, Nuggehalli, Chiasserini, & Rao, 2003). Special attention to the fairness aspect in communication scenarios is given in Aram, Singh, & Sarkar (2009) and Cai & Pooch (2004), among others. Moreover, contributions can be found in the literature, which deal with coalitional games that are of particular interest for the problem discussed in this chapter: In Saad, Hjørungnes, Han, Niyato, & Hossain (2011) the focus is on vehicular networks, in Mathur, Sankaranarayanan, & Mandayam (2008) a coalitional problem is proposed to study fairness and cooperation gains in virtual MIMO systems, in Han & Poor (2009) packet forwarding issues in ad-hoc networks are addressed, whereas in Shehory & Kraus (2009) a task allocation problem is studied.

In this chapter, game theoretic solutions for the wireless cooperative content sharing will be discussed in details. Particular focus will be first on the analysis of the fairness issue in cellular-Bluetooth and cellular-WLAN cooperative scenarios, where energy saving is the objective to be obtained by the end-users. In particular, a centralized approach will allow both to minimize the energy consumption of a cooperating group and to tune any proposed solution such that fairness is guaranteed and the solution is stable. The proposed model considers a transferable utility (TU in short) game, whereby nodes use side payments to fill the gap between an optimal cost distribution that minimizes the overall costs and a fair solution indicated by the game theoretic model. If the adopted TU model allows to minimize the energy consumption for the scenario, it requires additional efforts to handle side payments via credit/debit exchanges among nodes to reflect the fair solution.

An alternative effective solution to the problem will then be presented considering that the energy consumption parameter is, by nature, hardly "transferable" and a solution where no side payments are required is of practical interest. A framework is proposed to define the coalitions of cooperating players to be formed and select the game theoretic bargaining solution to adopt for each coalition.

In the last part of the chapter, a monetary cost analysis based on a TU-game model will be presented. This model will consider the overall service managing, where the benefits for end-users, service and network providers are all taken into account. The proposed model will allow a comparison on the performance and analytical features of the Shapley-value and Nucleolus solution concepts for the game.

BACKGROUND

The reference scenario for the solutions proposed in this chapter is a group of users equipped with a mobile device with multiple network connectivity possibilities, accessing some content over the Internet. For instance, it is very usual for groups of students attending aggregation places in a University campus to exchange and download lesson notes, books, and other contents. For these users it is common to exchange and download multimedia content, access a podcast service (i.e., music, videos), playing together videogames exploiting continuously updated information (maps, game plug-in etc.), and so on. Other interesting scenarios are a stadium where spectators sit on the stands while attending a sport event and watching relevant multimedia content, or open-space music concerts and cultural events when additional information can be made available to users attending the event. In all these cases, many users in mutual proximity are interested in the same contents and can exploit short-range links to cooperatively access the data. For these and similar cases, a cooperative content-sharing service has been proposed, based on a centralized architecture that relies on a coordinating entity, called *Cooperation server* as in Figure 1. This could be managed, for example, by a content provider selling web content (e.g., movies, music, ringtones, etc.) to the subscribers. This central entity will act as a service promoter and "guarantee" for the nodes that cooperation will be secure and fair. Moreover, it will also introduce important energy savings for the users, whereas the central entity guarantees that none of the users is misbehaving (which is typical of classic P2P systems). A high performing central entity can compute any solution in a fast and

Figure 1. Reference wireless cooperative scenario

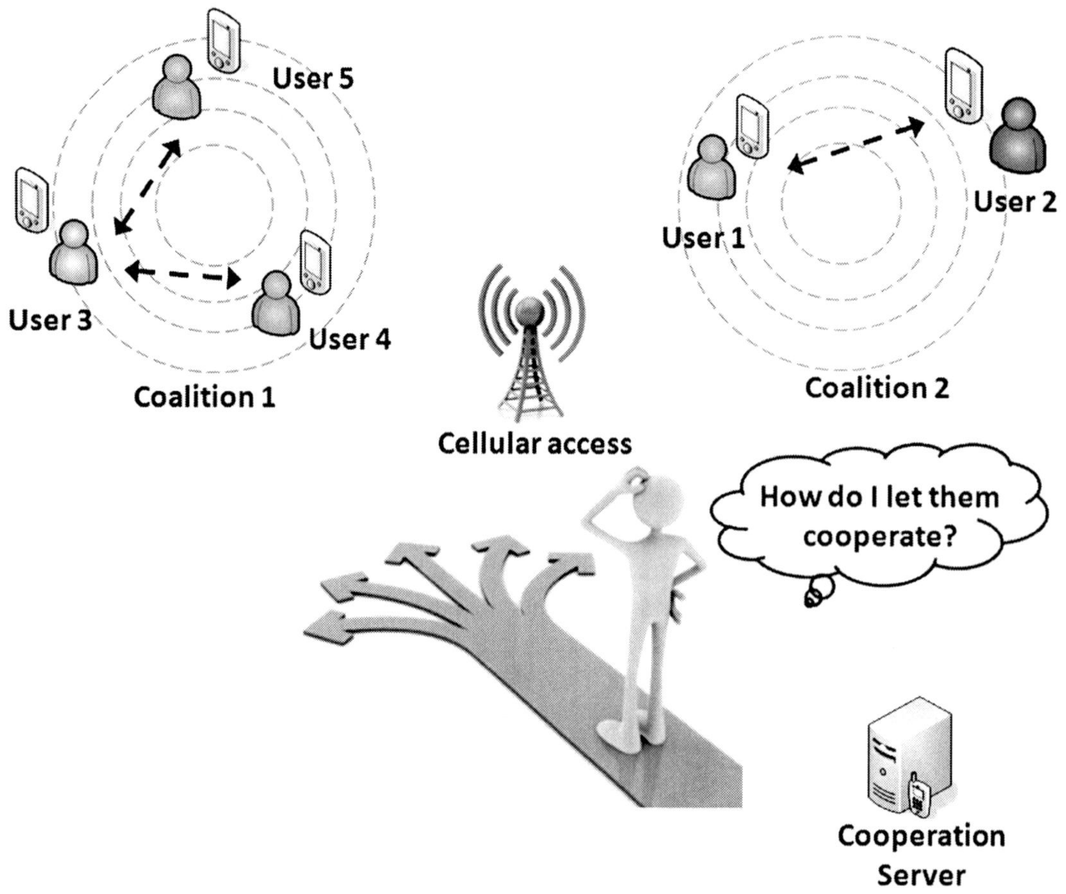

effective way without introducing further overhead and energy costs for the mobile devices. The server will be responsible of defining the group(s) of cooperating users, the so-called coalition(s) according to game theory, and orchestrating the cooperative data sharing by assigning tasks to each user.

The service provider managing and guaranteeing cooperative data-sharing is an attractive add-on service for both users and providers. On the one hand, any service provider has interest in this novel data sharing-service, because it can consider new business models for the service and, by offering group discounts for the contents, it can enlarge its subscribers basin. In doing so, the provider will still have full control over the content distribution and it will enlarge its profits. In fact, users will most probably prefer such a controlled, secure and fair file-sharing service instead of classic "illegal" and insecure peer-to-peer services. Moreover, a legal peer-to-peer service that introduces also energy consumption gains will attract and involve those users which otherwise would not access the content due to the high related costs. When thinking this service as an alternative to classical Peer to Peer systems causing profits losses for content providers, the sharing among users is now transparent and legal, and it is essentially under its control. On the other hand, users will likely subscribe to such an additional service in the hope of achieving significant savings in terms of energy consumption, of money (discounts on the cost of the content are offered to groups of cooperating users), and of reduced usage of cellular bandwidth (in general, costly). Moreover, for nodes with very slow cellular system equipment, this could be the only possibility to access services otherwise not attractive in a "stand-alone download" fashion.

A TRANSFERABLE-UTILITY COALITIONAL GAME FOR ENERGY SAVINGS ANALYSIS

The solutions presented in this section refer to the study case where the utility is defined in terms of the energy consumption for the user devices when taking part in a cooperative content sharing. The proposed framework foresees the implementation of the following steps: *(i)* Interested users, send the Cooperation server information about their position, equipment identifier, measured cellular link data rate, transmitting and receiving power consumption levels; *(ii)* the Cooperation server collects the requests during a given time interval, and uses linear-programming to compute the minimum cost (energy consumption) of any possible coalition and the corresponding data-shares for each cooperating device; *(iii)* the Cooperation server implements an node grouping policy exploiting the output of the energy minimization step; *(iv)* the Cooperation server applies a Transferable Utility (TU) game model to avoid an "unfair" cost distribution among users. If a fair and stable solution is found for a coalition, this is considered in the final configuration.

The Cooperation server notifies each device about: The coalition it belongs to, the exact data share to download, short-range information for data sharing in the coalition if needed, and the cost compensation according to the game theoretic solution.

The Cooperation server collects a number of candidates requiring the service within a limited period of time, before computing the solutions. Noteworthy, mobility and rapid scenario modifications are phenomena not of interest for this study (please refer to the scenario description). In those cases when a node is suddenly leaving the cooperative group, the Cooperation server, which has already computed any possible solution, can easily update the cooperation scheme (it repeats only the operations from step *(iii)* to the end). Should a new node ask for the service, then it will register to the Cooperation server, but it will only be considered for future sessions. The Cooperation server is responsible for compensating the gap between the "real" cost allocation determined by the energy minimization step, and the fair cost distribution given by the game-theoretic solution. Note that fairness is guaranteed in TU games thanks to the possibility of transferring utilities among players. However, energy cannot be transferred directly among nodes, so some different transferable "metrics" for utilities must be found. For example, the service provider could offer a credit/bonus compensation (with a credit unit proportional to an energy consumption unit) provided that a specific business model is agreed with cooperating customers. Such credits are associated with economic advantages for users, e.g., additional discounts to buy product licenses.

Game Theoretic Notions

It is natural to model the process of fairly assigning a cost to each player as a cooperative cost-game $G=\langle N,c\rangle$ with transferable utilities (TU), where N is the set of players and *c* is a characteristic (cost) function that models the feasible cost for every coalition $S \subseteq N$. Henceforth, components of any vector $x \in \mathbb{R}^{|N|}$ are one-to-one associated with players in N, so that x_i denotes the component associated with player $i \in N$. Moreover, for any vector $x \in \mathbb{R}^{|N|}$, x(S) is the value $\sum_{i \in S} x_i$, where $S \subseteq N$ is a coalition. A feasible payoff profile (or pre-imputation) of G is a vector $x \in \mathbb{R}^{|N|}$ such that $x(N)=c(N)$, usually called *efficiency* property. An imputation of G is a feasible cost profile $x \in \mathbb{R}^{|N|}$ such that $x_i \leq c(\{i\})$, for each $i \in N$. This condition is usually called *individual rationality* and states that cost allocations where a player experiences higher costs than what it obtains on its own, without cooperating with anyone else, are unacceptable. For the reference application, this entails that every player $i \in N$ should spend at most the energy it would spend when it is not cooperating and thus downloading the complete file over the cellular link; i.e., $c(\{i\}) = Enocoop_i$. The set of all imputations of the game G is denoted by X(G). An *outcome* for G is an imputation from X(G) that specifies the distribution of the cost to any player of the game. The third wished property for a cost allocation is that it is *stable* with respect to the possibility that subsets of players get incentive to deviate from it, by forming coalitions on their own. The set of such stable outcomes is known as the *core* of the game.

Definition 1: The core **C(G)** of a cost TU game $G = \langle N, c \rangle$ is the set of all imputations x such that, for each coalition $S \subseteq N, x(S) \leq c(S)$.

In words, the *core* is the set of all *imputations* that satisfy the cost upper-bound of all coalitions according to the cost function *c*. We say that if an imputation associated with a coalition is in the core then the coalition is *stable*; otherwise, we say it is unstable. Indeed, if $y \notin C(G)$, there exists some coalition S such that $y(S) > c(S)$. Therefore, players in the group S might leave the group N and download the desired file on their own at the total cost c(S), which is less than what requested according to the cost distribution y. In general, the core of a game may be empty as well as it may contain an infinite number of imputations. For more information on this subject, the reader is referred to (Osborne, 2004).

The last notion to be introduced is the definition of a *subadditive* game. In particular, a (cost) game $G = \langle N, c \rangle$ is said subadditive if, for every pair of coalitions S and T with $S \cap T = \varnothing, c(S \cup T) \leq c(S) + c(T)$. Intuitively, if the game is subadditive, it is always convenient for players to join larger coalitions. If this condition does not hold, it may be preferable for players to join smaller coalitions, thus forming a partition π of N, called *coalition structure* (Saad, Han, Debbah, Hjørungnes, & Basar, 2009). Note that standard game-theoretic solution concepts for coalitional structures, such as the core, consider stability conditions also on sets of players belonging to different elements of the partition π (Aumann & Dreze, 1974). However, such inter-structure conditions lead to high computational costs that are not suitable for the practical applications. Therefore, a simplified approach is applied where every set of players N_i of π induces a distinct (sub)game $\langle N_i, c_i \rangle$, where $c: 2^{N_i} \to \mathbb{R}$ is naturally defined as $c_i(T) = c(T), \forall T \subseteq N_i$.

Cooperative-Download Cost Game

The first step for modeling the TU coalitional game for assigning cost to nodes participating in a group download is to define the cost c(S) of any group $S \subseteq N$ of nodes willing to download together a given content. For any singleton coalition {i}, the feasible cost for player i alone is

$$c(\{i\}) = Enocoop_i = \frac{X}{R_{c_i}} \cdot P_{c_i},$$

where X is the file size in MB, P_{c_i} and R_{c_i} are respectively the power consumption (in W) and data rate (in byte/s) of the cellular link for node i. For any coalition $S \subseteq N$, with cardinality $|S|>1$, the cost is determined as follows:

$$c(S) = \min Ecoop_S = \min_{\alpha_i} \left\{ \sum_{i \in S} Ecoop_i(\alpha_i) \right\}$$
$$s.t. \sum_{i \in S} \alpha_i = 1$$
$$0 \leq \alpha_i \leq 1, \forall i \in S$$
$$Ecoop_i(\alpha_i) < Enocoop_i, \forall i \in S \quad (1)$$

where $Ecoop_s$ is the total energy consumption for coalition S and α_i is the file fraction assigned for download to node i. If the above linear-program has no solution, then define: $c(S) = \sum_{i \in S} c(\{i\})$. For the *grand coalition*, i.e., for S=N, the cost is just the solution of the above linear-programming optimization problem. In particular, if there is no solution then the cost c(N) is not defined and the game cannot be played by the available players. The optimization process identifies the best assignment of data portions to download, i.e., the one guaranteeing the lowest overall cost in terms of energy consumption for any coalition S. The problem constraints assure that the whole content is downloaded, suitably partitioned according to the values of variables α_i and the energy consumption for each node in cooperation does not exceed the consumption in the stand-alone case.

The energy consumption spent in a coalition S, $Ecoop_s$ is, by definition, the sum of the energy consumption of single nodes belonging to the coalition: $Ecoop_S = \sum_{i \in S} Ecoop_i$. The energy consumption for a cooperating node i, $Ecoop_i$ is a function of the assigned portion of file α_i, and is the sum of different terms that can be summarized as in equation (2):

$$Ecoop_i(\alpha_i) = Ecell_i(\alpha_i) + Esrtx_i(\alpha_i) + Esrrx_i(1 - \alpha_i) + Esridle_i + Egps \quad (2)$$

The first term is the energy spent to download the assigned data-share α_i over the cellular link; the second, the third and the fourth terms represent the energy consumption for transmitting and receiving data packets on the short-range link and for idle times on the short-range link, respectively; the fifth term accounts for the energy spent to localize the terminal. The exact value for the energy consumption due to GPS positioning can be assumed as equal to 13,32 J for all nodes (Paek, Kim, & Govindan, 2010). As for the other energy consumption terms, we assume that each node downloads the assigned data portion α_i over its cellular link and then begins to share it on the short-range link. Thus, the download energy consumption can be simply computed as the mean receiving power Pc_i multiplied by the assigned file fraction and divided by the mean cellular link data rate Rc_i for node i. This simplified computation avoids the need for a detailed link model of the cellular network. To define the remaining terms a specific model is required for the short-range communication, since technology-dependent features, number of nodes, and exchanged file fractions influence both energy consumption terms. This chapter will not report details on this modeling, but the interested reader can find a detailed model for the short-range energy consumption in a wireless cooperative setting in (Iera, Militano, Romeo, & Scarcello, 2011) and (Militano, Iera, Molinaro, & Scarcello, 2014) for Bluetooth and WLAN scenarios respectively.

Fairness and Coalition Structure Definition

To guarantee fairness in the cost allocation several criteria have been proposed in the literature; although all of them might be applied to the wireless cooperation paradigm under study, the results obtained in, where a comparison of different fairness criteria in terms of computational costs and stability features, has shown that for the considered application, the criterion offering the best performance is the *alternate cost avoided (ACA)* method . A common approach for cost allocation methods is to begin by charging every player a minimal cost, called *separable cost* (marginal contribution/cost), which can be associated with each player. This is the additional cost of including the player in a solution already designed for the other players. Thus, for a (cost) game $G = \langle N, c \rangle$, the separable cost $SC_i(c)$ of player $i \in N$ is formally defined as $SC_i(c) := c(N) - c(N \setminus \{i\})$.

As discussed in (Otten, 1993), when adopting cost allocation solutions the well-known *balancedness* conditions are reasonable assumptions about the underlying cost game:

$$SC_i(c) \le c(\{i\}), \forall i \in N, \sum_{i \in N} SC_i(c) \le c(N) \le \sum_{i \in N} c(\{i\}).$$

The first condition states that if $SC_i(c) > c(\{i\})$ for some $i \in N$, then it is not favourable to include player i in the joint cooperative solution. The second condition implies that after each player is charged his separable cost there is still a positive amount of cost remaining which should be allocated. This remaining cost is called the *nonseparable* cost and is $NSC_i(c) := c(N) - \sum_{i \in N} SC_i(c)$.

According to the ACA method, the *nonseparable* cost is divided in proportion to $c(\{i\}) - SC_i(c)$. Hence, the cost associated to each node $i \in N$ is:

$$ACA_i(c) := SC_i(c) + \frac{c(\{i\}) - SC_i(c)}{\sum_{j \in N} c(\{j\}) - SC_j(c)}, \forall i \in N \qquad (3)$$

where the number $SC_i(c)$ - $c(\{i\})$ represents the alternate cost avoided by including player i in the cooperative coalition.

Due to the physical constraints and the energy costs associated with communications on the short range and GPS positioning, the TU games associated with the proposed cooperative data sharing are *not subadditive* in general. Therefore, any realistic modeling of the application context has to deal with coalition structures, with players suitably partitioned in smaller coalitions where it is convenient for them to join the download-and-share process. To this end, a grouping policy is defined based on the so-called *maximum percentage gain sum algorithm*. Let $G_i(S)$ be the percentage gain each node experiences in the cooperative coalition S w.r.t. the stand-alone case, and let $TG(S)$ be the sum of the percentage gains experienced by the nodes in the coalition S. By ordering all the $TG(S)\ \forall S \subseteq N$ in a decreasing order, the topmost coalition in the list is the one guaranteeing the maximum percentage gain sum. The system computes the ACA solution for this coalition to introduce fairness in the cost allocation, and checks if it belongs to the core of the *subgame* induced by that set of players. If this is not the case, then the coalition is discarded and the subsequent coalition in the list (in decreasing order) is considered. This process is iterated until all nodes belong to a stable coalition. Nodes will then cooperate by downloading a certain amount of data as dictated by the energy minimization solution computed for the coalition they belong to. Eventually, the gap between the "real" cost allocation determined by the global optimization algorithm and the "fair" cost allocation obtained by the game theoretic solution is filled through a suitable compensation, e.g., in terms of credits/debits to be used for buying further products.

Performance Evaluation in WLAN and Bluetooth Settings

The proposed framework can be applied for any cooperative short-range link, given that a correct energy consumption modeling is used for the adopted link as shown in and for Bluetooth and WLAN settings respectively. Under these conditions, and the realistic power consumption values for the involved network interfaces, the main results that can be obtained are reported next.

Starting from the Bluetooth study case, Figure 2 considers a configuration with 8 devices, 4 of which use an EDGE (Enhanced Data rates for GSM Evolution) link (40kbps), while the remaining 4 use an UMTS (Universal Mobile Telecommunications System) link (340kbps). In this case a large coalition of 7 nodes is formed including all the EDGE and three UMTS nodes, while one UMTS node is left alone. In the large coalition, one out of the two UMTS nodes is elected as master of the Bluetooth piconet and the three UMTS nodes are in charge of downloading the whole file (a larger share is downloaded by the master). The EDGE nodes actually only receive data without downloading anything. It is manifest that the immediate energy consumption gain is not fairly distributed among the nodes, as the master node has a negligible gain, the other two UMTS nodes have a relatively low gain of about 25%, whereas EDGE nodes have a gain of 95%. This testifies to the need for a compensation algorithm. Following the compensation policy, enabled by the game theory application step, the UMTS nodes obtain a greater benefit, while the EDGE nodes slightly reduce their benefits. Side payments allow the UMTS nodes to obtain a percentage energy consumption gain of 45%, whereas EDGE nodes reduce their gain only to 93%. Moreover, compensation allows the UMTS node acting as a master to recover from the minor benefit obtained by operating according to the output of the energy minimization step.

A similar analysis can be repeated for other configurations with interesting results as the reader can find in (Iera, Militano, Romeo, & Scarcello, 2011). When moving the attention to the WLAN case, other interesting findings can be obtained. In particular, consider the case with eight players where half of the

Figure 2. Energy percentage gain in cooperation after side payments, four EDGE nodes (40kbps) - three UMTS nodes (340kbps) coalition

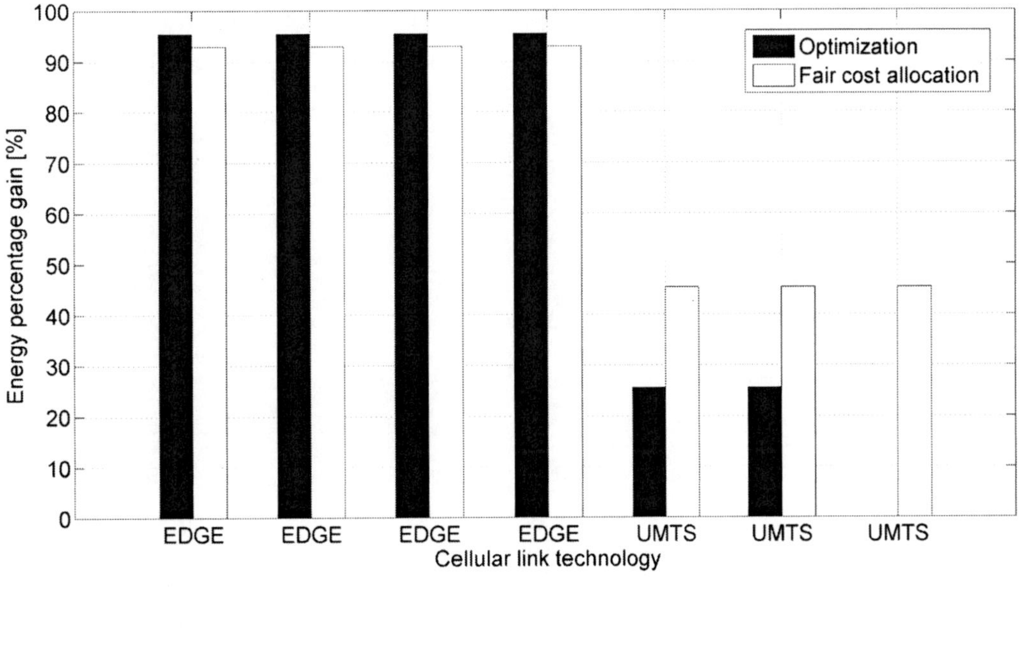

nodes has an available cellular data rate R_{c1}= 0.2Mbps, and the second half a cellular data rate R_{c2} either 1.2Mbps or 2.2Mbps. Tests are repeated in the coalition of eight nodes when changing the proportion of nodes with cellular data rate R_{c1} and R_{c2}. The analysis is also parameterized with regards to the Bit Error Rate (BER) value in the WLAN and aims at understanding whether a cooperative coalition can be formed in the presence of other combinations of heterogeneous nodes. In Figure 3 the average energy saving is represented for a BER set to either 10-8 or 10-3 and for a set of cases with different ratios of nodes with R_{c1} and nodes with R_{c2}. In the plot, the scenarios where the *grand coalition* is formed are highlighted. When this is not the case, usually two cooperative coalitions are formed of smaller sizes, and sometimes one of the nodes does not cooperate. As expected, the average energy saving is lower for higher BER values in the WLAN and for higher values of R_{c2}. Finally, with a greater number of nodes, out of the set of 8 nodes, transmitting at the higher cellular data rate R_{c2}, again the energy savings decrease.

The next analysis has the objective of highlighting how, in different network configurations, the proposed framework splits the nodes into different coalitions, i.e., it forms the coalition structure. Consider the scenarios where the BER in the WLAN is either 10^{-4}, 10^{-6}, or 10^{-8} and the number of nodes is 2, 4, 6, 8, 10, or 12. The attention is on two different cases to show how the solution depends on the adopted cellular link: A) half of the nodes in each of these scenarios uses cellular links with R_c= 0.2Mbps, while the second half uses higher rate cellular links with R_c = 2.2Mbps; B) half of the nodes in each of these scenarios uses still a cellular links with R_c = 0.2Mbps, while the second half uses cellular links the highest data rate considered so far, namely R_c = 4.2Mbps.

The plots in Figures 4 and 5 show the coalitions that are formed in the two proposed cases (in each studied case the different blocks identify the different coalitions). It can be observed that depending on

Figure 3. Average per node energy savings for heterogeneous scenarios with half of the nodes with cellular rate R_{c1} and half of the nodes with cellular rate R_{c2}; BER is set to 10^{-8}

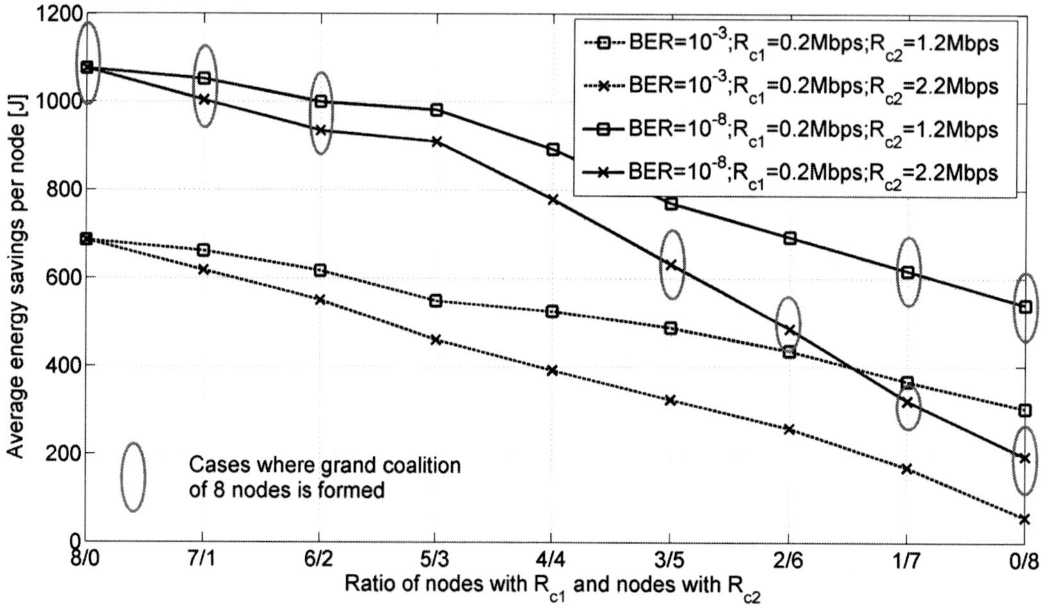

Figure 4. Coalition structure for different BER values and number of nodes, R_c = 0.2Mbps for half of the nodes and R_c = 2.2Mbps for the second half of the nodes.

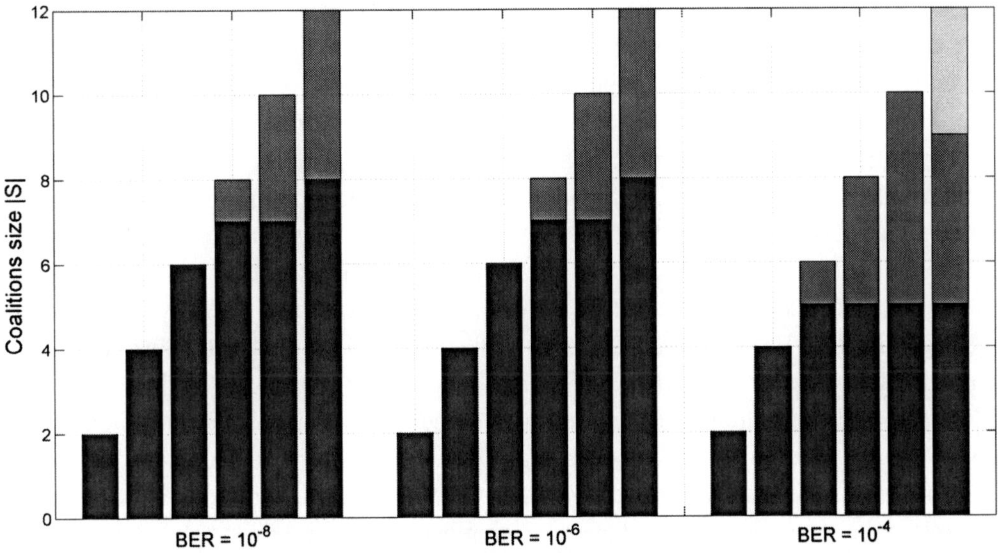

Figure 5. Coalition structure for different BER values and number of nodes, R_c = 0.2Mbps for half of the nodes and R_c = 4.2Mbps for the second half of the nodes.

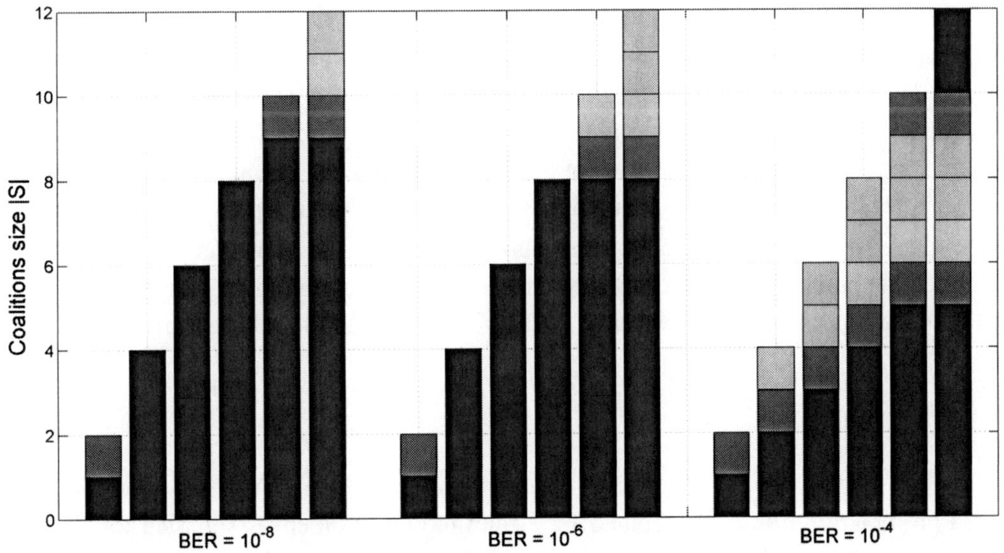

the cellular link for the nodes, on the BER value in the WLAN, and on the number of nodes in the coalition, different solutions exist. In particular, in the cases plotted in Figure 4 either the grand coalition or up to three distinct cooperating coalitions are formed (three coalitions are formed only with 12 nodes and 10-4). Instead, when higher cellular data links are involved, as for Figure 5, only one cooperating coalition is formed whenever possible, while the remaining nodes never cooperate with each other.

BARGAINING SOLUTIONS FOR ENERGY SAVINGS ANALYSIS

In this section, the focus is on game theoretic bargaining solutions as an alternative effective solution for the wireless cooperation with energy savings for the end-users still the objective of the cooperative content download (Militano & Iera, Adopting bargaining solutions for Bluetooth-based user cooperation, 2012). The main reason to investigate in the direction of bargaining solutions is the consideration that the energy consumption parameter is, by nature, hardly "transferable". The solution presented next, foresees still a central entity, but no side payments are required. In general, in bargaining problems it is not guaranteed that players converge to a solution autonomously. However, in centralized scenarios no explicit individual bargaining is necessary as the network will directly set the bargaining solution and the players are ready to accept them since they are equilibrium solutions given by a "central authority" (Fattahi & Paganini, 2005), (Militano, et al., 2014). The proposed solution will help in defining both the cooperative coalitions to be formed and the utility distribution among the nodes (Militano & Araniti, 2014).

Four game theoretic bargaining solutions are considered in this analysis. For the axiomatic definition the interested reader can refer to the literature, e.g., (Rubinstein, Safra, & Thomson, 1992), while here the basic formulations are briefly reported (Araniti, Condoluci, Iera, Militano, & Pansera, 2013). Let N={1,2,…,N} be the set of cooperative players. Each player is characterized by a performance function f_i defined on a convex closed and non-empty set $X \subset \mathbb{R}^N$, which is the set of game strategies for the N players. The performance functions define the performance vector $f(x) = (f_1(x),\ldots,f_n(x))$. The set $U = \{u \in \mathbb{R}^N : \exists x \in X, u = (f_1(x),\ldots,f_N(x))\}$ is the set of achievable performance values for the players, which in the reference problem is the energy consumption savings obtained by each cooperating player. Each player is also characterized by an initial minimum performance, the so-called disagreement point, the players require to enter the game, $d = (d_1, d_2, \ldots, d_N) \in \mathbb{R}^N$, such that $U_d = \{u \in U : u \geq d\} \neq \varnothing$. Let $X_d = \{x \in X : f(x) \geq d\}$ be the subset of strategies that enable the players to achieve at least their disagreement point performance. A bargaining problem is a couple (U,d) where U is the feasibility set, i.e., the closed, convex, bounded and non-empty set of payoffs which players can agree on after bargaining and $d = (d_1, d_2, \ldots, d_N) \in \mathbb{R}^N$ is the disagreement point. An agreement point is a possible outcome of a bargaining process where the utility obtained by every player has to be at least as much as the utility obtained at the disagreement point. Many solutions were proposed for a bargaining problem; the first considered in this study is the *Nash bargaining solution, NBS* (U,d) (Nash, 1950), which is given by:

$$\arg\max\left\{\prod_{i \in N}(u_i - d_i) \mid u \in U, u \geq d\right\}. \tag{4}$$

The idea promoted by the NBS is that after a minimal requirement is satisfied for all players, the remaining utility is allocated according to the conditions of each player.

The second one is the *Kalai-Smorodinsky solution, KSS(U,d)* (Kalai & Smorodinsky, 1975). According to KSS the utilities assigned to the players should be proportional to their maximum possible values. In practice, this solution assigns the maximal point of the feasible set on the segment connecting the disagreement point to the so-called *utopia point*. This ideal point, in which each player would get his maximum possible benefit, is defined as $a_i = \max\{u_i \in \mathbb{R} \mid u \in U, u \geq d\}, i \in N$. The *KSS(U,d)* is thus given by:

$$\arg\max\left\{\frac{u_1 - d_1}{a_1 - d_1} = \ldots = \frac{u_N - d_N}{a_N - d_N} \mid u \in U, u \geq d\right\}. \quad (5)$$

The third bargaining solution is the *egalitarian solution*, *ES*(*U*,*d*) (Roth, 1979), which assigns the point in the feasible set where all players achieve maximal equal increase in utility with respect to the disagreement point. It is given by:

$$\arg\max\{u_1 - d_1 = \ldots = u_N - d_N \mid u \in U, u \geq d\}. \quad (6)$$

Finally, the *utilitarian solution*, *US*(*U*,*d*) maximizes the sum of the utilities and can, therefore, also be seen as the global optimum solution for the problem, given by:

$$\arg\max\left\{\sum\nolimits_{i \in N} u_i \mid u \in U, u \geq d\right\}. \quad (7)$$

In the reference problem setting, the nodes registering for the cooperative content download will be the players of the game, the performance functions f_i are defined by the energy savings for the users. As for the X set for the bargaining problem, this is the set of file shares α that can be assigned for download to the single nodes. In the implementation of the bargaining solutions, further constraints will be set on these file shares, namely: $\sum\nolimits_{i \in N} \alpha_i = 1$ and $0 \leq \alpha_i \leq 1$, $i \in N$, so that the whole content will be available in the cooperative group and no multiple copies of the content will be downloaded by the nodes. These constraints will be equally applied to any possible coalition of cooperating nodes involving the whole set of nodes or also a subset of them. Under these conditions, the feasibility set U is convex, closed, and non-empty as required by the bargaining problem setting. Moreover, for the particular wireless cooperative problem, the disagreement point d is always set to zero, since no savings are obtained by the nodes when an agreement on cooperation is not reached (each node downloads the file independently over its cellular link). Finally, concerning the utopia point for the KSS, this is considered as the case where a node spends energy only to receive data over the short-range link.

Performance Evaluation

As a sample analysis, the attention is on a Bluetooth cooperative link which energy consumption model is detailed in (Militano & Araniti, Introducing Fairness-Efficiency Trade-off for Energy Savings in Wireless Cooperative Networks, 2014), showing the utility distribution for the cooperating nodes. A simple comparison is proposed among the different bargaining solutions for a sample case with only 2 cooperating nodes downloading a 100MB file, see Figure 6. The first node is equipped with a 3G cellular network interface with a low data rate of 200kbps, and the second node is equipped with a 3G network interface with a higher data rate of 400kbps (similar results can be observed in other configurations). As it can be observed, when the ES is adopted, both nodes experience the same utility; for the NBS node 1 is experiencing a higher utility than node 2 and a similar behavior is observed for the KSS but with higher emphasis, as this solution is weighted by the utopia point. Concerning the US, it clearly appears how node 2 is actually having no energy savings. This solution is offering the highest total savings (global optimum solution), but it guarantees savings only to the first of the cooperating nodes as the first node

Figure 6. Energy savings comparison for the four bargaining solutions

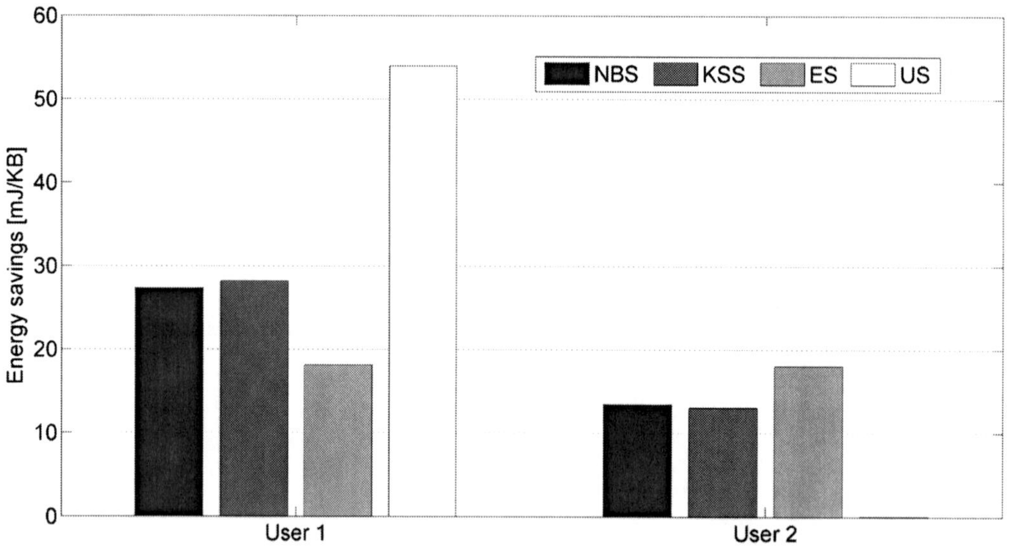

having a lower cellular data rate causes higher energy consumptions for the cellular data download. Consequently, although this solution is maximizing the energy savings in cooperation, it is clearly an unfair situation. This observation is clearly supported considering the FI corresponding to the energy savings allocation as the FI is equal to 0.89, 0.88, 1, 0.5 for the NBS, KSS, ES and US respectively.

A TRANSFERABLE-UTILITY COALITIONAL GAME FOR MONETARY COST ANALYSIS IN COOPERATION

The models presented so far, focused on the end-user side and the energy savings it can obtain from the cooperative content sharing. However, it is also interesting to analyze the network and service provider perspectives, and obtain a framework where providers may have economic benefits too. In fact, whenever any form of content sharing violates copyright constraints, this behavior may cause large profit losses to content and network providers. A successful reaction for providers is to promote themselves as an active part of the system (Militano, Iera, & Scarcello, 2013). The idea is that users in a group will be able to select a special option when they buy some product and get a group license that benefits of a high discount with respect to standard single-user licenses. Moreover, they are allowed to freely share the product within the wireless group by downloading only some parts of the product over the costly phone-link. The service coordinator will compute and assign to users the fractions of the product to be downloaded by each of them, by taking into account download costs, time and the energy consumption. The main advantage for providers would derive from the increased chances to involve in the business those nodes (the majority) inherently oriented to *"legality"* and not accessing classic Peer to Peer (P2P) services, but that avoid to access the provider downloading-service because prices are too high. Users instead, may save money because of the special group discount, so that the actual price to be paid could match the actual value/interest they assign to the product. Moreover, they have further pros in terms of energy/throughput/download cost, as well as the additional security factor guaranteed by a downloading-service under the direct control of providers (instead of possible illegal and uncontrolled alternatives).

Also in this case, a key feature that the provider shall promote to catch the interest of more users, is the implementation of a cost distribution among cooperating nodes, which is judged *fair* by all the players. This feature may be the only effective incentive to access the service since, in real environments, where nodes are rational and *selfish* players, it is difficult to reach such an agreement in cooperation. Moreover, any proposed cost allocation should also be tested according to the stability property, guaranteeing that no (sub)group of users has an incentive to leave the cooperation group and to form a different coalition. This way, the fare to be paid by each player (member of the group) takes into account his contribution to any possible coalition in terms of download costs of his terminal over the cellular link, possible agreements with the provider (e.g., frequent buyers), and so on. The service coordinator will act as an *"impartial"* entity guaranteeing that only solutions are applied that are judged fair, according to the proposed (publicly available) criterion. Moreover, since every player contributes to the process (at least allowing the group to obtain a higher multiple-license discount), the mechanism also avoids free riding, a typical behavior of pure peer-to-peer systems where some nodes get benefits from the community without giving anything back.

The Framework for the Service

The attention is on a content provider that offers the group-option to its clients, and plays the role of the Cooperation server. In order to reach its goal of selling licenses of its product to all interested users, the service coordinator implements a policy aiming at minimizing the overall monetary cost for the users, while allowing a cost distribution that is perceived to be a fair and stable one. Let N be a group of users interested in downloading some product from this provider. Nodes start the procedure by contacting the Cooperation server P and requesting the desired product with the cooperative option. Each node sends information about its device interface identifier, cellular link throughput, cost per second of the cellular link (or alternatively the cost per unit of data), information related to the short-range link they cooperate through (e.g., Bluetooth or WLAN link). For every node i, the Server P knows the monetary cost cd_i afforded to download the full product file from the provider, as well as the time required to complete such a download. Users may also specify some constraints over their contribution to the proposed P2P framework: any user i may impose a limit on the amount of data to be downloaded in the P2P application, expressed as a fraction $0 \leq \text{DownloadBound}_i \leq 1$ of the desired product file, which of course entails a bound $\text{DownloadBound}_i \bullet cd_i$ for the maximum download cost for player i.

While the short-link technology is available for free to all users in N, the energy consumption and the content transfer delay are considered as further parameters of interest in the final product-sharing phase over this inexpensive link. In particular, any user i may define a $\text{TimeConstraint}_i \geq 0$ to express the maximum time it is willing to spend to receive the wished content, in relation to the non-cooperative download. Similarly, any user i may define an $\text{EnergyConstraint}_i \geq 0$ to express the maximum energy it is willing to spend, compared to the non-cooperative download. In order to evaluate these constraints, the Cooperation server P should be able to estimate the time and the energy consumption for the cooperative content download for every node. This requirement can be fulfilled providing the service provider with a specific model which deals both with the short-range link and the cellular data-exchange.

A suitable pricing function $p(S)$ is defined by P for every non-empty group of users $S \subseteq N$, which may take into account frequent buyers, clients with high-feedbacks, and so on. To model the intended business strategy where large groups (buying many licenses) should be attracted, such a pricing function is required to be sub-additive, that is, $p(S \cup T) \leq p(S) + p(T)$, for every pairs of non-intersecting sub-groups S and T. Therefore, adding users to any group always leads to a lower price-per-license, so that larger groups have higher discounts, as expected.

The Cooperation server reports decisions to the interested users about "how" to cooperate. This task consists in defining the cooperative coalitions and assigning the portion α_i of content (file-share) that any node i in the cooperative framework shall download over its costly cellular link, and the total cost x_i for player i to get the desired product. Note that this cost includes the network cost, and thus player i actually pays to the content provider $x_i - \alpha_i \bullet cd_i$. Nodes will then proceed to download their assigned file-share, and then they exchange these parts of the file over the (free) short-range link. In general, more copies of the file may be downloaded by nodes of the group, the only constraint is that the fractions downloaded by peers cover at least a whole copy of the desired file. That is, $\sum_{i \in N} \alpha_i \geq 1$ must hold, i.e., non-overlapping file fractions (may) sum to more than 1.

Noteworthy, because of the group-discount modeled by the pricing function and the optimization performed to reduce the download cost (exploiting devices with lower bandwidth costs), the cost assigned to each node *i* will always be not higher than the cost that i would sustain for a stand-alone download (which is the sum of the content license cost $p(\{i\})$ and the download cost cd_i). In fact, the precise cost is determined according to a game theoretical solution-concept, that guarantees a fair cost allocation for all nodes and that is the key issue for the success of the proposed paradigm. In particular, the service coordinator needs to take into careful consideration the contribution of every node i in terms of price reduction and of device capability (bandwidth cost), to suitably evaluate how much discount it deserves.

Besides the service provider also the network provider (if it is distinct), that could oppose the proposed framework because it may reduce its profits on the costly cellular link, can be considered. In a simple approach, one may just assume that the network provider has a mutual business agreement with the service provider that covers somehow such a loss. In fact, if the network provider has some control power over the proposed protocol, we could imagine that, in the business agreement, the network provider may require a lower bound on its profit per transaction, with respect to the case of stand-alone downloads for all users. Formally, an additional parameter $0 \leq \text{GainConstraint} \leq 1$ is used to impose the constraint $\sum_{i \in N} \alpha_i \cdot cd_i \geq \sum_{i \in N} cd_i \cdot \text{GainConstraint}$ over the total download-cost of the group.

Game-Theoretic Solution Concepts for the P2P Framework

As for the energy saving analysis, a coalitional cost-game $G = \langle N, c \rangle$ with transferable utilities (TU), is considered where N is the set of players and is the characteristic cost function mapping a real value to every coalition $S \subseteq N$. In this case, the cost function models the fact that every coalition S has at its disposal the ability of each player $i \in S$ to download the file with some costs and some constraints, besides his contribution to determine the price $p(S)$ of the desired product. A fair price assignment should consider all possible contribution of players, trying to make everyone happy, and not only to meet the imposed constraints.

Every player $i \in N$ is associated with the two constants cd_i and DownloadBound_i, which model the cost for player *i* to download the full desired file from the network, and the maximum amount of data that i would like to download in the P2P process (expressed as a fraction of the whole file). Thus, $0 \leq \text{DownloadBound}_i \leq 1$, where 1 means that *i* sets no a-priori limits on his possible download, and 0 means that i will not download anything from the costly-link (therefore this player will just participate in the sharing phase over the short-range link). Therefore, by *individual rationality*, the total cost for every player i should not be larger than $c(\{i\}) = p(\{i\})$ + DownloadBound$_i$•cd_i. *Stability* is studied through the *core* which in general may be empty as well as it may contain an infinite number of imputations. However, an important class of cost games where the core is always non-empty is the class of concave games (dually, the convex games for value games). A game G is said concave if, for every pair of coalitions S and T, $c(S \cup T) + c(S \cap T) \leq c(S) + c(T)$. It can be shown that this holds if the cost function is *submodular*, that is, if

$$c(T \cup \{i\}) - c(T) \leq c(S \cup \{i\}) - c(S),$$

for each pair of coalitions $S \subseteq T \subseteq N \setminus \{i\}$. Nonetheless, even if the core is not empty, it remains the problem of choosing an *outcome* out of possibly infinite many candidates belonging to the core. Thus, solution concepts associated with unique profiles are usually desirable in applications. In particular, the *Shapley value* (Shapley, 1953) is one of the most used solution concepts in cost-sharing applications (see, e.g., (Misra, Ioannidis, Chaintreau, & Massouliè, 2010)).

Definition 2: The Shapley value (Shapley, 1953) of a cost TU game $G = \langle N, c \rangle$ is the pre-imputation of G assigning to every player $i \in N$ the following cost:

$$\phi_i(G) = \frac{1}{|N|!} \sum_{S \subseteq N \setminus \{i\}} |S|!(|N| - |S| - 1)![c(S \cup \{i\}) - c(S)]. \quad (11)$$

In words, the Shapley value assigns a cost to each player i taking into account his *"average marginal contribution"*, where the average is computed over all different sequences according to which the grand coalition could be built up from the empty coalition. This solution concept has also a nice axiomatic characterization supporting its notion of fairness (it is the unique pre-imputation that satisfies the Symmetry, Dummy Player, and Additivity axioms). It is known that, in any concave game, the Shapley value belongs to the core and thus it is a stable imputation. However, in the general case the Shapley value may be outside the core, even if the core is not empty. Thus, the Shapley value is not necessarily an imputation, and it may also violate the individual rationality condition.

Another approach to single out a fair outcome for TU games is based on the notion of *Nucleolus*, first introduced by Schmeidler (Schmeidler, 1969), and based on the lexicographical minimization of the maximum unhappiness of coalitions. Formally, given a cost game $G = \langle N, c \rangle$, a coalition $S \subseteq N$ and a cost allocation x, the *excess* (or unhappiness) of S w.r.t. x is equal to $e(S, x) = \sum_{i \in S} x_i - c(S)$. For any imputation x, define $\Theta(x)$ as the vector where the various excesses of all coalitions (but the empty one) are arranged in non increasing order:

$$\Theta(x) = (e(S_1, x), e(S_2, x), \ldots, e(S_{2^{|N|}-1}, x)). \quad (12)$$

Let $\Theta(x)[i]$ denote the i-th element of $\Theta(x)$. For a pair of imputations x and y, we say that $\Theta(x)$ is lexicographically smaller than $\Theta(y)$, denoted by $\Theta(x) \prec \Theta(y)$, if there exists a positive integer q such that $\Theta(x)[q] < \Theta(y)[q]$.

Definition 3: The Nucleolus $N(G)$ of a TU game G is the set $N(G) = \{x \in X(G) \mid \nexists y \in X(G)\, s.t. \Theta(y) \prec \Theta(x)\}$ (Schmeidler, 1969).

Therefore, this solution concept first cares about players that are less satisfied, then it cares about the less satisfied among the remaining coalitions, and so on for the rest of the coalitions, with the same approach. Interestingly, there is a unique point that leads to the lexicographically minimum vector of excesses, that is, $N(G)$ is in fact a singleton. Moreover, this unique outcome belongs to the core, whenever it is not empty, and usually offers a valid solution also for games with an empty core, so that the Nucleolus has been considered as one of the most interesting solutions to investigate (Han & Poor, 2009).

Modeling the Game

This section introduces a class of TU coalitional games the cooperative-download cost games (short: CDC games), defined by means of a suitable class of pricing functions determining the discount policy. Such games best model the problem of assigning costs to nodes participating in the mediated cooperative file-sharing process in a fair (and stable) way. In particular, in it is shown that these games *have always a non-empty core*, even if they are not concave. As a consequence, the Nucleolus of any CDC game is not only fair but also always a stable imputation (in that it belongs to the core of the game), thus meeting two important requirements for the service. The Shapley value instead, for the CDC game sometimes exhibits the undesired behavior of not being a stable allocation; see for more details and the formal demonstrations.

Formally, a CDC game G is a TU cost game $\langle N, c \rangle$, where N is the group of players interested in buying some product (or bundle of products) according to the mediated P2P download framework, and *c* is a cost function. The provider defines a suitable price function p(S) to determine how to charge the nodes of any coalition S for buying the |S| licenses for the product they are cooperatively downloading, and this function is assumed to be sub-additive, that is, for every pair of coalitions S and T with $S \cap T = \varnothing, p(S \cup T) \leq p(S) + p(T)$. In fact, this property formalizes the usual business logic *"the more licenses the more discount"*. With any coalition $S \subseteq N$, a cost is associated equal to:

$$\begin{aligned} & c(S) = \min\left\{p(S) + \sum\nolimits_{i \in S} \alpha_i \cdot cd_i\right\} \\ & s.t.: \sum\nolimits_{i \in S} \alpha_i \geq 1 \\ & \qquad 0 \leq \alpha_i \leq \text{DownloadBound}_i, \forall i \in S \\ & \qquad \sum\nolimits_{i \in S} \alpha_i \cdot cd_i \geq \left(\sum\nolimits_{i \in S} cd_i\right) \cdot \text{GainConstraint} \end{aligned} \tag{13}$$

In any optimal solution of the linear program above, the α_i control variable contains the (amount of) file fraction to be downloaded by player i, in order to minimize the total cost for the coalition S, while satisfying the problem constraints in (13). In particular, it is required that the whole file can be reconstructed after the download of these fractions (in fact, the constraint just checks the sizes of the fractions, as the actual subdivision of the file is performed later by the Cooperation server, respecting the information in the α_i values). Also, every player does not download more data than it is allowed by the DownloadBound_i parameter, thus meeting the individual rationality constraint, and the GainConstraint of the network provider should be satisfied, as well. Finally, the data fractions to be downloaded are required to meet also users' requirements in terms of energy consumption and time in cooperation which are limited by $\text{EnergyConstraint}_i$ and TimeConstraint_i.

It is required that the linear program (13) has a solution at least for the case $S=N$. Indeed, in this case, the above optimization procedure provides the cost of the grand-coalition, and thus the total cost associated with any pre-imputation of the game. An unfeasible linear program would mean that the constraints imposed by players and network provider are too restrictive for the characteristics of the involved players, and the cooperative download involving the whole group of players N cannot be executed. In such a situation the Cooperation server will enter in a further procedure of coalition partition to find alternative coalitions involving subsets of the nodes for which the constraints are met and the cooperative download can be executed. For any other coalition $S \subseteq N$ different from the grand coalition, it is not necessarily required that a feasible solution is found by the above linear program. In fact, in these cases can be associated to the specific coalition the cost $c(S) = p(S) + \sum_{i \in S} cd_i$, that is, the cost when the nodes in coalition S are not cooperating, i.e., the sum of the costs for players in S assuming that each of them downloads the whole file on her/his own.

Properties for the Cooperative-Download Cost Games

Recall that a cost TU game is said *sub-additive* if its cost function is sub-additive, that is, for every pair of coalitions S and T with $S \cap T = \varnothing, c(S \cup T) \leq c(S) + c(T)$. For sub-additive games usually the grand-coalition forms. Indeed, in such games adding players to any coalition is always cost-effective, which entails that the grand-coalition is the most convenient configuration. Intuitively, the CDC games are expected to have this property, since the business strategy of providers ("the more licenses the more discount") is modeled by the choice of sub-additive license (product) price functions. In it is demonstrated how any cooperative-download cost game is indeed sub-additive.

A further important property of the class of CDC games of highest interest, that is, the CDC games based on cardinality-discount cost-functions, called hereafter cardinality-discount CDC games. These games are based on pricing functions that depend only on the cardinality $|S|$ of the given coalition, and have the following form:

$$p(S) = \text{price} \cdot |S| \cdot \frac{100 - d(|S|)}{100}$$

where price is the basic price of the product, that should be paid by any single buyer not involved in any cooperative download, and $d(\bullet)$is a non-decreasing function over the naturals $[0,N]$ which determines the (percentage) discount to be applied to the group S, given the number of users $|S|$ belonging to it. No special assumption is required for the form of $d(\bullet)$, but for its codomain [0, 100] and the base cases $d(0) = d(1) = 0$ (no discount is given to the empty coalition or to single non cooperating nodes).

Observe that the sub-additivity property holds for this class of functions, as required for legal pricing functions for CDC games. However it is worthwhile noting that these functions are not concave in general, and hence the cardinality-discount CDC games based on them are not concave, too. Indeed, whether or not this property holds depends on the choice of the specific discount function $d(\bullet)$. Nevertheless, in (Militano, Iera, & Scarcello, 2013) it is proved that, no matter on the choice of the discount function, these games (even if non-concave) have a non-empty core, and thus its Nucleolus is always a stable imputation. Having a non-empty core and a stable Nucleolus is a remarkable property of cardinality-discount CDC games, as they seem quite appropriate to model the most frequent scenario for the proposed file sharing framework. On the contrary, the Shapley value is not necessarily in the core, if the game is not concave.

Cost of Product Licenses for the Nodes in the Coalition

The last phase of the proposed framework foresees the Cooperation server to actually assign to every node i the part of the file to be downloaded. At this step, the Nucleolus x has already been computed, and thus every player i knows his total cost x_i to get the product, comprising both the license cost and the download cost. Once the money issues are solved, there is the possibility to perform a final optimization step possibly focusing on different aspects. For instance, while meeting the constraints of the players and the total bandwidth cost to be paid to the network provider $c(N) - p(N)$, the actual file shares to be downloaded may be assigned to nodes in such a way that the download time over the cellular link is minimized, in order to obtain the desired content in a shorter time. Other alternatives can be proposed as well, for instance one could minimize the total energy consumed over the cellular link, or relax the constraints on the total bandwidth cost to be paid to the network provider when service and network provider are not distinct entities, and so on. The time optimization may be obtained by minimizing the maximum download-time over the participating nodes. To this end, the Cooperation server may perform the following linear-programming optimization:

$$\begin{aligned} &\min t_{\max} \\ &s.t.: \sum\nolimits_{i\in S} \alpha_i \geq 1 \\ &\quad t_{\max} \geq \alpha_i \cdot t_i, \forall i \in N \\ &\quad 0 \leq \alpha_i \leq \text{DownloadBound}_i, \forall i \in N \\ &\quad \sum\nolimits_{i\in N} \alpha_i \cdot cd_i = c(N) - p(N) \\ &\quad \text{EnergyCoop}_i(\alpha_i) \leq \text{EnergyNocoop}_i \times \text{EnergyConstraint}_i, \forall i \in N \\ &\quad \text{TimeCoop}_i(\alpha_i) \leq t_i \cdot \text{TimeConstraint}_i, \forall i \in N \end{aligned} \tag{14}$$

where t_{max} is the variable to be minimized, whose feasible (lowest) values are determined by the greatest product $\alpha_i \bullet t_i$, that is, by the slowest download. Then, from the values $\alpha_i^o \left(i \in \{1,\ldots,|N|\}\right)$ of variables in any optimal solution of the linear program (14), the Cooperation server computes the actual file parts, say β_i, that any player i has to download over the costly link. The transaction ends when every player i has paid its license fee $x_i - \alpha_i^o \cdot cd_i$, and all parts have been downloaded.

Note that the unique constraint $\sum\nolimits_{i\in N} \alpha_i cd_i = c(N) - p(N)$ suffices to deal with economic issues. For instance, it entails that the network provider's GainConstraint is fulfilled, because it is so in the solution leading to the computation of $c(N)$, which determines the network profit $c(N) - p(N)$. Further constraints to be considered for this final optimization are the constraints on the energy consumption and the time in the cooperative content download as defined by the players. The different constraints set by the users and/or by the network may cause the linear optimization not to find a valid solution for the grand coalition. Moreover, if the Shapley value is the chosen cost-allocation solution, then the property of being outside the (non-empty) core may be considered unacceptable.

To deal with these cases, the Cooperation server may be equipped with a strategy to find alternative groups of players to cooperate successfully. The proposed policy is to exclude iteratively some of the nodes until a first coalition is found where all the requirements are met. If more alternatives are meeting the constraints, then a further policy should be introduced to select the preferred coalition. In this framework, we adopt the well-known concept of maximizing the social welfare (Saad, Han, Debbah, Hjørungnes, & Basar, 2009), whereby the coalitions that maximize the total monetary savings (for themselves) are preferred. In more detail, if for the coalition involving *N* nodes a valid solution cannot be found, then smaller coalitions are considered. Whenever a coalition with the required properties is formed (clearly, it will be a maximal one), the process restarts with the excluded users, until a partition π of the players *N* in coalitions, also called a coalition structure (Saad, Han, Debbah, Hjørungnes, & Basar, 2009), is found.

Performance Evaluation

The numerical results presented next are related to Bluetooth as short-range link, but the overall framework can equally work with any other short-range technology. All presented cases assume that nodes have a time-based billing agreement with the provider expressed in terms of a given amount of generic Cost Units per second (CU). In particular, the cost per second is set to 0,05 cost units, the file size is 100 Mbyte, and the basic content price *price* is equal to 400 cost units. Concerning the power consumption on both the cellular link and the Bluetooth interface, the values used in the present paper result from measurement campaigns (conducted on N95 Nokia smartphones). In particular, different values of power consumption have been measured for a device connected to a 2.5G or to a beyond 2.5G system (400mW for 2.5G, 1400mW for beyond 2.5G and we assume 250kbps the maximum data rate value for a 2.5G connection). For the Bluetooth the adopted values are 178.2mW, 108mW and 59.4mW in transmission, reception and idle time respectively. Finally, the energy consumed by the nodes to gather their GPS positioning is set according to results in (Paek, Kim, & Govindan, 2010), whereby the energy consumption for a Nokia N95 smartphone is equal to 13,32 J in the best case.

It is assumed that the Cooperation server chooses a price function of the type described in previous sections. More specifically, cardinality-discount CDC games are considered whose discount functions are based on the functions proposed in (Kirkwood, 1997), suitably adapted to the scope. The discount proposed by the content provider w.r.t. to the normal basic price for different number of nodes is given in (15):

$$d_{\rho}\left(\left|S\right|\right) = Max_{discount} \frac{\exp\left[-\frac{\left|S\right|-1}{\rho}\right]-1}{\exp\left[-\frac{\left|S\right|_{\max}-1}{\rho}\right]-1}, with \rho \neq \infty \tag{15}$$

where $|S|$ and $|S|_{max}$ are respectively the cardinality of the coalition and the maximum cardinality, ρ determines the discount amount for either larger or smaller coalitions, $Max_{discount}$ represents the highest discount proposed by the provider which is set here to 50. It is easy to see from the formula that the highest 50% discount is offered when the coalition size $|S|$ is equal to $|S|_{max}$, while no discount is given to the non cooperating nodes when $|S|$ is equal to 1. Based on the above discounts the following class of cardinality-discount price-functions is obtained:

$$p_{\rho}(S) = \text{price} \cdot \left|S\right| \cdot \frac{100 - d_{\rho}\left(\left|S\right|\right)}{100}.$$

These price functions are used for the experiments focusing on the case with ρ=2, as this offers higher savings immediately, encouraging the users to join cooperating groups also of very small sizes. Interestingly, for positive ρ values and in particular for the chosen ρ=2 value, sub-modularity does not hold for such a function, because the marginal discount obtained by larger coalitions is smaller than the marginal discount obtained by smaller coalitions. Thus, the considered games are not concave, but as demonstrated the core is always not empty.

For simplicity in the presentation of the results and to better discuss the influence of this parameter, the analysis assumes all nodes having the same $DownloadBound_i$, hereafter just called DownloadBound. Several throughput distributions for the nodes can be considered. The interested reader can find a complete analysis in (Militano, Iera, & Scarcello, 2013), whereas here the main results are selected. They are characterized by half of the nodes with a cellular throughput of T=100 kbps, while the throughput of the remaining nodes equal to (T+δ) kbps, with δ variable.

The first results focus on the cost assignment to nodes, which is also equal to providers' profit. The cost assigned to each node is a combination of the content license cost and the cost deriving from the download of the assigned file-share over the cellular links. Of course, these values directly depend on DownloadBound and node throughput values, while also the GainConstraint of the network provider influences the final results. These dependencies are investigated next, while it is assumed the nodes have not set stringent constraints on the energy consumption and time delay in cooperation.

In Figure 7 it is shown how the file-shares to be downloaded by nodes, hence their download costs, change with the parameter DownloadBound and the GainConstraint; this behavior is highlighted in sample cases in which δ=100kbps. Considering the case with GainConstraint (in short GC) set to zero the influence of the DownloadBound (in short DB) can be analyzed, left side of the plot. To minimize the maximum download time according to the final optimization, the solution is to assign larger portions of data to be downloaded to the four nodes with higher cellular throughput (nodes 5-8 in the scenario). For any scenario with DownloadBound≥0.125 these four nodes will download 1/4 of the file each. When the nodes set more stringent constraints on the DownloadBound, also the other nodes are involved. For any value DownloadBound<0.125 (that is maximum 1/8 of the file) no feasible solution can be found, since the total number of nodes is 8. Note that similar considerations hold for different values of the total number of nodes in the coalition; obviously, these lead to different limiting values for the DownloadBound value. Moreover, it is worth commenting that in a more general system setting, where all values of nodes' cellular throughput are different, again to minimize the maximum download time, the solution is to assign a larger portion of file to download to the nodes with higher cellular throughput as far as the DownloadBound constraint allows it.

Figure 7. Influence of DownloadBound (DB) and GainConstraint (GC) on the file portions assigned for download to nodes; δ is set to 100 kbps

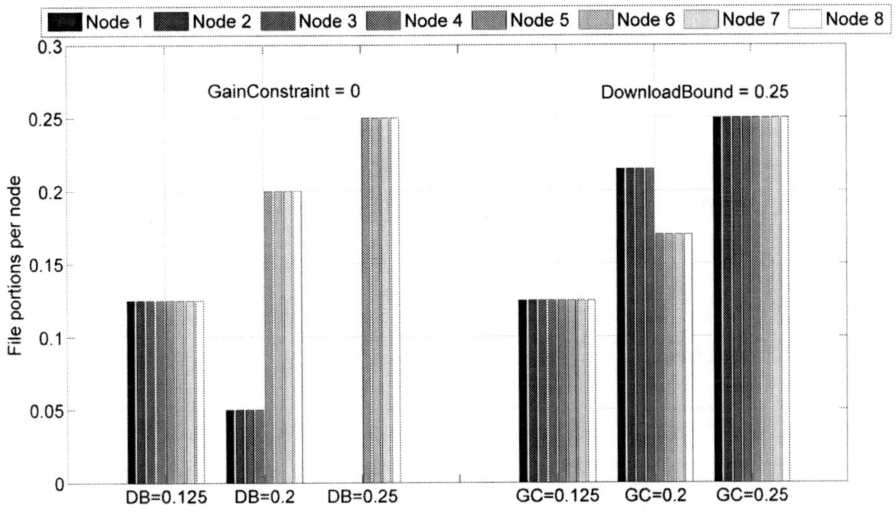

To assess the impact of the GainConstraint set by the network, the case is analyzed where DownloadBound is set to 0.25. Comparing the scenario with GainConstraint not set, the third case in the left side of the plot in Figure 7 and the scenario corresponding to the results reported on the right side of the plot in Figure 7, two main effects of an increase in the GainConstraint parameter emerge. First, the nodes with a less performing cellular link are now also involved in the content download over the cellular link. This is justified by the increase in the bandwidth income required by the network provider. When the GainConstraint increases even more, the second effect is that the cooperating cluster is forced to download more than one copy of the total file (e.g., 2 copies are downloaded when GC=0.25).

The last analysis is on the costs repartition among the nodes in the considered scenarios. In Figure 8 the cost for the content license and the content download is reported for the nodes in the sample sce-

Figure 8. Cost distribution among the nodes according to the Nucleolus solution for different values of DownloadBound and δ=100kbps, GainConstraint=0.

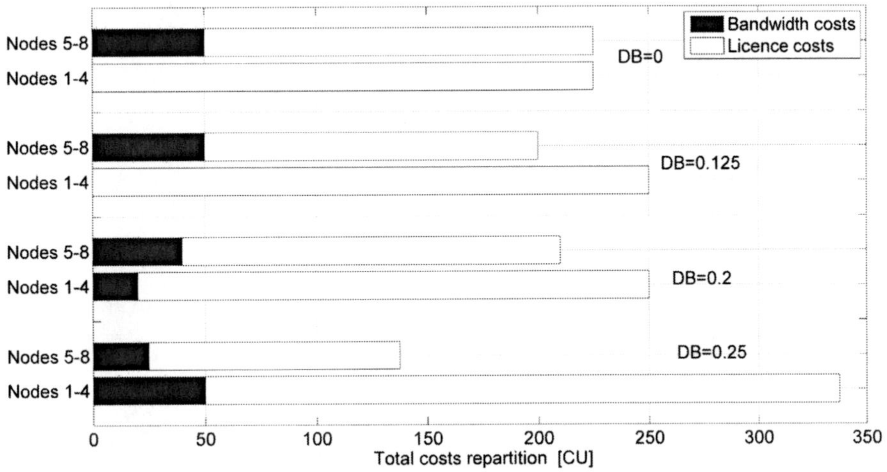

narios with δ=100kbps, GainConstraint=0 and the DownloadBound set to one of the following values: 0, 0.125, 0.2, 0.25. In Figure 9 instead, the same information is reported when GainConstraint assumes one of the following values 0, 0.125, 0.2, 0.25, while δ=100kbps and DownloadBound=0. What can be observed in both of the Figures is that a node having higher costs for the content download, will pay a smaller contribution for the content license. This observation has a general validity also in a system setting with more differentiated values of cellular throughput for the nodes.

It can also be observed that the cost for the content is often higher, for some nodes, than the cost given by the price function $p_\rho(S)$. The reason for this is that it has also to cover the costs for the node "more devoted to the content download".

FUTURE RESEARCH DIRECTIONS

The solutions proposed in this chapter give a clear and detailed description of the potentialities of game theoretic solutions for a wireless problem setting. However, other game theoretic solutions can be further investigated to propose alternative solutions and further improve the proposed wireless cooperative framework.

For instance, a first research direction is the analysis of distributed solutions where users find an agreement on cooperation in an autonomous way, without the need for a central authority. There is a large set of possible solutions to this issue which require a careful analysis. In particular, some directions to investigate could be classic Nash equilibrium solutions, bargaining processes and their convergence, the use of incentives for cooperation. Another interesting field of investigation could be the use of evolutionary game theory and the strategic alternatives in social behavior.

Another future research direction is the investigation of a broader framework where the end-users may not necessarily be interested in the content, but for economic benefits may offer their support for users that need help to reach their objectives. This aspect, which is quite interesting, however introduces further challenges related to a framework where a historical view of the service is needed. On the other

Figure 9. Cost distribution for the nodes according to the Nucleolus solution for different values of GainConstraint and δ=100kbps, DownloadBound=0.25.

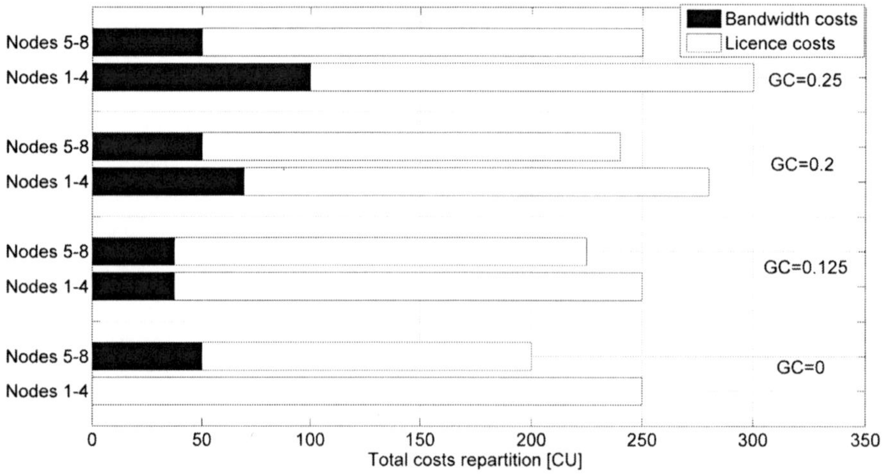

hand, privacy and security issues should be addressed to guarantee the service sustainability. In such a framework also mobility issues can be considered, where a user can join and leave a cooperating coalition over the time. This aspect has not yet been investigated and game theoretic tools might definitely be helpful for this further side of the problem.

CONCLUSION

In this chapter the potentialities of game theory have been described considering a set of solutions for the specific reference problem. The attention has been devoted to a wireless cooperative paradigm where users download portions of data of common interest over long-range cellular links while exchanging the downloaded portions over short-range radio links. A valid solution for the application should be based on fairness in the utility distribution among the involved nodes and should be stable. To this aim and to achieve an agreement among cooperating users this chapter described how game theory and its tools can be exploited. In particular, the analysis focused first on the fairness and the stability issues in cellular-Bluetooth and cellular-WLAN cooperative scenarios, where energy saving is the objective to be obtained by the end-users. The proposed model considered a transferable utility game, whereby nodes use side payments to fill the gap between an optimal cost distribution that minimizes the overall costs and a fair solution indicated by the game theoretic model. Moreover, the stability has been studied based on the core set solution. An alternative effective approach to the problem has then been presented based on a bargaining game modeling for the problem and four classic bargaining solutions have been evaluated. In the last part of the chapter, a monetary cost analysis has been presented based on a transferable utility game, where also network and service providers are modeled in the game. In particular, the Shapley-value and Nucleolus solution concepts have been proposed for the game and their properties have been discussed with reference to the stability of a solution. All the proposals have also been supported by a simulative analysis for an evaluation of the performance results for the proposals. The results obtained indeed testify the great potentialities of game theory for real wireless systems and the benefits it can introduce for practical implementation in real systems.

REFERENCES

Aram, A., Singh, C., & Sarkar, S. (2009). *Cooperative profit sharing in coalition based resource allocation in wireless networks*. IEEE INFOCOM. doi:10.1109/INFCOM.2009.5062136

Araniti, G., Condoluci, M., Iera, A., Militano, L., Pansera, B. A. (2013). Analytical modeling of bargaining solutions for multicast cellular services. *Electronic Journal of Differential Equations, 2013*(165), 1-10.

Aumann, R., & Dreze, J. (1974). Cooperative Games with Coalition Structures. *International Journal of Game Theory*, *3*(4), 217–237. doi:10.1007/BF01766876

Bhatia, R., Li, L. E., Luo, H., & Ramjee, R. (2006). ICAM: Integrated Cellular and Ad Hoc Multicast. *IEEE Transactions on Mobile Computing*, *5*(8), 1004–1015. doi:10.1109/TMC.2006.116

Cai, J., & Pooch, U. (2004). Allocate fair payoff for cooperation in wireless ad hoc networks using shapley value. *International Parallel and Distributed Processing Symposium*.

Campolo, C., Iera, A., Militano, L., & Molinaro, A. (2012). Scenario-adaptive and gain-aware content sharing policies for cooperative wireless environments. *Computer Communications*, *35*(10), 1259–1271. doi:10.1016/j.comcom.2012.03.017

Cavalcanti, D., Agrawal, D., Cordeiro, C., Xie, B., & Kumar, A. (2005). Issues in Integrating Cellular Networks, WLANs, and MANETs: A Futuristic Heterogeneous Wireless Network. *IEEE Wireless Communications Magazine*, *12*(3), 30–41. doi:10.1109/MWC.2005.1452852

Fattahi, A., & Paganini, F. (2005). New economic perspectives for resource allocation in wireless networks. *IEEE American Control Conference*. doi:10.1109/ACC.2005.1470595

Goemans, M., Li, L., Mirrokni, V., & Thottan, M. (2006). Market sharing games applied to content distribution in ad hoc networks. *IEEE Journal on Selected Areas in Communications*, *24*(5), 1020–1033. doi:10.1109/JSAC.2006.872884

Han, Z., Niyato, D., Saad, W., Basar, T., & Hjørungnes, A. (2011). *Game Theory in Wireless and Communication Networks: Theory, Models and Applications*. Cambridge, UK: Cambridge University Press. doi:10.1017/CBO9780511895043

Han, Z., & Poor, V. (2009). Coalition games with cooperative transmission: A cure for the curse of boundary nodes in selfish packet-forwarding wireless networks. *IEEE Transactions on Communications*, *57*(1), 203–213. doi:10.1109/TCOMM.2009.0901.060661

Iera, A., Militano, L., Romeo, L., & Scarcello, F. (2011). Fair cost allocation in cellular-bluetooth cooperation scenarios. *Wireless Communications. IEEE Transactions on*, *10*(8), 2566–2576.

Kalai, E., & Smorodinsky, M. (1975). Other Solutions to Nash's Bargaining Problem. *Econometrica*, *43*(3), 513–518. doi:10.2307/1914280

Kirkwood, C. (1997). *Notes on Attitude Toward Risk Taking and the Exponential Utility Function*. Working paper, Arizona State University.

Mathur, S., Sankaranarayanan, L., & Mandayam, N. (2008). Coalitions in cooperative wireless networks. [JSAC]. *IEEE Journal on Selected Areas in Communications*, *26*(7), 1104–1115. doi:10.1109/JSAC.2008.080908

Militano, L., & Araniti, G. (2014). Introducing Fairness-Efficiency Trade-off for Energy Savings in Wireless Cooperative Networks. *Wireless Personal Communications*, *76*(1), 3–21. doi:10.1007/s11277-013-1483-3

Militano, L., Condoluci, M., Araniti, G., & Iera, A. (2012). Bargaining solutions for multicast subgroup formation in LTE. *IEEE Vehicular Technology Conference (VTC Fall)*. doi:10.1109/VTCFall.2012.6399273

Militano, L., Condoluci, M., Araniti, G., & Iera, A. (2013). Multicast service delivery solutions in LTE-Advanced systems. *IEEE International Conference on Communications (ICC)*. doi:10.1109/ICC.2013.6655551

Militano, L., Fitzek, F., Iera, A., & Molinaro, A. (2010a). Network coding and evolutionary theory for performance enhancement in wireless cooperative clusters. *European Transactions on Telecommunications*, *21*(8), 725–737. doi:10.1002/ett.1435

Militano, L., Fitzek, F., Iera, A., & Molinaro, A. (2010b). A genetic algorithm for source election in cooperative clusters implementing network coding. *IEEE International Conference on Communications Workshops (ICC)*. doi:10.1109/ICCW.2010.5503955

Militano, L., Fitzek, F., Iera, A., & Molinaro, A. (2011). Group interactions in wireless cooperative networks. *IEEE 73rd Vehicular Technology Conference (VTC Spring)*.

Militano, L., & Iera, A. (2012). Adopting bargaining solutions for Bluetooth-based user cooperation. *IEEE 23rd International Symposium on Personal Indoor and Mobile Radio Communications (PIMRC)*.

Militano, L., Iera, A., Molinaro, A., & Scarcello, F. (2014). Energy saving analysis in cellular-WLAN cooperative scenarios. *IEEE Transactions on Vehicular Technology*, *63*(1), 478–484. doi:10.1109/TVT.2013.2270315

Militano, L., Iera, A., & Scarcello, F. (2013). A fair cooperative content-sharing service. *Computer Networks*, *57*(9), 1955–1973. doi:10.1016/j.comnet.2013.03.014

Militano, L., Molinaro, A., Iera, A., & Petkovics, Á. (2013). A Game Theoretic Approach to Guarantee Fairness in Cooperation Among Green Mobile Network Operators. [IJBDCN]. *International Journal of Business Data Communications and Networking*, *9*(3), 1–15.

Militano, L., Niyato, D., Condoluci, M., Araniti, G., Iera, A., & Molica Bisci, G. (2014). Radio Resource Management for Group-Oriented Services in LTE-A. *IEEE Transactions on Vehicular Technology*, 1. doi:10.1109/TVT.2014.2359686

Misra, V., Ioannidis, S., Chaintreau, A., & Massoulié, L. (2010). Incentivizing peerassisted services: A fluid Shapley value approach. *Performance Evaluation Review*, *38*(1), 215–226. doi:10.1145/1811099.1811064

Nash, J. (1950). The Bargaining Problem. *Econometrica*, 128–140.

Osborne, M. (2004). *An introduction to game theory*. Oxford, UK: Oxford University Press.

Otten, G. (1993). Characterizations of a game theoretical cost allocation method. *Zeitschrift fur Operations Research*, *38*(2), 175–185.

Paek, J., Kim, J., & Govindan, R. (2010). Energy-Efficient Rate-Adaptive GPS-based Positioning for Smartphones. *8th international conference on Mobile systems, applications, and services, MobiSys*.

Roth, A. (1979). *Axiomatic Models of Bargaining*. New York City: Springer. doi:10.1007/978-3-642-51570-5

Rubinstein, A., Safra, Z., & Thomson, W. (1992). On the Interpretation of the Nash Bargaining Solution and its Extension to Non-Expected Utility Preferences. *Econometrica*, *60*(5), 1171–1186. doi:10.2307/2951543

Saad, W., Han, Z., Debbah, M., Hjørungnes, A., Basar, T. (2009). Coalition game theory for communication networks: A tutorial. *IEEE Signal Processing Magazine, Special issue on Game Theory in Signal Processing and Communications, 26*(5), 77-97.

Saad, W., Hjørungnes, A., Han, Z., Niyato, D., & Hossain, E. (2011). Coalition formation games for distributed roadside units cooperation in vehicular network. [JSAC]. *IEEE Journal on Selected Areas in Communications*, *29*(1), 48–60. doi:10.1109/JSAC.2011.110106

Schmeidler, D. (1969). The Nucleolus of a Characteristic Function Game. *SIAM Journal on Applied Mathematics*, *17*(6), 1163–1170. doi:10.1137/0117107

Shapley, L. (1953). A Value for n-Person Games. In H. W. Kuhn & A. W. Tucker (Eds.), *Contributions to the Theory of Games II, Annals of Mathematics Studies* (pp. 307–317). Princeton, New Jersey, USA: Princeton University Press.

Shehory, O., & Kraus, S. (2009). Methods for task allocation via agent coalition formation. *IEEE Transactions on Communications*, *101*(1), 459–469.

Srinivasan, V., Nuggehalli, P., Chiasserini, C., & Rao, R. (2003). *Cooperation in wireless ad hoc networks*. IEEE INFOCOM. doi:10.1109/INFCOM.2003.1208918

Tang, A., & Andrew, L. (2008). Game theory for heterogeneous flow control. *Proceedings of 42nd Annual Conference on Information Sciences and Systems*. doi:10.1109/CISS.2008.4558494

Yenumula, B. (2009). A Game Theory Approach to Detect Malicious Nodes in Wireless Sensor Networks. *Third International Conference on Sensor Technologies and Applications*.

Zhang, H., & Fang, X. (2007). A Pricing and Game theory based Call Admission Control Scheme for CDMA Systems. *International conference on Wireless Communications, Networking and Mobile Computing*. doi:10.1109/WICOM.2007.1593

KEY TERMS AND DEFINITIONS

Content Sharing: The activity of users that share some data of common interest over a communication network.

Energy Efficiency: It is a measure of how good an approach in limiting the amount of energy required to provide a service.

Fairness: A solution or approach is fair if it is free from iniquity or injustice for the involved entities.

Short-Range Network: A network that allows devices to communicate over short-range links, with typical distance of tens of meters.

Stability: A solution or agreement among users is stable if there is no motivation for the involved users to change their behavior and leave the agreement.

Utility: A measure of how good a situation or solution is for the user.

Wireless Cooperative Networking: A framework where users cooperatively access some data using simultaneously multiple networks interfaces.

Chapter 16
Game Theoretic Analysis for Cooperative Video Transmission over Heterogeneous Devices:
Mobile Communication Networks and Wireless Local Area Networks as a Case Study

Wen Ji
Chinese Academy of Sciences, China

Yiqiang Chen
Chinese Academy of Sciences, China

Bo-Wei Chen
Princeton University, USA

Shaojie Kang
Chinese Academy of Sciences, China

Shuili Zhang
Chinese Academy of Sciences, China

ABSTRACT

This chapter presents a mechanism for cooperative video transmission based on game theory for heterogeneous devices during broadcasting. Broadcasting is a multipoint delivery of transmission that sends data from a source to multiple destinations. The terminal is involved in cooperative transmission when the station broadcasts video data. To enhance performance, the heterogeneity and forwarding capabilities should be considered. This work studies power control and allocation in a collaborative transmission based on game theory, which provides an effective strategy when network resources are limited. First, a novel power-allocation model of the base station (BS) based on noncooperative game theory and bidding is presented in this study. Additionally, we also propose a utility function of Signal-to-Noise Ratios (SNRs) along with Signal-to-Interference Ratio (SIRs). Subsequently, based on such noncooperative game theory with a utility function, the model of the power distribution of terminals in cooperative transmission can be built. Experiments on the System-in-the-Loop (SITL) mode in OPNETs have proven the correctness of the designed model and superiority, verifying the effectiveness of the proposed power-control idea.

DOI: 10.4018/978-1-4666-8642-7.ch016

INTRODUCTION

Previous studies showed that the resource allocation problem in cooperative networks could be divided into two types - Centralized (Nosratinia & Hunter, 2007; Ng & Yu, 2007) and distributed methods (Alamouti, 1998; Savazzi & Spagnolini, 2007; Laneman & Wornell, 2003; Abdelnasser & Hossain, 2013). Usually, centralized methods considered channel state information. However, such a method is not frequently used since channel conditions are fluctuating, and a large number of video packets are delivered during transmission. Compared with centralized methods, distributed approaches focused only on local information. Hence, the distributed approaches were frequently adopted in the wireless network research. For example, Nosratinia and Hunter (2007) devised a centralized method for maximizing the network bandwidth. Ng and Yu (2007) presented a distributed resource-allocation algorithm for networks of OFDMA (Orthogonal Frequency Division Multiple Access), which was designed for power/bandwidth allocation and the selection for relay nodes. Alamouti (1998) proposed distributive power allocation of multihop transmissions for a single user. (Liu, Tao, & Huang, 2013; Huang, Zhu, Mung, & Poor, 2008) presented a cooperative transmission scheme based on auction mechanisms. Typically, most researchers concentrated only on the power allocation of the terminal while maximizing the network benefit. However, cooperative transmission is important as it can reduce the burden of the base station (BS) and improves power utilization when limited resources are present. In view of this, this chapter studies the power-allocation strategy for terminals to collaboratively transmitting data. That is, video streaming is broadcast to heterogeneous terminals by a base station (BS). Moreover, terminals cooperate with each other during the proposed process. First, terminals are classified into different groups, and each terminal in the same group can co-work with another member while transmitting video streaming. For instance, if the terminals are divided into three parts "*A*, *B*, and *C*" based on three different transmission powers "p_A, p_B, and p_C," then they can receive video streams from the BS, which broadcasts video streaming with p_A, p_B, or p_C. The members in a group can be further divided into different subgroups. With Scalable Video Coding (SVC), the cooperative transmission process in groups can be described as follows. If the total type of terminals is one, terminals transmit and receive the same video layer. Otherwise, if more than one type exists, then terminals cooperatively transmit different layers. In this study, game theory is used to model and understand the cooperative behavior of mobile users because the related work on Device-to-Device (D2D) communication proved its effectiveness (Song, L., Niyato, D., Han, Z., & Hossain, E., 2014). Song (2014) presented an auction model and a coalitional game-theoretical model. They described advanced research on development of game-theoretical models and introduced many significant challenges for wireless-network resources allocation in D2D data transmission.

As for the Wireless Local Area Networks (WLANs), with the development of wireless network standards and transmission technologies, communications become more efficient and gradually enrich our lives. Besides the advanced wireless communication techniques, recent research focused on video encoding methods. In 2003, H.264 video coding standard was issued by International Telegraph Union — Telecommunication Standardization Sector (ITU-T). As one of the new generation of video coding standards, H.264 coding has its own incomparable superiority. This coding scheme takes advantage of Quality-of-Service (QoS) mechanism in order to preserve the most significant information and to decrease video distortion under congestion conditions (Choudhry, U. I., & Kim, J., 2005). The performance of wireless local network systems is based on the Quality of Service and the Signal-to-Noise Ratio (SNR). QoS is a necessary element for supporting a number of applications which utilize wireless network resources. These applications may be video streaming, voice over IP, video conferencing, etc.

(Chauhan, N. S., Kaur, L., & Walleh, A. S., 2011). As we know, an optimal resource and bandwidth allocation can optimize system performance (Song, W., & Zhuang, W., 2012). However, a dynamic environment usually exists in WLANs, such as bandwidth fluctuation, error rates, transmission delays, and so on (Ismail, M., & Zhuang, W., 2011). Therefore, it is difficult to maintain optimal performance for WLANs simply with a unique resource-allocation strategy. To find an optimal resource-allocation strategy, game theory can be applied into video transmissions in WLANs. The theory provides various mathematical methods to simulate and analyze the workstation behavior. The previous study (Song and Zhuang, 2012) has demonstrated that the game-theoretical model could be applied to radio resource allocation problems in networks. As conflicts between limited bandwidth and network resources frequently exist in WLANs, the maximization of network resources and the fairness of competition are the main targets in the development of wireless local networks (Wu, H., Xia, W., Li, J., & Shen, L., 2014). Recent research has proved that game-theoretical mechanisms for video transmission could improve the whole performance of WLANs. For example, an opportunistic distributed algorithm was presented in order to improve throughput metrics in ad-hoc networks (Giuliano, R., Mazzenga, F., Petracca, M., & Pomposini, R., 2012). This algorithm was implemented by the power-control strategy of data transmission. Giuliano and Mazzenga (2012) also researched the applicability to power allocation, medium access control, node participation, and so on. However, these recent works only analyzed the model adaptations that were related to different layers of the applied protocols in the network. The simulation of wireless communication network is still necessary for research. For video transmission, a fully functioning simulation toolbox named OPNET was imported. OPNET provides a typical simulation framework for video transmission systems, which consists of a model builder, simulation kernels, model modules, and a proper processor. OPNET has fulfilled such a framework and drawn much attention from researchers (Khan, T. A., Beg, M. T., & Khan, M. A., 2013). OPNET is widely used in many systems as it contains various modules for simulating real-world systems, especially for WLANs. This makes it easy to conduct research on video transmission for WLANs. Many researchers have successfully examined video transmission over WLANs with the help of OPENET. The actual responses of mobility can be simulated based on SVC that allows videos to be transmitted with mutable packet lengths under fluctuant channel conditions. To model different performance of WLANs for video transmission, communication lags, packet delay variation, and throughput were evaluated. Besides, mutable movement speeds of groups of workstations were also simulated (Singh, G., Grover, A., & Grover, N., 2013). Because a large number of simulation scenarios are built in OPNET, more and more developers can tune parameters of WLANs for each scenario.

The rest of this chapter is organized as follows. Section II introduces the cooperative video transmission over heterogeneous devices. Section III then describes the details of the proposed video transmission mechanism in WLANs. Next, Section IV summarizes the performance of the proposed mechanism and the analytical results. Conclusions are finally drawn in Section V.

COOPERATIVE VIDEO TRANSMISSION OVER HETEROGENEOUS DEVICES

This section introduces the proposed system, where the division of terminals is based on the video content and the distance between a terminal and a BS. In this study, two aspects are considered while terminals are divided into different types.

1. The distance between terminals and the BS is considered because it is efficient for a team to co-operate with each other when the members are close to each other.
2. Team division based on distance has no effect on cooperative transmission process, as this work focuses on single content broadcasting. It is unnecessary to consider the factor of contents.

The power-allocation model for the BS along with the cooperative transmission model is built based on game theory, respectively. Therefore, the transmission time for a BS to broadcast videos to the terminal members within each group by using fixed power is the same as that for terminals to transmit videos to each other. The proposed process is shown in Figure 1.

Figure 1. Workflow

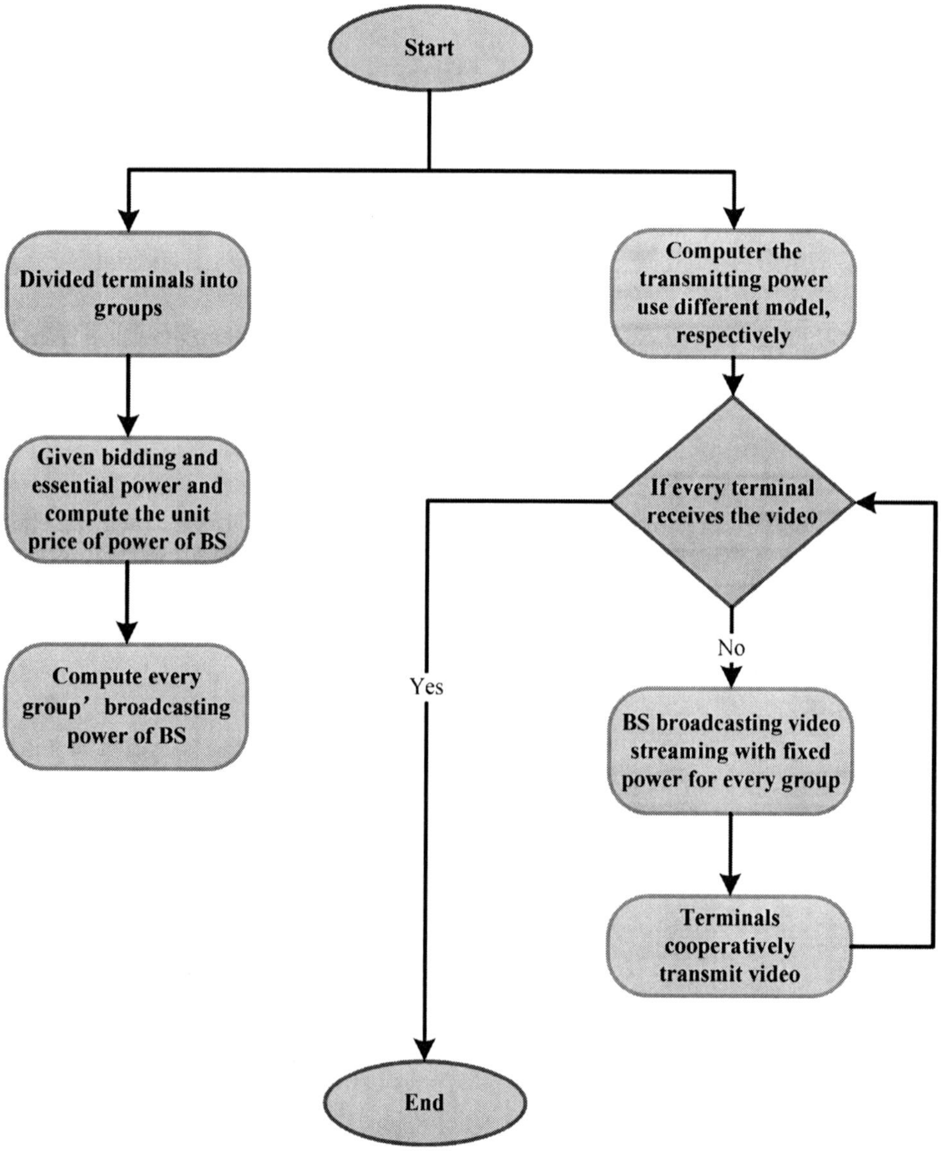

The workflow can be divided into two parts. One is power allocation for the BS, and the other is power allocation for terminals, which transmit video streams to the other terminals. As for the BS part, firstly we divide terminals into several groups. Then, each group gives its bidding price and essential power, followed by computation of power allocation for the BS based on the proposed model. If only one type of groups exists, we propose using the exchange transmission mechanism to compute power allocation for terminals. If more than one type is present, this study develops an SVC-based transmission mechanism.

In this work, we use $N=\{1,2,\ldots,n\}$ to represent the overall users, and they are divided into m cooperative groups denoted as $M=\{g_i|i=1,2,\ldots,m\}$ where g_i denotes the number of users in group i. Besides, $D=\{d_i|i=1,2,\ldots,m\}$ specifies the distance between the BS and group i. As more than one user exists in a group, such distance is computed as the distance between the BS and the farthest member, where the condition is $d_i \leq d_{i+1}$ and $i + 1 \leq m$. The whole arrangement of these teams is shown in Figure 2.

Figure 2. Layout of the cooperative model

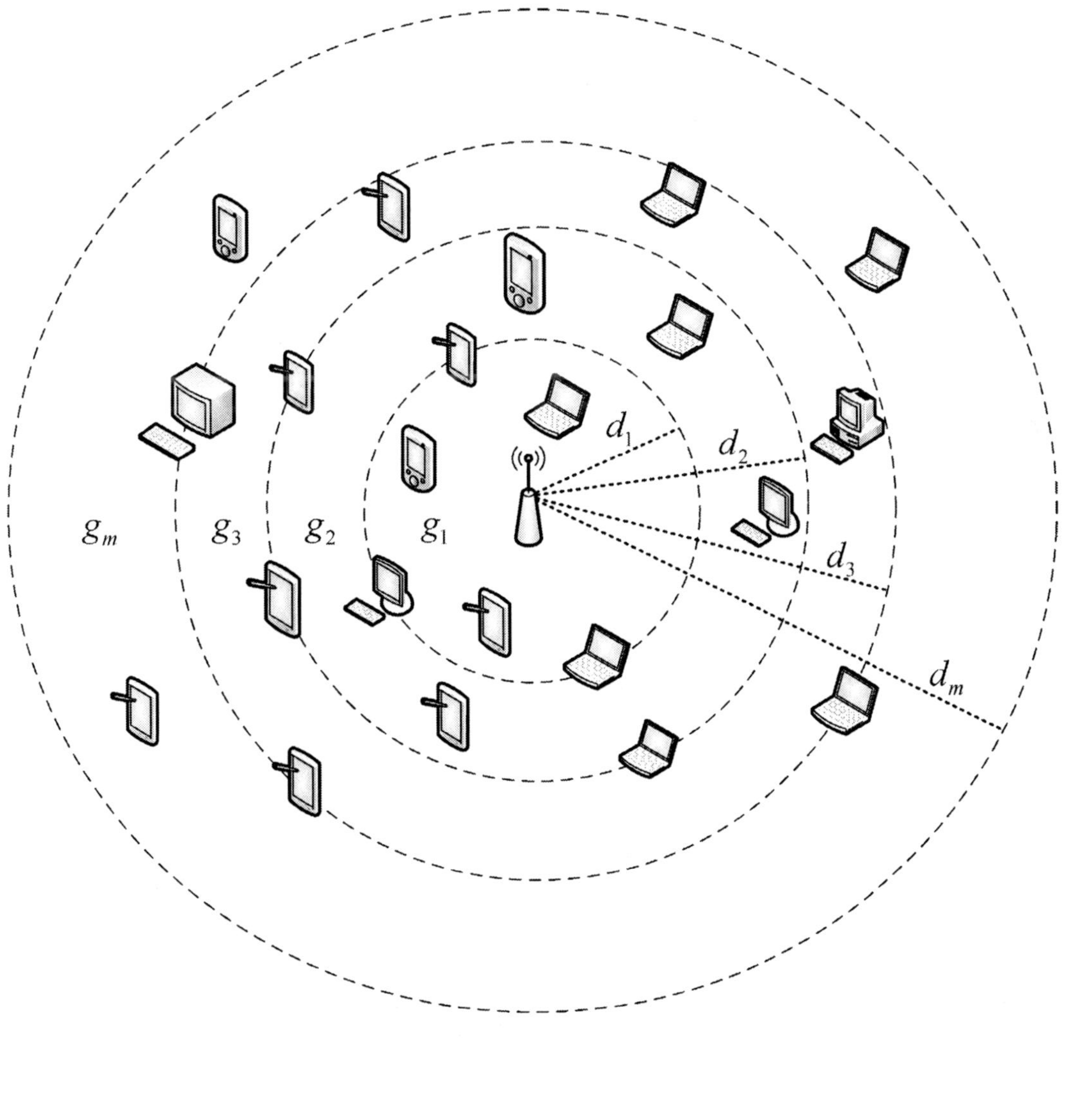

Punishment Mechanism

We hope to use game theory to restrain the greedy behavior of terminals for fairness and efficiency. The power of the BS is limited. However, it is easy for terminals if the power is provided for free, which causes a waste of energy and increases the BS burden. Thus we put the price on power, which means the more power a terminal demands, the more cost a terminal should pay for it.

The most terminals cannot pay for it if the price is too high, or we fail to restrain power if the price is too low. Thus, we need a suitable price for power. The normal network pricing mechanism is listed as following (Cao, Zhao, & Jing, 2012; Li, 2009).

1. Pricing model based on a specific user (Koriiis, Varvarigou, & Ahuja, 1998).
2. Negotiated pricing mechanism (Zhang & Li, 2003).
3. Pricing model based on auction (Cao, Shen, Milito, & Wirth, 2002; Liu, Tao, & Huang, 2013).

The bargain information is the auction method (Yaipairoj & Harmantzis, 2006; Sallent et al., 2006) including the unit price of the resources offered by network providers to the bidders. Namely, a network provider sets a unit price and offers resources to the bidders according to bargain information.

The pricing mechanism based on auction has been widely studied in the research, and recent reports have demonstrated its effectiveness and fairness (Cao et al., 2002; Liu et al., 2013). Therefore, this work followed the idea of pricing mechanism and applied it to power pricing with a modification to support the computation — The unit price of power generated by the network provider, the bidding prices offered by bidders, and power requirements given by the terminals.

As the power can be viewed as a type of commodities, the market pricing mechanism of commodities is complex (Zhang & Zhang, 2014). It needs to consider the cost of production, supply and demand, and transportation costs. To simplify this problem, only supply and demand is considered herein.

Suppose the terminals do not know pricing rules in advance, and assume x_i presents the bidding price of team i, where p_i indicates its essential power, where $p_1 \geq p_2 \geq \cdots \geq p_m$. Further suppose that P denotes the maximum power of the BS, and then power pricing model is expressed as follows.

- If $P < \sum_{i=1}^{m} p_i$, the pricing scheme is

$$x_0 = x_i, where\ i = \begin{cases} \sum_{i=j}^{k} p_i \leq P < \sum_{i=j}^{k+1} p_i, & k \in M \\ \min\{j\} \cap p_j \leq P \end{cases}.$$

- Otherwise, the power of the BS must meet all the demands from the terminals. The unit price of power is $x_0 = 0$.

Power Allocation of the Base Station

The utility function is not unique in game theory when we measure the loss and gain (Niyato & Hossain, 2009; Li, Guo, Li, & Zheng, 2009; Niyato & Hossain, 2008; Li, Zhang, Wang, Zhao, & Chen, 2013).

Usually, there are a variety of ways to construct a utility function for modeling users, such as currency measurement and network engineering indices (Scutari, Palomar, Facchinei, & Pang, 2010). In this study, we devise a utility function based on two aspects — i) "f(SNR)" for communication quality and ii) "$f(x_0,p_i)$" for punishment mechanism. Equation shows the proposed utility function.

$$u_i(p_i) = f(SNR) \bullet f(x_0,p_i)$$

The following content discusses in detail the aforementioned two aspects — f(SNR) and $f(x_0,p_i)$. The SNR and SIR reflect the quality of communication, and when they become larger, the transmission rate is higher. This enhances the quality of communication.

As the noise has an effect on the quality of video streaming during the broadcast, thus the noise needs to be modeled to guarantee the quality when the power of the BS is allocated. Generally, SNRs provide a feasible way to measure the noise. In this study, we use the Gaussian noise model to formulate the noise because such a model is well investigated. The model is defined as follows

$$SNR_i = \gamma_i = \frac{p_i h_i}{\sigma^2}$$

where p_i denotes the power of the BS when the BS transmits data to team i, and h_i denotes the channel gain during the transmission, and σ^2 is Gaussian noise power.

If the shadowing and fast fading is not considered, the channel gain model is defined as follows.

$$h_i = A / d_i^{\psi}$$

where A and ψ respectively denote the constant signal gain and the path attenuation factor. Thus, the SNR in becomes

$$\gamma_i = \frac{p_i A}{\sigma^2 d_i^{\psi}}.$$

Based on (Huang et al., 2008; Shah, Mandayam, & Goodman, 1998) and, the utility function becomes

$$u_i(p_i, \gamma_i) = f(SNR) = \frac{\alpha}{p_i}\left(1 - e^{-0.5 \cdot \gamma_i}\right)^L, \quad \alpha \neq 0$$

where α is a constant, and L represents packet length.

Regarding the punishment mechanism, as the groups, who require more power, need to pay more costs to the BS, the cost function is

$$f(x_0,p_i) = x_0 p_i.$$

Combining and yields the proposed SNR utility function,

$$u_i(p_i, \gamma_i) = \frac{\alpha}{p_i}(1 - e^{-0.5 \cdot \gamma_i})^L - x_0 p_i .$$

Notably, u_i=0 when p_i=0.

To sum up, the proposed SNR power-allocation model based on the punishment mechanism for the BS is shown as follows.

$$\arg\max_{p_i} \left\{ u_i(p_i) = \frac{\alpha}{p_i}\left(1 - e^{-0.5 \cdot \gamma_i}\right)^L - x_0 p_i \right\}$$
$$\text{s.t. } 0 \le p_i \le P$$

According to the aforementioned power distribution model, we can prove the existence and the uniqueness of Nash equilibrium. With Nash equilibrium, the power of a terminal can be calculated. The following content shows the steps.

Step 1: Every group generates the bid and essential power
Step 2: Calculate the price based on
Step 3: Calculate the power of a terminal by using

Cooperative Transmission Mechanism

This study assumes that when all the forwarding terminals deliver videos to the terminal that makes a request, the distance between all the terminals falls within the scope of transmission. Besides, all the terminals agree on cooperative transmission under the condition of the same network environment, namely, the channel condition of each terminal is consistent.

To design this cooperative transmission mechanism such that the terminals can receive the whole video by sharing the video streaming, which they have already received from the BS or the other terminals, with each other, the following problems should be considered:

- How do the terminals of the same type cooperate.
- How do the terminals of different types cooperate.
- How is the power allocated during cooperative transmission.

To overcome the aforementioned three problems, Scalable Video Coding (SVC) is introduced first herein because SVC can satisfy different requirements of all terminals corresponding to different layers (Hou, Wan, Zhan, & Chau, 2013; Schwarz, Marpe, & Wiegand, 2007; Li, Amon, Hutter, & Kaup, 2011).

During SVC transmission, the broadcasting system sends different coding layers to terminals based on their heterogeneous needs. However, every terminal still needs to receive basic (or basic enhancement) coding layer. Thus, the terminals, which need to receive the basic coding layer, can share the basic coding layer with each other. On the other hand, the terminals, which receive enhancement coding layers, do not take part in the process.

The following subsection discusses in detail the proposed cooperative transmission mechanism, including the exchange transmission mechanism for the same type of terminals and the SVC-based transmission for different types of terminals.

As the signals would interfere with each other when a terminal transmits video streaming to the other terminals of the same type, we use SIRs to measure such interference and the quality of video streaming by using the proposed SIR utility function.

Recall that a team is divided into several subgroups according to video frame packets. For example, suppose that the video frame is separated into four segments corresponding to the division of the team i. If the total type of terminals is one, each subgroup receives videos from the BS. Besides, terminals collaboratively transmit the video between subgroups as illustrated in Figure 3. Otherwise, if more than one type exists, the process is shown in Figure 4.

If the terminals of a group belong to the same type, the transmission exchange mechanism is shown as follows.

Assume that F denotes the length of video streaming, and the number of members of group i is g_i. Also assume the video is divided into y_i segments when group i receives video streams from the BS. There are g_i/y_i members in each group (where the remainder joins in the last group). It is not efficient

Figure 3. Collaborative signal interference

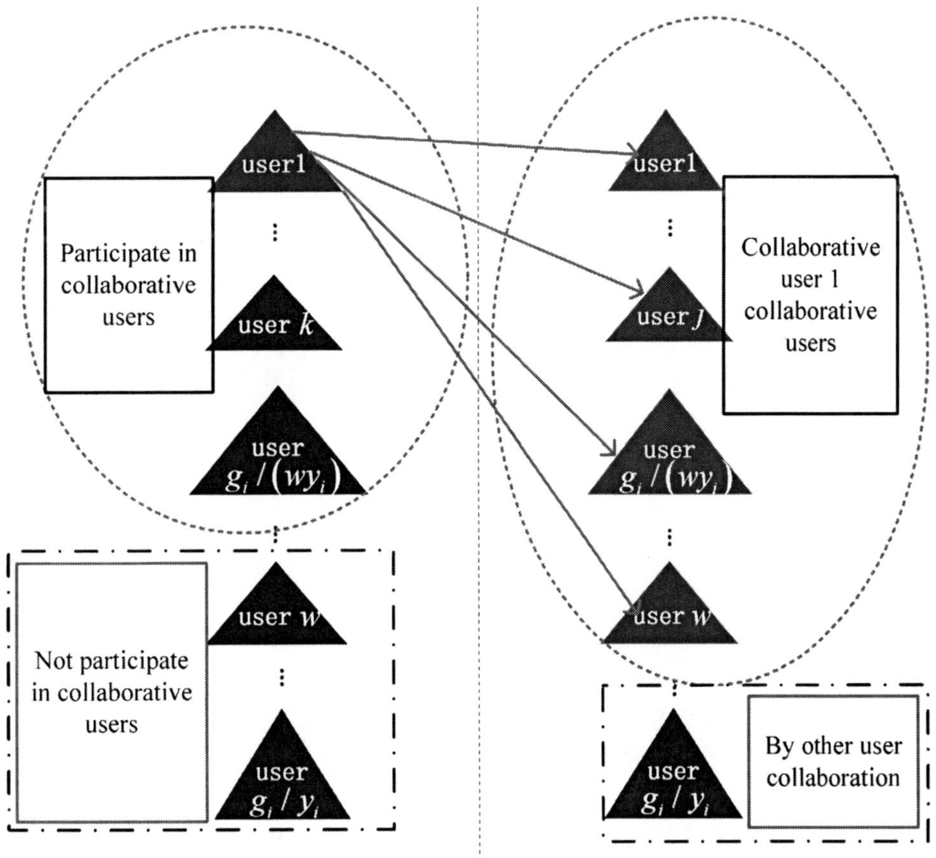

Figure 4. Signal interference of the cooperative mode in SVC video streams

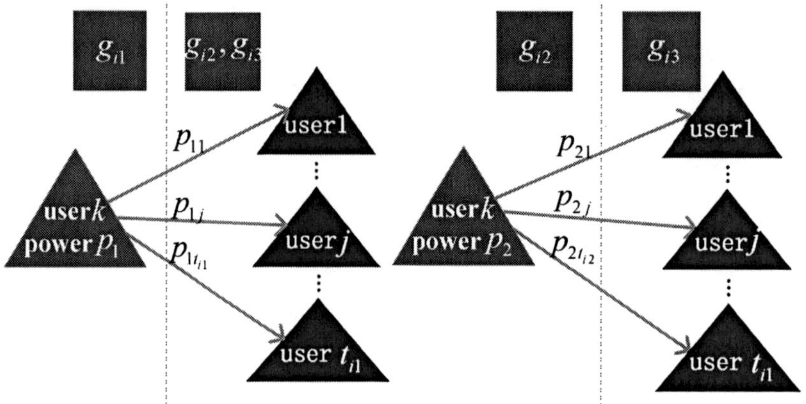

when every terminal participates in cooperative transmission. Thus, assume that no more than $g_i/(wy_i)$ members in each group take part in transmission. Also suppose the number of terminals equals w when these terminals co-work in the cooperative transmission. A base station broadcasts F/y_i video segments w times.

The SIR of terminal j is

$$SIR_{kj} = \gamma_{kj} = \frac{W}{R} \cdot \frac{h_{kj} p^u_{kj}}{\sum_{a=1 \cap a \neq j}^{ag_i} h_{ka} p^u_{ka} + \sigma^2}$$

where W denotes the network bandwidth, R represents the bit rate, and a denotes the terminal that receives video streams from terminal j .

The proposed power-allocation model for the homogeneous terminal collaborative transmission is then defined as follows.

$$\arg\max_{p^u_{kj}} \left\{ u_{kj}\left(p^u_{kj}\right) \right\} = \max \left\{ \frac{\alpha}{p^u_{kj}} \left(1 - e^{-0.5 \cdot \gamma_{kj}}\right)^L \right\}$$
$$\text{s.t.} \begin{cases} \sum p^u_{kj} \leq p^u_{k0} \\ p^u_{kj} \geq 0 \end{cases}$$

where γ_{kj} denotes the SIR of terminal j, and p^u_{kj} specifies the power of terminal k (where $k = 1,2,\ldots,w$) when it transmits a video layer to terminal j (where $j = 1,2,\ldots,y_i$). Besides, p^u_{k0} represents the transmission power that the terminal k can use.

On the other hand, if more than one type exists, the SVC-based transmission mechanism is used instead. The detail of the proposed SVC-based transmission mechanism is shown as follows.

This work assumes that there are three types of terminals, namely, D1, CIF, and QCIF. The corresponding video requests are QD1, QCIF, and QQCIF, respectively, where the quality is QD1 > QCIF > QQCIF. Furthermore, the QCIF terminals receive only the basic coding layer, and the CIF terminals receive the basic coding layer and the basic enhancement coding layer. The QCIF terminals receive the basic coding layer, the basic enhancement coding layer, and the enhancement video layer.

Suppose group i is divided into three subgroups, that is, subgroup QCIF, subgroup CIF, and subgroup D1, respectively denoted by g_{i1}, g_{i2}, and g_{i3}. Additionally, m_{i1}, m_{i2}, and m_{i3} represent the number of subgroups g_{i1}, g_{i2}, and g_{i3}, respectively.

The cooperative transmission mechanism works as follows. Subgroup g_{i1} receives the basic coding layer from the BS and then transmits it to subgroups g_{i2} and g_{i3}. Subgroup g_{i2} receives the basic enhancement coding layer from the BS and subsequently delivers it to subgroup g_{i3}. Subgroup g_{i3} only receives the enhancement video coding layer from the BS.

Further suppose that the cooperative task is equally assigned to every member within a subgroup. The terminals, with which every member of subgroup g_{i1} needs to cooperate, are defined as

$$t_{i1} = (m_{i2}+m_{i3})/m_{i1}.$$

Likewise, the terminals for subgroup g_{i2} are

$$t_{i2} = m_{i3}/m_{i2}.$$

As discussed above, more than one terminal may receive video streams from the same terminal. Thus, terminals may compete with each other, and this leads to unnecessary power consumption. To deal with such a problem, two power-allocation models of subgroups g_{i1} and g_{i2} are developed as follows.

Denote p_j^{ik} (where $k = 1,\ldots,m_{ij}$, and $j = 1$ or 2) as the power consumption of member k in subgroup g_{ij}. Also denote p_{jb}^{ik} as power that terminal b (where $b = 1,\ldots,t_{ij}$) consumes while receiving video streaming from terminal k in subgroup g_{ij}. Notably, terminal t_{ij} may interfere with the other terminals when the source terminal k transmits streams. Thus, we can define the SIR of terminal b as

$$SIR_b = \gamma_b = \frac{W}{R}\cdot\frac{h_{kb}p_{jb}^{ik}}{\sum_{a=1\cap a\neq b}^{t_{ij}} h_{ka}p_{ja}^{ik}+\sigma^2}.$$

Based on the aforementioned equations, the proposed power-allocation model for the heterogeneous terminal collaborative transmission is shown as follows.

$$\arg\max_{p_{jb}^{ik}}\left\{u_b\left(p_{jb}^{ik}\right)\right\} = \max\left\{\frac{\alpha}{p_{jb}^{ik}}\left(1-e^{-0.5\cdot\gamma_b}\right)^L\right\}$$
$$\text{s.t.}\begin{cases}\sum p_{jb}^{ik} \le p_j^{ik}\\ p_{jb}^{ik} \ge 0\\ b = 1,2,\cdots,t_{ij}\end{cases}.$$

VIDEO TRANSMISSION IN WLANs

This section discusses the above-mentioned mechanism realized in WLANs. Actually, it is not easy to obtain satisfactory QoS for real-time video transmission for workstations, especially for mobile nodes., WLANs are not eliable and stable due to the movement of mobile users and the fluctuating transmission environment. This study uses the above-mentioned proposed mechanism and a virtual WLAN environment to understand the cooperation among individuals and groups. The terrain modeling module and the SITL module were applied to simulate the activities of workstations.

Typically, it is not easy to transmit raw video-frames through WLANs because the storage space is limited, which results from the limitation of channel bandwidth. Therefore, the coding schemes are usually required to make full use of bandwidth efficiently (Abdelhadi, A., Gerstlauer, A., & Vishwanath, S., 2014).

Figure 5 reflects a typical system which includes 254 terminal workstations or mobile users. In this system these terminals can receive or deliver the video traffic flows and other regular traffic. Moreover, a video server in the figure serves as a video transmitter, and a router serves as an access point (AP).

Figure 5. Video transmission system

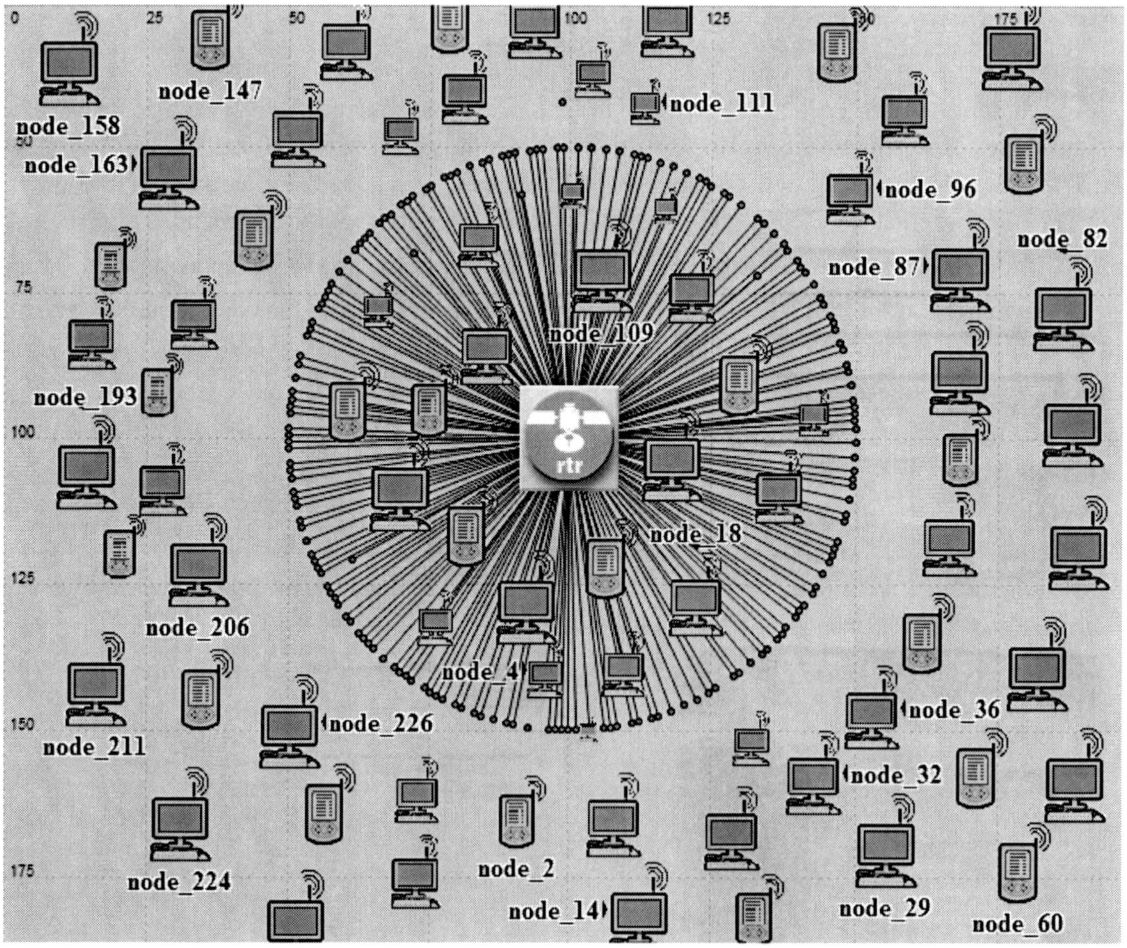

The modeling procedure is divided into the following steps according to (Khan, M. A., Khan, T. A., & Beg, M. T., 2012). Firstly, attributes of each workstation in the system are analyzed and configured. Secondly, SVC-format videos are processed and converted into a compatible OPNET format. Thirdly, for SVC formats, we utilize different video traffic for various packet sizes. Fourth, common traffic that does not fit SVC-formats is then modeled in the simulation. Finally, the mobility of the stations along movement paths at a proper speed is configured in the system.

To deal with limited-resource, fluctuating, delay, and limited-bandwidth problems, this study utilizes SVC standard instead of H.264 version as the empirical result of SVC shows satisfactory performance during video transmission.

System in the Loop

SITL module can help us make an interaction between real-world and virtual wireless networks for analyzing real-time SVC traffic. In the SITL model simulation, the video frames are converted into packets. These packets are exchanged through an external interface named SITL gateway, i.e., an Ethernet network adaptor to the virtual simulation environment (Babazadeh, D., Chenine, M., Zhu, K., & Nordström, L., 2012).

Three modeling scenarios in the SITL are listed as follows.

- **Sim-to-Sim:** Interaction between the simulated environment and another one.
- **Real-to-Real:** Interaction between the real-world network and another one.
- **Sim-to-Real:** Interaction between the real-world network and the simulated environment.

The latest OPNET simulation version supports many SITL module protocols, such as IP, ICMP, TCP, RIP, UDP, ICMP, OSPF, and so on. The researchers can develop their own translation method with C language when they need to use non-built-in protocols.

Figure 6 presents the visualization of video transmission environments, in which the virtual WLAN network is employed in OPNET programming environment. Based on this figure, observations show that two WLAN routers are employed in this virtual network. The right WLAN router works as an access point, and the left router serves as a receiver. They are both connected through the virtual link (SITL-virtual-eth and SITL-virtual-gateway) that provides an interface for the real network.

Game Theory for Video Transmission Systems

The WLAN is a wireless communication environment for multiusers. All the nodes share a certain number of network resources, such as channels and receiver power. In such a multiuser environment, all users can either cooperate or conflict with each other to reach their desired Quality of Services or bandwidth (Luo, H. C., Wu, E. K., & Chen, G. H., 2013). Under such a circumstance, proposing a method for allocating a limited resource to all nodes to obtain the optimum environment performance is a challenge. Game theory provides a feasible solution for that. Game theory has been widely used in the communication to improve power control, routing, medium access, and so on. This study utilizes game theory to understand CSMA to obtain the fairness while allocating the network resources to users

Figure 6. Situational WLANs in OPNET

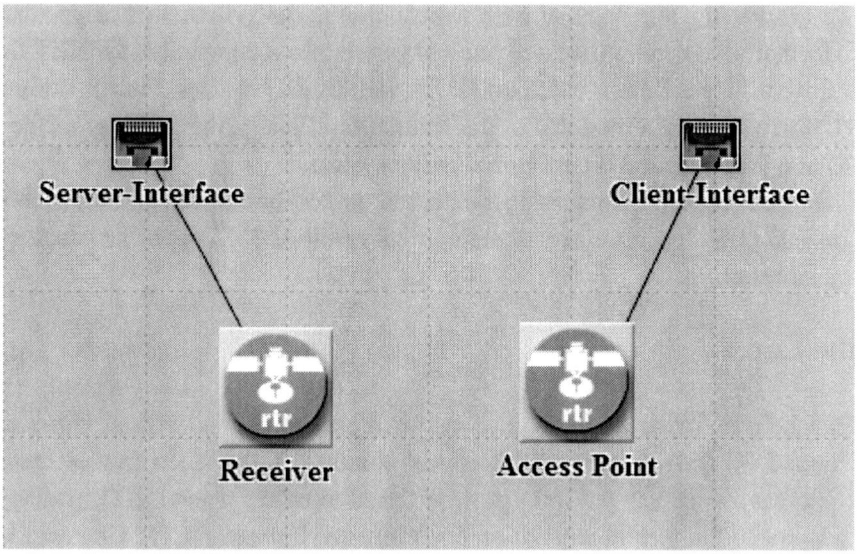

in a WLAN (Alvi, A. N., Naqvi, S. S., Bouk, S. H., Javaid, N., Qasim, U., & Khan, Z. A., 2012). We use game theory to distribute the channel bandwidth and to solve the problems of conflictions between nodes during transmission.

As the WLAN utilizes CSMA/CA schedules, users compete with each other to send data packets over the allocated channel. When a collision occurs, the receiver node then reaches a low Signal-to-Interference-plus-Noise Ratio (SINR) and the process cannot proceed in the right way.

To maximize the total system performance, each user of the wireless local area network should compromise with others by using a cooperative scheme (Rozario, F., Han, Z., & Niyato, D., 2011). Besides, the user has to work synchronically. To such an end, this work uses Nash bargaining game, for it is considered as a cooperative game, and each user can increase the profit. Nash bargaining game can be expressed as follows (for simplicity, the mathematic symbols are reused and are irrelevant to the previous sections).

$$s^* = \arg\max_{s} \prod_{i \in \prod} \left(u_i\left(s_i\right) - u_i^d \right)$$

where $u_i(\bullet)$ represents the payoff function of user i, and s_i is the resource competition strategy of user i. Besides, s^* is the solution to Nash equilibrium for all the users, and u_i^d denotes the competition point, where user i gives up cooperating with the other users in the same competing system.

As CSMA/CA strategy is a game of choosing the appropriate back-off windows to achieve the Nash equilibrium, all the users may configure their back-off windows at the minimum level due to the selfishness. The throughput of WLANs then becomes low when all the users are selfish. Therefore, to solve this problem, cooperative game models are utilized in this study. In this situation, users cooperate with each other to avoid penalty (Li, Z., & Shen, H., 2012). Table 1 summarizes typical game-theoretic schemes for SVC video transmission over CSMA/CA scheme.

Table 1. CSMA/CA Game Strategies

Model	Symmetric Strategy	Schedule
Repeated game	No	CRISP cooperative mechanism
Dynamic game	Yes	N/A
	Yes	Best-response and gradient-update algorithms
	Yes	Dynamic Bayesian learning mechanism
Cooperative game	No	Round-based distributed algorithm and coordination algorithm
	No	A distributed method to update user information

In this chapter, we use the dynamic game theory to simulate the user activities. User i either sends the packets or stops based on the wireless channel gain hi and the user type. Furthermore, M steps are related to M back-off slots in a dynamic game based on Bayesian theory. At the step $m \in \{1,2,\ldots,M\}$, when user i sends the video traffic packets successfully, it can be modeled by the pay-off function $\mu_i(h_i)$-$c_i(h_i)$, in which μ_i represents the utility function, and $c_i(h_i)$ represents the cost function. Contrarily, when the transmission fails, the user should be panelized by -$c_i(h_i)$. If no data is sent, the step of the competition system reaches $m + 1$. The game finishes as more than one sending procedure exists at any step m. Each user maximizes its expectation through all the steps in order to achieve the optimal Nash equilibrium.

To maximize the payoffs of each user, we consider only the symmetric strategy with $s_1^* = \ldots = s_i^* = \ldots = s_M^* = p^*$, where p^* represents the probability of sending data. The conflicting probability of package sending p and collision q in a channel is

$$n = f(p,q) = 1 + \frac{\log(1-q)}{\log(1-p)}.$$

The optimal solution to the game is based on the number of users n who face conflictions. The sending probability varies according to the following equation, in which only confliction parameter $CW_{\min}$ is considered.

$$p = \frac{2\left(1-2q\right)}{(1-2q)(CW_{\min}+1)+q.CW_{\min}.(1-(2q)^n)}$$

SIMULATION RESULTS

Cooperative Video Transmission over Heterogeneous Devices

Suppose that the maximum transmission power of the base station is 10 mW. In total, ten teams are present, and the market power pricing is 0.5 RMB/mW, where the abbreviation mW means milliwatt.

Table 2 gives the experimental data of these ten teams. The experimental data in the table is generated by a random function. This study examines a condition, where the behavior of each team is rational, and the price is not over high. Besides, the range of the bidding price is 0.5 RMB/mW.

Suppose the bidding behavior and the power consumption form normal distribution. We use a normal random function to generate the experimental data for bidding behavior. The mean of the distribution for bidding behavior is 2.5, and the standard deviation is 1.0. Furthermore, the mean of the power consumption is 5.0 with the standard deviation equal to 0.1.

By using, we can calculate the unit price of power, and the simulation results are displayed in Figure 7.

1. Utility Function of Power and SNRs

Figure 8 and Figure 9 respectively display the utility functions, where the vertical axis is the output of the function, whereas the horizontal axis specifies power consumption and SNRs.

From Figure 8, we observe that the utility function diminishes as power consumption deceases if a fixed SNR is specified. Besides, the downtrend of the function with the proposed punishment mechanism is greater than that of the original function, and the magnitude of the decrease becomes higher with the increase of power.

Table 2. Experimental data of the pricing model

No.		1	2	3	4	5	6	7	8	9	10
1	**Bidding price**	3.0	4.3	0.2	3.4	2.8	1.2	2.1	2.8	6.1	5.3
	Power (mW)	4.325	6.515	5.365	4.97	5.355	4.9	4.940	5.745	5.705	5.710
2	**Bidding price**	3.2	1.3	3.2	4.1	3.0	3.5	3.2	2.2	2.8	1.7
	Power (mW)	5.445	4.425	4.465	4.595	3.530	5.720	5.165	4.625	5.685	4.145
3	**Bidding price**	2.4	2.3	2.8	2.8	1.6	2.5	2.3	3.1	3.6	3.6
	Power (mW)	4.570	5.040	4.395	4.445	4.995	5.765	4.615	5.185	4.885	5.560
4	**Bidding price**	1.4	2.5	3.1	3.6	4.0	2.6	1.0	1.8	1.4	4.9
	Power (mW)	4.690	5.375	4.905	5.445	4.620	4.300	4.290	5.245	4.910	4.900
5	**Bidding price**	3.9	2.8	2.7	4.1	1.7	3.2	3.3	2.3	2.7	1.3
	Power (mW)	4.425	5.050	5.360	6.295	4.665	5.095	4.960	4.035	4.780	4.105
6	**Bidding price**	3.3	1.6	2.6	2.0	2.8	1.9	3.0	3.2	4.2	2.3
	Power (mW)	3.930	4.580	5.675	4.465	5.480	5.060	5.720	4.020	4.900	4.395
7	**Bidding price**	5.4	3.3	3.9	1.4	2.0	2.2	3.6	2.2	3.2	0.4
	Power (mW)	4.825	4.590	4.210	5.255	5.140	5.015	4.335	5.565	5.175	4.850
8	**Bidding price**	2.5	2.2	0.7	2.2	1.7	1.5	1.3	2.0	0.5	3.5
	Power (mW)	5.260	4.990	4.985	4.600	5.510	4.935	4.645	5.675	4.890	4.705
9	**Bidding price**	2.2	1.7	1.4	5.0	4.2	2.8	1.2	1.6	2.3	3.3
	Power (mW)	4.335	3.835	4.275	5.165	5.195	5.225	4.935	5.090	4.760	5.430
10	**Bidding price**	1.1	3.0	1.7	2.2	3.1	3.5	1.4	3.8	3.2	2.4
	Power (mW)	4.900	4.890	4.850	5.010	5.025	5.415	5.765	5.235	4.895	5.315

Figure 7. Unit price of power fluctuations

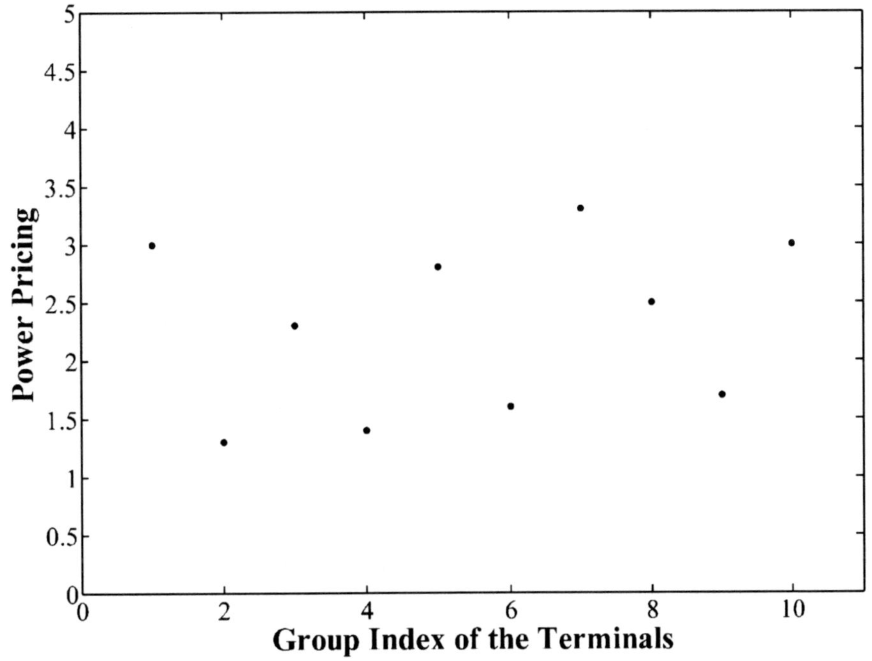

Figure 8. Output of the utility function changes with power when the SNR is fixed

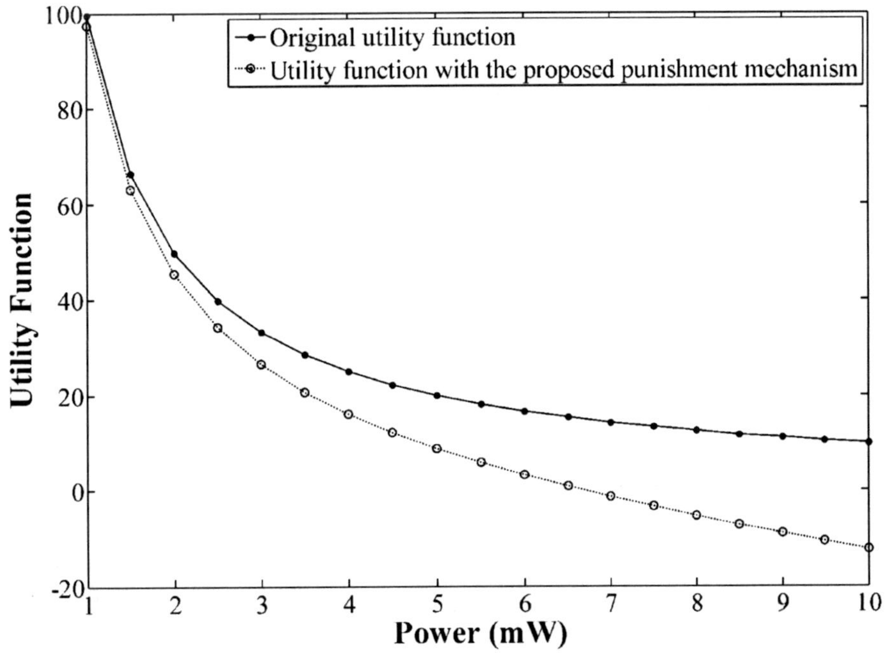

Figure 9. Output of the utility function changes with SNRs when the power is fixed

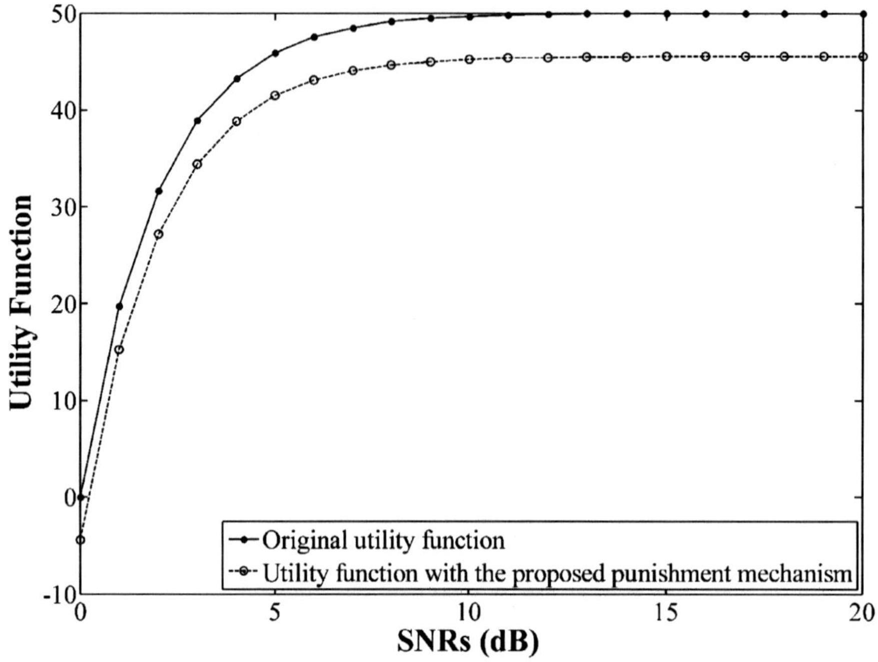

As shown in Figure 9, we observe that the utility function rises as SNRs grow gradually while power is fixed. Besides, the uptrend of the function with the proposed punishment mechanism is greater than that of the original function. Like Figure 8, the magnitude of the increase becomes higher with the increase of SNRs.

2. Power Allocation Model of SNRs for the BS

To evaluate the proposed model, the terminals are divided into ten groups, and the distance between the BS and the groups 1–10 is respectively {100, 200,..., 1000} meters. The remaining parameters are shown in Table 3.

The data of power allocation of the BS for ten groups is listed in Table 4. Then, calculation of the power allocation by using and generates the result listed in Table 5.

Table 3. Simulation Parameters

Parameter	Value
Power of Gaussian noise σ^2	5×10^{-12} (W)
Path attenuation factor ψ	3
Constant signal gain A	0.097
Constant parameter α	100
Packet length L	80

Table 4. Pricing settings of the BS for different groups

No.	1	2	3	4	5	6	7	8	9	10
Price	1.3	0.8	1.9	2.2	2.9	0.8	3.0	3.5	2.6	3.2
Power (mW)	3.485	4.375	4.525	4.630	4.745	4.840	5.005	5.195	4.770	5.620

Table 5. Power of the BS for different groups based on pricing

No.	1	2	3	4	5	6	7	8	9	10
Power (mW)	0.755	1.050	1.315	2.075	4.000	4.930	5.055	6.660	7.750	8.860

As the power changes with the distance, once the distance increases, the transmission power increases. Figure 10 shows the result.

As shown in Figure 10, the curve reveals that the relationship between the power and the distance is not linear. This also demonstrates that the distance is not the only factor that has an effect on the quality of videos. In addition, it shows that the utility function cannot constrain the behavior of terminals. In other words, terminals far away from the BS still require high power.

Figure 10. Power consumption versus the distance between the terminals and the BS

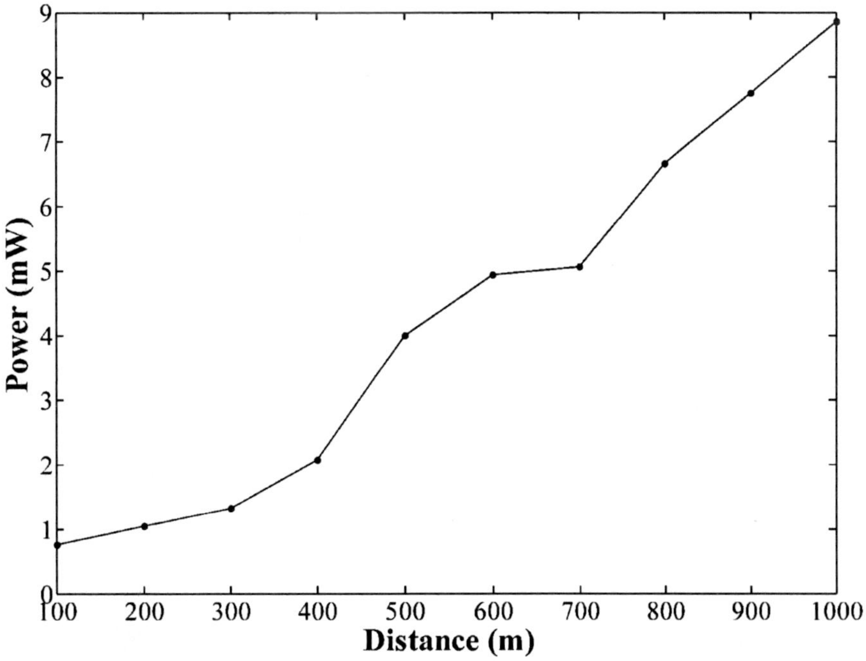

3. Cooperative Transmission Mechanism

A. Exchange Transmission Mechanism

For the exchange transmission mechanism, we use simple two subgroups cooperating with each other as the test subjects in the experiment. Suppose all the transmission members are of the same type, and they are divided into two subgroups. Each subgroup contains 20 terminals, or members. The distance from the BS to the two subgroups is 50 meters and 90 meters, respectively. Also suppose that four terminals take part in cooperative transmission in each subgroup, and every terminal of the four terminals cooperates with the other three terminals.

The terminals are numbered from 1 to 40, where the four terminals in the cooperative mode among the first subgroup use labels 1–4, whereas those in the second subgroup are labeled from 21 to 24. The cooperative transmission is illustrated in Figure 11.

Assume that the limited power of the eight terminals using the cooperative mode is {0.710, 0.950, 1.010, 0.690, 0.820, 0.940, 1.120, 1.200} milliwatt. The limited power refers to the maximum constrained power which a terminal outputs.

Let D_1 denote the distance matrix, of which the element represents the distance between the four cooperative terminals (i.e., terminals 1–4) in subgroup 1 and all the terminals (i.e., terminals 21–40) in subgroup 2. Moreover, D_2 denotes the distance between the four cooperative terminals (i.e., terminals 21–24) in subgroup 2 and the terminals (i.e., terminals 1–20) in subgroup 1. Therefore,

Figure 11. Two-subgroup cooperative transmission mechanism. The first group contains terminals 1–20, whereas the second group consists of terminals 21–40

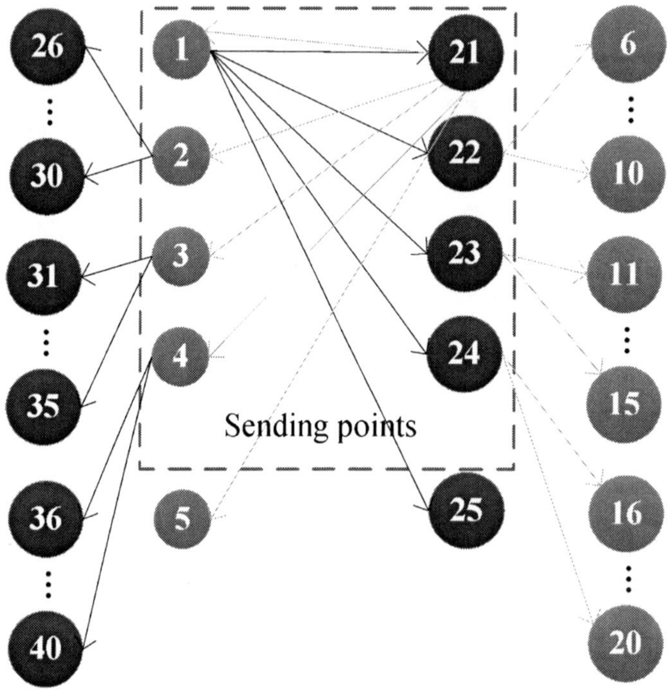

$$D_1 = \begin{bmatrix} (1,21) & (1,22) & (1,23) & (1,24) & (1,25) \\ (2,26) & (2,27) & (2,28) & (2,29) & (2,30) \\ (3,31) & (3,32) & (3,33) & (3,34) & (3,35) \\ (4,36) & (4,37) & (4,38) & (4,39) & (4,40) \end{bmatrix} = \begin{bmatrix} 30 & 25 & 50 & 13 & 41 \\ 14 & 5 & 52 & 44 & 12 \\ 19 & 11 & 20 & 45 & 21 \\ 10 & 21 & 34 & 33 & 10 \end{bmatrix}$$

and

$$D_2 = \begin{bmatrix} 30 & 35 & 9 & 33 & 61 \\ 23 & 52 & 12 & 24 & 9 \\ 49 & 17 & 31 & 50 & 15 \\ 30 & 11 & 34 & 23 & 20 \end{bmatrix}.$$

The power-allocation results are shown in Table 6.

Table 6 shows the power allocation between the eight sending terminals of two subgroups (i.e., the first column) and the other receiving terminals. Examining the allocated power and the distance indicates that when the distance is larger, the transmission power becomes greater to overcome attenuation due to the fact that it is far away from the sending terminal.

Table 6. Power allocation of the exchange transmission mechanism

No.	Limited Power (mW)	No./Power (mW)*				
1	0.710	21	22	23	24	25
		0.134	0.112	0.223	0.058	0.183
2	0.950	26	27	28	29	30
		0.105	0.037	0.389	0.329	0.090
3	1.010	31	32	33	34	35
		0.165	0.096	0.174	0.392	0.183
4	0.690	36	37	38	39	40
		0.064	0.134	0.217	0.211	0.064
21	0.820	1	2	3	4	5
		0.146	0.171	0.044	0.161	0.298
22	0.940	6	7	8	9	10
		0.180	0.407	0.094	0.188	0.071
23	1.120	11	12	13	14	15
		0.339	0.118	0.214	0.346	0.104
24	1.200	16	17	18	19	20
		0.305	0.112	0.346	0.234	0.203

*The upper part of the third column specifies the terminal label

b. SVC-Based Transmission Mechanism

For the SVC-based transmission mechanism, suppose that the numbers of terminals using the QCIF, CIF, and D1 mode are respectively 8, 8, and 20. Besides, four QCIF members serve as sending terminals, each of which transmits the video coding layer to the other seven terminals, i.e., (8 CIF + 20 D1) / 4 QCIF = 7. Moreover, four CIF terminals work as sending points, each of which also transmits the video coding layer to the other five terminals, i.e., 20 D1 / 4 CIF = 5.

The limited power of the sending terminals is defined as follows. The vector shown in contains the power of four cooperative terminals in the QCIF transmission mode, whereas the vector in is in the CIF transmission mode.

[1.92, 2.12, 1.89, 2.49] and [4.20, 4.61, 6.10, 5.90].

Distance matrixes are listed as follows, respectively

$$D_1 = \begin{bmatrix} 10 & 21 & 12 & 17 & 31 & 20 & 9 \\ 23 & 8 & 10 & 13 & 15 & 28 & 30 \\ 45 & 47 & 50 & 32 & 17 & 9 & 10 \\ 31 & 17 & 24 & 9 & 7 & 18 & 22 \end{bmatrix}$$

where the row denotes four QCIF cooperative members, and the column specifies the other seven terminals as mentioned earlier.

$$D_2 = \begin{bmatrix} 53 & 32 & 16 & 23 & 51 \\ 32 & 12 & 32 & 14 & 29 \\ 39 & 7 & 13 & 10 & 21 \\ 20 & 31 & 14 & 13 & 20 \end{bmatrix}.$$

Likewise, the row denotes four CIF cooperative members, and the column specifies the other five terminals. The power-allocation results based on – are shown in Table 7 and Table 8. The first column in Table 7 specifies the four cooperative terminals in the first subgroup, and that in Table 8 represents the four cooperative terminals in the second subgroup.

Table 7. Power allocation based on the SVC transmission mechanism for subgroup QCIF

Limited Power (mW)	Power Allocation (mW)						
1.920	0.160	0.336	0.192	0.272	0.496	0.320	0.144
2.120	0.384	0.134	0.167	0.217	0.250	0.467	0.501
1.890	0.405	0.423	0.450	0.288	0.153	0.081	0.090
2.490	0.603	0.331	0.467	0.175	0.136	0.350	0.428

Table 8. Power allocation based on the SVC transmission mechanism for subgroup CIF

Limited Power (mW)	Power Allocation (mW)				
4.200	1.272	0.768	0.384	0.552	1.224
4.610	1.240	0.465	1.240	0.542	1.123
6.100	2.643	0.474	0.881	0.678	1.423
5.900	1.204	1.866	0.843	0.783	1.204

Video Transmission in WLANs

This study used the real-to-real model provided in the SITL module for simulation. The WLAN and the SVC video transmission, where communication between real devices travels via a simulated network, are adopted. In the simulation network illustrated in Figure 12, we used two computers (i.e., Clients 1 and 2), with the video playback software installed on both operating systems. Both computers were connected to the Ethernet Network Interface Card (NIC) in the simulation computer over a hub. The topology of the simulation environment is the same as that in Figure 6. The IP addresses of Clients 1 and 2 respectively

Figure 12. SITL Model in OPNET

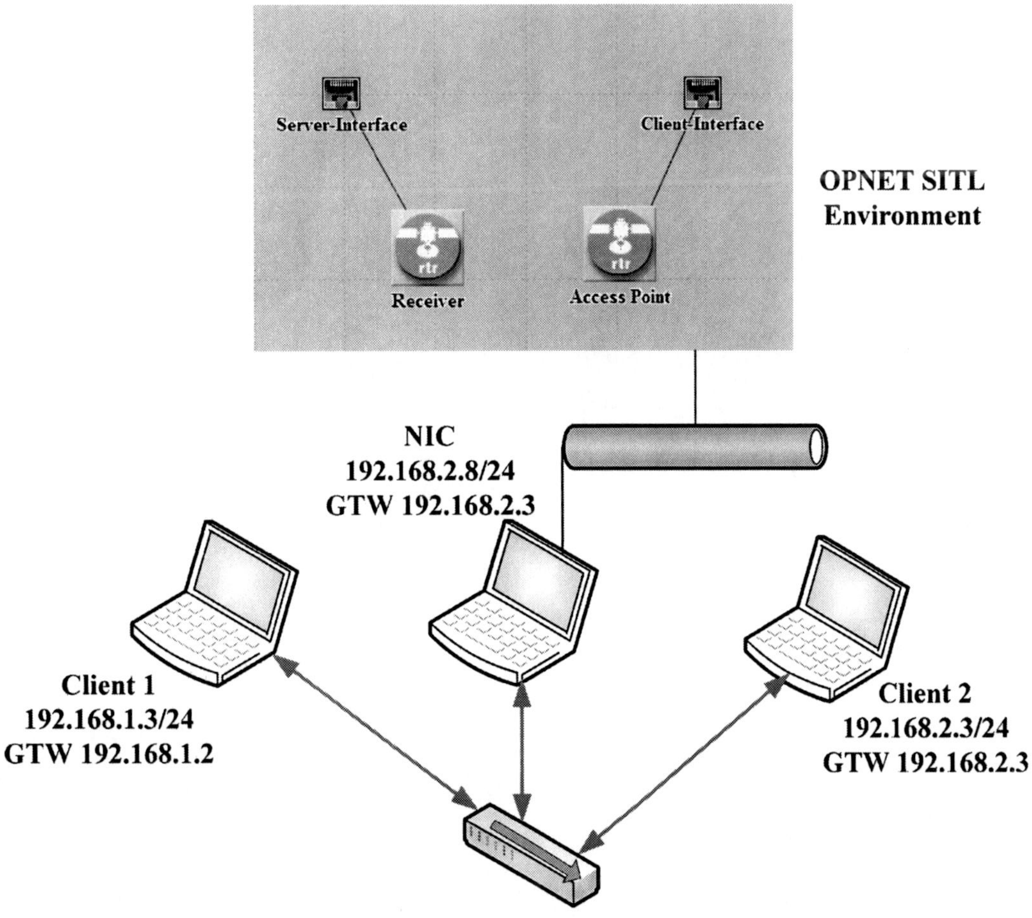

belonging to different and the same logical subnets were considered in the simulation. In the latter case, the communication between them went directly over a hub instead of through the simulation process. To support such a case, where both external computers had IP addresses from the same subnet, more than one network interface card was installed (i.e., one NIC for each external computer).

Real-time videos are received by OPNET through SITL gateways, which also have a user-defined packet filter. When the filter is applied, only desired packets can enter the simulation process from the real network. SITL gateways also translate real packets into appropriate simulation packets. In our simulation, we simulate WLAN communication devices by using two wireless routers, each connected to its own SITL gateway.

By using both wireless routers, we can define appropriate wireless parameters, such as the channel frequencies, transmission power, protocol types, modulations, antenna altitude, and emissive diagrams.

Figure 13 shows the position of the transmitter with a defined trajectory of movement and the receiver across a virtual terrain. The inner structure for transmitters and receivers were the same in both cases. Because the mobile users are not stable, it is not easy to predict the efficiency of network resource allocation in WLANs. By simulating the movement of mobile users, the performance of the network can be made more efficient and accurate (Manoj, P., 2011).

Figure 13. Movement of nodes

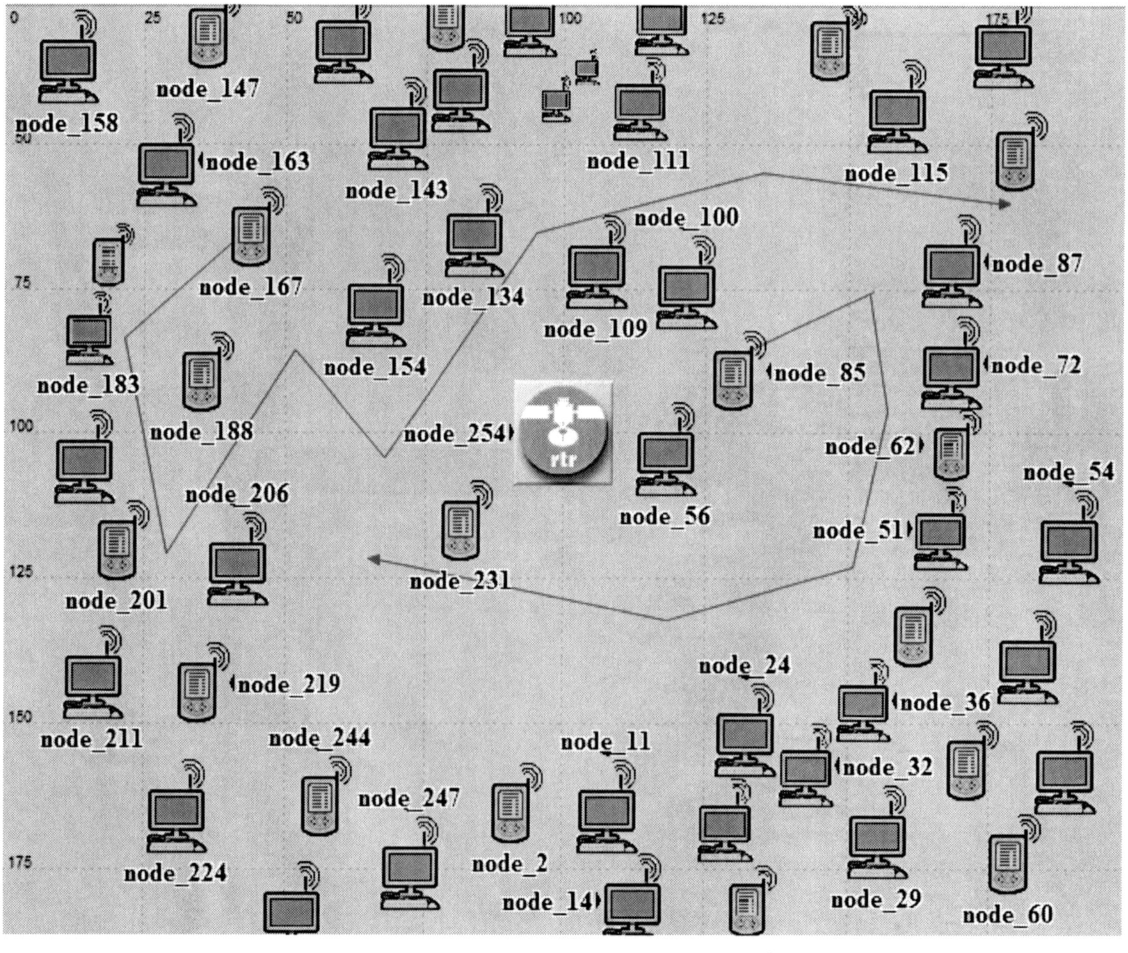

Figure 14 shows that the transmission delay becomes shorter when game theory is applied, whereas the transmission delay without the game-theoretical model is longer. It implies that our game-theoretical model can help group cooperation work better.

Figure 15 reveals that the whole duration of transmitting a short video clip is quite low. The horizontal axis is the video size, and the vertical axis reflects the time expense. As the figure shows, when a 4Mb-size video segment is transmitted, the transmission time expense with game theory is only 2.5 minutes, while the transmission time expense without game theory is as high as 6.5 minutes. As for the aspect of transmission time expense, the SITL module and the game theory both make a contribution to the group cooperation in the whole WLAN transmission.

CONCLUSION

This work presents a collaborative transmission mechanism for heterogeneous terminals based on the game theory. In this study, a power-allocation strategy is designed to fulfill collaborative transmission between terminals because controlling the distribution of power is the focus of this research. To this end, three utility functions are proposed — 1) SNR power-allocation model based on the punishment mechanism for the BS, 2) power-allocation model for homogenous terminal collaborative transmission, and 3) power-allocation model for heterogeneous terminal collaborative transmission. The first

Figure 14. Video transmission delay

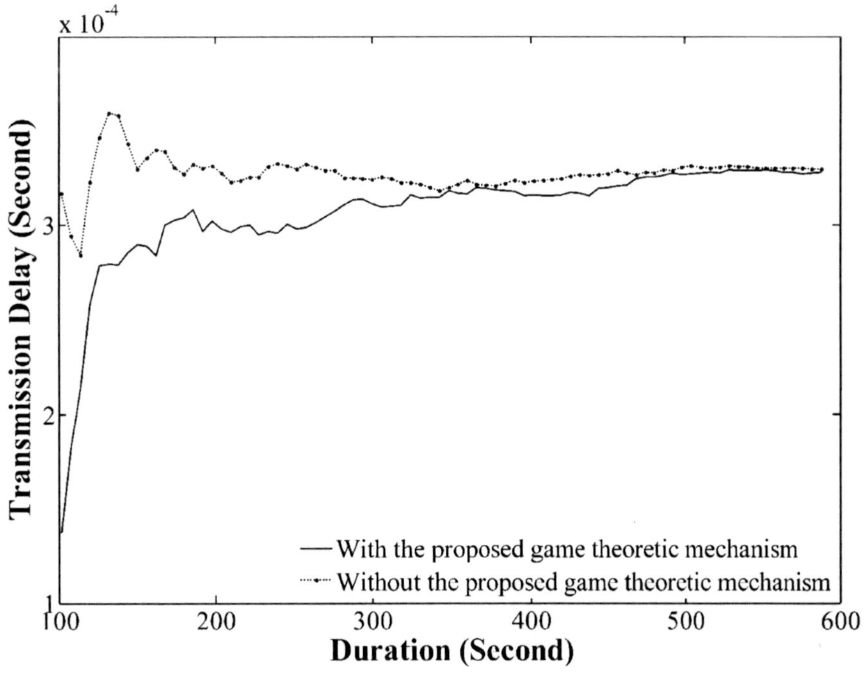

Figure 15. Transmission time when the video size changes

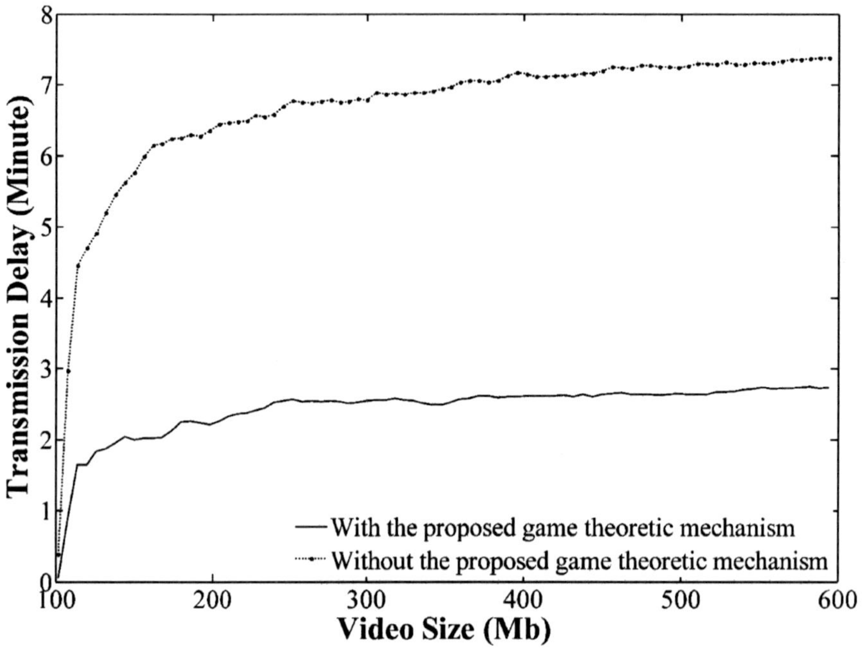

focuses on the power of the base station based on SNR distribution, whereas the last two highlight the distribution of the transmission power of a terminal based on SIRs. Experiments on ten terminals and one base station were carried out for evaluation. The experimental results verified the importance of the cooperative transmission and the distribution of power. The analysis showed that the proposed utility functions realize a collaborative transmission mechanism for the BS and heterogeneous terminals. Moreover, the study also examines the possibilities and limitations of the video transmission for OPNET environments. Simulations of the game theory and OPNET simulation show that the resource allocation strategy converges to Nash equilibrium when decisions are updated according to a better response dynamic. The proposed functions indeed saved power and enhanced video streaming, thereby proving the effectiveness of the proposed idea.

REFERENCES

Abdelhadi, A., Gerstlauer, A., & Vishwanath, S. (2014). *Real-time rate-distortion optimized streaming of wireless video.* Retrieved June 07, 2004, from http://arxiv.org/abs/1406.1915

Abdelnasser, A., & Hossain, E. (2013). Subchannel and power allocation schemes for clustered femtocells in two-tier OFDMA HetNet. In *Proceedings of IEEE International Conference on Communications Workshops* (pp. 1129-1133). New York: IEEE. doi:10.1109/ICCW.2013.6649406

Alamouti, S. M. (1998). A simple transmitter diversity scheme for wireless communications. *IEEE Journal on Selected Areas in Communications*, *1*(8), 1451–1458. doi:10.1109/49.730453

Alvi, A. N., Naqvi, S. S., Bouk, S. H., Javaid, N., Qasim, U., & Khan, Z. A. (2012). Evaluation of slotted CSMA/CA of IEEE 802.15.4. In *Proceedings of the 2012 Seventh International Conference on Broadband, Wireless Computing, Communication and Applications* (pp. 391-396). New York, NJ: IEEE. doi:10.1109/BWCCA.2012.69

Babazadeh, D., Chenine, M., Zhu, K., & Nordström, L. (2012). Real-Time smart grid application testing using OPNET SITL, In *Proceedings of OPNETWORK* (pp. 1586-1572). New York, NY: Riverbed Technology.

Cao, Q., Zhao, H. V., & Jing, Y. (2012). Power allocation and pricing in multi-user relay networks using Stackelberg and bargaining games. *IEEE Transactions on Vehicular Technology*, *61*(7), 3177–3190. doi:10.1109/TVT.2012.2204076

Cao, X.-R., Shen, H.-X., Milito, R., & Wirth, P. (2002). Internet pricing with a game theoretical approach concepts and examples. *IEEE/ACM Transactions on Networking*, *10*(2), 208–216. doi:10.1109/90.993302

Chauhan, N. S., Kaur, L., & Walleh, A. S. (2011). Evaluation of QoS for different voice schemes in wireless LAN using OPNET modular. *International Journal on Recent Trends in Engineering & Technology*, *5*(1), 110–114.

Choudhry, U. I., & Kim, J. (2005). Performance evaluation of H. 264 mapping strategies over IEEE 802.11e WLAN for robust video streaming. In *Proceedings of the 6th Pacific-Rim Conference on Advances in Multimedia Information Processing* (pp. 818-829). Berlin, Heidelberg, Germany: Springer-Verlag. doi:10.1007/11582267_71

Giuliano, R., Mazzenga, F., Petracca, M., & Pomposini, R. (2012). Performance evaluation of an opportunistic distributed power control procedure for wireless multiple access. In *Proceedings of the 2012 5th International Symposium on Communications Control and Signal Processing* (pp. 1-4). New York, NJ: IEEE. doi:10.1109/ISCCSP.2012.6217767

Hou, J. H., Wan, S., Zhan, M., & Chau, L.-P. (2013). Consistent video quality control in scalable video coding using dependent distortion quantization model. *IEEE Transactions on Broadcasting*, *59*(4), 717–724. doi:10.1109/TBC.2013.2289632

Huang, J., Zhu, H., Mung, C., & Poor, H. V. (2008). Auction-based resource allocation for cooperative communications. *IEEE Journal on Selected Areas in Communications*, *26*(7), 1226–1237. doi:10.1109/JSAC.2008.080919

Ismail, M., & Zhuang, W. (2011). A distributed resource allocation algorithm in heterogeneous wireless access medium. In *Proceedings of the IEEE International Conference on Communications* (pp. 1-8). New York, NJ: IEEE.

Khan, M. A., Khan, T. A., & Beg, M. T. (2012). RTS/CTS mechanism of MAC layer IEEE 802.11 WLAN in presence of hidden nodes. *International Journal of Engineering and Innovative Technology*, *2*(5), 235–239.

Khan, T. A., Beg, M. T., & Khan, M. A. (2013). Performance analysis of WLAN using OPNET. *International Journal of Innovative Technology and Exploring Engineering*, *2*(5), 2278–2375.

Koriiis, Y. A., Varvarigou, T. A., & Ahuja, S. R. (1998). Incentive-compatible pricing strategies in noncooperative networks. In *Proceedings of 1998 IEEE International Conference on Computer Communications* (pp. 439-446). New York: IEEE.

Laneman, J. N., & Wornell, G. W. (2003). Distributed space-time-coded protocols for exploiting cooperative diversity in wireless networks. *IEEE Transactions on Information Theory*, *49*(10), 2415–2425. doi:10.1109/TIT.2003.817829

Li, F., Tolley, D., Padhy, N. P., & Wang, J. (2009). Framework for assessing the economic efficiencies of long-run network pricing models. *IEEE Transactions on Power Systems*, *24*(4), 1641–1648. doi:10.1109/TPWRS.2009.2030283

Li, X., Amon, P., Hutter, A., & Kaup, A. (2011). Performance analysis of linter-layer prediction in scalable video coding. *IEEE Transactions on Broadcasting*, *57*(9), 66–74. doi:10.1109/TBC.2010.2082370

Li, Y., Guo, X., Li, Y., & Zheng, X. (2009). A utility-based network selection mechanism in heterogeneous wireless networks. In *Proceedings of International Conference on Wireless Network and Information Systems* (pp. 201-204). New York: IEEE. doi:10.1109/WNIS.2009.94

Li, Y., Zhang, Z., Wang, C., Zhao, W., & Chen, H.-H. (2013). Blind cooperative communications for multihop ad hoc wireless networks. *IEEE Transactions on Vehicular Technology*, *62*(7), 3110–3122. doi:10.1109/TVT.2013.2256475

Li, Z., & Shen, H. (2012). Game-theoretic analysis of cooperation incentive strategies in mobile ad hoc networks. *IEEE Transactions on Mobile Computing*, *11*(8), 1287–1303. doi:10.1109/TMC.2011.151

Liu, Y., Tao, M., & Huang, J. (2013). An auction approach to distributed power allocation for multiuser cooperative networks. *IEEE Transactions on Wireless Communications*, *12*(1), 237–247. doi:10.1109/TWC.2012.121112.112284

Luo, H. C., Wu, E. K., & Chen, G. H. (2013). A transmission power/rate control scheme in CSMA/CA-based wireless ad hoc networks. *IEEE Transactions on Vehicular Technology*, *62*(1), 427–431. doi:10.1109/TVT.2012.2219567

Manoj, P. (2011). *Effect of real video streaming over virtual MANET environment*. (Master thesis). Rochester Institute of Technology, Rochester, NY.

Ng, T., & Yu, W. (2007). Joint optimization of relay strategies and resource allocations in cooperative cellular networks. *IEEE Journal on Selected Areas in Communications*, *25*(2), 328–339. doi:10.1109/JSAC.2007.070209

Niyato, D., & Hossain, E. (2008). A game theoretic analysis of service competition and pricing in heterogeneous wireless access networks. *IEEE Transactions on Wireless Communications*, *7*(12), 5150–5155. doi:10.1109/T-WC.2008.070609

Niyato, D., & Hossain, E. (2009). Dynamic of network selection in heterogeneous wireless networks: An evolutionary game approach. *IEEE Transactions on Vehicular Technology*, *58*(4), 2008–2017. doi:10.1109/TVT.2008.2004588

Nosratinia, A., & Hunter, T. E. (2007). Grouping and partner selection in cooperative wireless networks. *IEEE Journal on Selected Areas in Communications*, *25*(2), 369–378. doi:10.1109/JSAC.2007.070212

Rozario, F., Han, Z., & Niyato, D. (2011). Optimization of non-cooperative P2P network from the game theory point of view. In *Proceedings of Wireless Communications and Networking Conference* (pp. 868-873). New York: IEEE.

Sallent, O., Perez-Romero, J., Agusti, R., Giupponi, L., Kloeck, C., Martoyo, I., & Luo, J. et al. (2006). Resource auctioning mechanisms in heterogeneous wireless access networks. In *Proceedings of the 63rd IEEE Vehicular Technology Conference* (pp. 52-56). New York: IEEE. doi:10.1109/VETECS.2006.1682774

Savazzi, S., & Spagnolini, U. (2007). Energy aware power allocation strategies for multihop-cooperative transmission schemes. *IEEE Journal on Selected Areas in Communications*, *25*(2), 318–327. doi:10.1109/JSAC.2007.070208

Schwarz, H., Marpe, D., & Wiegand, T. (2007). Overview of the scalable video coding extension of the H.264/AVC standard. *IEEE Transactions on Circuits and Systems for Video Technology*, *17*(9), 1103–1120. doi:10.1109/TCSVT.2007.905532

Scutari, G., Palomar, D. P., Facchinei, F., & Pang, J. S. (2010). Convex optimization, game theory, and variational inequality theory. *IEEE Signal Processing Magazine*, *27*(3), 35–49. doi:10.1109/MSP.2010.936021

Shah, V., Mandayam, N. B., & Goodman, D. J. (1998). Power control for wireless data based on utility and pricing. In *Proceedings of IEEE Personal, Indoor and Mobile Radio Communications Conference*. (pp. 1427-1432). New York: IEEE. doi:10.1109/PIMRC.1998.731433

Singh, G., Grover, A., & Grover, N. (2013, July). Simulation and analysis: The effect of mobility on IPTV (VOD) over Wi-Max using OPNET. In *Proceedings of the 7th Asia Modelling Symposium* (pp. 267-272). New York: IEEE. doi:10.1109/AMS.2013.47

Song, L., Niyato, D., Han, Z., & Hossain, E. (2014). Game-theoretic resource allocation methods for device-to-device communication. *IEEE Wireless Communications*, *21*(3), 136–144. doi:10.1109/MWC.2014.6845058

Song, W., & Zhuang, W. (2012). Resource allocation for integrated voice and data services. In *Interworking of wireless LANs and cellular networks* (pp. 11–27). New York, NY: Springer. doi:10.1007/978-1-4614-4379-7_2

Wu, H., Xia, W., Li, J., & Shen, L. (2014). Max-min fair resource allocation in multi-cell ultra-high throughput WLAN. In *Proceedings of the 2014 International Conference on Wireless Communications and Signal Processing* (pp. 1-7). New York: IEEE. doi:10.1109/WCSP.2014.6992093

Yaipairoj, S., & Harmantzis, F. C. (2006). Auction-based congestion pricing for wireless data services. In *Proceedings of IEEE Int. Communications* (pp. 1059-1064). New York: IEEE. doi:10.1109/ICC.2006.254887

Zhang, C. R., & Zhang, E. (2014). Optimized bidding algorithm of real time bidding in online ads auction. In *Proceedings of International Conference on Management Science and Engineering* (pp. 33-42). New York: IEEE. doi:10.1109/ICMSE.2014.6930205

Zhang, L., & Li, D. (2003). Resource allocation models based on pricing for multiservice WCDMA networks. In *Proceedings of 14th IEEE Int. Symp. Personal, Indoor and Mobile Radio Communication* (pp. 1466-1470). New York: IEEE. doi:10.1109/PIMRC.2003.1260356

KEY TERMS AND DEFINITIONS

Broadcasting: The distribution/transmission of audio or video content to terminals.
Cooperative Transmission: Joint data transmission by terminals and base stations.
Game Theory: The study of strategic decision making.
Power Allocation: The strategy of resource distribution used by a base station.
Video Transmission: The distribution/transmission of audio visual content to terminals.

Compilation of References

3GPP R2-130416. (2013). *Small cell challenges and benefits of dual connectivity*. Ericsson.

3GPP R2-131678. (2013). *Further discussions on UL/DL split*. Ericsson.

3GPP R2-131902. (2013). *On the potential UL/DL imbalance problem*. Nokia Corporation.

3GPP R2-131958. (2013). *Intra-frequency small cell enhancements*. Alcatel Lucent.

3GPP R2-132042. (2013). *Study of Small Cell Enhancement Gains for Dual Connectivity Scenario*. NEC Corporation.

3GPP Technical Specification TS 36.814v9.0.0. (2012). *Evolved universal terrestrial radio access (E-UTRA) and evolved universal terrestrial radio access network (E-UTRAN); Further advancements for E-UTRA physical layer aspects*. Retrieved from http://www.3gpp.org/ftp/Specs/html-info/36814.htm

3GPP TR 36.814. (2013). *E-UTRA; Further Advancements for E-UTRA Physical Layer Aspects, v. 11.1.0*. Academic Press.

3GPP TR 36.839 V 11.1.0. (2012). *Mobility Enhancements in Heterogeneous Networks*. Academic Press.

3GPP TR 36.842. (2013). *Small Cell enhancements for E-UTRA and E-UTRAN-Higher layer aspects, v12.0.0*. Academic Press.

3GPP TR 36.842. (2013). *Study on Small Cell Enhancements for E-UTRA and E-UTRAN - Higher layer aspects, v0.4.1*. Academic Press.

3GPP TR 36.872. (2013). *Small cell enhancements for E-UTRA and E-UTRAN-Physical layer aspects, v12.1.0*. Academic Press.

3GPP TR 36.902. (2011). *Self-configuring and self-optimizing network (SON) use cases and solutions. v9.3.1*. Academic Press.

3GPP TR 36.932. (2013). *Scenarios and requirements for small cell enhancements for E-UTRA and E-UTRAN, v12.1.0*. Academic Press.

3GPP TR36.872. (2013). *Small cell enhancements for E-UTRA and E-UTRAN Physical layer aspects (Release 12)*. Academic Press.

3GPP TS 32.500. (2011). *Self-organizing networks (SON); concepts and requirements. v11.1.0*. Academic Press.

3GPP TS 36.101. (2014). *Evolved Universal Terrestrial Radio Access (E-UTRA); User Equipment (UE) radio transmission and reception, v12.4.0*. Academic Press.

3GPP TS 36.104. (2014). *Evolved Universal Terrestrial Radio Access (E-UTRA); Base Station (BS) radio transmission and reception, v12.4.0*. Academic Press.

3GPP TS 36.133. (2014). *Evolved Universal Terrestrial Radio Access (E-UTRA); Requirements for support of radio resource management, v12.4.0.* Academic Press.

3GPP TS 36.213. (2014). *Evolved Universal Terrestrial Radio Access (E-UTRA); Physical layer procedures, v12.2.0.* Academic Press.

3GPP. (2001). *3GPP TS 23.236, 3rd Generation Partnership Project; Shared Networks: An operator alternative to reduce initial investments, increase coverage and reduce Time To Market for WCDMA by sharing infrastructure.* Author.

3GPP. (2002). *3GPP TS 23.236. 3rd Generation Partnership Project; Technical Specification Group Services and System Aspects; Intra-domain connection of Radio Access Network (RAN) nodes to multiple Core Network (CN) nodes. Release 5. Technical Report.* Author.

3GPP. (2010). *3GPP TR 32.826 V10.0.0. 3rd Generation Partnership Project; Technical Specification Group Services and System Aspects; Telecommunication management; Study on Energy Savings Management (ESM); Release 10, Technical Report.* Author.

3rd Generation Partnership Project. (2012). *TR 22.803 V0. 3.0, 3rd Generation Partnership Project; Technical Specification Group SA; Feasibility Study for Proximity Services (ProSe) (Release 12), May, 2012.* Academic Press.

Abdelhadi, A., Gerstlauer, A., & Vishwanath, S. (2014). *Real-time rate-distortion optimized streaming of wireless video.* Retrieved June 07, 2004, from http://arxiv.org/abs/1406.1915

Abdelkader, M., Hamdi, M., & Boudriga, N. (2011). Game theory and iteratedvoronoi diagrams for self-organized wireless sensor networks. In *International Wireless Communications and Mobile Computing Conference (IWCMC)* (pp. 1250–1255). IEEE.

Abdelnasser, A., & Hossain, E. (2013). Subchannel and power allocation schemes for clustered femtocells in two-tier OFDMA HetNet. In *Proceedings of IEEE International Conference on Communications Workshops* (pp. 1129-1133). New York: IEEE. doi:10.1109/ICCW.2013.6649406

Ahmadi, H., Macaluso, I., & Dasilva, L. A. (2013). Carrier aggregation as a repeated game: Learning algorithms for efficient convergence to a Nash equilibrium. In *Proceedings of the 2013 IEEE Global Communications Conference (GLOBECOM)*, Atlanta, GA: IEEE. doi:10.1109/GLOCOM.2013.6831243

Ahmed, M., Peng, M., Ahmad, I., Li, Y., & Abana, M. (2013). Stackelberg game based optimized power allocation scheme for two-tier femtocell network. *Wireless Communications and Signal Processing (WCSP), 2013 International Conference on* (Vol.411, pp.1-6). IEEE.

Ahmed, M., Peng, M., Ahmad, I., Li, Y., & Abana, M. (2013, October). Stackelberg game based optimized power allocation scheme for two-tier femtocell network. In *Proceedings of Wireless Communications & Signal Processing (WCSP)* (pp. 1-6). IEEE. doi:10.1109/WCSP.2013.6677160

Ahmed, M., Wang, W., Zhang, B., Sun, C., Ahmad, I., Rehman, W. U., & Yan, S. et al. (2015). A dynamic coalition formation framework for interference management in dense small cell networks. *Transactions on Emerging Telecommunications Technologies*, *26*(4), 718–731. doi:10.1002/ett.2901

Ai, X., Srinivasan, V., & Tham, C. (2008). Optimality and complexity of pure Nash equilibria in the coverage game. *IEEE Journal on Selected Areas in Communications*, *26*(7), 117–1182. doi:10.1109/JSAC.2008.080914

Akbari Torkestani, J., & Meybodi, M. R. (2010a). An intelligent backbone formation algorithm for wireless ad hoc networks based on distributed learning automata. *Computer Networks*, *54*(5), 826–843. doi:10.1016/j.comnet.2009.10.007

Akbari Torkestani, J., & Meybodi, M. R. (2010b). Mobility-based multicast routing algorithm for wireless mobile Ad-hoc networks: A learning automata approach. *Computer Communications*, *33*(6), 721–735. doi:10.1016/j.comcom.2009.11.019

Akyildiz, I., Lee, W., Vuran, M., & Mohanty, S. (2006). Next generation/dynamic spectrum access/cognitive radio wireless networks: A survey. *Computer Networks*, *50*(13), 2127–2159. doi:10.1016/j.comnet.2006.05.001

Alamouti, S. M. (1998). A simple transmitter diversity scheme for wireless communications. *IEEE Journal on Selected Areas in Communications*, *1*(8), 1451–1458. doi:10.1109/49.730453

Ali, S. H., & Leung, V. C. M. (2009, August). Dynamic frequency allocation in fractional frequency reused OFDMA networks. *IEEE Transactions on Wireless Communications*, *8*(8), 4286–4295. doi:10.1109/TWC.2009.081146

Al-Rubaye, S., Al-Dulaimi, A., & Cosmas, J. (2011). Cognitive femtocell. *IEEE Vehicular Technology Magazine*, *6*(1), 44–51. doi:10.1109/MVT.2010.939902

Al-Tous, H., & Barhumi, I. (2014). (in press). Resource Allocation for Multiple-Users AF-OFDMA Systems Using the Auction Framework. *DOI.*

Alvi, A. N., Naqvi, S. S., Bouk, S. H., Javaid, N., Qasim, U., & Khan, Z. A. (2012). Evaluation of slotted CSMA/CA of IEEE 802.15.4. In *Proceedings of the 2012 Seventh International Conference on Broadband, Wireless Computing, Communication and Applications* (pp. 391-396). New York, NJ: IEEE. doi:10.1109/BWCCA.2012.69

Andrews, J. G. (2013). Seven ways that HetNets are a cellular paradigm shift. *Communications Magazine, IEEE*, *51*(3), 136–144. doi:10.1109/MCOM.2013.6476878

Apt, K., & Radzik, T. (2006). *Stable partitions in coalitional games.* Retrieved from arXiv:cs/0605132v1 [cs.GT].

Apt, K., & Witzel, A. (2007). *A generic approach to coalition formation.* Retrieved from arXiv preprint arXiv:0709.0435.

Apt, K., & Witzel, A. (2009). A generic approach to coalition formation (extended version). *International Journal of Game Theory*, *11*(3), 347–367. doi:10.1142/S0219198909002352

Aram, A., Singh, C., & Sarkar, S. (2009). *Cooperative profit sharing in coalition based resource allocation in wireless networks.* IEEE INFOCOM. doi:10.1109/INFCOM.2009.5062136

Araniti, G., Condoluci, M., Iera, A., Militano, L., Pansera, B. A. (2013). Analytical modeling of bargaining solutions for multicast cellular services. *Electronic Journal of Differential Equations, 2013*(165), 1-10.

Arnold, T., & Schwalbe, U. (2002). Dynamic coalition formation and the core. *Journal of Economic Behavior & Organization*, *49*(3), 363–380. doi:10.1016/S0167-2681(02)00015-X

Ashraf, I., Claussen, H., & Ho, L. T. W. (2010). Distributed radio coverage optimization in enterprise femtocell networks. In *Proceeding of 2010 IEEE International Conference in Communication (ICC).* Cape Town: IEEE. doi:10.1109/ICC.2010.5502072

Assad, M. (Sep.2008). Optimal fractional frequency reuse (FFR) in multicellular OFDMA system. In *Proc. IEEE Vehicular Technology Conf.*, 1-5. doi:10.1109/VETECF.2008.381

Aumann, R. J. (1987). Correlated equilibrium as an expression of Bayesian rationality. *Econometrica*, *55*(1), 1–18. doi:10.2307/1911154

Aumann, R., & Drèze, J. (1974). Cooperative games with coalition structures. *International Journal of Game Theory*, *3*(4), 217–237. doi:10.1007/BF01766876

Ausubel, L. M. (2004). An efficient ascending-bid auction for multiple objects. *The American Economic Review, 94*(5), 1452–1475. doi:10.1257/0002828043052330

Babazadeh, D., Chenine, M., Zhu, K., & Nordström, L. (2012). Real-Time smart grid application testing using OPNET SITL, In *Proceedings of OPNETWORK* (pp. 1586-1572). New York, NY: Riverbed Technology.

Bacci, G., Bulzomato, A., & Luise, M. (2011). Uplink power control and subcarrier assignment for an OFDMA multicellular network based on game theory.*International Conference Perfect Evaluation methodology Tools (Value-Tools).* doi:10.4108/icst.valuetools.2011.245691

Bahl, P., Chandra, R., Moscibroda, T., Murty, R., & Welsh, M. (2009). *White space networking with wi-fi like connectivity* (pp. 27–38). SIGCOMM. doi:10.1145/1592568.1592573

Balasubramaniam, S., Botvich, D., Donnelly, W., Foghlu, M., & Strassner, J. (2007). *Bio-inspired framework for autonomic communication systems* (pp. 3–19). Advances in Biologically Inspired Information Systems.

Ba, T., & Srikant, R. (2002). A Stackelberg network game with a large number of followers. *Journal of Optimization Theory and Applications, 115*(3), 479–490. doi:10.1023/A:1021294828483

Beckman, C., & Smith, G. (2005). Shared networks: Making wireless communication affordable. *IEEE Wireless Communications, 12*(2), 78–85. doi:10.1109/MWC.2005.1421931

Belmega, E. V., Lasaulce, S., Debbah, M., & Hjorungnes, A. (2010). Learning distributed power allocation policies in MIMO channels. In *Proceedings of the European Signal Processing Conference (EUSIPCO-2010).* Aalborg, Denmark: EURASIP.

Bennis, M., Perlaza, S. M., & Debbah, M. (2012). Learning coarse correlated equilibrium in two-tier wireless networks. *InProceedings of the IEEE International Conference on Communications* (pp. 1592-1596). Ottawa, Canada: IEEE.

Bhatia, R., Li, L. E., Luo, H., & Ramjee, R. (2006). ICAM: Integrated Cellular and Ad Hoc Multicast. *IEEE Transactions on Mobile Computing, 5*(8), 1004–1015. doi:10.1109/TMC.2006.116

Bkassiny, M., Li, Y., & Jayaweera, S. K. (2013). A survey on machine-learning techniques in cognitive radios. *IEEE Communications Surveys and Tutorials, 15*(3), 1136–1159. doi:10.1109/SURV.2012.100412.00017

Bogomonlaia, A., & Jackson, M. (2002). The stability of hedonic coalition structures. *Games and Economic Behavior, 38*(2), 201–230. doi:10.1006/game.2001.0877

Bousia, A., Kartsakli, E., Alonso, L., & Verikoukis, C. (2012b). Energy efficient base station maximization switch off scheme for LTE-Advanced. *Proceedings from CAMAD'12: The IEEE 17th International Workshop on Computer Aided Modeling and Design of Communication Links and Networks.* Barcelona, Spain.

Bousia, A., Katsakli, E., Antonopoulos, A., Alonso, L., & Verikoukis, C. (2013). Game Theoretic Approach for Switching Off Base Stations in Multi-Operator Environments, *Proceedings from ICC'13:The IEEE International Conference on Communications.* Budapest, Hungary. doi:10.1109/ICC.2013.6655262

Bousia, A., Antonopoulos, A., Alonso, L., & Verikoukis, C. (2012). "Green" distance-aware base station sleeping algorithm in LTE-Advanced. *Proceedings from ICC'12:The IEEE International Conference on Communications.* Ottawa, ON, Canada.

Bousia, A., Antonopoulos, A., Alonso, L., & Verikoukis, C. (2012). Green Distance-Aware Base Station Sleeping Algorithm in LTE-Advanced. In *Proceedings of the IEEE International Conference on Communications (ICC 2012)* (pp. 1347 – 1351). Ottawa, Canada: IEEE. doi:10.1109/ICC.2012.6364240

Bousia, A., Kartsakli, E., Alonso, L., & Verikoukis, C. (2012a). Dynamic energy efficient distance-aware base station switch on/off scheme for LTE-Advanced. *Proceedings from GLOBECOM'12:The IEEE Global Communications Conference*. Anaheim, CA, USA.

Boyd, S., & Vandenberghe, L. (2004). *Convex optimization*. Cambridge: Cambridge university press. doi:10.1017/CBO9780511804441

Branzei, R., Dimitrov, D., & Tijs, S. (2005). *Models in Cooperative Game Theory: Crisp, Fuzzy and Multichoice Games*. New York, NY: Springer.

Brunner, M., Eggert, L., Felis, S., Schmid, S., & Zimmermann, K. (2011). Autonomicwireless network management. IFIP Lecture Notes in Computer Science (LNCS), 3854(3854):57–70.

Budapest, Z. H. M., Xue, R., Yu, H., Luo, H., & Chen, W. (2013, June). Secrecy capacity optimization in coordinated multipoint processing. In *Proceedings of International Conference on Communications (ICC '13)*. Budapest, Hungary: IEEE.

Buzzi, S., Colavolpe, G., Saturnino, D., & Zappone, A. (2012). Potential games for energy-efficient power control and subcarrier allocation in uplink multicell OFDMA systems. *IEEE Journal of Selected Topics in Signal Processing*, *6*(2), 89–103.

Cai, J., & Pooch, U. (2004). Allocate fair payoff for cooperation in wireless ad hoc networks using shapley value.*International Parallel and Distributed Processing Symposium*.

Campolo, C., Iera, A., Militano, L., & Molinaro, A. (2012). Scenario-adaptive and gain-aware content sharing policies for cooperative wireless environments. *Computer Communications*, *35*(10), 1259–1271. doi:10.1016/j.comcom.2012.03.017

Cao, Q., Zhao, H. V., & Jing, Y. (2012). Power allocation and pricing in multi-user relay networks using Stackelberg and bargaining games. *IEEE Transactions on Vehicular Technology*, *61*(7), 3177–3190. doi:10.1109/TVT.2012.2204076

Cao, X.-R., Shen, H.-X., Milito, R., & Wirth, P. (2002). Internet pricing with a game theoretical approach concepts and examples. *IEEE/ACM Transactions on Networking*, *10*(2), 208–216. doi:10.1109/90.993302

Cao, Y., & Wei, J. (2012, October). Distributed coalition formation for selfish relays and eavesdroppers in wireless networks: A job-hopping game. In *Proceedings of Wireless Communications & Signal Processing (WCSP '12)*. Huangshan, China: IEEE. doi:10.1109/WCSP.2012.6542873

Carmichael, F. (2005). *A Guide to Game Theory*. Harlow, UK: Prentice Hall.

Cavalcanti, D., Agrawal, D., Cordeiro, C., Xie, B., & Kumar, A. (2005). Issues in Integrating Cellular Networks, WLANs, and MANETs: A Futuristic Heterogeneous Wireless Network. *IEEE Wireless Communications Magazine*, *12*(3), 30–41. doi:10.1109/MWC.2005.1452852

Cesa-Bianchi, N., & Lugosi, G. (2006). *Prediction, learning, and games*. Cambridge, UK: Cambridge University Press. doi:10.1017/CBO9780511546921

Chai, B., Deng, R., Cheng, P., & Chen, J. (2012, December). Energy-efficient power allocation in cognitive sensor networks: A game theoretic approach. In *Proceedings of Global Communications Conference (GLOBECOM)* (pp. 416-421). IEEE.

Chandrasekhar, V., Andrews, J. G., & Gatherer, A. (2008). Femtocell networks: A survey. *IEEE Communications Magazine*, *46*(9), 59–67. doi:10.1109/MCOM.2008.4623708

Chandrasekhar, V., Andrews, J. G., Muharemovic, T., Shen, Z., & Gatherer, A. (2009). Power control in two-tier femtocell networks. *IEEE Transactions on Wireless Communications*, *8*(7), 4316–4328. doi:10.1109/TWC.2009.081386

Chandrasekhar, V., Andrews, J. G., Muharemovict, T., Shen, Z., & Gatherer, A. (2009). Power control in two-tier femtocell networks. *IEEE Transactions on Wireless Communications*, *8*(8), 4316–4328.

Chang, R. Y., Tao, Z., Zhang, J., & Kuo, C. (Jun.2009). A graph approach to dynamic fractional frequency reuse (FFR) in multi-cell OFDMA networks. In *Proc. IEEE Intl. Conf. on Communications*, Dresden, Germany, 1-6. doi:10.1109/ICC.2009.5198612

Charalambous, C. D., Djouadi, S. M., & Denic, S. Z. (2005). Stochastic power control for wireless networks via SDEs: Probabilistic QoS measures. *IEEE Transactions on Information Theory*, *51*(12), 4396–4401.

Chauhan, N. S., Kaur, L., & Walleh, A. S. (2011). Evaluation of QoS for different voice schemes in wireless LAN using OPNET modular. *International Journal on Recent Trends in Engineering & Technology*, *5*(1), 110–114.

Chaves, F. D. S., de Oliveira Neto, R. A., & Cavalcanti, F. R. P. (2006, September). Energy efficiency improved game-theoretic distributed power control algorithm for best effort services in wireless communications systems. In *Proceedings of Telecommunications Symposium, 2006 International* (pp. 874-879). IEEE. doi:10.1109/ITS.2006.4433394

Chen, T., & Zhang, H. Zhao, & Chen, X. (2010). Towards green wireless access networks. *Proceeding from CHINA-COM'10: The 5th International ICST Conference on Communications and Networking in China*. Beijing, China.

Chen, X., & Huang, J. (2012a). Game theoretic analysis of distributed spectrum sharing with database. *In the 32nd International Conference on Distributed Computing Systems (ICDCS).*

Chen, X., Zhang, H., Chen, T., & Lasanen, M. (2013, June). Improving energy efficiency in green femtocell networks: a hierarchical reinforcement learning framework. In *Proceedings of Communications (ICC)* (pp. 2241-2245). IEEE. doi:10.1109/ICC.2013.6654861

Chen, Y., Zhang, J., Zhang, Q., & Jia, J. (2012). A reverse auction framework for access permission transaction to promote hybrid access in femtocell network. In *Proceedings ofIEEE International Conference on Computer Communications (INFOCOM)*. IEEE.

Chen, J., Yu, Q., Chai, B., Sun, Y., Fan, Y., & Shen, X. S. (2015). Dynamic channel assignment for wireless sensor networks: A regret matching based approach. *IEEE Transactions on Parallel and Distributed Systems*, *26*(1), 95–106. doi:10.1109/TPDS.2014.2307868

Chen, M., Liew, S., Shao, Z., & Kai, C. (2010). Markov approximation for combinatorial network optimization. In *Proceedings of IEEE INFOCOM* (pp. 1–9). IEEE.

Chen, X., & Huang, J. (2012b). Spatial spectrum access game: Nash equilibria and distributed learning.*InACM International Symposium on Mobile Ad Hoc Networking and Computing (MobiHoc)*. doi:10.1145/2248371.2248401

Chen, X., & Huang, J. (2013a). Database-assisted distributed spectrum sharing. *IEEE Journal on Selected Areas in Communications*, *31*(11), 2349–2361. doi:10.1109/JSAC.2013.131110

Chen, X., & Huang, J. (2013b). Distributed spectrum access with spatial reuse. *IEEE Journal on Selected Areas in Communications*, *31*(3), 593–603. doi:10.1109/JSAC.2013.130323

Chen, X., & Huang, J. (2013c). Evolutionarily stable spectrum access. *IEEE Transactions on Mobile Computing*, *12*(7), 1281–1293. doi:10.1109/TMC.2012.94

Chen, X., & Huang, J. (2014a). Imitation-based social spectrum sharing. *IEEE Transactions on Mobile Computing*.

Chen, X., & Huang, J. (2014b). Spatial spectrum access game. *IEEE Transactions on Mobile Computing*.

Chen, Y., Xu, X., & Lei, Q. (2013). Joint Subcarriers and Power Allocation with Imperfect Spectrum Sensing for Cognitive D2D Wireless Multicast. *Transactions on Internet and Information Systems (Seoul)*, *7*(7), 1533–1546. doi:10.3837/tiis.2013.07.001

Chen, Y., Zhang, J., & Zhang, Q. (2012). Utility-Aware Refunding Framework for Hybrid Access Femtocell Network. *IEEE Transactions on Wireless Communications*, *11*(5), 1688–1697. doi:10.1109/TWC.2012.031212.110002

Choi, D., Monajemi, P., Shinjae, K., & Villasenor, J. (2008). Dealing with Loud Neighbors: The Benefits and Tradeoffs of Adaptive Femtocell Access. In *Proceedings ofIEEE Global Communications Conference (GLOBECOM)*. IEEE. doi:10.1109/GLOCOM.2008.ECP.540

Choudhry, U. I., & Kim, J. (2005). Performance evaluation of H. 264 mapping strategies over IEEE 802.11e WLAN for robust video streaming. In *Proceedings of the 6th Pacific-Rim Conference on Advances in Multimedia Information Processing* (pp. 818-829). Berlin, Heidelberg, Germany: Springer-Verlag. doi:10.1007/11582267_71

Cisco (2012, February 14). *Cisco Visual Networking Index: Global Mobile Data, Traffic Forecast Update, 2011-2016.* Retrieved from: http://www.cisco.com/en/US/solutions/collateral/ns341/ns525/ns537/ns705/ns827/white_paper_c11-481360_ns827_Networking_Solutions_White_Paper.html

Clancy, C., Hecker, J., Stuntebeck, E., & O'Shea, T. (2007). Applications of machine learning to cognitive radio networks. *IEEE Wireless Communications*, *14*(4), 47–52. doi:10.1109/MWC.2007.4300983

Claussen, H. (2007). Performance of macro-and co-channel femtocells in a hierarchical cell structure. In *Proceeding of IEEE 18th Annual International Symposium on Personal, Indoor and Mobile Radio Communications (PIMRC2007)*. Athens: IEEE. doi:10.1109/PIMRC.2007.4394515

Claussen, H., Ashraf, I., & Ho, L. T. (2010). Dynamic idle mode procedures for femtocells. *Bell Labs Technical Journal*, *15*(2), 95–116. doi:10.1002/bltj.20443

Claussen, H., Ho, L. T. W., & Samuel, L. G. (2008). Self-optimization of coverage for femtocell deployments. In *Proceeding of Wireless Telecommunications Symposium 2008 (WTS2008)*, Pomona, CA: IEEE doi:10.1109/WTS.2008.4547576

Correia, L., Zeller, D., Blume, O., Ferling, D., Jading, Y., Godor, I., & Van der Perre, L. et al. (2012). Challenges and enabling technologies for energy aware mobile radio networks. *IEEE Communications Magazine*, *48*(11), 66–72. doi:10.1109/MCOM.2010.5621969

Cui, S., Goldsmith, A. J., & Bahai, A. (2004). Energy-efficiency of MIMO and cooperative MIMO techniques in sensor networks. *IEEE Journal on Selected Areas in Communications*, *22*(6), 1089–1098. doi:10.1109/JSAC.2004.830916

Cui, T., Chen, L., & Low, S. H. (2008). A game-theoretic framework for medium access control. *IEEE Journal on Selected Areas in Communications*, *26*(7), 1116–1127. doi:10.1109/JSAC.2008.080909

Damnjanovic, A., Montojo, J., Wei, Y., Ji, T., Luo, T., Vajapeyam, M., & Malladi, D. et al. (2011). A survey on 3GPP heterogeneous networks. *IEEE Wireless Communications*, *18*(3), 10–21. doi:10.1109/MWC.2011.5876496

Debreu, G. (1952). A social equilibrium existence theorem. National Academy of Sciences, 38, 886–893.

Deb, S., Srinivasan, V., & Maheshwari, R. (2009). *Dynamic spectrum accessing DTV whitespaces: design rules, architecture and algorithms* (pp. 1–12). MOBICOM.

Dong, L. (2007). Supermodel game Nash equilibrium stability. Wuhan University of Science and Technology, 20(1), 55-57.

Dong, L., Han, Z., Petropulu, A. P., & Poor, H. V. (2008, September). Secure wireless communications via cooperation. In *Proceedings of* 46th *Annual Allerton Conference on Communication, Control, and Computing*. Urbana-Champaign, IL: IEEE. doi:10.1109/ALLERTON.2008.4797687

Dong, L., Han, Z., Petropulu, A. P., & Poor, H. V. (2010). Improving wireless physical layer security via cooperating Relays. *IEEE Transactions on Signal Processing*, *58*(3), 1875–1888. doi:10.1109/TSP.2009.2038412

Doppler, K., Rinne, M., Wijting, C., Ribeiro, C., & Hugl, K. (2009). Device-to-device communication as an underlay to LTE-advanced networks. *IEEE Communications Magazine*, *47*(12), 42–49. doi:10.1109/MCOM.2009.5350367

Dou yinan, the research of the classification of network users according to the multi-dimensional characteristics. (2010). Beijing: Beijing University of Posts and Telecommunications.

Douros, V. G., Toumpis, S., & Polyzos, G. C. (2012). Power control under best response dynamics for interference mitigation in a two-tier femtocell network. In *Proceeding of 2012 10th International Symposium on Modeling and Optimization in Mobile, Ad Hoc and Wireless Networks (WiOpt)*. Paderborn: IEEE.

Dreze, J., & Greenberg, J. (1980). Hedonic coalitions: Optimality and stability. *Econoemetrica*, *48*(4), 987–1003. doi:10.2307/1912943

Duan, L., Gao, L., & Huang, J. (2014). Cooperative spectrum sharing: A contract-based approach. *IEEE Transactions on Mobile Computing*, *13*(1), 174–187. doi:10.1109/TMC.2012.231

Elayoubi, S. E., Ben Haddada, O., & Fourestie, B. (2008, May). Performance evaluation of frequency planning schemes in OFDM-based networks. *IEEE Transactions on Wireless Communications*, *7*(5), 1623–1633. doi:10.1109/TWC.2008.060458.

ElSawy, H., Hossain, E., & Haenggi, M. (2013). Stochastic geometry for modeling, analysis, and design of multi-tier and cognitive cellular wireless networks: A survey. *IEEE Communications Surveys and Tutorials*, *13*(3), 996–1019. doi:10.1109/SURV.2013.052213.00000

Eun-Kyu, L., Gerla, M., & Oh, S. Y. (2012). Physical layer security in wireless smart grid. *IEEE Communications Magazine*, *50*(8), 46–52. doi:10.1109/MCOM.2012.6257526

Fang, L., & Zhang, X. (Oct.2008). Optimal fractional frequency reuse in OFDMA based wireless networks. In *Proc. Intl. Conf. on Wireless Communications, Networking and Mobile Computing*, 1-4. doi:10.1109/WiCom.2008.166

Fattahi, A., & Paganini, F. (2005). New economic perspectives for resource allocation in wireless networks.*IEEE American Control Conference*. doi:10.1109/ACC.2005.1470595

FCC. (2010, September 23). *Second memorandum opinion and order*. FCC.

Feng, X., Zhang, J., & Zhang, Q. (2011). Database-assisted multi-AP network on TV white spaces: Architecture, spectrum allocation and AP discovery. In *IEEE Symposia on New Frontiers in Dynamic Spectrum Access Netoworks*. DySpan. doi:10.1109/DYSPAN.2011.5936215

Feng, X., Zhang, Q., & Zhang, J. (2014). A hybrid pricing framework for TV white space database. *IEEE Transactions on Wireless Communications*, *13*(5).

Fettweis, G., & Zimmermann, E. (2008). ICT Energy Consumption - Trends and Challenges, *Proceedings from WPMC'08: The 11th International Symposium on Wireless Personal Multimedia Communications*. Lapland, Filand.

Friedman, J. W., & Mezzetti, C. (2001). Learning in games by random sampling. *Journal of Economic Theory*, *98*(1), 55–84. doi:10.1006/jeth.2000.2694

Frisanco, T., Tafertshofer, P., Laurin, P., & Ang, R. (2008). Infrastructure Sharing and Shared Operations for Mobile Network Operators-From a Deployment and Operations View, *Proceeding from NOMS'08:The IEEE Network Operations and Management Symposium*. Salvadro, Bahia.

Fu, W., Yao, R., Gao, F., Li, J. C., & Lei, M. (2013). *Robust null-space based interference avoiding scheme for D2D communication underlaying cellular networks.* Paper presented at IEEE Wireless Communications and Networking Conference (WCNC), Shanghai, China. doi:10.1109/WCNC.2013.6555244

Fudenberg, D., & Tirole, J. (1994). *Game Theory.* Cambridge, MA: MIT Press.

Fu, S., Wu, B., Wen, H., Ho, P.-H., & Feng, G. (2014). Transmission Scheduling and Game Theoretical Power Allocation for Interference Coordination in CoMP. *IEEE Transactions on Wireless Communications, 13*(1), 112–123. doi:10.1109/TWC.2013.120113.121646

Gao, L., Huang, J., Chen, Y.-J., & Shou, B. (2013). An integrated contract and auction design for secondary spectrum trading. *IEEE Journal on Selected Areas in Communications, 31*(3), 581–592. doi:10.1109/JSAC.2013.130322

Garcia, V., Zhou, Y. Q., & Shi, J. L. (2014). Coordinated Multipoint Transmission in Dense Cellular Networks with User-Centric Adaptive Clustering. *IEEE Transactions on Wireless Communications, 13*(8), 4297–4308. doi:10.1109/TWC.2014.2316500

Gavrilovska, L., Atanasovski, V., Macaluso, I., & DaSilva, L. (2013). Learning and reasoning in cognitive radio networks. *IEEE Communications Surveys and Tutorials, 15*(4), 1761–1777. doi:10.1109/SURV.2013.030713.00113

Geraci, G., Singh, S., & Andrews, J. G. (2014). Secrecy Rates in Broadcast Channels with Confidential Messages and External Eavesdroppers. *IEEE Transactions on Wireless Communications, 13*(5), 2931–2943. doi:10.1109/TWC.2014.041014.131101

Gharehshiran, O. N., & Krishnamurthy, V. (2010). Coalition formation for bearings-only localization in sensor networks—a cooperative game approach. *IEEE Transactions on Signal Processing, 58*(8), 4322–4338. doi:10.1109/TSP.2010.2049201

Ghimire, J., & Rosenberg, C. (2013). Resource allocation, transmission coordination and user association in heterogeneous networks: A flow-based unified approach. *IEEE Transactions on Wireless Communications, 12*(3), 1340–1351. doi:10.1109/TWC.2013.013013.120940

Ghosh, A., Mangalvedhe, N., Ratasuk, R., Mondal, B., Cudak, M., Visotsky, E., & Novlan, T. D. et al. (2012). Heterogeneous cellular networks: From theory to practice. *IEEE Communications Magazine, 50*(6), 54–64. doi:10.1109/MCOM.2012.6211486

Giuliano, R., Mazzenga, F., Petracca, M., & Pomposini, R. (2012). Performance evaluation of an opportunistic distributed power control procedure for wireless multiple access. In *Proceedings of the 2012 5th International Symposium on Communications Control and Signal Processing* (pp. 1-4). New York, NJ: IEEE. doi:10.1109/ISCCSP.2012.6217767

Glicksberg, L. (1952). A further generalization of the Kakutani fixed point theorem with application to Nash equilibrium points. *American Mathematical Society, 3*(1), 170-174.

Goel, S., & Negi, R. (2008). Guaranteeing secrecy using artificial noise. *IEEE Transactions on Wireless Communications, 7*(6), 2180–2189. doi:10.1109/TWC.2008.060848

Goemans, M., Li, L., Mirrokni, V., & Thottan, M. (2006). Market sharing games applied to content distribution in ad hoc networks. *IEEE Journal on Selected Areas in Communications, 24*(5), 1020–1033. doi:10.1109/JSAC.2006.872884

Goldsmith, A. (2005). *Wireless Communications.* New York, NY: Cambridge University Press. doi:10.1017/CBO9780511841224

Grandhi, S. A., Vijayan, R., Goodman, D. J., & Zander, J. (1993). Centralized power control in cellular radio systems. *IEEE Transactions on Vehicular Technology, 42*(4), 466–468. doi:10.1109/25.260766

Gunduz, D., Yener, A., Goldsmith, A., & Poor, V. (2013). The multi-way relay channel. *IEEE Transactions on Information Theory*, *59*(1), 51–63. doi:10.1109/TIT.2012.2219156

Gunes, M., Sorges, U., & Bouazizi, I. (2002). Ara-the ant-colony based routing algorithm for MANENTs. In *International Conference on Parallel Processing Workshops*, pages 79–85. IEEE. doi:10.1109/ICPPW.2002.1039715

Gupta, P., & Kumar, P. R. (2000). The capacity of wireless networks. *IEEE Transactions on Information Theory*, *46*(2), 388–404. doi:10.1109/18.825799

Guruacharya, S., Niyato, D., Bennis, M., & Kim, D. I. (2013). Dynamic coalition formation for network MIMO in small cell networks. *IEEE Transactions on Wireless Communications*, *12*(10), 5360–5372. doi:10.1109/TWC.2013.090513.130516

Guruacharya, S., Niyato, D., Kim, D. I., & Hossain, E. (2013). Hierarchical competition for downlink power allocation in ofdma femtocell networks. *IEEE Transactions on Wireless Communications*, *12*(4), 1543–1553. doi:10.1109/TWC.2013.022213.120016

Han, Z., Pandana, C., & Liu, K. J. R. (2007). Distributive opportunistic spectrum access for cognitive radio using correlated equilibrium and no-regret learning. In *Proceedings of the 2007 IEEE Wireless Communications and Networking Conference (WCNC)*. Hong Kong, China: IEEE. doi:10.1109/WCNC.2007.8

Han, K., Choi, S., Kang, D. H., & Choi, S. (2012). Game based self-organizing scheme for femto-cell networks. *Lecture Notes of the Institute for Computer Sciences. Social Informatics and Telecommunications Engineering*, *75*, 57–75.

Han, M.-H., Kim, B.-G., & Lee, J.-W. (2012). Subchannel and Transmission Mode Scheduling for D2D Communication in OFDMA Networks. In *Proceedings of the IEEE Vehicular Technology Conference (VTC Fall 2012)*. Quebec City, Canada: IEEE. doi:10.1109/VTCFall.2012.6399205

Hansly, S., & Mathar, R. (2002). On the Optimal Base-Station Density for CDMA Cellular Networks. *IEEE Transactions on Communications*, *50*(8), 1274–1281. doi:10.1109/TCOMM.2002.801523

Han, T., & Ansari, N. (2013). On Greening Cellular Networks via Multicell Cooperation. *IEEE Wireless Communications*, *20*(1), 82–89. doi:10.1109/MWC.2013.6472203

Han, Z., Ji, Z., & Liu, K. J. R. (2007, August). Non-Cooperative Resource Competition Game by Virtual Referee in Multi-Cell OFDMA Networks. *IEEE Journal on Selected Areas in Communications*, *25*(6), 1079–1090. doi:10.1109/JSAC.2007.070803

Han, Z., Niyato, D., & Saad, W. (2012). *Game theory in wireless and communication networks*. Cambridge, United Kingdom: Cambridge University Press.

Han, Z., Niyato, D., Saad, W., Basar, T., & Hjørungnes, A. (2011). *Game Theory in Wireless and Communication Networks: Theory, Models and Applications*. Cambridge, UK: Cambridge University Press. doi:10.1017/CBO9780511895043

Han, Z., & Poor, V. (2009). Coalition games with cooperative transmission: A cure for the curse of boundary nodes in selfish packet-forwarding wireless networks. *IEEE Transactions on Communications*, *57*(1), 203–213. doi:10.1109/TCOMM.2009.0901.060661

Hanzo, L., Haas, H., Imre, S., O'Brien, D., Rupp, M., & Gyongyosi, L. (2012). Wireless myths, realities, and futures: from 3G/4G to optical and quantum wireless. *Proceedings of the IEEE, 100*(Special Centennial Issue), 1853-1888.

Hao, X., Cheung, M. H., Wong, V. W. S., & Leung, V. C. M. (2012). Hedonic coalition formation game for cooperative spectrum sensing and channel access in cognitive radio networks. *IEEE Transactions on Wireless Communications*, *11*(11), 3968–3979. doi:10.1109/TWC.2012.092412.111833

Hart, S. (2004). Adaptive heuristics: a little rationality goes a long way. *Walras–Bowley Lecture slides*. Retrieved from http://www.ma.huji.ac.il/~hart/papers/wb.pdf

Hart, S., & Mas-Colell, A. (2000). A simple adaptive procedure leading to correlated equilibrium. *Econometrica, 68*(5), 1127–1150. doi:10.1111/1468-0262.00153

Hart, S., & Mas-Colell, A. (2001). *A reinforcement procedure leading to correlated equilibrium*. New York: Springer. doi:10.1007/978-3-662-04623-4_12

Hart, S., & Schmeidler, D. (1989). Existence of correlated equilibria. *Mathematics of Operations Research, 14*(1), 18–25. doi:10.1287/moor.14.1.18

Hasan, Z., Boostanimehr, H., & Bhargava, V. K. (2011). Green Cellular Networks: A Survey, Some Research Issues and Challenges. *IEEE Communications Surveys and Tutorials, 13*(4), 524–540. doi:10.1109/SURV.2011.092311.00031

Hasbani, G., El-Darwiche, B., Mourad, M., & Chanab, L. A. (2007). *Telecom Infrastructure sharing-Regulatory Enablers and economic benefits. Perspective report*. Booz Allen Hamilton.

Haykin, S. (2005). Cognitive radio: Brain-empowered wireless communications. *IEEE Journal on Selected Areas in Communications, 23*(2), 201–220. doi:10.1109/JSAC.2004.839380

He, G., Debbah, M., & Altman, E. (2010). A Bayesian Game-Theoretic Approach for Distributed Resource Allocation in Fading Multiple Access Channels. *EURASIP Journal on Wireless Communications and Networking, 2010*(1), 8:1–8:12.

He, G., Lasaulce, S., & Hayel, Y. (2011). Stackelberg games for energy-efficient power control in wireless networks. *2011. Proceedings - IEEE INFOCOM, 2*(3), 591–595.

Helmy, A., Musavian, L., & Le-Ngoc, T. (2013). *Energy-efficient power adaptation over a frequency-selective fading channel with delay and power constraints*. Academic Press.

Hkimi, M., Langar, R., Secci, S., Boutaba, R., & Pujolle, G. (2013, June). Demands rescaling for resource and power allocation in cooperative femtocell networks. In *Proceedings of Communications (ICC)* (pp. 6235-6239). IEEE. doi:10.1109/ICC.2013.6655605

Ho, L. T. W., Ashraf, I., & Claussen, H. (2009). Evolving femtocell coverage optimization algorithms using genetic programming. In *Proceeding of IEEE 20th Annual International Symposium on Personal, Indoor and Mobile Radio Communications (PIMRC2009)*. Tokyo: IEEE. doi:10.1109/PIMRC.2009.5450062

Hong, E. J., Yun, S. Y., & Cho, D. H. (2009). Decentralized power control scheme in femtocell networks: A game theoretic approach. In *Proceeding of IEEE 20th Annual International Symposium on Personal, Indoor and Mobile Radio Communications (PIMRC2009)*. Tokyo: IEEE. doi:10.1109/PIMRC.2009.5449782

Hossain, E., Niyato, D., & Han, Z. (2009). *Dynamic spectrum access and management in cognitive radio networks*. Cambrige University Press. doi:10.1017/CBO9780511609909

Hou, J. H., Wan, S., Zhan, M., & Chau, L.-P. (2013). Consistent video quality control in scalable video coding using dependent distortion quantization model. *IEEE Transactions on Broadcasting, 59*(4), 717–724. doi:10.1109/TBC.2013.2289632

Hoydis, J., & Debbah, M. (2010). Green, Cost-Effective, Flexible, Small Cell Networks. *IEEE ComSoc MMTC E-Letter Special Issue on. Multimedia Over Femto Cells, 5*(5), 23–26.

Hoydis, J., Kobayashi, M., & Debbah, M. (2011). Green small-cell networks—a cost- and energy-efficient way of meeting the future traffic demands. *IEEE Vehicular Technology Magazine, 6*(1), 37–43. doi:10.1109/MVT.2010.939904

Huang, H., Sun, Y., Li, X., Chen, S., Xiao, M., & Huang, L. (2014). (in press). Truthful Auction Mechanisms with Performance Guarantee in Secondary Spectrum. *IEEE Transactions on Mobile Computing*. doi:10.1109/TMC.2014.2343976

Huang, J. (2013). Market mechanisms for cooperative spectrum trading with incomplete network information. *IEEE Communications Magazine*, *51*(10), 201–207. doi:10.1109/MCOM.2013.6619584

Huang, J. W., & Krishnamurthy, V. (2011). Cognitive base stations in LTE/3GPP femtocells: A correlated equilibrium game-theoretic approach. *IEEE Transactions on Communications*, *59*(12), 3485–3493. doi:10.1109/TCOMM.2011.093011.100693

Huang, J., Berry, R. A., & Honig, M. L. (2006). Auction-based spectrum sharing. *ACM/Springer. Mobile Networks and Applications*, *11*(3), 405–418. doi:10.1007/s11036-006-5192-y

Huang, J., Zhu, H., Mung, C., & Poor, H. V. (2008). Auction-based resource allocation for cooperative communications. *IEEE Journal on Selected Areas in Communications*, *26*(7), 1226–1237. doi:10.1109/JSAC.2008.080919

Huang, L., Zhou, Y. Q., Han, X., Wang, Y. Y., Qian, M. L., & Shi, J. L. (2013). Coverage optimization for small cell clusters using game theory. In *Proceeding of 2013 IEEE Wireless Communication and Networking Conference (WCNC)*. Shanghai: IEEE. doi:10.1109/WCNC.2013.6554917

Huang, L., Zhou, Y. Q., Hu, J. L., Han, X., & Shi, J. L. (2012). *Coverage optimization for femtocell clusters using modified particle swarm optimization*. In *Proceeding of 2012 IEEE International Conference in Communication (ICC)*. Ottawa: IEEE. doi:10.1109/ICC.2012.6364494

Hui, D. S. W., Lau, V. K. N., & Lam, W. H. (June2006). Cross layer designs for OFDMA wireless system with heterogeneous delay requirements, In *Proc. of the IEEE International Conference on Communications*, *12*, 5325-5330.

IEEE 802.16ppc-10/0008r3. (2010). *Study report on hierarchical networks*. IEEE.

Iera, A., Militano, L., Romeo, L., & Scarcello, F. (2011). Fair cost allocation in cellular-bluetooth cooperation scenarios. *Wireless Communications. IEEE Transactions on*, *10*(8), 2566–2576.

IMT Vision. (2013). Framework and overall objectives of the future development of IMT for 2020 and beyond. *Working document toward preliminary draft new recommendation ITU-R M.[IMT.Vision]*, Working Party 5D/TEMP/126 Rev. 1.

Ismail, M., & Zhuang, W. (2011). A distributed resource allocation algorithm in heterogeneous wireless access medium. In *Proceedings of the IEEE International Conference on Communications* (pp. 1-8). New York, NJ: IEEE.

Janis, P., Koivunen, V., Ribeiro, C. B., Doppler, K., & Hugl, K. (2009). *Interference-Avoiding MIMO Schemes for Device-to-Device Radio Underlaying Cellular Networks*. Paper presented at IEEE Personal, Indoor and Mobile Radio Communications (PIMRC), Tokyo, Japan. doi:10.1109/PIMRC.2009.5450284

Jayaweera, S. K., & Tinaming, L. (2009). Dynamic spectrum leasing in cognitive radio networks via primary-secondary user power control games. *IEEE Transactions on Wireless Communications*, *8*(6), 3300–3310. doi:10.1109/TWC.2009.081230

Jo, H. S., Mun, C., Moon, J., & Yook, J. G. (2010). Self-optimized coverage coordination in femtocell networks. *IEEE Transactions on Wireless Communications*, *9*(10), 2977–2982. doi:10.1109/TWC.2010.090210.081313

Joshi, T., Ahuja, D., Singh, D., & Agrawal, D. P. (2008). SARA: Stochastic automata rate adaptation for IEEE 802.11 networks. *IEEE Transactions on Parallel and Distributed Systems*, *19*(11), 1579–1590. doi:10.1109/TPDS.2007.70814

Kalai, E., & Smorodinsky, M. (1975). Other Solutions to Nash's Bargaining Problem. *Econometrica*, *43*(3), 513–518. doi:10.2307/1914280

Kameda, H., & Altman, E. (2008). Inefficient noncooperation in networking games of common-pool resources. *IEEE Journal on Selected Areas in Communications*, *26*(7), 1260–1268. doi:10.1109/JSAC.2008.080922

Kang, X., Liang, Y.-C., Garg, H. K., & Zhang, L. (2008). Sensing-based spectrum sharing in cognitive radio networks. In *Proceedings of the IEEE Global Telecommunications Conference*. New Orleans, LA: IEEE. doi:10.1109/GLOCOM.2008.ECP.844

Kang, X., Zhang, R., & Motani, M. (2012). Price-based resource allocation for spectrum-sharing femtocell networks: A stackelberg game approach. *IEEE Journal on Selected Areas in Communications*, *30*(3), 538–549. doi:10.1109/JSAC.2012.120404

Kannan, R., & Iyengar, S. S. (2004). Game-theoretic models for reliable path-length and energy-constrained routing with data aggregation in wireless sensor networks. *IEEE Journal on Selected Areas in Communications*, *22*(6), 1141–1150. doi:10.1109/JSAC.2004.830937

Kashyap, A., Basar, T., & Srikant, R. (2004). Correlated jamming on MIMO Gaussian fading channels. *IEEE Transactions on Information Theory*, *50*(9), 2119–2123. doi:10.1109/TIT.2004.833358

Katti, S., Gollakota, S., & Katabi, D. (2007). Embracing wireless interference: analog network coding. In *Proceedings of ACM SIGCOMM*. ACM. doi:10.1145/1282380.1282425

Kemeny, J. G., & Snell, J. L. (1976). *Finite Markov Chains*. New York, NY: Springer.

Kennedy, J., & Eberhart, R. (1995). *Particle swarm optimization*. Paper presented at the IEEE International Conference on Neural Networks, Australia.

Khan, A., Kellerer, W., Kozu, K., & Yabusaki, M. (2011). Network sharing in the next mobile network: TCO reduction, management flexibility, and operational independence. *IEEE Communications Magazine*, *49*(10), 134–142. doi:10.1109/MCOM.2011.6035827

Khan, M. A., Khan, T. A., & Beg, M. T. (2012). RTS/CTS mechanism of MAC layer IEEE 802.11 WLAN in presence of hidden nodes. *International Journal of Engineering and Innovative Technology*, *2*(5), 235–239.

Khan, M. A., Tembine, H., & Vasilakos, A. V. (2012). Game Dynamics and Cost of Learning in Heterogeneous 4G Networks. *IEEE Journal on Selected Areas in Communications*, *30*(1), 198–213. doi:10.1109/JSAC.2012.120118

Khan, T. A., Beg, M. T., & Khan, M. A. (2013). Performance analysis of WLAN using OPNET. *International Journal of Innovative Technology and Exploring Engineering*, *2*(5), 2278–2375.

Khoshkholgh, M. G., Navaie, K., & Yanikomeroglu, H. (2010). Access strategies for spectrum sharing in fading environment: Overlay, underlay, and mixed. *IEEE Transactions on Mobile Computing*, *9*(12), 1780–1793. doi:10.1109/TMC.2010.57

Kirkwood, C. (1997). *Notes on Attitude Toward Risk Taking and the Exponential Utility Function*. Working paper, Arizona State University.

Ko, C.-H., & Wei, H.-Y. (2011). On-Demand Resource-Sharing Mechanism Design in Two-Tier OFDMA Femtocell Networks. *IEEE Transactions on Vehicular Technology*, *60*(3), 1059–1071. doi:10.1109/TVT.2011.2106171

Koczy, L. A. (2006). A recursive core for partition function form games. *Meteor Research Memorandum*, *63*(1), 41-51.

Koriiis, Y. A., Varvarigou, T. A., & Ahuja, S. R. (1998). Incentive-compatible pricing strategies in noncooperative networks. In *Proceedings of 1998 IEEE International Conference on Computer Communications* (pp. 439-446). New York: IEEE.

Koudouridis, G. P., & Li, H. (2012). Distributed Power On-Off Optimisation for Heterogeneous Networks. *Proceedings from CAMAD'12: The IEEE 17th International Workshop on Computer Aided Modeling and Design of Communication Links and Networks*. Barcelona, Spain.

Krikidis, I., Thompson, J. S., & McLaughlin, S. (2009). Relay selection for secure cooperative networks with jamming. *IEEE Transactions on Wireless Communications*, *8*(10), 5003–5011. doi:10.1109/TWC.2009.090323

Krishnamurthy, V., Maskery, M., & Yin, G. (2008). Decentralized adaptive filtering algorithms for sensor activation in an unattended ground sensor network. *IEEE Transactions on Signal Processing*, *56*(12), 6086–6101. doi:10.1109/TSP.2008.929664

Krishna, V. (2009). *Auction Theory*. New York, NY: Academic Press.

Kruger, B., & Dressler, F. (2005). Molecular processes as a basis for autonomous networking. *IPSI Transactions on Advances Research: Issues in Computer Science and Engineering*, *1*(1), 43–50.

Lai, W. S., Lee, S. C., Chiang, M. E., & Lee, T. S. (2012).Low complexity eigenvalue-based cooperative sensing in cognitive femtocell. In *Proceedings of the 9th IEEE Vehicular Technology Society Asia Pacific Wireless Communications Symposium*, Kyoto, Japan: IEEE.

Laneman, J. N., & Wornell, G. W. (2003). Distributed space-time-coded protocols for exploiting cooperative diversity in wireless networks. *IEEE Transactions on Information Theory*, *49*(10), 2415–2425. doi:10.1109/TIT.2003.817829

Lan, Y., & Harada, A. (2012). Interference Analysis and Performance Evaluation on the Coexistence of Macro and Micro/Pico Cells in LTE Networks. In *Proceedings of the IEEE Vehicular Technology Conference (VTC Spring 2012)*. Yokohama, Japan: IEEE. doi:10.1109/VETECS.2012.6239907

Lasaulce, S., & Tembine, H. (2011). *Game Theory and Learning for Wireless Networks: Fundamentals and Applications* (1st ed., pp. 153–180). Waltham, MA: Elsevier. doi:10.1016/B978-0-12-384698-3.00006-2

Le Treust, M., Tembine, H., Lassaulce, S., & Debbah, M. (2010). Coverage games in small cell networks. In Proceeding of 2010 Future Network and Mobile Summit. Florence: IEEE.

Le, L. B. (2012). *Fair resource allocation for device-to-device communications in wireless cellular networks*. Paper presented at IEEE Global Communications Conference (GLOBECOM), Anaheim, California, USA.

Lee, D. H., Choi, K. W., Jeon, W. S., & Jeong, D. G. (2013). Resource Allocation Scheme for Device-to-Device Communication for Maximizing Spatial Reuse. In *Proceedings of the IEEE Wireless Communications and Networking Conference (WCNC 2013)*. Shanghai, China: IEEE. doi:10.1109/WCNC.2013.6554548

Lee, C., & Suzuki, J. (2009). An immunologically-inspired autonomic framework for self-organizing and evolvable network applications.[TAAS]. *ACM Transactions on Autonomous and Adaptive Systems*, *4*(4), 22. doi:10.1145/1636665.1636668

Leibner, T. (2004). *Network and Infrastructure Sharing in 2G networks*. Munich: Siemens.

Leung-Yan-Cheong, S. K., & Hellman, M. E. (1978). The Gaussian wiretap channel. *IEEE Transactions on Information Theory*, *24*(4), 451–456. doi:10.1109/TIT.1978.1055917

Li, W., Zhang, H., Zheng, W., Su, T., & Wen, X. (2012, December). Energy-efficient power allocation with dual-utility in two-tier OFDMA femtocell networks. In Proceedings of Globecom Workshops (GC Wkshps) (pp. 535-540). IEEE. doi:10.1109/GLOCOMW.2012.6477630

Liang, C., & Dandekar, K. R. (2007). Power management in MIMO ad hoc networks: A game-theoretic approach. *IEEE Transactions on Wireless Communications*, *6*(4), 1164–1170. doi:10.1109/TWC.2007.348307

Liang, Y., Poor, H. V., & Shamai, S. (2008). Secure communication over fading channels. *IEEE Transactions on Information Theory*, *54*(6), 2470–2492. doi:10.1109/TIT.2008.921678

Liang, Y., Poor, H. V., & Ying, L. (2011). Secure Communications Over Wireless Broadcast Networks: Stability and Utility Maximization. *IEEE Transactions on Information Forensics and Security*, *6*(3), 682–692. doi:10.1109/TIFS.2011.2158311

Li, F., Tolley, D., Padhy, N. P., & Wang, J. (2009). Framework for assessing the economic efficiencies of long-run network pricing models. *IEEE Transactions on Power Systems*, *24*(4), 1641–1648. doi:10.1109/TPWRS.2009.2030283

Li, J., Petropulu, A. P., & Weber, S. (2011). On cooperative relaying schemes for wireless physical layer security. *IEEE Transactions on Signal Processing*, *59*(10), 4985–4997. doi:10.1109/TSP.2011.2159598

Lim, G., & Cimini, L. J. (2012). Energy-efficient cooperative relaying in heterogeneous radio access networks. *IEEE Communications Letters*, *1*(5), 476–479. doi:10.1109/WCL.2012.070312.120366

Li, S., Huang, J., & Li, S.-Y. R. (2013). Dynamic profit maximization of cognitive mobile virtual network operator. *IEEE Transactions on Mobile Computing*.

Liu, J., Zheng, W., Li, W., Wang, X., Xie, Y., & Wen, X. (2013, April). Distributed uplink power control for two-tier femtocell networks via convex pricing. In *Proceedings ofWireless Communications and Networking Conference (WCNC)* (pp. 458-463). IEEE.

Liu, Y., Tao, M., & Huang, J. (2013). An auction approach to distributed power allocation for multiuser cooperative networks. *IEEE Transactions on Wireless Communications*, *12*(1), 237–247. doi:10.1109/TWC.2012.121112.112284

Li, X. F., Qian, L. J., & Kataria, D. (2009). Downlink power control in co-channel macrocell femtocell overlay. In *Proceeding of 43rd Annual Conference on Information Sciences and Systems, 2009 (CISS 2009)*. Baltimore, MD: IEEE. doi:10.1109/CISS.2009.5054750

Li, X., Amon, P., Hutter, A., & Kaup, A. (2011). Performance analysis of linter-layer prediction in scalable video coding. *IEEE Transactions on Broadcasting*, *57*(9), 66–74. doi:10.1109/TBC.2010.2082370

Li, Y. Y., Macuha, M., Sousa, E. S., Sato, T., & Nanri, M. (2009). Cognitive interference management in 3G femtocells. In *Proceeding of IEEE 20th Annual International Symposium on Personal, Indoor and Mobile Radio Communications (PIMRC2009)*. Tokyo: IEEE. doi:10.1109/PIMRC.2009.5450030

Li, Y., Guo, X., Li, Y., & Zheng, X. (2009). A utility-based network selection mechanism in heterogeneous wireless networks. In *Proceedings of International Conference on Wireless Network and Information Systems* (pp. 201-204). New York: IEEE. doi:10.1109/WNIS.2009.94

Li, Y., Zhang, Z., Wang, C., Zhao, W., & Chen, H.-H. (2013). Blind cooperative communications for multihop ad hoc wireless networks. *IEEE Transactions on Vehicular Technology*, *62*(7), 3110–3122. doi:10.1109/TVT.2013.2256475

Li, Z., & Shen, H. (2012). Game-theoretic analysis of cooperation incentive strategies in mobile ad hoc networks. *IEEE Transactions on Mobile Computing*, *11*(8), 1287–1303. doi:10.1109/TMC.2011.151

Lopez-Perez, D., Guvenc, I., de la Roche, G., Kountouris, M., Quek, T. Q. S., & Zhang, J. (2011). Enhanced inter-cell interference coordination challenges in heterogeneous networks. *IEEE Wireless Communications*, *18*(3), 22–30. doi:10.1109/MWC.2011.5876497

Lu, Y., He, H., Wang, J., & Li, S. (2009). Energy-efficient dynamic spectrum access using no-regret learning. In *Proceedings of the 7th International Conference on Information, Communications and Signal Processing* (ICICS 2009). Beijing, China: Springer.

Luo, H. C., Wu, E. K., & Chen, G. H. (2013). A transmission power/rate control scheme in CSMA/CA-based wireless ad hoc networks. *IEEE Transactions on Vehicular Technology*, *62*(1), 427–431. doi:10.1109/TVT.2012.2219567

Luo, Y., Gao, L., & Huang, J. (2012). Spectrum broker by geo-locationdatabase. In *IEEE Global Communications Conference (GLOBECOM)*, pages 5427–5432. IEEE. doi:10.1109/GLOCOM.2012.6503984

Luo, Y., Gao, L., & Huang, J. (2014a). Price and inventory competitionin oligopoly TV white space markets. *IEEE Journal on Selected Areas in Communications.*

Luo, Y., Gao, L., & Huang, J. (2014b). Trade information, not spectrum: A novel TV white space information market model. In *International Symposium on Modeling and Optimization in Mobile, Ad Hoc, and Wireless Networks (WiOpt)*, pages 405–412. IEEE. doi:10.1109/WIOPT.2014.6850326

Ma, C., Yue, J., Yu, H., Luo, H., Zhou, W., & Sun, X. (2013). An Interference Coordination Mechanism Based on Resource Allocation for Network Controlled Device-to-Device Communication. In *Proceedings of the IEEE/CIC International Conference on Communications in China - Workshops (CIC/ICCC 2013).* Xi'an, China: IEEE.

Ma, Y., Lv, T., & Lu, Y. (2013, April). Efficient power control in heterogeneous Femto-Macro cell networks. In *Proceedings ofWireless Communications and Networking Conference (WCNC)* (pp. 4215-4219). IEEE.

MacKenzie, A. B., & DaSilva, L. A. (2007). *Game theory for wireless engineers.* Morgan & Claypool Publisher.

Manoj, P. (2011). *Effect of real video streaming over virtual MANET environment.* (Master thesis). Rochester Institute of Technology, Rochester, NY.

Marden, J. R., Arslan, G., & Shamma, J. S. (2009). Cooperative control and potential games. *IEEE Transactions on Systems, Man, and Cybernetics. Part B, Cybernetics, 39*(6), 1393–1407. doi:10.1109/TSMCB.2009.2017273 PMID:19369160

Marden, J. R., & Shamma, J. S. (2012). Revisiting log-linear learning: Asynchrony, completeness and payoff-based implementation. *Games and Economic Behavior, 75*(2), 788–808. doi:10.1016/j.geb.2012.03.006

Marden, J. R., Young, H. P., & Pao, L. Y. (2012). Achieving pareto optimality through distributed learning. In *Proceedings of the 51th IEEE Conference on Decision and Control (CDC2012).*Hawaii, HI: IEEE.

Marsan, M. A., Chiaraviglio, L., Ciullo, D., & Meo, M. (2009). Optimal energy savings in cellular access networks, *Proceedings from ICC Workshops '09:The IEEE International Conference on Communications Workshops.* Dresden, Germany.

Marsan, M. A., & Meo, M. (2009). Energy efficient management of two cellular access networks. *Performance Evaluation Review, 37*(4), 69–73. doi:10.1145/1773394.1773406

Marsan, M. A., & Meo, M. (2011). Energy efficient wireless Internet access with cooperative cellular networks. *Elsevier Journal Computer Networks, 55*(2), 386–398. doi:10.1016/j.comnet.2010.10.017

Maskery, M., & Krishnamurthy, V. (2007a). Decentralized activation in a ZigBee-enabled unattended ground sensor network: a correlated equilibrium game theoretic analysis. In *Proceedings of the 2007 IEEE International Conference on Communications (ICC),*Glasgow, UK: IEEE. doi:10.1109/ICC.2007.645

Maskery, M., & Krishnamurthy, V. (2007b). Decentralized adaptation in sensor networks: Analysis and application of regret-based algorithms. In *Proceedings of the 46th IEEE Conference on Decision and Control(CDC).*New Orleans, LA: IEEE. doi:10.1109/CDC.2007.4434388

Maskery, M., Krishnamurthy, V., & Zhao, Q. (2009). Decentralized dynamic spectrum access for cognitive radios: Cooperative design of a non-cooperative game. *IEEE Transactions on Communications, 57*(2), 459–469. doi:10.1109/TCOMM.2009.02.070158

Massey, J. L. (1988). An introduction to contemporary cryptology. *Proceedings of the IEEE*, *76*(5), 533–549. doi:10.1109/5.4440

Mathur, S., Sankaranarayanan, L., & Mandayam, N. (2008). Coalitions in cooperative wireless networks. *IEEE Journal on Selected Areas in Communications*, *26*(9), 1104–1115. doi:10.1109/JSAC.2008.080908

McHenry, M. A., McCloskey, D., Roberson, D., & MacDonald, J. T. (2005). *Spectrum occupancy measurements. Technical report*. Shared Spectrum Company.

Meddour, D. E., Rasheed, T., & Gourhant, Y. (2011). On the role of infrastructure sharing for mobile network operators in emerging markets. *Elsevier Computer Networks.*, *55*(7), 1567–1591. doi:10.1016/j.comnet.2011.01.023

Mehbodniya, A., & Aiffissa, S. (2012). Game-theoretic approach for interference management in heterogeneous multimedia wireless personal area networks. *IET Communications*, *6*(15), 2278–2286. doi:10.1049/iet-com.2012.0297

Menon, R., MacKenzie, A., Buehrer, R., & Reed, J. (2009). Interference avoidance in networks with distributed receivers. *IEEE Transactions on Communications*, *8*(10), 5225–5237.

Mériaux, F., Valentin, S., Lasaulce, S., & Kieffer, M. (2012, October). An energy-efficient power allocation game with selfish channel state reporting in cellular networks. In *Proceedings of Performance Evaluation Methodologies and Tools (VALUETOOLS)* (pp. 290-297). IEEE. doi:10.4108/valuetools.2012.250288

Miao, G., Himayat, N., Li, G., & Talwar, S. (2011). Distributed interference-aware energy-efficient power optimization. *IEEE Transactions on Wireless Communications*, *10*(4), 1323–1333. doi:10.1109/TWC.2011.021611.101376

Militano, L., & Iera, A. (2012). Adopting bargaining solutions for Bluetooth-based user cooperation. *IEEE 23rd International Symposium on Personal Indoor and Mobile Radio Communications (PIMRC).*

Militano, L., Fitzek, F., Iera, A., & Molinaro, A. (2011). Group interactions in wireless cooperative networks. *IEEE 73rd Vehicular Technology Conference (VTC Spring).*

Militano, L., & Araniti, G. (2014). Introducing Fairness-Efficiency Trade-off for Energy Savings in Wireless Cooperative Networks. *Wireless Personal Communications*, *76*(1), 3–21. doi:10.1007/s11277-013-1483-3

Militano, L., Condoluci, M., Araniti, G., & Iera, A. (2012). Bargaining solutions for multicast subgroup formation in LTE.*IEEE Vehicular Technology Conference (VTC Fall).* doi:10.1109/VTCFall.2012.6399273

Militano, L., Condoluci, M., Araniti, G., & Iera, A. (2013). Multicast service delivery solutions in LTE-Advanced systems.*IEEE International Conference on Communications (ICC).* doi:10.1109/ICC.2013.6655551

Militano, L., Fitzek, F., Iera, A., & Molinaro, A. (2010a). Network coding and evolutionary theory for performance enhancement in wireless cooperative clusters. *European Transactions on Telecommunications*, *21*(8), 725–737. doi:10.1002/ett.1435

Militano, L., Fitzek, F., Iera, A., & Molinaro, A. (2010b). A genetic algorithm for source election in cooperative clusters implementing network coding.*IEEE International Conference on Communications Workshops (ICC).* doi:10.1109/ICCW.2010.5503955

Militano, L., Iera, A., Molinaro, A., & Scarcello, F. (2014). Energy saving analysis in cellular-WLAN cooperative scenarios. *IEEE Transactions on Vehicular Technology*, *63*(1), 478–484. doi:10.1109/TVT.2013.2270315

Militano, L., Iera, A., & Scarcello, F. (2013). A fair cooperative content-sharing service. *Computer Networks*, *57*(9), 1955–1973. doi:10.1016/j.comnet.2013.03.014

Militano, L., Molinaro, A., Iera, A., & Petkovics, Á. (2013). A Game Theoretic Approach to Guarantee Fairness in Cooperation Among Green Mobile Network Operators.[IJBDCN]. *International Journal of Business Data Communications and Networking*, *9*(3), 1–15.

Militano, L., Niyato, D., Condoluci, M., Araniti, G., Iera, A., & Molica Bisci, G. (2014). Radio Resource Management for Group-Oriented Services in LTE-A. *IEEE Transactions on Vehicular Technology*, 1. doi:10.1109/TVT.2014.2359686

Min, H., Lee, J., Park, S., & Hong, D. (2011). Capacity enhancement using an interference limited area for device-to-device uplink underlaying cellular networks. *IEEE Transactions on Wireless Communications*, *10*(12), 3995–4000. doi:10.1109/TWC.2011.100611.101684

Mishra, A., Banerjee, S., & Arbaugh, W. (2005). Weighted coloring based channel assignment for WLANs. *Mobile Computing and Communications Review*, *9*(3), 19–31. doi:10.1145/1094549.1094554

Mishra, A., Rozner, E., Banerjee, S., & Arbaugh, W. (2005). Exploiting partially overlapping channels in wireless networks: Turning a peril into an advantage. In *Proceedings of the 5th ACM SIGCOMM conference on Internet Measurement.*Philadelphia, PA: ACM. doi:10.1145/1330107.1330144

Misra, S., Tiwari, V., & Obaidat, M. S. (2009). Lacas: Learning automata-based congestion avoidance scheme for healthcare wireless sensor networks. *IEEE Journal on Selected Areas in Communications*, *27*(4), 466–479. doi:10.1109/JSAC.2009.090510

Misra, V., Ioannidis, S., Chaintreau, A., & Massouliè, L. (2010). Incentivizing peerassisted services: A fluid Shapley value approach. *Performance Evaluation Review*, *38*(1), 215–226. doi:10.1145/1811099.1811064

Mitola, J., & Maguire, G. Q. Jr. (1999). Cognitive radio: Making software radios more personal. *IEEE Personal Communications*, *6*(4), 13–18. doi:10.1109/98.788210

Monderer, D., & Shapley, L. S. (1996a). Potential games. *Games and Economic Behavior*, *14*(1), 124–143. doi:10.1006/game.1996.0044

Monderer, D., & Shapley, L. S. (1996b). Fictitious play property for games with identical interests. *Journal of Economic Theory*, *1*(1), 258–265. doi:10.1006/jeth.1996.0014

Moretti, M., Todini, A., Baiocchi, A., & Dainelli, G. (2011, May). A Layered Architecture for Fair Resource Allocation in Multicellular Multicarrier Systems. *IEEE Transactions on Vehicular Technology*, *60*(4), 1788–1798. doi:10.1109/TVT.2011.2119501

Mukherjee, A., & Swindlehurst, A. L. (2013). Jamming Games in the MIMO Wiretap Channel with an Active Eavesdropper. *IEEE Transactions on Signal Processing*, *61*(1), 82–91. doi:10.1109/TSP.2012.2222386

Mukherjee, S. (2012). Distribution of downlink SINR in heterogeneous cellular networks. *IEEE Journal on Selected Areas in Communications*, *30*(3), 575–585. doi:10.1109/JSAC.2012.120407

Murty, R., Chandra, R., Moscibroda, T., & Bahl, P. (2011). Senseless: A database-driven white spaces network. In IEEE Symposia on New Frontiers in Dynamic Spectrum Access Netoworks (DySpan). doi:10.1109/DYSPAN.2011.5936197

Mustika, I. W., Yamamoto, K., Murata, H., & Yoshida, S. (2011) Potential Game Approach for Self-Organized Interference Management in Closed Access Femtocell Networks. In *Proceedings of IEEE 73rd Vehicular Technology Conference (VTC Spring)*. IEEE.

Myerson, R. (1991). *Game theory: analysis of conflict*. Cambridge, UK: Harvard University Press.

Myerson, R. (1997). *Game Theory: Analysis of Conflict*. Cambridge, MA: Harvard University Press.

Myung, H. G., & Goodman, D. J. (2008). *Single Carrier FDMA: A New Air Interface for Long Term Evolution*. Chichester, UK: John Wiley and Sons. doi:10.1002/9780470758717

Nash, J. (1950). The Bargaining Problem. *Econometrica*, 128–140.

Nash, J. F. (1950). Equilibrium points in n-person games. (1950). *Proceedings of the National Academy of Sciences of the United States of America*, *36*(1), 48–49. doi:10.1073/pnas.36.1.48 PMID:16588946

Negi, R., & Goel, S. (2005, September). Secret communication using artificial noise. In *Proceedings of Vehicular Technology Conference (VTC '05)*. Tokyo, Japan: IEEE.

Newton, J. (2010). Non-cooperative convergence to the core in Nash demand games without random errors or convexity assumptions. *Journal of Economic Theory*, *24*(6), 1223–1235.

Ng, T., & Yu, W. (2007). Joint optimization of relay strategies and resource allocations in cooperative cellular networks. *IEEE Journal on Selected Areas in Communications*, *25*(2), 328–339. doi:10.1109/JSAC.2007.070209

Nicopolitidis, P., Papadimitriou, G. I., & Pomportsis, A. S. (2002). Using learning automata for adaptive push-based data broadcasting in asymmetric wireless environments. *IEEE Transactions on Vehicular Technology*, *51*(6), 1652–1660. doi:10.1109/TVT.2002.802978

Nicopolitidis, P., Papadimitriou, G. I., & Pomportsis, A. S. (2003). Learning automata-based polling protocols for wireless LANs. *IEEE Transactions on Communications*, *51*(3), 453–463. doi:10.1109/TCOMM.2003.809788

Nicopolitidis, P., Papadimitriou, G. I., & Pomportsis, A. S. (2004). Distributed protocols for ad hoc wireless LANs: A learning-automata-based approach. *Ad Hoc Networks*, *2*(4), 419–431. doi:10.1016/j.adhoc.2003.09.004

Nie, P., & Zhang, P. (2008,). *A Note on Stackelberg Games*. Paper presented at IEEE Control and Decision Conference, Yantai, China.

Nie, N., & Comaniciu, C. (2006). Adaptive channel allocation spectrum etiquette for cognitive radio networks. *Mobile Networks and Applications*, *11*(6), 779–797. doi:10.1007/s11036-006-0049-y

Niu, Z., Wu, Y., Gong, J., & Yang, Z. (2010). Cell Zooming for Cost-Efficient Green Cellular Networks. *IEEE Communications Magazine*, *48*(11), 74–79. doi:10.1109/MCOM.2010.5621970

Niu, Z., Zhou, S., Hua, Y., Zhang, Q., & Cao, D. (2002). Energy-aware network planning for wireless cellular system with inter-cell cooperation. *IEEE Transactions on Wireless Communications*, *11*(4), 1412–1423. doi:10.1109/TWC.2012.021412.110147

Niyato, D., & Hossain, E. (2008). A game theoretic analysis of service competition and pricing in heterogeneous wireless access networks. *IEEE Transactions on Wireless Communications*, *7*(12), 5150–5155. doi:10.1109/T-WC.2008.070609

Niyato, D., & Hossain, E. (2008). A Non-cooperative game-theoretic framework for radio resource management in 4G heterogeneous wireless access networks. *IEEE Transactions on Mobile Computing*, *7*(3), 332–345. doi:10.1109/TMC.2007.70727

Niyato, D., & Hossain, E. (2009). Dynamic of network selection in heterogeneous wireless networks: An evolutionary game approach. *IEEE Transactions on Vehicular Technology*, *58*(4), 2008–2017. doi:10.1109/TVT.2008.2004588

Nokia. (2001). *Nokia launches Multi-Operator Radio Access Network for controlled 3G network sharing*. Press release (23.05.2001).

Nosratinia, A., & Hunter, T. E. (2007). Grouping and partner selection in cooperative wireless networks. *IEEE Journal on Selected Areas in Communications*, *25*(2), 369–378. doi:10.1109/JSAC.2007.070212

Obaid, N., & Czylwik, A. (2013, September). The impact of deploying pico base stations on capacity and energy efficiency of heterogeneous cellular networks. In *Proceedings of Personal Indoor and Mobile Radio Communications (PIMRC)* (pp. 1904-1908). IEEE. doi:10.1109/PIMRC.2013.6666454

Obaidat, M. S., Papadimitriou, G. I., & Pomportsis, A. S. (2002). Guest editorial learning automata: Theory, paradigms, and applications. *IEEE Transactions on Systems, Man, and Cybernetics. Part B, Cybernetics*, *32*(6), 706–709. doi:10.1109/TSMCB.2002.1049604 PMID:18244877

Oh, E., Krishnamachari, B., Liu, X., & Niu, Z. (2011). Towards dynamic energy-efficient operation of cellular network infrastructure. *IEEE Communications Magazine*, *49*(6), 56–61. doi:10.1109/MCOM.2011.5783985

Osborne, M. (2004). *An introduction to game theory*. Oxford, UK: Oxford University Press.

Osborne, M. J., & Rubinstein, A. (1994). *A course in game theory*. Cambridge, MA: MIT press.

Osborne, M. J., & Rubinstein, A. (1994). *A Course in Game Theory*. Cambridge, MA: MIT Press.

Otten, G. (1993). Characterizations of a game theoretical cost allocation method. *Zeitschrift fur Operations Research*, *38*(2), 175–185.

Paek, J., Kim, J., & Govindan, R. (2010). Energy-Efficient Rate-Adaptive GPS-based Positioning for Smartphones. *8th international conference on Mobile systems, applications, and services, MobiSys.*

Pan, Y., Nix, A., & Beach, M. (2008, September). A game theoretic approach to distributed resource allocation for OFDMA-based relaying networks. In *Proceedings of Personal, Indoor and Mobile Radio Communications, 2008* (pp. 1-5). IEEE.

Pantisano, F., Bennis, M., Saad, W., Debbah, M., & Latva-aho, M. (2013). Interference alignment for cooperative femtocell networks: A game-theoretic approach. *IEEE Transactions on Mobile Computing*, *12*(11), 2233–2246. doi:10.1109/TMC.2012.196

Pantisano, F., Bennis, M., Saad, W., Verdone, R., & Latva-Aho, M. (2011). Coalition formation games for femtocell interference management: A recursive core approach.*Wireless Communications and Networking Conference (WCNC)*, 2011IEEE (Vol.34, pp.1161-1166). IEEE. doi:10.1109/WCNC.2011.5779295

Parkvall, S., Furuskar, A., & Dahlman, E. (2011). Evolution of LTE toward IMT-advanced. *IEEE Communications Magazine*, *49*(2), 84–91. doi:10.1109/MCOM.2011.5706315

Phunchongharn, P., Hossain, E., & Kim, D. I. (2013). Resource allocation for device-to-device communications underlaying LTE-advanced networks. *IEEE Wireless Communications*, *20*(4), 91–100. doi:10.1109/MWC.2013.6590055

Pradelski, B. S., & Young, H. P. (2012). Learning efficient Nash equilibria in distributed systems. *Games and Economic Behavior*, *75*(2), 882–897. doi:10.1016/j.geb.2012.02.017

Quek, T., Roche, G., Guvenc, I., & Kountouris, M. (2013). *Small cell networks: Deployment, PHY techniques, and resource Allocation*. Cambridge, UK: Cambridge University Press. doi:10.1017/CBO9781139061421

Rahman, M., & Yanikomeroglu, H. (2010, April). A downlink dynamic interference avoidance scheme with inter-cell coordination. *IEEE Transactions on Wireless Communications*, *9*(4), 1414–1425. doi:10.1109/TWC.2010.04.090256

Rasti, M., Sharafat, A. R., & Zander, J. (2011). Pareto and energy-efficient distributed power control with feasibility check in wireless networks. *IEEE Transactions on Information Theory*, *57*(1), 245–255.

Ray, D. (2007). *A Game-Theoretic Perspective on Coalition Formation*. New York, NY: Oxford University Press. doi:10.1093/acprof:oso/9780199207954.001.0001

Ren, S., Park, J., & van der Schaar, M. (2013). Entry and Spectrum Sharing Scheme Selection in Femtocell Communications Markets. *IEEE/ACM Transactions on Networking*, *21*(1), 218–232. doi:10.1109/TNET.2012.2198073

Richter, F., Fehske, A. J., & Fettweis, G. P. (2009). Energy efficiency aspects of base station deployment strategies in cellular networks. *Proceedings from VTC Fall'09: The IEEE 70th Vehicular Technology Conference Fall*. Anchorage, AK. doi:10.1109/VETECF.2009.5379031

Rogers, A., David, E., & Jennings, N. (2005). Self-organized routing for wireless microsensor networks. *IEEE Transactions on Systems, Man, and Cybernetics. Part A, Systems and Humans*, *35*(3), 349–359. doi:10.1109/TSMCA.2005.846382

Rose, L., Le Martret, C. J., & Debbah, M. (2013). Joint channel and power allocation in tactical cognitive networks: Enhanced trial and error. In *Proceedings of the 2013 Military Communications and Information Systems Conference (MilCIS)*. Canberra, Australia: IEEE.

Rose, L., Perlaza, S. M., Le Martret, C. J., & Debbah, M. (2013). Achieving Pareto optimal equilibria in energy efficient clustered ad hoc networks. In *Proceedings of the 2013 IEEE International Conference on Communications (ICC2013)*. Budapest, Hungary: IEEE. doi:10.1109/ICC.2013.6654723

Rose, L., Lasaulce, S., Perlaza, S. M., & Debbah, M. (2011). Learning equilibria with partial information in decentralized wireless networks. *IEEE Communications Magazine*, *49*(8), 136–142. doi:10.1109/MCOM.2011.5978427

Rose, L., Perlaza, S. M., Debbah, M., & Le Martret, C. J. (2012). Distributed power allocation with SINR constraints using trial and error learning. In *Proceedings of the 2012 Wireless Communications and Networking Conference (WCNC)*. Paris, France: IEEE doi:10.1109/WCNC.2012.6214083

Rose, L., Perlaza, S. M., Le Martret, C. J., & Debbah, M. (2014). Self-organization in decentralized networks: A trial and error learning approach. *IEEE Transactions on Wireless Communications*, *13*(1), 268–279. doi:10.1109/TWC.2013.112613.130405

Rosen, J. B. (1965). Existence and uniqueness of equilibrium points for concave n-person games. *Econometrica*, *33*(3), 520–534. doi:10.2307/1911749

Rosenthal, R. W. (1973). A Class of Games Possessing Pure-Strategy Nash Equilibria. *International Journal of Game Theory*, *2*(1), 65–67. doi:10.1007/BF01737559

Rota, G. C. (1964). The number of partitions of a set. *The American Mathematical Monthly*, *71*(5), 498–504. doi:10.2307/2312585

Roth, A. (1979). *Axiomatic Models of Bargaining*. New York City: Springer. doi:10.1007/978-3-642-51570-5

Rozario, F., Han, Z., & Niyato, D. (2011). Optimization of non-cooperative P2P network from the game theory point of view. In *Proceedings of Wireless Communications and Networking Conference* (pp. 868-873). New York: IEEE.

Rubinstein, A., Safra, Z., & Thomson, W. (1992). On the Interpretation of the Nash Bargaining Solution and its Extension to Non-Expected Utility Preferences. *Econometrica*, *60*(5), 1171–1186. doi:10.2307/2951543

Saad, W., Han, Z., Debbah, M., & Hjørungnes, A. (2009). Coalitional games for distributed collaborative spectrum sensing in cognitive radio networks. In *Proceddings of the 28th Conference on Computer Communications (IEEE INFOCOM '09)*. Rio de Janeiro, Brazil: IEEE. doi:10.1109/INFCOM.2009.5062135

Saad, W., Han, Z., Debbah, M., & Hjørungnes, A. (2009). Physical layer security: Coalitional games for distributed cooperation. In *Proceedings of the 7th Intl.Symposium on Modeling and Optimization in Mobile, Ad Hoc, and Wireless Networks (WiOpt '09)*. Seoul, Korea: IEEE.

Saad, W., Han, Z., Debbah, M., Hjørungnes, A., Basar, T. (2009). Coalition game theory for communication networks: A tutorial. *IEEE Signal Processing Magazine, Special issue on Game Theory in Signal Processing and Communications, 26*(5), 77-97.

Saad, W., Han, Z., Baar, T., Debbah, M., & Hjørungnes, A. (2011). Hedonic coalition formation for distributed task allocation among wireless agents. *IEEE Transactions on Mobile Computing, 10*(9), 1327–1344. doi:10.1109/TMC.2010.242

Saad, W., Han, Z., Debbah, M., & Hjørungnes, A. (2009). A distributed coalition formation framework for fair user cooperation in wireless networks. *IEEE Transactions on Wireless Communications, 8*(9), 4580–4593. doi:10.1109/TWC.2009.080522

Saad, W., Han, Z., Debbah, M., Hjørungnes, A., & Baar, T. (2009). Coalitional game theory for communication networks. *IEEE Signal Processing Magazine, 26*(5), 77–97. doi:10.1109/MSP.2009.000000

Saad, W., Hjørungnes, A., Han, Z., Niyato, D., & Hossain, E. (2011). Coalition formation games for distributed roadside units cooperation in vehicular network.[JSAC]. *IEEE Journal on Selected Areas in Communications, 29*(1), 48–60. doi:10.1109/JSAC.2011.110106

Salem, M., Adinoyi, A., Rahman, M., Yanikomeroglu, H., Falconer, D., Kim, Y.-D., & Cheong, Y.-C. et al. (2010). An Overview of Radio Resource Management in Relay-Enhanced OFDMA-Based Networks. *IEEE Communications Surveys and Tutorials, 12*(3), 422–438. doi:10.1109/SURV.2010.032210.00071

Sallent, O., Perez-Romero, J., Agusti, R., Giupponi, L., Kloeck, C., Martoyo, I., & Luo, J. et al. (2006). Resource auctioning mechanisms in heterogeneous wireless access networks. In *Proceedings of the 63rd IEEE Vehicular Technology Conference* (pp. 52-56). New York: IEEE. doi:10.1109/VETECS.2006.1682774

Sandholm, T., Larson, K., Anderson, M., Shehory, O., & Tohme, F. (1999). Coalition structure generation with worst case guarantees. *Artificial Intelligence, 111*(1-2), 209–238. doi:10.1016/S0004-3702(99)00036-3

Sandvine. (2011). *Intelligent Broadband Networks Global Internet phenomena report*. Author.

Saraydar, C. U., Mandayam, N. B., & Goodman, D. J. (2002). Efficient power control via pricing in wireless data networks. *IEEE Transactions on Communications, 50*(2), 291–303. doi:10.1109/26.983324

Sastry, P. S. S., Phansalkar, V. V., & Thathachar, M. (1994). Decentralized learning of Nash equilibria in multi-person stochastic games with incomplete information. *IEEE Transactions on Systems, Man, and Cybernetics, 24*(5), 769–777. doi:10.1109/21.293490

Savazzi, S., & Spagnolini, U. (2007). Energy aware power allocation strategies for multihop-cooperative transmission schemes. *IEEE Journal on Selected Areas in Communications, 25*(2), 318–327. doi:10.1109/JSAC.2007.070208

Schmeidler, D. (1969). The Nucleolus of a Characteristic Function Game. *SIAM Journal on Applied Mathematics, 17*(6), 1163–1170. doi:10.1137/0117107

Schneier, B. (1998). Cryptographic design vulnerabilities. *IEEE Computer, 31*(9), 26–33. doi:10.1109/2.708447

Schwarz, H., Marpe, D., & Wiegand, T. (2007). Overview of the scalable video coding extension of the H.264/AVC standard. *IEEE Transactions on Circuits and Systems for Video Technology, 17*(9), 1103–1120. doi:10.1109/TCSVT.2007.905532

Scutari, G., Barbarossa, S., & Palomar, D. P. (2006). Potential games: A framework for vector power control problems with coupled constraints. In *Proceedings of IEEE ICASSP*. IEEE. doi:10.1109/ICASSP.2006.1660950

Scutari, G., Palomar, D. P., & Barbarossa, S. (2008, July). Asynchronous iterative water-filling for Gaussian frequency-selective interference channels. *IEEE Transactions on Information Theory, 54*(7), 2868–2878. doi:10.1109/TIT.2008.924723

Scutari, G., Palomar, D. P., Facchinei, F., & Pang, J. S. (2010). Convex optimization, game theory, and variational inequality theory. *IEEE Signal Processing Magazine*, *27*(3), 35–49. doi:10.1109/MSP.2010.936021

Shah, V., Mandayam, N. B., & Goodman, D. J. (1998). Power control for wireless data based on utility and pricing. In *Proceedings of IEEE Personal, Indoor and Mobile Radio Communications Conference*. (pp. 1427-1432). New York: IEEE. doi:10.1109/PIMRC.1998.731433

Shapley, L. (1953). A Value for n-Person Games. In H. W. Kuhn & A. W. Tucker (Eds.), *Contributions to the Theory of Games II, Annals of Mathematics Studies* (pp. 307–317). Princeton, New Jersey, USA: Princeton University Press.

Shehory, O., & Kraus, S. (2009). Methods for task allocation via agent coalition formation. *IEEE Transactions on Communications*, *101*(1), 459–469.

Sheng, M., Xu, C., Wang, X., Zhang, Y., Han, W., & Li, J. (2014). Utility-Based Resource Allocation for Multi-Channel Decentralized Networks. *IEEE Transactions on Communications*, *62*(10), 3610–3620. doi:10.1109/TCOMM.2014.2357028

Shetty, N., Parekh, S., & Walrand, J. (2009). Economics of Femtocells. In *Proceedings ofIEEE Global Telecommunications Conference (Globecom)*. IEEE.

Shi, Y., & Eberhart, R. (1998). A modified particle swarm optimizer. In *Evolutionary Computation Proceedings, 1998. The 1998 IEEE International Conference on IEEE World Congress on Computational Intelligence*. Anchorage: IEEE doi:10.1109/ICEC.1998.699146

Shum, K. W., Leung, K.-K., & Sung, C. W. (2007). Convergence of iterative water-filling algorithm for Gaussian interference channels. *IEEE Journal on Selected Areas in Communications*, *25*(6), 1091–1100. doi:10.1109/JSAC.2007.070804

Singh, M., & Chaporkar, P. (2013). An efficient and decentralised user association scheme for multiple technology networks. In *Proceedings of the 2013 11th International Symposium on Modeling & Optimization in Mobile, Ad Hoc & Wireless Networks (WiOpt)*. Tsukuba Science City, Japan: IEEE.

Singh, G., Grover, A., & Grover, N. (2013, July). Simulation and analysis: The effect of mobility on IPTV (VOD) over Wi-Max using OPNET. In *Proceedings of the 7th Asia Modelling Symposium* (pp. 267-272). New York: IEEE. doi:10.1109/AMS.2013.47

Siomina, I., Varbrand, P., & Yuan, D. (2006). Automated optimization of services coverage and base station antenna configuration in UMTS networks. *IEEE Wireless Communications*, *13*(6), 16–25. doi:10.1109/MWC.2006.275194

Small Cell Forum. (2011). Retrieved June 15, 2013, from http://www.smallcells.org

Song, Y., Wong, S. H., & Lee, K.-W. (2011). Optimal gateway selection in multi-domain wireless networks: a potential game perspective. In *Proceedings of the* 17th annual international conference on Mobile computing and networking (MOBICOM). Beijing, China: ACM.

Song, L., Niyato, D., Han, Z., & Hossain, E. (2014). Game-theoretic resource allocation methods for device-to-device communication. *IEEE Wireless Communications*, *21*(3), 136–144. doi:10.1109/MWC.2014.6845058

Song, W., & Zhuang, W. (2012). Resource allocation for integrated voice and data services. In *Interworking of wireless LANs and cellular networks* (pp. 11–27). New York, NY: Springer. doi:10.1007/978-1-4614-4379-7_2

Son, K., Chong, S., & de Veciana, G. (2009, July). Dynamic association for load balancing and interference avoidance in multi-cell networks. *IEEE Transactions on Wireless Communications*, *8*(7), 3566–3576. doi:10.1109/TWC.2009.071140

Srinivasan, V., Nuggehalli, P., Chiasserini, C., & Rao, R. (2003). *Cooperation in wireless ad hoc networks*. IEEE INFOCOM. doi:10.1109/INFCOM.2003.1208918

Stiakogiannakis, & Kaklamani. (Jul. 2011). A Combinatorial Auction Based Subcarrier Allocation Algorithm for Multiuser OFDMA. *IEEE VTC Spring*, 1-5.

Stoltz, G., & Lugosi, G. (2007). Learning correlated equilibria in games with compact sets of strategies. *Games and Economic Behavior*, *59*(1), 187–208. doi:10.1016/j.geb.2006.04.007

Stolyar & Viswanathan. (2009). Self-organizing dynamic fractional frequency reuse for best-effort traffic through distributed inter-cell coordination. In *Proceedings of IEEE INFOCOM* (pp. 1287-1295). IEEE.

Sun, Y., Zhang, B., Peng, M., & Sun, C. (2014). Interference coordination in small cell networks using coalition formation game. *Communications in China (ICCC), 2014 IEEE/CIC International Conference on* (pp.642-646). IEEE.

Sun, C., Peng, M., Sun, Y., & Li, Y. (2014). Distributed power control for device-to-device network using stackelberg game.*Wireless Communications and Networking Conference (WCNC)*, 2014IEEE (pp.1344-1349). IEEE. doi:10.1109/WCNC.2014.6952365

Sun, C., Peng, M., Zhang, B., Sun, Y., Li, Y., & Wang, C. (2014). A coalitional game scheme for cooperative interference management in cloud radio access networks. *Transactions on Emerging Telecommunications Technologies*, *25*(9), 954–964. doi:10.1002/ett.2838

Sun, G., Feng, X., Tian, X., Gan, X., Xu, Y., Wang, X., & Guizani, M. (2014, June). Coalitional Double Auction for Spatial Spectrum Allocation in Cognitive Radio Networks. *IEEE Transactions on Wireless Communications*, *13*(6), 3196–3206. doi:10.1109/TWC.2013.042114.130952

Suryaprakash, V., Fonseca dos Santos, A., Fehske, A. J., & Fettweis, G. P. (2012). Energy Consumption Analysis of Wireless Networks using Stochastic Deployment Models. *Proceedings from GLOBECOM'12:The IEEE Global Communications Conference*. Anaheim, CA. doi:10.1109/GLOCOM.2012.6503603

Swarm Intelligence. (n.d.). Retrieved June 15, 2013, from http:// www.swarmintelligence.org

Tan, C. K., Sim, M. L., & Chuah, T. C. (2008). Game theoretic approach for channel assignment and power control with no-internal-regret learning in wireless ad hoc networks. *IET Communications*, *2*(9), 1159–1169. doi:10.1049/iet-com:20070547

Tang, H., Zhu, C., & Ding, Z. (2013). *Cooperative MIMO precoding for D2D underlay in cellular networks*. Paper presented at IEEE International Conference on Communications (ICC), Budapest, Hungary. doi:10.1109/ICC.2013.6655469

Tang, A., & Andrew, L. (2008). Game theory for heterogeneous flow control.*Proceedings of 42nd Annual Conference on Information Sciences and Systems*. doi:10.1109/CISS.2008.4558494

Teng, F., Guo, D., Honig, M. L., Xiao, W., & Liu, J. (2012). *Power Control Based on Interference Pricing in Hybrid D2D and Cellular Networks*. Paper presented at IEEE Global Communications Conference (GLOBECOM), Anaheim, CA. doi:10.1109/GLOCOMW.2012.6477655

Thathachar, M., & Sastry, P. S. (2002). Varieties of learning automata: An overview. *IEEE Transactions on Systems, Man, and Cybernetics. Part B, Cybernetics*, *32*(6), 711–722. doi:10.1109/TSMCB.2002.1049606 PMID:18244878

The 25th China Internet Development Statistics Report. (2012). Beijing: China.

The 3rd Generation Partnership Project. (2010). *Telecommunication Management; Self-Organizing Networks (SON); Concepts and requirements* (3GPP TS 32.500 V10.0.0). Retrieved June 15, 2013, from www.qtc.jp/3GPP/Specs/32500-a00.pdf

Third Generation Partnership Project (3GPP). (2006). *3GPP TR 25.814 3GPP TSG RAN Physical Layer Aspects for Evolved UTRA, v7.1.0,2006*. Author.

Third Generation Partnership Project (3GPP). (2014a). *3GPP TS 36.211 3GPP TSG RAN Evolved Universal Terrestrial Radio Access (E-UTRA) Physical Channels and Modulation, version 12.3.0, Release 12,2014*. Author.

Third Generation Partnership Project (3GPP). (2014b). *3GPP TS 36.213 3GPP TSG RAN Evolved Universal Terrestrial Radio Access (E-UTRA) Physical layer procedures, version 12.3.0, Release 12,2014*. Author.

Tian, Y., Liu, H., & Li, Z. (2012, December). Power Allocation Algorithm Based on Non-cooperative Game Theory for LTE Downlink. In *Proceedings of Web Intelligence and Intelligent Agent Technology (WI-IAT)* (Vol. 1, pp. 643-647). IEEE. doi:10.1109/WI-IAT.2012.81

Tombaz, S., Vastberg, A., & Zander, J. (2011). Energy- and cost-efficient ultra-high-capacity wireless access. *IEEE Wireless Communications.*, *18*(5), 18–24. doi:10.1109/MWC.2011.6056688

Torkestani, J. A., & Meybodi, M. R. (2011). Weighted Steiner connected dominating set and its application to multicast routing in wireless MANETs. *Wireless Personal Communications*, *60*(2), 145–169. doi:10.1007/s11277-010-9936-4

Tse, D., & Viswanath, P. (2005). *Fundamentals of Wireless Communications*. Cambridge, UK: Cambridge University Press. doi:10.1017/CBO9780511807213

Tseng, Y., & Huang, C. (2012). Analysis of Femto Base Station Network Deployment. *IEEE Transactions on Vehicular Technology*, *50*(2), 748–757. doi:10.1109/TVT.2011.2176355

Tsiropoulou, E. E., Katsinis, G. K., Vamvakas, P., & Papavassiliou, S. (2013, September). Efficient uplink power control in multi-service two-tier femtocell networks via a game theoretic approach. In *Proceedings of Computer Aided Modeling and Design of Communication Links and Networks (CAMAD)* (pp. 104-108). IEEE. doi:10.1109/CAMAD.2013.6708098

UMTS Forum. (2011). *Mobile traffic forecasts 2010-2020 report*. Author.

Uykan, Z., & Jantti, R. (2014). Joint optimization of transmission-order selection and channel allocation for bidirectional wireless links—part I: Game theoretic analysis. *IEEE Transactions on Wireless Communications*, *13*(7), 4003–4013. doi:10.1109/TWC.2014.2314633

Varma, V. S., Elayoubi, S. E., Lasaulce, S., & Debbah, M. (2012, October). A flow level perspective on base station power allocation in green networks. In *Proceedings of Performance Evaluation Methodologies and Tools (VALUETOOLS)* (pp. 83-88). IEEE. doi:10.4108/valuetools.2012.250249

Varma, V. S., Mhiri, M., Treust, M. L., Lasaulce, S., & Samet, A. (2013, July). On the benefits of repeated game models for green cross-layer power control in small cells. In *Proceedings of Communications and Networking (BlackSeaCom)* (pp. 137-141). IEEE. doi:10.1109/BlackSeaCom.2013.6623397

Verbeeck, K., & Nowe, A. (2002). Colonies of learning automata. *IEEE Transactions on Systems, Man, and Cybernetics. Part B, Cybernetics*, *32*(6), 772–780. doi:10.1109/TSMCB.2002.1049611 PMID:18244883

Von Stackelberg, H. (1952). *The Theory of the Market Economy*. Oxford, UK: Oxford University Press.

Wang, F., Song, L., Han, Z., Zhao, Q., & Wang, X. (May, 2013). *Joint scheduling and resource allocation for Device-to-device underlay communication*. Paper presented at IEEE Wireless Communications and Networking Conference (WCNC), Shanghai, China. doi:10.1109/WCNC.2013.6554552

Wang, A., Cai, Y., Yang, W., & Hou, Z. (2013). A Stackelberg security game with cooperative jamming over a multiuser OFDMA network. In *Proceedings of Wireless Communications and Networking Conference (WCNC '13)*. Shanghai, China: IEEE. doi:10.1109/WCNC.2013.6555246

Wang, B., Han, Z., & Liu, K. (2009). Distributed relay selection and power control for multiuser cooperative communication networks using stackelberg game. *IEEE Transactions on Mobile Computing*, *8*(7), 975–990. doi:10.1109/TMC.2008.153

Wang, B., & Liu, K. J. R. (2011). Advances in cognitive radio networks: A survey. *IEEE Journal of Selected Topics in Signal Processing*, *5*(1), 5–23. doi:10.1109/JSTSP.2010.2093210

Wang, C.-Y., & Wei, H.-Y. (2013). Profit maximization in femtocell service with contract design. *IEEE Transactions on Wireless Communications*, *12*(5), 1978–1988. doi:10.1109/TWC.2013.032513.111789

Wang, D., Wang, X., & Zhao, Y. (2012). An Interference Coordination Scheme for Device-to-Device Multicast in Cellular Networks. In *Proceedings of the IEEE Vehicular Technology Conference (VTC Fall 2012)*. Quebec City, Canada: IEEE. doi:10.1109/VTCFall.2012.6398945

Wang, H., & Chu, X. (2012). Distance-Constrained Resource-Sharing Criteria for Device-to-Device Communications Underlaying Cellular Networks. *Electronics Letters*, *48*(9), 528–530. doi:10.1049/el.2012.0451

Wang, J., Zhu, D., Zhao, C., Li, J. C., & Lei, M. (2013). Resource sharing of underlaying device-to-device and uplink cellular communications. *IEEE Communications Letters*, *17*(6), 1148–1151. doi:10.1109/LCOMM.2013.042313.130239

Wen, S., Zhu, X., Lin, Z., Zhang, X., & Yang, D. (2013). *Distributed resource management for Device-to-Device (D2D) communication underlay cellular networks*. Paper presented at IEEE Personal Indoor and Mobile Radio Communications (PIMRC), London, UK. doi:10.1109/PIMRC.2013.6666402

WINNER+/CELTIC Deliverable CELTIC/CP5-026 D1.3. (2008). *Innovative Concepts in Peer-to-Peer and Network Coding*.

Wu, D., Cai, Y., Zhou, L., & Wang, J. (2012). A cooperative communication scheme based on coalition formation game in clustered wireless sensor networks. *IEEE Transactions on Wireless Communications*, *1*(3), 1190–1200. doi:10.1109/TWC.2012.012712.111049

Wu, D., Cai, Y., Zhou, L., Zheng, Z., & Zheng, B. (2013). Cooperative strategies for energy-aware ad hoc networks: A correlated-equilibrium game-theoretical approach. *IEEE Transactions on Vehicular Technology*, *62*(5), 2303–2314. doi:10.1109/TVT.2012.2235189

Wu, D., Wang, J., Hu, R. Q., Cai, Y., & Zhou, L. (2014). Energy-efficient resource sharing for mobile device-to-device multimedia communications. *IEEE Transactions on Vehicular Technology*, *63*(5), 2093–2103. doi:10.1109/TVT.2014.2311580

Wu, D., Zhou, L., & Cai, Y. (2012). Energy-efficient resource allocation for uplink orthogonal frequency division multiple access systems using correlated equilibrium. *IET Communications*, *6*(6), 659–667. doi:10.1049/iet-com.2011.0625

Wu, H., Xia, W., Li, J., & Shen, L. (2014). Max-min fair resource allocation in multi-cell ultra-high throughput WLAN. In *Proceedings of the 2014 International Conference on Wireless Communications and Signal Processing* (pp. 1-7). New York: IEEE. doi:10.1109/WCSP.2014.6992093

Wu, Q., Ding, G., Xu, Y., Feng, S., Du, Z., Wang, J., & Long, K. (2014). Cognitive Internet of Things: A New Paradigm Beyond Connection. *IEEE Internet of Things Journal*, *1*(2), 129–143. doi:10.1109/JIOT.2014.2311513

Wu, Q., Xu, Y., Wang, J., Shen, L., Zheng, J., & Anpalagan, A. (2013). Distributed channel selection in time-varying radio environment: Interference mitigation game with uncoupled stochastic learning. *IEEE Transactions on Vehicular Technology*, *62*(9), 4524–4538. doi:10.1109/TVT.2013.2269152

Wu, Y., Zhu, Q., Huang, J., & Tsang, D. H.-K. (2014). *Revenue sharing based resource allocation for dynamic spectrum access networks*. IEEE Journalon Selected Areas in Communications.

Wyner, A. D. (1975). The wire-tap channel. *The Bell System Technical Journal*, *54*(8), 1355–1387. doi:10.1002/j.1538-7305.1975.tb02040.x

Xiang, L., Pantisano, F., Verdone, R., Ge, X., & Chen, M. (2011). Adaptive Traffic Load-Balancing for Green Cellular Networks. In *Proceedings of the IEEE International Symposium on Personal Indoor and Mobile Radio Communications (PIMRC 2011)*. Toronto, Canada: IEEE. doi:10.1109/PIMRC.2011.6139995

Xiao, M. B., & Shroff, N. B. (2001). Utility-Based power control in cellular wireless systems. In Proceedings of IEEE INFOCOM (pp. 412-421). IEEE.

Xing, Y., & Chandramouli, R. (2008). Stochastic learning solution for distributed discrete power control game in wireless data networks. *IEEE/ACM Transactions on Networking*, *16*(4), 932–944. doi:10.1109/TNET.2007.911424

Xing, Y., Mathur, C. N., Haleem, M. A., Chandramouli, R., & Subbalakshmi, K. (2007). Dynamic spectrum access with QoS and interference temperature constraints. *IEEE Transactions on Mobile Computing*, *6*(4), 423–433. doi:10.1109/TMC.2007.50

Xu, C., Song, L., Han, Z., Zhao, Q., Wang, X., Cheng, X., & Jiao, B. (2013). Efficiency resource allocation for device-to-device underlay communication systems: A reverse iterative combinatorial auction based approach. *IEEE Journal on Selected Areas in Communications*, *31*(9), 348–358. doi:10.1109/JSAC.2013.SUP.0513031

Xu, J., Wang, J., Zhu, Y., Yang, Y., & Zheng, X. (2014). Cooperative distributed optimization for the hyper-dense small cell deployment. *IEEE Communications Magazine*, *52*(5), 61–67. doi:10.1109/MCOM.2014.6815894

Xu, S. Y., Wang, H. M., Chen, T., Peng, T., & Kwak, K. S. (2012). Device-to-Device Communication Underlaying Cellular Networks: Connection Establishment and Interference Avoidance. *Transactions on Internet and Information Systems (Seoul)*, *6*(1), 203–228.

Xu, Y., Anpalagan, A., Wu, Q., Shen, L., Gao, Z., & Wang, J. (2013). Decision-theoretic distributed channel selection for opportunistic spectrum access: Strategies, challenges and solutions. *IEEE Communications Surveys and Tutorials*, *15*(4), 1689–1713. doi:10.1109/SURV.2013.030713.00189

Xu, Y., Wang, J., Wu, Q., Anpalagan, A., & Yao, Y. (2012). Opportunistic spectrum access in cognitive radio networks: Global optimization using local interaction games. *IEEE Journal of Selected Topics in Signal Processing*, *6*(2), 180–194. doi:10.1109/JSTSP.2011.2176916

Xu, Y., Wang, J., Wu, Q., Anpalagan, A., & Yao, Y.-D. (2012b). Opportunistic spectrum access in unknown dynamic environment: A game-theoretic stochastic learning solution. *IEEE Transactions on Wireless Communications*, *11*(4), 1380–1391. doi:10.1109/TWC.2012.020812.110025

Xu, Y., Wang, J., Wu, Q., Du, Z., Shen, L., & Anpalagan, A. (2015). A game-theoretic perspective on self-organizing optimization for cognitive small cells. *IEEE Communications Magazine*, *53*(7), 100–108. doi:10.1109/MCOM.2015.7158272

Xu, Y., Wu, Q., Shen, L., Wang, J., & Anpalagan, A. (2013). Opportunistic spectrum access with spatial reuse: Graphical game and uncoupled learning solutions. *IEEE Transactions on Wireless Communications*, *12*(10), 4814–4826. doi:10.1109/TWC.2013.092013.120862

Xu, Y., Wu, Q., Wang, J., Shen, L., & Anpalagan, A. (2013). Opportunistic spectrum access using partially overlapping channels: Graphical game and uncoupled learning. *IEEE Transactions on Communications*, *61*(9), 3906–3918. doi:10.1109/TCOMM.2013.072913.120881

Xu, Y., Wu, Q., Wang, J., Shen, L., & Anpalagan, A. (in press). Robust multiuser sequential channel sensing and access in dynamic cognitive radio networks: Potential games and stochastic learning. *IEEE Transactions on Vehicular Technology*.

Xu, Z., Li, G. Y., Yang, C., & Zhu, X. (2012). Throughput and optimal threshold for FFR schemes in OFDMA cellular networks. *IEEE Transactions on Wireless Communications*, *11*(8), 2776–2785.

Yaacoub, E. E., & Dawy, Z. (2012). *Resource Allocation in Uplink OFDMA Wireless Systems*. Hoboken, NJ: Wiley & Sons. doi:10.1002/9781118189627

Yaacoub, E., & Abu-Dayya, A. (2012). Multihop routing for energy efficiency in wireless sensor networks. In M. Matin (Ed.), *Wireless Sensor Networks - Technology and Protocols* (pp. 165–188). Rijeka, Croatia: InTech Academic Publishers. doi:10.5772/39221

Yaacoub, E., Ghazzai, H., Alouini, M.-S., & Abu-Dayya, A. (2013). Achieving Energy Efficiency in LTE with Joint D2D Communications and Green Networking Techniques. In *Proceedings of the IEEE International Wireless Communications and Mobile Computing Conference (IWCMC 2013)*. Cagliari, Italy: IEEE. doi:10.1109/IWCMC.2013.6583571

Yaipairoj, S., & Harmantzis, F. C. (2006). Auction-based congestion pricing for wireless data services. In *Proceedings of IEEE Int. Communications* (pp. 1059-1064). New York: IEEE. doi:10.1109/ICC.2006.254887

Yang, S., Chi, Z., & Yuguang, F. (2010). Stochastic Traffic Engineering in Multihop Cognitive Wireless Mesh Networks. *IEEE Transactions on Mobile Computing*, *9*(3), 305–316. doi:10.1109/TMC.2009.111

Yan, Y., Huang, J., & Wang, J. (2013). Dynamic bargaining for relay-based cooperative spectrum sharing. *IEEE Journal on Selected Areas in Communications*, *31*(8), 1480–1493. doi:10.1109/JSAC.2013.130812

Yates, R. D. (1995). A framework for uplink power control in cellular radio systems. *IEEE Journal on Selected Areas in Communications*, *13*(7), 1341–1347.

Yates, R. D. (1995). A framework for uplink power control in cellular radio systems. *IEEE Journal on Selected Areas in Communications*, *13*(7), 1341–1347. doi:10.1109/49.414651

Yenumula, B. (2009). A Game Theory Approach to Detect Malicious Nodes in Wireless Sensor Networks.*Third International Conference on Sensor Technologies and Applications.*

Yi, Y., Zhang, J., Zhang, Q., & Jiang, T. (2012). Spectrum leasing to femto service provider with hybrid access. In *Proceedings ofIEEE International Conference on Computer Communications (INFOCOM)*. IEEE.

Yilmaz, E., Zakhour, R., Gesbert, D., & Knopp, R. (2010). Multi-pair two-way relay channel with multiple antenna relay station. In *Proceedings ofIEEE International Conference on Communications (ICC)*. IEEE. doi:10.1109/ICC.2010.5502396

Yingmin, W. (2010). *TD-LTE technology principle and system design*. Beijing: People's Posts and Telecommunications Press.

Young, H. P. (2001). *Individual strategy and social structure: An evolutionary theory of institutions*. Princeton, NJ: Princeton University Press.

Young, H. P. (2009). Learning by trial and error. *Games and Economic Behavior*, *65*(2), 626–643. doi:10.1016/j.geb.2008.02.011

Yuan, W., Leung, H., Cheng, W., & Chen, S. (2011). Optimizing voting rule for cooperative spectrum sensing through learning automata. *IEEE Transactions on Vehicular Technology*, *60*(7), 3253–3264. doi:10.1109/TVT.2011.2161001

Yu, C., Doppler, K., Ribeiro, C. B., & Tirkkonen, O. (2011). Resource sharing optimization for device-to-device communication underlaying cellular networks. *IEEE Transactions on Wireless Communications*, *10*(8), 2752–2763. doi:10.1109/TWC.2011.060811.102120

Yu, Q., Chen, J., Sun, Y., Fan, Y., & Shen, X. (2010). Regret matching based channel assignment for wireless sensor networks. In *Proceedings of the 2010 IEEE International Conference on Communications (ICC).*Cape Town, South Africa: IEEE. doi:10.1109/ICC.2010.5502547

Yu, W., Ginis, G., & Cioffi, J. M. (2002). Distributed multiuser power control for digital subscriber lines. *IEEE Journal on Selected Areas in Communications*, *20*(5), 1105–1115. doi:10.1109/JSAC.2002.1007390

Zander, J. (1992). Performance of optimum transmitter power control in cellular radio systems. *IEEE Transactions on Vehicular Technology*, *41*(1), 57–62.

Zappone, A., Alfano, G., Buzzi, S., & Meo, M. (2011, September). Energy-efficient non-cooperative resource allocation in multi-cell OFDMA systems with multiple base station antennas. In *Proceedings ofOnline Conference on Green Communications (GreenCom)* (pp. 82-87). IEEE. doi:10.1109/GreenCom.2011.6082512

Zemlianov, A., & de Veciana, G. (2005). Cooperation and decision-making in a wireless multi-provider setting. In *Proceedings of the 24th Annual Joint Conference of the IEEE Computer and Communication Societies* (vol. 1, 386-397). Miami, FL: IEEE. doi:10.1109/INFCOM.2005.1497908

Zeng, Y., & Liang, Y.-C. (2009). Eigenvalue-based spectrum sensing algorithms for cognitive radio. *IEEE Transactions on Communications*, *57*(6), 1784–1793. doi:10.1109/TCOMM.2009.06.070402

Zhang, G., & Zhang, H. (2008, November). Adapative resource allocation for downlink OFDMA networks using cooperative game theory. In *Proceedings of Communication Systems, 2008* (pp. 98-103). IEEE.

Zhang, C. R., & Zhang, E. (2014). Optimized bidding algorithm of real time bidding in online ads auction. In *Proceedings of International Conference on Management Science and Engineering* (pp. 33-42). New York: IEEE. doi:10.1109/ICMSE.2014.6930205

Zhang, H., Chu, X., & Wen, X. (2013). *Energy Efficient Power Control in Femtocells with Interference Pricing. In 4G Femtocells* (pp. 51–61). Springer New York.

Zhang, H., & Fang, X. (2007). A Pricing and Game theory based Call Admission Control Scheme for CDMA Systems.*International conference on Wireless Communications, Networking and Mobile Computing*. doi:10.1109/WICOM.2007.1593

Zhang, J., Zhang, R., Li, G., & Hanzo, L. (2012). Remote Coalition Network Elements for Base Station Cooperation Aided Multicell Processing. *IEEE Transactions on Vehicular Technology*, *61*(3), 1406–1415. doi:10.1109/TVT.2012.2186600

Zhang, L., & Li, D. (2003). Resource allocation models based on pricing for multiservice WCDMA networks. In *Proceedings of 14th IEEE Int. Symp. Personal, Indoor and Mobile Radio Communication* (pp. 1466-1470). New York: IEEE. doi:10.1109/PIMRC.2003.1260356

Zhang, R., Song, L., Han, Z., & Jiao, B. (2011, April). Improve physical layer security in cooperative wireless network using distributed auction games. In *Proceedings of Conference on Computer Communications Workshops (INFOCOM WKSHPS '11).*Shanghai, China: IEEE.

Zhang, R., Song, L., Han, Z., & Jiao, B. (2011, June). Distributed coalition formation of relay and friendly jammers for secure cooperative networks. In *Proceedings of International Conference on Communications (ICC '11)*. Milan, Italy: IEEE. doi:10.1109/icc.2011.5962513

Zhang, S., Liew, S., & Lam, P. P. (2006). Hot topic: physical-layer network coding. In *Proceedings of ACM MobiCom* (pp. 358-365). ACM.

Zhang, Z., Song, L., Han, Z., Saad, W., & Lu, Z. (2013). Overlapping coalition formation games for cooperative interference management in small cell networks. *Wireless Communications and Networking Conference (WCNC)* (pp.643-648). IEEE. doi:10.1109/WCNC.2013.6554639

Zhao, X., & Yang, X. (2012). Downlink Ergodic Capacity Analysis for Wireless Networks with Cooperative Distributed Antenna Systems. In *Proceedings of the IEEE International Conference on Communications (ICC 2012)*. Ottawa, Canada: IEEE. doi:10.1109/ICC.2012.6364775

Zheng, J., Cai, Y., Liu, Y., Xu, Y., Duan, B., & Shen, X. (2014). Optimal power allocation and user scheduling in multicell networks: Base station cooperation using a game-theoretic approach. *IEEE Transactions on Wireless Communications*, *13*(12), 6928–6942. doi:10.1109/TWC.2014.2334673

Zhong, W., Chen, G., Jin, S., & Wong, K. K. (2014). Relay selection and discrete power control for cognitive relay networks via potential game. *IEEE Transactions on Signal Processing*, *62*(20), 5411–5424. doi:10.1109/TSP.2014.2347261

Zhong, W., Xu, Y., Tao, M., & Cai, Y. (2010). Game theoretic multimode precoding strategy selection for MIMO multiple access channels. *IEEE Signal Processing Letters*, *17*(6), 563–566. doi:10.1109/LSP.2010.2047315

Zhong, W., Xu, Y., & Tianfield, H. (2011). Game-theoretic opportunistic spectrum sharing strategy selection for cognitive MIMO multiple access channels. *IEEE Transactions on Signal Processing*, *59*(6), 2745–2759. doi:10.1109/TSP.2011.2121063

Zhou, P., Liu, W., Yuan, W., & Cheng, W. (2008, March). Energy-efficient joint power and rate control via pricing in wireless data networks. In *Proceedings of Wireless Communications and Networking Conference, 2008* (pp. 1091-1096). IEEE. doi:10.1109/WCNC.2008.197

Zulhasnine, M., Huang, C., & Srinivasan, A. (2010). *Efficient resource allocation for device-to-device communication underlaying LTE network*. Paper presented at IEEE Wireless and Mobile Computing, Networking and Communications (WiMob), Niagara, Canada. doi:10.1109/WIMOB.2010.5645039

About the Contributors

Chungang Yang received the Bachelor and Doctoral Degrees in the Department of Electrical and Information Engineering, Xidian University, Xi'an, China in 2006 and 2011, respectively. Between Sep. 2010 and March 2011, he held the visiting scholar position in the department of electrical and computer engineering at Michigan Technological University. Since July 2011, he has been with Xidian University as an assistant professor, where now he works as associated professor. Dr. Yang is a member of IEEE. His research interests include communication theory, information theory and their applications in practical wireless communications and networks with emphasis on Cognitive, Heterogeneous, Cooperative Wireless Cellular Networks, and Cloud Computing Networks; Advanced, flexible, and autonomous multiple resource management and interference management, and mathematical modeling, analysis and design of both spectral and energy efficiency; Application and development of game theory and intelligent computing in wireless communications community, and network economics.

Jiandong Li was graduated from Xidian University with Bachelor Degree, Master Degree and Ph. D in Communications and Electronic System respectively in 1982, 1985 and 1991. He is with Xidian University from 1985, an associate professor from 1990 to 1994, professor from 1994, Ph.D student supervisor from 1995, and Dean of School of Telecommunication Engineering, Xidian University from 1997 respectively. Now, he also serves as Executive Vice Dean Graduate School, Xidian University. Dr. Li is a senior member of IEEE and China Institute of Electronics (CIE) and the fellow of China Institute of Communication (CIC). He was the member of PCN specialist group for China 863 Communication high technology program between Jan. 1993 and Oct. 1994 and from 1999 to 2000. He is also the member of Communication specialist group for The Ministry of Information Industry. His current research interest and projects are funded by the 863 High Tech Project, NSFC, National Science Fund for Distinguished Young Scholars, TRAPOYT, MOE and MOI.

Manzoor Ahmed received his B.E. degree from Balochistan University of Engineering and Technology Khuzdar, Pakistan in 1996 and M.S. degree from Balochistan University of Information Technology Engineering and Management Sciences in 2010. He worked as Divisional Engineer in National Telecommunication Corporation, and serving there since 2000. Currently, he is pursuing Ph.D. degree with the Key Laboratory of Universal Wireless Communications at Beijing University of Posts and Telecommunications, Beijing, China since 2011. His research interests include the optimization, resource allocation and interference management using game theory in Heterogeneous small cell networks and in C-RAN etc.

Seyed Alavi received his B.S. and M.S. degrees in Electrical Engineering form K. N. Toosi and Sharif University of Technology, Tehran, Iran, in 2004 and 2006, respectively. He was a teaching assistant in Sharif University of technology from 2004 to 2006. He received his Ph.D. at Illinois Institute of Technology, Chicago in 2012. His research interests include wireless communication networks, resource allocation in multi user, multi carrier networks, the application of optimization and game theory to network design and analysis and admission control.

Luis Alonso obtained the PhD from UPC (Barcelona) in 2001 and got a permanent tenured position at the same University becoming an Associate Professor in 2006. He has been co-founder of the Wireless Communications and Technologies Research Group (WiComTec), to which currently belongs. His current research interests are within the field of medium access protocols, radio resource management, cross-layer optimization, cooperative transmissions, cognitive radio and QoS features for all kind of wireless communications systems. He is author of more than forty research papers in international journals and magazines, one book, twelve chapters of books and more than one hundred papers in international congresses and symposiums. He has received several best paper awards.

Mohamed-Slim Alouini was born in Tunis, Tunisia. He received the Ph.D. degree in Electrical Engineering from the California Institute of Technology (Caltech), Pasadena, CA, USA, in 1998. He served as a faculty member in the University of Minnesota, Minneapolis, MN, USA, then in the Texas A&M University at Qatar, Education City, Doha, Qatar before joining King Abdullah University of Science and Technology (KAUST), Thuwal, Makkah Province, Saudi Arabia as a Professor of Electrical Engineering in 2009. His current research interests include the modeling, design, and performance analysis of wireless communication systems.

Angelos Antonopoulos (Thessaloniki, 1983) received his PhD degree (Cum Laude) from the Signal Theory and Communications (TSC) Department of the Technical University of Catalonia (UPC) in December 2012, while he holds an MSc degree from the Information and Communication Systems Engineering Department (ICSDE) of the University of the Aegean (2007). His main research interests include cooperative communications, MAC protocols, network coding and energy efficient network planning. He has participated in several European and Spanish national projects (e.g., GREENET, Green-T, CO2GREEN, etc.) and has served as an expert evaluator of research projects funded by the Romanian Government through the National Council for Scientific Research. He has been granted three annual scholarships by the Greek State Scholarships Foundation (IKY) and, recently, he has been awarded the First Polytechnic Graduates Prize by the Technical Chamber of Greece (TEE-TGC).

Giuseppe Araniti is an Assistant Professor of Telecommunications at the University Mediterranea of Reggio Calabria, Italy. From the same University he received the Laurea (2000) and the Ph.D. degree (2004) in Electronic Engineering. His major area of research includes Personal Communications Systems, Enhanced Wireless and Satellite Systems, Traffic and Radio Resource Management, Multicast and Broadcast Services, device-to-device and machine type communications over 4G/5G cellular networks.

Mehdi Bennis received his M.Sc. degree from the Ecole Polytechnique Federale de Lausanne (EPFL), Switzerland and the Eurecom Institute, France in 2002. He obtained his Ph.D. degree in electrical engineering December 2009 on spectrum sharing for future mobile cellular systems. His main research interests are in radio resource management, heterogeneous networks, game theory and machine learning. Dr. Bennis was the co-PI of the Broadband Evolved FEMTO (FP7-BeFEMTO) project, and is currently the PI of the European CELTIC project SHARING (2013-2016). He has published more than 100 research papers in international conferences, journals, book chapters and patents. Dr. Bennis is currently an editor for the IEEE Transaction of Wireless Communications. Dr. Bennis co-chaired a dozen of IEEE workshops including the four consecutive editions of the international workshop on small cell wireless networks (SmallNets) held in conjunction with IEEE ICC 2012-2015. Recently, Dr. Bennis gave tutorial presentations at IEEE PIMRC 2012, IEEE GLOBECOM 2012, IEEE DySPAN 2014, IEEE WCNC 2014 and IEEE GLOBECOM 2014, IEEE ICC 2015. Dr. Bennis has also given keynote speeches at IEEE ICT 2013, and IEEE WCNC 2014 CLEEN WS. Finally, Dr. Bennis was a guest editor on 4 special issues (forthcoming JSAC SI on Hetnets, IEEE Communication magazine on full duplexing, big data and UAVs). Dr. Bennis is the winner of the 2015 IEEE Communications Society Fred W. Ellersick Prize.

Alexandra Bousia received her B.S. and M.S. degree in Computer & Communication Engineering from the Department of Computer & Communication Engineering in University of Thessaly, in 2008 and 2009, respectively. She is currently working towards the PhD degree in Wireless Communications at the Technical University of Catalonia (UPC) and she is working as a Marie Curie researched in the same department. Since 2011, she has been involved in several national and European research funded projects (Greenet, CO2Green, Green-T, Geocom, AGAUR). Her research interests include wireless networks, MAC protocols, energy efficient communication protocols and RRM algorithms.

Yueming Cai received his B.S. degree in Physics from Xiamen University, Xiamen, China in 1982, the M.S. degree in Micro-electronics Engineering and the Ph.D. degree in Communications and Information Systems both from Southeast University, Nanjing, China in 1988 and 1996 respectively. His current research interests include MIMO systems, OFDM systems, signal processing in communications, cooperative communications and wireless sensor networks.

Tsung-Hui Chang received the B.S. degree in electrical engineering and the Ph.D. degree in communications engineering from the National Tsing Hua University (NTHU), Hsinchu, Taiwan, in 2003 and 2008, respectively. He was an exchange Ph.D. student of University of Minnesota, Minnea polis, MN, a visiting scholar of The Chinese University of Hong Kong, Hong Kong, and a Postdoctoral Researcher with the Institute of Communications Engineering, NTHU. He currently worksasaPostdoc toral Researcher with the Depart-ment of Electrical and Computer Engineering, University of California, Davis. His research interests are widely in wireless communications, digital signal pro-cessing, and convex optimization and its applications.

Xu Chen obtained the Ph.D. degree in Information Engineering from The Chinese University of Hong Kong, Hong Kong, China, in 2012. Before that, he received the B.Eng. degree in Electronic Engineering from South China University of Technology, Guangzhou, China, in 2008. From August 2012 to February 2014, Dr. Xu Chen was a Postdoctoral Research Fellow in Arizona State University, USA. Since April 2014, he is a Humboldt Scholar Fellow at the Computer Networks Group of University of Goettingen, Germany. Dr. Xu Chen is the recipient of the 2014 Hong Kong Young Scientist Runner-Up Award, the Best Paper Runner-up Award of IEEE International Conference on Computer Communications (INFOCOM) 2014 and the Honorable Mention Award of IEEE international conference on Intelligence and Security Informatics (ISI) 2010.

Bo-Wei Chen served as a volunteer in IEEE Region 10 during 2010-2012. He currently works as a postdoctoral research fellow at Princeton University, USA. He used to serve as a guest editor in several journals, e.g., IEEE/ACM, Springer, and Elsevier. Moreover, he also hosted many international conferences. Dr. Chen published many SCI journals, including high-quality papers in IEEE/ACM Transactions. His research interest covers machine learning, audiovisual signal processing, and human-machine interactions.

Yiqiang Chen received the B.A.Sc. and M.A. degrees from the University of Xiangtan, Xiangtan City, China, in 1996 and 1999, respectively, and the Ph.D. degree from the Institute of Computing Technology (ICT), Chinese Academy of Sciences (CAS), Beijing, China, in 2002. In 2004, he was a Visiting Scholar Researcher at the Department of Computer Science, Hong Kong University of Science and Technology (HKUST), Hong Kong. Currently, he is a Professor and Director of the pervasive computing research center at ICT, CAS. His research interests include artificial intelligence, pervasive computing, and human computer interface.

Rui Gao received her Master degrees of Information and Communication Engineering in Beijing University of Posts and Telecommunications (BUPT) in 2015. Her research interests focus on key technologies in small cell enhancement such as power control in small cell network, small cell on/off, offloading between small cells and macro cells.

Hakim Ghazzai was born in Tunisia. He received the Dipl. Ing. degree (with highest distinction) in telecommunication engineering and the Master's degree in high-rate transmission systems from Ecole Supérieure des Communications de Tunis, Tunis, Tunisia, in 2010 and 2011, respectively. He obtained his Ph.D. degree in electrical engineering from King Abdullah University of Science and Technology, Thuwal, Saudi Arabia, in 2015. Afterwards, he joined the Qatar Mobility Innovations Center where he is currently a research scientist. His general research interests include the intersection of wireless networks and green communications and its optimization.

Zhu Han received the B.S. degree in electronic engineering from Tsinghua University, in 1997, and the M.S. and Ph.D. degrees in electrical engineering from the University of Maryland, College Park, in 1999 and 2003, respectively. Currently, he is a Full Professor in Electrical and Computer Engineering Department at University of Houston, Texas. His research interests include wireless resource allocation and management, wireless communications and networking, game theory, wireless multimedia, and security. Dr. Han is an NSF CAREER award recipient 2010, and he is an Associate Editor of IEEE Transactions on Wireless Communications since 2010. Dr. Han is the winner of the 2011 IEEE Communications Society Fred W. Ellersick Prize, and he is the coauthor for the papers that won eight IEEE best conference paper awards. Dr. Han is an IEEE Fellow since 2014, and IEEE distinguished lecturer since 2015.

Jianwei Huang is an Associate Professor and Director of the Network Communications and Economics Lab (ncel.ie.cuhk.edu.hk) in the Department of Information Engineering at the Chinese University of Hong Kong. He received his Ph.D. degree from Northwestern University in 2005. Dr. Huang is the co-recipient of 8 Best Paper Awards, including IEEE Marconi Prize Paper Award in Wireless Communications in 2011. He has co-authored three books: Wireless Network Pricing, Monotonic Optimization in Communication and Networking Systems, Cognitive Mobile Virtual Network Operator Games, Social

Cognitive Radio Networks, and ESI Highly Cited Papers in the field of Computer Science. He has served as the Editor of IEEE Journal on Selected Areas in Communications - Cognitive Radio Series and IEEE Transactions on Wireless Communications, and Guest Editor of IEEE Journal on Selected Areas in Communications and IEEE Communications Magazine. He has served as Associate Editor-in-Chief of IEEE Communications Society Technology News and Chair of IEEE Communications Society Multimedia Communications Technical Committee. He is a Distinguished Lecturer of IEEE ComSoc since 2015.

Kaizhi Huang, born in 1973, received the Ph.D. degree in communication and information system from Tsinghua University, Beijing, China. Currently, she is a professor and doctoral supervisor of National Digital Switching System Engineering & Technological R&D Center, China. She is serving as a leader of wireless mobile communication innovation technology team, Henan Prov., China. Her research interests include wireless communications and networks, wireless physical layer security.

Liang Huang received his B.S. degree in Computer Science from Chong Qing University of Post and Telecommunications in 2004, M.S. degree in Computer Science from Beijing University of Posts and Telecommunications in 2007 and Ph.D. degree in Computer Architecture from the Institute of Computing Technology, Chinese Academy of Sciences in 2013. From 2007 to 2009, he worked as a telecommunication R&D engineer in Agilent Technologies. In 2013, he joined the National Computer Network Emergency Response Technical Team, Coordination Center of China, as a principle engineer. His research interest includes radio resource management, heterogeneous network, centralized Radio Access Network and mobile internet.

Antonio Iera graduated in Computer Engineering at the University of Calabria, Italy, in 1991 and received a Master Diploma in Information Technology from CEFRIEL/Politecnico di Milano, Italy, in 1992 and a Ph.D. degree from the University of Calabria in 1996. Since 1997 he has been with the University of Reggio Calabria and currently holds the position of full professor of Telecommunications and Director of the Laboratory for Advanced Research into Telecommunication Systems (www.arts.unirc.it). IEEE Senior Member since 2007. His research interests include, next generation mobile and wireless systems, RFID systems, and Internet of Things.

Wen Ji is an Associate Professor with the Institute of Computing Technology (ICT), Chinese Academy of Sciences (CAS), Beijing, China. She was a Post Doctoral Fellow from 2007 to 2009, and was an Assistant Professor from 2009 to 2010 at ICT, CAS. Her research areas include video communication and networking, video coding, channel coding, information theory, optimization, network economics, and pervasive computing.

Shaojie Kang is a M.D. student with the Institute of Computing Technology, Chinese Academy of Sciences, China. His research interest includes video transmission, Wireless video communication, pervasive computing, communication, and networking.

Elli Kartsakli received her Ph.D. in Wireless Telecommunications from the Technical University of Catalonia (UPC) in February 2012. She holds a degree in Electrical and Computer Engineering from the National Technical University of Athens, Greece (2003) and an M.Sc. in Mobile and Satellite Communications from the University of Surrey, UK (2004). She is currently a Post-Doctoral Researcher in

UPC and has participated in several national and European projects (GREENET, WSN4QoL, etc.). Her primary research interests include wireless networking, channel access protocols and energy efficient communication protocols.

Wei-Sheng Lai received the Bachelor degrees in electrical engineering from National Taipei University of Technology, Taipei, Taiwan, in 2002 and the Master degrees in electrical engineering from National Sun Yat-Sen University, Kaohsiung, Taiwan, in 2004. Currently, he is a Ph.D. candidate in the department of communication engineering, National Chiao Tung University, Hsinchu, Taiwan. Simultaneously, he is also a R&D Engineer at Taoyuan. His research interests encompasses the wireless communication and network, signal processing and moving target estimation. Currently, his research topic includes deployment of small cell networks, resource allocation design and game-theoretic models for communication networks.

Ta-Sung Lee received the B.S. degree from National Taiwan University in 1983, the M.S. degree from University of Wisconsin, Madison, in 1987, and the Ph.D. degree form Purdue University, W. Lafayette, IN, in 1989, all in electrical engineering. In 1990, he joined the Faculty of National Chiao Tung University (NCTU), Hsinchu, Taiwan, where he holds a position as Professor of Department of Electrical and Computer Engineering. From 1999 to 2001, he was Director of Communications & Computer Continuing Education Program, NCTU. From 2005 to 2007, he was Chairman of Department of Communication Engineering, and from 2007 to 2008, he was Dean of Student Affairs of NCTU. From 2008 to 2010, he was Commissioner of the National Communications Commission (NCC), a regulatory agency of Taiwan similar to the FCC, and responsible for the strategic planning, policy making and technical regulation for the telecommunications and broadcasting services in Taiwan. He was Vice Chairman and Chairman of IEEE Communications Society Taipei Chapter for 2005-2008, a Board Member of IEEE Taipei Section for 2007-2010, an Associate Editor of IEEE Transactions on Signal Processing for 2009-2013, and IEEE Signal Processing Society Regional Director-at-Large for R10 for 2011-2013. He is currently an Associate Editor of Journal of Signal Processing Systems. He has been Chairman of Telecom Technology Center, a government funded agency for telecommunications R&D, since 2013. Dr. Lee is actively involved in research and development in signal processing and system design for wireless communications. He has won several awards for his research, engineering and education contributions; these include National Science Council (NSC) Research Award, 1999 Young Electrical Engineer Award of the Chinese Institute of Electrical Engineering (CIEE), 2011 Distinguished Electrical Engineering Professor Award of CIEE, NCTU Distinguished Scholar Award, and NCTU Teaching Award.

Mingliang Li born in 1988, received his B.E. in Henan University of China in 2010, and received his M.E. in National Digital Switching System Engineering&Technological R&D Center of China (NDSC) in 2013 and then became a Ph.D student in NDSC. His research area is wireless physical layer security.

Mingyang Li received his B.S. degree of Electronic Information Science and Technology in Beijing University of Posts and Telecommunications (BUPT) in 2013. He is currently a Master candidate at BUPT. His research interests focus on wireless networking theory and key technologies, including heterogeneous networking, radio resource management and mobility.

Leonardo Militano is currently an Assistant Professor at the Mediterranea University of Reggio Calabria, Italy. He received a Ms. degree in Telecommunications Engineering in 2006 and his Ph.D in Telecommunications Engineering in 2010 from the University of Reggio Calabria. He has been a visiting Ph.D student at the Mobile Device group at University of Aalborg, Denmark. His major areas of research are wireless systems, radio resource management, device-to-device communications and user cooperation.

Antonella Molinaro is an Associate Professor of Telecommunications at University Mediterranea of Reggio Calabria, Italy. Before her current appointment, she worked at the University of Messina (1998–2001) and the University of Calabria (2001–2004) as an Assistant Professor, and at Polytechnic of Milano (1997) as a research assistant. She worked with Telesoft S.p.A. in Rome as a network designer (1992–1993), and with Siemens AG, Munich, Germany (1994–1995) as an EU Fellow within the RACE II ATDMA (Advanced-TDMA Mobile Access) project. She received the Laurea Degree in Computer Engineering from University of Calabria in 1991, a post-laurea master diploma in Information Technology from CEFRIEL/Politecnico di Milano in 1992, and a Ph.D. degree in Multimedia Technology and Communication Systems in 1996. She has served in the Technical Programme Committee, Steering Committee, and Advisory Board of several IEEE conferences. Her recent research activity mainly focuses on architectures and protocols design and optimization for wireless networking, multihop communications, wireless ad hoc networks, vehicular networks, information-centric networking for the future Internet.

Mugen Peng received the B.S. degree in electronics engineering in 2000 from Nanjing University of Posts & Telecommunications (NJUPT), and a PhD degree from the Beijing University of Posts & Telecommunications (BUPT) in 2005. After the PhD graduation, he joined in BUPT, and has become a full professor with the school of information and communication engineering in BUPT since Oct. 2012. From Dec. 2013 to now, he is also an academic visiting fellow in Princeton University, USA. He is leading a research group focusing on wireless transmission and networking technologies in the Key Laboratory of Universal Wireless Communications (Ministry of Education) at BUPT, China. Prof. Peng is currently on the Editorial/Associate Editorial Board of IEEE Communications Mag- azine, IEEE Access, International Journal of Antennas and Propagation (IJAP), China Communi- cation, and International Journal of Communications System (IJCS). He has been the guest leading editor for the special issues in IEEE Wireless Communications, IJAP, and the International Journal of Distributed Sensor Networks (IJDSN). He was the guest editor of IET Communications. He is serving as the track chair for GameNets 2014, WCSP 2013 and ICCT 2011, the workshop co-chair of ChinaCom 2012, and the leading chair for SON-HetNet 2013 in IEEE PIMRC 2013 and Green HetNet 2013 in IEEE GreenCom2013. He has served as the reviewers for major IEEE journals and conferences in the areas of communications.

Francesco Scarcello received the PhD degree in Computer Science from the University of Calabria in 1997. He is a professor of computer science (SSD ING-INF/05) at the University of Calabria. His research interests are computational complexity, graph and hypergraph theory, constraint satisfaction, logic programming, knowledge representation, non-monotonic reasoning, and database theory. He has extensively published in all these areas in leading conferences and journals. In particular, in database theory he solved some open problems on the complexity of acyclic conjunctive queries (Journal of the ACM 48(3):431-498,2001), and on some techniques to identify tractable classes of queries. From these studies the idea of a new notion of degree of cyclicity for hypergraphs, called hypertree-width came out. It turned out that this notion is the best technique for identifying tractable classes of hard problems, in that it is able to single out the largest islands of tractability among all other methods previously known in the literature (Artif. Intell. 124(2): 243-282, 2000, J. Comput. Syst. Sci. 64(3): 579-627, 2002, J. Comput. Syst. Sci. 66(4): 775-808, 2003, J. Comput. Syst. Sci. 73(3): 475-506, 2007). He participated in a number of national and international projects dealing with database and knowledge representation systems. In particular, the notions and the algorithms proposed in his Phd thesis have been used for developing DLV, a widely used knowledge-base management system based on disjunctive logic programming (Inf. Comput. 135(2): 69-112, 1997, ACM Trans. Comput. Log. 7(3): 499-562, 2006). Professor Scarcello holds his current position since 2001. Before that, he was recipient of two grants from the Italian National Research Council (CNR) until 1999, when he became an assistant professor at the University of Calabria. Also, during these years, he has been visiting the Department of Information Systems at the Vienna University of Technology. Professor Scarcello serves on program committees and as a reviewer for many international conferences and journals. Moreover, it is a member of the Editorial Board of the Artificial Intelligence journal (Elsevier). He is co-recipient of the 2008 IJCAI-JAIR Best Paper Prize, awarded to an outstanding paper published in the Journal of Artificial Intelligence Research in the preceding five calendar years. He is co-recipient of the 2009 ACM PODS Alberto O. Mendelzon Test-of-Time Award, awarded every year to a paper published in the proceedings of the ACM Symposium on Principles of Database Systems (PODS) ten years prior that had the most impact in terms of research, methodology, or transfer to practice over the intervening decade.

Jinglin Shi received his Ph.D. degree from Beijing Institute of Technology. He currently serves as the director of Wireless Communication Technology Research Center in Institute of Computing Technology, Chinese Academy of Science. He is also a visiting professor of Beijing University of Posts and Telecommunications, University of Sydney, University of Wollongong and Macquarie University. His research interests focuses on wireless communication systems. As a team leader, he led the development of TD-SCDMA, WiMAX, and LTE protocol stack systems successfully. He is currently responsible for several national projects in broadband wireless communication, including TDD-LTE baseband chip design and radio resource management techniques for IMT-A. Prof. Shi has published 2 books and over 100 papers in telecommunications journals and conference proceedings and has more than 30 patents granted. He was the General Co-chair of ChinaCom2012, and the TPC member of IEEE WCNC, ICC, AusWireless 2006, 7th IEEE ISCIT'07, and ChinaCom 2007/2009.

Chengdan Sun received the B.S. degree from the school of Electronics and Information Engineering, Tongji University, China, in 2012. He is currently working toward the Master's degree with the Key Laboratory of Universal Wireless Communications (Ministry of Education), BUPT. His current research interests are interference coordination, game theory, and resource allocation.

Yaohua Sun received his bachelor degree in Telecommunications Engineering with Management from Beijing University of Posts and Telecommunications (BUPT) in 2014 with first class honors, and also got BUPT Excellent Project Award. During the undergraduate study, he won the national scholarship and the Scholarship of Tang Jun and Sun Chunlan. He is currently a Ph.D. student in the Key Laboratory of Universal Wireless Communications (Ministry of Education) at BUPT. His main research interests include coalition game, Stackelberg game, contract theory, matching theory and their applications to resource management in cloud-radio access networks/heterogeneous-cloud radio access networks and fog computing based radio access networks.

Meixia Tao received the B.S. degree in electronic engineering from Fudan University, Shanghai, China, in 1999, and the Ph.D. degree in electrical and electronic engineering from Hong Kong University of Science and Technology in 2003. She is currently a Professor with the Department of Electronic Engineering, Shanghai Jiao Tong University, China. Her current research interests include cooperative communications, wireless resource allocation, MIMO techniques, and physical layer security. Dr. Tao is a member of the Executive Editorial Committee of the IEEE Transactions on Wireless Communications since Jan. 2015. She serves as an Editor for theIEEE Transactions on Communications and the IEEE Wireless Communications Letters. Dr. Tao is the recipient of the IEEE Heinrich Hertz Award for Best Communications Letters in 2013, the IEEE ComSoc Asia-Pacific Outstanding Young Researcher Award in 2009, and the International Conference on Wireless Communications and Signal Processing (WCSP) Best Paper Award in 2012.

Lin Tian is an assistant researcher of Wireless Communication Technology Research Center, Institute of Computing Technology, Chinese Academy of Sciences. She received her B.S and M.S. degree from Beihang University, Beijing, China in 2002 and 2005 respectively. In Apr. 2012, she received the Ph.D. degree from Institute of Computing Technology, Chinese Academy of Sciences. Her research interests include wireless resource management and multimedia multicast scheme in next-generation mobile communication systems. She has published more than twenty research papers in IEEE journals and international conferences. She is also the inventor of more than ten Chinese patents and pending applications. She currently serves as the Publication Chair of the 7th International ICST Conference on Communications and Networking in China, and as reviewers for numerous referred journals and international conferences.

Athanasios V. Vasilakos is recently Professor with the Lulea University of Technology,Sweden . He served or is serving as an Editor for many technical journals, such as the IEEE TRANSACTIONS ON NETWORK AND SERVICE MANAGEMENT; IEEE TRANSACTIONS ON CLOUD COMPUTING, IEEE TRANSACTIONS ON INFORMATION FORENSICS AND SECURITY, IEEE TRANSACTIONS ON CYBERNETICS; IEEE TRANSACTIONS ON NANOBIOSCIENCE; IEEE TRANSACTIONS ON INFORMATION TECHNOLOGY IN BIOMEDICINE; ACM Transactions on Autonomous and Adaptive Systems; the IEEE JOURNAL ON SELECTED AREAS IN COMMUNICATIONS . He is also General Chair of the European Alliances for Innovation(www.eai.eu).

Christos Verikoukis got his Ph.D. from the Technical University of Catalonia in 2000. He is currently a Senior Researcher at CTTC and an adjunct professor at UB. He has published 60 journal papers and over 130 conference papers. He is also co-author in 2 books, 14 chapters in different books and in 2 patents. He has supervised 15 Ph.D. students and 5 Post Docs researchers since 2004. Dr. Verikoukis has participated in more than 30 competitive projects while he has served as the Principal investigator in national projects in Greece and Spain as well as the technical manager in Marie-Curie and Celtic projects. He was General Chair of the 17th and 18th IEEE Workshop on Computer-Aided Modeling, Analysis and Design of Communication Links and Networks (CAMAD), and the TPC Co-Chair of the 15th IEEE International Conference on E-health Networking, Application & Services (Healthcom). He is currently officer (Secretary) of the IEEE ComSoc Technical Committee on Communication Systems Integration and Modeling (CSIM). Dr. Verikoukis received the best paper award of the Communication QoS, Reliability & Modeling Symposium (CQRM) symposium in the IEEE ICC11 conference and the EURASIP 2013 Best Paper Award for the Journal on Advances in Signal Processing.

Chih-Yu Wang received the B.S. and Ph.D. degrees in electrical engineering and communication engineering from National Taiwan University (NTU), Taipei, Taiwan, in 2007 and 2013, respectively. He has been a visiting student in University of Maryland, College Park in 2011. He is currently an Assistant Research Fellow with the Research Center for Information Technology Innovation, Academia Sinica, Taipei, Taiwan. His research interests include game theory, wireless communications, social networks, and data science.

Jiaheng Wang received the B.E. and M.S. degrees from Southeast University, Nanjing, China, in 2001 and 2006, respectively, and the Ph.D. degree in electrical engineering from The Hong Kong University of Science and Technology, Kowloon, Hong Kong, in 2010. He is currently an Associate Professor with the National Mobile Communications Research Laboratory (NCRL), Southeast University. From 2010 to 2011, he was with the Signal Processing Laboratory, ACCESS Linnaeus Center, KTH Royal Institute of Technology, Stockholm, Sweden. He also held a visiting position at the Department of Computer and Information Science, University of Macau, Macau. His research interests mainly include optimization in signal processing, wireless communications, and networks. Dr. Wang serves as an Associate Editor for the IEEE Signal Processing Letters. He is a recipient of the Humboldt Fellowship for Experienced Researchers, and a recipient of the Best Paper Award in WCSP 2014.

Jinlong Wang received the B.S. degree in mobile communications, M.S. degree and Ph.D. degree in communications engineering and information systems from Institute of Communications Engineering, Nanjing, China, in 1983, 1986 and 1992, respectively. Since 1979, Dr. Wang has been with the PLA University of Science and Technology, where he is currently a Full Professor. His current research interests are the broad area of digital communications systems with emphasis on cooperative communication, adaptive coding and modulation, multiple-input-multiple-output systems, soft defined radio, cognitive radio, green wireless communications, and game theory. He has published over 100 papers in refereed mainstream journals and reputed international conferences and has been granted over 20 patents in his research areas. He is an IEEE Senior Member.

Yunting Wang, the postgraduate student of the School of Information and Communication Engineering, Beijing University of Posts and Telecommunications.

Hung-Yu Wei received the B.S. degree in electrical engineering from National Taiwan University in 1999. He received the M.S. and the Ph.D. degree in electrical engineering from Columbia University in 2001 and 2005 respectively. He was a summer intern at Telcordia Applied Research in 2000 and 2001. He was with NEC Labs America from 2003 to 2005. He joined Department of Electrical Engineering at the National Taiwan University in July 2005. His research interests include wireless mesh networks, mobility management in mobile Internet, sensor networks, cross-layer design and optimization in wireless multimedia communications, and game theoretical models for communications networks. He received NTU Excellent Teaching Award in 2008. He also received Recruiting Outstanding Young Scholar Award from the Foundation for the Advancement of Outstanding Scholarship in 2006, K. T. Li Young Researcher Award from ACM Taipei/Taiwan Chapter and The Institute of Information and Computing Machinery in 2012, Ministry of Science and Technology Research Project for Excellent Young Scholars in 2014, and Excellent Young Engineer Award from the Chinese Institute of Electrical Engineering in 2014. He was a consulting member of Acts and Regulation Committee of National Communications Commission during 2008~2009. He has been actively participating in the IEEE 802.16 and 3GPP standardization, and was a voting member of the IEEE 802.16 working group.

Dan Wu received her B.S., M.S., Ph.D. degrees in Institute of Communications Engineering, PLA University of Science and Technology, Nanjing, China in 2006, 2009, 2012 respectively. Currently she is a postdoctoral researcher in Institute of Communications Engineering, PLA University of Science and Technology, Nanjing, China. Her research interests are mainly in resource allocation and management, game theory, cooperative communications and wireless sensor networks.

Qihui Wu received his B.S. degree in communications engineering, M.S. degree and Ph.D. degree in communications and information system from Institute of Communications Engineering, Nanjing, China, in 1994, 1997 and 2000, respectively. He is currently a full professor at the PLA University of Science and Technology, China. His current research interests are soft defined radio and wireless communication systems, cognitive radio networks, Internet of Things, and spectrum management and optimization for 5G networks. He has published over 100 papers in refereed mainstream journals and reputed international conferences and has been granted several patents in his research areas. He is an IEEE Senior Member.

Chunmei Xia is a master candidate student in Beijing Jiaotong University. Her research interset includes radio resource allocation in D2D systems.

Shaoyi Xu received her Ph. D degree from Inha University, South Korea in 2007. After that, she worked in Nokia Research Center in Nokia China Ltd. as a senior researcher and a postdoctor. From 2009 to now, she worked in Beijing Jiaotong University. From Feb. 2013 to Feb. 2014, she is a visiting scholar in Department of Electrical and Computer Engineering, University of California, USA. She has obtained more than 20 international and Chinese patents and published more than 70 journal papers and conference papers. Her research interests include D2D communication, cooperative communication in high-speed environments, cognitive radio and UWB, etc.

Xiaodong Xu received his Ph.D. degree of Circuit and System in Beijing University of Posts and Telecommunications (BUPT) in 2007. He is currently an associate professor at BUPT. His research interests focus on wireless networking theory and key technologies, including coordinated distributed cellular network architecture, heterogeneous networking, interference mitigation, radio resource management and Software Defined Network. He has participated in several research projects such as Natural and Science Foundation of China, 863 and National Major Projects. He also works with the 3GPP stardards. He is the author or co-author of 90 papers and has more than 30 patents pended.

Yuhua Xu received his B.S. degree in Communications Engineering, and Ph.D. degree in Communications and Information Systems from College of Communications Engineering, PLA University of Science and Technology, in 2006 and 2014 respectively. He has been with College of Communications Engineering, PLA University of Science and Technology since 2012, and currently as an Assistant Professor. His research interests focus on opportunistic spectrum access, learning theory, game theory, and distributed optimization techniques for wireless communications. He has published several papers in international conferences and reputed journals in his research area. He is an Editor for the KSII Transactions on Internet and Information Systems. In 2011 and 2012, he was awarded Certificate of Appreciation as Exemplary Reviewer for the IEEE Communications Letters.

Elias Yaacoub received the B.E. degree in Electrical Engineering from the Lebanese University in 2002, the M.E. degree in Computer and Communications Engineering from the American University of Beirut (AUB) in 2005, and the PhD degree in Electrical and Computer Engineering from AUB in 2010. He was a Research Assistant at the American University of Beirut from 2004 to 2005, and in the Munich University of Technology in Spring 2005. From 2005 to 2007, he worked as a Telecommunications Engineer with Dar Al-Handasah, Shair and Partners. After obtaining his PhD, he worked as a research scientist in the Qatar mobility Innovations Center (QMIC) till December 2014, when he joined Strategic Decisions Group (SDG) in Beirut as a Consultant. He has more than 130 publications in the field of wireless communications. His research interests include wireless communications, green cellular networks, vehicular communications, sensor networks, antenna theory, and bioinformatics.

Shuili Zhang is a M.D. student with the Institute of Computing Technology, Chinese Academy of Sciences, China. Her research interest includes video transmission, wireless video communication, pervasive computing, communication, and networking.

Hua Zhao, born in 1979, received the master degree in communication and information system from Zhengzhou University. Currently, she is a lecturer of National Digital Switching System Engineering Technology Research Center. Her research interests include wireless mobile communications and wireless physical layer security.

Yiqing Zhou received the B.S. degree in communication and information engineering and the M.S. degree in signal and information processing from the Southeast University, China, in 1997 and 2000, respectively. In February 2004, she received the Ph.D. degree in electrical and electronic engineering from the University of Hong Kong, Hong Kong. Now she is a professor supported by the 100-Talent Program in Wireless Communication Research Center, Institute of Computing Technology, Chinese Academy of Sciences. Dr. Zhou has published over 80 papers and book chapter in the areas of wireless mobile communications. She is a the associate/guest editor for IEEE Trans. Vehicular Technology, IEEE JSAC (Special issue on ``Broadband Wireless Communication for High Speed Vehicles'' and ``Virtual MIMO''), WCMC, ETT and JCST. She is the symposia co-chair of IEEE ICC2015, the symposium co-chair of ICC2014, TPC co-chair of ChinaCom2012, the tutorial co-chair of IEEE WCNC2013 and ICCC2014, and the workshop co-chair of SmartGridCom2012 and GlobeCom2011. Dr. Zhou also served as the member-at-large for IEEE NIC in 2013.

Wei Zhong currently holds a position as a lecturer at the College of Communications Engineering, PLAUST, Nanjing, China. His current research interests include game theory, signal processing, wireless communication, machine learning, and computing.

Zhou Zhong, born in 1982, received the PH.D. degree in communication and information system from the PLA Information Engineering University, Zhengzhou, China. Currently, he is a lecturer of National Digital Switching System Engineering & Technological Research Center. His research interests include signal processing, channel coding and physical layer security.

Chi Zhou received two B.S. degrees in both Automation and Business Administration from Tsinghua University, China, in 1997. She received the M.S. and Ph.D. degrees in Electrical and Computer Engineering from Northwestern University, Evanston, IL, in 2000 and 2002, respectively. She is currently serving as an Associate Professor in the Department of Electrical and Computer Engineering, Illinois Institute of Technology, Chicago, IL. Her primary research interests include wireless sensor networks, demand response for smart grid, scheduling for OFMA/MIMO systems, and network coding for wireless mesh networks.

Index

I

J

L

M

N

P

Q

R

S

T

U

V

W

CPSIA information can be obtained at www.ICGtesting.com
Printed in the USA
BVOW04*0013010416

442546BV00010B/28/P

9 781466 686427